third edition
Genetics

Robert F. Weaver
University of Kansas

Philip W. Hedrick
Arizona State University

WCB **Wm. C. Brown Publishers**

Dubuque, IA   Bogotá   Buenos Aires   Caracas   Chicago   Guilford, CT   London
Madrid   Mexico City   Seoul   Singapore   Sydney   Taipei   Tokyo   Toronto

Project Team

Editor *Elizabeth Sievers*
Developmental Editor *Terrance Stanton*
Production Editor *Carla D. Kipper*
Marketing Manager *Patrick E. Reidy*
Designer *K. Wayne Harms*
Art Editor *Jodi K. Banowetz*
Photo Editor *Nicole Widmyer*
Permissions Editor *Karen L. Storlie*
Advertising Projects Coordinator *Leslie Dague*

Wm. C. Brown Publishers

President and Chief Executive Officer *Beverly Kolz*
Vice President, Director of Editorial *Kevin Kane*
Vice President, Sales and Market Expansion *Virginia S. Moffat*
Vice President, Director of Production *Colleen A. Yonda*
Director of Marketing *Craig S. Marty*
National Sales Manager *Douglas J. DiNardo*
Executive Editor *James M. Smith*
Advertising Manager *Janelle Keeffer*
Production Editorial Manager *Renée Menne*
Publishing Services Manager *Karen J. Slaght*
Royalty/Permissions Manager *Connie Allendorf*

Copyedited by Julie Bach

Cover Photo © Douglas Struthers/Tony Stone Images

Freelance Permissions Editor *Karen Dorman*

The credits section for this book begins on page 627
and is considered an extension of the copyright page.

Library of Congress Catalog Card Number: 96–83917

ISBN 0–697–16000–9

Printed in the United States of America
2460 Kerper Boulevard, Dubuque, IA 52001

10 9 8 7 6 5 4 3 2 1

BRIEF CONTENTS

CONTENTS

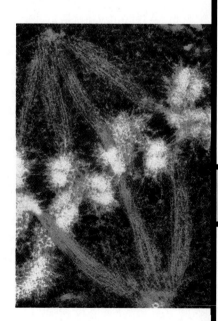

CHAPTER 4

Chromosomes and Heredity   73

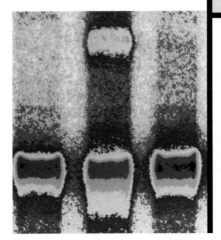

CHAPTER 5

Genetic Linkage   104

PART 2   MOLECULAR GENETICS   129

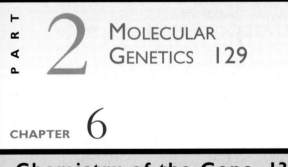

CHAPTER 6

Chemistry of the Gene   130

CHAPTER 14

Developmental Genetics 369

CHAPTER 15

Gene Cloning and Manipulation 415

CHAPTER 16

Applications of Gene Cloning and Advanced Genetic Mapping 448

CHAPTER 20

Extensions and Applications of Population Genetics   557

PREFACE

This textbook is intended for a first college course in genetics. In writing the book, we have faced the age-old question of organization: Which to present first, Mendelian genetics or molecular genetics? The traditional organization of a genetics text is historical—Mendelian, or transmission, genetics first, followed by molecular genetics. However, because molecular approaches to genetics have become more sophisticated and the body of knowledge about molecular genetics has grown explosively, it is easy to make a case for starting with molecular aspects. Some have followed a middle road and mingled the two topics.

Although modern genetics is a wonderfully fruitful blend of classical and molecular techniques, we think it makes sense to separate these two approaches when presenting them to students for the first time. Moreover, we have chosen to follow the traditional historical order. To us, this seems to allow for a more logical progression of ideas, moving from early, relatively simple concepts of heredity to later, more complex explanations at the molecular level. Besides that, we enjoy teaching genetics because it is like telling a story, and it doesn't seem quite right to start the story in the middle.

In spite of our own prejudices, we have tried to make this book easy to use no matter which order of topics an instructor prefers. Accordingly, we have organized this third edition of our book in the following way:

Chapter 1 is a historical introduction to genetics. It provides the bare essentials of both major branches of the discipline. After reading this chapter, students should have some appreciation for the molecular character of genes as they study the subsequent chapters on transmission genetics. If, on the other hand, an instructor chooses to start with molecular genetics, chapter 1 introduces students to the basics of heredity, providing a context for the molecules. In either case a brief review of basic concepts, such as cell cycle, mitosis, meiosis, and cell structure has been added in chapter 1 to assure that all students have at least this prerequisite knowledge.

Chapters 2 through 5 constitute a unit on transmission genetics. In chapter 2 we use Mendel's experiments to introduce the principles of segregation and independent assortment. At the same time, since it is impossible to discuss these topics without dealing with probabilities, we provide a brief introduction to statistics. In chapter 3 we show that simple Mendelian analysis does not always work, and introduce concepts such as penetrance, expressivity, and epistasis that illustrate this point. Chapter 4 deals with chromosomes: their behavior and the fact that genes reside on chromosomes. The chromosome concept leads naturally to the idea of linkage, the subject of chapter 5. Here, students learn about linkage and recombination and how to construct linkage maps with eukaryotes.

Each chapter in this first part of the book includes a discussion of the application of genetics to human problems. For example, in chapter 2 students learn to apply the concepts of segregation and independent assortment to pedigree analysis of human genetic diseases. In chapter 4 they see how human chromosomal abnormalities can have severe consequences.

The central part of the book, and its major theme, is a thirteen-chapter sequence emphasizing molecular genetics. We believe that the direction of modern genetics research demands that the heart of a modern genetics text be molecular.

Chapters 6 through 11 present the fundamentals of molecular genetics. Chapter 6 introduces the structure and properties of DNA and a minimal outline of gene function. In the first and second editions, chapter 7 consisted of a much more complete introduction to gene function, including gene replication, transcription, translation, and mutation. The following five chapters then expanded on these themes. As a result, some closely related concepts were found in separate chapters—some in chapter 7; others in chapter 11, for example. This "jumping" around to cover some topics in full caused a problem for students. To solve this problem, we have taken the material in chapter 7, "An Introduction to Gene Function," and distributed it among the appropriate subsequent chapters in this new edition. Chapter 6, "Chemistry of the Gene," is now followed by five chapters (7–11) that present the concepts of gene function. Chapter 7 describes the replication and recombination of genes. Chapters 8, 9, and 10 describe the process of gene expression. In particular, chapters 8 and 9 cover transcription in prokaryotes and eukaryotes, respectively, and chapter 10 outlines the process of translation. Finally, chapter 11 deals with another characteristic of genes, their capacity for mutation.

Chapters 12 through 18 further illustrate specific aspects of the functions of genes. Chapter 12 is an extension of the material in chapter 11; it details the structure and function of transposable elements in both prokaryotes and eukaryotes and shows how these mobile elements can participate in mutagenesis. Chapter 13 is a treatment of the transmission genetics of bacteria and phages. It might reasonably be placed in the first part of the book, but since so much is known about the molecular basis of prokaryotic genetics, it seems to fit better in the molecular genetics section. Also, students can better appreciate the analysis of mutations in bacterial operons after they have learned how operons work (in chapter 8) and the chemical nature of mutations (in chapter 11). Chapter 14 explores the ways that genes govern the development of organisms. The intricate and fascinating development of the fruit fly is the major paradigm for this chapter.

Chapters 15 through 17 present practical applications of molecular genetics. Chapter 15 covers gene cloning and manipulation and shows how one branch of genetics has evolved into a kind of engineering. Chapter 16 is a new chapter, titled "Applications of Gene Cloning and Advanced Genetic Mapping," which expands upon the information presented in chapter 15. It begins with uses of cloned genes in various fields, followed by techniques used to map and sequence the human genome, which leads to discussions of cloning genes responsible for human diseases. The chapter concludes with the subject of DNA typing in the courtroom. Chapter 17 is about the molecular genetics of cancer, with special emphasis on the behavior of oncogenes. Our experience has been that students have fun with these topics, but since they need to master the basics first, these chapters come near the end of this section. Chapter 18 deals with the genetics of mitochondria and chloroplasts from both classical and molecular perspectives.

The book concludes with a two-chapter sequence on population genetics. Chapter 19 deals with introductory population genetics, including selection, genetic drift, inbreeding, mutation, and gene flow. Chapter 20 integrates these factors to discuss human diseases, molecular evolution, conservation genetics, and pesticide resistance.

The most obvious change from the second edition is in organization and updating of information. The material from the original chapter 7 has been split up and incorporated into the five chapters dealing with DNA replication, transcription, translation and mutation (chapters 7–11). A new chapter, chapter 16, has been added to this edition to reflect the explosive advancements seen in applications of gene cloning and genetic mapping. Chapter 14, "Developmental Genetics," has been heavily revised, especially in the area of *Drosophila* development, but you will also see extensive updating and new information in almost every chapter in this edition.

This edition also features a greater emphasis on problem solving. We have expanded worked-out problems within the text to give students more guidance in how to solve important types of problems, and have included a greater variety of problems at the end of each chapter. In most cases, at least one end-of-chapter problem tests the skill taught by each worked-out problem within the same chapter.

The most important underlying theme of this book is its experimental approach. Genetics, like any science, is not a collection of facts, but a way of posing and answering questions about nature. Wherever possible, we have included the experimental rationale and results that have led to our present understanding of genetics. Furthermore, many of the end-of-chapter questions require students to examine and interpret hypothetical experimental results, or to design experiments to test hypotheses. We know that genetics, especially molecular genetics, can be bewildering to college students. However, we have found that if given a clear understanding of the way geneticists practice their science, students can master the subject much more easily. That is the goal of this book.

LEARNING AIDS FOR THE STUDENT

To achieve success in the study of genetics, students must understand the material presented, utilize the text as a tool for learning, and enjoy reading the text. Therefore, we have included many aids to make your study of genetics efficient and enjoyable. Each chapter contains the following:

CHAPTER OPENING QUOTATION

The chapter opening quotation is designed to spark interest and provide a perspective on the chapter's contents.

CHAPTER OPENING LEARNING OBJECTIVES

The chapter opening learning objectives were written to help the student preview the basic concepts in the chapter.

BOLDFACED KEY TERMS

Important terms are emphasized and clearly defined when they are first presented.

SOFTWARE CORRELATIONS

Selected sections of text are cross-referenced to WCB software/multimedia products, including the *Explorations in Cell Biology and Genetics* CD-ROM by George Johnson. Icons appear by section headings to signal a software application. In addition, an Appendix provides a description of each *Explorations* module that is cited, along with "Questions to Explore."

Genetic Inheritance: Peas and Drosophila Software (Mac).

Explorations in Cell Biology and Genetics CD-ROM.

BOX 7.2 Telomeres, the Hayflick Limit, and Cancer

Everyone knows that organisms, including humans, are mortal. But biologists used to assume that cells cultured from humans were immortal. Each individual cell would ultimately die, of course, but the cell *line* would go on dividing indefinitely. Then in the 1960s Leonard Hayflick discovered that ordinary human cells are not immortal. They can be grown in culture for a finite period of time—about 50 generations. Then they enter a period of senescence, and finally they all die. This ceiling on the lifetime of normal cells is known as the Hayflick limit. But cancer cells do not obey any such limit. They do go on dividing generation after generation indefinitely.

BOXED READINGS

Most chapters contain a boxed reading. These readings cover a variety of interesting topics, including some of the ethical issues involved with recent developments in molecular biology and gene therapy. A portion of box 7.2, is shown above.

SECTION SUMMARIES

These summaries help the student review and understand the section's basic concepts before continuing on to the next section of the chapter. For example:

The ends of eukaryotic chromosomes are formed by an enzyme called telomerase, which adds DNA to the 3′-ends of the chromosomes according to the instructions of its own RNA template. This renewal of chromosome ends counteracts the tendency of chromosomes to lose telomeric DNA with every cell division.

END-OF-CHAPTER WORKED-OUT PROBLEMS

To help students develop their analytical, quantitative, and problem-solving skills, example problems and detailed solutions are provided. Look for this heading at the end of the chapter:

WORKED-OUT PROBLEMS

END-OF-CHAPTER SUMMARY

This narrative section summarizes the important basic concepts of the chapter.

END-OF-CHAPTER PROBLEMS AND QUESTIONS

The problems and questions are written to aid students in their understanding of basic genetic concepts and processes. They also help students test and develop their analytical problem-solving skills. Selected End-of-Chapter questions are cross-referenced to genetic software published by Wm. C. Brown Publishers. Product information is available under "Supplementary Materials." This section is identified by the following heading:

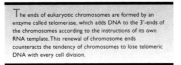

PROBLEMS AND QUESTIONS

Problems and Questions with Links to Software

Mendel's Peas: 6, 7, 11, 12, 13, 20

ANSWERS TO END-OF-CHAPTER PROBLEMS AND QUESTIONS (APPENDIX A)

The answers to all of the end-of-chapter problems and questions are provided to help students immediately evaluate their problem-solving skills.

SELECTED READINGS FOR EACH CHAPTER (APPENDIX B)

References for each chapter are provided to help students investigate and study important topics.

GLOSSARY

Definitions of boldfaced terms are assembled in a glossary at the end of this textbook.

INDEX

An easy-to-use, comprehensive index is provided to help students locate information on topics they need to review.

NEW STUDENT SUPPLEMENT

A text-specific *Student Study Guide,* titled "A Synopsis of Genetics," is available with the third edition of *Genetics.* The study guide is written by Roger Denome of Stonehill College and supports the student in understanding the key concepts presented in the text. More than just an outline or set of questions, it brings a distinctive approach, complete with its own illustrations, and designed to zero in on and clarify major concepts. (ISBN 29086–7)

SUPPLEMENTARY MATERIALS

1. An *Instructor's Manual with Test Item File (IM/TIF)* is available free to all adopters of Weaver/Hedrick, *Genetics,* third edition. This IM/TIF consists of two sections. The first section is a list of transparencies that accompany the text and thought questions and problems for each text chapter. The second section (TIF) contains objective questions pertinent to the material found in each text chapter. These include multiple choice, true/false, fill-in-the-blank, and short-answer questions.

2. *Computerized Testing and Classroom Management Software* is offered free upon request to adopters of this textbook. The TIF from the *Instructor's Manual* is included in the database of this computerized testing software. The software also provides a grade-recording program. The software requires no programming experience and is available in IBM and Macintosh formats. Diskettes are available through your local WCB sales representative.

3. A set of *100 full-color acetate transparencies*, based on illustrations taken from the textbook, may be used to supplement classroom lectures and is available free to adopters.

4. *Gene Game Software*, by Bill Sofer of Rutgers University, linked to this text, is a Macintosh software game that gives students the opportunity to test their logic and critical thinking skills while attempting to clone a fictitious "Fountain of Youth" gene. Contact your local WCB sales representative for more information on this fun, interactive software game. (ISBN 24893)

5. *Genetic Inheritance: Peas and Drosophila Software,* linked directly to this text, is an affordable software program, developed at Purdue University, that can be packaged with this text. The software allows students to simulate hundreds of genetic crosses right at their computers to gain valuable practice in the quantitative aspects of genetics. The pea experiments investigate Mendel's theories of dominance, segregation, independent assortment, and how numbers of offspring affect test results. Once students master the concepts illustrated with the basic pea crosses, they can challenge themselves with the *Drosophila* experiments, which also explore the concepts of monohybrid, dihybrid, and trihybrid crosses, as well as the determination of linkage, map distances, and gene order on chromosomes. This software can also be packaged with the text. (Mac ISBN 28861–7), (Windows ISBN 35225-0)

6. *Explorations in Genetics CD-ROM,* linked directly to this text, is an interactive multimedia program developed by George B. Johnson, Washington University–St. Louis, and WCB. It calls on users to manipulate genetic variables and examine how they impact the results as they explore such modules as Constructing a Genetic Map, DNA Fingerprinting: You Be the Judge, and Gene Regulation. The CD-ROM is compatible with both Windows and Macintosh systems. (ISBN 28862–5)

7. *GenPak: Computer-Assisted Academic Programs in Genetics,* by Tully Turney, Hampden-Sydney College, is a Macintosh hypercard program featuring interactive, problem-solving exercises in Mendelian, molecular, and population genetics. (ISBN 17370–4)

8. The new fourth edition of *Laboratory Manual of Genetics,* by A. M. Winchester and P. J. Wejksnora, University of Wisconsin–Milwaukee, is an up-to-date, practical manual. It features classical and molecular biology exercises that give students the opportunity to apply the scientific method to "real"—not simulated—lab investigations. (ISBN 12287–5)

9. *How Scientists Think,* by George B. Johnson, is a concise, illustrated book that presents discussions of twenty-one classic genetics and molecular biology experiments. It is an intriguing way to foster critical thinking and reinforce the scientific method in a genetics course. (ISBN 27875–1)

10. *Case Workbook in Human Genetics,* by Ricki Lewis, SUNY–Albany, includes thought-provoking case studies in human genetics, with many examples gleaned from the author's experiences as a practicing genetics counselor. (ISBN 22287–X)

11. *Genetics Problem-Solving Guide,* by William Wellnitz, is a companion guide that systematically walks students through the logical steps involved in solving genetics problems. (ISBN 13739–2)

12. *Compendium of Problems in Genetics,* by John Kuspira and Ramesh Bhambhani, University of Alberta, includes logical, illustrated exercises—including many based on actual experimental data from classic papers—for students at basic and advanced levels. (ISBN 16734–8)

REVIEWERS

We are indebted to the following reviewers of our manuscript for countless helpful suggestions.

David Evans
Penn College/PSU

Gunther Schlager
University of Kansas

Norman Johnson
University of Texas

Harvey Friedman
University of Missouri

Patrick Galliart
North Iowa Area Community College

Ernest Hannig
University of Texas at Dallas

Richard Kowles
St. Mary's University of Minnesota

Alvan Karlin
University of Arkansas at Little Rock

Paul Madtes Jr.
Mount Vernon Nazarene College

Louise A. Paquin
Western Maryland College

Gerald T. Schlink
Missouri Southern State College

Daphne Foreman
Macalester College

Arden Campbell
Iowa State University

Robert J. Wiggers
Stephen F. Austin State University

Gaston Griggs
John Brown University

Robin L. Bennett
University of Washington School of Medicine

James L. Eliason
Manhattan College

Gregory J. Podgorski
Utah State University

Rosemary H. Ford
Washington College

David Foltz
Louisiana State University

Susan J. Karcher
Purdue University

Larry R. Eckoat
Penn State Erie

Laura Adamkewicz
George Mason University

D. H. "Denny" Crews Jr.
Louisiana State University

Robert Winning
Eastern Michigan University

Edmund Zimmerer
Murray State University

John Belote
Syracuse University

Kim O'Neill
Brigham Young University

Roger Shanks
University of Illinois

Thomas King
Central Connecticut State University

Mary Ann Polasek
Cardinal Stritch College

DuWayne Englert
Southern Illinois University at Carbondale

Wade Hazel
DePauw University

Lee Ehrman
State University of New York at Purchase

Martin LaBar
Southern Wesleyan University

Charles H. Green
Rowan College

Carl Luciano
Indiana University of Pennsylvania

Ronald K. Hodgson
Central Michigan University

Ming Y. Zheng
Houghton College

James C. Gibson
Chadron State College

Jun Tsuji
Siena Heights College

Paul F. Lurquin
Washington State University

Paul Goldstein
University of Texas–El Paso

Louis Levinger
York College

Michael H. Perlin
University of Louisville

Michele Morek
Brescia College

I. R. Schmoyer
Muhlenberg College

James Courtright
Marquette University

S. Cecilia Agnes Mulrennan
Regis College

Harry Wistrand
Agnes Scott College

Lance Urven
University of Wisconsin–Whitewater

John Rinehart
Eastern Oregon State College

James Curry
Franklin College of Indiana

W. H. Stone
Trinity University

David Parker
North Virginia Community College

William Ettinger
Gonzaga University

Margaret Hollingsworth
SUNY at Buffalo

Matthew M. White
Ohio University

David Wofford
University of Florida

Darryl Grennell
Alcorn State University

Tully Turney
Hampden–Sydney College

Mitrick Johns
Northern Illinois University

Lori R. Scott
Augustana College

William Steinhart
Bowdoin College

Dwayne Wise
Mississippi State University

Ron L. Frank
University of Missouri–Rolla

James Robinson
University of Texas at Arlington

Thomas Wolf
Washburn University

Tim Mullican
Dakota Wesleyan University

Mark Holland
Salisbury State University

David S. Dennison
Dartmouth College

John Harris
Tennessee Technology University

Sandra Michael
Binghamton University

Joan Redd
Walla Walla College

Charles Stufflebeam
Southwest Missouri State

Eliot Krause
Seton Hall University

David Campbell
California State Polytechnic University

Monte Turner
University of Akron

Judith M. Johnston
Community College of Philadelphia

Charles A. Owens
King College

Chris L. Kapicka
Northwest Nazarene College

Charles Herr
Eastern Washington University

Joel Chandlee
University of Rhode Island

Thomas Eveland
Luzerne County Community College

Albert C. Jensen
Central Florida Community College

Susan R. Capasso
Middlesex Community-Technical College

Douglas K. Walton
College of St. Scholastica

Bruce D. Parker
Utah Valley State College

Michael McCann
St. Joseph's University

Stephen Kucera
University of Tampa

Robert McGuire
University of Montevallo

Peter T. Rowley
University of Rochester

Bonnie S. Wood
University of Maine at Presque Isle

Peter Luykx
University of Miami

Patricia Dorn
Loyola University

Kiran Misra
Edinboro University of Pennsylvania

James Tan
Valparaiso University

David S. Durica
University of Oklahoma

Sister Anne M. Donigan
Gwynedd-Mercy College

James F. Hancock
Michigan State University

Fred Allendorf
University of Montana

Todd Stanislav
Xavier University of Louisiana

Daryl Sas
Geneva College

Shelley Jansky
University of Wisconsin–Stevens Point

Richard Kliman
Radford University

Jeffrey Byrd
St. Mary's College of Maryland

Kenneth Wireman
Southeastern College

Charles Paulson
University of Arizona

Joyce L. Azevedo
Southern Adventist University

Ann Zeeh
The College of Saint Rose

Anna W. Berkovitz
Purdue University

Michael Myszewski
Drake University

Stephen S. Daggett
Avila College

Walter Kaczmarczyk
West Virginia University

Sheryl K. Ayala
Marion College of Fond Du Lac

Kathleen Triman
Franklin & Marshall College

Frank J. Doe
University of Dallas

Robert D. Seager
University of Northern Iowa

Michael F. Tibbetts
Bard College

Philip M. Meneely
Haverford College

Robert A. Paoletti
King's College

Mark Kaminski
Western States Chiropractic College

David Scott
South Carolina State University

Virginia Wootten
Massachusetts College of Pharmacy

Jacquelyn K. Beals
Mary Baldwin College

Amelia J. Ahern-Rindell
Weber State University

Paul E. Bibbins Jr.
Kentucky State University

Jacob Varkey
Humboldt State University

Judith Van Houten
University of Vermont

Carla Barnwell
University of Illinois, Urbana

Case Okubo
Florida International University

Ammini Moorthy
Wagner College

Ian Boussy
Loyola University

Ernie Hannig
University of Texas–Dallas

Florian Muckenthaler
Bridgewater State College

Patrick Calie
Eastern Kentucky University

Deborah Clark
Middle Tennessee State University

Delmar Vander Zee
Dordt College

Bob Cohen
University of Kansas

Doug Ruden
University of Kansas

Karl J. Fryxell
University of California, Riverside

Chuck Staben
University of Kentucky

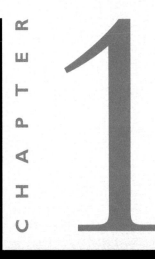

CHAPTER 1

An Introduction

Learning Objectives

This chapter will introduce:

1. The three branches of genetics.
2. The fundamentals of transmission genetics.
3. The fundamentals of molecular genetics.

I am the family face;
Flesh perishes. I live on,
Projecting trait and trace
Through time to times anon,
And leaping from place to place
Over oblivion.

Thomas Hardy

English novelist

What is the most fundamental life science? Different biologists give different answers, but one of our students recently asserted that it is genetics. And it is hard to argue with her, because the things we study in biology, including most of the physical characteristics, behavior, and capabilities of living things, are inherited. Therefore, to understand an organism fully, we must understand its heredity. The word "genetics" comes from the Latin *genesis*, which means birth. Thus, genetics is the study of birth, or, more broadly, the study of heredity.

Modern geneticists take three major approaches to this study, and these define the three main branches of contemporary genetics: (1) transmission genetics, so called because it seeks to explain the pattern of transmission of traits from one generation to the next; (2) molecular genetics, the study of the activities of genes, the molecules that carry genetic information from generation to generation; and (3) population genetics, the study of the variation of genes between and within populations.

THE THREE BRANCHES OF GENETICS

The first branch of genetics relies mainly on the same kind of experimental approach used by Gregor Mendel, an Austrian monk, in the middle of the nineteenth century. Organisms with differing traits, or **phenotypes,** are mated, and the transmission of these traits to the next generation is observed. We can then mate these progeny organisms with others of the same or different phenotypes (or even with themselves, as with Mendel's peas) and again observe the transmission of traits. Because Mendel pioneered this approach to genetics, we frequently call it **Mendelian,** or **classical, genetics.** Since it deals with the transmission of genes between generations we also call it **transmission genetics.**

The second branch, **molecular genetics,** approaches the subject from its chemical foundation: molecules. Instead of examining phenotypic characters, molecular geneticists examine the genes themselves. They are concerned with the molecules that compose genes, the molecules that control genes, and the molecules that are the products of genes. The molecular geneticist's job has become much easier (and more fun) since we learned how to clone genes in the 1970s. With these gene-cloning techniques, we can produce a gene in large quantities in pure form and study its structure and function.

The third branch, **population genetics,** examines the extent of genetic variation within and among populations. Geneticists traditionally studied this variation on the phenotypic level, but population geneticists now focus increasingly on molecular variation in a population. Much of population genetics study is aimed at understanding how the observed genetic variation evolved.

Population genetics techniques are also useful in describing the genetic differences between species and in learning about the process of species formation. Remember that a **species** is a reproductively isolated group, such as *Zea mays* (corn) or *Drosophila melanogaster* (fruit fly). A species name has two parts. The first part, which is capitalized (e.g., *Drosophila*), is the **genus,** and may encompass several species. The second part, which starts with a lowercase letter, designates a particular species within the genus.

A great deal of genetics research is now done at the molecular level. However, the other branches of genetics are in no danger of disappearing. Although molecular techniques can uncover new genes, they frequently cannot reveal a gene's function. A more fruitful approach is this: A researcher first discovers an abnormal organism with interesting characteristics. The abnormality in this organism is caused by a mutated, or altered, gene, so the **mutant** is the "red flag" that leads us to a new gene. Once the gene has been identified this way, we can begin to use molecular techniques to examine its structure and function in detail and perhaps to understand the evolutionary significance of variation in the gene. Actually, it is misleading to put too much emphasis on the divisions within the field of genetics. Most geneticists now approach their subject in several ways, using methods from more than one of the three branches.

Genetics teachers face an organization problem: Do we start by talking about the molecular aspects of the gene, so that later, when we show how genes are transmitted from generation to generation, students can appreciate fully what this means? Do we talk about transmission genetics first and only later explain molecular genetics? Or do we try to deal with both at the same time?

In this book, we have decided to use the second approach: transmission genetics first, then molecular genetics, and finally population genetics. This feels right to us for two reasons. First, transmission genetics is in some ways easiest to understand. It deals with the inheritance of characteristics that we can frequently see and therefore understand more readily than abstract concepts like molecules.

A second reason for treating transmission genetics first is that it developed first historically. We like a good story, and the history of genetics is a very good one. By adopting a historical approach, we get a chance to retell that story.

No matter which part of genetics we cover first, we always wish our students had prior exposure to the other, because it would make their understanding so much richer. This problem is impossible to resolve completely, but we propose to compromise by providing in this introductory chapter a brief overview of transmission and molecular genetics, so that at least the most fundamental concepts of these branches of genetics will be familiar as we look at each in detail in later chapters. Since you do not need an appreciation of population genetics in order to understand the other two branches, we will not discuss it further in this chapter. In keeping with the basic plan of this book, we will present our overview of transmission and molecular genetics in historical fashion.

FIGURE 1.1 Gregor Mendel.

Courtesy of the Department of Library Services, American Museum of Natural History, negative 219467.

TRANSMISSION GENETICS

In 1865, Gregor Mendel (figure 1.1) published his findings on the inheritance of seven different traits in the garden pea. Prior to this time, scientists thought inheritance occurred through a blending of each trait of the parents in the offspring. Mendel concluded instead that inheritance is particulate. That is, each parent contributes particles, or genetic units, to the offspring. We now call these particles **genes.** Furthermore, by carefully counting the number of progeny plants having a given **phenotype,** or observable characteristic (e.g., yellow seeds, white flowers), Mendel was able to make some important generalizations. The word "phenotype," by the way, comes from the same Greek root as "phenomenon," meaning "appearance"; thus, a tall pea plant exhibits the tall phenotype, or appearance. "Phenotype" can also refer to the whole set of observable characteristics of an organism.

MENDEL'S LAWS OF INHERITANCE

Mendel saw that a gene can exist in several different forms called **alleles.** For example, the pea can have either yellow or green seeds. One allele of the gene for seed color gives rise to yellow seeds, the other to green. Moreover, one allele can be **dominant** over the other, **recessive** allele. In this case, yellow is dominant. Mendel showed this when he mated a green-seeded pea with a yellow-seeded pea. All of the progeny in the first filial generation (F_1) had yellow seeds. However, when these F_1 yellow peas were allowed to self-fertilize, some green-seeded peas reappeared. The

ratio of yellow to green seeds in the second filial generation (F_2) was very close to 3:1.

The term "filial" comes from the Latin: *filius* = son; *filia* = daughter. Therefore, the first filial generation (F_1) contains the offspring (sons and daughters) of the original parents. The second filial generation (F_2) is the offspring of the F_1 individuals.

Mendel concluded that the green allele must have been preserved in the F_1 generation, even though it did not affect the seed color of those peas. His explanation was that each parent plant carried two copies of the gene; that is, the parents were **diploid,** at least for the characteristics he was studying. According to this concept, **homozygotes** have two copies of the same allele, either two yellow alleles or two green alleles. **Heterozygotes** have one copy of each allele. The two parents in the first mating above were homozygotes; the resulting F_1 peas were all heterozygotes. Further, Mendel reasoned that sex cells contain only one copy of the gene; that is, they are **haploid.** Homozygotes can therefore produce sex cells, or **gametes,** that have only one allele, but heterozygotes can produce gametes having either allele.

This explains what happened in the matings of yellow with green peas. The yellow parent contributed a gamete with a gene for yellow seeds; the green parent, a gamete with a gene for green seeds. Therefore, all the F_1 peas got one yellow and one green allele. They had not lost the green allele at all, but since yellow is dominant, all the peas were yellow. However, when these heterozygous peas were self-fertilized, they produced yellow and green gametes in equal numbers, and this allowed the green phenotype to reappear.

Here is how that happened. Assume that we have two sacks, each containing equal numbers of green and yellow marbles. If we take one marble at a time out of each sack and pair it with a marble from the other sack, we will wind up with the following results: ¼ of the pairs will be yellow/yellow; ¼ will be green/green; and the remaining ½ will be yellow/green. The yellow and green alleles for peas work the same way. Recalling that yellow is dominant, you can see that only ¼ of the progeny (the green/green ones) will be green. The other ¾ will be yellow because they have at least one yellow allele. Hence, there is a 3:1 ratio of yellow to green peas in the second (F_2) generation.

Mendel also found that the genes for the seven different characteristics he chose to study operate independently of one another. Therefore, combinations of alleles of two different genes (for example, yellow or green peas with round or wrinkled seeds, where yellow and round are dominant and green and wrinkled are recessive) gave ratios of 9:3:3:1 for yellow/round, yellow/wrinkled, green/round, and green/wrinkled, respectively.

Genes can exist in several different forms, or alleles. One allele can be dominant over another, so heterozygotes having two different alleles of one gene will generally exhibit the characteristic dictated by the dominant allele. The recessive allele is not lost; it can still exert its influence when paired with another recessive allele in a homozygote.

THE CHROMOSOME THEORY OF INHERITANCE

Other scientists either did not know about or uniformly ignored the implications of Mendel's work until 1900 when three botanists, who had arrived at similar conclusions independently, rediscovered it. After that time, most geneticists accepted the particulate nature of genes, and the field of genetics began to blossom. One factor that made it easier for geneticists to accept Mendel's ideas was a growing understanding of the nature of chromosomes, which had begun in the latter half of the nineteenth century. Mendel had predicted that gametes would contain only one allele of each gene instead of two. If chromosomes carry the genes, their numbers should also be reduced by half in the gametes—and they are. Chromosomes therefore appeared to be the discrete physical entities that carry the genes.

This notion that chromosomes carry genes is the **chromosome theory of inheritance.** It was a crucial new step in genetic thinking. No longer were genes disembodied factors; now they were observable objects in the cell nucleus. (See figure B1.1 to review key structures in the cell.) Some geneticists, particularly Thomas Hunt Morgan (figure 1.2), remained skeptical of this idea. Ironically, in 1910 Morgan himself provided the first definitive evidence for the chromosome theory.

Morgan worked with the fruit fly (*Drosophila melanogaster*), which was in many respects a much more convenient organism than the garden pea for genetic studies because of its small size, short generation time, and large number of offspring. When he mated red-eyed flies (dominant) with white-eyed flies (recessive), most but not all of the F_1 progeny were red-eyed. Furthermore, when Morgan mated the red-eyed males of the F_1 generation with their red-eyed sisters, they produced about ¼ white-eyed males, but no white-eyed females. In other words, the eye color phenotype was sex linked. It was transmitted along with sex in these experiments. How could this be? (See figure B1.2 to review

FIGURE 1.2 Thomas Hunt Morgan.
Courtesy of the National Library of Medicine.

the cell cycle and the process of mitosis, which explains how traits are transmitted from parent to daughter cells.)

We now realize that sex and eye color are transmitted together because the genes governing these characteristics are located on the same chromosome—the X chromosome. (Most chromosomes, called **autosomes,** occur in pairs in a given individual, but the X chromosome is an example of a **sex chromosome,** of which the female fly has two copies and the male has one.) However, Morgan was reluctant to draw this conclusion until he observed the same sex linkage with two more

REVIEW BOX I.I

Cell Structure

Before we proceed with our discussion of transmission genetics and molecular genetics, it is important to review cellular structures that are of particular interest in the study of genetics (figure B1.1).

The **nucleus** is a discrete structure within the cell that contains the genetic material, DNA. The DNA is usually present in the cell as a network of fibers called **chromatin.** During mitosis, the DNA molecules form the condensed chromosome structure by coiling and supercoiling around specialized proteins.

Other structures that are of particular importance to geneticists are the centriole, the endoplasmic reticulum, the Golgi apparatus, ribosomes, and the mitochondrion. The **centriole** is a cylindrical structure composed of microtubules. Paired centrioles organize the formation of spindle fibers during mitosis. The

endoplasmic reticulum (ER) is a network of membranes throughout the cell; cells have both smooth and rough ER. The rough ER is studded with ribosomes, giving it a rough appearance. The bound ribosomes of the ER synthesize proteins that are destined for secretion from the cell, or that are to be incorporated into the membrane or into specific vacuoles. The **Golgi apparatus** is a continuation of the membrane network of the endoplasmic reticulum. **Ribosomes** that are free in the cytosol synthesize proteins that remain in the cell and are not transported through the endoplasmic reticulum and Golgi apparatus. The **mitochondrion** is the site of energy production in the cell. The mitochondrion consists of a double membrane system. The outer membrane is smooth and the inner membrane has convoluted folds. The mitochondrion can divide independently of the cell and contains its own circular, double-stranded DNA.

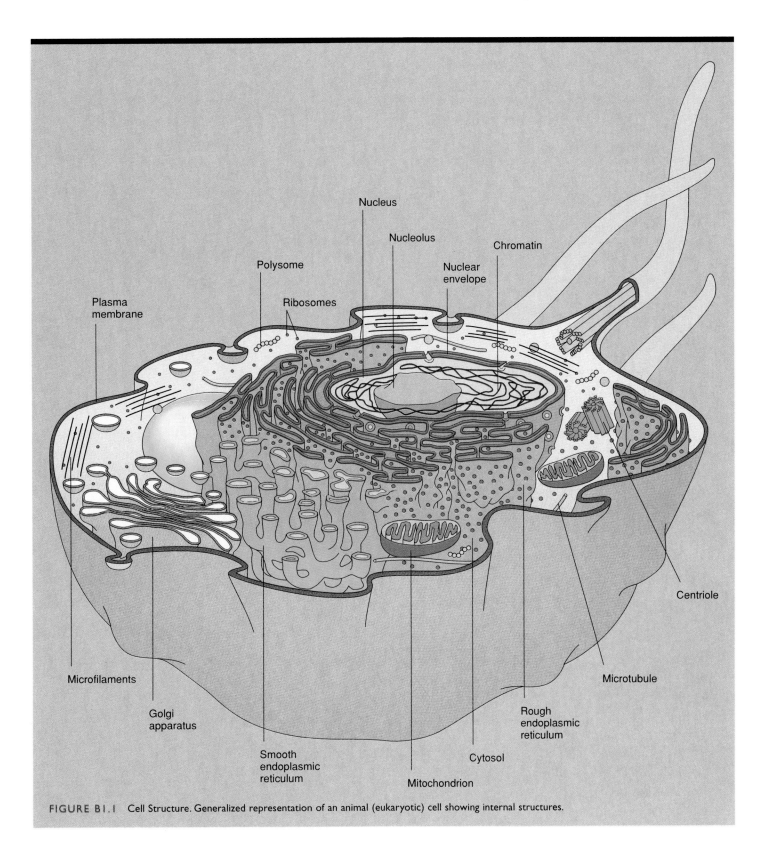

FIGURE B1.1   Cell Structure. Generalized representation of an animal (eukaryotic) cell showing internal structures.

Cell Cycle and Mitosis

Cell Cycle

Cellular reproduction is a cyclic process in which daughter cells are produced through nuclear division (mitosis) and cellular division (cytokinesis). Mitosis and cytokinesis are part of the growth–division cycle called the **cell cycle** (figure B1.2a). Mitosis, which is divided into four stages (see figure B1.2b), is a relatively small part of the total cell cycle; the majority of the time, a cell is in a growth stage called **interphase.** Interphase is divided into three parts, G_1, S, and G_2. The first gap phase, G_1, follows mitosis and is a period of growth and metabolic activity. The S phase follows G_1 and is a period of DNA synthesis in which the DNA is replicated. Another gap phase, G_2, follows DNA synthesis and precedes the next mitotic division. Certain mature cell types do not continue to divide but remain in interphase (in G_0).

Mitosis

Mitosis is the period of time in the cell cycle in which the nucleus divides and gives rise to two daughter cells that have chromosome numbers identical to that of the original cell (figure B1.2b). Mitosis can be divided into four distinct stages (**prophase, metaphase, anaphase,** and **telophase**)

characterized by the appearance and orientation of the homologous pairs of chromosomes.

Prophase   At the end of the G_2 interphase, the chromosomes have replicated, and each has two sister chromatids attached to each other at the centromere. Prophase indicates the beginning of mitosis, when the replicated chromosomes coil and condense. The centrioles divide and the spindle apparatus also appears.

Metaphase   During metaphase, the spindle fibers from the spindle apparatus attach to the centromeres of the chromosomes. The chromosomes gradually migrate to the midline of the cell oriented between the two centrioles.

Anaphase   Anaphase begins with the separation of the centromeres and the contraction of the spindle fibers pulling the centromeres toward the centrioles. As a result, the sister chromatids, now called daughter chromosomes, separate, and the homologous chromosomes move in opposite directions toward the centrioles.

Telophase   The two sets of chromosomes reach opposite poles of the cell, where they begin to uncoil. Telophase ends with cytokinesis (cellular division) and the formation of two daughter cells.

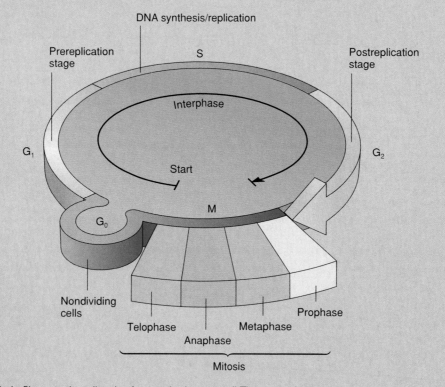
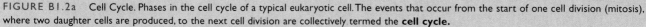

FIGURE B1.2a   Cell Cycle. Phases in the cell cycle of a typical eukaryotic cell. The events that occur from the start of one cell division (mitosis), where two daughter cells are produced, to the next cell division are collectively termed the **cell cycle.**

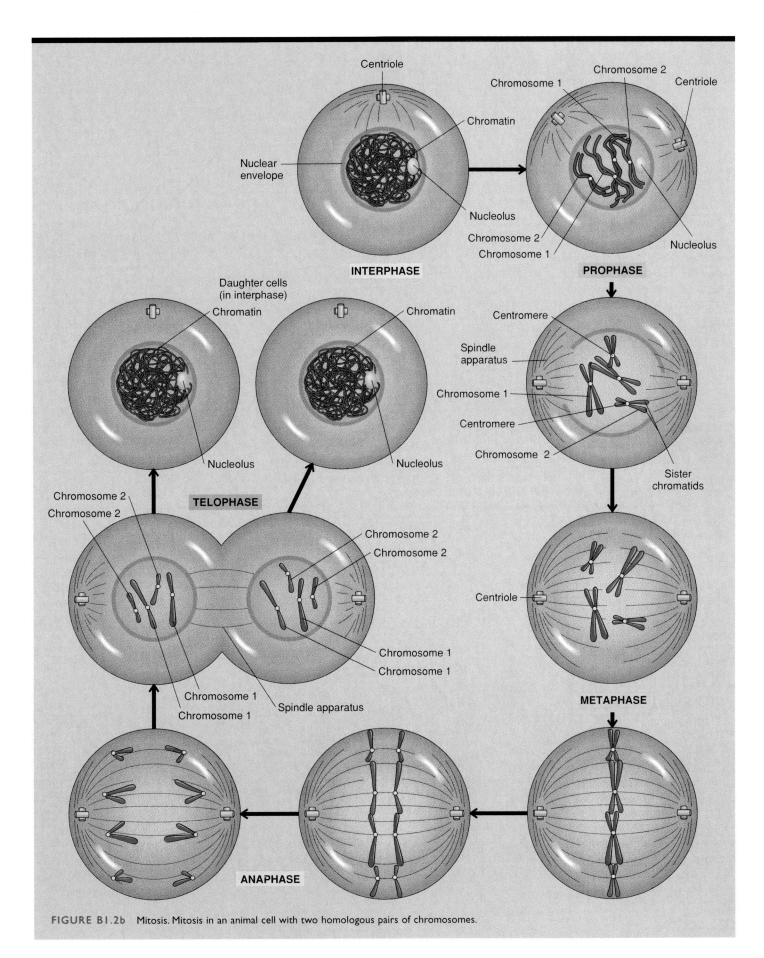

FIGURE B1.2b Mitosis. Mitosis in an animal cell with two homologous pairs of chromosomes.

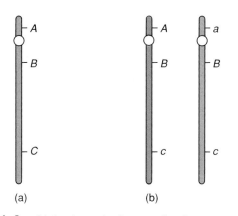

FIGURE 1.3 (a) A schematic diagram of a chromosome, indicating the positions of three genes—*A, B,* and *C.* (b) A schematic diagram of a diploid pair of chromosomes, indicating the positions of the three genes—*A, B,* and *C*—on each, and the genotype (*A* or *a*; *B* or *b*; and *C* or *c*) at each locus.

phenotypes, miniature wing and yellow body, also in 1910. That was enough to convince him of the validity of the chromosome theory of inheritance.

Before we leave this topic, let us reiterate two crucial points. First, every gene has its place, or **locus,** on a chromosome. Figure 1.3 depicts a hypothetical chromosome and the positions of three of its genes, called *A, B,* and *C.* Second, diploid organisms such as human beings contain two copies of all chromosomes (except sex chromosomes). That means that they have two copies of most genes, and that these copies can be the same alleles, in which case the organism is **homozygous,** or different alleles, in which case it is **heterozygous.** For example, figure 1.3*b* shows a diploid pair of chromosomes with different alleles at one locus (*Aa*) and the same alleles at the other two loci (*BB* and *cc*). The **genotype,** or allelic constitution, of this organism with respect to these three genes is *AaBBcc.* Since this organism has two different alleles (*A* and *a*) in its two chromosomes at the *A* locus, it is heterozygous at that locus (Greek: *hetero* = different). Since it has the same, dominant *B* allele in both chromosomes at the *B* locus, it is homozygous dominant at that locus (Greek: *homo* = same). And since it has the same, recessive *c* allele in both chromosomes at the *C* locus, it is homozygous recessive there. Finally, because the *A* allele is dominant over the *a* allele, the phenotype of this organism would be the dominant phenotype at the *A* and *B* loci, and the recessive phenotype at the *C* locus.

This discussion of varying phenotypes in *Drosophila* gives us an opportunity to introduce another important genetic concept: **wild-type** versus **mutant.** The wild-type phenotype is the most common, or at least the generally accepted standard phenotype of an organism. To avoid the mistaken impression that a wild organism is automatically a wild-type, some geneticists prefer the term **standard type.** In *Drosophila,* red eyes and full-size wings are wild-type. Mutations in the *white* and *miniature* genes result in mutant flies with white eyes and miniature wings, respectively. Mutant alleles are usually recessive, as in these two examples, but not always.

FIGURE 1.4 Recombination in *Drosophila.* The two X chromosomes of the female are shown schematically. One of them (red) carries two wild-type genes: (*m⁺*), which results in normal wings, and (*w⁺*), which gives red eyes. The other (blue) carries two mutant genes: *miniature (m)* and *white (w).* During egg formation, a recombination, or cross over, indicated by the X, occurs between these two genes on the two chromosomes. The result is two recombinant chromosomes with mixtures of the two parental genes. One is *m⁺ w,* the other is *m w⁺.*

GENETIC RECOMBINATION AND MAPPING

It is easy to understand that genes on separate chromosomes behave independently in genetic experiments, and that genes on the same chromosome—like the genes for miniature wing (*miniature*) and white eye (*white*)—behave as if they are linked. However, genes on the same chromosome do not always show perfect genetic linkage. In fact, Morgan discovered this when he examined the behavior of the sex-linked genes he had found. For example, although *white* and *miniature* are both on the X chromosome, they remain linked in offspring only 65.5% of the time. The other offspring have a new combination of alleles not seen in the parents and are therefore called **recombinants.**

How are these recombinants produced? The answer was already apparent by 1910, because microscopic examination of chromosomes during meiosis (gamete formation) had shown crossing over between **homologous** chromosomes (chromosomes carrying the same genes or alleles of the same genes). This resulted in the exchange of genes between the two homologous chromosomes. In the example above, during formation of eggs in the female, an X chromosome bearing the *white* and *miniature* alleles experienced crossing over with a chromosome bearing the red-eye and normal-wing alleles (figure 1.4). Because the crossing over event occurred between these two genes, it brought together the *white* and normal-wing alleles on one chromosome and the red and *miniature* alleles on the other. Because it produced a new combination of alleles, we call this process **recombination.** (See figure B1.3 to review the process of meiosis and the occurrence of crossing over.)

Morgan assumed that genes are arranged in a linear fashion on chromosomes, like beads on a string. This, together with his awareness of recombination, led him to propose that the farther apart two genes are on a chromosome, the more likely they are to recombine. This makes sense because there is simply more room between widely spaced genes for crossing over to occur. A. H. Sturtevant extended this hypothesis to predict that there should be a mathematical relationship between the distance separating two genes on a chromosome and the frequency of recombination between these two genes. Strurtevant collected data on recombination in the fruit fly that supported his hypothesis. This established the rationale for **genetic mapping** techniques still in use today. By the 1930s, other investigators found that the same rules applied to other **eukaryotes** (nucleus-containing organisms), including the mold *Neurospora,* the garden pea, maize (corn), and even human beings. These rules also apply to **prokaryotes,** organisms that have no nuclei.

PHYSICAL EVIDENCE FOR RECOMBINATION

Barbara McClintock (figure 1.5) and Harriet Creighton provided a direct physical demonstration of recombination in 1931. By examining maize chromosomes microscopically, they could detect recombinations between two easily identifiable features of a particular chromosome (a knob at one end and a long extension at the other). Furthermore, whenever this physical recombination occurred, they could also detect recombination genetically. Thus, there is a direct relationship between a region of a chromosome and a gene. Shortly after McClintock and Creighton performed this work on maize, C. Stern observed the same phenomenon in *Drosophila.* So recombination could be detected both physically and genetically in animals as well as plants.

FIGURE 1.5   Barbara McClintock.
The Bettmann Archive.

The chromosome theory of inheritance holds that genes are arranged in linear fashion on chromosomes. The reason that certain traits tend to be inherited together is that the genes governing these traits are on the same chromosome. However, recombination between two chromosomes during meiosis can scramble the parental alleles to give nonparental combinations. The farther apart two genes are on a chromosome the more likely such recombination between them will be.

MOLECULAR GENETICS

Studies such as those just discussed tell us important things about the transmission of genes and even about how to map genes on chromosomes, but they do not tell us what genes are made of or how they work. This has been the province of molecular genetics, which also happens to have its roots in Mendel's era.

THE DISCOVERY OF DNA

In 1869, Friedrich Miescher (figure 1.6) discovered in the cell nucleus a mixture of compounds that he called nuclein. The major component of nuclein is **deoxyribonucleic acid (DNA).**

REVIEW BOX 1.3

Meiosis

The process of meiosis involves two cell divisions that result in daughter cells (gametes) with half the parental number of chromosomes. If the number of chromosomes were not reduced prior to fertilization, the chromosome number would double in each generation.

Meiosis also allows the exchange of genetic material on homologous chromosomes through crossing over, which occurs at the end of prophase I. Crossing over results in new combinations of alleles of different genes on a chromosome.

Meiosis, like mitosis, can be divided into distinct phases (figure B1.3). **Meiosis I** includes prophase I (where crossing over

occurs), metaphase I, anaphase I (unlike in mitosis, homologous pairs of chromosomes, not sister chromatids, separate to the opposite poles of the cell), and telophase I. **Meiosis II** begins with prophase II, continues on to metaphase II (the centromeres attach to the spindle apparatus), anaphase II (as in mitosis, the centromeres separate and the chromosomes begin to migrate to opposite poles), and telophase II, which completes meiosis, resulting in four daughter cells with half the number of chromosomes as the parental cell. As shown in figure B1.3, the parental cell contains four chromosomes (four homologous pairs), and the resulting four daughter cells have two chromosomes each.

the probability of either of two mutually exclusive events occurring is the sum of their individual probabilities. Thus, using the coin example again, if we want to know the total probability of obtaining two heads *or* two tails when two coins are flipped:

$$P(\text{two heads or two tails}) = P(\text{two heads}) + P(\text{two tails})$$
$$= P(\text{head}) \times P(\text{head}) + P(\text{tail}) \times P(\text{tail})$$
$$= \frac{1}{2} \times \frac{1}{2} + \frac{1}{2} \times \frac{1}{2}$$
$$= \frac{1}{4} + \frac{1}{4}$$
$$= \frac{1}{2}$$

In this case, notice that the events—for example, two heads—are actually themselves composed of two coin flips.

Thinking again of the segregation probabilities of progeny from an F_1 cross, the probability of an *RR* offspring is $\frac{1}{4}$, that of an *Rr* offspring is $\frac{1}{2}$, and that of an *rr* offspring is $\frac{1}{4}$. Because both genotypes *RR* and *Rr* are dominant, we may want to know the probability of obtaining either *RR* or *Rr*. Using the sum rule:

$$P(RR \text{ or } Rr) = P(RR) + P(Rr)$$
$$= \frac{1}{4} + \frac{1}{2}$$
$$= \frac{3}{4}$$

The product rule and the sum rule can be extended to more than two events. For example, the probability of simultaneously obtaining heads on each of four coins, using the product rule, is $\frac{1}{2} \times \frac{1}{2} \times \frac{1}{2} \times \frac{1}{2} = (\frac{1}{2})^4 = \frac{1}{16}$. In a standard card deck with thirteen types of cards (aces, twos, threes . . . kings), the probability of drawing a particular type of card (ignoring different suits) is $\frac{1}{13}$. Using the sum rule, the probability of drawing a face card (jack, queen, or king) is $\frac{1}{13} + \frac{1}{13} + \frac{1}{13} = \frac{3}{13}$.

CONDITIONAL PROBABILITY ✗

As we have just discussed, more than two outcomes or events may be possible. For example, a fair, six-sided die should have an equal probability of landing with a 1, 2, 3, 4, 5, or 6 showing. In other words, the probability of each of these outcomes is $\frac{1}{6}$. Obviously, half the time the die should land with an even number (2, 4, or 6) showing. We can ask, given that an even number is rolled, what is the probability of its being a 2? *A probability that is contingent on given circumstances is called a* **conditional probability.** In other words, the probability that we will obtain a 2, given that an even number is rolled on the die, is:

$$P(2 \mid \text{even number}) = \frac{P(2)}{P(\text{even})}$$

$$P(2 \mid \text{even number}) = \frac{\frac{1}{6}}{\frac{1}{2}}$$

$$= \frac{1}{3}$$

(The vertical line means "given.") In other words, the total probability of a general event, an even number in this case, is subdivided into more specific events.

When Mendel examined his F_2 progeny, he used conditional probability to describe the proportion of heterozygotes among the dominant progeny. In this case, the probability of heterozygotes among dominant progeny is:

$$P(\text{heterozygous} \mid \text{dominant}) = \frac{P(\text{heterozygous})}{P(\text{dominant})}$$

$$P(\text{heterozygous} \mid \text{dominant}) = \frac{\frac{1}{2}}{\frac{3}{4}}$$

$$= \frac{2}{3}$$

In other words, two-thirds of the dominant progeny are expected to be heterozygotes (and one-third are expected to be homozygotes). We will use this same logic later to calculate the probability that normal individuals who have a sibling with a recessive disease are themselves **carriers** of the disease allele; that is, that they are heterozygotes.

BINOMIAL PROBABILITY ✗

Finally, consider the probability that in a group of a given size, say a sibship (a group of brothers and sisters), a certain number of individuals will be of one type and the remainder of another type. To calculate these probabilities, we can use an expansion of the product rule. First, we calculate the probability of the occurrence of a particular sequence of events when there are three different events and only two alternatives for each event. As an example, imagine flipping a coin three consecutive times. The following eight sequences are possible:

	HHT	TTH	
HHH	HTH	THT	TTT
	THH	HTT	

There could be three consecutive heads and no tails, as in the first column; two heads and one tail (second column); one head and two tails (third column); or three tails and no heads (fourth column). In other words, if the order is ignored, four outcomes are possible.

Now, assuming again that the probability of heads is $\frac{1}{2}$ (and that the probability of tails is $\frac{1}{2}$), the probability of their coming up in any one of the orders shown in the second column is, by the product rule:

$$P(HTH) = P(H) \times P(T) \times P(H)$$
$$= \frac{1}{2} \times \frac{1}{2} \times \frac{1}{2}$$
$$= \frac{1}{8}$$

However, to determine the probability of two heads and one tail, independent of order, we have to use the sum rule to get the total probability of the three sequences that give this constitution, or:

$$P(\text{two heads, one tail}) = P(HHT) + P(HTH) + P(THH)$$
$$= \frac{1}{8} + \frac{1}{8} + \frac{1}{8}$$
$$= \frac{3}{8}$$

Thus, the probability of three heads is ⅛; the probability of two heads (and one tail) is ⅜; the probability of one head (and two tails) is ⅜; and the probability of no heads (and three tails) is ⅛. The sum of all these probabilities is ⅛ + ⅜ + ⅜ + ⅛ = 1, as expected.

Now let us generalize the approach we have just developed. First, assume there are two possible alternatives, say heads or tails, having probabilities p and q, respectively. We know that p + q = 1 because there are only two alternatives. If N different events occur (N being equal to three in this example because three coins were flipped), what is the probability of x events of the first type and N − x of the second type? In the previous example, x = 2 (two heads) and N − x = 1 (one tail)? In general, the **binomial formula** (binomial because there are only two different types of events) gives these probabilities:

$$P(x \text{ type one events out of N total events}) = Cp^x q^{N-x}$$

In this formula, C is the binomial coefficient and indicates the number of sequences that give the same overall constitution. In the current example, C = 3.

The binomial coefficient for most of the examples we will consider can be obtained using the following formula:

$$C = \frac{N!}{x!\,(N-x)!}$$

where the symbol ! indicates a factorial. By definition, 0! = 1, 1! = 1, 2! = 2 × 1 = 2, 3! = 3 × 2 × 1 = 6, and so on. For example, if x = 2 and N = 3, then

$$C = \frac{3!}{2!\,(3-2)!} = \frac{3 \times 2 \times 1}{2 \times 1 \times 1} = 3$$

This is the same answer we found previously by writing out all the possibilities.

Another way to obtain the binomial coefficient is to use Pascal's triangle (figure 2.9). In this scheme, a given number is the sum of the two numbers above it as indicated by the two arrows. This relationship was named after a seventeenth-century French mathematician and philosopher. It is useful if you forget the formula for the binomial coefficient. Any given number in this array is obtained by summing the two numbers directly above. As an example, let us look at N = 3, row 4 in figure 2.9. The first number in this row, C = 1, is the coefficient for all events of one type (for example, three heads), x = 3; the second number is for x = 2 (two heads), giving C = 3 as before; the third number is for x = 1 (one head, giving C = 3; and the fourth number is for x = 0 (no heads), giving C = 1.

We will use the binomial formula to calculate the probability of a particular number of boys and girls in a sibship. Assuming there are four children, what is the probability that four will be girls? In other words, what is the probability that x = 4 and N − x = 0? Assuming that the probability of a girl is ½ and the probability of a boy is also ½, then:

$$P(\text{four girls of four children}) = C(\tfrac{1}{2})^4(\tfrac{1}{2})^0$$

From the formula for the binomial coefficient:

$$C = \frac{4!}{4!\,0!} = 1$$

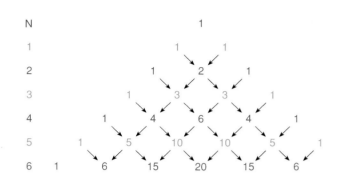

FIGURE 2.9   Pascal's triangle showing the size of the binomial coefficient (C) for different N values, the number of different events. The numbers in a horizontal row indicate C for x = N, N − 1, N − 2, . . . , etc.

Or, looking at figure 2.9, where N = 4 and x = 4, then C = 1. Given that any number to the zero power is 1:

$$P(\text{four girls of four children}) = (1)(\tfrac{1}{2})^4(1)$$
$$= \tfrac{1}{16}$$

Now let us look at a more complicated situation and calculate the probability of two girls in a sibship of size four. Again, using the formula where x = 2 and N − x = 2:

$$P(\text{two girls of four children}) = C(\tfrac{1}{2})^2(\tfrac{1}{2})^2$$

In this case:

$$C = \frac{4!}{2!\,2!} = 6$$

so that:

$$P(\text{two girls of four children}) = 6(\tfrac{1}{2})^2(\tfrac{1}{2})^2$$
$$= \tfrac{3}{8}$$

C = 6 here because there are six orders, all of which have two girls and two boys: BBGG, BGBG. BGGB, GBBG, GBGB, and GGBB.

We can apply these probability concepts to other situations. For example, a number of genes affect the coats of domestic cats. One gene causes white spotting, primarily on the chest of the cat, where *S,* the white-spotting allele is dominant over *s* for no white spotting. Another gene affects hair length, where *L,* the short-hair allele, is dominant over *l,* the long-hair allele. Figure 2.10 shows a cat with white spotting (either genotype *SS* or *Ss*) and short hair (genotype *LL* or *Ll*). Assuming that two cats, both heterozygous at these genes, mate, what proportion of their progeny will exhibit the different phenotypes? Figure 2.11 gives the expected frequency of the progeny, after segregation and independent assortment. Most of the time (⁹⁄₁₆), the offspring will have the same phenotype as their parents—white spotting and short hair—but ¹⁄₁₆ of the time, the offspring will be different from their parents for both characters; that is, they will have no white spotting and long hair.

We can apply probability concepts to the following example: Given these same two cats as parents and a sibship of four kittens, what is the probability that all the kittens will have long

FIGURE 2.10 An orange cat with white spotting and short hair.
© H. Reinhard/OKAPIA/Photo Researchers, Inc.

hair (*ll*) and white spotting (*S–*)? Because there is only one order in which this can occur and the coat patterns of the different kittens are independent (given these same parental genotypes), then:

$$P(4 \text{ of } 4 \text{ long-hair, white-spotted kittens}) = [P(ll\ S–)]^4$$
$$= (\tfrac{3}{16})^4$$
$$= 0.0012$$

In this case, the probability is quite low, with such a litter occurring about one in a thousand times. Perhaps the male parent was misidentified and actually had long hair!

Several basic theories of probability are necessary to understand Mendelian genetics: 1) the product rule, which states that the joint probability of two events occurring is the product of their individual probabilities; 2) the sum rule, which states that the probability of either one of two mutually exclusive events occurring is the sum of their individual probabilities; 3) the principle of conditional probability, which is the probability of a specific event divided by the probability of a more general event; and 4) the binomial formula, which states the probability of x events of a particular type out of a total of N events.

FORKED-LINE APPROACH

The approach just explained is the best way to visualize the probability of different types of progeny for two genes. For three

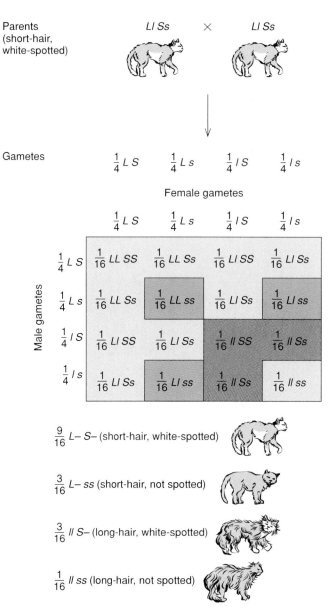

FIGURE 2.11 A cross between two cats doubly heterozygous at the long-hair and white-spotted genes, giving the expected proportions of progeny.

or more genes, the **forked-line approach,** in which the probabilities for each category are determined and combined one gene at a time, is quite useful.

The forked-line approach is based on the assumption of independent assortment of all the genes involved. An example is given in figure 2.12 for the garden pea dihybrid cross discussed previously. First, the progeny are separated into proportions as to round or wrinkled (¾ to ¼) at the first gene. Within each of these categories, they are then separated into proportions at the second gene (¾ yellow to ¼ green). To determine the number in any combined phenotypic class, the probabilities are multiplied over the genes. For example, the expected proportion of wrinkled, yellow offspring is ¼ × ¾ = ³⁄₁₆.

This same approach can be used to calculate proportions for more than two genes. Let us consider cat coat color again and introduce a third gene, *D*, where *dd* homozygotes have dilute (or light) color and *D*– is nondilute or full color. (The difference between gray and black cats is caused by the difference between dilute and nondilute.) Given a cross between two cats that are *Ll ss Dd* and *Ll SS dd*, we can write out all the progeny types using the forked-line approach (figure 2.13). We can then ask, what is the probability of a long-hair, white-spotted, nondilute offspring from this mating? This probability, given in the third row of figure 2.13, is ⅛. Or, assuming that these genes assort independently, we can use the product rule to write:

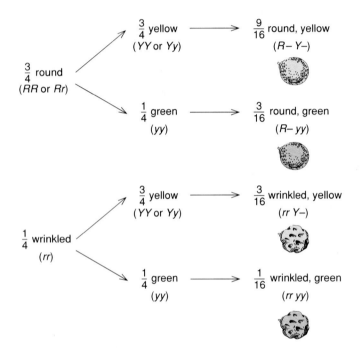

FIGURE 2.12   The forked-line approach for progeny of a dihybrid cross in the garden pea.

P(long-hair, white-spotted,

nondilute offspring) = P(*ll S– D–*)

$$= P(ll) \times P(S–) \times P(D–)$$
$$= \tfrac{1}{4} \times 1 \times \tfrac{1}{2}$$
$$= \tfrac{1}{8}$$

We could also ask from the same parental cross, what is the probability of an offspring of genotype *Ll Ss Dd*? In this case:

$$P(Ll\ Ss\ Dd) = P(Ll) \times P(Ss) \times P(Dd)$$
$$= \tfrac{1}{2} \times 1 \times \tfrac{1}{2}$$
$$= \tfrac{1}{4}$$

The forked-line approach can be extended to calculate the probability of genotypes or phenotypes in offspring for any number of genes as long as they assort independently.

Often the alleles at a gene are symbolized by a capital letter for the dominant allele and a lowercase letter for the recessive allele, as in alleles *R* and *r*. However, to describe all the mutants that have been discovered, the twenty-six letters of the alphabet are not enough. For example, the recessive mutant in *Drosophila melanogaster* that has small, underdeveloped wings is called *vestigial* and is symbolized by *vg*. The normal allele at this gene, instead of being indicated by a capital letter, is symbolized either by vg^+ or just by +, both of which mean the **wild-type,** or nonmutant allele. Heterozygotes at this locus having normal wings are indicated as either vg^+vg or +*vg*. Mutants themselves may be dominant, such as the curly-wing mutant *Cy* in *Drosophila*, as shown in figure 2.14*a*, where the first letter of the symbol is capitalized to indicate its dominance. Heterozygotes *Cy*+ have curly wings because *Cy* is dominant over the wild-type allele +.

Let us use these symbols to illustrate the forked-line approach. (Notice that to avoid confusion when using these symbols we will leave a space between the letters indicating the genotypes for different genes.) If we have the cross +*vg Cy*+ *sese* × +*vg* + + +*se* (*se* indicates the eye color mutant *sepia*), what proportion of progeny would we expect to be curly-winged—that is, mutant at the second locus and wild-type at the other

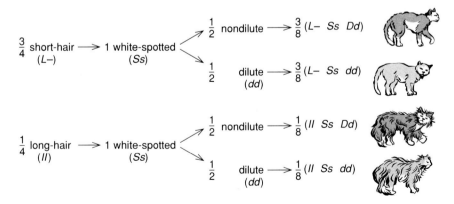

FIGURE 2.13   The four progeny categories and their probabilities for a cross between cats of genotypes *Ll ss Dd* and *Ll SS dd*.

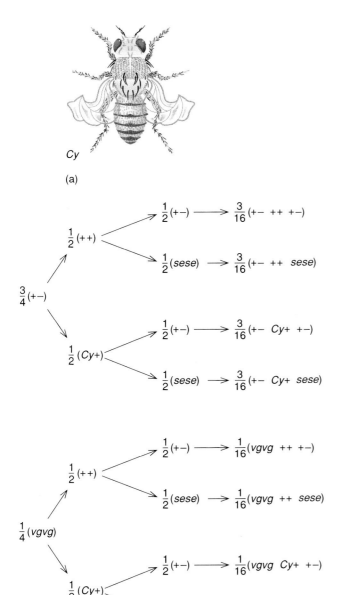

FIGURE 2.14 (*a*) Curly-wing mutant *Drosophila melanogaster*, *Cy*. (*b*) The eight progeny categories and their probabilities for the cross between *Drosophila* with genotypes *+vg Cy+ sese* and *+vg ++ +se*.

two genes? Figure 2.14*b* shows the eight possible genotypes and their probabilities in the progeny from this cross. The particular category we are interested in is shown in row three of the far right column. To calculate this:

$$P(+- Cy+ +-) = P(+-) \times P(Cy+) \times P(+-)$$
$$= \tfrac{3}{4} \times \tfrac{1}{2} \times \tfrac{1}{2}$$
$$= \tfrac{3}{16}$$

The forked-line approach, which is based on the product rule, can be used to determine the probability of various offspring types for any number of genes, assuming they assort independently.

GOODNESS OF FIT

How can we use predictions based on probabilities to understand the transmission of genes from one generation to the next? Remember that Mendel's F_2 data given in table 2.1 very closely fit 3:1 expectations. But what if the numbers had not been so close to the expected ratio? Would this have been due to chance or because the underlying hypothesis, the principle of segregation, was incorrect? It is useful to measure the "goodness of fit" to find out whether the observed numbers are adequately close to the numbers expected according to a particular hypothesis. Mendel himself did not use any statistical tests (they weren't developed until decades later), but such tests have helped identify exceptions to Mendel's principles in other studies.

A test commonly used to measure goodness of fit is the **chi-square test.** In this statistical test, the *observed numbers* in each category, say dominant and recessive progeny, are compared to the *expected numbers* under a certain hypothesis, say Mendel's principle of segregation. Notice that in this test we are not using the observed and expected frequencies but the observed and expected numbers. If the observed number in one category is larger than the expected number, the opposite will be true in another category because the total number over all classes for both observed and expected numbers is the same. To obtain a test that measures the difference between observed and expected numbers, irrespective of the sign of this difference, we use the square of the difference between observed and expected numbers. This value is then standardized by dividing it by the expected number in a category, making the formula for chi-square:

$$\chi^2 = \Sigma \frac{(O - E)^2}{E}$$

where Σ is the Greek letter sigma indicating the sum over all categories, χ is the Greek letter chi, and O and E are the observed and expected numbers.

This statistical technique can be illustrated using the data from one of Mendel's backcross experiments. In this experiment, he crossed F_1 violet-flowered plants (Vv) with white-flowered plants (vv). Of 166 progeny, 85 had violet flowers and 81 had white flowers. According to the principle of segregation, 0.5 should be violet and 0.5 should be white. Or, using numbers, 166(0.5) = 83 would be expected in each class. Notice that the sum of the expected numbers must always equal the sum of the observed numbers; the sum here is 166. With these observed and expected numbers, we can calculate the chi-square value as

Table 2.4   Chi-Square Values of Mendel's Backcross Data

Flower Color	Observed Numbers	Expected Numbers	$\dfrac{(O-E)^2}{E}$
Violet	85	166(0.5) = 83	$\dfrac{2^2}{83} = 0.05$
White	81	166(0.5) = 83	$\dfrac{(-2)^2}{83} = 0.05$
	$\overline{166}$	$\overline{166}$	$\chi^2 = 0.10$

Table 2.5   Probabilities of Different Theoretical Chi-Square Values for Given Degrees of Freedom[1]

Degrees of Freedom	Probability				
	0.9	0.5	0.1	0.05★	0.01★★
1	0.02	0.46	2.71	3.84	6.64
2	0.21	1.39	4.60	5.99	9.21
3	0.58	2.37	6.25	7.82	11.34
4	1.06	3.86	7.78	9.49	13.28
5	1.61	4.35	9.24	11.07	15.09
6	2.20	5.35	10.64	12.59	16.81
7	2.83	6.35	12.02	14.07	18.48
8	3.49	7.34	13.36	15.51	20.09
9	4.17	8.34	14.68	16.92	21.07
10	4.86	9.34	15.98	18.31	23.21

[1]Probabilities of 0.05 and 0.01 are known as the statistically significant and highly significant levels and are indicated by ★ and ★★, respectively.

in table 2.4. For the violet and white categories, the contributions to chi-square are $(2)^2/83$ and $(-2)^2/83$, respectively, both of which equal 0.05. Notice that if we had not squared O − E, then these two classes would exactly cancel out each other. Summing over both the values, the chi-square is 0.10. Obviously, this is a small value, indicating the similarity of the observed and expected numbers, but how do we know how large or small it really is?

The significance of a chi-square value can be determined by comparing it to a theoretical value from a chi-square table that has the same degrees of freedom. The term **degrees of freedom** refers to an integer that is generally equal to the number of categories in the data, minus one. In other words, the degrees of freedom indicate the number of categories (phenotypic classes) that are independent of each other. For example, if there are two categories, there is only one degree of freedom. This constraint occurs because, given the total number (say 166 as previously) and the number in one class (say 85), the number in the other class (81) is fixed.

Assuming that the Mendelian segregation hypothesis we are examining is correct, if we drew many, many samples and calculated a chi-square value for each, we would obtain a theoretical distribution of chi-square values. Due to chance, the chi-square would be large in a small proportion of the cases because the observed numbers were quite different from the expected numbers. However, most of the time the observed numbers would be close to those expected, making the chi-square small.

Instead of giving the complete theoretical distribution of chi-square values, we will provide a few important values (table 2.5). These critical chi-square values are exceeded by chance only rarely, given that the underlying hypothesis is correct. There are two different critical theoretical values, 0.05 and 0.01, to which the observed (or calculated) chi-square is generally compared. If the calculated chi-square value exceeds the theoretical values given under 0.05 and 0.01 (for given degrees of freedom), the observations are said to be *statistically significant* or *highly statistically significant,* respectively. Such high chi-square values suggest that the hypothesis used to determine the expected numbers is probably incorrect. The values labeled 0.05 are equaled or exceeded by chance only one in twenty times (0.05) even

though the underlying hypothesis is correct. In other words, by chance, the observed numbers could be quite different from those expected one in twenty times, making the chi-square value this large or larger. If the observed and expected numbers are very different, the chi-square may equal or exceed the numbers in the last column, a value that occurs by chance only one in one hundred times, given a correct hypothesis.

In the example of violet and white flowers, there are two categories, so there is only one degree of freedom. The calculated or observed chi-square value, 0.10, is substantially less than the critical value needed for statistical significance. (With one degree of freedom, it is 3.84.) Therefore, we can say that the observed chi-square value is not significantly different from the expected chi-square value. In other words, the observed numbers are *consistent* with the numbers expected under the principle of segregation. Notice that we did not say this proves the principle of segregation. One cannot prove a hypothesis by such a statistical test. However, because many such crosses have been evaluated and are consistent with the principle of segregation, we can feel confident that this process is responsible for the inheritance of these genes.

Let us go back to the data from Mendel's dihybrid cross and ask whether the observed phenotypic numbers are consistent with the hypothesis of independent assortment. We can use the chi-square test, but first we need to calculate the expected numbers in each category. There were 556 progeny, so if we multiply 556 by $\frac{9}{16}$, $\frac{3}{16}$, $\frac{3}{16}$, and $\frac{1}{16}$, we obtain the expected numbers in the four classes (table 2.6). Using the chi-square formula, and then summing over the four categories, the chi-square value is 0.48. Again, like the single-gene example, this is a very small chi-square value. However, in this case, there are four categories, making the degrees of freedom (number of categories minus one) equal to 3. Looking at table 2.5, we find that the calculated chi-square value would have to be equal to, or greater than 7.82

Table 2.6 Chi-Square of a Dihybrid Cross

Seed Type	Observed Numbers	Expected Numbers	$\dfrac{(O-E)^2}{E}$
Round, yellow	315	$556(\%) = 312.8$	$\dfrac{(2.2)^2}{312.8} = 0.02$
Wrinkled, yellow	101	$566(\%) = 104.2$	$\dfrac{(-3.2)^2}{104.2} = 0.10$
Round, green	108	$556(\%) = 104.2$	$\dfrac{(3.8)^2}{104.2} = 0.14$
Wrinkled, green	32	$556(\%) = 34.8$	$\dfrac{(-2.8)^2}{34.8} = 0.22$
	$\overline{556}$	$\overline{556}$	$\chi^2 = \overline{0.48}$

for the observed values in order to be significantly different from expectations with 3 degrees of freedom. Obviously, 0.48 is much less than 7.82, so we can conclude that the observed numbers are consistent with the expected numbers under the hypothesis of independent assortment.

Remember that when calculating chi-square values, observed and expected numbers, not frequencies, are used. Also, as a rule of thumb for using a chi-square test, the expected numbers in a category should be five or greater. When the expected number is less than five, a modification of the formula should be used, or else the observed chi-square value may be somewhat inflated.

An interesting historical note is that all of Mendel's results fit expectations very closely and consequently give very low chi-square values. For instance, in the dihybrid example, given a correct hypothesis, the chi-square value should be less than 0.58 only 10% of the time. However, the observed value was 0.48, less than 0.58. Several statisticians have suggested that Mendel (or some of his student helpers) biased the results to conform to expectations of the principles of segregation and independent assortment. We will probably never know whether this is true, but it should not detract from the genius of Mendel's insights into inheritance. We should note too that the workers who repeated Mendel's experiments also obtained results extremely close to expectations (see table 2.2).

A useful statistical test of the goodness of fit between observed and expected numbers is the chi-square. Chi-square values larger than a critical value indicate that the underlying hypothesis used to generate the expected numbers may be incorrect. All of Mendel's results fit expectations very closely and thus yield very low chi-square values.

MENDELIAN INHERITANCE IN HUMANS

Appendix C, Module 12

The same patterns of inheritance occur in humans as in peas, *Drosophila,* or cats. At this point 6,678 single traits and diseases for humans are categorized in the 1994 edition of Victor McKusick's *Mendelian Inheritance in Man.* There is an extensive worldwide research effort to locate human disease genes, and a new gene is described at least once a week in the scientific literature. Determining the mode of inheritance is somewhat more difficult in humans because the sibship sizes are relatively small; that is, there are generally fewer than ten children in a family, whereas hundreds of seeds are produced from a given cross in peas. However, using Mendel's ratios, the same inheritance patterns can be seen even in the pedigrees of small families.

Before looking at Mendelian inheritance in humans, we must introduce the basic symbols used in human pedigree analysis. As shown in figure 2.15, a female or a male in a pedigree chart is indicated by a circle or a square, respectively. Diamonds are used for individuals whose sex has not been determined or in cases where sex is unimportant for genetic analysis. When an individual has a given trait or genetic disease, the symbol is filled in or is a different color. Mating individuals are joined by a single horizontal line. A mating between relatives, called **consanguineous**

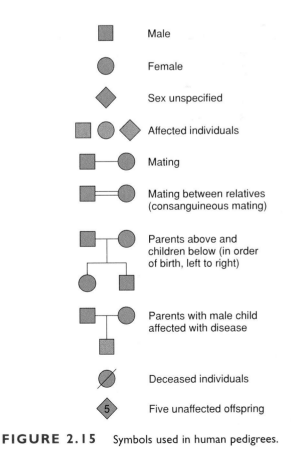

FIGURE 2.15 Symbols used in human pedigrees.

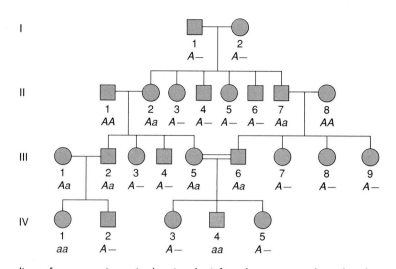

FIGURE 2.16   A human pedigree for a recessive trait, showing the inferred genotypes, where *A* and *a* are the dominant recessive alleles, respectively. *A*— indicates that the state of the second allele is not known.

(Latin: *consanguineus* = of the same blood), is indicated by a double horizontal line. Siblings are joined by a line below their parents and are listed in order of birth, from left to right. A pedigree chart begins with the oldest generation at the top; successive generations are identified by Roman numerals (figure 2.16).

RECESSIVE TRAITS

Let us examine a pedigree exemplifying the inheritance pattern for a recessive trait (figure 2.16). Two recessive diseases in humans are **albinism,** in which individuals lack the pigment melanin (figure 2.17), and **cystic fibrosis,** a severe disorder that affects the production of mucus in the pancreas and lungs. (McKusick lists a total of 1,730 entries for recessive loci.) Cystic fibrosis is the most common recessive disease among Caucasians, occurring in about one in 2,500 births. In the pedigree in figure 2.16, two individuals, IV-1 and IV-4, have the recessive trait or disease. Notice that the parents of the affected individuals, III-1, III-2, III-5, and III-6, are themselves unaffected, a typical observation in pedigrees for recessive traits. Obviously, the trait appeared because the parents are heterozygotes whose recessive alleles were masked by the dominant alleles (or carriers). Notice also that either males or females can have the recessive phenotype.

In a large proportion of recessive disease cases, the parents of the affected individual are relatives. For example, the parents of individual IV-4 are first cousins; that is, III-5 and III-6 have the same grandparents, I-1 and I-2. (We will discuss the basis for the increase in recessive diseases from consanguineous matings in chapters 16 and 17.) The other grandparents of III-5 and III-6, as well as some other relatives, are omitted in order to simplify the pedigree.

Several other aspects of recessive pedigrees are important. First, if both parents are affected or have the recessive genotype (not shown in figure 2.16), all progeny should be affected. Second, among the sibs of an affected individual, given that both

FIGURE 2.17   A Hopi Indian girl with albinism.
Field Museum of Natural History, Chicago, Neg. 118.

parents are unaffected (heterozygotes), approximately ¼ should be affected, according to the principle of Mendelian segregation. Finally, the probability that an additional child born to this family would have the trait is also ¼. The attributes of pedigrees for recessive diseases are summarized in table 2.7. However, because human families are usually small, these proportions are generally seen only when there are a large number of sibships.

Calvin and Hobbes

by Bill Watterson

Table 2.7 Patterns for Human Genetic Diseases

Recessive Diseases

(a) Parents are generally unaffected.
(b) Approximately one-quarter of the siblings of an affected individual are affected (assuming both parents are carriers).
(c) The probability that an additional child will be affected is ¼.
(d) Recessive traits often result from consanguineous mating.
(e) Two affected parents cannot have unaffected offspring.

Dominant Diseases

(a) Trait occurs every generation (at least one parent is affected).
(b) When one parent is affected, approximately one-half of progeny are affected.
(c) The probability that an additional child will be affected is ½.
(d) Two unaffected parents do not produce affected offspring.
(e) Two affected parents can produce unaffected offspring.

In any given sibship, the numbers of different types of offspring may not be close to Mendelian ratios. To understand this, we can use the expression for binomial probabilities given previously. For example, let us assume we have a sibship of size six in which two individuals are affected. We will designate one affected individual as the **proband** (or index case), the individual who was first identified as having the recessive trait or disease. What is the probability that one of the five remaining sibs will also be affected? Given the Mendelian probabilities of segregation, let us call ¾ dominant (p) and ¼ recessive (q). Thus:

$$P(4 \text{ unaffected, } 1 \text{ affected offspring}) = C(\tfrac{3}{4})^4(\tfrac{1}{4})^1$$

In this case, C = 5, thus:

$$P(4A-, 1aa) = 5(\tfrac{3}{4})^4(\tfrac{1}{4}) = 0.396$$

This makes the probability of such a sibship quite large. In fact, it is the most likely type of sibship containing five sibs. The next

most probable sibships are $P(5A-, 0aa) = 0.237$ and $P(3A-, 2aa) = 0.264$.

Two other points should be made about recessive trait pedigrees before we continue. First, note that unrelated individuals who do not have affected offspring, such as individuals II-1, and II-8 in figure 2.16, are assumed to be homozygous for the dominant allele. Such individuals could possibly be heterozygotes, but because most recessive alleles are rare, it is assumed that unrelated individuals with the dominant phenotype are homozygotes.

Second, given an affected offspring in a family, we know that the unaffected parents are heterozygous (barring mutation, which is extremely rare). Using segregation proportions, we can then calculate the conditional probability that an unaffected sib in such a family is heterozygous. The probability that a sib is heterozygous is ½ and the probability that a sib is unaffected is ¾, so:

$$P(\text{Heterozygous} \mid \text{unaffected}) = \frac{P(\text{heterozygous})}{P(\text{unaffected})}$$

$$= \frac{\tfrac{1}{2}}{\tfrac{3}{4}}$$

$$= \tfrac{2}{3}$$

In other words, two-thirds of the time, unaffected sibs of individuals with a recessive disease are carriers (heterozygotes) for the disease allele.

DOMINANT TRAITS

Now let us examine a pedigree that illustrates the inheritance pattern for a dominant trait. **Achondroplasia,** a common type of dwarfism (figure 2.18), and **Huntington's disease,** a degenerative neurological disorder, are examples of dominant disorders. (McKusick lists a total of 4,458 entries for dominant loci.) For dominant traits, at least one of the parents of an affected individual is also affected, and the trait is present in every generation, as shown in the pedigree in figure 2.19. This pattern results because an offspring must obtain the dominant allele from one of its parents. As a result, unaffected individuals do not produce affected offspring (unless there is a rare new mutation). Notice also that either males or females can have the dominant phenotype. Unlike a recessive trait, two affected parents can produce an unaffected offspring, because both parents may be heterozygotes and consequently have recessive alleles. As expected according to Mendelian segregation, when one parent is affected, half the sibs of the proband are expected to be affected. Given such a family, the probability that an additional child will be affected is ½. (See table 2.7 for a summary about dominant diseases in pedigrees.) Again, because of the small sibship size in humans, sibships in which the proportion of affected individuals

is not ½ occur with probabilities that can be calculated using the binomial formula.

The pedigree patterns listed in table 2.7 are not completely true for some diseases, and there are other types of inheritance patterns as well. We will discuss several such situations in chapter 3 and illustrate how they may affect pedigrees. Furthermore, the pedigrees given in figures 2.16 and 2.19 are large ones, allowing the mode of inheritance to be determined with reasonable ease. In many pedigrees having fewer individuals, the mode of inheritance may not be clear.

Mendelian principles apply to traits in humans, but because human families are relatively small, determining modes of inheritance can be difficult. However, some general patterns have been discerned regarding the passing on of recessive and dominant traits.

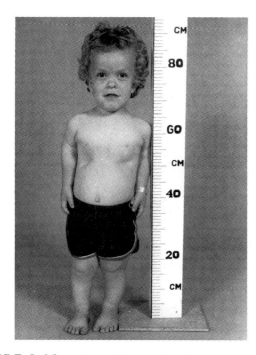

FIGURE 2.18   A boy with achondroplasia, a common type of dwarfism.

Courtesy Dr. Judith Hall.

WORKED-OUT PROBLEMS

PROBLEM 1

A number of researchers repeated Mendel's crosses in order to confirm his genetic principles. For example, both Correns and Tschermak crossed yellow and green varieties of peas and self-fertilized the subsequent F_1 progeny to get F_2 progeny arrays. Correns observed 1,394 yellow and 453 green, while Tschermak observed 3,580 yellow and 1,190 green. Are these values consistent with the principle of segregation? (Calculate χ^2 values to determine consistency.)

SOLUTION

Set up a table with the observed and expected numbers for the two crosses, then calculate the χ^2 values:

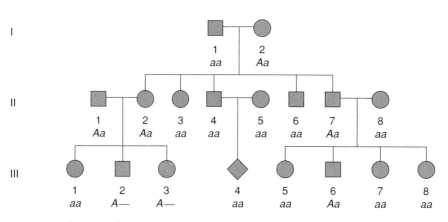

FIGURE 2.19   A human pedigree for a dominant trait, showing the inferred genotypes, where A and a are the dominant and recessive alleles, respectively. A— indicates that the second allele is not known.

		Observed Numbers	Expected Numbers	$\dfrac{(O-E)^2}{E}$
Correns	Yellow	1,394	$1,847 \times \frac{3}{4} = 1,385.25$	0.055
	Green	453	$1,847 \times \frac{1}{4} = 461.75$	0.166
		1,847	1,847	$\chi^2 = 0.221$
Tschermak	Yellow	3,580	$4,770 \times \frac{3}{4} = 3,577.5$	0.002
	Green	1,190	$4,770 \times \frac{1}{4} = 1,192.5$	0.032
		4,770	4,770	$\chi^2 = 0.074$

In this case, there is 1 degree of freedom (2 categories minus 1 = 1 degree of freedom). As a result, the χ^2 values are not significant because both are less than 3.84 (see table 2.5), making the observations consistent with the principle of segregation.

PROBLEM 2

Using the same traits Mendel used in his experiments to demonstrate independent assortment, assume that a true-breeding pea plant with round, yellow seeds was crossed to one with wrinkled, green seeds. One of the progeny from this cross was then crossed back to plants from the wrinkled, green line. From this mating, the progeny types were 42 round, yellow; 45 round, green; 46 wrinkled, yellow; and 45 wrinkled green. What are the genotypes of all the phenotypes, parents, and progeny? What are the expected numbers of the four progeny types assuming independent assortment? What is the χ^2 value resulting from this progeny array, and is it statistically significant?

SOLUTION

Parents	Round, yellow $RR\,YY \times$ Wrinkled, green $rr\,yy$			
Backcross	*Phenotype*	*Genotype*	*Observed*	*Expected*
progeny	Round, yellow	$Rr\,Yy$	42	44.5
	Round, green	$Rr\,yy$	45	44.5
	Wrinkled, yellow	$rr\,Yy$	46	44.5
	Wrinkled, green	$rr\,yy$	45	44.5

$$\chi^2 = \frac{(42-44.5)^2}{44.5} + \frac{(45-44.5)^2}{44.5} + \frac{(46-44.5)^2}{44.5} + \frac{(45-44.5)^2}{44.5}$$

$$= 0.14 + 0.01 + 0.05 + 0.01$$

$$= 0.21$$

With three degrees of freedom, the number of classes minus one, this value is not statistically significant. That is, 0.21 is less than 7.82, the value in table 2.5 that it would have to exceed to be statistically significant.

PROBLEM 3

A couple had a child with cystic fibrosis, a recessive disease. If we assume that C is the normal allele at this locus and c is the allele that causes cystic fibrosis as a homozygote, what are the genotypes of the unaffected parents and the child? What is the probability that the couple's second child would not have cystic fibrosis?

Treatments are now available for individuals with cystic fibrosis, so that this couple's child could grow up and have children. If we assume that the affected child has children with an individual who does not have the cystic fibrosis allele (as is true for about 95% of all people), what is the chance that these children would be affected?

SOLUTION

The parents are both heterozygotes, Cc, and the child is cc. The probability that the second child would be unaffected, that is, either CC or Cc, is 0.75. If the grown-up affected child with the genotype cc mates with an unaffected person with the genotype CC, then all of their children would be Cc, unaffected heterozygotes.

SUMMARY

Mendel first demonstrated that inheritance is based on the segregation of alleles at a gene. Through self-fertilization of F_1 garden pea plants obtained from crosses between plants with different characteristics, he observed a 3:1 ratio of dominant to recessive phenotypes in the F_2 progeny. The principle of segregation he developed from these experiments states that heterozygous individuals produce equal proportions of gametes containing the two alleles. Mendel's second principle, that of independent assortment, states that alleles at different genes assort independently of each other, resulting in a 9:3:3:1 ratio in F_2 progeny from a dihybrid cross.

Several basic theories of probability are useful in understanding Mendelian genetics. These are: (1) the sum rule, which states that the probability of either of two mutually exclusive events occurring is the sum of their individual probabilities; (2) the product rule, which states that the joint probability of two events occurring is the product of their individual probabilities; (3) the theory of conditional probability, which is the probability of a specific event divided by the probability of a more general event; and (4) the binomial formula, which figures the probability of x events of a particular type out of a total N.

A useful statistical test of the goodness of fit between the numbers observed and the numbers expected according to Mendel's principles is the chi-square. Chi-square values larger than a critical value indicate that the underlying hypothesis used to generate the expected numbers may be incorrect.

Mendelian principles apply to traits in humans. Human pedigrees show certain patterns that allow us to differentiate recessive and dominant traits.

PROBLEMS AND QUESTIONS

Problems and Questions with Links to Software

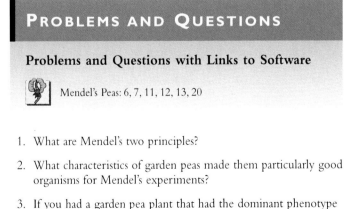

Mendel's Peas: 6, 7, 11, 12, 13, 20

1. What are Mendel's two principles?

2. What characteristics of garden peas made them particularly good organisms for Mendel's experiments?

3. If you had a garden pea plant that had the dominant phenotype R, what cross would you make to determine if it has genotype RR or Rr? Explain your answer.

4. Assume that you have a large container filled with 500 red marbles and 500 blue marbles and that the marbles are thoroughly mixed. (*a*) If you reached in without looking and picked out one marble, what is the probability that it would be red? (*b*) If you put that marble back and picked out two more in the same manner, what is the probability that one would be red and one would be blue?

5. Maple syrup urine disease is a rare disease that runs in families. It is so called because a metabolic error causes the urine of affected babies to have an odor like maple syrup. The babies die at an early age if not treated. The parents are almost never affected with these symptoms. If we assume that the disease is caused by a single gene, is it a recessive or a dominant gene? Explain your answer.

6. The F_2 numbers of the dominant and recessive phenotypes for Mendel's seven different traits are given in table 2.1. Calculate the chi-square values for axial versus terminal flowers and for tall versus short stems. Are these values consistent with Mendelian expectations using the chi-square test?

7. If the F_1 yellow-seeded plants from Mendel's experiments (see table 2.1) were crossed to green-seeded plants, what proportion of the progeny would be expected to be yellow? If F_2 yellow-seeded plants from self-fertilized F_1 plants (table 2.1) were crossed to green plants, what proportion of the plants would produce only yellow progeny?

8. A student has a penny, a nickel, a dime, and a quarter. She flips them all simultaneously and checks for heads or tails. What is the probability that all four coins will come up heads? She again flips all four coins. What is the probability that she will get four heads both times? What probability rule did you use to determine your answer?

9. The first child of two normally pigmented parents has albinism, a recessive trait that results from a lack of the pigment melanin. (*a*) Given that the normal allele is *A* and the albino allele is *a,* draw this pedigree and label both the phenotypes and the genotypes. (*b*) What is the probability that a second child from these parents will be an albino? What is the probability that the second child will be a carrier of the albino allele? Given that the second child is unaffected, what is the probability that he or she is a carrier?

10. Assuming that the probability of producing a female child is 0.5, what is the probability of having three girls and one boy in a family of four? Calculate this value both by writing out all the possible birth orders that can lead to such a family and by using the binomial formula. The actual proportion of girls at birth is about 0.49 in several human populations. Given this value, what is the probability of a family having three girls and one boy?

11. Mendel crossed purebred wrinkled, green-seeded plants with purebred round, yellow-seeded plants. Then he crossed the F_1 progeny to purebred wrinkled, green-seeded plants and observed 31 plants with round, yellow seeds, 26 with round, green seeds, 27 with wrinkled, yellow seeds, and 26 with wrinkled, green seeds. (*a*) Assuming independent assortment, what proportion of these types would be expected? (*b*) Calculate the appropriate chi-square and determine whether it is consistent with Mendelian expectations.

12. Using the forked-line approach and given the cross between two cats with the genotypes *ll Ss dd* and *Ll Ss Dd,* what is the probability of a cat having the genotype *ll ss dd?* What is the probability of a cat having the short-hair, white-spotted, nondilute phenotype?

13. Using the forked-line approach in a trihybrid cross involving three traits, where the parents are both *Aa Bb Cc,* what is the probability of their producing an offspring recessive for all three traits? What is the probability of producing an offspring with the phenotype *A– bb C–?*

14. Given the pedigree shown here for a rare genetic disease, would you conclude that the trait is recessive or dominant? Why? What do you expect the genotypes of individuals II-3, II-4, and III-2 to be, assuming *A* and *a* are the dominant and recessive alleles, respectively? Given only this information, what is the probability that I-2 is heterozygous? What is the probability that III-1 will have the trait?

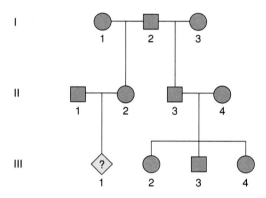

15. Given the following pedigree for a rare genetic disease, would you conclude that the trait is recessive or dominant? Why? What do you expect the genotypes of II-1, II-2, and III-3 to be, assuming that *B* and *b* are the dominant and recessive alleles, respectively? What is the probability that a child of II-3 will be affected? Why?

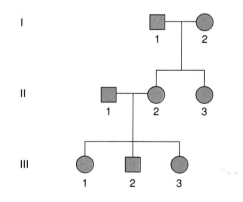

16. A woman with the rare recessive disease phenylketonuria (PKU), who had been treated with a diet having low levels of the amino acid phenylalanine, was told that it was unlikely her children would inherit PKU because her husband did not have it. However, her first child had PKU. What is the most likely explanation? Assuming this explanation is true, what would be the probability of her second child having PKU?

17. In an examination of Mendel's principles, a strain of light brown mice was crossed to a strain of dark brown mice. All F_1 were dark brown. In the F_2, 42 were dark brown and 15 were light brown. Is this consistent with the principle of segregation? (Use a chi-square to check.)

18. The 42 dark brown F_2 mice from question 17 were crossed to light brown mice. Thirty of these mice produced litters composed of both dark and light brown mice, while twelve produced only dark brown mice. Is this consistent with the principle of segregation? (Use a chi-square to check.)

19. What is the probability that, if three identical coins were flipped, all would end up heads? What is the probability that the three coins would *not* be either all heads or all tails?

20. Assume that a cross was made between fruit flies of genotype *AAbb* and those of genotype *aaBB*. (*a*) Give the Punnett square for the expected F_2 progeny types. (*b*) What proportions of *A–B–, A–bb, aaB–,* and *aabb* progeny do you expect in the F_2?

21. In the recessive trait pedigree in figure 2.16, what is the probability that individual IV-2 is a carrier (heterozygote)? What is the probability that a third child of III-1 and III-2 would have the recessive disease?

22. In the dominant trait pedigree in figure 2.19, what is the probability that individual III-2 would be *AA?* If III-1 and III-6 produce an offspring, what is the probability that he or she will have the dominant trait?

23. In a common kind of congenital deaf-mutism, one or both parents are usually affected. Do you think it is a recessive or a dominant trait? What other criteria would you use to distinguish between these two types of deaf-mutism?

24. In the F_2 progeny from two purebred lines of rats, 50 were doubly dominant, *A–B–*. If these were tested by crossing to *aabb* rats, how many would you expect to be double heterozygotes?

25. Assume that one-twelfth of all individuals are born each month. What is the probability that two particular individuals were both born in April? What is the probability that two particular individuals were both born in the same month?

26. A geneticist crossed true-breeding garden peas that were tall (having long stems) with ones that were short (having short stems). All the F_1 offspring were tall. The F_1 were then allowed to self-fertilize. In the F_2 progeny, 153 were tall and 48 were short. Assume that the genotype for the short plants was *aa* and that *A* allele is associated with the tall phenotype. (*a*) What are the genotypes of the parents, the F_1 and the F_2 for the various phenotypes? (*b*) What are the observed proportions of tall to short plants in the F_2 progeny?

27. Some of the F_1 seeds in the experiment discussed in question 26 were crossed to individuals of the short true-breeding line. What are the genotypes in these parents and the expected proportions of genotypes and phenotypes in the progeny?

28. The cob of corn in figure 2.8 represents the F_2 progeny array for two genes. (*a*) First, at the color gene, an *Rr* (purple) plant was self-fertilized to produce an F_2 array expected to be ¾ purple (*RR* or *Rr*) and ¼ yellow (*rr*). Count all the kernels you can see in the photo and calculate the purple to yellow ratio. (*b*) Second, at the second gene, which influences kernel shape, an *Ss* (normal) plant was self-fertilized to produce an F_2 array expected to be ¾ normal (*SS* or *Ss*) to ¼ shrunken (*ss*). Again, count all of the kernels you can see in the photo to calculate normal to shrunken ratio. (*c*) Finally, count all the kernels, categorizing them by both color and shape. Determine the ratio of purple-normal, purple-shrunken, yellow-normal, and yellow-shrunken phenotypes. Do the ratios of (*a*) and (*b*) come close to that expected for segregation? Does the ratio for (*c*) approximate that expected for independent assortment? Calculate the chi-square values for all three.

29. Given that the parent is genotype *RR,* what is the probability of an *R* gamete? Given that the parent is *Rr,* what is the probability of an *R* gamete? Given that there is a mating between two individuals both with genotype *Rr,* what is the probability that an offspring will be *RR?*

30. (*a*) Given that one parent is *RrYy,* what is the probability of the gametes *RY, Ry, rY,* and *ry?* (*b*) If the other parent is *rryy,* what is the probability that an offspring will be *RrYy? Rryy? RRYy?*

Answers appear at end of book.

Extensions and Applications of Mendelian Genetics

The phenotype of an individual is what is perceived by observation: the organism's structures and functions—in short, what a living being appears to be to our sense organs. . . . The genotype is the sum total of the hereditary materials received by an individual from its parents and other ancestors.

Theodosius Dobzhansky

American geneticist

Learning Objectives

In this chapter you will learn:

1. The differences between complete dominance, incomplete dominance, and codominance.
2. The effects of pleiotropy and penetrance.
3. How there can be multiple alleles at a gene.
4. The pattern of inheritance of sex-linked genes.
5. The patterns of inheritance when multiple genes affect a trait.
6. The genetic basis of quantitative traits.
7. How to estimate heritability.
8. The basis of genetic counseling.
9. The basis of paternity exclusion.

Τhe elegance of Mendel's experiments was partly due to the complete consistency between his observations and the hypotheses he developed. However, after Mendel's work was rediscovered, it became clear that the simple Mendelian model did not adequately predict experimental observations in all situations. In this chapter, we will examine some of the more complex phenomena that make it necessary to adopt an expanded view of simple transmission, or Mendelian, genetics. In addition, we will introduce the topics of genetic counseling and paternity analysis, which utilize the principles of Mendelian genetics. In chapter 5 we will discuss genetic linkage, which can also be viewed as an extension of Mendelian principles.

DOMINANCE

All seven traits that Mendel selected had one allele dominant over the other, so that when he examined the F_2 progeny, he found a 3:1 dominant to recessive ratio of the phenotypes. (As in chapter 2, we will use the terms *dominant* and *recessive* to refer to alleles as well as to traits.) In fact, Mendel suggested that dominance was a general characteristic of traits, similar to the principles of segregation and independent assortment. However, shortly after 1900, several different patterns of dominance were documented.

COMPLETE DOMINANCE

In most situations, the normal, or wild-type, allele is completely dominant over mutant alleles. The basis for this will be more clear when we later examine the biochemical products of genes and find that one functioning copy of a gene is generally enough to produce sufficient gene product for a wild-type phenotype. For example, the albino allele in humans does not produce functioning gene product, and when an individual is homozygous for this allele, the biochemical pathway for the pigment melanin is blocked. However, one copy of the wild-type allele, as is present in heterozygotes, allows sufficient production of the pigment melanin so that normal pigmentation results.

In some instances, the mutant allele is dominant over the wild-type; that is, the wild type is recessive. There are 4,458 entries representing dominant loci in McKusick's 1994 catalog of human traits, many of them rare diseases. For example, the most common type of dwarfism, achondroplasia, is dominant, so that heterozygotes exhibit the mutant phenotype. In a subsequent section of this chapter, we will discuss some aspects of Huntington's disease, a fatal neurological disorder that is also dominant.

INCOMPLETE DOMINANCE

When Carl Correns, one of the rediscoverers of Mendel's principles, experimentally crossed red-flowered four-o'clocks with white-flowered ones, the F_1 plants were unlike either parent, being an intermediate pink instead (figure 3.1). When he

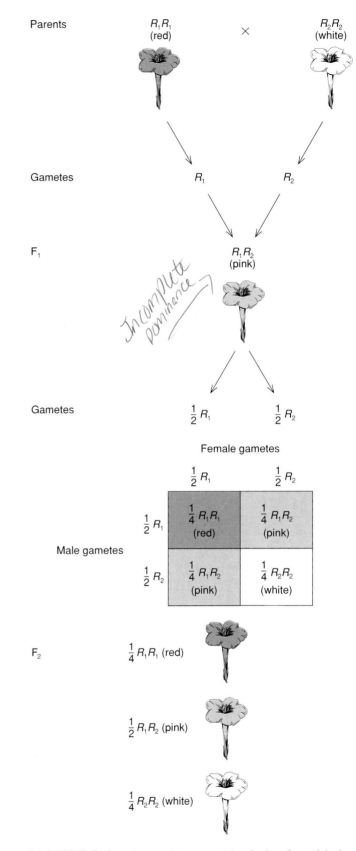

FIGURE 3.1 A cross between red and white four-o'clocks, with the F_2 progeny showing incomplete dominance.

allowed pink F_1 plants to self-fertilize, the F_2 ratio of phenotypes was 1 red : 2 pink : 1 white, or ¼, ½, ¼. Although at first glance these results appear to deviate from Mendel's, what Correns actually observed in the F_2 was a 1:2:1 phenotypic ratio corresponding to the 1:2:1 genotypic ratio expected, based on Mendelian principles. Because neither parental phenotype was fully expressed in the F_1, the trait is said to exhibit **incomplete dominance;** that is, all genotypes have different phenotypes, with the heterozygote's phenotype intermediate between the two homozygotes. The notation in figure 3.1 indicates this lack of complete dominance by using capital letters with different subscripts, R_1 for the red allele and R_2 for the white.

The type of dominance for a given genetic variant is not always easy to designate, however. For example, in what appears to be complete dominance, the observable phenotype of the heterozygote is the same as that of the dominant homozygote, but may show incomplete dominance on a biochemical level. For example, RR homozygotes and Rr heterozygotes at the gene affecting seed shape in garden peas both have the same round seed phenotype. However, these genotypes differ biochemically: the heterozygote has starch grains that are intermediate between the two homozygotes in type and amount. For this reason, we must be specific about the level of the trait—for example, gross phenotypic or biochemical—we are describing.

CODOMINANCE

In some cases, the traits associated with both alleles are observable in the heterozygote. Such alleles are said to exhibit **codominance.** For example, individuals heterozygous for **red blood cell antigens,** genetically determined chemicals on the membranes of red blood cells, often exhibit properties of both alleles. A well-known case is type AB in the ABO blood group system, which has the antigens from two different alleles (see the discussion of multiple alleles later in this chapter for more about ABO). Another blood group system, called MN, has two alleles, generally denoted as L^M and L^N, and three genotypes, $L^M L^M$, $L^M L^N$, and $L^N L^N$. (Sometimes these genotypes are called *MM, MN,* and *NN,* respectively.) Table 3.1 gives the six possible mating combinations between these genotypes, assuming

reciprocal matings are equivalent. The expected proportions of genotypes in the offspring are also shown, and it can be seen that they follow the Mendelian segregation pattern. In chapters 16 and 17, we will discuss naturally occurring allelic forms that determine some biochemical properties of proteins, most of which are also codominant.

Although the example of flower color in four-o'clocks illustrates incomplete dominance on a gross phenotypic scale, close examination of the pink flowers of heterozygotes show that both red plastids and white plastids are present. (Plastids are a type of cell organelle.) In other words, for this trait we find incomplete dominance on the gross phenotypic level and codominance on a subcellular level.

Dominance occurs in different forms. It may be complete, as found by Mendel, where either the mutant or the wild-type allele is dominant. It may be incomplete, where the heterozygote has a phenotype in between the two homozygotes. Or two traits may be codominant, where both alleles are expressed in the heterozygote.

LETHALS

The expectations of segregation given in chapter 2 are true for zygotes (newly fertilized eggs). However, for many traits, the zygotic ratio is modified by differential mortality (that is, survival) as the progeny develop. For example, the initial 3:1 dominant to recessive ratio expected in progeny from two heterozygotes for the sickle-cell allele, an allele that causes sickle-cell disease when homozygous, changes to much less than 25% anemics in adults because of the higher mortality for the sickle-cell homozygotes.

Humans who suffer from genetic diseases have higher mortality than normal individuals in nearly all cases. In other organisms as well, genetic conditions may result in high mortality; for example, genes can cause a lack of chlorophyll in plants resulting in no photosynthesis and mortality of all such seedlings. When a genetic defect causes 100% mortality, it is termed a **lethal allele.** Lethals are generally recessive, resulting in the death of the recessive homozygote. However, some lethals are dominant, as in Huntington's disease, where the heterozygote is affected, although generally not until later in life.

In 1904, shortly after the rediscovery of Mendel's principles, the French geneticist Lucien Cuénot, carrying out experimental crosses on coat color in mice, obtained results that were not consistent with Mendelian predictions. He observed from his experiments that the yellow body color allele was dominant, but that crosses between two yellow mice yielded approximately a 2:1 ratio of yellow to wild-type, rather than the expected 3:1. Furthermore, when he crossed yellow individuals to the recessive wild-type (agouti-colored mice), Cuénot found that all

---

Table 3.1 Expected Proportions of Offspring for the MN Blood Group System

Parents	Offspring		
	$L^M L^M$	$L^M L^N$	$L^N L^N$
$L^M L^M \times L^M L^M$	1	—	—
$L^M L^M \times L^M L^N$	½	½	—
$L^M L^M \times L^N L^N$	—	1	—
$L^M L^N \times L^M L^N$	¼	½	¼
$L^M L^N \times L^N L^N$	—	½	½
$L^N L^N \times L^N L^N$	—	—	1

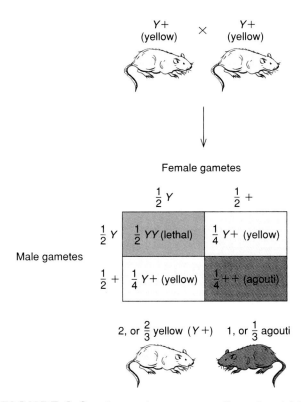

FIGURE 3.2 A cross between two yellow mice, yielding a 2:1 ratio of yellow to agouti-colored mice in the offspring because mice with genotype YY die.

FIGURE 3.3 A tailless Manx cat.
© Robert Pearcy/Animals Animals/Earth Scenes.

yellow mice produced wild-type progeny. He concluded that all yellow mice were heterozygotes and that there were no yellow homozygotes as would have been expected. Later it was suggested that homozygosity for yellow is lethal, and that these individuals died in utero, a fact eventually validated by the histological observation that approximately ¼ of the embryos from yellow × yellow crosses failed to develop.

The explanation for this situation is given in figure 3.2, where Y indicates the dominant, yellow mutant, and + indicates the recessive, wild-type allele. If one were to calculate a chi-square using the principle of segregation expectations for such data, it would give a statistically significant value, suggesting that the observations were not consistent with the Mendelian hypothesis. However, the principle of segregation obviously works properly in this case. The observed ratio of phenotypes differs from expectations because yellow homozygotes die before they can be counted. Many dominant disease alleles in humans, such as achondroplastic dwarfism, also appear to be lethals when they are homozygous. So are many dominant mutants in *Drosophila*, such as *Curly* wing, which we discussed in the last chapter.

The lack of a tail (taillessness) in the Manx cat (figure 3.3) is another trait caused by an allele that has a dominant effect in heterozygotes and is a lethal in homozygotes. This trait is thought to have first occurred as a mutation among domestic cats on the Isle of Man in 1935. The Manx and normal alleles are denoted by M and m, respectively, so that Manx individuals are the heterozygous genotype Mm. True-breeding Manx breeds are not possible because matings between Manx cats, $Mm \times Mm$, produce about ⅓ normal-tailed progeny.

Many genes have alleles that affect the rate of mortality but are not lethals. In general, these are termed **deleterious, or detrimental, alleles.** They result in various levels of mortality, ranging from a few percent more then the wild-type genotype to nearly lethal (almost 100% mortality). Natural selection (selection that does not result from human intervention), which we will discuss in some detail in chapter 16, results in the lower representation in the progeny of genotypes (and their alleles) having higher mortality than of genotypes with lower mortality.

PLEIOTROPY

The genes we have been discussing are generally well known because they substantially affect an obvious trait such as flower color or seed shape. However, many of these genes may also have secondary, or related, effects. For example, the yellow coat color in mice discussed previously is an allele that affects more than one character; that is, it not only produces yellow body color in heterozygotes but affects survival, causing lethality in homozygotes. Another example of multiple effects we mentioned is the gene affecting seed shape in garden peas; this gene also affects starch grain morphology. In fact, many genes affect more than one trait. Mendel himself noticed that the gene causing the flower colors violet and white also influenced seed color and caused the presence or absence of colored areas on the leaves. The phenomenon in which a single gene affects two or more characteristics is called **pleiotropy.**

Many genetic diseases in humans are caused by genes that have pleiotropic effects or a syndrome of diagnostic characteristics. One such disease is phenylketonuria (PKU), which occurs in individuals homozygous for a defective recessive allele. These

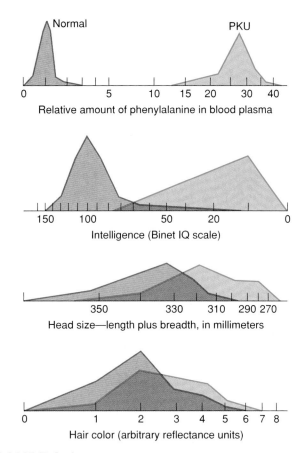

Relative amount of phenylalanine in blood plasma

Intelligence (Binet IQ scale)

Head size—length plus breadth, in millimeters

Hair color (arbitrary reflectance units)

FIGURE 3.4   The values of four different characteristics in PKU and normal individuals, illustrating the pleiotropic effects of this disease. The purple regions show the overlap in the two distributions.

people lack the enzyme necessary for the normal metabolism of the amino acid phenylalanine to the next biochemical product. As a result, when normal and PKU individuals are compared, the level of phenylalanine is much higher in the diseased group (figure 3.4). In addition, this basic biochemical difference results in a suite of other changes in untreated PKU patients, including lower IQ, smaller head size, and somewhat lighter hair. All of these pleiotropic effects in PKU can be understood as a consequence of the basic biochemical defect. For example, in PKU patients, a toxic compound accumulates in the head, causing brain damage and leading to a lower IQ and a smaller head size.

Many mutations of *Drosophila* have multiple phenotypic effects. New mutants are generally recognized by a gross change in the phenotype, such as white eyes or vestigial wings. Invariably, after detailed inspection, less extreme variations in body morphology or other characteristics are found. For example, the *w* (or *white*) allele that results in white-eyed flies also causes many changes in the fly's internal organs, and the *vg* (or *vestigial*) allele that results in vestigial, undeveloped wings also alters its halteres (balancers) and modifies the fly's bristles.

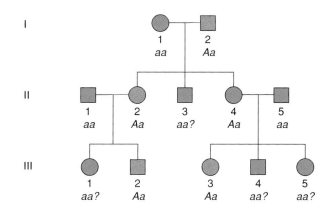

FIGURE 3.5   A pedigree illustrating the pattern of phenotypes that can occur with an incompletely penetrant dominant allele. As before, individuals indicated in purple are affected. Individual II-4 has genotype *Aa* but is not affected by the disease.

PENETRANCE AND EXPRESSIVITY

All of the genes we have considered to this point have a definite genotype-phenotype relationship; for example, a pea plant with genotype *rr* is always wrinkled. In fact, all of the genes Mendel examined, all blood group types, and many human diseases show such an absolute pattern. However, for some genes, a given genotype may or may not show a given phenotype. This phenomenon is described as the **penetrance** of a gene. The level of penetrance can be calculated as the proportion of individuals with a given genotype who exhibit a particular phenotype. When all individuals of a particular genotype have the same phenotype, the gene shows complete penetrance and the level of penetrance is defined as 1.0. Otherwise, the gene is termed incompletely penetrant and the particular level of penetrance can be calculated. For example, if there are eight individuals of a particular genotype and five of them express a diseased phenotype, the level of penetrance is ⅝ = 0.625.

The presence of incomplete penetrance at a gene may cause a phenotype to skip a generation in a pedigree; that is, a dominant trait present in a given generation may skip the first progeny generation but appear again in the second generation. Such a case is illustrated in figure 3.5, a human pedigree where unaffected individual II-4 is both the daughter and mother of affected individuals. This indicates that she has the genotype *Aa,* but because of incomplete penetrance, she does not show the dominant phenotype. For example, the dominant form of retinoblastoma, a disease that causes malignant eye tumors in children, is only about 90% penetrant.

In addition, given that a particular genotype exhibits the expected phenotype, the level of expression, or **expressivity,** may vary. For example, although a gene causes a detectable disease in most individuals with a given genotype, some may be much more severely affected than others. In other words, individuals of a given genotype may be unaffected, may have intermediate symptoms, or may be severely affected. Figure 3.6 gives

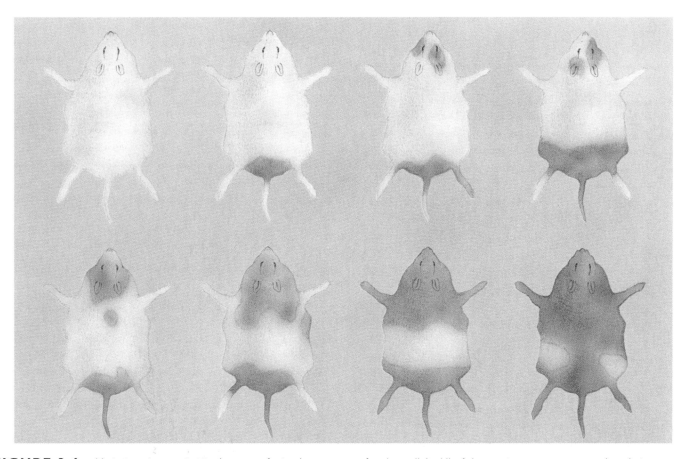

FIGURE 3.6 Variations in spotting in the coat of mice homozygous for the s allele. All of these mice are genotype ss, but their phenotype varies greatly because of variable expressivity.

the variation seen in spotting in the coats of mice homozygous for the *s* allele. Notice that the phenotype may range from virtually no spotting (bottom right) to completely white.

Huntington's disease in humans is a trait that shows variable expressivity in its age of onset. The disease generally occurs in early middle age, with the average age of onset in the thirties, but can occasionally occur very early (at younger than ten years) or in an older individual (figure 3.7). Some people with the disease genotype die of other causes before they express the disease, resulting in incomplete penetrance as well as variable expressivity for this gene.

What are the causes of incomplete penetrance and variable expressivity? In particular instances, it is often impossible to determine causation, but both environmental factors and other genes are known to influence the penetrance or expressivity of a gene. For example, alleles of a number of genes in the fruit fly *Drosophila melanogaster* affect viability and may be lethal at elevated temperatures (say, greater than 28°C) but have little or no effect at lower temperatures. Such genes, known as **conditional lethals,** generally show their lethal effects in extreme environmental situations.

Temperature can also have a dramatic effect on the phenotypes of plants. For example, primroses have red flowers when grown at 24°C but white flowers when grown above 32°C. The

coats of Siamese cats and Himalayan rabbits are darker in color at the paws, ears, and nose because of the lower body temperature at these extremities (figure 3.8). If the animal is placed in a warmer temperature, the new growth of fur at these points is light in color.

In addition, alleles at other genes may influence the penetrance and expressivity of a gene. For example, the level of expression of a trait is generally more similar among relatives than among unrelated individuals, given that the relatives and unrelated individuals are raised in fairly similar environments. Such genes that have a secondary effect on a trait are called **modifier genes** and can substantially influence a phenotype. In many animals, the dilute allele that reduces the intensity of pigmentation when homozygous (such as from black to gray in house cats) is an example of a modifier gene. The extent of taillessness in Manx cats varies from having no tail vertebrae to having a small number of fused tail vertebrae, apparently due to the influence of modifying genes on the expression of the dominant Manx mutant. Mutants of *D. melanogaster* that have been kept in laboratory culture for many years sometimes do not have as extreme a phenotype as when first observed, presumably because of the increased frequency of alleles that modify the mutant phenotype.

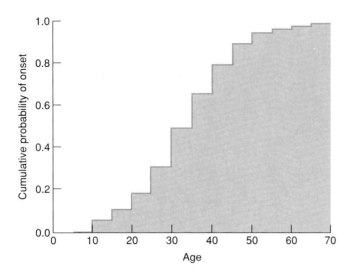

FIGURE 3.7 The cumulative age of onset of Huntington's disease in a human population. All individuals are the diseased genotype *Hh,* but some show signs of the disease much earlier than others.

FIGURE 3.8 The color of Siamese cats is darkened at the paws, ears, and nose because of a lower temperature at these extremities.

© Mary Eleanor Browning/Photo Researchers, Inc.

A related phenomenon, called a **phenocopy,** occurs when environmental factors induce a particular abnormal phenotype that would usually be genetically determined. In this case, an extreme environment may disrupt normal development in a pattern similar to that of a mutant gene. One tragic example occurred in western Europe in the late 1950s and early 1960s when a number of babies were born with severely shortened limbs, an abnormality that resembled a rare recessive genetic disorder. Only after an intensive investigation was it shown that the defect had been environmentally induced by the sleeping pill medication thalidomide, which the mothers of the affected individuals had taken during early pregnancy.

Some alleles of genes may be lethal. Pleiotropy occurs when one gene affects several traits. Some genotypes may not always have the same phenotype, a phenomenon resulting from incomplete penetrance or variable expressivity of the genotype.

MULTIPLE ALLELES

All the genes we have discussed so far (and all the traits Mendel studied) have had only two segregating alleles. However, when detailed analysis of a gene is undertaken in a population, generally more than two alleles are discovered. Although some of these alleles may be rare, the existence of more than two alleles per gene (**multiple alleles**) is relatively frequent in a population.

One commonly known genetic variation resulting from multiple alleles is the ABO blood group system in humans. (There are closely related systems in other mammals such as chimpanzees and cattle.) The different ABO blood group types were discovered in 1900, although their genetic basis was not uncovered through pedigree examination until 1925. In the ABO system, there are three major types of alleles (these classes themselves being heterogeneous), which are generally identified as I^A, I^B, and I^O (or sometimes as A, B, and O). If we assume that female and male gametes can contain one of the three allelic types, then the different genotypes and phenotypes shown in figure 3.9 are possible. Notice that there are six different genotypes—that is, six different ways the three alleles can be combined two at a time.

In general, the number of genotypes possible in a population of a diploid organism with "n" different alleles is found by adding $1 + 2 + 3 + \ldots + n$, where . . . indicates all the integers between 3 and n, or by using the formula $n(n + 1)/2$ (table 3.2). Both of these expressions give the number of different ways n

Female gametes

		I^A	I^B	I^O
Male gametes	I^A	$I^A I^A$ (A)	$I^A I^B$ (AB)	$I^A I^O$ (A)
	I^B	$I^A I^B$ (AB)	$I^B I^B$ (B)	$I^B I^O$ (B)
	I^O	$I^A I^O$ (A)	$I^B I^O$ (B)	$I^O I^O$ (O)

FIGURE 3.9 The six genotypes produced by the three alleles in the ABO blood group system. The four corresponding phenotypes are in parentheses.

Table 3.2 Numbers of Homozygotes, Heterozygotes, and Genotypes, Given Multiple Alleles

Alleles	Homozygotes	Heterozygotes	Genotypes
1	1	0	1
2	2	1	3
3	3	3	6
4	4	6	10
5	5	10	15
6	6	15	21
•	•	•	•
•	•	•	•
•	•	•	•
n	n	$\dfrac{n(n-1)}{2}$	$\dfrac{n(n+1)}{2}$

Recipient

Donor Genotype	Phenotype	A (anti-B)	B (anti-A)	AB	O (anti-A and anti-B)
$I^A I^A, I^A I^O$	A	+	A, anti-A	+	A, anti-A
$I^B I^B, I^B I^O$	B	B, anti-B	+	+	B, anti-B
$I^A I^B$	AB	B, anti-B	A, anti-A	+	A, anti-A B, anti-B
$I^O I^O$	O	+	+	+	+

FIGURE 3.10 The potential for successful blood transfusions, indicated by a + for different ABO donor-recipient combinations. A, anti-A, and B, anti-B, indicate a reaction between A antigens and anti-A antibodies and between B antigens and anti-B antibodies that would result in unsuccessful transfusions.

objects (or alleles) can be combined two at a time when order is not important. For the ABO system with three alleles (n = 3), the number of genotypes is either $1 + 2 + 3 = 6$ or $(3)(4)/2 = 6$, as given in figure 3.9. If there were four alleles, there would be $(4)(5)/2 = 10$ possible genotypes. In addition, the number of homozygotes is equal to the number of alleles (n), and the number of heterozygotes is equal to $n(n-1)/2$ (table 3.2). For example, the ten genotypes resulting from four alleles are composed of four homozygotes and $(4)(3)/2 = 6$ heterozygotes.

Because allele I^O of the ABO system is recessive to both alleles I^A and I^B, there are only four different phenotypes: type A (either genotype $I^A I^A$ or $I^A I^O$), type B ($I^B I^B$ or $I^B I^O$), type AB ($I^A I^B$), and type O ($I^O I^O$). Alleles I^A and I^B are codominant, illustrating that different dominance relationships are possible among the alleles at a given gene.

Differences among ABO blood group types occur because of the presence of different antigens and antibodies. **Antigens** of the ABO blood group system are molecules composed of a protein-sugar combination that occur on the surface of red blood cells. Individuals with alleles I^A, I^B, and I^O have A, B, and neither A nor B antigens, respectively, on these cells. The A and B antigens differ in the type of terminal sugar on the molecules. The antigen from the I^O allele actually lacks this terminal sugar because when this allele is present, no enzyme to add the terminal sugar is produced.

Antibodies are substances that can be produced by the immune system in large amounts as a response to specific external antigens. (Antibodies are not produced against self antigens.) For example, if cells from a type A individual are injected into rabbits, the immune system of the rabbits will produce anti-A antibodies. In humans, individuals of type A have anti-B antibodies; type B have anti-A antibodies; type AB have neither of these antibodies; and type O have both kinds of antibodies. The presence of antibodies that recognize A and B antigens in individuals lacking these antigens appears to be the result of immune

responses to substances produced by bacteria that are similar to these antigens.

In order to transfuse blood successfully from one individual to another, differences between individuals in the ABO blood group system must be taken into account. As an example of transfusion problems that can result from the ABO system, if red blood cells with B antigens (from type B individuals) are transfused to type A individuals, the type B cells will be attacked and eliminated by the anti-B antibodies present in the type A individuals. All combinations of donor and recipient phenotypes are given in figure 3.10. Notice that type O individuals are universal donors; that is, type O blood, because it does not have antigens comparable to the other types, is accepted by all individuals. On the other hand, type AB individuals are called universal recipients, because they do not have anti-A or anti-B antibodies and can therefore accept A, B, AB, or O blood. Type O recipients given any blood but O will suffer dire consequences because O individuals have both anti-A and anti-B antibodies. Type AB individuals can donate blood only to type AB individuals because other blood types have anti-A, anti-B, or both antibodies.

Some genes may have a series of alleles showing sequential dominance; for instance, a number of alleles can affect color in mammals, insects, or mollusks. In some parts of England, the peppered moth, *Biston betularia,* has at least three such alleles. In chapter 16 we will discuss the evolution of these color types, a phenomenon that has been associated with industrial pollution in England. At this locus, the allele that was characteristic of the species before industrial pollution, the pale typical allele (*m*), is recessive. A second allele, the darkly mottled insularia (*M'*), is dominant to the typical. The third allele, the nearly black melanic (*M*), is dominant to both of the other alleles. In other words, there are six genotypes but only three phenotypes: (1) the melanic phenotype, having three genotypes—homozygous melanic (*MM*) and heterozygous between melanic and the other two alleles (*MM'* and *Mm*); (2) the insularia

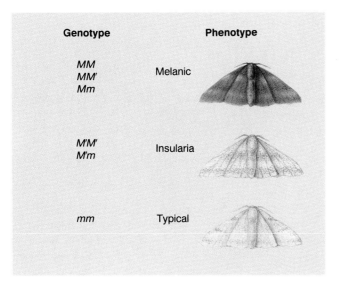

FIGURE 3.11   The six possible genotypes of the peppered moth from all diploid combinations of three alleles are listed in the left column. The resulting phenotypes are shown in the far right column.

	Recipient		
Donor	A1 A28	A2 A10	A1 A9
A2 A2	A2	+	A2
A1 A28	+	A1, A28	A28
A9 A11	A9, A11	A9, A11	A11
A9 A9	A9	A9	+

FIGURE 3.12   The potential for an organ transplant from four donor individuals to three recipients based on *HLA-A*. The + indicates a match. *A2, A9,* and so on indicate the antigens in the donor that would cause rejection.

phenotype, composed of the insularia homozygote ($M'M'$) and the insularia-typical heterozygote ($M'm$); and (3) the typical phenotype, all homozygous recessive typical (*mm*) (figure 3.11). As an example of the types of mating possible, assume that two heterozygotes, a melanic *Mm* and an insularia $M'm$, were crossed. The expected progeny genotypes are one-quarter each MM', *Mm*, $M'm$, and *mm*. However, because of dominance, half the progeny should be the melanic phenotype (MM' and *Mm* genotypes), one-quarter insularia ($M'm$), and one-quarter typical (*mm*).

Some genes, such as the human genes for HLA (human leukocyte antigen, which determines antigens that are on the surfaces of nearly all cells), may have many alleles. For example, gene *HLA-B* has more than thirty different antigenically defined alleles in some populations. These alleles are denoted by A1, A2, A3, A11, and so on for *HLA-A*, and by B5, B7, B8, B12, and so on for *HLA-B*. As a result of this diversity, an extraordinary number of genotypes generally occur at this gene in a population. Again, we can use the formulas in table 3.2 to calculate the number of possible genotypes for genes with different numbers of alleles. *HLA-B,* with thirty alleles, potentially has 30 different homozygotes, 435 different heterozygotes [(30)(29)/2], and a total of 465 different genotypes [(30)(31)/2], truly an extensive array of genotypic variation.

HLA genes are important in determining the potential success of organ transplants and are related to susceptibility to a number of autoimmune diseases, such as insulin dependent diabetes and rheumatoid arthritis. For these genes, all the alleles are codominant, so the number of genotypes equals the number of phenotypes. Matching the genotypes of *HLA-B* (and the *HLA-A* gene, which has nearly twenty different alleles) in the donor and recipient of a kidney increases the likelihood of a successful

transplant. Figure 3.12 gives the *HLA-A* genotype for three potential recipients and four potential donors. Only where there is a + is the donor compatible with the recipient. For all other combinations, one or more antigens present in the donor are not present in the recipient, and the likelihood of a successful transplant is much lower. Because there are so many different genotypes, donors and recipients generally do not match. Because relatives are more likely to share HLA types, often close relatives are examined for HLA to find a suitable organ donor. Even with a suitable donor for the major HLA types, it is still necessary to use immunosuppressant drugs to reduce the organ recipient's immune response to foreign antigens from other unmatched genes and thus achieve a successful transplant.

Genes may have more than two alleles. In diploid organisms, many different genotypes can exist when there are multiple alleles.

SEX-LINKED GENES

Appendix C, Module 12

In chapter 1 we mentioned that most genes are on autosomes—the chromosomes for which each diploid individual has a homologous pair. However, shortly after the rediscovery of Mendel's principles, some experiments with *Drosophila* yielded progeny that differed from Mendelian predictions in the phenotypic proportions of males and females. These initially perplexing results were explained by assuming that the genes in question were not on autosomes but on the X chromosome, which is present in two copies in the female *Drosophila* and in one copy in the male. (These experiments will be discussed in chapter 4.)

Mammals, some insects, and a few plants have this kind of sex chromosome system in which one sex has two homologous chromosomes and the other has two different kinds of chromosomes. (In chapter 4 we will also discuss sex determination.) In mammals, females have two X chromosomes and are designated XX. Males have one X chromosome and one Y chromosome and are designated XY. If we examine the inheritance of genes on these chromosomes, we find patterns different from the autosomal genes examined by Mendel, but the basis of this inheritance is consistent with Mendelian principles.

Let us first examine traits determined by genes on the X chromosome, or **X-linked traits.** McKusick lists 161 identified loci on the human X chromosome. One example is the inheritance of the common type of color blindness, which is caused by a recessive allele located on the X chromosome. We will indicate X chromosomes having the two alternative alleles as X^C and X^c, the normal and color-blind alleles, respectively. In females there are three different genotypes ($X^C X^C$, $X^C X^c$, and $X^c X^c$) in which only the last is color-blind because the normal allele is dominant in heterozygotes. Males have only one X and a Y. Because Y chromosomes do not carry the color blindness gene (the Y has only a few genes allelic to those on the X), the two types of males are $X^C Y$ and $X^c Y$—normal and color-blind, respectively. (Another way to indicate X-linked inheritance is to leave out the X symbol and write the X-linked alleles as we symbolized alleles previously. In other words, a female heterozygous for color blindness would be Cc, and a color-blind male would be cY.)

Figure 3.13 shows a test for color blindness. People who have good color vision will see large numbers in both diagrams, while color-blind individuals will not be able to detect the numbers in part (*b*) of the figure. (We should caution that this color vision test

is only part of a complete test and that people with some color vision may not be able to distinguish the numbers in part (*b*).)

In females, segregation for X-linked genes is the same as that for autosomal genes, while in males, half of the gametes carry the X and half carry the Y. To illustrate this, let us examine the progeny from a cross between a woman heterozygous for color blindness and a man with normal sight (figure 3.14).

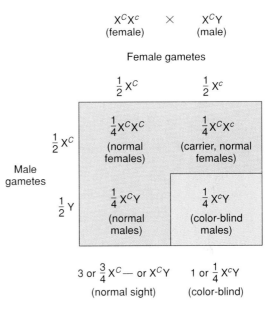

FIGURE 3.14 Probable outcome of a mating between a female heterozygous for color blindness and a male with normal sight.

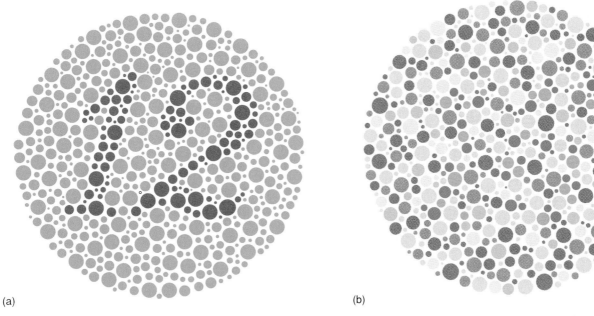

(a)   (b)

FIGURE 3.13 Plates to test for color blindness. All individuals with normal sight will read (*a*) as "12" and (*b*) as "16." Color-blind people will not be able to read (*b*) at all.

The above has been reproduced from *Ishihara's Tests for Colour Blindness* published by Kanehara & Co., Ltd., Tokyo, Japan, but tests for color blindness cannot be conducted with this material. For accurate testing, the original plates should be used.

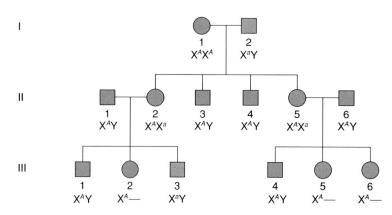

FIGURE 3.15   A pedigree illustrating the pattern of inheritance expected for an X-linked recessive gene.

Notice that all male progeny receive a Y chromosome from their father. We would expect half the progeny to be female, all normal-sighted (although half of them are homozygotes and half heterozygotes), and half the progeny to be male (half color-blind and half normal-sighted). The ratio of normal to affected progeny is still 3:1, but unlike the F_2 progeny of monohybrid cross for an autosomal gene, all the recessive types are males. Obviously, the number of progeny from any one mating in humans is small, but if a number of such matings are lumped together, these expectations are appropriate.

Pedigrees for X-linked traits have particular patterns that generally allow them to be distinguished from pedigrees for autosomal traits. Figure 3.15 shows a pedigree for an X-linked recessive. First, the X-linked recessive affects many more males than females because males need only one copy of the defective allele to express the trait, while females need two. When the allele is rare, it is unlikely that homozygous females will be encountered. Second, the X-linked recessive trait skips generations because males can receive an X-linked allele only from their mother and, as just suggested, very few females exhibit X-linked recessive traits.

Several common types of the recessive genetic disease hemophilia, in which blood does not clot normally, are also determined by genes on the X chromosome. As a result, most individuals with X-linked forms of hemophilia are males who have received the hemophilia allele from their unaffected carrier mothers. Hemophilia afflicted a number of the descendants of Queen Victoria of England, suggesting that she was heterozygous for this allele. Figure 3.16 shows the family of Czar Nicholas II of Russia, whose wife Alexandra was a granddaughter of Victoria. Their son Alexis, pictured in front, had hemophilia and was executed, along with the rest of the family, at age 14.

X-linked dominant traits have a different pattern of inheritance than X-linked recessive traits, as shown in figure 3.17. Notice in this pedigree that both males and females are affected by the trait (sometimes slightly more females are affected), but that the patterns of affected offspring produced by affected individuals differ between the two sexes. Irrespective of sex, approximately half the offspring of affected females, such as I-1 in figure 3.17, are affected, while for

FIGURE 3.16   A photograph of Czar Nicholas II, the Czarina Alexandra, their four daughters, all with normal blood, and their son, Alexis, who suffered from hemophilia.
The Bettmann Archive.

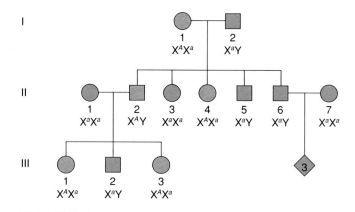

FIGURE 3.17   A pedigree illustrating the pattern of inheritance expected for an X-linked dominant gene.

FIGURE 3.18 Men with hairy ear rims, a putative Y-linked trait.
Courtesy of Dr. S. D. Sigamony.

affected males, such as II-2, all female offspring and no male offspring are affected.

The last type of sex-linked inheritance we will mention is that expected of genes on the Y chromosome. Although there is evidence for Y-linked inheritance for only a few genes in humans (McKusick lists fewer than twenty) and those are genes that determine maleness or are homologous to genes on the X chromosome, other organisms do carry a number of genes on the Y chromosome. One putative Y-linked gene in humans results in hairy ear rims (figure 3.18). However, determining the precise mode of inheritance of this trait is complicated by a variable age of expression. The characteristics of Y-linked traits are straightforward: (1) They are expressed only in males, and (2) they are always passed from father to son. An interesting analogy to Y-linked inheritance is the traditional transmission of family names in English-speaking (and many other) societies. A summary of the characteristics of pedigrees for the different types of sex-linked traits is given in table 3.3.

Some traits are expressed differently in the two sexes but are not sex-linked. First, there are **sex-limited** traits, characters expressed only in one sex. Many such traits are related to reproductive characters, such as milk yield in dairy cattle, egg laying in chickens, or oviposition behavior in insects.

Second, other traits are **sex-influenced;** that is, the genotypes determined by autosomal genes are expressed differently in the two sexes. For example, the genes determining the presence of horns in sheep are expressed differently in males and females. In some sheep breeds, such as the Suffolk, neither sex has horns (they are called polled), while in other breeds, such as the Dorset Horn, both sexes have horns, although the horns of the male are much larger than those of the female (table 3.4). When these two breeds are crossed, the hybrids, which are heterozygous for the autosomal *H* gene, are horned if male but polled if female. In the heterozygotes, the presence of male or female hormones determines whether or not horns develop. It is thought that premature pattern baldness in

Table 3.3 General Patterns of Sex-Linked Inheritance

X-linked Recessive

(a) Usually more males than females are affected.
(b) No offspring of an affected male are affected, making the trait skip generations in the pedigree, always with an unaffected female in the intermediate generation. (An exception to this pattern occurs in the rare instance when the affected male mates with a female carrier, producing affected female offspring.)

X-linked Dominant

(a) Affected males produce all affected female offspring and no affected male offspring.
(b) Approximately half the offspring of affected females are affected, regardless of their sex.

Y-linked Inheritance

(a) Trait is always passed from father to son.
(b) Only males are affected.

Table 3.4 Example of a Sex-Influenced Trait in Sheep

	HH (Suffolk)	*Hh* (Cross)	*hh* (Dorset Horn)
Males	Polled	Horned	Horned
Females	Polled	Polled	Horned

humans (occurring before age thirty-five) follows this same pattern of inheritance; that is, baldness is dominant in male heterozygotes and recessive in female heterozygotes.

> Genes on the X and Y chromosomes have different patterns of inheritance than autosomal genes. X-linked recessive traits occur almost entirely in males and generally skip generations in families.

MULTIPLE GENES AND EPISTASIS

In our discussion of penetrance and expressivity, we mentioned that the phenotype resulting from the genotype at one gene can be altered by alleles at other genes called modifier genes. In fact, many traits are affected by several genes, as we observed by a number of scientists shortly after the rediscovery of Mendel's principles. (Remember that when Mendel examined two genes in his dihybrid crosses, he used genes that affected different traits, not the same trait.)

For a number of traits, different genes affecting the same trait interact so that the phenotype expressed is a function of the particular combination of alleles present at the different genes. The general term for interactions of alleles at different loci to influence a trait is **epistasis** (Greek: *epi* + *stasis* = standing upon). Note that epistasis involves two (or more) genes and differs from dominance, which occurs when the phenotype is determined by different alleles at the same locus.

One of the first observations of epistasis resulted from a cross between two varieties of white-flowered sweet peas, which was done by early geneticists William Bateson and R. C. Punnett. When they crossed the two white-flowered varieties, the F_1 flowers were unexpectedly purple (figure 3.19). When the F_1 individuals were allowed to self-fertilize, 382 purple and 269 white flowers resulted, which is close to a ratio of 9 purple to 7 white in the F_2 progeny. The explanation for this ratio, confirmed by other crosses, was that homozygosity of recessive alleles of two different genes could result in white flowers. When both genes had one or more dominant alleles, the phenotype was purple. Because nine of the sixteen categories in figure 3.19 are like this, there is a 9:7 ratio in the F_2 progeny.

It is useful to understand the biochemical basis of these phenotypes. Pigmentation in the flowers of sweet peas results from chemicals called *anthocyanins*. Production of anthocyanins occurs through a series of metabolic steps catalyzed by enzymes that are themselves products of genes. Whenever a step in this synthetic process is blocked by the absence of a functional enzyme, pigmentation does not occur. Although the exact details of pigmentation synthesis in sweet peas are not known, the general pattern is illustrated in figure 3.20, where the enzyme products of gene C and gene P are both necessary for the production of anthocyanins. When the genotype is cc, the

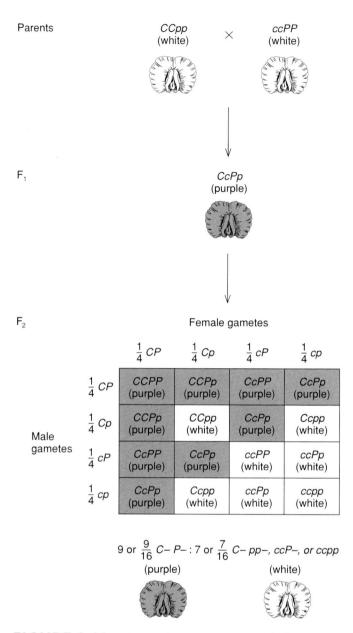

FIGURE 3.19   A cross between two types of white sweet peas, showing the 9:7 ratio that results in the F_2 generation.

first enzyme is not produced; consequently, reaction to the intermediate product is stopped. Likewise, if the genotype is pp, the lack of the second enzyme halts the second metabolic step. In other words, if the genotype is either cc or pp, the synthetic pathway is blocked and no pigments are produced, resulting in a white flower. Such an interaction of genes to jointly produce a specific gene product is a type of epistasis called **complementary gene action.**

As we will discuss later, many recessive diseases in humans are caused by metabolic errors, often called inborn errors of metabolism. If two individuals with metabolic errors for the same trait (but in different steps of a biochemical pathway) mate, their offspring should have a normal phenotype. For example,

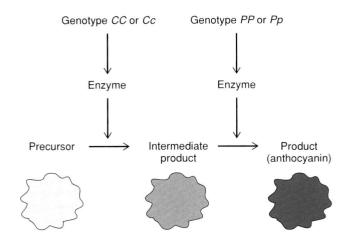

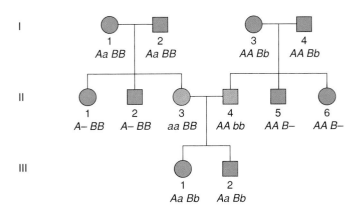

FIGURE 3.20 A diagram illustrating the relationship between genes *C* and *P* and the production of anthocyanins in sweet peas.

FIGURE 3.21 A pedigree illustrating a recessive disease caused by either of two genes.

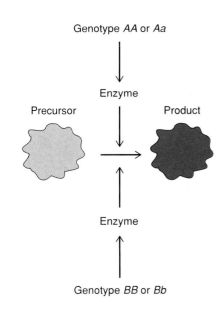

FIGURE 3.22 A diagram illustrating duplicate gene action, in which the enzyme from either gene *A* or gene *B* can catalyze the biochemical reaction.

	A– B–	*A– bb*	*aa B–*	*aa bb*
Dominance with no epistasis	9	3	3	1
Complementary gene action	9	7		
Duplicate gene action	15			1
Dominance with epistasis of *aa* over *B–*	9	3	4	
Dominance with epistasis of *A–* over *bb*	12		3	1

FIGURE 3.23 Examples of epistatic F$_2$ ratios from the cross *Aa Bb × Aa Bb*.

this is what happens when two individuals with deaf-mutism caused by different recessive genes mate and have offspring. In the pedigree in figure 3.21, II-3 and II-4 are affected with a recessive disease because they are homozygous at different loci. Note that the pedigree is different from that of a single recessive gene because no progeny, instead of all progeny, from a mating of two affected individuals are affected.

The 9:7 ratio is only one type of F$_2$ pattern caused by epistasis. In other situations, the F$_2$ ratios may be 9:3:4, 13:3, or 15:1. Notice that the different categories of a 9:3:3:1 ratio can be combined to obtain these ratios. The biochemical basis for the 15:1 ratio can be understood by examining one step of a pathway in which a dominant allele at either of two genes is enough to produce enzymes for catalysis of a given reaction (figure 3.22). Only when there is a double recessive, say *aabb,* is the pathway blocked and the aberrant phenotype expressed. When either of two genes can function to produce the dominant phenotype, this type of epistasis is called **duplicate gene action.** Presumably, the two different genes produce similar gene products, and one of them may have arisen by duplication from the

other gene (see chapter 4). A classic example of duplicate gene action occurs in shepherd's purse, a widespread weed. Round seed pods result when dominant alleles are present at either of two genes, while narrow seed pods are produced only in the absence of dominant alleles at both loci.

Let us diagrammatically summarize the types of epistasis we have discussed. Where there is dominance at both loci and no epistasis, there are four phenotypes, with a 9:3:3:1 F$_2$ ratio as shown in row 1 of figure 3.23. The two types of epistasis we have mentioned, complementary and duplicate gene action, are given next, with their 9:7 and 15:1 ratios, respectively. Two other types of epistasis, in which there are three phenotypic classes, are given in the last rows. Can you find any other phenotypic ratios that are possible by combining the different classes—for example, 12:4 or 9:6:1?

Coat color in mammals is a trait that is affected by a number of different genes. In fact, at least 63 different genes with 148 different alleles affect coat color in mice. Some of the variations in mouse coat color are shown in figure 3.24. We have already

FIGURE 3.24   Mice showing black and brown phenotypes.
© Runk/Schoenberger/Grant Heilman Photography, Inc.

Table 3.5   Genotypes and Resulting Coat Color in Mice[1]

A and B Loci		B and C Loci	
Genotype	Phenotype	Genotype	Phenotype
A– B–	Agouti	B– C–	Black
A– bb	Cinnamon	bb C–	Brown
aa B–	Black	B– cc	Albino
aa bb	Brown	bb cc	Albino

[1]A = agouti; B = brown; C = albino

mentioned two coat-color genes, the yellow gene in mice and the dilute gene in house cats. Other coat-color genes interact in a number of different ways, three of which we will discuss here to illustrate epistatic interaction.

First, the dominant agouti color is controlled by the *A* allele at the *A* gene and results in an overall gray or "salt and pepper" appearance if you look closely. It is the common color of many mammals in natural populations. In actual fact, the individual hairs of agouti animals have a band of yellow, giving an overall brindled grayish color to the coat. When the genotype is the recessive non-agouti (*aa*), the hair shaft is all one solid color.

Second, the alleles at the *B*, or brown coat color, gene result in different colors, depending upon the genotype at the *A* gene (table 3.5). When the *B* allele is present, the phenotype is agouti if the genotype is *AA* or *Aa*, but black if the genotype is *aa*. On the other hand, when the genotype at the *B* gene is *bb*, the phenotype is cinnamon-colored for genotype *A*– and brown for *aa*.

Finally, a third gene, called *C*, allows the formation of coat pigments. (The symbol *C* is used for this albino gene because *A* already represents the agouti gene.) For the *C* gene, the recessive genotype, *cc* prevents the formation of color from other genes (table 3.5). For example, the genotype *bb cc* is albino

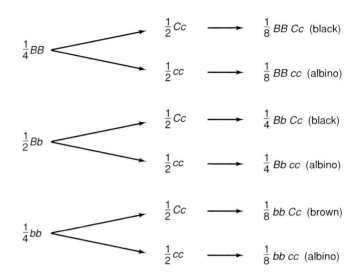

FIGURE 3.25   The expected results of a cross between mouse coat-color genotypes *Bb Cc* and *Bb cc*.

(no pigment) because the *cc* genotype inhibits the formation of the brown color. If two mice of genotype *Bb Cc* are crossed, the 9:3:4 phenotypic ratio in the fourth row of figure 3.23 results. What phenotypic proportions would you expect if *Bb Cc* and *Bb cc* were crossed? Using the forked-line approach (figure 3.25), you can see that the expected ratio of phenotypes in the progeny is ⅜ black, ⅛ brown, and ½ albino or a 6:2:8 ratio.

> More than one gene can affect a single trait. The interaction of two genes to determine a phenotype is called epistasis. Epistatic interaction can result in F$_2$ ratios different from 9:3:3:1.

GENOTYPE-PHENOTYPE RELATIONSHIPS

For the traits that Mendel examined, a given genotype always resulted in a certain phenotype. However, in other situations, the phenotype of a given genotype varies, depending upon the penetrance and expressivity of that genotype. Often such variable expression results from environmental factors, and the exact phenotype caused by a given genotype-environment combination is not easily predicted.

In an ideal situation, the extent of the effect of the environment on a given genotype can be determined by looking at the phenotype of genetically identical individuals raised in different environments. In humans, identical, or monozygotic, twins separated at birth provide such a test situation. However, the best indications of environmental effects come from studies

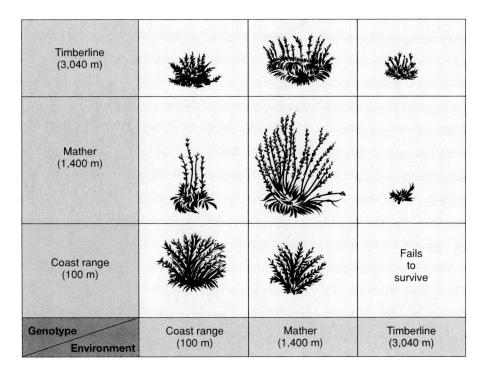

FIGURE 3.26 The phenotypes of *Potentilla* plants coming from three altitudes, and each cloned and grown at three different altitudes.

of species that can be raised in large numbers in controlled environments.

Working with several plant species, J. Clausen and his coworkers performed a classic example of such a study in the 1940s. To determine the effect of the environment on phenotypes, these researchers grew **clones** (genetically identical individuals) from cuttings at different altitudes. For example, they collected *Potentilla glandulosa*, a member of the rose family, from three locations: the Pacific coast range at an altitude of 100 meters; midway up the Sierras at 1,400 meters (Mather); and high in the Sierras at 3,040 meters (Timberline). From these samples, they generated clones by vegetative propagation and grew them at the three different locations. In other words, genotypes collected from the three locations were grown in each of the three environmental conditions (100 m, 1,400 m, and 3,040 m).

Figure 3.26 shows the results of these experiments. Each row represents a genotype from the different areas, and each column indicates the environment in which these plants were grown. From this, it appears that both genetic and environmental factors are important in determining plant growth. For example, the coast range genotypes grow well at both low and intermediate altitudes but cannot survive at the high altitude (bottom row), indicating the importance of environmental factors. On the other hand, the timberline genotype (top row) grows much better than the other genotypes at the high altitude (far right column), indicating the importance of genetic factors.

The way the phenotype for a given genotype changes according to its environment is called the **norm of reaction.** Figure 3.27 gives norms of reaction similar to those of the low- and middle-altitude genotypes in figure 3.26. Genotype *A*, like

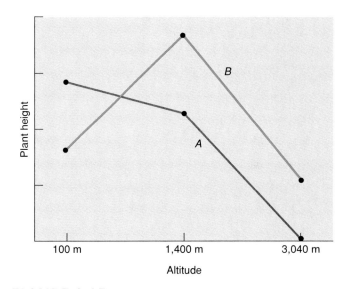

FIGURE 3.27 The norms of reaction for two different genotypes: *A*, the low-altitude (coast range) clone, and *B*, the middle-altitude (Mather) clone.

the low-altitude clone, does best at the low end of the environmental spectrum and worst at the high end. But genotype *B*, like the middle-altitude clone, does best at intermediate altitude levels, falling off in phenotypic value in both the low- and high-altitude environments.

Norms of reaction may differ greatly among genotypes for traits involving behavior. Although we have not discussed behavior as a phenotype until now, a number of mutant genes

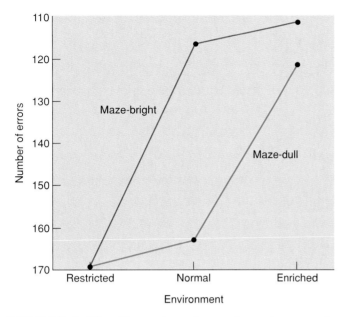

FIGURE 3.28   The numbers of errors in running a maze for two strains of rats, maze-bright and maze-dull, when the two strains are raised in three different environments.
Source: Data from Cooper and Zubek, in *Canadian Journal of Psychology* 12:159–64, 1958.

do affect behavior. For example, individuals with untreated PKU have lower than normal IQs, a behavioral trait that can be measured using IQ tests.

The norm of reaction for a behavioral trait can be shown by comparing two strains of rats and their ability to complete a maze in three different environments: restricted, normal, and enriched. In this experiment, the environments are ranked according to the amount of visual stimulus present during the maturation of the rats. The two strains of rats are selected through a number of generations for their increased or decreased maze-learning ability in a normal environment. The lines are therefore called maze-bright and maze-dull, respectively. In the normal environment, animals from the maze-bright strain make an average of fifty fewer errors than animals from the maze-dull strain, reflecting the effectiveness of selection in causing genetic differences between the strains (figure 3.28). However, if the two strains of rats are raised in either restricted or enriched environments, the two strains make a similar number of errors, primarily because the maze-bright strain cannot run the maze well under restricted conditions and the maze-dull strain can perform well under enriched conditions. The implications of these observations are intriguing; they suggest that some differences in behavior may be present only in a narrow range of environments and may not be generalizable to a diversity of environments.

The relationship of genotype to phenotype may be complicated. The norm of reaction over different environments is one way to quantify this relationship.

QUANTITATIVE TRAITS

For many traits, the genetic basis is not known precisely, and it may not be feasible, or even possible, to determine the number of genes and their individual effects on a particular trait. The phenotypic value of traits such as size, weight, or shape can generally be measured on a quantitative scale; hence, these are called **quantitative traits.** Although this phenotypic value may result from some general underlying genetic determination, for most quantitative traits there is no precise relationship between the phenotypic value and a particular genotype. However, in recent years using molecular marker techniques (see chapter 16) genes having large effects on particular traits, called QTLs or quantitative trait loci, have been identified for complex human diseases, yield in crop plants, and a number of other characters. In general, quantitative traits have these characteristics: (1) They have a continuous distribution; (2) they appear to be affected by many genes; that is, they are **polygenic;** and (3) they are affected to varying degrees by environmental factors.

One of the first good documentations of a polygenic trait was by Swedish geneticist Herman Nilsson-Ehle, who showed in 1909 that the color of wheat kernels was affected by several genes. To understand his findings, we will examine the genetic basis of his crosses, where color differences are determined by two genes. Figure 3.29 gives the F_2 results of a cross between wheat plants with red kernels and those with white kernels. The F_1 plants are all intermediate, having pink kernels, as expected with incomplete dominance. However, instead of just three categories in the F_2, there are five different phenotypes because there are two extra classes, light red and light pink, intermediate between the parents and the F_1. The explanation for the ratios of color phenotypes seen in the F_2 (1:4:6:4:1) is that two genes equally affect this trait, so that red occurs with four red alleles (indicated by the subscript 1) present $(A_1A_1 \ B_1B_1)$; light red occurs with three red alleles present at either the A or the B gene $(A_1A_1 \ B_1B_2$ and $A_1A_2 \ B_1B_1)$; and so on. In addition, a third gene affects wheat kernel color (not shown in figure 3.27). All three genes influence color independently of each other; that is, the phenotype is the sum of the individual effects of the different genes.

To illustrate in general the polygenic nature of quantitative traits, let us assume that we can visualize the influence of multiple genes affecting a trait by comparing the phenotypic distribution that results when one, two, or four genes, each with two alleles, determine the phenotypic value of a trait (figure 3.30). For simplicity, we assume that heterozygotes are exactly intermediate, that the two alleles at each locus are equal in frequency, and that the environment has no effect on the trait. The frequency of the polygenic genotypes is calculated as the product of the frequencies of the single-locus genotypes. In fact, we have already shown examples of one-gene and two-gene cases in figures 3.1 and 3.29, respectively. From figure 3.30, it is obvious that as the number of genes affecting the trait increases from one to four, the distribution changes from one of discrete

Parents $A_1A_1\,B_1B_1$ $\times$ $A_2A_2\,B_2B_2$
(red)   (white)

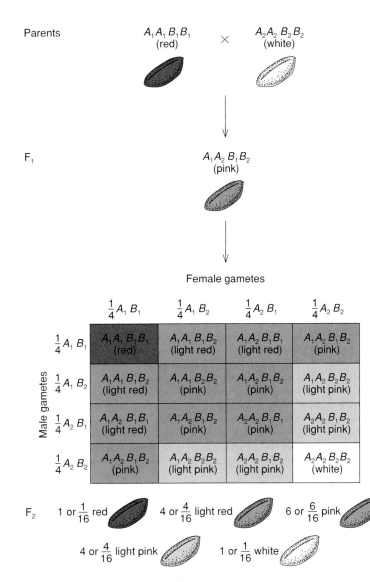

F$_1$ $A_1A_2\,B_1B_2$
(pink)

FIGURE 3.29   A cross of red- and white-kerneled wheat differing at two genes, showing the five types of F$_2$ progeny.

classes to one in which there is no discontinuity. In fact, the exact genotype-phenotype correspondence for all classes disappears when there are only two genes. That is, when two genes affect a trait, the second lowest phenotypic value can result from either genotype $A_1A_1\,B_1B_2$ or genotype $A_1A_2\,B_1B_1$.

Most quantitative traits are influenced by the environment. For example, it is obvious that adult size in most organisms is affected by the amount of available resources. Thus, food availability will determine size in an inbred line of mice, just as light and soil differences will result in phenotypic variation in height among the clones of a fruit tree.

Typically, quantitative traits such as size and weight have a continuous distribution in a population, given that the method of measurement is sufficiently accurate. The distribution of such traits often closely approximates a bell-shaped, or normal, distribution, with a high proportion of the individuals having an intermediate phenotypic value. Both the polygenic character of

such traits and the influence of the environment help determine the shape of such a distribution. A classic example of a normal distribution curve is shown in figure 3.31, depicting a group of 175 male students arranged by height. In this case, the students were put into discrete groups for every inch increment in height for ease of measurement even though a trait such as height is really continuous.

MEAN AND VARIANCE

How can we evaluate the phenotypic variation for height in the group of students (or the phenotypic variation in any organism)? To assess the attributes of populations, it is necessary to use some descriptive statistics. Here we will briefly discuss how to calculate the **mean** ($\bar{x}$), a measure of the average of a group of values, and the **variance,** a measure of the variability around that central value.

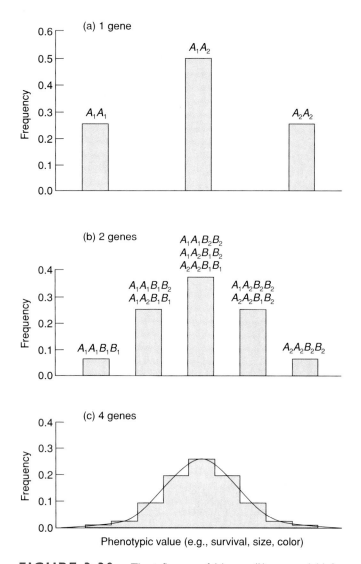

FIGURE 3.30   The influence of (*a*) one, (*b*) two, and (*c*) four genes on the occurrence of a quantitative trait and the resulting phenotypic distribution.

Let us consider the heights of a sample of n different plants. If we symbolize the height of plant i as x_i, we can calculate the mean plant height ($\bar{x}$) as:

$$\bar{x} = \frac{1}{n} \sum_{i=1}^{n} x_i$$

where Σ (Greek letter sigma) indicates the summation of all n plant heights in the sample. For example, if there are five plants in the sample and their heights in centimeters are $x_1 = 28$, $x_2 = 30$, $x_3 = 27$, $x_4 = 32$, and $x_5 = 33$, then:

$$\begin{aligned}\bar{x} &= \tfrac{1}{5}(x_1 + x_2 + x_3 + x_4 + x_5) \\ &= \tfrac{1}{5}(28 + 30 + 27 + 32 + 33) \\ &= 30\end{aligned}$$

In other words, the mean plant height in this sample of individuals is 30 centimeters.

Where a sample of individuals has a given mean, different amounts of dispersion, or variation, may exist around that mean. Thus, there may be no dispersion if all plants are exactly the same height, or there may be an extreme amount of dispersion if there are equal numbers of small and large individuals. The variance (V_x) is a measure of the dispersion around the mean and is calculated as:

$$V_x = \frac{1}{n-1} \sum_{i=1}^{n} \left(x_i - \bar{x} \right)^2$$

In other words, the variance of x is the sum of the squared deviations of individual values from the mean, divided by n − 1. The variance for the data given here is:

$$\begin{aligned}V_x &= \tfrac{1}{4}[(x_1 - \bar{x})^2 + (x_2 - \bar{x})^2 + (x_3 - \bar{x})^2 + (x_4 - \bar{x})^2 \\ &\quad + (x_5 - \bar{x})^2] \\ &= \tfrac{1}{4}[(28 - 30)^2 + (30 - 30)^2 + (27 - 30)^2 + (32 - 30)^2 \\ &\quad + (33 - 30)^2] \\ &= 6.5\end{aligned}$$

For the group of students pictured in figure 3.31, the mean height is 67.3 inches and the variance is 7.3.

The variance is not on the same scale as the mean because the variance is a squared value. Thus the square root of the variance, or the **standard deviation,** is often used to measure the dispersion around the mean. For example, the standard deviation for the sample of students is:

$$\begin{aligned}(V)^{1/2} &= (7.3)^{1/2} \\ &= 2.7\end{aligned}$$

(The exponent ½ indicates the square root.) If the distribution of observations is bell-shaped, or normal, approximately 95% of the sample should be within two standard deviations of the mean. In this case, two standard deviations is 5.4; therefore, 67.3 − 5.4 = 61.9 is two standard deviations less than the mean, and 67.3 + 5.4 = 72.7 is two standard deviations greater. Looking at figure 3.31, we can see that a very high proportion of the students are within these limits. In fact, 168 out of 175 students, or 96%, are within these limits. Only two are below 61.9 inches in height and only five are above 72.7 inches.

However, some polygenic traits have only discrete classes, such as the number of bird eggs in a clutch, the number of scales on a reptile, the number of petals or leaves on a plant, or the number of vibrissae (whiskers) on the face of a mouse. Such discontinuous traits, sometimes called **countable,** or **meristic, traits,** are also assumed to be controlled by many genes and affected by environmental factors. Even more extreme, there may be only two phenotypic categories for a discontinuous trait, such as the presence/absence situation that occurs for certain disease states (diseased/not diseased) or for reproductive states (fertile/sterile). In such cases, the presence of one state or the other is the result of an underlying genetic determination of liability or susceptibility that has a continuous distribution.

The presence of a genetic disease in a particular individual is determined by a combination of the individual's underlying genetic liability and his or her particular environment. Figure 3.32 shows

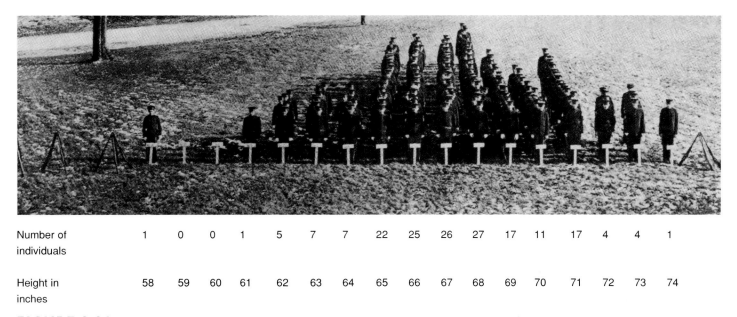

Number of individuals	1	0	0	1	5	7	7	22	25	26	27	17	11	17	4	4	1
Height in inches	58	59	60	61	62	63	64	65	66	67	68	69	70	71	72	73	74

FIGURE 3.31 Frequency distribution of heights of male students at Connecticut Agricultural College.
Library of Congress.

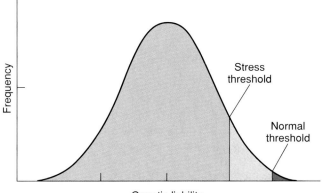
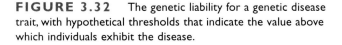

FIGURE 3.32 The genetic liability for a genetic disease trait, with hypothetical thresholds that indicate the value above which individuals exhibit the disease.

the hypothetical threshold in a normal environment for a disease such as diabetes. Here, a relatively small percentage of the population exceeds the threshold, and thus exhibits diabetes (red region). Given a different environment, in which obesity, alcoholism, or other factors increase the likelihood of diabetes, the threshold is moved to the left (indicated as the stress threshold). This shift results in a higher incidence of the disease, because some individuals, represented by the pink region, would not contract the disease in a normal environment but become afflicted under stress conditions.

Hip dysplasia in dogs, a condition that eventually leads to degenerative joint disease, appears to be determined by a combination of underlying genetic liability (that is, susceptibility) and environmental factors. This trait occurs frequently in many large dog breeds, such as German shepherds and Labrador retrievers, despite substantial efforts to eliminate it. The lack of simple inheritance as well as ambiguity in diagnosis (some dogs considered dysplastic by radiographs are clinically unaffected) has made it difficult to reduce the incidence of this condition.

It is important to realize that the genetic determination of quantitative traits may vary considerably among populations. Such a varied response has occurred in populations of insects exposed to various pesticides and may be primarily due to a single gene (or to a number of genes) that confer resistance to a particular insecticide. In addition, the single genes and the method by which they confer resistance may differ among populations. For example, resistance to the insecticide DDT may occur because particular genes influence either detoxification of the insecticide, storage of DDT in fat bodies, or resistance to its diffusion into the circulatory system. (A more detailed discussion of pesticide resistance is given in chapter 17.)

It has been generally believed, and recent molecular genetics evidence has verified, that the genes affecting quantitative traits are essentially like the genes (or are the same genes) that determine single-gene qualitative traits such as color, enzyme, or shape differences. In other words, there is no reason to assume that a different type of gene accounts for quantitative genetic variation; in fact, some genes known for their qualitative effects also affect quantitative traits. Genes affecting quantitative traits follow Mendelian patterns of inheritance, may have multiple alleles, and show various patterns of dominance and epistasis.

Quantitative traits are affected by many genes as well as by the environment. Polygenic traits generally have a continuous phenotype distribution in a population, but they may occur in discrete classes, such as diseased or not diseased.

MODEL OF QUANTITATIVE TRAITS

In order to understand and examine the importance of quantitative traits, we need to construct a model that will allow us to break down phenotypic values into genetic and environmental components. We can accomplish this in a simple manner by symbolizing the phenotypic value (P) for a genotype (i) in a particular environment (j) as:

$$P_{ij} = G_i + E_j$$

G_i is the genetic contribution of genotype i to the phenotype, and E_j is the deviation resulting from environment j. E_j may be either positive or negative, depending on the effect of environment j. This simple model illustrates the basic genetic-environment or nature-nurture dichotomy often discussed in relation to human intelligence and other traits.

In many cases, different populations have different mean phenotypic values. It is generally difficult to determine whether such a difference is the result of genetic factors, environmental factors, or a combination of both. However, if the populations can be grown · in the same environment, called a **common garden experiment** for plants and animals, we can make some inferences. For example, if individuals from different populations retain phenotypic differences in a common environment, this is consistent with the hypothesis that the populations are genetically different. But if the phenotypic differences disappear in the common environment, this generally indicates that environmental factors are important in determining phenotypic differences. For the *Potentilla* plants shown in figure 3.26, both genetic and environmental factors were important.

Different genotypes may interact differently with their environment in producing a phenotype. If such specific interactions occur between genotypes and environments, the basic model just presented can be expanded to include a term for **genotype-environment interaction,** with the phenotypic value becoming:

$$P_{ij} = G_i + E_j + GE_{ij}$$

where GE_{ij} measures the interaction between genotype i and environment j. As with the environmental deviation, the genotype-environment interaction may be either positive or negative.

Maze-learning ability in the rat experiment (figure 3.28) is an example of genotype-environment interaction. To further illustrate the complexity introduced by genotype-environment interaction, consider the simple situation of two genotypes in two environments. It is possible that a reversal of the order of the phenotypic values of two genotypes may occur in two different environments, as with a domesticated breed and wild-progenitor of a particular species, say pigs or sheep, raised either in association with humans or in the natural environment. The wild progenitor will survive and reproduce better in its natural situation, while the domesticated breed will survive and reproduce better in association with humans (figure 3.33). As a result, the phenotypic value of a trait like survival is primarily the result of the interaction between the specific genotype and the specific environment and cannot be predicted without information

about both the genotype and the environment. Only when the norms of reaction are parallel or close to parallel for different genotypes or breeds is there no genotype-environment interaction. If the range of environments examined is limited, then it is probably less likely that genotype-environment interaction will be important in influencing phenotypic variation.

The genetic model can be more precisely defined by subdividing the genotypic value into two components, with the value of genotype i being:

$$G_i = A_i + D_i$$

where A_i and D_i are the additive and dominant components, respectively. If the heterozygote is exactly intermediate between the homozygotes, the contribution of the dominant component is zero; that is, all the genetic component is additive. For example, the genetic variation in kernel color in wheat that we discussed earlier is due to the additive component. In other words, the phenotypic value can be symbolized as:

$$P_{ij} = A_i + D_i + E_j$$

assuming that the genotype-environment interaction is negligible.

As pointed out earlier, there is no exact genotype-to-phenotype correspondence for quantitative traits. Therefore, a more practical way to examine quantitative traits is to partition the variation in a population for a particular phenotype into that caused by genetic factors and that caused by environmental factors. If the basic model given above is used to examine the phenotypic variance, then:

$$V_P = V_G + V_E$$

where V_P, V_G and V_E are the phenotypic, genetic, and environmental variances, respectively.

Adopting the same approach as before, if we use the expression with separate additive and dominant components to examine the phenotypic variance, it becomes:

$$V_P = V_A + V_D + V_E$$

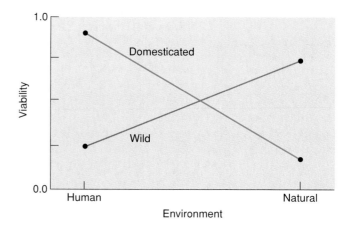

FIGURE 3.33   The viability of domesticated and wild animals in human and natural environments, illustrating the effects of genotype–environment interaction.

Separated at birth, the Mallifert twins meet accidentally.

Drawing by Chas. Addams; © 1981 The New Yorker Magazine, Inc. Reprinted by permission.

where V_A and V_D are the genetic variances due to additive and dominant effects respectively. If this expression is divided by V_P, we can obtain the proportion of phenotypic variance due to the different components. In particular, the ratio of the additive genetic variance to the phenotypic variance is known as the **heritability** (h^2):

$$h^2 = \frac{V_A}{V_P}$$

The magnitude of the heritability is important in determining the rate and amount of response to directional selection (that is, selecting for an extreme phenotype) and is often estimated by plant and animal breeders prior to initiating a selection program to improve yield or other traits.

> Quantitative traits can be understood by partitioning out the genetic, environmental, and genotype-environment effects. The heritability is the ratio of the additive genetic variance to the phenotypic variance.

ESTIMATING GENETIC VARIANCE AND HERITABILITY

Several different approaches can be used to estimate the amount of genetic variance and heritability. In some organisms, experimental manipulation, or breeding, can either reduce or eliminate the environmental or genetic variance components. Any remaining variance must therefore be due to the component not eliminated. In other words, if $V_G = 0$, then $V_P = V_E$, or if $V_E = 0$, then $V_P = V_G$. For example, the genetic variation is zero in inbred lines, clones, identical twins, and repeated measurements on the same individual, so that any phenotypic variation must be caused by environmental differences. Environmental variance may approach zero if all important environmental factors, such as food, moisture, and temperature, are known and carefully controlled.

Identical (monozygotic) twins have the same genotype; thus any differences between the twins should be the result of environmental factors. In fact, identical twins raised together share similar environments as well as identical genotypes. One way to determine the extent of similarity is to measure the **correlation** between individuals for different traits. The highest value a correlation can be is 1.0, where all the paired individuals have the same phenotypic value. If the traits are uncorrelated among paired individuals, the correlation is 0.0. The first column of table 3.6 gives the correlation for four different traits in monozygotic twins raised together. The second column gives the correlation for the same traits in dizygotic (nonidentical) twins of the same sex. The difference between these values should be the result of the lower genetic similarity of dizygotic twins, who share (on the average) only half their genes in common. One way to estimate the heritability is from the expression:

$$h^2 = 2(r_M - r_D)$$

where r_M and r_D are the correlations between monozygotic and dizygotic twins, respectively. Using this formula, the heritability for the four traits in table 3.6 varies from 0.98 to 0.16. In other words, most of the phenotypic variation in IQ and social maturity scores appears to be environmental, while most of the variation for fingerprint-ridge count (see figure 3.36) and weight appears to be genetic.

The response of a quantitative trait to artificial selection depends upon the heritability of the trait under selection. This is quite obvious when h^2 is either 0 or 1. When $h^2 = 0$, none of the phenotypic variation is from the additive genetic variance component; as a result, selecting even extreme individuals as

Table 3.6 Correlations of Four Traits Between Human Twins

	Correlations		Heritability
	Monozygotic Twins (r_M)	Dizygotic Twins (r_D)	
Fingerprint-ridge count	0.96	0.47	0.98
Height	0.90	0.57	0.66
IQ score	0.83	0.66	0.34
Social maturity score	0.97	0.89	0.16

Parent generation

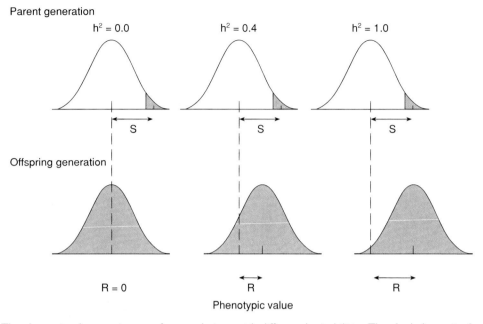

Phenotypic value

FIGURE 3.34   The change in phenotypic mean for populations with different heritabilities. The shaded area in the parent generation indicates those selected parents. The tic marks indicate mean values of all the parents, the selected parents, and the offspring.

parents would not change the phenotypic mean (figure 3.34). On the other hand, if $h^2 = 1$, all the phenotypic variation is from the additive genetic variance component, and phenotypically extreme individuals are that way because of their genotypes. The difference in the parental mean and the offspring mean is called the **response (R)** and can be expressed as:

$$R = h^2S$$

where S is the **selectional differential,** the difference in the mean of the selected parents and the mean of all the individuals in the parental population. This equation may be rearranged so that an estimate of heritability is:

$$h^2 = \frac{R}{S}$$

This expression is called the **realized heritability** because it is the ratio of observed response over total response possible.

Perhaps the best way to appreciate this affect is to examine the expected change in phenotypic mean from one generation of selection for traits with different heritabilities (figure 3.34). Notice that even when the selected individuals are quite extreme (blue shaded region at top of figure), if the heritability is low, the mean of the offspring is similar to that of all the parents (response is low). On the other hand, if the heritability is high, the offspring are much more like the selected parents than the distribution of all parents.

Many directional (or artificial) selection experiments give evidence of a change in phenotypic values over ten to forty generations, often reaching a maximum, or plateau, before the termination of the experiment. Evidence indicates that selection response may continue for many generations, given an adequate

initial sample of genetic variation, a reasonable population size during the selection process, and an effective selection scheme. (See chapter 17 for further discussion.)

One of the longest selection experiments ever conducted, for oil and protein percentage in corn, is still running at the University of Illinois. High and low directional selection has been carried out for nearly 100 plant generations, and progress is continuing, although large changes in oil and protein percentage have already been made. The data for the high oil and low oil lines for the first 76 generations are given in figure 3.35. The percentage of oil has increased nearly four-fold in the high line and has been reduced to below 1% in the low line.

Perhaps the most important approach used to measure components of genetic variation and heritability is based on the phenotypic resemblance between relatives. If genetic factors are important, one would expect the offspring of parents with high phenotypic values to have high phenotypic values, and the offspring of parents with low phenotypic values to have low phenotypic values. Obviously, close relatives share more genes with one another than with nonrelatives or distant relatives, and thus should have greater phenotypic resemblance. For example, monozygotic twins share all of their genes; siblings share one-half of their genes (on the average); and first cousins share one-eighth of their genes. As a result, phenotypic values between close relatives should be more highly correlated than those between distant relatives. Using this approach, estimates of heritability for a number of traits in different organisms are given in table 3.7.

Fingerprints have been widely used for identification purposes because the fingerprint patterns of an individual are unique. A general way of classifying fingerprints is to count the number of ridges, the raised areas of skin on the fingertips. The three basic configurations of ridges are known as the arch, the

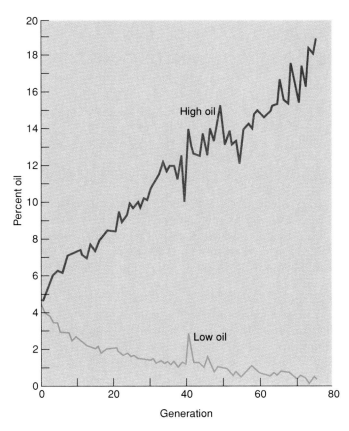

FIGURE 3.35 The percent of high and low oil content in long-term selection experiments in corn.

From *Proceedings of the International Conference on Quantitative Genetics,* August 16–21, 1976, edited by W. Pollak, O. Kempthorne, and T. B. Bailey, Jr. Copyright © 1977 by Iowa State University Press, Ames, IA. Reprinted by permission.

Table 3.7 Approximate Values of Heritability of Different Traits in Five Organisms

Organism	Trait	Heritability
Swine	Back fat thickness	0.55
	Body weight	0.3
	Litter size	0.15
Poultry	Egg weight	0.6
	Body weight	0.2
	Viability	0.1
Mice	Tail length	0.6
	Body weight	0.35
	Litter size	0.15
Drosophila melanogaster	Bristle number	0.5
	Body size	0.4
	Egg production	0.2
Corn	Plant height	0.70
	Yield	0.25
	Ear length	0.17

Source: Data from D. S. Falconer, *Introduction to Quantitative Genetics,* Ronald Press, New York.

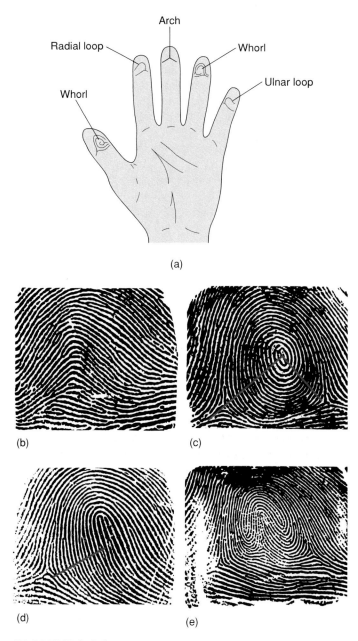

FIGURE 3.36 The position of fingerprints on a hand (*a*) and examples of the basic fingerprint types: (*b*) arch; (*c*) whorl; (*d*) loop; and (*e*) combination.

loop, and the whorl (figure 3.36). Fingerprint-ridge count seems to be almost entirely genetic, with heritability estimates near unity (see table 3.6) and with little noticeable influence from the environment.

RACE AND INTELLIGENCE

In the late nineteenth century, some British scientists suggested that the high reproductive rate among the "lower" classes was leading to an overall deterioration of the average intelligence of the British population. The discussion that followed became known as the nature-nurture controversy; that is, given real differences in intellectual ability among social classes, were these differences due to genetic or environmental factors? The controversy was revived again in the late 1960s when it was claimed that the difference in IQ scores among racial groups was genetically determined, and in particular, that lower IQ scores among African-Americans were the result of genetic factors. The recent publication of the book *The Bell Curve* has again generated controversy about the role of genetics in determining intelligence.

Many researchers have suggested that environmental factors have a substantial effect on IQ scores, citing a range of factors that includes nutrition (prenatal, during childhood, and at the time of testing), socioeconomic status of parents, and educational and testing experience. One concern in comparing the IQ scores of blacks and whites is that these groups differ not only ethnically but also in education, upbringing, and other socioeconomic factors. An obvious way to evaluate the effect of socioeconomic factors is to compare blacks and whites within the same socioeconomic category.

Table 3.8 gives the IQ scores of African-American and Caucasian seven-year-olds, some from an upper socioeconomic background and some from a lower socioeconomic background. Notice first that both groups from the high socioeconomic background score higher than the groups from the low socioeconomic background, suggesting that socioeconomic background has a substantial influence on IQ test scores. Second, the difference between racial groups from the same socioeconomic background is quite small, only 4.2 points in the high group and only 4.1 points in the low group. These differences are in fact much smaller than any of the standard deviations, indicating that the means of the two racial categories are not statistically different within each socioeconomic group. It appears that the differences seen between whites and blacks are probably due in large part to socioeconomic differences between the two groups.

Genetic variance can be estimated by examining identical twins. Realized heritability can be estimated by determining the rate of response from selection. Intelligence is affected by a number of environmental factors; thus any racial differences in IQ are generally confounded with environmental differences.

Table 3.8   The Effect of Socioeconomic Background on IQ Scores

Socioeconomic Background	African-American	Caucasian
High	100.0 (12.6)★	104.2 (13.4)
Low	91.2 (12.1)	95.3 (12.9)

★The numbers in parentheses are standard deviations.

GENETIC COUNSELING

The importance of genetic disease has increased substantially in the last several decades as the incidence of infectious disease has declined. For example, infectious diseases such as smallpox and diphtheria used to be primary causes of infant mortality, but genetic disease is now a leading factor. Some of these diseases follow a single-gene pattern of inheritance; others are polygenic, with significant environmental components; and still others are the result of aberrant chromosomal numbers or types (see chapter 4).

After a child is born with a genetic disease, the parents may seek advice concerning the probability of a second child having the same disease. Such **genetic counseling** is offered at a number of medical centers. Once the disease is properly diagnosed and its mode of inheritance known, the **recurrence risk,** or probability that the condition will recur in a given family, can be calculated.

A large proportion of known genetic diseases are inherited as autosomal recessives. Many of these are **inborn errors of metabolism**—that is, enzymatic defects that result in the buildup or absence of an important biochemical product. Most of these diseases are quite rare, generally occurring in fewer than 1 in 10,000 individuals. However, because there are a large number of such diseases, their cumulative incidence is substantial.

The probability of a genetic disease occurring in a child depends on whether the parents or other relatives are affected or unaffected, and on whether other offspring have already been affected. Table 3.9 shows the risk of producing affected children for parents with different disease states when a previous child has been affected. These probabilities follow the pattern of inheritance for recessive traits discussed in chapter 2.

Genetic counseling can also be useful when parents strongly wish to have children but there is a probability that their children will inherit a genetic disease. For example, if both parents are known carriers of a recessive disease allele (heterozygotes can be detected for sickle-cell disease, Tay-Sachs, and a number of other recessive diseases), the probability of a sibship of size N with no affected progeny is $(\frac{3}{4})^N$. With a small family of one or two children, the chances of having no affected offspring are 0.75 and 0.5625, respectively, a risk that some people find acceptable.

Table 3.9   Probability of Affected Offspring for an Autosomal Recessive[1]

Parents	Second Child Affected
Both unaffected	¼
One affected	½
Both unaffected; one has affected sib	¼
Both affected	1

[1]Given that one child is already affected.

Table 3.10 Empirical Risks for Congenital Pyloric Stenosis in an English Population

Number of Affected Sibs	Neither Parent Affected	Father Affected	Mother Affected	Both Parents Affected
0	0.003	0.037	0.051	0.298
1	0.030	0.115	0.135	0.365
2	0.119	0.216	0.231	0.431

Source: Data from C. Bonaiti-Pellie and C. Smith, "Risk Tables for Genetic Counseling in Some Common Congenital Malformations," in *J. Med. Genet.* 11:374:376, 1974.

A number of diseases, such as diabetes, cleft palate, and spina bifida, have a strong genetic component but do not have a single-gene mode of inheritance. For such diseases, it is assumed that the genetic liability to the disease is polygenically controlled and that, consequently, close relatives of affected individuals are more likely to be affected. Obviously, the closer the affected relative and the more relatives are affected, the higher the genetic liability in an individual.

For such diseases, **empirical recurrence risks** are often calculated from a large number of case studies. These values may be figured where both parents are affected, where one parent is affected, where one sib is affected, or in other situations. Empirical risk values can be used only as a guideline, since specific genetic or environmental factors may either increase or reduce the risk in a particular family.

Pyloric stenosis is a disease in which the pyloric valve, the sphincter valve between the stomach and the small intestine, is not completely opened. Before corrective surgery was introduced in the 1920s, most affected individuals died in infancy because little or no food could pass on to the small intestine. With corrective surgery, individuals born with pyloric stenosis have normal digestion and generally no other related problems. Risk values for this disease are given in table 3.10 with respect to the disease state of the parents and the number of affected sibs. The risk, as intuitively expected, increases as either the number of affected sibs or the number of affected parents increases. For example, with no affected relatives, the risk is only 0.3%, while with two affected sibs and two affected parents, the risk is greater than 40%.

Genetic counseling is used to determine the risks of producing genetically defective infants. Mendelian inheritance probabilities are used to predict single-gene traits, while empirical risk values are used for polygenic diseases.

PATERNITY EXCLUSION

In cases where the father of a child is unknown or disputed, blood groups, enzymes, or HLA types have been used to determine the probability of parentage. In the past, such data were admitted in U.S. courts only when they ruled out an individual as a parent. But in recent years, because of better genetic analysis, courts will often accept information that indicates the probability of parenthood.

When a locus has only a few allelic forms in a population, it is difficult to exclude an individual as a parent, because many individuals have the same alleles, making them all possible parents. As an example, mother-child genotype combinations and the excluded paternal genotypes for the MN blood group locus are listed in table 3.11. Given the genotype of the mother and child, one may exclude certain genotypes as the father. For example, if the genotypes of both mother and child are $L^M L^M$, the father could be either $L^M L^M$ or $L^M L^N$, and only the paternal genotype $L^N L^N$ can be excluded. Notice that only one genotype at most can be excluded and that when both the mother and child are heterozygotes, no paternal genotypes can be excluded.

Even with a large number of loci, the probability of exclusion is not as high as one might wish. The use of HLA **haplotypes** (all the HLA alleles at different genes on a single chromosome) and DNA markers, both of which can have many alleles at a locus, may resolve this difficulty. To illustrate, figure 3.37 gives the HLA haplotypes of a mother, a child, and two possible fathers. For example, the haplotype A1 B17 is on one of the chromosomes in the putative father on the left. (These haplotypes stay intact from generation to generation because the HLA genes are close to each other on the chromosome, that is, they are tightly linked. See chapter 5.) From this information, it is apparent that the mother produced an A1 B8 gamete, and therefore the father

Table 3.11 Excluded Paternity Types, Given Genotypes of Mother and Child

Mother	Child	Excluded Paternity Types
$L^M L^M$	$L^M L^M$	$L^N L^N$
	$L^M L^N$	$L^M L^M$
$L^M L^N$	$L^M L^M$	$L^N L^N$
	$L^M L^N$	—
	$L^N L^N$	$L^M L^M$
$L^N L^N$	$L^M L^N$	$L^N L^N$
	$L^N L^N$	$L^M L^M$

BOX

3.1 Grandpaternity Exclusion

Genetic information has long been used to attempt to identify parents. Recently, the same approach has been extended to grandparents and other relatives. A major impetus for this work was the disappearance of at least nine thousand Argentinians between 1975 and 1983. Among the missing were a number of children who had been abducted by the military and police or who had been born to kidnapped women in captivity. For example, Liliana Pereyra (figure B3.1) was abducted when she was five months pregnant and kept alive in a torture center until she gave birth to a boy in February 1978. Her son has not been found and may have been kidnapped by the military.

It appears that many of the Argentinian children are now living with military couples who claim to be their biological parents but may actually be the kidnappers. If all four grandparents are alive, and if the child has an allele that none of them have, grandpaternity can be excluded. But if the alleles in the child are also present in the grandparents, the probability of grandpaternity can be calculated. For example, if the shared alleles are rare in the population, the probability of grandpaternity is quite high.

Figure B3.2 is a pedigree in which a child's HLA types were found in the putative grandparents. In this case, Paula Eva Logares, an Argentinian kidnap victim, was raised by a policeman and his wife. One of her HLA haplotypes, A2 B5, was present in her paternal grandfather. The reconstructed genotype of her maternal grandparent (from his other children) contains the other haplotype, A1 B5. It is very likely that these haplotypes were passed on to her from her deceased parents. These data and probability principles make it 99.9% certain that Paula is the grandchild of the Logares and Grinspon families.

FIGURE B3.1   Liliana Pereyra, an Argentinian kidnap victim.
From Dr. Jared M. Diamond. *Nature* 327:552, 1987. Macmillan Magazines Limited.

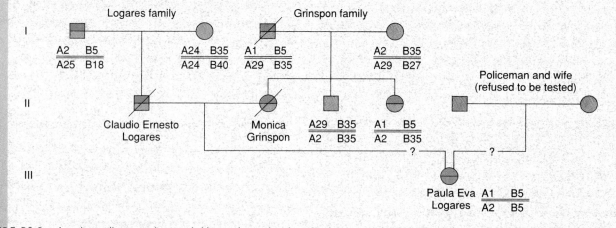
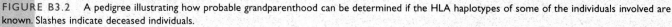

FIGURE B3.2   A pedigree illustrating how probable grandparenthood can be determined if the HLA haplotypes of some of the individuals involved are known. Slashes indicate deceased individuals.

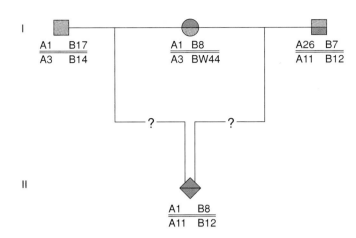

FIGURE 3.37 The HLA haplotypes and genotypes of a mother, child, and two putative fathers.

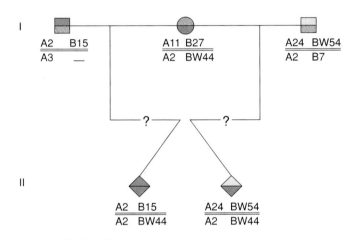

FIGURE 3.38 The HLA haplotypes and genotypes of a mother, dizygotic twins, and the putative fathers of the twins.

produced an A11 B12 gamete. Putative father 1 does not have this haplotype, so he can be excluded. Putative father 2 does, and if there is no other possible male parent, this would indicate that he is the father. If there were other possible fathers, one of those other men could have been the father, so it cannot be stated with complete assurance that putative father 2 is the father.

In an actual case that proved the usefulness of the HLA system, two different men were identified as the fathers of a set of dizygotic twins. The genotypes of the individuals are given in figure 3.38. The twins both received an A2 BW44 gamete from their mother, but twin 1 received an A2 B15 gamete from putative father 1 and twin 2 received an A24 BW54 gamete from putative father 2. This was not detectable by looking at the ABO blood group system because all of the individuals were type A.

Recently, the identification of a number of loci, each with many alleles, has allowed the development of **DNA fingerprints.** The principles involved are the same as discussed before,

except that nearly every individual has a different DNA pattern for these genes. We will discuss some details of the molecular basis of this procedure in chapter 15.

Paternity exclusion based on genetic variants is used when the father of a given child is unknown. Generally, blood group or HLA alleles are used, but recently, DNA fingerprints have improved the power of this technique.

WORKED-OUT PROBLEMS

PROBLEM 1

A woman with blood type A and a man with blood type B had children that are blood types A, B, AB, and O. What are the genotypes of the parents?

SOLUTION

Without any knowledge about the offspring, the mother could be either I^AI^A or I^AI^O, and the father could be either I^BI^B or I^BI^O. There are then four possible mating types, with the following possible progeny phenotypes:

Mother	Father	Offspring
I^AI^A	I^BI^B	All AB
I^AI^A	I^BI^O	½AB, ½A
I^AI^O	I^BI^B	½AB, ½B
I^AI^O	I^BI^O	¼AB, ¼A, ¼B, ¼O

In other words, only if the mother is I^AI^O and the father I^BI^O can they have progeny of all four types.

PROBLEM 2

Figure 3.21 gives the pattern of inheritance that might be expected for a disease that is caused by two different genes, as for example several types of deaf-mutism in humans. Assume that individuals II-3 (who is deaf) and I-3 in this pedigree have a child. What are the possible genotypes from this mating and the probability that the child will be deaf?

Although individuals III-1 and III-2 in this pedigree have normal hearing, their parents were both deaf. What proportion of the progeny of a mating between two individuals with the genotypes of these individuals would be expected to have normal hearing?

SOLUTION

In the mating between II-3 and I-3, one would expect 0.5 of the progeny to be *Aa BB* and 0.5 to be *Aa Bb.* None of these children would be deaf.

There are two ways to calculate this proportion. First, in this mating between individuals of genotypes *Aa Bb* and *Aa Bb*, it is expected that ⅛₆ would be *A– bb,* ⅛₆ would be *aa B–,* and ⅛₆ would be *aa bb,* all of which would be deaf. Overall then, 1 − ⁷⁄₁₆ or ⁹⁄₁₆ of the progeny would be expected to have normal hearing. Another way to calculate this is to determine that ⁹⁄₁₆ would be expected to be *A– B–,* all of which have normal hearing.

PROBLEM 3

For many traits, the genetic contribution is determined by many genes, and the only way to characterize them is to estimate the proportion of phenotypic variation that is genetic. If such a quantitative trait, say yield in corn, is selected so that the response is 1.4 grams per plant from a selection differential of 3.1 grams per plant, what is the estimated heritability? Do you think that subsequent selection in future generations would be successful in further changing the yield in corn?

SOLUTION

The heritability is $h^2 = \dfrac{R}{S} = \dfrac{1.4}{3.1} = 0.45.$ Because nearly half the phenotypic variation for this trait is genetic, one would expect that future generations of selection would significantly increase the yield further.

SUMMARY

Dominance can occur in different forms: (1) complete, as found by Mendel, where either the mutant or the wild-type allele is dominant; (2) incomplete, with the heterozygote having a phenotype in between the two homozygotes; or (3) codominant, in which both alleles are expressed in the heterozygote.

Many variations on Mendel's original findings have been discovered. Some genes may be lethal. Pleiotropy occurs when one gene affects several traits. Some genotypes may not always have the same phenotype, a phenomenon resulting from incomplete penetrance or variable expressivity of the genotype. Genes may have more than two alleles. In diploid organisms, many different genotypes can exist when there are multiple alleles. Genes on the X and Y chromosomes have different patterns of inheritance than do autosomal genes. X-linked recessive traits occur almost entirely in males and generally skip generations in pedigrees.

More than one gene can affect a single trait. When two genes interact to determine a phenotype, this phenomenon is called *epistasis.* Epistatic interaction can result in F_2 ratios different from 9:3:3:1. The relationship of the genotype to the phenotype may be complicated. The norm of reaction over different environments is one way to quantify this relationship. Quantitative traits are affected by many genes and are influenced by the environment. Polygenic traits may occur in a continuous phenotypic distribution in a population or in discrete classes, such as diseased or not diseased.

Quantitative traits can be understood by partitioning out the genetic, environmental, and genotype-environment effects. Heritability is the ratio of the additive genetic variance to the phenotypic variance. Genetic variance can be estimated by examining identical twins. Realized heritability can be estimated by determining the rate of response from selection. Intelligence is affected by a number of environmental factors, and any racial differences in IQ are generally confounded with environmental differences.

Genetic counseling is employed to determine the risks of producing genetically defective infants. The probabilities from Mendelian inheritance are used for single-gene traits, while empirical risk values are used for quantitative traits. Paternity exclusion, based on genetic variants, is applied when the father of a given child is unknown. Generally, blood group or HLA alleles are used, but recently, DNA fingerprints have improved the power of this technique.

PROBLEMS AND QUESTIONS

Problems and Questions with Links to Software

 Drosophila Genetics: 8, 15, 16, 17, 18, 19, 20, 26

1. The seven traits that Mendel used in his experiments were all completely dominant. Although this was crucial in his ability to learn about particulate inheritance, we now know that most traits are not completely dominant. What other kinds of dominance are there? Give examples of each.

2. What are penetrance and expressivity? Give examples and discuss them to show that you understand the difference between these concepts.

3. (*a*) If a gene has three alleles, how many genotypes are possible? (*b*) If another gene has five alleles, how many homozygotes and how many heterozygotes are possible?

4. (*a*) What are the differences between sex-linked, sex-limited, and sex-influenced traits? (*b*) Describe how to determine whether a trait is sex-linked or sex-limited.

5. If in his experiments Mendel had used traits that were polygenically determined, what outcome do you think he would have observed in his F_2 progeny?

6. One experiment to elucidate the inheritance of flower color in four-o'clocks crossed two plants with pink flowers. In the progeny from this cross, there were 42 plants with red flowers, 86 with pink flowers, and 39 with white flowers. Using a chi-square test, determine whether those numbers are consistent with Mendelian expectations.

7. A series of matings between individuals where one parent was blood group phenotype N and the other parent was phenotype MN yielded 84 type MN progeny and 91 type N progeny. Is this consistent with expectations? (Use chi-square to test.)

8. Write out all the possible mating types for the ABO blood group system. (Given four phenotypes, there are ten different mating types if it is assumed that reciprocal matings are

equivalent.) Specifying genotypes, write out the four mating types designated phenotypically as A × B and the expected proportion of progeny types from each.

9. Trying to understand the inheritance of the dominant yellow gene in mice, researchers mated two yellow heterozygous mice. A typical result was 56 yellow progeny to 31 wild-type. Use a chi-square test to determine if this is consistent with 3:1 Mendelian predictions for a dominant gene. Are these data consistent statistically with the lethal model of Cuénot?

10. (*a*) If two Manx cats are mated (all Manx are genotype *Mm*), what proportion of the progeny should be tailless? (*b*) If a Manx male cat mates with a female cat with a normal tail, what proportion of the progeny should have a tail?

11. Some researchers feel that most genes are pleiotropic for many traits. If you discovered a new genetic disease, how would you go about documenting its pleiotropic effects?

12. Draw a pedigree for a trait such as Huntington's disease, illustrating variable expressivity. Assuming *H* and *h* are the dominant and recessive alleles, respectively, give the genotypes and phenotypes for three generations (one set of grandparents in the first generation, at least three progeny, and at least five grandchildren).

13. (*a*) Describe how you would document the importance of environmental factors, such as temperature and diet, on the expression of a recessive mutation in *Drosophila*. (*b*) How would you document the effect of modifier genes on the expression of the same mutation?

14. Which blood group type of the ABO system is known as the universal donor? The universal recipient? Explain your answers.

15. A female *Biston betularia* moth having the typical phenotype is mated to a male melanic. If half the progeny are melanic and half are insularia, what are the genotypes of the two parents?

16. (*a*) Given that two plants with genotypes B_2B_4 and B_1B_5 are mated, what types of progeny (and in what proportions) would you expect? (*b*) For the same gene, if a progeny array from a single mating has equal numbers of B_1B_2 and B_2B_4 individuals (and no other progeny), what are the parents' genotypes?

17. A woman who is blood type A and a man who is blood type B have a child who is blood type O. What are the genotypes of the three individuals? What is the probability that a second child would again be blood type O? Blood type AB?

18. If a color-blind man and a woman with normal color vision have a female child who is color-blind, what are the genotypes of the parents and the child? What proportion of male children from such a mating would be expected to have normal vision?

19. The pedigree in figure 3.17 shows the pattern of inheritance for an X-linked dominant gene. If individuals III-3 and II-6 in that pedigree mate and have children, what would be the proportion of individuals in the four classes: male, unaffected; male, affected; female, unaffected; and female, affected?

20. (*a*) If female *Drosophila* homozygous for the white allele (*ww*) affecting eye color are mated to red-eyed males (*w⁺Y*), what types of F_1 progeny could you expect? (Give both phenotype and sex.) (*b*) F_1 males and females were mated, and in the F_2 progeny there were 40 white-eyed females, 42 red-eyed females, 39 white-eyed males, and 44 red-eyed males. Using a chi-square test, determine whether these results are consistent with the expectations for X-linked genes.

21. Using the forked-line approach, calculate the proportions of the five phenotypes in a cross between two wheat plants with pink kernels and genotype $A_1A_2B_1B_2$. (See figure 3.29 for information on this cross and another way of calculating the proportions.)

22. Assume there is a third gene affecting color in wheat so that the F_1 plants are $A_1A_2B_1B_2C_1C_2$. (*a*) What are expected proportions of the F_2 genotypes using the forked-line approach? (*b*) Assume that gene *C* contributes equally to the other genes for color and that the color of the F_2 can be divided into seven categories: 0 (no "2" alleles, $A_1A_1B_1B_1C_1C_1$); 1 (one "2" allele, $A_1A_2B_1B_1C_1C_1$ or $A_1A_1B_1B_1C_1C_2$); on through to 6 (no "1" alleles, $A_2A_2B_2B_2C_2C_2$). Draw a graph showing the seven phenotypic categories equally spaced on the horizontal axis and their frequencies as obtained from the F_2 progeny on the vertical axis. (*c*) How does this compare to the distribution expected if only one gene or two genes affect color?

23. What is the difference between dominance and epistasis? Why didn't any pairs of genes used by Mendel show epistasis?

24. Bateson and Punnett observed 382 purple-flowered and 269 white-flowered sweet peas in an F_2. Are these numbers consistent with the 9:7 ratio they predicted, using a chi-square test?

25. If the purple F_1 in question 24 were crossed to one of the parental varieties, what proportions of white-flowered progeny would you expect?

26. Assume a mouse from a true-breeding line of albino mice with genotype *BBcc* was crossed to a mouse from a true-breeding brown line with genotype *bbCC*. (*a*) What are the genotype and phenotype of the F_1 mice? (*b*) Assume two F_1 mice were crossed, what are the proportions of the different phenotypes in the F_2 progeny and the genotypes in each color class?

27. Two types of epistasis discussed were duplicate and complementary gene action. How would you distinguish between these types of epistasis in determining purple and white flower color if you could set up any crosses necessary?

28. Graph the effect of the environmental temperature on the phenotype for plant height when the environment affects two genotypes, a short and a tall, equally. Assume that as temperature increases, plant height increases. Next, assume that at high temperature, the tall genotype plant is inhibited in growth so that it has a lower height than the short genotype. Draw the norms of reaction of the two genotypes in this case.

Chromosomes and Heredity

Learning Objectives

In this chapter you will learn:

1. How different chromosomes are characterized.
2. How chromosome number is maintained during the process of mitosis.
3. How chromosome number is reduced from two sets (diploid) to one set (haploid) during the process of meiosis.
4. That genes and chromosomes exhibit parallel behavior.
5. The four types of changes in chromosomal structure: duplication, deletion, inversion, and translocation.
6. The two ways in which chromosome number may vary: polyploidy and aneuploidy.
7. That sex determination is generally associated with the chromosomes or specific genes, although the mechanism varies with the species.

I may finally call attention to the probability that the association of paternal and maternal chromosomes in pairs and their subsequent separation during the reducing division . . . constitute the physical basis of the Mendelian law of heredity.

Walter Sutton

American geneticist

Although chromosomes had been noticed as cell structures in the nineteenth century, it was only after the rediscovery of Mendel's principles that their fundamental role in heredity was established. In 1902, Walter Sutton in the United States and Theodore Boveri in Germany both suggested that genes might be on chromosomes. They based their hypotheses on the analogous behavior of genes and chromosomes. In fact, the behavior of chromosomes in meiosis, as we will see later, provides a mechanism for the Mendelian principles of segregation and independent assortment.

SEXUAL REPRODUCTION

Before we describe the behavior of chromosomes during cell division, let us briefly discuss sexual reproduction and the overall structure of chromosomes. Reproduction in eukaryotes in general is either asexual or sexual. **Asexual reproduction** occurs when a single individual produces new individuals identical to itself. This is a common mode of reproduction in plants and simple animals. Asexual reproduction using cuttings is the common method of propagating many horticultural plants. **Parthenogenetic** female organisms (such as some aphids) can also produce offspring without fertilization. All progeny produced asexually are genetically identical to their parent (barring a rare mutation).

Sexual reproduction occurs when individuals produce male and female sex cells, or gametes, that in turn unite to form a **zygote,** a single cell from which a new individual develops. Sexual reproduction occurs in nearly all types of organisms, including most simple animals (figure 4.1), plants, and even bacteria (see chapter 13). Usually the male and female gametes come from different individuals, ensuring that the progeny are different from either parent. An important exception occurs in self-fertilization, as in Mendel's garden peas, where the same individual produces both male and female gametes. In the following discussion, we will concentrate on sexual reproduction and its genetic consequences.

The general scheme of reproduction and growth in a sexual organism is shown in figure 4.2. Let us examine two zygotes, one of which will mature into a female and the other into a male. Each zygote has received a complement of chromosomes of number n in the gamete from each parent, making the total number of chromosomes in each zygote 2n. The chromosomal number in the gamete is called the **haploid** (n) number, and the chromosomal number in the zygote is called the **diploid** (2n) number. The n different chromosomes within each haploid set each have different genes. Each haploid set has one copy of each type of chromosome. The two chromosomes of a particular type present in a diploid set are called **homologous chromosomes.**

The zygote increases in cell number by the process of **mitosis** and cell division. In this process, the chromosomes are

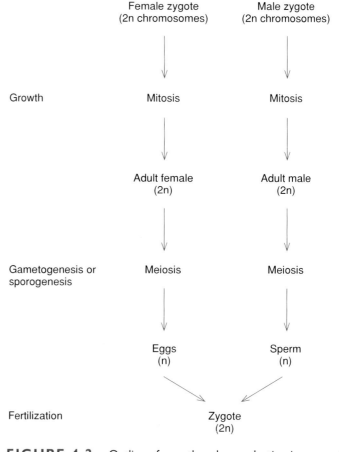

FIGURE 4.1   Sexual reproduction in barnacles, as shown by two adjacent individuals mating.
© Heather Angel/Biofotos.

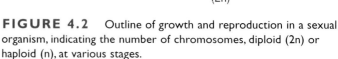

FIGURE 4.2   Outline of growth and reproduction in a sexual organism, indicating the number of chromosomes, diploid (2n) or haploid (n), at various stages.

relocated so that they remain at the 2n number in the daughter cells. This type of cell division continues until reproductive maturity, at which time some cells undergo **meiosis.** Meiosis is a second type of cell division that produces female and male gametes, each having a chromosomal number of n. This process is known as **gametogenesis** in animals and as **sporogenesis** in plants. The female gametes (eggs) and the male gametes (sperm) then unite to form progeny zygotes by the process of **fertilization.** These new zygotes mature into either males or females, and the whole cycle starts over.

Each species has a characteristic number of chromosomes; that is, virtually all individuals of a given species have the same chromosome number. Generally, this number is expressed as the diploid, or 2n, number. Table 4.1 lists a number of organisms used for genetic research, including corn, with a diploid number of 20 chromosomes, and pink bread mold, with a haploid number of 7 (for this organism, the haploid phase is the major one). The fruit fly has a 2n number of only 8, while carp (and some ferns) have over 100 chromosomes. Notice that closely related species, such as humans and chimpanzees or horses and donkeys,

have similar chromosome numbers. In Hymenoptera (bees and wasps), the females are diploid (2n) and the males are haploid (n). The term **haplodiploid** is used to describe such organisms. Certain species have some very small chromosomes, called **supernumerary,** or **B, chromosomes,** that can vary in number among individuals. These chromosomes are generally thought to be devoid of important genes, although in some instances supernumerary chromosomes affect fertility. Other species, including many birds, have a number of very small chromosomes called **microchromosomes.** Although microchromosomes are hard to count exactly because of their number and small size, they are thought to carry genes and to be constant in number, just as the larger chromosomes are.

CHROMOSOMAL MORPHOLOGY

In order to further understand chromosomes and their function, we need to be able to discriminate among different chromosomes. First, chromosomes differ greatly in size. Between organisms the size difference can be over 100-fold, while within a species some chromosomes are often ten times as large as others. In a species **karyotype,** a pictorial or photographic representation of all the different chromosomes in a cell of an individual, chromosomes are usually ordered by size and numbered from largest to smallest.

Second, chromosomes may differ in the position of the **centromere,** the place on the chromosome where spindle fibers are attached during cell division. The centromere is identified as a large constriction where the chromosome appears to be pinched. (The term **kinetochore** is used to designate a structure around the centromeric region.) In general, if the centromere is near the middle, the chromosome is **metacentric;** if the centromere is toward one end, the chromosome is **acrocentric** (or **submetacentric**); and if the centromere is very near the end, the chromosome is **telocentric** (figure 4.3). The

---

Table 4.1 Some Organisms and Their Diploid Chromosome Numbers

Common Name (Scientific Name)	Diploid Chromosome Number
Mammals	
Human (*Homo sapiens*)	46
Chimpanzee (*Pan troglodytes*)	48
Dog (*Canis familiaris*)	78
Cat (*Felis catus*)	38
Horse (*Equus caballus*)	64
Donkey (*Equus asinus*)	62
House mouse (*Mus domesticus*)	40
Other Animals	
Chicken (*Gallus domesticus*)	~78
Leopard frog (*Rana pipiens*)	26
Carp (*Cyprinus carpio*)	104
Fruit fly (*Drosophila melanogaster*)	8
Plants	
Corn (*Zea mays*)	20
Tobacco (*Nicotiana tabacum*)	48
Pine (*Pinus* species)	24
Garden pea (*Pisum sativum*)	14
Fungi	
Baker's yeast (*Saccharomyces cerevisiae*)	17★
Black bread mold (*Aspergillus nidulans*)	8★
Pink bread mold (*Neurospora crassa*)	7★

★Haploid number

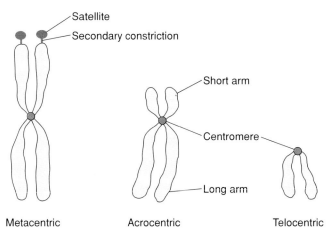

FIGURE 4.3 The three major types of chromosomes as they appear in human karyotypes: metacentric, acrocentric, and telocentric. Satellites may be present on any of the chromosome types.

Table 4.2 Karyotypes of Eight Common Domestic Mammals

		Autosomal Pairs		Sex Chromosomes	
	Diploid number (2n)	*Number of metacentrics*	*Number of acrocentrics or telocentrics*	*X*	*Y*
Cat (*Felis catus*)	38	16	2	M	M
Dog (*Canis familiaris*)	78	0	38	M	A
Pig (*Sus scrofa*)	38	12	6	M	M
Goat (*Capra bircus*)	60	0	29	A	M
Sheep (*Ovis aries*)	54	3	23	A	M
Cow (*Bos taurus*)	60	0	29	M	M
Horse (*Equus caballus*)	64	13	18	M	A
Donkey (*Equus asinus*)	62	24	6	M	A

M = metacentric
A = acrocentric
Source: Data from Ohno, "Veterinary Medical Cytogenetics," *Advances in Veterinary Science and Comparative Medicine* 12:1–31, 1968; and F. Eldridge and W. F. Blazak, "Horse, Ass, and Mule Chromosomes," *Journal of Heredity* 67:361–67, 1976.

centromere divides the chromosome into two **arms,** so that, for example, an acrocentric chromosome has one short and one long arm, while a metacentric chromosome has arms of equal length. All house mouse chromosomes are telocentric, while human chromosomes include both metacentrics and acrocentrics, but no telocentrics. Table 4.2 describes the karyotypes of eight common domestic animals to illustrate the variation in gross chromosomal morphology that occurs among these species. In addition to a centromere, there may also be a **secondary constriction** on particular chromosomes. Often the part of the chromosome located distal to a secondary constriction is a chromosomal **satellite.** The secondary constrictions usually contain genes that help organize or form the **nucleolus,** a structure that is apparent in various stages of cell division. The ends of a chromosome are called **telomeres** and often have DNA sequences different from the rest of the chromosome.

Third, chromosomes may be identified by regions that stain in a particular manner when treated with various chemicals. In fact, the term *chromosome* literally means "colored body." Several different chemical techniques are used to identify certain chromosomal regions by staining them so that they form chromosomal **bands.** For example, darker bands are generally found near the centromeres or on the ends (telomeres) of the chromosomes, while other regions do not stain as strongly. The positions of the dark-staining or **heterochromatic,** regions (also called *heterochromatin*) and the light-staining, or **euchromatic,** regions (called *euchromatin*) generally remain constant in different cells or individuals of a given species and are therefore useful in identifying particular chromosomes.

Euchromatic regions often undergo a regular cycle of contraction and extension. The two types of heterochromatin are **constitutive** and **facultative.** Constitutive heterochromatin is a permanent part of the genome and is not convertible to euchromatin. On the other hand, facultative heterochromatin consists of euchromatin that takes on the staining and compactness characteristics of heterochromatin during some phase of development. Evidence suggests that the constitutive heterochromatin is genetically inactive and that the euchromatin contains most of the genes. We should note that the G, Q, and R bands (Giemsa, quinacrine, and reverse bands, respectively), which are chromosomal bands that are observed using different chemical techniques, are not associated with centromeric heterochromatin.

In humans (and other species), a system is used for identifying chromosomes based on chromosomal size, position of the centromere, and banding patterns (figure 4.4). First, the nonsex chromosomes (autosomes) are numbered 1 to 22 on the basis of length, with the X and Y chromosomes identified separately. Second, the short arm of a chromosome is indicated by p and the long arm by q based on the position of the centromere. Finally, each arm is first divided into regions and then identified by bands. For example, 9q34 indicates the long arm of chromosome 9, region 3, and band 4. This happens to be the location of the gene for ABO blood group types.

In 1956, Joe-Hin Tjio and Albert Levan determined the correct diploid number of human chromosomes (46). (The number had been erroneously thought to be 48 for over thirty years.) Since then, knowledge of human cytogenetics has increased remarkably. These advances resulted mainly from the development of new techniques—ways to culture dividing cells, to swell cells using low osmotic solutions, and to break open cells and spread out the chromosomes. Various new staining techniques have allowed each chromosome, and even small parts of each chromosome, to be identified. Using such biochemical techniques, it is now possible to locate particular genes to certain positions on chromosomes.

Figure 4.5 gives a schematic representation of the X chromosome in *Drosophila melanogaster,* an organism in which a detailed banding pattern is known. The chromosomes in the

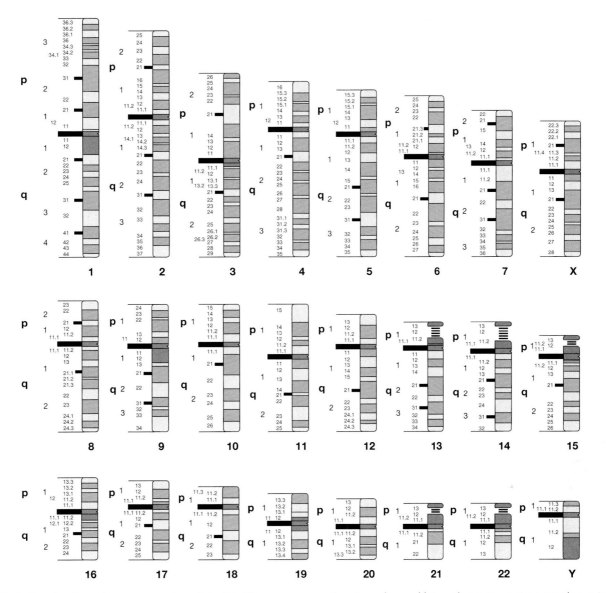

FIGURE 4.4 A schematic representation of the 24 different human chromosomes, in which the heavy black bar indicates the centromeric region, and p and q indicate the short and long arms, respectively. The numbers alongside the chromosomes indicate the regions in each arm. Heterochromatic regions are shown in blue (mainly around centromeres and in satellites), while yellow and green indicate other banded or nonbanded regions.

salivary glands of *Drosophila* and other dipteran insects are extremely large, thus easier to study than other chromosomes. These giant chromosomes, called **polytene chromosomes,** are extended over one hundred times the length of regular chromosomes and include many parallel replicates of the same chromosome held tightly together. More than five thousand bands have been identified in *D. melanogaster,* and more than one thousand of them are on the X chromosome alone. Notice in figure 4.5 the variety in the size of the bands and in their position in relation to each other, allowing identification of particular chromosomal regions. This chromosome is telocentric, having the centromere in region 20, on the far right of the drawing. The gene that causes the white-eyed phenotype, which we will discuss later, is near the left end of the chromosome in region 3, at the other end from the centromere. This was the first chromosome on which the location of genes was mapped, as we will discuss in the next chapter.

The diploid number of chromosomes varies for different species. Specific chromosomes can be identified according to size, position of the centromere, and banding patterns.

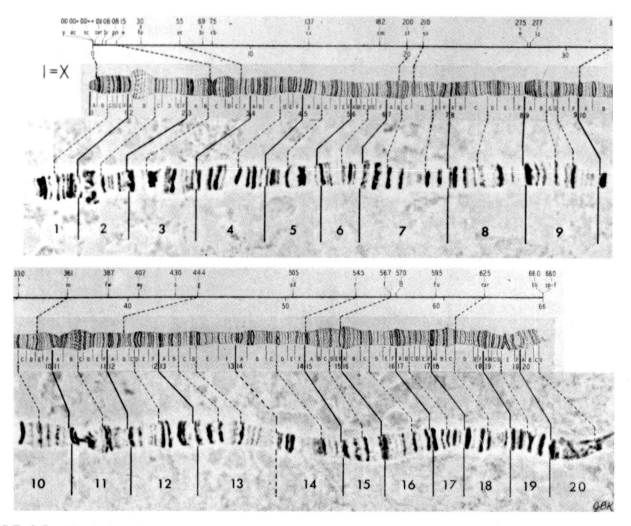

FIGURE 4.5   The chromosome map (top) and a photographic map (bottom) of the X chromosome of *D. melanogaster*. It has been arbitrarily divided into twenty numbered regions, each beginning with a dark band.

From *Modern Genetics* by Ayala and Kiger, p. 141, Benjamin/Cummings Publishers, 1980. (From *Handbook of Genetics*, Vol. 3, by R. C. King, Plenum Press, New York, 1975.)

Photo: From Lefevre, *The Genetics and Biology of Drosophila*, 1:31–66, 1976. Academic Press. [Ashburner and Novitski, editors].

Mitosis

In order to understand genetics, we must understand the process by which hereditary information is passed from cell to cell and from parents to offspring. Although nineteenth century researchers documented the behavior of chromosomes in cell division, they did not realize the genetic implications of chromosomal duplication and movement. Only after the rediscovery of Mendel's principles was the connection made between the earlier chromosomal observations and the principles of segregation and independent assortment.

We will first trace the steps of mitosis, the process of division in which daughter cells are produced that have chromosomal numbers identical to the parental cell. Mitosis is part of the total **cell cycle** for cells undergoing division. In figure 4.6 notice that the period of mitosis (designated M) is a relatively

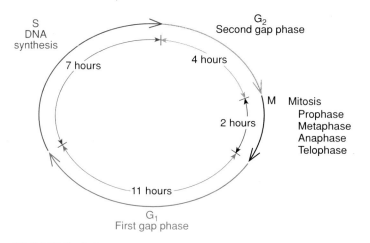

FIGURE 4.6   A schematic representation of the cell cycle for human white blood cells, showing the time involved for each phase.

4.1

Recently, there have been many exciting, technical developments for classification of specific chromosomes. For example, with the identification of specific markers on each human chromosome, it has been possible to identify each chromosome and to artificially color each chromosome differently (termed spectral karyotyping). Figure B4.1 shows all 22 human autosome pairs in decreasing size and the two sex chromosomes, X and Y, each with a different color. Such techniques can be used to identify individuals that have extra or missing chromosomes (or parts of chromosomes) as discussed later in the chapter. In addition, they can be used to determine the structure of chromosomes in related species as discussed on page 99. As an example, figure B4.2 compares chromosomes of a gibbon, an ape from Asia, to that of humans.

The upper row gives examples of four chromosomes that are unchanged in structure between the two species: chromosomes 12, 16, X, and Y in the gibbon are identical in structure to chromosomes 1, 8, X, and Y in humans. The middle row gives gibbon chromosomes that are composed of translocations of whole arms from the human chromosomes. For example, gibbon chromosome 5 is composed of parts of human chromosomes 1 and 13. In the bottom row, chromosomes with more complicated patterns are shown where, as an example, gibbon chromosome 9 is made up of parts of human chromosome 13, 1, 4, and 10.

Figures B4.1 and B4.2: From E. Schröck, et al., "Multicolor Spectral Karyotyping of Human Chromosomes," *Science* 273:495, 26 July 1996. © American Association for the Advancement of Science.

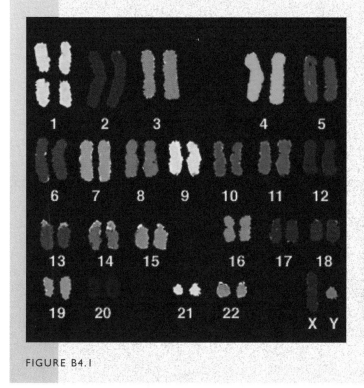

FIGURE B4.1

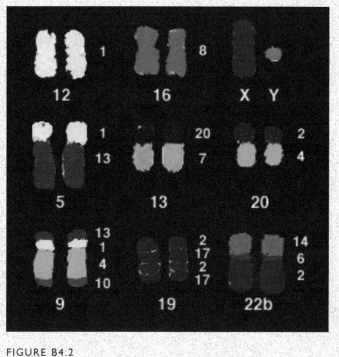

FIGURE B4.2

small part (2 hours out of 24) of the total cell cycle. The times given in figure 4.6 for each stage are for dividing human white blood cells. The duration of the different stages in the cell cycle varies considerably, depending upon the organism, cell type, temperature, and other factors.

The remainder of the cell cycle, called **interphase,** is divided into three parts, termed G_1, S, and G_2. Interphase is generally a time of high metabolic activity, during which DNA, the hereditary material, is synthesized and replicated. Some mature cell types, such as some nerve cells, remain in interphase and never divide further. After a given mitosis is

completed, there is an initial time gap, called G_1, in which cells grow, metabolize, and perform their designated functions for the organism. Following the G_1 is a period of DNA synthesis, designated S, in which the DNA is replicated (we will discuss this process in detail in chapter 7). After synthesis, another gap phase, G_2, occurs, followed by the next mitotic division. When an organism is growing, this cycle repeats itself many times, eventually resulting in an individual with billions of cells.

Mitosis itself can be divided into four different stages, listed in chronological sequence as: **prophase, metaphase, anaphase,**

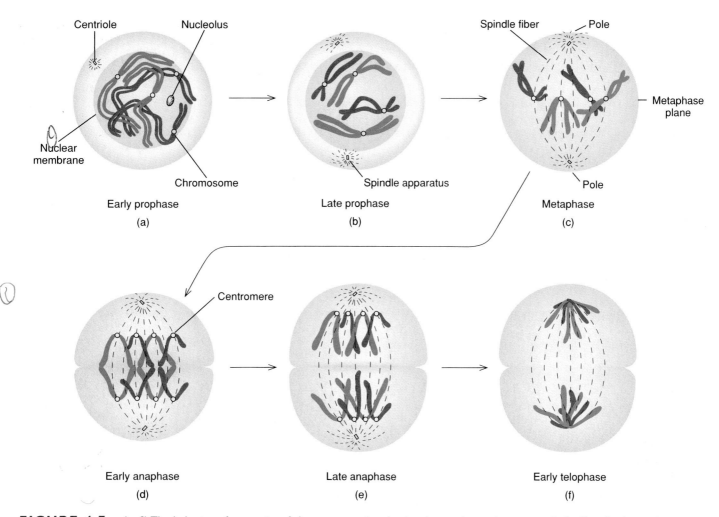

FIGURE 4.7   (a–f) The behavior of two pairs of chromosomes in mitosis, where pair one is acrocentric (red) and pair two is metacentric (blue).

and **telophase.** These stages actually merge one into another, but each has definite characteristics, particularly in relation to chromosomal behavior, that we can use to identify them. Let us go through these phases to understand the elements of mitosis and cell division and how cells produce identical daughter cells (figure 4.7 and review box figure 1.2B).

PROPHASE

At the beginning of mitosis (the end of G_2 interphase), the chromosomes coil and condense, thus becoming visible under the light microscope. The chromosomes have replicated and each has two sister **chromatids** at this stage (figure 4.7a). The two sister chromatids are attached to each other at the centromere. The cromatids are identical and are the result of the DNA replication in the earlier S phase of the cell cycle. Because the two sister chromatids are attached at the centromeric region, they are considered one chromosome at this stage.

Also during prophase, the **nucleolus,** the largest cell organelle besides the nucleus, usually disappears and the nuclear

membrane begins to break down (figure 4.7b). The **centriole,** an organelle important in organizing cell division, divides. Around it a new structure, the spindle apparatus, which is composed of a series of **spindle fibers** that stretch to the poles of the cell, also appears (figure 4.8). Some spindle fibers attach directly to the kinetochore. The spindle structure is fundamental to the proper movement of the chromosomes to the two poles in the latter part of mitosis.

METAPHASE

The nuclear membrane completely disappears during metaphase, leaving the chromosomes free in the cytoplasm (figure 4.7c). The spindle fibers attach to the centromeres of the chromosomes, and the chromosomes gradually change their orientation, moving to the **metaphase** (or **equatorial**) **plane,** an imaginary region midway between the two poles. Generally, the chromosomes are most condensed at this stage of mitosis, making it a good stage for preparation of karyotypes. Because sister chromatids are attached at the centromeres, a chromosome at

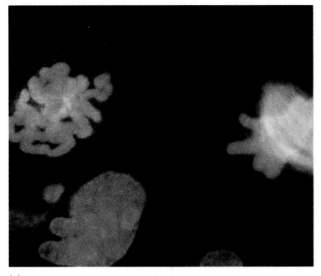

(a)

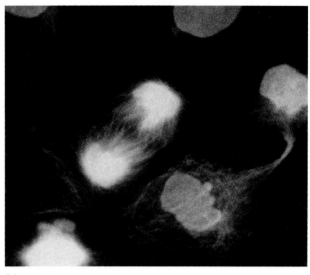

(b)

FIGURE 4.8 Photo (*a*) shows two cells, one in prophase and the other in metaphase. Photo (*b*) shows one cell in anaphase and another in very late telophase. The images in both slides are stained for immunofluorescene microscopy. The green color is from an antibody directed against microtubules, which compose the spindle fibers, and the red is a chemical called propidium iodide, which specifically stains chromosomes.

Courtesy of Mark S. Ladinsky and J. Richard McIntosh, University of Colorado, Boulder.

metaphase when it is spread out in a karyotype appears either as an X or a V, depending upon whether it is metacentric or telocentric, respectively. The acrocentric is intermediate between these shapes.

ANAPHASE

Anaphase, nearly always the shortest period of mitosis, begins with the separation of the centromeres (figure 4.7*d* and figure B1.2). As a result, each sister chromatid is now called a daughter chromosome. The spindle fibers contract and cause the movement of the centromeres to the poles, with the two identical (homologous) daughter chromosomes moving to opposite poles (figure 4.7*e* and figure B1.2). A chromosome appears as a V, J, or I at this stage, depending upon whether it is metacentric, acrocentric, or telocentric, respectively.

TELOPHASE

In telophase, the last stage of mitosis, the two sets of daughter chromosomes have reached opposite poles (figure 4.7*f* and review box figure 1.2B) where they begin to uncoil. The nuclear membrane reappears and encloses the chromosomes, and the nucleoli are re-formed.

Next, in animals, the cell membrane develops a furrow; in plants, a cell plate develops that divides the cell in two. This splits the two sets of daughter chromosomes and the cytoplasm between the two daughter cells. The daughter cells have chromosome numbers identical to the original cell.

> Mitosis, cell division that results in the production of two identical daughter nuclei, is divided into four stages: prophase, metaphase, anaphase, and telophase.

MEIOSIS

Before describing meiosis, we will summarize why it is fundamental for maintaining the genetic state of an organism and for generating new genetic types. First, meiosis allows the conservation of chromosome number in sexually reproducing species. If not reduced via meiosis prior to fertilization, the chromosome number would be doubled. Second, meiosis permits the different maternal and paternal chromosomes to randomly assort into each gamete. And finally, the presence of crossing over in meiosis I allows the exchange of genetic material on homologous chromosomes and can result in new combinations of alleles at different genes on a chromosome. (See chapter 5 for a discussion of crossing over.)

Meiosis produces cells with half the parental number of chromosomes through a process involving two cell divisions, meiosis I and meiosis II. **Meiosis I** is also called **reductional division,** because the number of chromosomes is reduced to the haploid number. In meiosis I, the sister chromatids remain attached to each other while the homologous chromosomes separate. **Meiosis II** is called **equational division** and is much like mitosis in that the centromeres on sister chromatids separate and the chromosome number remains unchanged (figure 4.9 and figure B1.3a).

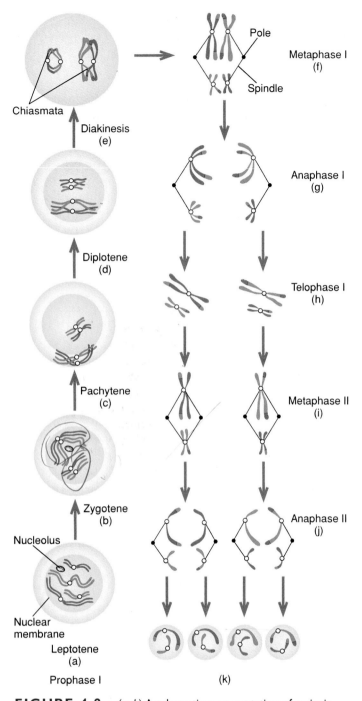

FIGURE 4.9 (a–k) A schematic representation of meiosis for two chromosomes. Meiosis I results in the separation of the homologous chromosomes, and meiosis II results in the separation of duplicated chromosomes. Here the two larger red and blue chromosomes are homologous, as are the two smaller red and blue chromosomes.

MEIOSIS I

As with mitosis, both meiotic divisions can be divided into four stages, based on the position of the chromosomes and other characteristics. The interphase period before meiosis I is similar to that in mitosis, having an S period for DNA synthesis.

Prophase I

The most complex phase of the whole process of meiosis, **prophase I,** is itself broken down into five different parts (figure 4.9*a–e* and figure B1.3a). The first stage, called the **leptotene** (thin-thread) stage, is marked by the appearance of the chromosomes as long threads when seen through a light microscope. This stage is much like the start of prophase in mitosis.

However in the next stage, the **zygotene** (joined-thread) stage, unlike in mitosis, homologous chromosomes pair side by side and gene by gene with each other. The process of the lateral association of homologues, called **synapsis,** occurs during zygotene and is a key difference between meiosis and mitosis. When the two homologous chromosomes (the large red and blue pair or the small red and blue pair), consisting of four chromatids, are paired, the structure is called a **bivalent.** The next, or third, stage of prophase I, the **pachytene** (thick-thread) stage, consists of a shortening and thickening of the bivalents and is the stage during which synapsis is complete. During this stage, portions of homologous chromosomes may be exchanged, a phenomenon called **crossing over** or **recombination.** The structure that allows crossing over is the **synaptonemal complex,** a structure made of protein and DNA that is found between the synapsed homologues. We will discuss the consequences of recombination in chapter 5.

In the fourth stage, the **diplotene** (double-thread) stage, the homologues begin to separate as the synaptonemal complex breaks down, particularly in the region on either side of the centromere. The sister chromatids remain attached at the centromeric region. In addition, the homologous chromosomes generally have one or two areas, called **chiasmata,** in which they are still close or in contact. Chiasmata are the physical evidence that recombination occurred earlier when the homologues were synapsed. Generally, there is at least one chiasma per chromosome arm, but in long chromosomes, there may be several.

The last stage of prophase I, **diakinesis,** is characterized by shortened chromosomes and the terminalization of the chiasmata; that is, the chiasmata appear to move to the ends of the chromosomes. As in mitosis, at the end of prophase, the nucleolus and the nuclear membrane disappear.

Metaphase I

In **metaphase I** (figure 4.9*f* and figure B1.3a), the chromosomes have aligned at the equatorial plane as homologous pairs and become attached to the spindle fibers. The bivalents orient so that one centromere is on each side of the metaphase plane, with the chiasmata lined up along it. As a result, metaphase I is different from mitotic metaphase, in which the homologous chromosomes were not associated as bivalents.

Anaphase I

At the stage known as **anaphase I** (figure 4.9*g* and figure B1.3a), the two sister centromeres in homologous chromosomes of each bivalent move apart, so that one of each homologous chromosome pair goes toward each pole. As the homologues move apart, the chiasmata completely terminalize. In anaphase I, unlike mitotic anaphase, the sister chromatids stay together, joined at the centromere.

Telophase I

In most organisms, after the chromosomes have migrated to the poles, the nuclear membrane forms around them and the cell divides into two daughter cells, just as in mitosis. However, the exact cytological details of **telophase I** (figure 4.9*h* and figure B1.3a) vary, particularly in plants.

MEIOSIS II

Between meiosis I and meiosis II there is usually only a short interphase called **interkinesis.** No DNA synthesis takes place during this period; at this point, each cell contains only a haploid complement (n) of chromosomes. (Note that each chromosome contains two sister chromatids.) The next stage is **prophase II,** which unlike prophase I is quick, uncomplicated, and generally similar to mitotic prophase. In **metaphase II** (figure 4.9*i* and figure B1.3a), the centromeres attach to spindle fibers and migrate to the metaphase plate. At the start of **anaphase II** (figure 4.9*j*), as in mitotic anaphase, the chromosomes first separate at the centromeres when they divide. One centromere begins to move with one daughter chromosome to one pole, while the other centromere and chromosome go to the opposite pole. There are n chromosomes at each pole (each chromosome is composed of a single chromatid) when the nuclear membrane forms around them in **telophase II.**

Before we continue, it is useful to compare the changes in chromosome number and amount of DNA that occur during the mitotic and meiotic cycles (see figure B1.3a). First, as we have already discussed, homologous chromosomes synapse in meiosis I, allowing recombination and precise separation of homologues, but this does not occur in mitosis. Second, at the beginning of both mitosis and meiosis, there are 2n chromosomes in each cell (figure 4.10*a*). Each chromosome is composed of two sister chromatids, so that the amount of DNA is four times the haploid amount (indicated as 4C in figure 4.10*b*). At the end of mitosis, there are still 2n chromosomes, but the amount of DNA has been reduced to 2C per cell. At the end of meiosis I (indicated by A-I, or anaphase I, in figure 4.10), the number of chromosomes has been reduced to n and the amount of DNA to 2C. As in mitosis, meiosis II does not change the number of chromosomes, but the amount of DNA is halved to C as the sister chromatids separate.

SPERMATOGENESIS AND OOGENESIS

The sequence of nuclear events in meiosis that we have described is similar in the production of both male and female gametes; however, there are cytoplasmic differences (figure 4.11). The production of male gametes in animals, termed **spermatogenesis,** takes place in the **testes,** the male reproductive organs. The process begins with the growth of an undifferentiated diploid cell called a spermatogonium, which then differentiates into a primary spermatocyte. This cell, in turn, undergoes the first meiotic division (the reductional division), becoming two haploid secondary spermatocytes. After meiosis II, these cells each divide into two haploid spermatids. The last step is the differentiation of the spermatids into new sperm cells with long tails (figure 4.12).

Oogenesis, the production of female gametes in animals, occurs in the **ovaries,** the female reproductive organs. The process is parallel to that of spermatogenesis in that it also begins

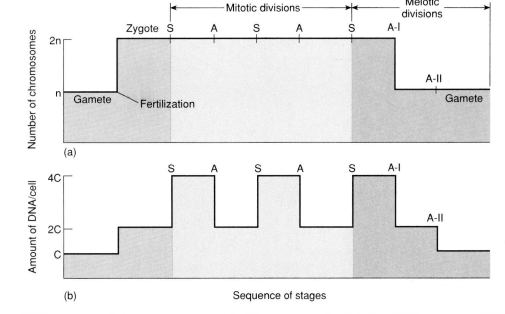

FIGURE 4.10   (*a*) The numbers of chromosomes per cell in different stages of cell division. (*b*) The amount of DNA per cell at the various stages of cell division. S indicates DNA synthesis and A indicates anaphase.

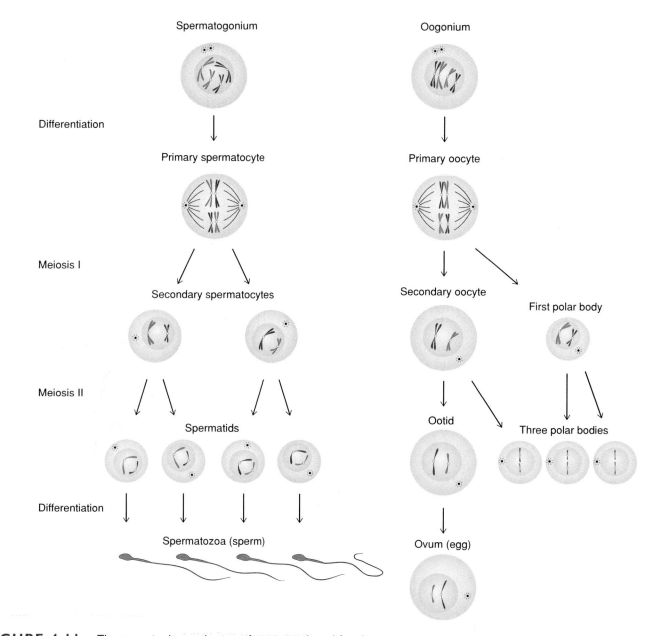

FIGURE 4.11 The stages in the production of mature male and female gametes.

with a diploid cell, or oogonium. However, the cytoplasm is concentrated into only one of the cell products, one of the secondary oocytes, while the other cell products (first and second polar bodies) receive only a small amount of cytoplasm. Meiosis takes place near the cell membrane, allowing the formation of polar bodies with little cytoplasm. The polar bodies remain on the surface of the egg where they eventually disintegrate. Therefore, only one mature haploid egg is produced in oogenesis, while four mature haploid sperm are produced in spermatogenesis. The concentration of cytoplasm in one egg cell provides nourishment for the developing embryo after fertilization.

Spermatogenesis is continuous in adult animals that reproduce all year round, such as humans, and seasonal in animals that breed seasonally. The capacity for sperm production is extremely high in most animals, with adult males producing many millions of sperm in their lifetimes. On the other hand, egg production in females is generally much lower; for example, a human female can produce only around five hundred mature eggs in her lifetime. All these eggs develop into primary ooctyes before the female is born and are arrested in prophase I until puberty, when they resume meiosis and begin maturing into eggs one by one and are ovulated each month. Many other eggs never mature and are reabsorbed.

In higher plants, the formation of male gametes is known as **microsporogenesis.** It is similar to spermatogenesis in animals in that meiosis gives rise to four haploid cells, all of an equal size and all potentially functional. **Megasporogenesis,**

FIGURE 4.12 Sea urchin sperm on a sea urchin egg. Only one sperm will fertilize this egg.
© Francis Leroy, Biocosmos/SPL/Photo Researchers, Inc.

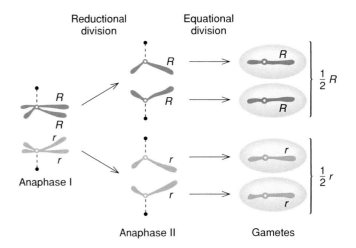

FIGURE 4.13 The chromosomal basis of Mendel's principle of segregation operating in an *Rr* heterozygote. As the diagram shows, ½ *R* gametes and ½ *r* gametes are produced.

the formation of eggs in plants, is also similar to the meiotic process described for animals. However, the products vary considerably among species and are complicated by double fertilization. (It is best to consult a botany text for more details.)

> Meiosis is composed of one replication of DNA and two cell divisions. Meiosis I, the reductional division, results in the separation of homologous chromosomes and the halving of chromosome number. Meiosis II results in the separation of the sister chromatids, with no reduction in chromosome number. In animals, mature male and female gametes are produced by spermatogenesis and oogenesis, respectively.

GENES AND CHROMOSOMES

As we mentioned before, several early researchers pointed out the parallel between the behavior of genes and chromosomes. Now that we have discussed meiosis, we can more precisely show how the behavior of genes and chromosomes is related. For example, figure 4.13 illustrates the principle of segregation in a heterozygote, *Rr* (see also review box figure 1.3A). In reductional division (meiosis I), the two homologues containing different alleles separate into two different cells. Each of these cells then divides into two cells, which contain the same alleles as the cell produced from meiosis I. As a result, half the gametes are *R* and half are *r*. In other words, equal segregation of alleles into gametes is really the result of reductional division in meiosis.

The chromosomal explanation for independent assortment is given in figure 4.14 and review box figure 1.3A. Let us examine gamete production in the F₁ double heterozygote *Rr Yy*, where the two genes are on different chromosomes. Two outcomes are equally likely: Either the chromosomes containing *R* and *Y* go to the same pole and the ones containing *r* and *y* go to the other pole, or the *R* and *y* go to one pole and the *r* and *Y* go to the other pole. These two alternatives are equally possible because they result from the chance placement of the centromeres on one side or the other of the metaphase I plane. Each of the four types of cells divides in meiosis II, so that overall, the four gametic types occur in equal proportions. Therefore, the principle of independent assortment also generally results from the behavior of chromosomes in meiosis I. (We should note that this assumes that no recombination takes place between the loci and their centromeres. See chapter 5 for a discussion of recombination.)

The parallel behavior of genes and chromosomes was strong support that genes actually resided in chromosomes, but it was not direct or definitive evidence that specific genes and specific chromosomes were connected. The first evidence of this sort was provided by the experiments of Thomas H. Morgan, who in 1910 described X-linked inheritance in *Drosophila*. Morgan had begun to maintain *Drosophila* cultures shortly after the turn of the century. A momentous event occurred when a white-eyed male appeared in a culture of all wild-type (red-eyed) flies. Using this mutant, Morgan experimentally confirmed that genes are on chromosomes.

Morgan carried out several experiments designed to understand the observed association between the white-eyed mutant and its sex. For example, he crossed red-eyed females and white-eyed males to produce F₁ progeny that were all red-eyed. However, when he crossed the F₁ individuals, the F₂ had the expected 3:1 ratio, but all the white-eyed flies were males (figure 4.15). Furthermore, the reciprocal cross,

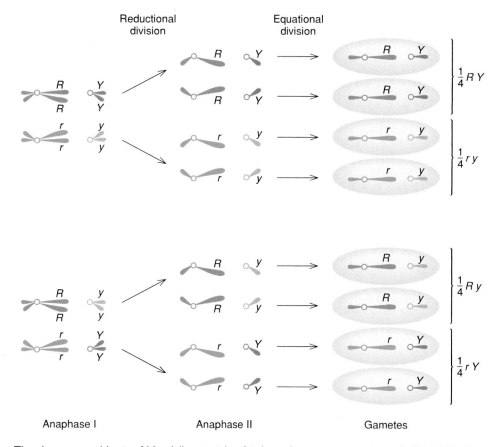

Reductional division

Equational division

$\frac{1}{4} R Y$

$\frac{1}{4} r y$

$\frac{1}{4} R y$

$\frac{1}{4} r Y$

Anaphase I   Anaphase II   Gametes

FIGURE 4.14   The chromosomal basis of Mendel's principle of independent assortment in an *Rr Yy* double heterozygote. Equal proportions of the four types of gametes are produced. The top and bottom halves of the figure are two equally likely outcomes of independent assortment.

white-eyed females and red-eyed males, gave different results (figure 4.16), with half the F_1 being white-eyed (all of which were males) and half red-eyed (all females). The F_2 progeny from a cross of these F_1 individuals were half white-eyed and half wild-type.

It was known that male and female *Drosophila* differ with respect to their chromosomes: females have two X chromosomes and males have one X and one Y. Morgan suggested that the results of his crosses could be explained if the eye color gene was on the X chromosome and not on the Y chromosome. Obviously, the explanation that the white-eyed mutant was X-linked was consistent with the observations from his experimental crosses.

Although the association of the *white* allele with the X chromosome appeared to be convincing proof that genes were on chromosomes, another experiment was necessary to exclude all other explanations. Calvin Bridges, a student of Morgan, provided the crucial evidence in 1916, also using the *white* gene. When white-eyed females and red-eyed males were crossed, as in figure 4.17, approximately one in two thousand F_1 offspring had unexpected eye color; that is, they were either white-eyed females or red-eyed males. Bridges suggested that

these exceptional flies were the product of an abnormal separation of chromosomes in meiosis called **nondisjunction.** In this case, nondisjunction occurred in females such that the two X chromosomes did not separate properly; instead both went to the same pole, resulting in eggs with either two X chromosomes or no X chromosomes (see figure 4.17).

If nondisjunction did occur, and assuming normal meiosis in males, some offspring would be $X^W X^W Y$, with three sex chromosomes. These would be the unexpected white-eyed females, because there is no wild-type allele. Other flies (the red-eyed males) would be X^+, with only one sex chromosome. (We will discuss the chromosomal basis of sex determination in *Drosophila* later in this chapter.) In this explanation, the white-eyed females inherit both of their X chromosomes from their mother, and the red-eyed males inherit their X chromosome from their father. Bridges corroborated his hypothesis by examining the chromosomes of the unexpected flies. He found that the exceptional females were XXY and the exceptional males were only X. In other words, Bridges's discovery that an abnormal genetic condition and an abnormal chromosomal condition directly corresponded gave convincing proof that genes were indeed on chromosomes.

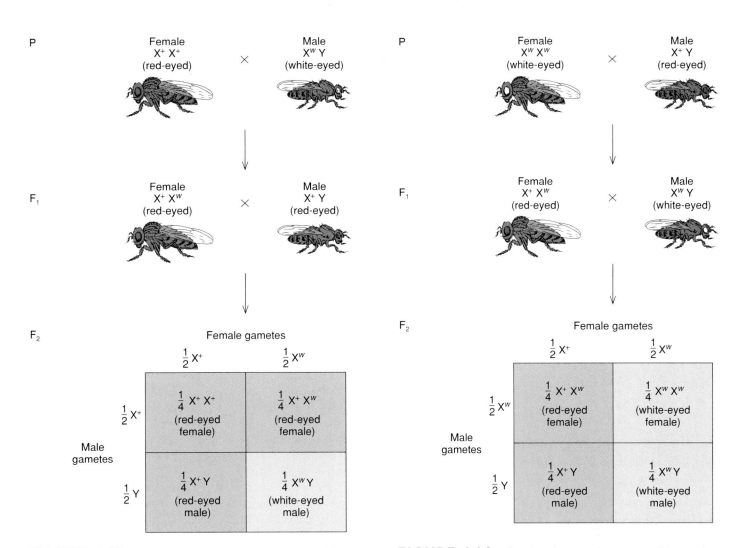

FIGURE 4.15 Results of a cross between a red-eyed female and a white-eyed male to produce an F$_2$ phenotypic ratio that is sex dependent; that is, all white-eyed flies are male.

FIGURE 4.16 Results of a cross between a white-eyed female and a red-eyed male to produce an F$_2$ that is different from the reciprocal cross in figure 4.15.

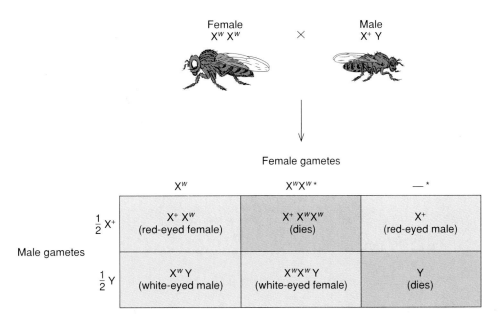

FIGURE 4.17 The results of a cross between white-eyed females and red-eyed males. The * indicates female gametes produced by rare nondisjunction events.

The parallel behavior of genes and chromosomes provides a mechanistic explanation for the principles of segregation and independent assortment. The observation of this parallel behavior was also the first indication that genes were on chromosomes. Definitive proof resulted from experiments with X-linked genes in *Drosophila* and from the confirmation that individuals with exceptional eye color were the result of meiotic mistakes that caused abnormal chromosomal numbers.

CHROMOSOMAL CHANGES

Until now, we have been discussing individuals with a normal set of chromosomes. However, an organism's chromosomal constitution may change, or mutate, resulting in differences among individuals of a given species. Chromosomes can change in two basic ways—by altering their structure or by altering their number (table 4.3). Both types of alterations may have consequences besides their immediate effect on the chromosomes. For example, individuals heterozygous for chromosomes with different structures often have lowered fertility, and individuals with altered numbers of chromosomes may be inviable or sterile.

Table 4.3   Types of Chromosomal Changes

Changes in Structure
Duplications
Deletions (or deficiencies)
Inversions
Translocations

Changes in Number
Euploidy level
Autopolyploids
Allopolyploids
Aneuploidy
Monosomies (2n − 1)
Trisomies (2n + 1)

CHANGES IN CHROMOSOMAL STRUCTURE

The four possible types of changes in chromosomal structure are duplications, deletions (or deficiencies), inversions, and translocations. When breaks occur in chromosomes, any two broken chromosomal ends may reunite. Chromosomal changes are often the result of chromosomal breaks in which the same broken ends do not reunite. Generally, such chromosomal mutations occur infrequently, but some researchers have estimated that more than one in a thousand new gametes may have some type of chromosomal mutation. Note that these chromosomal changes result from chromosomal breaks and abnormal rejoining. The ensuing consequences, which we will discuss in the following section, are the result of the gene-for-gene synapsis of normal and abnormal chromosomes and the subsequent problems that occur in anaphases I and II.

Duplications

When a chromosomal segment is represented twice, it is called a **duplication.** We can categorize duplications by the position and order of the duplicated region (figure 4.18). First, the duplication may be adjacent to the original chromosomal region. When this occurs, the order may either be the same as the original order, called a **tandem duplication,** or the opposite order, called a **reverse duplication.** Second, the duplicated region may not be adjacent to the original segment, resulting in a **displaced duplication.** In this case, the displaced duplication may still be on the same chromosome, as illustrated in figure 4.18, or it may be on another chromosome.

One possible way a tandem duplication can be generated is illustrated in figure 4.19. It is assumed that the homologues overlap and that there are simultaneous breaks in the two homologous chromosomes at *different* points. If the different homologues reunite, one chromosome will have a tandem duplication and the other a deletion of the duplicated region (figure 4.19c). In this case, the duplication and the deletion are reciprocals of each other. In chapter 5, we will discuss another mechanism (unequal crossing over) by which small duplications (and deletions) can be generated.

When an individual is heterozygous for a duplication and a normal chromosome, the duplicated region does not have a homologous segment to pair with in meiosis I. As a result, a loop of the duplicated region may develop (figure 4.20a). In some cases, part of the chromosome may bend back and join

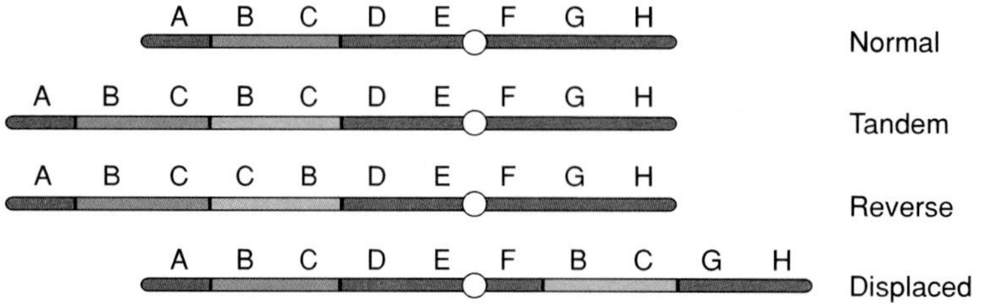

FIGURE 4.18   A normal chromosome (top), followed by three types of duplication that can result, depending on the position and order of the duplicated segment (shown in green).

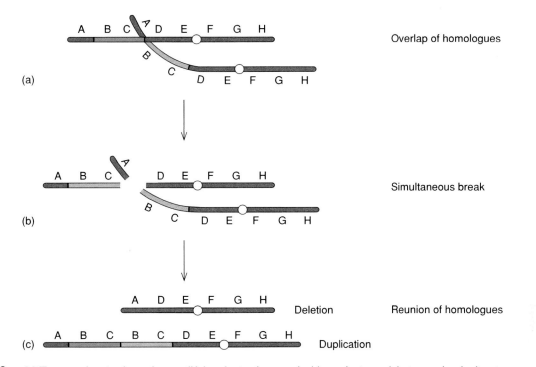

FIGURE 4.19 (*a*) Two overlapping homologues (*b*) break simultaneously, (*c*) producing a deletion and a duplication.

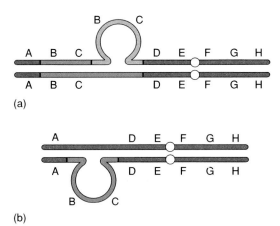

FIGURE 4.20 Pairing of chromosomes in individuals heterokaryotypic for (*a*) duplication or (*b*) deletion.

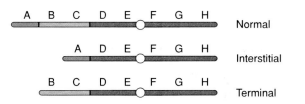

FIGURE 4.21 A normal chromosome (top), followed by two types of deletions that may result, depending on the positions of the deleted segment.

with the duplicated sequence on the same chromosome. Individuals having two chromosomal types such as this are called **heterokaryotypic.**

Individuals that are either heterozygous or homozygous for small duplicated segments may be viable, although they often exhibit some phenotypic effects (see the discussion of the *Bar* eye mutations in *Drosophila* in chapter 5). If individuals are viable, there is a potential for further evolutionary change in these extra genes. In fact, it is thought that this happened with the different globin genes, the genes that code for the components of the protein hemoglobin. These genes may have descended from an ancestral gene was that duplicated, and then the duplicate copies diverged in their function.

Deletions

A missing chromosomal segment is termed a **deletion** or a **deficiency.** We show one way a deletion can occur in figure 4.19. This type of deletion, where an internal part of the chromosome is missing, is called an **interstitial deletion.** If there is only one break and the homologue fails to rejoin, a **terminal deletion** can occur. In this case, the tip of the chromosome is usually lost in cell division because it does not have centromere. Interstitial and terminal deletions are illustrated in figure 4.21.

When deletions are homozygous, they are often lethal, because essential genes are missing. Even when heterozygous, lethals can cause abnormal development. A well-known example in humans is the deletion of a substantial part of the short arm of chromosome 5 (5p), which when heterozygous causes the *cri du chat* (cry-of-the-cat) syndrome. Infants with this syndrome generally have a characteristic high-pitched, catlike cry as well as microcephaly (small heads) and severe mental retardation. Generally, they die in infancy or early childhood.

In addition, deletion heterozygotes usually show abnormal chromosomal pairing in meiosis. Because the normal chromosome does not have a homologous region to pair with, a deletion loop is formed (figure 4.20b). This phenomenon may be seen in meiotic chromosomes or in the polytene chromosomes of *Drosophila* and a few other organisms.

Several other characteristics are useful in identifying deletions. First, deletions, unlike other mutations, generally do not revert, or mutate back, to the wild-type chromosome. Second, in deletion heterozygotes, recessive alleles on the normal chromosome are expressed because the deletion chromosome is missing the homologous region. Expression of recessive alleles in such cases, called **pseudodominance,** is useful in defining the length of the deleted segment. For example, let us assume that genes *B* and *C* were deleted on one chromosome as in figure 4.20b. If we have wild-type chromosomes with recessive mutants at different genes, these should be expressed if they are in the deleted region. Figure 4.22 illustrates that the deletion covers genes *B* and *C* but not gene *D*. As will be discussed in chapter 5, deletions can be used to map the sequence of genes on the chromosome.

> Duplications occur when a chromosomal segment is present more than once; deletions occur when a segment is missing. Both can result in phenotypic changes or lethality, although deletions often have greater average effects. In meiotic chromosomes, duplication and deletion heterozygotes often form loops.

Inversions

Most of the homologous chromosomes in a population have genes in the same sequence. However, in some instances the sequence may differ on different chromosomes. Alterations in

the sequence of genes, called **inversions,** may be of two different kinds relative to the position of the centromere. If the inverted segment does not contain the centromere, it is called a **paracentric inversion** (Greek: *para* = next to), while if the inversion spans the centromere, it is called a **pericentric inversion** (Greek: *peri* = around).

Inversions can be generated by a simultaneous break at two points in a chromosome, followed by an incorrect reunion. For example, figure 4.23 illustrates how a looped chromosome may break and rejoin to generate either a paracentric or a pericentric inversion. Note in this example that the paracentric inversion results in an acrocentric chromosome just like the noninverted chromosome, but that the pericentric inversion results in a metacentric chromosome, because the position of the centromere has been changed.

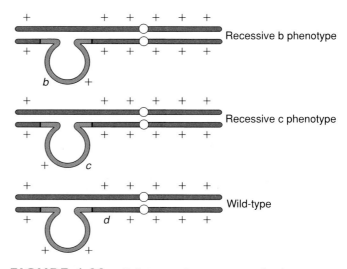
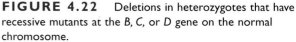

FIGURE 4.22   Deletions in heterozygotes that have recessive mutants at the *B, C,* or *D* gene on the normal chromosome.

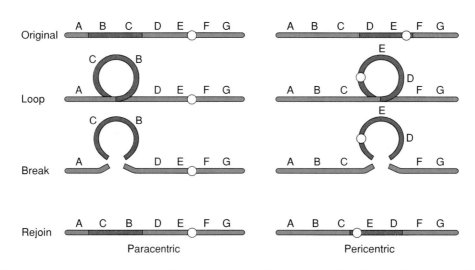

FIGURE 4.23   A mechanism for the generation of paracentric and pericentric inversions by two chromosomal breaks.

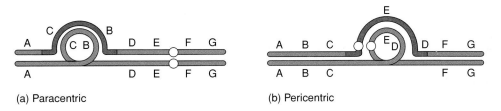

FIGURE 4.24   Loop formed in chromosomal heterozygotes for (a) a paracentric inversion and (b) a pericentric inversion.

Individuals heterozygous for an inversion can be recognized by the presence of inversion loops in meiotic pachytene chromosomes. The results of meiosis in inversion heterozygotes will be discussed in chapter 5. These structures occur because of the affinity of the two homologues. The only way the two homologues can pair is if one twists on itself and makes a loop while the other makes a loop without a twist. Examples of such loops for the two types of inversions are given in figure 4.24. These loops can best be seen in the polytene chromosomes of organisms such as *Drosophila pseudoobscura*. Both a single inversion example, Arrowhead-Standard, and a multiple inversion example, Chiricahua-Standard, are shown in figure 4.25. We will mention these sequences again when we discuss population genetics in chapter 16.

> Inversions occur when the sequence of genes on a chromosome is reversed.

Translocations

A **translocation** is the movement (by breaking and rejoining) of a chromosomal segment from one chromosome to another, nonhomologous chromosome. Figure 4.26 illustrates two types of translocations, an **interstitial translocation,** involving the

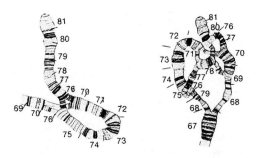

FIGURE 4.25   Inversion heterokaryotypes in *Drosophila pseudoobscura.* The Arrowhead-Standard heterokaryotype (left) illustrates a single inversion, while the Chiricahua-Standard heterokaryotype (right) shows a multiple inversion.
From Elliot B. Spiess, *Genetics,* 23:28, 1938. John Wiley and Sons, Inc.

one-way movement of a segment, and the more common **reciprocal translocation,** involving a two-way exchange of chromosomal segments. If two of the segments that join in a reciprocal translocation are large and the other two are small, the smaller translocated chromosomes are often lost. In this case, the number of chromosomes is reduced by the chromosomal exchange. Obviously, translocations can change both the size of chromosomes and the position of the centromere.

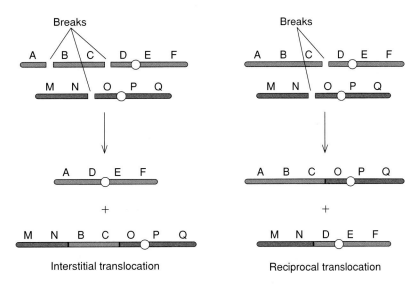

FIGURE 4.26   The most common types of events leading to two types of translocations. An interstitial translocation involves the one-way movement of a segment, while a reciprocal translocation involves a two-way exchange of chromosomal segments.

Even though chromosomal segments have been exchanged between chromosomes in a reciprocal translocation, the affinity of the homologous regions results in pairing during meiosis I. If nearly equal parts of chromosomes are exchanged or not exchanged, the paired chromosomes in a translocation heterozygote have a cross appearance in metaphase I (figure 4.27). During anaphase I, two major types of segregation occur: one in which alternate centromeres go to the same pole and one in which adjacent centromeres go to the same pole (adjacent-1). The third type of segregation pictured in figure 4.27 (adjacent-2) is rare, because it requires that homologous centromeres go to the same pole.

When alternate centromeres go to the same pole, the chromosomes often form a figure eight shape in early anaphase I. (Remember that the chiasmata terminalize so that the homologous chromosome ends remain together.) The products of this event, which is known as **alternate segregation,** are balanced so that each gamete has a full complement of chromosomes, either two untranslocated or two balanced translocated. On the other hand, when adjacent centromeres segregate together, **adjacent segregation,** the chromosomes appear as a ring at metaphase I. When this occurs, the products are unbalanced, resulting in duplications and deletions in the gametes.

Some plants, and also a few animals, have a series of reciprocal translocations, so that chromosomal heterozygotes also have nearly all the chromosomes associated in a large ring (or rings) in meiosis (figure 4.28). However, at anaphase these chromosomes may undergo an orderly alternate segregation, producing only zygotes with a balanced chromosomal complement.

Although translocations can result in normal chromosomes, they can also cause several human diseases. For example, about 5% of individuals with **Down syndrome** have one parent who is heterozygous for a translocation (figure 4.29). In this instance, chromosome 14 is translocated onto chromosome 21. Half of the time, the heterozygote produces either the normal set or a balanced translocated set of chromosomes, making the progeny either normal or translocation heterokaryotypes, respectively. The other half of the time, unbalanced chromosomes are produced, either a 14 without the translocated 21 segment or a translocated 14 with the attached 21 plus a normal 21. In the first case, offspring get only one 21 chromosome, a lethal chromosomal component. In the second instance, three 21 chromosomes are received, resulting in Down syndrome. Overall then, approximately one-third of the live births from such a translocation heterokaryotype can be expected to have Down syndrome. In actual fact, the proportion is less than this, primarily because some Down individuals do not survive gestation.

Note that this cause of Down syndrome has implications for genetic counseling. First, Down syndrome could recur in children of a translocation heterokaryotype, whereas normally Down syndrome does not recur in sibs (see discussion of trisomy later in this chapter). Second, half of the phenotypically normal sibs of Down individuals are themselves translocation heterokaryotypes, and therefore could produce Down progeny.

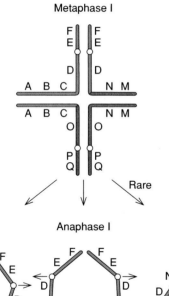

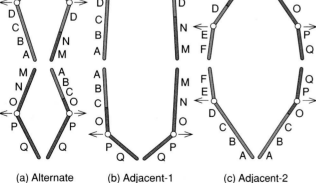

FIGURE 4.27   A heterozygote for a reciprocal translocation and the outcomes of meiotic division. (*a*) Alternate segregation produces balanced chromosomes. (*b*) and (*c*) Adjacent-1 and adjacent-2 segregation results in duplication and deletions.

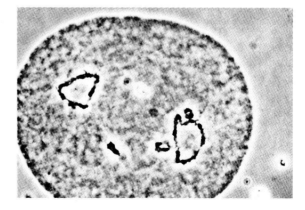

FIGURE 4.28   Cell of *Clarkia,* an annual plant, with two rings of six at about four o'clock and ten o'clock and three smaller bivalents.

Courtesy of William L. Bloom, University of Kansas.

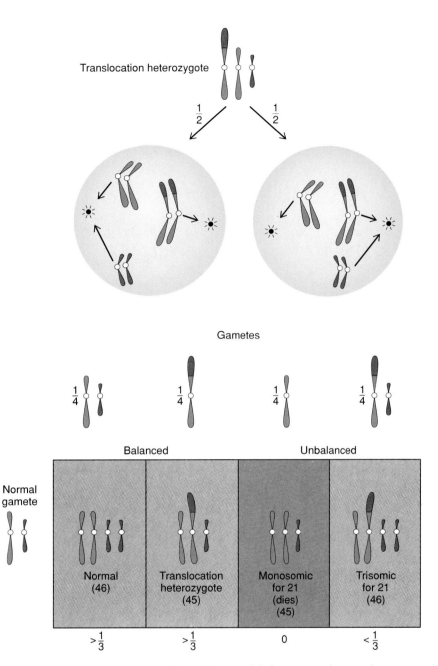

Gametes

FIGURE 4.29 Production of individuals with Down syndrome (trisomy 21) from a translocation heterozygote. Chromosome 14 is indicated in blue and chromosome 21 in red.

Translocations are the result of the movement of a chromosomal segment to a nonhomologous chromosome. Normal meiotic products result when there is alternate segregation in translocation heterokaryotypes, but otherwise, duplications and deficiencies occur.

CHANGES IN CHROMOSOMAL NUMBER

The numbers of chromosomes may vary in two basic ways: **euploid** variants, in which the number of chromosomal sets differ, and **aneuploid** variants, in which the number of a particular chromosome is not diploid. As one might expect, changes in chromosome number, either euploid or aneuploid, generally have an even greater effect on survival than do changes in chromosome structure. In fact, in humans, more than half of the spontaneous abortions that occur in the first three months of pregnancy involve fetuses with aneuploidy, polyploidy, or other large chromosomal aberrations.

Before we discuss variants in chromosome number caused by increased or decreased amounts of genetic material, let us consider the other mechanisms by which chromosome number can change. As already mentioned, if reciprocal translocations

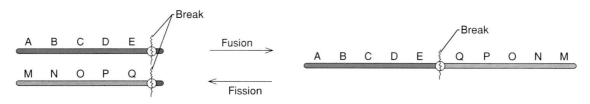

FIGURE 4.30   Centric fusion occurs when two nonhomologous acrocentric chromosomes join to form a metacentric chromosome. Centric fission occurs when the metacentric breaks to form two telocentric chromosomes.

occur between two acrocentric chromosomes such that the large segments reattach, the result is a large metacentric chromosome and a small chromosome that may be lost during cell division. In other words, two acrocentric chromosomes may be combined into one metacentric chromosome. This is a mechanism for **chromosomal** (or **centric**) **fusion,** in which two nonhomologous chromosomes are joined at their centromeres to form a single metacentric (figure 4.30). On the other hand, a chromosome may split at its centromere perpendicular to the length of the chromosome, resulting in two smaller chromosomes, a change termed **chromosomal** (or **centric**) **fission.** It is generally thought that chromosomal fusions are much more common than chromosomal fissions. Later in this chapter, we will give an example of a fusion that results in the 46 human chromosomes.

Polyploidy (Euploidy Variation)

Organisms with three or more complete sets of chromosomes are called **polyploids.** If we let the haploid number of chromosomes be x, then organisms with three chromosome sets have 3x chromosomes and are called triploids; those with 4x chromosomes are tetraploids; those with 6x chromosomes are hexaploids; and so on. Earlier in this chapter, we used the symbol n to refer to the haploid number of chromosomes; here, x will stand for the haploid number. However, for organisms that are regularly polyploid, such as many plants, x usually refers to the number of chromosomes in a set and n to the number in a gamete. Thus, in a hexaploid organism with 60 chromosomes, 6x = 2n = 60, so that x = 10 and n = 30.

Polyploidy is relatively common in plants but rare in most animals, occurring only in certain beetles, earthworms, salamanders, fishes, and a few other organisms. On the other hand, nearly half of all flowering plants are polyploids, as are many important crops. For example, potatoes are tetraploid (4x = 48), bread wheat is hexaploid (6x = 42), and strawberries are octoploid (8x = 56).

Polyploidy is less frequent in animals than in plants for several reasons. First, sex determination is often more sensitive to polyploidy in animals than in plants. Second, plants can often self-fertilize, so a single new polyploid plant with an even number of chromosomal sets (tetraploid, hexaploid, etc.) can still reproduce. Finally, plants generally hybridize more easily with other related species, an important attribute, because the different sets of chromosomes in a polyploid often have different origins.

We can distinguish two types of polyploids: those that receive all their chromosomal sets from the same species, **autopolyploids,** and those that obtain their chromosomal sets from different species, **allopolyploids.** For example, if an unreduced, or diploid pollen grain from a diploid organism fertilizes a diploid egg of the same species, the offspring are **autotetraploids,** or *AAAA,* where *A* indicates a complete chromosomal set, or **genome,** of type *A.* On the other hand, if diploid pollen of one species fertilizes a diploid egg of another, related species, the offspring are **allotetraploids,** or *AABB,* where *B* indicates a genome from the second species. All the chromosomal sets in an autopolyploid are homologous, just as they are in a diploid. But in allopolyploids, the different chromosomal sets generally vary somewhat and are called **homeologous,** or partially homologous.

Polyploids occur naturally, but in very low frequency, when a cell undergoes abnormal mitosis or meiosis. For example, if during mitosis all chromosomes go to one pole, that cell will have an autotetraploid chromosome number. If abnormal meiosis takes place, an unreduced gamete may result, having 2n chromosomes. However, in most situations this diploid gamete would combine with a normal haploid gamete and produce a triploid. Polyploids can also be produced artificially using colchicine, a chemical that interferes with the formation of spindle fibers. As a result of colchicine application, chromosomes do not move to the poles, and autotetraploids are often formed.

Autopolyploids   Triploid organisms are usually autopolyploids (*AAA*) that result from fertilization involving a haploid and a diploid gamete. They are normally sterile because the probability of producing balanced gametes is quite low. In meiosis, the three homologues may pair and form a trivalent, or two may pair as a bivalent, leaving the third chromosome unpaired. Gametes are equally likely to have one or two homologues of a given chromosome (figure 4.31). However, because the behavior of nonhomologous chromosomes is independent, the probability of a gamete having exactly n chromosomes is $(\frac{1}{2})^n$ (using the product rule), and the probability of a gamete having exactly 2n chromosomes is also $(\frac{1}{2})^n$. All other gametes will be unbalanced and generally nonfunctional, as will the zygotes containing them. For example, most bananas are triploids; they produce unbalanced gametes, and as a result are seedless (they are propagated by cuttings). Other polyploids with odd numbers of chromosome sets (for example, pentaploids) are also usually sterile.

Polyploids are often larger (that is, they produce larger fruit) than related diploids; thus, many food crops are autotetraploids or other types of polyploids. Autotetraploids may have normal meiosis if they form either bivalents or quadrivalents only. However, if the four homologues form a trivalent and a univalent, gametes will generally have either too many or too few chromosomes (figure 4.32).

Allopolyploids Most naturally occurring polyploids are allopolyploids, and they may result in a new species. For example, the bread wheat *Triticum aestivum* is an allohexaploid with 42 chromosomes. By examining wild related species, it appears that bread wheat is descended from three different diploid ancestors, each of which contributed two sets of chromosomes (in this case designated as *AABBDD*). Pairing occurs only between the homologous sets, so that meiosis is normal and results in balanced gametes of n = 21.

A particularly interesting allotetraploid was developed by a Russian, G. Karpechenko, in 1928, when he crossed the cabbage and the radish, both of which have a diploid chromosome number of 18. He wanted to produce a hybrid having the leaves of a cabbage and the root of a radish. When he finally recovered seeds from an artificial hybrid, he planted them and found that they had

36 chromosomes. However, instead of the traits he had hoped for, this hybrid had the leaves of a radish and the roots of a cabbage!

A haploid pollen grain with genome *A* may pollinate a flower of a species with genome *B*, resulting in a sterile hybrid of genome constitution *AB*. If mitotic failure subsequently takes place in one branch, *AABB* cells may be produced. If these are self-fertile, an allopolyploid has been formed. Plant breeders use colchicine on sterile hybrids to produce allopolyploids in much the same manner.

Polyploids, species with more than two chromosomal sets, are most common in plants. They may be either autopolyploids, whose chromosomal sets come from the same species, or allopolyploids, whose chromosomal sets come from different species.

Aneuploidy

We have already discussed the observation by Bridges that *Drosophila* occasionally produce individuals having either one sex chromosome (for example, an X) or three sex chromosomes (for example, XXY). The cause of such aneuploidy is nondisjunction; that is, two homologous chromosomes fail to separate properly during meiosis or mitosis. Nondisjunction in meiosis itself is thought to result from improper pairing of homologues in early meiosis so that the centromeres are not on opposite sides of the metaphase plane, or from failure of chiasma formation. As a result, both chromosomes may go to the same pole, leaving one daughter cell with an extra chromosome and the other daughter cell with no chromosome. When these gametes are fertilized by a normal gamete, they either have an extra chromosome, 2n + 1, termed **trisomy**, or are missing a chromosome, 2n − 1, termed **monosomy**. Nondisjunction is most common in meiosis I, but it can occur in meiosis II as well (figure 4.33). Nondisjunction can also take place in mitosis, resulting in mosaics for normal and aneuploid cells. Other combinations of extra

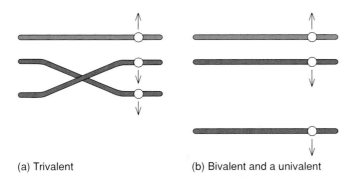

(a) Trivalent   (b) Bivalent and a univalent

FIGURE 4.31 Meiotic pairing and the direction of homologue movement in triploids when (*a*) a trivalent or (*b*) a bivalent and a univalent are formed.

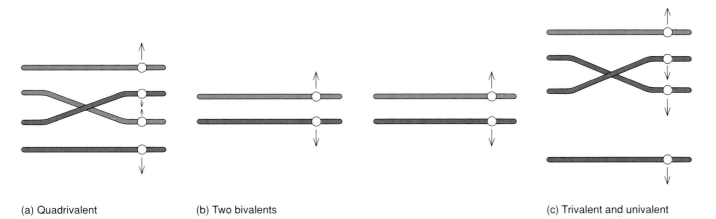

(a) Quadrivalent   (b) Two bivalents   (c) Trivalent and univalent

FIGURE 4.32 Meiotic pairing and the direction of homologue movement in tetraploids for (*a*) a quadrivalent and (*b*) two bivalents. One possible outcome for a trivalent and a univalent is shown in (*c*).

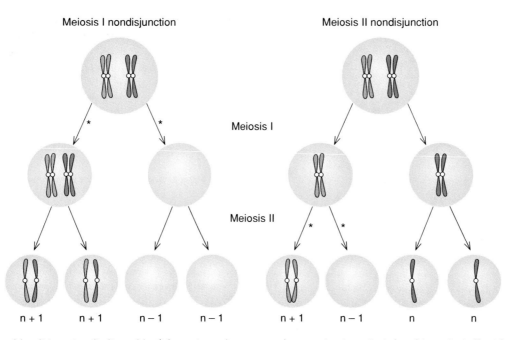

FIGURE 4.33   Nondisjunction (indicated by * for a given chromosome) occurring in meiosis I and in meiosis II, with the resulting gametes.

chromosomes are possible, the most important being a tetrasomic with 2n + 2 chromosomes and a nullisomic with 2n − 2 chromosomes, in which no copies of a particular homologue exist.

Trisomics are known in many different species. They are viable in many plants, but are less frequently viable in animals. For example, among the aneuploids that have been most thoroughly studied are those in the Jimson weed, or thorn apple, *Datura stramonium*. A series of *Datura* mutants with strange properties, studied by Alfred Blakeslee around 1920, turned out to be trisomics for different chromosomes. In fact, a trisomic for each of the twelve different chromosomes was found, and each had a particular phenotype. The effects on the appearance of the seed capsule were quite different for trisomies of the different chromosomes, suggesting that different chromosomes have different hereditary effects on this trait (figure 4.34).

Trisomics have been investigated in crop plants such as corn, rice, and wheat in an effort to identify the chromosomes carrying different genes. Crosses involving plants with trisomic chromosomes give unusual segregation ratios. For example, if a homozygous dominant trisomy, *AAA* (the *A* symbol again indicates a dominant allele), is crossed to a recessive diploid, *aa*, half the progeny are trisomic *AAa* and half are diploid *Aa*. When the trisomic progeny are backcrossed to *aa* individuals, approximately one-sixth of the progeny are recessive *aa* (figure 4.35). If the gene had been on a chromosome that was not trisomic, the F_1 would be *Aa*, and one-half, not one-sixth, of the backcross progeny would be homozygous recessive (*aa*).

In animals, trisomies and other aneuploid chromosomal complements are more unusual. From analysis of the chromosomal constitution of spontaneous abortions in humans, it appears that nearly all monosomics and many trisomics are fetal

lethals. However, several trisomies that sometimes come to full term compose a substantial part of congenitally abnormal births. One of the most common is Down syndrome, trisomy of chromosome 21, with a frequency of one in seven hundred live births (table 4.4). Down syndrome, first described nearly 150 years ago, is generally characterized by mental retardation, distinctive palm prints, and a common facial appearance. In general, mortality is higher than normal: the average life span is the middle teens to the forties, depending upon the country, but some individuals live much longer. People with Down syndrome generally have a positive disposition, and some are able to be partially independent (figure 4.36).

The chromosomal basis of Down syndrome was first discovered in 1959, shortly after the correct human diploid number was determined. Detailed banding of human chromosomes has shown that Down syndrome actually results from a trisomy of the smallest chromosome, which is actually chromosome 22. However, because Down syndrome is known so prevalently as trisomy 21, this association was not changed, and the smallest chromosome is still called chromosome 21. The current nomenclature to indicate an individual with trisomy 21 is 47, +21, in which 47 indicates the total number of chromosomes and +21 indicates that there are three, rather than two, copies of chromosome 21. We discussed previously how trisomy 21 can be generated from reciprocal translocation heterokaryotypes, but most commonly (in about 95% of cases), it results from meiotic nondisjunction. The other autosomal trisomies are much rarer, mostly because they are not viable as fetuses.

Nondisjunction of the sex chromosomes in humans is the source of several conditions. Table 4.5 shows the gametes and zygotes resulting from normal meiosis and from nondisjunction. Notice that four common viable, but abnormal chromosomal types, XO, XXX, XXY, and XYY, are produced through nondisjunction. The symbol O here indicates the lack of a sex chromosome in a gamete or zygote.

Klinefelter syndrome, XXY (or 47, XXY), occurs fairly frequently and generally results in a relatively mild abnormality. These individuals are sterile males with some female characteristics. Individuals with **Turner syndrome** (XO or 45, X) are sterile females, short in stature, with some neck webbing. The frequency of XYY (or 47, XXY) is about one per one thousand males, but such males do not appear to have any congenital problems. At one point, it was suggested that XYY individuals had criminal tendencies, but further study indicates minimal

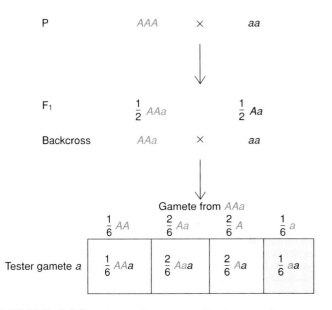

FIGURE 4.35 A cross between an F₁ trisomy and a recessive, giving ⅙ recessive progeny.

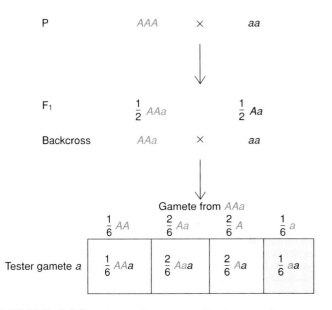

FIGURE 4.34 The normal fruit from *Datura* (top) and the twelve different types of trisomies, each having a different appearance and name.

Syndrome	Sex	Chromosomes	Frequency		Expected Viability and Fertility
			Abortuses	*Births*	
Down	M or F	Trisomy 21 (47, +21)	1/40	1/700	15 years, rarely reproductive
Patau	M or F	Trisomy 13 (47, +13)	1/33	1/15,000	<6 months
Edward	M or F	Trisomy 18 (47, +18)	1/200	1/5,000	<1 year
Turner	F	XO (45, X)	1/18	1/5,000	Sterile
Metafemale	F	XXX (47, XXX)	0	1/700	Nearly all sterile
Klinefelter	M	XXY (47, XXY)	0	1/2,000	Sterile
XYY	M	XYY (47, XYY)	?	1/2,000	Normal

Table 4.4 Frequency and Effects of the Most Common Aneuploidies in Humans

FIGURE 4.36   An individual with Down syndrome.
© M. Coleman/Visuals Unlimited.

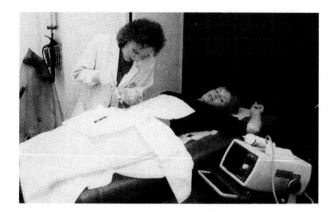

FIGURE 4.37   The procedure for amniocentesis.
© Eric Kroll.

correlation with behavior, if any. The frequency of XYY (or 47, XYY) individuals in prisons is significantly higher than that of the general population; however, less than 5% of all XYY individuals are actually institutionalized.

Abnormal chromosome numbers in a fetus can be diagnosed using **amniocentesis** (figure 4.37). In this procedure, a sample of fluid is withdrawn from the amniotic sac with a needle. The fetal cells contained in this fluid are cultured for two to three weeks. Dividing cells are then stained, and the chromosomes are examined and counted to check for chromosomal abnormalities.

The X chromosome is different from the other chromosomes in that only one is active in a given cell. Normal males have only one X, which is active in all cells. In normal females, only one X is active in a given cell and the other X is heterochromatinized, or mostly inactive. The mostly inactivated X forms a structure called a **Barr body** (named after its discoverer, Murray Barr) that can be identified in a cell (figure 4.38). Therefore, normal males and XO individuals have no Barr bodies; normal females and XXY individuals have one; XXX individuals have two; and so on. In other words, by counting the number of Barr bodies in a cell, chromosomal abnormalities involving the X chromosome can be determined.

The incidence of Down syndrome, and to some extent other aneuploidies, increases with the age of the mother (figure 4.39). The incidence of Down for mothers of age forty-five is nearly 50-fold that for teenage mothers. Although the exact mechanism for this increase is unknown, it appears to be related to the difference in gametogenesis between females and males. In females, oocytes are formed before birth and held in a resting stage (actually prophase of meiosis I) until just before ovulation. In older mothers, an oocyte may remain at this stage for over forty years, during which time it may be affected by environmental factors that may cause a nondisjunction.

Table 4.5   Gametes and Zygotes Resulting from Normal Meiosis and Nondisjunction

		Egg		
			Nondisjunction	
			(Meiosis I or Meiosis II)	
	Sperm	*Normal Meiosis*		
		X	XX	O
Normal Meiosis	X	XX	XXX	XO
	Y	XY	XXY	YO★
Nondisjunction in Meiosis I	XY	XXY		
	O	XO		
Nondisjunction in Meiosis II	XX	XXX	Extremely rare	
	YY	XYY		
	O	XO		

★Inviable zygotes.

Aneuploidy occurs when there are extra or missing chromosomes compared to the normal complement. Cases of trisomy, one extra chromosome, are well known in plants and in humans cause a number of disorders.

CHROMOSOMES IN DIFFERENT SPECIES

Appendix C, Module 10

As we discussed earlier in this chapter, variation in chromosomal structure or number may occur within a species. When the

structures of chromosomes in related species are compared, they are often quite similar. Early studies of this sort compared the banding patterns of different related *Drosophila* species and found a high degree of homology. For example, species such as *D. melanogaster* and *D. simulans,* which are virtually the same except in the morphology of the male genitalia, have nearly identical banding sequences, differing only in a few inversions.

Recently developed high-resolution techniques for chromosomal banding permit the comparison of chromosomes in many species. Much research has focused on human chromosomes and on those of the related great apes. Humans have a diploid chromosome complement of 46, while chimpanzees, gorillas, and orangutans have 48. The difference in number is explained by a centric fusion of two nonhomologous chromosomes in the great apes to form human chromosome 2 (figure 4.40). Notice that the

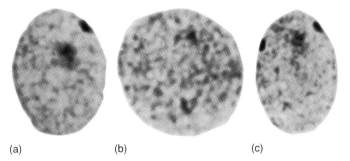

FIGURE 4.38 Nuclei from cells of (*a*) a normal female (XX) with one Barr body, (*b*) a normal male (XY) with no Barr bodies, and (*c*) an XXX female with two Barr bodies.

Dr. Dorothy Warburton/Peter Arnold, Inc.

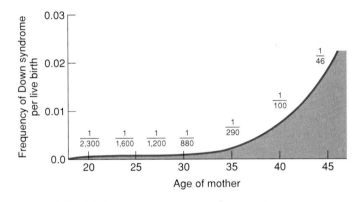

FIGURE 4.39 The frequency of Down syndrome in children as it relates to mothers of different ages.

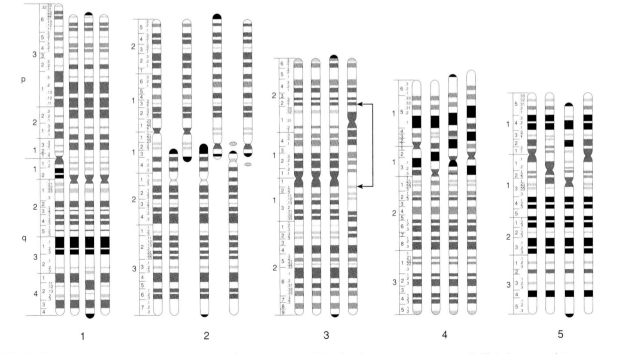

FIGURE 4.40 Comparison of late-prophase banding sequences of the five largest chromosomes (1–5) in humans, chimpanzees, gorillas, and orangutans (left to right).

two acrocentric chromosomes in the three great apes that are homologous to the metacentric human chromosome 2 have identical banding patterns, except in the fusion region and in the terminal heterochromatin.

A number of other structural differences exist in the chromosomes among these four species. For example, the banding patterns in chromosome 3 in humans, chimpanzees, and gorillas are virtually identical, while the patterns in orangutans differ by two inversions. The break points of the larger of these inversions are indicated by the arrows in figure 4.40, showing that the inversion is pericentric and changes the chromosome from a metacentric to an acrocentric. Evidence indicates only one translocation difference among these species. It occurs between chromosomes 5 and 17 in the gorilla and is not present in the other species. Overall, the banding patterns for humans are nearly identical to chimpanzees for 13 chromosomes, to gorillas for 9 chromosomes, and to orangutans for 8 chromosomes. Almost all the same bands are present in these species, suggesting that the primary chromosomal differences among them are in the order of genes, not in the presence or absence of genes.

CHROMOSOMES AND SEX DETERMINATION

In the past, literally hundreds of unscientific theories have attempted to explain sex determination by relating it to phases of the moon, the time of day of fertilization, and other factors. Actually, there is no universal mode of sex determination, but rather many different modes, some of which are environmental. For example, certain reptile eggs that are exposed to high temperatures produce mostly males in particular species and mostly females in others. The sex of some fish is affected by social dominance, and certain plants produce different sexes depending on day length or other factors that affect growth rate.

However, in most organisms, sex is determined by chromosomal differences and the genes on them (table 4.6). The first example of chromosomal determination of sex was documented in 1905 for the bug *Protenor*, in which it was discovered that females have two X chromosomes and males only one. Male gametes were found to be equally likely to have an X or no X chromosome, and female gametes were found to always carry one X, so that equal numbers of progeny are XX (female) and X (male).

Table 4.6   Different Types of Chromosomal Sex Determination

Organisms	Female	Male
Most mammals, some insects, some plants	XX	XY
Protenor, some other insects	XX	X
Birds, most reptiles, moths	ZW	ZZ
Hymenoptera	Diploid	Haploid

Shortly thereafter, several investigators established the presence of a second sex chromosome, the Y chromosome, in other organisms. In these species, the presence of a Y along with an X results in a male, while two copies of the X produce a female. Again, male gametes are of two types, so that there are equal numbers of X-carrying and Y-carrying sperm or pollen. The sex having both types of sex chromosomes, the XY, is termed the *heterogametic sex,* while the XX is called the *homogametic sex.* In some organisms, such as birds and some reptiles, males are homogametic and females are heterogametic. To avoid confusion, these chromosomes are often termed Z and W; thus, homogametic males are ZZ and heterogametic females are ZW. As a result, the females produce two types of gametes, while the male gametes all contain one Z chromosome.

In chickens, it is important to identify the sex of young birds so that, for example, males will not be raised in an egg laying operation. It is difficult and expensive to determine the sex of young birds, and several systems utilizing the sex chromosome difference between males and females are often used. For example, one gene on the Z chromosome has a recessive allele that determines gold feather color (either Z^sZ^s or Z^sW), and a dominant allele that determines silver feather color (Z^SZ^S, Z^SZ^s or Z^SW). As a result, the cross given in figure 4.41 produces all silver males and all gold females, allowing easy sex determination of day-old chicks. Could the opposite cross, $Z^SZ^S \times ZW$, yield chicks of different colors that could be used to determine their sex?

In nearly all mammals, the presence of a Y chromosome is necessary for the development of the male phenotype. For

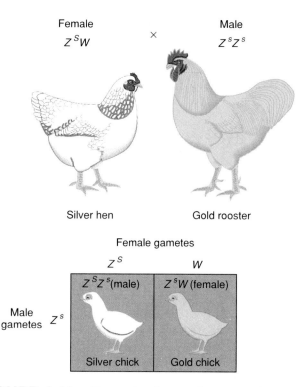

FIGURE 4.41   The results of a cross between a silver hen and a gold rooster to produce all male silver chicks and all female gold chicks.

example, humans that are 45, X (Turner syndrome) are phenotypically female. Furthermore, individuals that are 47, XXY (Klinefelter syndrome) are phenotypically male, even though they have two X chromosomes. Thus, it appears that the Y chromosome produces a gene product that acts as a necessary switch to begin development toward maleness. Although this switch gene, known as the SRY gene, is on the Y chromosome, the rest of the genes involved in the production of sexual phenotypes apparently are on other chromosomes.

Recently, the location of this switch gene on the Y has been identified by examining rare, sex-reversed individuals—that is, XX males and XY females. The XX males actually have a piece of the Y that contains the male-determining gene, and the XY females lack the same region of the Y. In *Drosophila*, there is an autosomal mutant, *tra* (transformer), which when homozygous in XX individuals causes them to be males, not females as expected. The opposite situation is known in humans where individuals with a mutation on the X chromosome have a condition called testicular feminization. In this case, the XY individuals are phenotypically female, having breasts and a vagina, but they are sterile. This mutation is apparently related to a hormone receptor that prevents these individuals from responding to the presence of testosterone; consequently, they do not develop secondary male characteristics.

Aside from this switch function to maleness, there appear to be few active genes on the Y chromosome in humans, according to the following evidence. First, much of the Y chromosome is heterochromatic. Second, the length of the long arm of the Y is quite variable. (Interestingly, the long arm is substantially longer in Japanese males than in males of most other populations, but this variation appears to have no effect on fertility or other phenotypic characteristics.) And finally, there is also extensive variation in the size and intensity of banding on the long arm of the Y in humans without apparent phenotypic effect. These facts support the hypothesis that much of the Y chromosome is inert and does not produce functional gene products.

Although XX individuals in *Drosophila* are female and XY individuals are male, the presence of the Y is not necessary for the male phenotype. For example, the exceptional phenotypes observed by Bridges (see figure 4.17) were red-eyed XO males and white-eyed XXY females. Later, Bridges developed triploid strains of *Drosophila*. When he examined individuals with different numbers of X chromosomes, he found that the sexual phenotype in *Drosophila* was a function of the ratio of the number of X chromosomes (X) to the number of sets of autosomes (A). For example, normal females have two X chromosomes and two autosomal sets, making the ratio $X/A = 2/2 = 1$ (table 4.7). Normal males have one X and two autosomal sets, so that $X/A = 1/2 = 0.5$. Likewise, his exceptional males and females still had ratios of 0.5 and 1.0, respectively.

When Bridges examined flies with three X chromosomes and two autosomal sets, $X/A = 3/2 = 1.5$, he found that they were sterile females (sometimes called **metafemales**), while flies with one X and three autosomal sets, $X/A = 1/3 = 0.33$, were sterile males **(metamales).** Finally, flies with two X

Table 4.7 Sexual Phenotypes of *D. melanogaster*[1]

Number X Chromosomes	Number Autosomal Sets (A)	X/A Ratio	Sexual Phenotype
3	2	1.5	Metafemale
2	2	1.0	Normal female
2	3	0.67	Intersex
1	2	0.5	Normal male
1	3	0.33	Metamale

[1]Expressed as a function of the ratio of the number of X chromosomes to sets of autosomes.

chromosomes and three autosomal sets, $X/A = 2/3 = 0.67$, had phenotypic characteristics of both sexes and were termed **intersexes.** Based on these findings, Bridges proposed that, overall, the X chromosomes were female-determining and the autosomes were male-determining. In other words, when the X/A ratio was 1 or greater, flies would have the female phenotype, and when the X/A ratio was 0.5 or less, flies would have the male phenotype. However, we should note that even though the Y is not necessary for the male phenotype in *Drosophila*, it is necessary for fertility; that is XO males and metamales are sterile.

Chromosomes in related species are generally quite similar. In most organisms, chromosomal differences are involved in sex determination. The Y chromosome determines maleness in nearly all mammals. In *Drosophila*, the relative proportion of X chromosome to autosomes determines sex.

WORKED-OUT PROBLEMS

PROBLEM 1

How many copies of a given gene does an autohexaploid have? Assuming there are two types of alleles at a gene, *B* and *b*, what are the different possible genotypes?

SOLUTION

Each gene is represented in each chromosomal set, so there would be six copies of each gene. There are seven different possible genotypes: two homozygotes *BBBBBB* and *bbbbbb*, and five types of heterozygotes, *BBBBBb*, *BBBBbb*, *BBBbbb*, *BBbbbb*, and *Bbbbbb*.

PROBLEM 2

Genes on the X chromosome in *Drosophila melanogaster* were used to show that genes are on chromosomes. In one cross, as outlined in figure 4.16, white-eyed females were crossed with red-eyed males and the F_1 progeny of this cross were mated to produce an F_2. In one such array, 40 red-eyed females, 37 white-eyed females, 42 red-eyed males, and 35 white-eyed males were produced. Seeing this array of genotypes and knowing the behavior of the X chromosome, why would you conclude that the white gene is on the X chromosome?

Given this array of phenotypes and the mode of inheritance given in figure 4.16, are the results consistent with the expectations? Use a chi-square test.

SOLUTION

Several observations from this cross point to the fact that genes are on chromosomes. For example, the F_1 males have a Y that they must have received from their father, because their mother does not have a Y chromosome. The males have a white-eye allele (because their phenotype is white) that they must have gotten from their mother. This suggests that the X chromosome that they do have is from their mother and contains the white allele. Likewise, all of the phenotype and sex combinations of the progeny in the F_2 can be explained by assuming that the white allele is on the X chromosome.

The chi-square test can be calculated as follows where the expected number for all classes is 155/4 or ¼ of the total number of progeny.

Phenotype	Sex	Observed	Expected	$(O - E)^2/E$
red-eyed	female	40	38.75	0.04
white-eyed	female	37	38.75	0.08
red-eyed	male	43	38.75	0.47
white-eyed	male	35	38.75	0.36
		155	155	$\chi^2 = 0.95$

With three degrees of freedom, the number of classes minus one, this value is not significant. In other words, these results are consistent with the hypothesis that the white allele is on the X chromosome.

SUMMARY

The diploid number of chromosomes is different in different species. Specific chromosomes can be identified by size, the position of the centromere, and banding patterns. Mitosis, nuclear division that results in the production of two identical daughter nuclei, is divided into four stages: prophase, metaphase, anaphase, and telophase. Meiosis is composed of one replication of DNA and two cell divisions. Meiosis I, the reductional division, results in the separation of homologous chromosomes or a halving of chromosome number. Meiosis II is similar to mitosis and results in the separation of the sister chromatids with no reduction in chromosome number. In animals, mature male and female gametes are produced by spermatogenesis and oogenesis, respectively.

The parallel behavior of genes and chromosomes provides a physical explanation of the principles of segregation and independent assortment. The first indication that genes are on chromosomes came from the parallel behavior of genes and chromosomes. Definitive

proof resulted from experiments using X-linked genes in *Drosophila* and the confirmation that individuals with exceptional eye color were the result of meiotic mistakes that produced abnormal chromosomal numbers.

Duplications and deletions occur when a chromosomal segment is present more than once or is missing. Both can result in phenotype changes or lethality, although deletions often have larger effects. In meiotic chromosomes, duplication and deletion heterozygotes often form loops. Inversions occur when the sequence of genes on a chromosome is reversed. When recombination takes place in the inverted region of inversion heterokaryotypes, duplications and deficiencies result in the meiotic products. Translocations are the result of the movement of a chromosomal segment to a homologous chromosome. Normal meiotic products result when there is alternate segregation in translocation heterokaryotypes, but otherwise duplications and deficiencies occur.

Polyploids, species with more than two chromosomal sets, are most common in plants. They may be either autopolyploids, in which all chromosome sets are from the same species, or allopolyploids, in which the chromosomal sets come from different species. Aneuploidy occurs when there are extra or missing chromosomes compared to the normal complement. Cases of trisomy, one extra chromosome, are well known in plants and in humans cause a number of disorders.

Chromosomes in related species are generally quite similar. In most organisms, chromosomal differences are involved in sex determination. In nearly all mammals, the Y chromosome determines maleness. In *Drosophila,* the relative dose of X chromosome to autosomes determines sex.

PROBLEMS AND QUESTIONS

Problems and Questions with Links to Software

Drosophila Genetics: 3

1. What are the three major morphological classes of chromosomes and how are they distinguished?

2. What are the four phases of the cell cycle, starting with the S phase? What takes place during the S phase?

3. A cross between red-eyed F_1 (heterozygous) female fruit flies and red-eyed male fruit flies yielded 94 red-eyed and 30 white-eyed progeny. Does this fit with expectations? (Use a chi-square test.) Of the red-eyed flies, 56 were females. Does this fit with expectations? (Again, use a chi-square test.)

4. In the cross shown in figure 4.17 involving a group of 4,240 flies, 3 red-eyed males and 2 white-eyed females were observed. From these data, what would be the estimated rate of nondisjunction?

5. Explain why the number of chromosomes per cell (n level) and the amount of DNA per cell (C level) are not synchronous in every stage of cell division.

6. What are two major differences between spermatogenesis and oogenesis in humans and most other mammals?

7. Differentiate between the behavior of chromosomes in metaphase of mitosis, meiosis I, and meiosis II.

8. (*a*) Given a heterozygote for albinism, *Aa,* what genotypes of sperm would you expect after meiosis? (*b*) Diagram the chromosomal behavior as in figure 4.13. (*c*) Extend this analysis to two genes, *Aa,* and another recessive disease heterozygote, *Bb.* Illustrate how chromosomal behavior explains independent assortment by drawing a diagram as in figure 4.14.

9. What are the four types of changes that can take place in chromosome structure? Describe briefly what happens in each.

10. Figure 4.19 shows how a tandem duplication may be generated. (*a*) Using the same logic, illustrate how a reverse duplication may be generated. (*b*) Show how a displaced duplication may be generated.

11. Let us assume that we are trying to identify the genes involved in a deletion. Crosses to lines homozygous for recessive mutants at genes *e* and *f* gave offspring that were mutant, while crosses to lines homozygous for recessive mutants at genes *g* and *h* yielded wild-type progeny. Which genes are missing in the deletion?

12. What is the difference between pericentric and paracentric inversions? What are the differences in meiosis and in the meiotic products from heterokaryotypes for paracentric and pericentric inversions?

13. Assume that the breaks for the reciprocal translocation in figure 4.27 were between E and F and between P and Q. (*a*) What do the chromosomes with the reciprocal translocation look like? (*b*) Diagram metaphase I and alternate segregation for a heterokaryotype for this reciprocal translocation.

14. Name four common aneuploidy conditions in humans and describe the chromosomal constitutions that are responsible.

15. Some Down syndrome cases are the result of a translocation of chromosome 21 onto chromosome 14. Approximately one-third of the offspring of a translocation heterozygote are expected to have Down syndrome. Explain why. In fact, only about one-sixth of the offspring of translocation heterozygotes actually do have Down. Explain why.

16. (*a*) Differentiate between the meiotic products expected from nondisjunction in meiosis I and meiosis II. (*b*) Assuming nondisjunction in meiosis I for the sex chromosomes in a male, what progeny genotypes would you expect? (*c*) What progeny genotypes would you expect from nondisjunction in meiosis II in males or in females?

17. In an effort to determine on which chromosome a gene is located, a homozygous recessive mutant was crossed to trisomies of two different chromosomes. A backcross of the trisomy F_1 to the recessive homozygote gave 42 wild-type and 38 mutant progeny for chromosome 5 and 56 wild-type and 12 mutant progeny for chromosome 6. On which chromosome do you think the gene was located?

18. How many copies of a gene does an autotetraploid have? Assuming that there are two types of alleles at a gene, *A* and *a,* give the five possible genotypes. If *A* is dominant, how many possible phenotypes would you expect?

19. Given an autotriploid with 3x = 12, what is the probability of producing a gamete having exactly one copy of each of the four homologues? Having exactly two copies of each homologue?

20. In figure 4.40, identify the smaller inversion on chromosome 3 that differs between orangutans and the other great apes. Identify the pericentric inversion that differentiates chromosome 4 for humans and chimpanzees.

21. In chickens, males are the homogametic sex, *ZZ,* and females are the heterogametic sex, *ZW.* In a cross between $Z^A W$ and $Z^A Z^a$, where the superscripts indicate the dominant and recessive alleles, what proportions of progeny would you expect? Give both genotype and sex.

22. Given the rules developed by Bridges to determine sex in *Drosophila,* what would be the sex of an individual with three X chromosomes and three autosomal sets? Of an individual with two X chromosomes and four autosomal sets? Explain why.

23. Using diagrams, show the differences between prophase in mitosis and in meiosis I.

24. Why might allotetraploids have higher rates of growth than either of the parental diploids? If the two sets of chromosomes from different species in an allotetraploid do not pair, would you expect balanced gametes?

25. Two related species of mice have different numbers of chromosomes. Species 1 has twenty pairs of metacentrics, and species 2 has twenty-three pairs of chromosomes, seventeen of which are metacentric and six acrocentric. How many chromosomal fissions would be necessary to produce a karyotype of species 2 from species 1?

Answers appear at end of book.

5

Genetic Linkage

T he results are a simple mechanical result of the location of the materials in the chromosomes, and of the method of union of homologous chromosomes, and the proportions that result are not so much the expression of a numerical system as the relative location of the factors in the chromosomes.

Thomas Hunt Morgan
American geneticist

Learning Objectives

In this chapter you will learn:

1 That genes on the same chromosome usually do not follow Mendel's principle of independent assortment.

2 That by estimating the rate of recombination between genes on the same chromosome, a linkage map of genes on the same chromosome can be constructed.

3 How crosses involving three linked genes can be used to order genes on a chromosome.

4 That the recombination frequency is generally reduced in a region where there has already been recombination.

5 That linkage in humans has traditionally been determined from particular matings and somatic cell hybridization— and that recently, DNA markers have made mapping in humans more feasible.

6 That crossing over takes place in the four-stranded stage of meiosis I.

7 That there is generally a good correspondence between genetic map distance and physical DNA distance, with important exceptions.

FIGURE 5.1 Thomas Hunt Morgan, a leading geneticist of the early twentieth century, in his fly lab.
Lowe Memorial Library, Columbia University.

The principles of segregation and independent assortment are consistent with the fact that genes are on chromosomes and with the behavior of chromosomes in meiosis. We have assumed that the different genes we followed were on different chromosomes. However, shortly after the rediscovery of Mendel's principles, Thomas Hunt Morgan (figure 5.1) realized that there were more genes than chromosomes in *Drosophila*. In fact, we know today that thousands of genes exist on each of the three major *Drosophila* chromosomes. As a result, the pattern of inheritance predicted by the principle of independent assortment does not always apply to genes on the same chromosome. In this chapter, we will examine the pattern of inheritance for two or more genes on the same chromosome. We will also describe how genes can be organized into genetic maps, showing their positions on the various chromosomes.

Because of the human genome project, a recent international effort to map all the genes in the human genome, a great resurgence of interest and advancement in the techniques used to map genes has occurred (see chapter 15). However, many of the fundamental ideas are based on the classic experiments that we will discuss in this chapter.

LINKAGE

Appendix C, Module 11

In the first decade of this century, William Bateson and R. C. Punnet investigated the inheritance patterns for various traits in the sweet pea. In one study, they examined the simultaneous inheritance of flower color, dominant purple (*P*) versus recessive

Table 5.1 F$_2$ Progeny of a Dihybrid Cross in Sweet Peas

Phenotype	Numbers		
	Observed	Expected	$(O - E)^2/E$
Purple, long	296	240.2	13.1
Purple, round	19	80.1	46.5
Red, long	27	80.1	35.1
Red, round	85	26.7	127.3
	427	427.1	$\chi^2 = 222.0$

red (*p*), and pollen shape, dominant long (*L*) versus recessive round (*l*). They carried out a dihybrid cross as Mendel had done for traits in the garden pea; that is, they crossed *PP LL* with *pp ll*, self-fertilized the F$_1$ (*Pp Ll*), and counted the numbers of different types of F$_2$ plants. But instead of the expected 9:3:3:1 ratio under independent assortment, they obtained the results given in table 5.1. Comparing the observed numbers to those expected under a 9:3:3:1 ratio, they found many more phenotypes like the original parents and many fewer single dominant or nonparental types. In this case, there are 3 degrees of freedom, making the chi-square value of 222.0 for the data in table 5.1 highly significant and inconsistent with the hypothesis of independent assortment (see table 2.5).

Bateson and Punnett knew that sometimes F$_2$ ratios were a modification of the 9:3:3:1 expectations because of epistasis, but the F$_2$ numbers they observed in this experiment did not fit a modified Mendelian ratio either (although they did try to fit it to a 7:1:1:7 ratio). Besides, the two genes affected quite different traits, unlike the epistatic examples with modified F$_2$ ratios. What could be the cause of the strange ratios they had found in F$_2$ progeny?

Bateson and Punnett suggested that because the two parental phenotypes were in excess in the F$_2$ progeny, there might be a physical connection between the parental alleles—the dominant alleles from one parent and the recessive alleles from the other parent. This phenomenon, which they termed **coupling,** interfered with independent assortment. On the other hand, they attributed the low numbers of the F$_2$ types, those having a dominant phenotype at one gene and a recessive at the other, to some innate negative affinity of dominant and recessive alleles, which they termed **repulsion.**

A PHYSICAL EXPLANATION

A physical explanation for Bateson and Punnett's observations came later from the research of Morgan, using two genes in *Drosophila*. One gene affected eye color (recessive purple, *pr*, and dominant wild-type red, *pr$^+$*, alleles), and the other gene affected wing development (recessive vestigial, *vg*, and dominant normal, *vg$^+$*, alleles). Instead of examining the F$_2$ progeny of a dihybrid cross, Morgan used a testcross for both genes simultaneously, as outlined in figure 5.2. Note that both wild-type alleles were

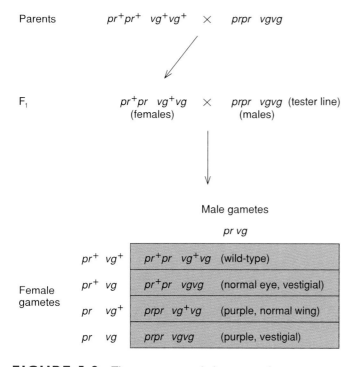

FIGURE 5.2   The genotypes and phenotypes from a two-gene testcross using the *vestigial* and *purple* mutants.

Table 5.2   Results of Morgan's First Testcross

		Numbers		
Phenotype	*Gamete*	*Observed*	*Expected*	*(O − E)²/E*
Wild-type	*pr⁺ vg⁺*	1,339	709.75	558
Normal, vestigial	*pr⁺ vg*	151	709.75	440
Purple, normal	*pr vg⁺*	154	709.75	435
Purple, vestigial	*pr vg*	1,195	709.75	332
		2,839	2,839.00	χ² = 1,765

originally in one parent of the F₁ and that the recessives were in the other parent. The value of such a cross is that only one parent, in this case the female, is heterozygous for both genes, and therefore, segregation and independent assortment would be expected only in the production of the gametes from that sex.

If the two genes assorted independently in Morgan's testcross, then the four types of female gametes, and consequently the four different phenotypes, should be equally frequent. However, as shown in table 5.2, the numbers of the parental phenotypes were many times those of the nonparental phenotypes. The chi-square value for these data is very large (with 3 degrees of freedom, a value greater than 11.34 is highly significant), indicating that the independent assortment hypothesis does not hold up. Although the four gametic numbers were not those expected from independent assortment, the two parental gametes, *pr⁺ vg⁺* and *pr vg*, were nearly equal in number, and the two nonparental

Table 5.3   Results of Morgan's Second Testcross

		Numbers		
Phenotype	*Gamete*	*Observed*	*Expected*	*(O − E)²/E*
Wild-type	*pr⁺ vg⁺*	157	583.75	312
Normal, vestigial	*pr⁺ vg*	965	583.75	249
Purple, normal	*pr vg⁺*	1,067	583.75	400
Purple, vestigial	*pr vg*	146	583.75	328
		2,335	2,335.00	χ² = 1,289

gametes, *pr⁺ vg* and *pr vg⁺*, were also nearly equal in number, but much less frequent than the parental gametes. Today we generally consider gametes with either two wild-type or two mutant alleles *coupling gametes,* and those with one wild-type and one mutant allele *repulsion gametes.* In other words, the two parental gametes in table 5.2 were in coupling, and the two nonparental gametes were in repulsion.

In a second experiment, Morgan crossed two different genotypes, each of which was homozygous for a wild-type allele at one gene and homozygous for a recessive allele at another gene (*pr⁺pr⁺ vgvg × prpr vg⁺vg⁺*). The F₁ flies, double heterozygotes, were then crossed to the tester line *prpr vgvg.* The results of this cross were quite different from the first cross (table 5.3). In this case, the progeny numbers of the single dominants were much larger than expected, while those of the double dominants and double recessives were much lower than expected. In other words, there were more repulsion types than coupling types. Note that in this cross, the parental gametes were in repulsion and the nonparental gametes were in coupling.

To explain his observations, Morgan suggested that the genes having the alleles for these traits are on the same chromosome. As a result, alleles at these genes tend to remain associated between generations because they are physically linked to each other. A schematic representation of this explanation is given in figure 5.3, where the wild-type alleles are on the blue chromosomes and the recessive alleles are on the red chromosomes. The F₁ flies have one of each type of chromosome. However, instead of these chromosomes always staying intact during the production of gametes, in some chromosomes recombination occurs between the chromosomes. As a consequence, these gametes contain nonparental chromosomes composed of the *pr⁺* and *vg* alleles or the *pr* and *vg⁺* alleles. A similar explanation is true for the cross discussed in table 5.3, except that the parental chromosomes have one dominant and one recessive allele, rather than two dominants or two recessives.

The intact original chromosomes are often called **parental chromosomal types,** while the new combinations are called **nonparental chromosomal types.** As we mentioned in chapter 4, the physical exchange through which new chromosomal

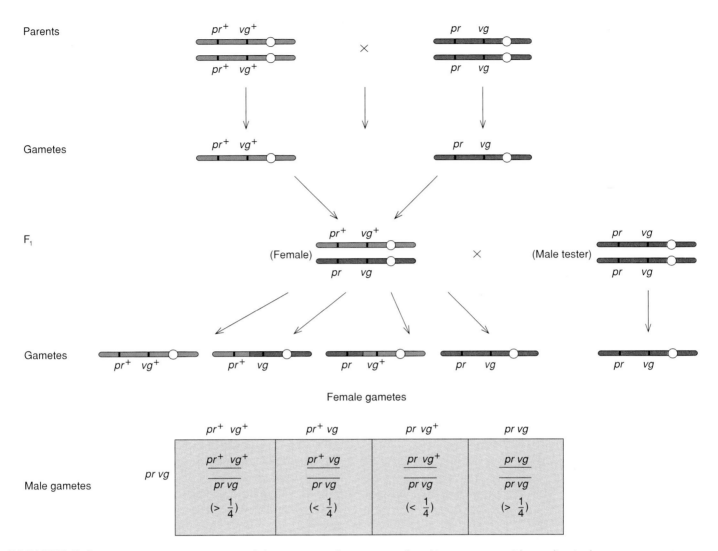

FIGURE 5.3 A schematic representation of the gametes and progeny produced in a testcross, with coupling in the parents.

types are formed is termed **crossing over,** or **recombination,** and is often indicated in diagrams by an ✕ joining homologous chromosomes. The other common names for parental and nonparental chromosomal types are **nonrecombinant** and **recombinant,** respectively, reflecting the process involved. We will discuss the biochemical mechanisms of recombination in a subsequent chapter. At this time, we can envision that the homologous chromosomes sometimes are broken but then reattach to the other member of the homologous pair. The farther apart two genes are on a chromosome, the more likely it is that such breakage–reunion between them will occur.

If a breakage and reunion were to take place between homologous chromosomes, one might expect some cytological manifestation of the event. In fact, it was observed in meiosis that sometimes nonsister chromatids appear to be attached to each other, forming a cross configuration. As we mentioned in chapter 4, this structure (termed a **chiasma**) is a cytological observation resulting from recombination.

Genes that are on the same chromosome are said to be **linked,** and the general phenomenon in which genes occur on the same chromosome is called **linkage.** (As we will see later in the chapter, if genes are far apart on a chromosome, they may actually assort independently.) We can refer to alleles that are linked on the same chromosome as being in either the coupling phase—that is, both dominants of $pr^+ vg^+$ or both recessives as $pr\ vg$—or in the repulsion phase, $pr^+ vg$ or $pr\ vg^+$. Genotypes containing the two different coupling chromosomes are called coupling double heterozygotes and can be indicated by $pr^+ vg^+/pr\ vg$. Genotypes with two different repulsion chromosomes are called repulsion double heterozygotes and can be expressed as $pr^+ vg/pr\ vg^+$. Note that the terms *coupling* and *repulsion* came from traits with complete dominance in which the dominant alleles were on the same chromosome. In other words, the definition is based on the phase in these initial crosses, an artifact that becomes apparent when we examine traits having intermediate dominance or codominance.

Shortly after the rediscovery of Mendel's principles, exceptions to the principle of independent assortment were discovered. The tendency for the alleles at different genes on parental chromosomes to remain together is explained by linkage of genes on a chromosome.

RATE OF RECOMBINATION

When two genes are on different chromosomes, all four gametic types (or phenotypes) are represented equally in the progeny of a testcross. In other words, independent assortment results in equal numbers of parental and nonparental gametes or offspring. However, when two genes are linked on a chromosome, the proportion of recombinant gametes is generally less than 50%, the specific amount depending upon the physical proximity of the two genes on the chromosome. As an extreme, if the two genes are very, very close to each other, the number of recombinant gametes or offspring should be close to zero. We can measure the **rate of recombination** (r) as the proportion of recombinant offspring observed in a testcross or in other crosses. For example, in table 5.2, 305 (151 + 154) or 2,839 progeny were recombinant, making the rate of recombination 305/2,839 = 0.107, or 10.7%. This value suggests that the genes are relatively close to each other but not exactly at the same location on the chromosome.

A general way to express the proportions of recombinant and nonrecombinant gametes is given in table 5.4. It is assumed that the parent is doubly heterozygous, either in coupling (second column) or in repulsion (third column). For a coupling parent, a proportion (r) of the gametes $A\ b$ and $a\ B$ are recombinant. Because we expect equal proportions of these two recombinant classes, the probability of each class is $(\frac{1}{2})$r. Likewise, the proportion of nonrecombinant gametes, $1 - r$, is equally divided between $A\ B$ and $a\ b$. The same logic applies in determining the expected proportion of gametes for the repulsion genotype, although the recombinant and nonrecombinant gamete categories are reversed.

THE LINKAGE MAP

Alfred Sturtevant, a student of Morgan's, was assigned the project of consolidating the early information on crossing over in *D. melanogaster*. He conceived an approach in which the recombination data would be used to describe the physical relationship of genes on a chromosome in a linear arrangement called a **linkage map**, or **genetic map.** A genetic map uses the frequency of recombinant gametes (or offspring) from a testcross, or from other crosses, as a measure of the distance between two genes. For example, 10.7% of the offspring from the cross in table 5.2 were recombinants between genes *pr* and *vg*, making the **map distance** between them 10.7 **map units.** Morgan and Sturtevant suggested that distance along the chromosome be measured in units determined by the percentage of recombination. In fact, later studies have shown that genetic distance measured by this statistical approach is generally similar to cytologically or biochemically measured distances on a chromosome.

Morgan and his students discovered a number of genes in *D. melanogaster* that showed X-linked inheritance. When they crossed strains differing in two of these genes and examined their progeny, they found substantial variation in the frequency of recombination for different gene pairs (table 5.5). For example, out of 21,736 potentially recombinant gametes between the genes for yellow and white, only 214, or 1%, were recombinants. On the other hand, 32.2% of the gametes for the genes for yellow and vermilion were recombinants.

Using the data in table 5.5, Sturtevant determined the physical relationship of these five genes and suggested that they formed a linear array. First, *y* and *w* had very few recombinant gametes, indicating that they were close to each other. Second, the next lowest level of recombination between *w* and another gene was with *v*, which had 0.297, or 29.7%, recombinants. But because the frequency of recombination between *y* and *v* is

Table 5.4   Proportions of Gametes[1]

Gamete	Parental Genotype	
	Coupling $A\ B/a\ b$	Repulsion $A\ b/a\ B$
$A\ B$	$\frac{1}{2}(1 - r)$	$(\frac{1}{2})$r
$A\ b$	$(\frac{1}{2})$r	$\frac{1}{2}(1 - r)$
$a\ B$	$(\frac{1}{2})$r	$\frac{1}{2}(1 - r)$
$a\ b$	$\frac{1}{2}(1 - r)$	$(\frac{1}{2})$r
	1	1

[1]r indicates the rate of recombination.

Table 5.5   Recombination Frequency for Five X-Linked Gene Pairs

Genes	Recombination Frequency
yellow (y)—white (w)	$\frac{214}{21,736} = 0.010$
yellow (y)—vermilion (v)	$\frac{1,464}{4,551} = 0.322$
white (w)—vermilion (v)	$\frac{471}{1,584} = 0.297$
vermilion (v)—miniature (m)	$\frac{17}{573} = 0.030$
white (w)—miniature (m)	$\frac{2,062}{6,116} = 0.337$
white (w)—rudimentary (r)	$\frac{406}{898} = 0.452$
vermilion (v)—rudimentary (r)	$\frac{109}{405} = 0.269$

THE FAR SIDE   By GARY LARSON

"Hey! What's this Drosophila melanogaster doing in my soup?"

Table 5.6	Chromosomal Locations for the Seven Traits in Mendel's Peas
Character	**Chromosome**
Seed color (yellow—green)	1
Flowers (violet—white)	1
Pods (smooth—wrinkled)	4
Flowers (axial—terminal)	4
Stem (tall—short)	4
Pods (green—yellow)	5
Seed form (round—wrinkled)	7

greater than that between w and v, the sequence of these three genes must be y-w-v, with w in between y and v. Next, the gene closest to v is m, only 3 map units away. It must be to the right of v, because the distance w-m is approximately the sum of v-m and w-v. Finally, r is loosely linked to v but shows no evidence of linkage to w in genetic crosses. As a result, r must be to the right of v, making the array of five genes y-w-v-m-r.

Sturtevant constructed a map beginning with y at 0.0 on the left side and using the frequencies of recombination between adjacent genes (figure 5.4). That is, he put w at 1.0, v at 30.7 (1.0 + 29.7), m at 33.7 (30.7 + 3.0), and r at 57.6 (30.7 + 26.9). Notice in the figure that the positions for w, v, m, and r today are slightly different from those given by Sturtevant because many intermediate genes have since been used to map them more accurately.

As indicated here, a genetic map can be constructed by estimating the recombination frequency of all pairs of genes. However, a more efficient approach, as we will discuss later, is to use three genes simultaneously. Using more marker genes also allows identification of cases with more than one crossover.

Notice that although both the *white* and *rudimentary* loci are X-linked, the data in table 5.5 indicate a very high recombination frequency between them, nearly that which would be

found if the genes were on different chromosomes. This suggests that genes far apart on the same chromosome would not show evidence of linkage in genetic crosses. Genes on the same chromosome, whether or not they show linkage in genetic crosses, are **syntenic.**

People have often wondered why Mendel did not observe linkage in his dihybrid crosses. It turns out that, although he picked seven traits and there are seven chromosomes in garden peas, two of the traits he studied are determined by genes on chromosome 1 and three by genes on chromosome 4 (table 5.6). Thus, out of the twenty-one different possible pairs of genes (the number of combinations of seven genes taken two at a time), four pairs of genes are syntenic. However, three of these pairs are far apart on the chromosome and show no indication of linkage in hundreds of crosses. The remaining pair of genes, containing the alleles that determine pod morphology and stem length, should have shown linkage in genetic crosses but were not investigated by Mendel. In other words, because only one of twenty-one pairs of traits would have shown linkage and Mendel examined only a few pairs, it is not surprising (and perhaps fortunate) that he did not observe patterns differing from the expectations of independent assortment.

The rate of recombination estimates the linkage of genes and ranges from 0 for very tightly linked loci to ½ for very loosely linked loci (or those on different chromosomes). The relationship of genes on a chromosome can be shown on a genetic map.

	y w		v	m		r	Centromere
Sturtevant	0.0 1.0		30.7	33.7		57.6	
Today	0.0 1.5		33.0	36.1		54.5	67.7

FIGURE 5.4   The five X-linked genes studied by Sturtevant, showing the positions he gave and those recognized today.

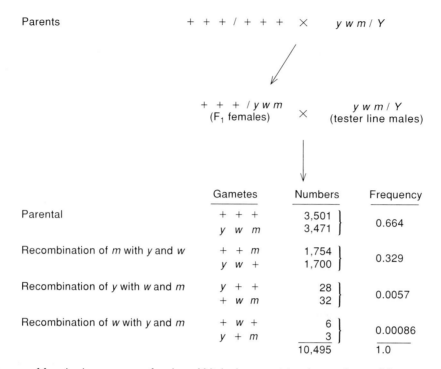

	Gametes	Numbers	Frequency
Parental	+ + + y w m	3,501 ⎫ 3,471 ⎭	0.664
Recombination of *m* with *y* and *w*	+ + m y w +	1,754 ⎫ 1,700 ⎭	0.329
Recombination of *y* with *w* and *m*	y + + + w m	28 ⎫ 32 ⎭	0.0057
Recombination of *w* with *y* and *m*	+ w + y + m	6 ⎫ 3 ⎭	0.00086
		10,495	1.0

FIGURE 5.5   A testcross of females heterozygous for three X-linked genes, giving the number and frequency of different progeny types.

THREE-POINT CROSSES

Although the linearity of the chromosome was apparent from the data for pairs of X-linked genes, complete substantiation of this model came by simultaneously crossing individuals differing by three different genes. One test again used the three X-linked genes in *Drosophila*—*y, w,* and *m.* In this experiment, a backcross of a female in coupling for all three genes was done (figure 5.5), and the gametes in the progeny were scored as receiving either an X chromosome with *y, w,* and *m* or a Y chromosome from the male tester parent. In the progeny gametes, those produced from the F_1 triple heterozygote female, eight types are possible, two of which are parental and six of which are recombinant.

Assuming that the three genes involved are arranged in a linear fashion, there are three possible orders: *y-w-m, w-y-m,* and *y-m-w,* each having a different gene in the center. Two of the recombined classes can be generated by a single recombination event, but the third must be the product of two crossovers involving the same chromatids. Because the likelihood of such a double crossover should be small (on the order of the product of the probability of two different crossover events), the smallest recombined class should represent double recombinants. The smallest observed recombinant class in figure 5.5 is that composed of the complementary gametes + *w* + and *y* + *m.* Using the three different gene orders, double recombination can produce these rare gametes only if the order of the genes is *y-w-m* (figure 5.6). Double recombination for the other two gene orders does not result in + *w* + and *y* + *m* as the rarest class but produces the other, more common recombinant classes.

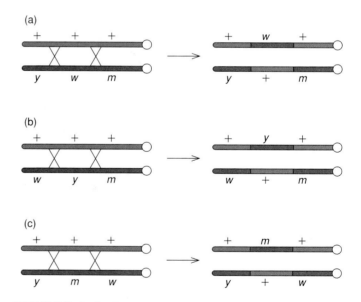

FIGURE 5.6   The three possible orders of genes *y, w,* and *m* with gametes produced from double recombination. Order (*a*) with gene *w* in the center is consistent with the data in figure 5.5. The X symbol indicates the position of crossing over.

With three linked genes, there are three different recombination frequencies between pairs of loci: between *w* and *m,* between *y* and *w,* and between *y* and *m.* From figure 5.5, the total frequency of recombination between *w* and *m* is the sum of the two recombinant classes that have recombined in this region, or

0.329 + 0.00086 = 0.330. (Double crossover gametes must have had recombination between w and m as well.) Likewise, for y and w the frequency of recombination is the sum 0.0057 + 0.00086 = 0.0066. Between y and m, the frequency is 0.329 + 0.0057 = 0.335. (Another way to ascertain the order of the genes is to note that the largest of these recombination values must be between the outside genes, y and m in this case.)

Given that the order of the genes is y-w-m, a double recombinant must involve both an exchange between loci y and w and one between w and m. Using the recombination frequencies just calculated, the expected frequency of a double recombination event (using the product rule) should be (0.0066)(0.330) = 0.00218. The observed frequency of double recombination is 0.00086, which is somewhat lower than that expected (see further discussion later in this chapter). The map distance between y and m is then calculated as the sum of the recombination between y and w plus that for w and m, or 0.006 + 0.330 = 0.337. Notice that the same information is obtained from one three-point cross as from three crosses involving different pairs of genes. In addition, three-point crosses allow us to identify gametes produced by double recombination.

Using such crosses and related techniques, researchers have constructed a genetic map of *Drosophila melanogaster* that is one of the most complete for any eukaryote. Some of this information is given in figure 5.7, listing the positions of some of the well-known genes on the four different chromosomes. Besides the X chromosome, which is 70 map units long, *Drosophila* has three autosomes. The two large autosomes have 108 and 106 map units, while the small fourth chromosome has only 3 map units. The total length of these chromosomes has been determined by the summation of empirically mapped intervals, as discussed previously. Overall then, the *Drosophila* genome has 287 map units. The linkage maps of other organisms used in genetic research, such as *Neurospora,* corn, and mice, are also well known.

Three-point crosses are used to determine the order of genes on a chromosome.

INTERFERENCE

Recombination frequency is defined as the proportion (or probability) of exchange between two linked genes. As we have suggested, more than one recombinational exchange may occur on a chromosome. Therefore, it is important to know whether these recombinations are independent of each other or whether they somehow influence each other. If different recombinant events are independent, the probability of two such events occurring in the same gamete should be equal to the product of the probability of the separate events. For the y, w, and m chromosomes discussed previously, the observed frequency of double crossovers was 0.00086, somewhat lower than the 0.00218 proportion expected. Fewer

double crossovers than expected are generally found in other organisms as well.

A standard way of describing the difference between the observed and expected numbers of double crossovers was first used by H. J. Muller. The method divides the observed numbers by the expected numbers (or proportions) of double crossovers and subtracts it from unity (one). This value gives a measure of the **interference (I)** of one crossover on another. This expression is:

$$I = 1 - \frac{\text{Observed frequency double recombinants}}{\text{Expected frequency double recombinants}}$$

Thus, for the y, w, and m data, I = 1 − 0.00086/0.00218 = 1 − 0.372 = 0.628. Positive values such as this one indicate interference between recombination events, while values near zero suggest no interference—that is, that different recombinant events are independent of each other. For closely linked genes, the measure of interference may often be near one, while for widely separated genes on the same chromosome, I is often near zero.

One theory to explain the infrequency of double crossovers is that the high values of interference for closely linked genes appear to be associated with a physical inflexibility of the chromatids, which keeps them from bending enough for two crossovers to occur in a very short region. However, interference may occur in genes up to 30 map units apart, indicating that other factors must be important as well. Interestingly, the centromere for metacentric chromosomes in *Drosophila* (and perhaps other organisms) acts as a barrier for interference. In other words, crossing over in one arm of a metacentric chromosome has no inhibitory effect on recombination in the other arm.

We should also note that when genes are far apart on a chromosome, the observed frequency of recombination may be less than that expected from knowledge of map distances. For example, we know that genes w and r are 53 map units apart on the X chromosome. First, the maximum proportion of recombinant gametes when the genes are on different chromosomes is 0.5, based on the principle of independent assortment, so we would not expect 0.53 recombinant gametes. Second, multiple even numbers of recombination will reduce the observed recombination frequency. Figure 5.8 illustrates that both double and quadruple crossovers appear as nonrecombinant gametes when there are no intermediate marker genes. As a result, the observed proportion of recombinant gametes for genes 50 map units apart is generally less than 0.5. Remember that the proportion of recombinants between w and r observed by Sturtevant was 0.452. Only when genes are 70 to 80 map units apart does the observed frequency of recombinants approach 0.5.

The frequency of double recombination between closely linked genes is often lower than that expected, a phenomenon called interference.

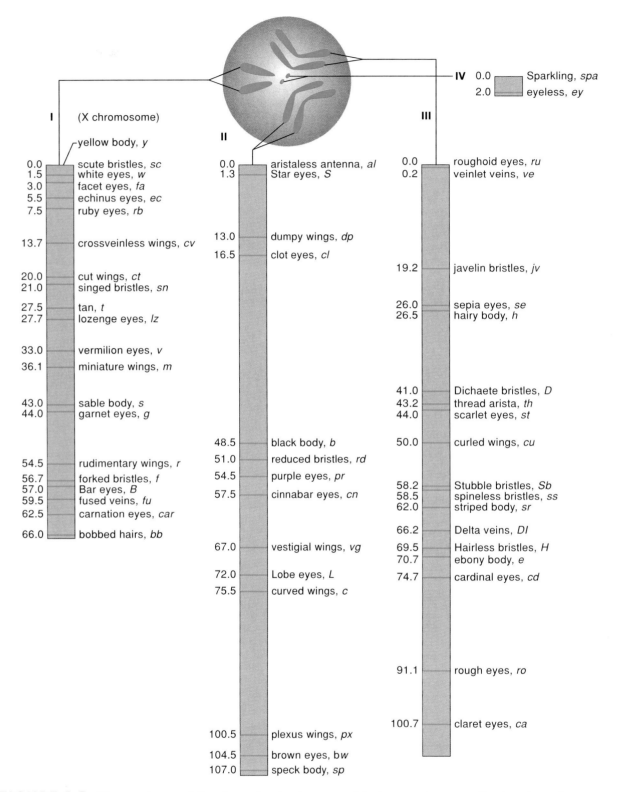

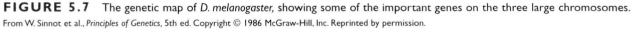

FIGURE 5.7   The genetic map of *D. melanogaster,* showing some of the important genes on the three large chromosomes.

From W. Sinnot et al., *Principles of Genetics,* 5th ed. Copyright © 1986 McGraw-Hill, Inc. Reprinted by permission.

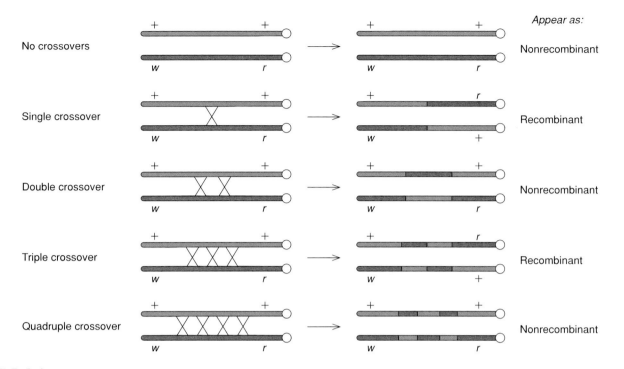

FIGURE 5.8 The types of gametes produced by different numbers of crossovers between two genes.

LINKAGE IN HUMANS

Planned genetic crosses are ethically not possible in humans, and the number of progeny from any one mating is small. As a result, until the recent introduction of molecular techniques, linkage had been determined for only a few human genes, mainly genes on the X chromosome. By studying informative matings (particular pedigrees segregating for the two genes of interest), tight linkage was established only between the ABO blood group system and the nail-patella syndrome (a condition causing abnormal and underdeveloped fingernails and small or absent kneecaps), between the β- and δ-globin genes, and between a few other gene pairs.

Mapping the X chromosome is made easier because recombination between X chromosomes takes place only in females (remember that males have only one X chromosome), and potential recombinant gametes can be directly identified in the male progeny. In other words, if the female parent is a double heterozygote with known gametic types, and if a sizable number of male offspring are available for study, the recombination frequency can be estimated.

An example of linkage is given in figure 5.9 for the X-linked traits color blindness and hemophilia. In this case, a woman (II-1) was unaffected by either recessive trait but produced three sons who had both hemophilia and color blindness, one son who was just color blind, and two who were normal. Because her father (I-2) was unaffected by both traits, the woman must have inherited an X chromosome with normal alleles from him and an X with both recessive alleles from her mother. In other words, she was a double heterozygote in

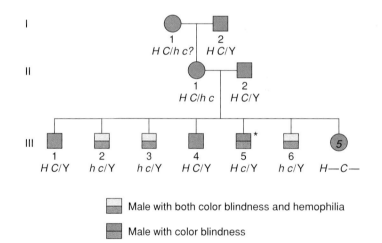

FIGURE 5.9 In this pedigree, a female (II-1) heterozygous for two X-linked genes has five nonrecombinant sons (1, 2, 3, 4, and 6) and one recombinant son (5, denoted by an asterisk). In addition, there are five unaffected daughters.

coupling phase (*H C/h c*). Therefore, the five sons who were either normal (*H C/Y*) or affected with both traits (*h c/Y*) received a nonrecombinant gamete from her, while the sixth son (III-5) received a recombinant gamete, *H c*. The frequency of recombination in this pedigree is then ⅙ = 0.167. Of course, to obtain a good estimate of the map distance between these genes, many such families must be examined, and the results from different families must then be combined using various statistical techniques. Generally, pedigrees segregating for these two traits do not have recombinants because the genes are closely linked

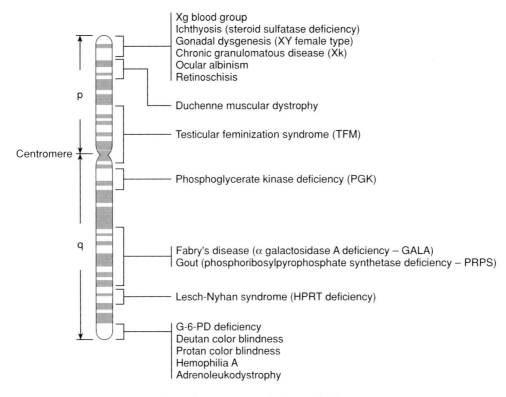

Xg blood group
Ichthyosis (steroid sulfatase deficiency)
Gonadal dysgenesis (XY female type)
Chronic granulomatous disease (Xk)
Ocular albinism
Retinoschisis

Duchenne muscular dystrophy

Testicular feminization syndrome (TFM)

Phosphoglycerate kinase deficiency (PGK)

Fabry's disease (α galactosidase A deficiency – GALA)
Gout (phosphoribosylpyrophosphate synthetase deficiency – PRPS)

Lesch-Nyhan syndrome (HPRT deficiency)

G-6-PD deficiency
Deutan color blindness
Protan color blindness
Hemophilia A
Adrenoleukodystrophy

FIGURE 5.10   The relative positions of a number of the genes on the human X chromosome, where p and q indicate short and long arms, respectively.

From Strickberger, Monroe W., *Genetics*, 3/e © 1985. Reprinted by permission of Prentice-Hall, Inc., Upper Saddle River, NJ.

at the end of the long arm of the X chromosome. These genes and a number of other X-linked genes, along with their relative map positions, are shown in figure 5.10.

Interestingly, the recombination rate for autosomal genes in males and females is often different. In fact, British evolutionary biologist J. B. S. Haldane suggested that in organisms with a chromosomal mechanism of sex determination, recombination is generally higher in the homogametic sex. In mammals, the homogametic sex is usually the female. And in both humans and mice, the recombination rate is higher in females than in males—about 50% to 100% higher for the human gene pairs examined. In *Drosophila,* which also has homogametic females, no recombination occurs in males; but in the silk moth *Bombyx mori,* which has heterogametic females, the females show no recombination.

Over the past few years, there has been a tremendous effort to map the human genome. An estimate of the total map length for humans can be obtained by observing the average number of chiasmata in meiosis I. The estimated mean number of chiasmata in male meiosis is about fifty per cell, or slightly more than two per bivalent. Assuming that a chiasma occurs approximately every 50 map units, the total map distance is around 2,000 to 2,500 map units, spread over the twenty-three human chromosomes—nearly ten times the map length of the *Drosophila* genome.

For many human genes, it is not known on which chromosome a gene is located, much less what linkage may exist

within the chromosome. In some cases, chromosomal abnormalities can give a clue to the location of genes. For example, trisomies are used to locate genes in a number of crops (see figure 4.35), and in theory, trisomy 21 (Down syndrome) individuals should have a different level of enzymes produced by genes on the affected chromosome. However, individuals with Down syndrome exhibit many abnormalities, so it is not clear whether any biochemical differences are due to three doses of a gene or to the abnormal physiological condition of these individuals. Likewise, individuals with deletions might also have lower levels of proteins produced by genes in the deleted region.

In both *Drosophila* and prokaryotes (see chapter 13), **deletion mapping** is a technique through which genes can be located to a particular chromosome segment (see also figure 4.22). For example, assume an individual is heterozygous for a deletion on a given chromosome that has a number of codominant genes (blood group, enzyme, or DNA variants). If different alleles are present at each gene, each genotype should be recognizable as a heterozygote (figure 5.11a). However, when a gene is in the deleted region, a heterozygote should always appear as a homozygote (figure 5.11b). To distinguish deletion heterozygotes from true homozygotes, some pedigree information is necessary. For example, if a deletion heterozygote B_2—was mated with a B_1B_2, there could be B_1—offspring appearing as B_1B_1 homozygotes. But if the parent really was a B_2B_2 homozygote and was mated with a B_1B_2, the offspring should be half

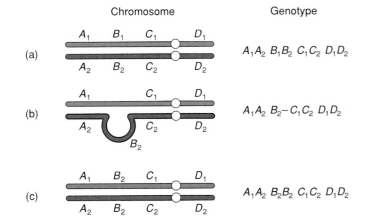

FIGURE 5.11   (a) A normal heterozygote for codominant alleles; (b) a heterozygote missing the gene B region; and (c) an individual homozygous for B_2.

B_1B_2 and half B_2B_2, with no individuals that are B_1 (and appear as B_1B_1 homozygotes).

SOMATIC CELL HYBRIDIZATION

Until recently, it was extremely difficult to locate human genes to specific chromosomes. One ingenious approach researchers used is **somatic cell hybridization.** In this technique, cells from two different species are fused in the laboratory, using a fusion agent such as Sendai virus. The Sendai virus has several points of attachment to its host cell, a characteristic that allows it to serve as a vehicle to fuse two quite different host cells. The chromosomes of the two species in the fused cells and the genetic trait (or traits) of interest are both monitored over time; human cells are usually hybridized with mouse or hamster cells. Eventually, the human chromosomes are more or less randomly lost, so that different cell lines lose different human chromosomes. The chromosome that contains the gene of interest is determined by noting which cell lines still carry the trait and finding a human chromosome that they have in common.

The first gene to be located in this way codes for thymidine kinase, an enzyme important in DNA replication. Mouse cells deficient for this enzyme (symbolized by Tk^-) were hybridized with human cells having the gene for the enzyme (Tk^+). These hybrid cells were then grown on a medium that required the cells to produce thymidine kinase in order to survive. Figure 5.12 shows the chromosomes found for five cell lines that were able to produce thymidine kinase. Notice that in these cell lines, human chromosome 17 is the only chromosome present in all lines. Therefore, these data indicate that the thymidine kinase gene must be on chromosome 17.

This experimental technique can be extended in several ways. For example, some cell lines also have deletions of a particular chromosomal segment. Hybrid cells with a deleted long arm of human chromosome 17 are Tk^-, indicating that the TK

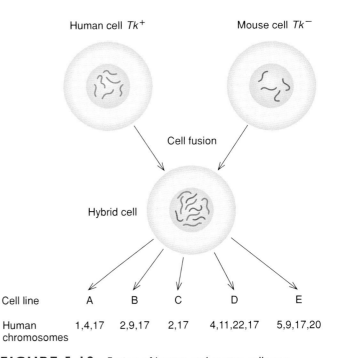

FIGURE 5.12   Fusion of human and mouse cells as a technique to identify the chromosomal location of the gene for thymidine kinase. The only human chromosome the five cell lines that produce thymidine kinase have in common is 17, demonstrating that the Tk gene must be on this chromosome.

gene must be on the long arm of chromosome 17. Furthermore, if several genetic traits are monitored simultaneously, the linkage relationship of the genes can be determined by the joint occurrence or absence of their protein products.

Studies locating genes to particular chromosomes in different organisms have shown that genes in related species are often linked in the same pattern. For example, the same genes appear on the X chromosome among related mammals. Other genes, such as TK, are on the same chromosome in all the great apes, supporting the evidence from chromosomal banding that indicates a high genetic homology for these species (see figure 4.40). In other words, if either the linkage relationship between two genes or the location of a gene on a chromosome is established in one species, similar species may have the same relationship. Knowledge of this sort could help shortcut the task of determining linkage patterns in each species. For example, the chromosomal location of the albino gene in humans is not known. However, the homologous albino genes in mice and cats are on regions with banding patterns like those on human chromosome 11p.

In the past, linkage of genes in humans generally was determined by accumulating information from informative matings. However, the technique of somatic cell hybridization with mouse cells did allow many more human genes to be assigned to particular chromosomes.

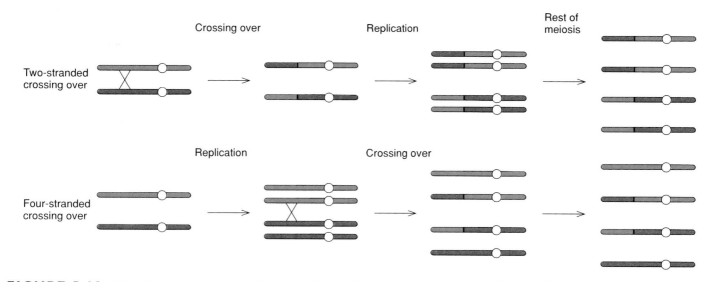

FIGURE 5.13   The different consequences of two-stranded and four-stranded crossing over. Four-stranded crossing over produces both parental and recombinant gametes, while two-stranded crossing over produces only recombinant gametes.

FOUR-STRANDED CROSSING OVER

Morgan's theories about the linkage of genes and crossing over were supported by extensive data. However, it was not clear initially whether recombination took place between chromosomes before they replicated (at the two-stranded stage) or after replication (at the four-stranded stage). If crossing over took place before replication, all four chromatids would be recombinants, while if it took place after replication, only two of the four chromatids would be recombinants (figure 5.13).

To distinguish between these alternatives, a genetic system was necessary in which all of the genetic products from meiosis could be identified. The first such system was developed in *Drosophila*, utilizing the **attached X,** a complex chromosome that has two X chromosomes attached at its centromere. As a result of this attachment, gametes from females having the attached X chromosome and a separate Y chromosome ($\overline{XX}$Y) contain either two X chromosomes or no X chromosomes (just a Y). An example of a cross in which both X chromosomes carry the wild-type allele at the white locus is given in figure 5.14. $\overline{XX}$Y females with the wild-type allele on both X chromosomes are crossed to normal white-eyed males. The viable offspring are identical to the parents, because the YY flies always die and the $\overline{XX}$X flies usually die.

When the alleles are different on the two attached X chromosomes, two-stranded and four-stranded crossing over can be differentiated. It is useful to use a mutant that is identifiable as a heterozygote in this experiment. For example, the *Bar* mutant reduces the number of eye facets from the normal number of approximately 780 to about 360 in the heterozygotes and about 70 in the homozygote *BB* (figure 5.15).

The top row of figure 5.16 indicates the types of chromosomes obtained when no crossing over occurs. In this case, all the

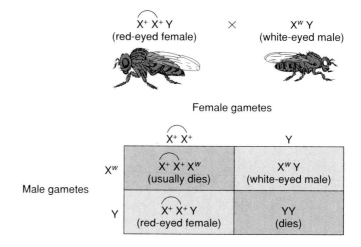

FIGURE 5.14   A cross between an attached X wild-type female and a white-eyed male. Notice that the surviving progeny are exactly like their parents.

gametes with X chromosomes have the heterozygous attached X and therefore the heterozygous *Bar* phenotype. Next, if crossing over takes place at the two-stranded stage (before replication), it will involve both strands. Therefore, when replication occurs later, all gametes are again heterozygous. In other words, the observed outcomes for no crossing over and two-stranded crossing over are the same.

Recombination at the four-stranded stage yields different results in the progeny. In this instance, only two of the four strands participate in recombination, so that new types are produced. With recombination at the four-stranded stage, both wild-type and *BB* homozygotes should result. In fact, this is the observation from actual experiments, indicating that

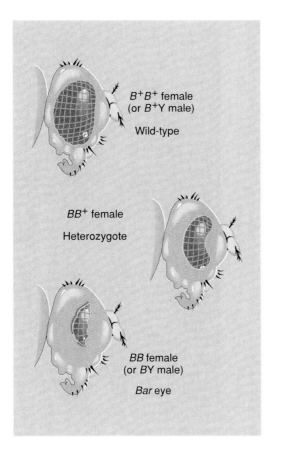

FIGURE 5.15 The phenotypes of wild-type, heterozygous *Bar* eye, and homozygous *Bar* eye.

recombination takes place after replication—that is, at the four-stranded stage.

TETRAD ANALYSIS

Confirmation of four-stranded crossing over was also obtained in *Neurospora crassa,* a fungus with the useful genetic attribute of having its meiotic products arranged in linear fashion. The life cycle of *Neurospora* is given in figure 5.17, illustrating mating and the production of a cell with the diploid chromosome number, which then undergoes meiosis. The meiotic products, called ascospores, are aligned in an ascus (or sac) so that the four on the left half descend from one type of segregants in meiosis I, while the four on the right half descend from the complementary segregants. In addition, within each half of the ascospore, one-half are products from half of meiosis II and the other half from the complementary half. After meiosis is completed, one mitotic division occurs, so that each pair of ascospores is descended from one of the products of meiosis II. These four products of meiosis, which are linearly arranged in the four pairs of ascospores, are called a **tetrad.**

When there is no crossing over in a heterozygote, one-half of the ascus has one allele and the other half has the second allele (figure 5.18). All the ascospores on one-half of the ascus have chromosomes with one of the centromeres and the other half

have the chromosomes with the other centromere. However, when crossing over does occur between the gene and the centromere, the pairs of the ascospores alternate as to which alleles they contain. This observation could be true only if recombination took place at the four-stranded stage. Two-stranded recombination would give results indistinguishable from no crossing over, assuming that there were no other marker genes.

Because the centromeres in the top half of the ascospores in an ascus are descended from one centromere and the bottom half from the other centromere, the proportion of recombinant asci can be used to determine the rate of recombination between the centromere and a given gene. For example, if 84% of a group of scored asci are like the top half of figure 5.18 (all *A* on one side and all *a* on the other side of the ascus) and the remaining 16% are like the bottom half of figure 5.18 (having both *A* and *a* in the top half and both *A* and *a* on the bottom half), the rate of recombination is 0.16.

Other experiments in *Neurospora* have increased our understanding of the fundamental features of meiotic recombination. For example, the three basic ways that nonsister chromatids can have double crossing over are distinguishable in *Neurospora* (figure 5.19). First, if there are three genes on the chromosome and both crossovers involve the same chromatids, let us say the center ones in figure 5.19a, then four types of ascospores are produced, with the parental ones on the outside. This is the type of double recombination event we assumed in our earlier discussion.

Second, different nonsister chromatids may be involved in the two recombination events (figure 5.19b). And finally, one chromatid could be involved in both events, while the second chromatid is different in the two crossovers (figure 5.19c). Notice that in this instance, one parental gametic type (*ab*) is on the outside of the ascus and the other (*AB*) is on the inside. An important observation from these results is that the products of recombination are reciprocal.

> The fact that crossing over takes place in the four-stranded stage of meiosis I was first demonstrated using attached X chromosomes in *Drosophila.* Confirmation of this theory was obtained from studies of the fungus *Neurospora.*

INVERSIONS AND RECOMBINATION

When individuals are heterozygous for chromosomal variants, then the products of meiosis may not be viable, a phenomenon that we mentioned in the last chapter. For example, when an individual is heterozygous for an inversion or a translocation, the resulting gametes may have either duplications or deletions. We can understand these problems for inversions by using the information we have just developed on four-stranded recombination. Remember from the last chapter that inversions may either include the centromere (pericentric) or may not include the centromere (paracentric). When an individual is an inversion heterozygote for either type of inversion and recombination

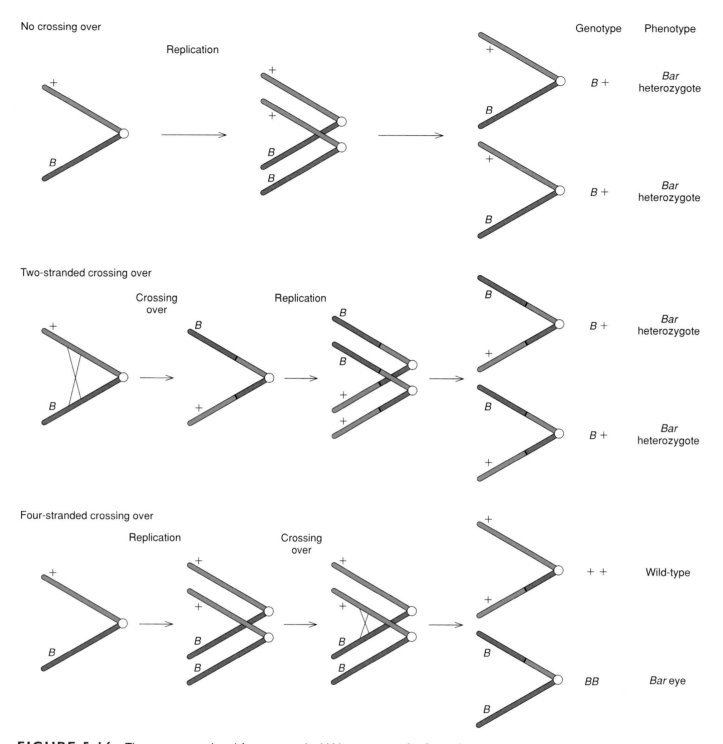

FIGURE 5.16   The gametes produced from an attached X heterozygous for *Bar* with no crossing over (top), followed by the results expected from two- and four-stranded crossing over. In reality, wild-type and *Bar* eye progeny are observed, indicating that crossing over takes place at the four-stranded stage.

occurs only outside the inversion loop, the products of meiosis have the same genes and the same gene sequence as the parental chromosomes. However, if there is crossing over within the inversion loop, the chromosomes may be abnormal. Figure 5.20*a* shows the products of a recombination between genes *B* and *C*

within a paracentric inversion loop. At anaphase I, the homologous chromosomes pull apart so that one chromatid involved in recombination is **acentric** (having no centromere) and the other is **dicentric** (having two centromeres). The acentric fragment usually does not go to a pole and becomes lost, while the

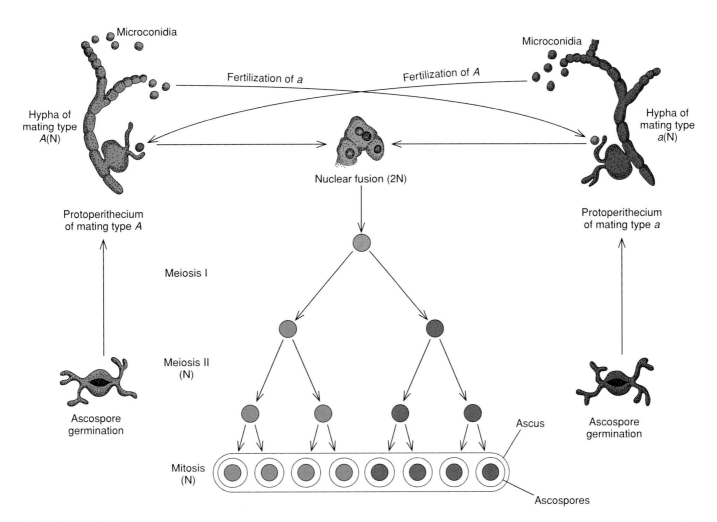

FIGURE 5.17 The production of ascospores following meiosis in *Neurospora crassa*. The ascospores are in a linear array such that each half is descended from the segregants at meiosis I. Each quarter is descended from chromatids associated with a particular centromere.

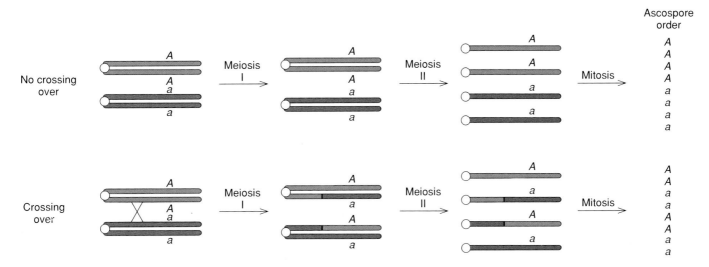

FIGURE 5.18 The ascospore order, illustrating the difference between no crossing over and a single crossover.

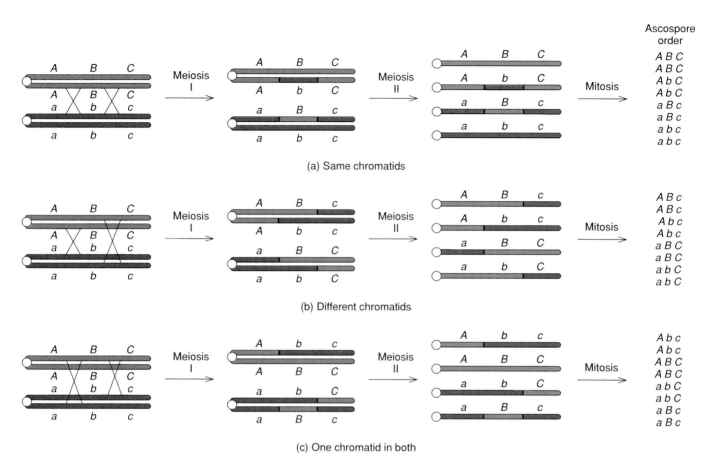

(a) Same chromatids

(b) Different chromatids

(c) One chromatid in both

FIGURE 5.19   The ascospore order that results from the three different types of double crossovers: (a) Involving the same chromatids; (b) involving different chromatids; and (c) involving one chromatid in both events.

dicentric fragment breaks more or less at random between the two centromeres. When the centromere splits in meiosis II, the result is two balanced chromosomes—one with the normal sequence and one with the inverted sequence—and two chromosomes with either a deletion or a deletion and a duplication. In this example, one abnormal chromosome (*EFG*) is missing more than half the total chromosome, while the other is missing only the *A* region but has a duplication of the *D* region. Zygotes formed from gametes containing these abnormal chromosomes would in all probability be lethal.

Recombination within a pericentric inversion loop can also result in abnormal chromosomes. As an example, assume that recombination takes place between genes *D* and *E* (figure 5.20*b*). Again there are two balanced chromosomes, but in this case the other pair has both a duplication and a deficiency. Zygotes with these unbalanced chromosomes would also likely be lethal.

As we have just seen, when recombination takes place within an inversion, the resulting chromosomes have either a deletion or a deletion and a duplication, and they are consequently often lethal. As a result, there appears to be a suppression of recombination in inverted regions, because no recombinant products are observed. Furthermore, in inversion heterokaryotypes, pairing difficulties often reduce recombination near and in the inverted region.

> When recombination takes place in the inverted region of inversion heterokaryotypes, duplications and deficiencies show up in the meiotic products.

UNEQUAL CROSSING OVER

Duplication and deletion of a relatively small amount of genetic material can result from a mistake related to recombination. Chromatids are strongly attracted to their homologous counterparts in prophase I, and they pair gene by gene, as discussed earlier. Occasionally, however, pairing does not occur precisely as it should. In particular, when regions already have duplications or repeated sequences, homologous chromosomes sometimes mispair. If recombination occurs within this mispaired region, duplications or deletions can result.

The classic demonstration of this phenomenon is for the *Bar* mutant in *Drosophila melanogaster*. In stocks homozygous for *Bar*, approximately one of 1,600 new chromosomes is either wild-type or a more extreme variant having only twenty-five eye facets, called *Ultrabar* (*B^u*). It was shown by Sturtevant

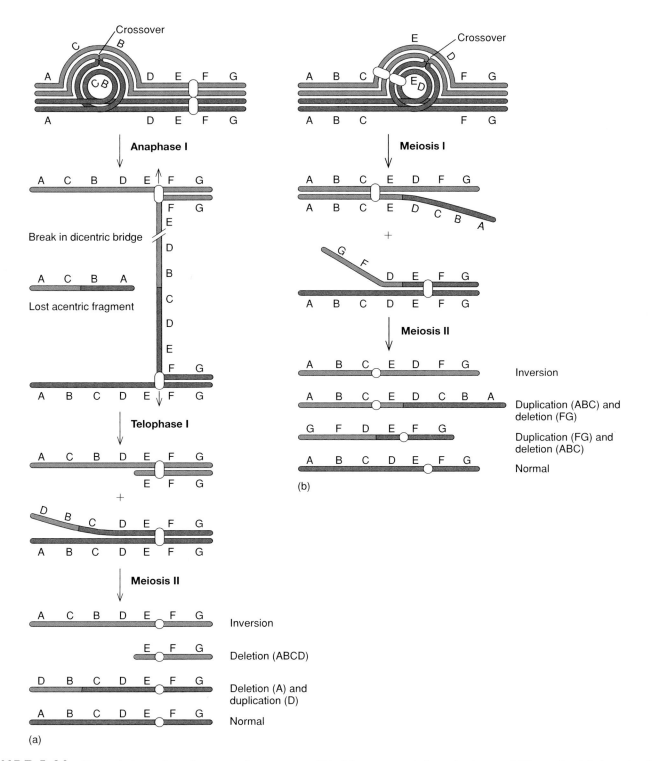

FIGURE 5.20   The meiotic products from a single crossover within (*a*) a paracentric inversion loop and (*b*) a pericentric inversion loop.

and Morgan that these are not new mutants but the products of recombination.

To understand the origin of *Ultrabar*, Sturtevant and Morgan constructed marked chromosomes that had the *Bar* mutations, as well as other mutations, on either side of the *Bar* region. In particular, these chromosomes had the recessive mutation for forked bristles (*f*) on one side and the recessive mutation causing fusion of two wing veins (*fu*) on the other side (figure 5.21). The map positions of these genes are 56.7 for *f*, 57.0 for *B*, and 59.5 for *fu*; thus, a relatively small region of about 3 map units on the X chromosome can be examined. Females homozygous at *Bar*, but heterozygous (and in coupling) at these other genes, were

Normal pairing and normal crossing over

Mispairing and unequal crossing over

Number of progeny

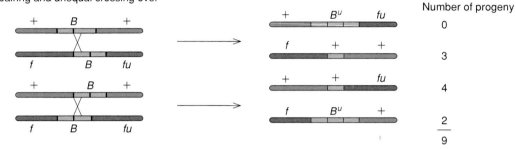

FIGURE 5.21 (*top*) The results of normal crossing over. (*bottom*) The results of mispairing and unequal crossing over that produce both wild-type and *Ultrabar* chromosomes. The number of progeny are those few abnormal progeny observed out of 20,000.

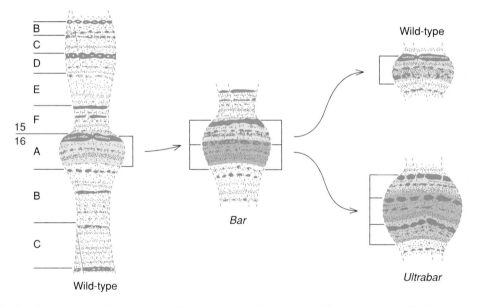

FIGURE 5.22 The banding patterns of homozygous *Bar* eye mutants. Note that wild-type, *Bar*, and *Ultrabar* have one, two, and three copies of band 16A, respectively.

crossed with males having the triply mutant chromosome, so that any recombinant chromosomes could be identified. Of the approximately 20,000 progeny examined, most chromosomes were nonrecombinant; that is, their chromosomes were either + *B* + or *f B fu*, because the marker loci were so tightly linked to *B*. Of the recombinant chromosomes, nearly all had *Bar* eyes, but seven had the wild-type allele and two had *Ultrabar*. In fact, all of the chromosomes that did not have the *Bar* allele were recombinant, three being *f* + +, four + +*fu*, and two *f B*ᵘ +.

Confirmation that duplication of the region was responsible for the *Bar* phenotype was obtained by examining in detail the banding pattern of the X chromosome. *Bar* chromosomes have a duplication of region 16A (figure 5.22), and the subsequent unequal recombinants have either a single copy of the genetic material in this region (wild-type) or three copies (B^u).

In some cases, related genes may be closely linked, such as the different globin genes or the genes of the major histocompatibility complex (MHC) in mammals. (The MHC in humans is called the HLA region.) It is thought that linked genes of similar structure and function are originally generated by duplication resulting from unequal crossing over. A group of related genes is called a **multigene family** (see box 5.1).

5.1 Multigene Families

Many genes are now known to exist as multigene families. Most of these families have probably arisen as a result of unequal crossing over and subsequent tandem duplication. Generally, the individual genes in a multigene family retain identical or similar functions. However, if the duplication occurred far back in the evolutionary past, the different individual genes in the family may have developed somewhat different functions.

A particularly striking multigene family is the cluster of linked genes comprising the **homeotic** genes in the flour beetle *Tribolium*. Different homeotic genes direct cells in different body segments of insects to differentiate into the appropriate anatomical structures (see figure B5.1). Five of these genes have been mapped to a chromosomal segment of 2.6 map units. In addition, the order of the genes is identical to the order of the body segments they affect (see figure B5.2).

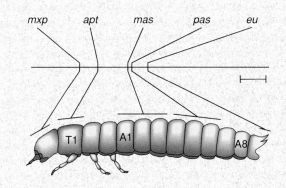

FIGURE B5.2 Five genes in *Tribolium*, showing the body segments they affect. (— = one map unit length.)
Source: Richard W. Beeman, USDA.

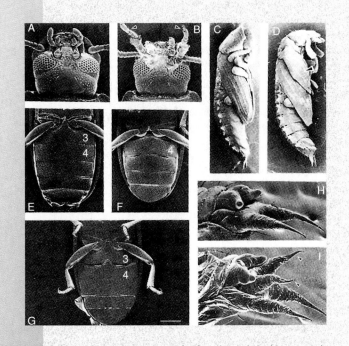

FIGURE B5.1 Normal phenotypes for five closely linked homeotic loci in a flour beetle are shown in (*a*), (*c*), (*e*), and (*h*); mutant phenotypes are shown in (*b*), (*d*), (*f*), (*g*), and (*i*). The five genes that cause the mutant phenotypes are shown in figure B5.2.
Courtesy of Richard W. Beeman, USDA.

In some cases, unequal crossing over occurs, and this can increase the amount of DNA by generating duplications.

MAP DISTANCE AND PHYSICAL DISTANCE

As we have mentioned before, the map distance between two genes estimated by genetic crosses generally corresponds with the physical distance, or length of DNA, that separates the genes. One major exception is the centromeric region, which has no recombination but can contain a sizable length of DNA. Furthermore,

it appears that recombination may often be localized into particular regions, and that it is not completely random over the length of the chromosome.

Comparing genetic and physical maps, figure 5.23 gives both the estimated map distance and the number of kilobases in the DNA (physical distance) for the major HLA genes. For example, genes *B* and *C* are close physically and are also only 0.1 map unit apart. On the other hand, genes *DP* and *DR* are physically closer than genes *C* and *A*, but the map distance is greater. This suggests that there may be higher than expected recombination between *DP* and *DR*.

Recently, a substantial effort has been focused toward obtaining a genetic map of the human genome (see figure 5.24 and chapter 15 for up-to-date details). Most of the initial landmarks on this map are DNA variants called **restriction fragment**

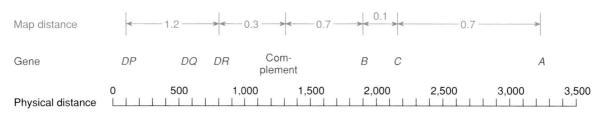

FIGURE 5.23   Comparison of map distance (given in map units) and physical distance (given in kilobases) for the HLA region in humans.

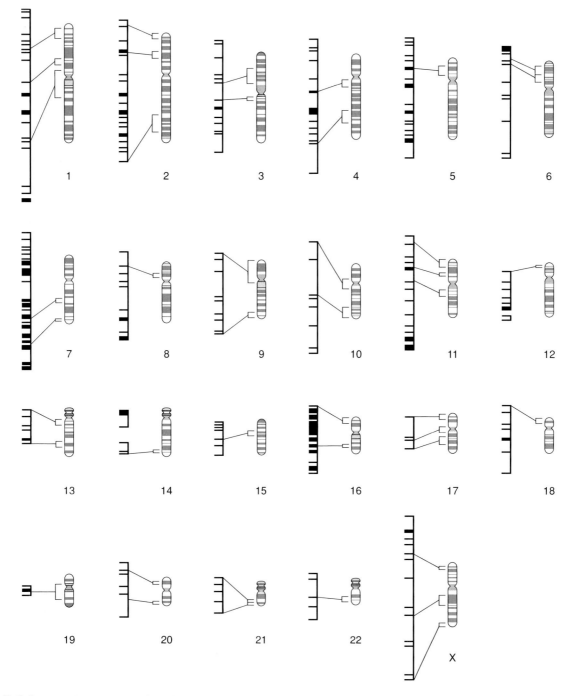

FIGURE 5.24   A preliminary map of the human genome (much more detailed maps are now available) in which the left-hand diagram for each chromosome indicates the relative map distance between different RFLPs. The right-hand chromosome gives the major banding patterns along with the general location of some RFLP sites.

length polymorphisms (RFLPs); we will discuss RFLPs more fully in chapter 15. Notice that some chromosomes, such as 1 and 7, are well covered by these genetic variants, while others, such as 14 and 19, have very few markers. Using these and other molecular markers, there are presently thousands of markers spread throughout the human genome. Adjacent to the genetic map of each chromosome is the physical map based on chromosome banding studies. This new map has great potential for helping us determine the location of genes that cause particular diseases and subsequently examine the DNA to see what has caused the defect.

The genetic map distance and physical DNA distance generally correspond quite well. Recently, DNA variants have been employed to map the human genome.

WORKED-OUT PROBLEMS

PROBLEM 1

A wild-type strain of flies was crossed to another strain homozygous for the recessive X-linked mutants m, r, and v. The F_1 females were backcrossed to recessive strain males, and the following numbers of progeny were observed.

+	+	+	331
m	+	+	2
+	r	+	73
+	+	v	14
m	r	+	10
m	+	v	81
+	r	v	0
m	r	v	309
			820

What is the gene order and the map distance between these mutants?

SOLUTION

There are three possible orders of the genes: m-r-v, m-v-r, and r-m-v. The two smallest progeny classes must be progeny produced by double recombination; that is, classes m + +, and + rv. The only way that these progeny can be produced from + m + r + v females by double recombination is if gene m is in the center, the gene order v-m-r. (See figure 5.6 as an example for genes y, w, and m.)

To calculate the map distance, rewrite the gene order for these progeny with m in the middle, count up all the progeny with recombination in a given region, and divide by the total number of progeny:

Genotype			Numbers	Region of recombination
+	+	+	331	None
+	m	+	2	v-m, m-r
+	+	r	73	m-r
v	+	+	14	v-m
+	m	r	10	v-m
v	m	+	81	m-r
v	+	r	0	v-m, m-r
v	m	r	309	None
			820	

For the region between v and m, the rate of recombination is:

$$v - m = (2 + 14 + 10 + 0)/820 = 0.032$$

For the region between m and r, the rate of recombination is:

$$m - r = (2 + 73 + 81 + 0)/820 = 0.190$$

In other words, the numbers of map units between v and m is 3.2; between m and r, 19; and between v and r, 22.2. The map units between v and m, as well as between m and r are calculated by multiplying the rates of recombination by 100.

PROBLEM 2

Let us assume that we have scored a number of *Neurospora* asci for the alleles at two loci that were heterozygous before meiosis. The results of this cross follow, first for the A gene only.

Ascus type	Number
AAAA aaaa	56
AAaa AAaa	8
AAaa aaAA	11
aaAA aaAA	12
aaAA AAaa	7
aaaa AAAA	62

What is the rate of recombination between the centromere and the A gene?

The types of asci when they are scored for both the A and B genes are given below. In this case, the two gametes that were fused at the beginning of meiosis were A B and a b. What is the rate of recombination between these two loci?

Ascus Type			
Top half	AB ab	0.5 AB, 0.5 Ab	0.5 ab, 0.5 aB
Bottom half	ab AB	0.5 ab, 0.5 aB	0.5 AB, 0.5 Ab
Number	54	51	23 28

SOLUTION

The rate of recombination between the centromere and the A gene is the number of recombinant asci divided by the total number of recombinant asci scored. In this data set, the middle four types were recombinant, i.e., not all A or not all a on one side of the ascus, so the rate of recombination is $(8 + 11 + 12 + 7)/156 = 0.244$.

The rate of recombination between loci A and B is calculated in a similar manner. In this case the first two columns were nonrecombinant asci, so the rate of recombination is the number in the last two columns divided by the total number of asci scored or $(23 + 28)/156 = 0.327$.

PROBLEM 3

A researcher wanted to find out the map distance between two loci in rats, which we will indicate as *C* and *D*. She crossed inbred lines that were *CC DD* and *cc dd* (Cross 1) and then backcrossed the progeny of this cross to inbred line *cc dd* and scored the progeny as recombinant or nonrecombinant. In another cross between inbred lines *CC dd* and *cc DD* (Cross 2), she also crossed the progeny back to *cc dd* line and scored the progeny. Given the data below from these crosses, what is the estimated rate of recombination between these two loci?

Numbers

Progeny type	Cross 1	Cross 2
Cc Dd	43	12
Cc dd	8	59
cc Dd	11	62
cc dd	51	15

SOLUTION

Because the only recombinant gametes in Cross 1 are *C d* and *c D* (the parental gametes are *C D* and *c d*), the number of recombinant gametes from this cross is $8 + 11 = 19$, i.e., the sum of the number of the *Cc dd* or *cc Dd* progeny. Likewise, the number of recombinant gametes from Cross 2 is $12 + 15 = 27$, i.e., the number of *Cc dd* or *cc Dd* progeny (the parental gametes in this cross are *C d* and *c D*). Therefore, the total number of recombinant gametes is $19 + 27 = 46$. The total number of progeny (or gametes) scored is $113 + 148 = 261$. The overall rate of recombination is then $46/261 = 0.176$.

SUMMARY

In the early 1900s, exceptions to Mendel's principle of independent assortment were discovered—namely, that alleles at different genes on parental chromosomes tend to remain together because of the distance between genes on a chromosome. The rate of recombination estimates the linkage of genes and ranges from 0 for extremely tightly linked loci to ½ for very loosely linked loci (or for those on different chromosomes). A genetic map illustrates the relationship of genes on a chromosome. Three-point crosses are used to determine the order of genes on a chromosome. In a phenomenon known as interference, the frequency of double recombination between closely linked genes is depressed compared to the frequency expected.

Linkage of genes in humans has been difficult to determine. Originally, it was done by accumulating information from informative matings, but another technique using somatic cell hybridization with mouse cells has located many more human genes on particular chromosomes. The fact that crossing over takes place in the four-stranded stage of meiosis I was first demonstrated in a study of attached X chromosomes in *Drosophila*. The theory was later confirmed using the fungus *Neurospora*.

In some cases, crossing over occurs unequally, thus generating duplications and increasing the amount of DNA. Generally, there is a close correspondence between the genetic map distance and the physical DNA distance. DNA variants are now being used to map the human genome.

PROBLEMS AND QUESTIONS

Problems and Questions with Links to Software

 Drosophila Genetics: 2, 3, 4, 6, 7, 11, 25

1. Which principle of Mendel's does linkage violate? Explain, in a sentence, how linkage causes a result different from this principle's expectations.

2. In a cross between two F_1 garden peas having genotype *Pp Ll*, the progeny types were 142 purple, long; 8 purple, round; 11 red, long; and 47 red, round. (This is a progeny array similar to that shown in table 5.1.) Does this fit 9:3:3:1 expectations, using a chi-square test?

3. Using a testcross, a recombination rate of 0.114 was found for the genes for purple and long. In the cross by Bateson and Punnett, what proportions of the four gametes would you expect from F_1 parents?

4. Using the recombination rate in question 3, what proportions of the four phenotypes would you expect in the F_2 progeny? Use a chi-square test to see if the observed data in table 5.1 are consistent with this explanation.

5. Why did Mendel not discover linkage as perhaps a third principle?

6. Calculate the rate of recombination between gene *pr* and *vg* using the data from Morgan's second testcross (table 5.3). Compare your results to the value calculated from the first testcross in which the parents were in coupling.

7. Show diagrammatically (using chromosomes as in figure 5.3) how gametes for the F_2 are produced when the original parents are *prpr vg⁺vg⁺* and *pr⁺pr⁺vgvg*.

8. In a series of testcrosses, the rate of recombination between genes *a* and *c* is 0.1; between *a* and *b*, 0.27; and between *b* and *c*, 0.2. Determine the order of the three genes. Explain why the sum of rates of recombination for the two closest pairs of genes is not equal to the rate between the two genes farthest apart.

9. Another gene, *d*, had a rate of recombination of 0.15 with gene *b* and a rate of 0.14 with gene *a*. Where would you place this gene on the map constructed for question 8? What cross would you make to confirm your map?

10. Why were three-point crosses a valuable approach to learning about the nature of the linkage of genes?

11. Assume a three-point cross in which the $F_1 kk^+ ll^+ mm^+$ was crossed to *kkllmm* and the offspring had the following genotypes:

$kk^+ ll^+ mm^+$?	621
$kk^+ ll^+ mm$	3
$kk^+ llmm^+$?	64
$kkll^+ mm^+$	103
$kk^+ llmm$?	109
$kkll^+ mm$ -	57
$kkllmm^+$	7
$kkllmm$	608

   Construct a genetic map of these three loci, including their order and the map distance between them.

12. Using the data in problem 11, calculate the expected proportion of double recombinants. What is the value of interference (I) estimated from these data? Explain what this level of I means.

13. Diagram the types of gametes you would expect from a $y+/+w$ female, given the following situations between these loci: no crossovers, one crossover, two crossovers, three crossovers, and four crossovers.

14. Why do you think it was much harder to determine linkage patterns in humans than in other species such as *Drosophila* or corn?

15. In a large family, the mother is heterozygous for both hemophilia and color blindness with repulsion gametes, or genotype *H c/h C*. Two of her sons have hemophilia, two other sons are color blind, and the fifth son is normal. Following the style of figure 5.9, give the genotypes for the sons. What is the estimated rate of recombination between the genes in this family?

16. Several families having a female parent heterozygous for both hemophilia and color blindness were gathered. In these families, the following numbers of recombinants and sons were observed: 1 of 5 sons, 2 of 8, 1 of 4, 1 of 5, and 1 of 6. What is the estimated rate of recombination, using all these data? In fact, the actual rate is much less. Can you give a possible explanation?

17. Assume that a deletion heterozygote (A_1-) was mated with an $A_2 A_2$ individual. What type of progeny, and in what proportions, would you expect? Would this differ from the results expected if the first parent were $A_1 A_1$?

18. Using somatic cell hybridization, a number of cells having a particular human enzyme were generated. Five of these cell lines had human chromosomes: (A) 4, 9, 11; (B) 2, 7, 11, 21; (C) 4, 9, 10, 11, 21; (D) 1, 2, 11, 18; and (E) 4, 11, 15. On which chromosome is the gene for this enzyme?

19. Discuss the difference between two-stranded and four-stranded crossing over. What would be the proportions (parental and nonparental) of the resulting gametes in two-stranded crossing over?

20. Give the progeny genotypes and phenotypes of a cross between $\overline{X^w X^+} Y$ and $X^+ Y$ where the two X chromosomes in the female parent are attached. What types of progeny survive? Why can't this cross be used to differentiate between two- and four-stranded crossing over?

21. Assume that *Neurospora* with allele *D* are crossed to those with allele *d*. If there were no crossing over between the centromere and this gene, what order of ascospores would you expect? If there was one crossover, what order would you expect?

22. Assume that we have two genes, *E* and *F*, which we know are on a particular *Neurospora* chromosome. If we cross *EF* with *ef* and examine the resulting asci, most of them contain the following sequence of ascospores: *EF, EF, EF, EF, ef, ef, ef, ef.* However, some are *EF, EF, eF, eF, Ef, Ef, ef, ef,* or *eF, eF, EF, EF, ef, ef, Ef, Ef.* What is the order of these genes and the centromere?

23. Sturtevant and Morgan used genes *f* (56.7) and *fu* (59.5) as markers in a cross to examine *Ultrabar*. What proportion of recombinant progeny would you expect between these markers? Sturtevant and Morgan found 9 out of approximately 20,000 progeny that were recombinant for the markers and had either wild-type or *Ultrabar* alleles. Assuming that mispairing takes place one out of ten times for *Bar* homozygotes, what is the estimated size of the region that can produce duplications or deficiencies?

24. In this chapter, we have been focusing on genetic or map distance between genes. Generally, map distance is similar to physical distance on the chromosome, but in some cases it is not. Suggest reasons why map and physical distance may not be exactly concordant.

25. Assume that two of the genes Mendel picked for his studies in garden peas were tightly linked. How might this have influenced the principle of independent assortment?

Answers appear at end of book.

MOLECULAR GENETICS

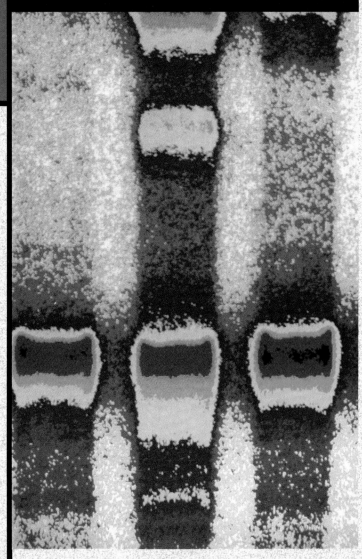

False-color scan of an RNA blot. Scientists at Xoma Corporation used gel electrophoresis to separate different-sized RNA molecules in three samples. Each RNA forms a horizontal band in the vertical gel lines. The workers blotted the RNAs to a membrane and treated them with radioactive DNA molecules that would bind only to specific RNAs. A phosphorimager collected beta rays from the radioactive DNA. In this image, made from a laser scan of the phosphorimager plate, the colors indicate intensity of the radioactivity, and therefore the amount of RNA in each horizontal band.
© Jay Fries/The Image Bank.

Chemistry of the Gene

Must we geneticists become bacteriologists, physiological chemists and physicists, simultaneously with being zoologists and botanists? Let us hope so.

H. J. Muller

American geneticist

Learning Objectives

In this chapter you will learn:

1. That most genes are made of a chainlike substance called DNA, and that the others are composed of a similar substance called RNA.

2. The chemical nature of DNA and RNA.

3. That most genes contain two strands of DNA wound around each other in a double helix.

4. How the two strands in a DNA double helix can be separated and put back together again.

5. That DNA comes in a variety of sizes and shapes.

Here is a fact of life: At its most fundamental level, life is chemistry. All living things, including you, are made of molecules. Moreover, your life is sustained by thousands of chemical reactions occurring constantly in your body; without chemical reactions, your life and all other life would be impossible. Therefore, in order to have a fundamental understanding of life, we must know something about its chemistry.

This is equally true of genetics: At its most fundamental level genetics is chemistry. One can perform genetic experiments using whole organisms, mating them and monitoring the phenotypes of their offspring to determine the nature and organization of genes. In fact, the history of genetics is full of such experimentation. Although this approach has been, and continues to be, very fruitful, it does not tell the whole story. That is because genes themselves are molecules—extremely large, *informational* molecules that contain the instructions for making other biomolecules. To understand genetics in detail, we must understand the structure and function of genes as molecules. This chapter introduces the molecular structure of genes.

THE NATURE OF GENETIC MATERIAL

The studies that have revealed the chemistry of genes began in Tübingen, Germany, in 1869. There, Friedrich Miescher isolated nuclei from pus cells (white blood cells) in waste surgical bandages. He found that these nuclei contained a novel phosphorus-bearing substance that he named *nuclein*. Nuclein is mostly **chromatin,** a complex of **deoxyribonucleic acid (DNA)** and chromosomal proteins.

By the end of the nineteenth century, both DNA and **ribonucleic acid (RNA)** had been separated from the protein that clings to them in the cell. This allowed more detailed chemical analysis of these **nucleic acids.** (Notice that the term *nucleic acid* and its derivatives, *DNA* and *RNA,* come directly from Miescher's term *nuclein.*) By the beginning of the 1930s, P. Levene, W. Jacobs, and others had demonstrated that RNA is composed of a sugar (ribose) plus four nitrogen-containing bases, and that DNA contains a different sugar (deoxyribose) plus four bases. They discovered that each base is coupled with a sugar-phosphate to form a nucleotide. Furthermore, the four bases in DNA or RNA seemed to be present in roughly equal quantities. We will return to the chemical structures of DNA and RNA later in the chapter. First, let us examine the evidence that genes are made of DNA (or sometimes RNA).

TRANSFORMATION IN BACTERIA

Frederick Griffith laid the foundation for the identification of DNA as the genetic material in 1928 with his experiments on **transformation** in the bacterium pneumococcus, now known

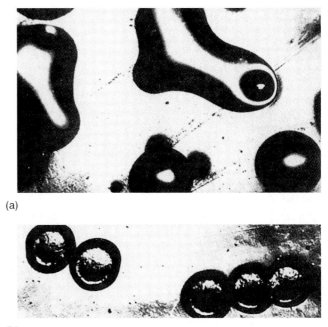

(a)

(b)

FIGURE 6.1 Variants of *Streptococcus pneumoniae.* (*a*) The large, glossy colonies contain smooth (S) virulent bacteria; (*b*) the small, mottled colonies are composed of rough (R) avirulent bacteria.

Photographs by Harriett Ephrussi-Taylor.

as *Streptococcus pneumoniae.* The wild-type organism is a spherical cell surrounded by a mucous coat called a capsule. The cells form large, glistening colonies, characterized as smooth (S) (figure 6.1*a*). These cells are **virulent**—capable of causing lethal infections upon injection into mice. A certain mutant strain of *S. pneumoniae* has lost the ability to form a capsule. As a result, it grows as small, rough (R) colonies (figure 6.1*b*). More importantly, it is **avirulent;** since it has no protective coat, it is engulfed by the host's white blood cells before it can proliferate enough to do any damage.

The key finding of Griffith's work was that heat-killed virulent colonies of *S. pneumoniae* could **transform** avirulent cells to virulent ones. Neither the heat-killed virulent bacteria nor the live avirulent ones by themselves could cause a lethal infection. Together, however, they were deadly. Somehow the virulent trait passed from the dead cells to the live, avirulent ones. This transformation phenomenon is illustrated in figure 6.2. Transformation was not transient; the ability to make a capsule and therefore to kill host animals, once conferred upon the avirulent bacteria, was passed to their descendants as a heritable trait. In other words, the gene for virulence, missing in the avirulent cells, was somehow restored during transformation. This meant that the transforming substance in the heat-killed bacteria was probably the gene for virulence itself. The missing piece of the puzzle was the chemical nature of the transforming substance. Whoever discovered this would reveal the nature of genes.

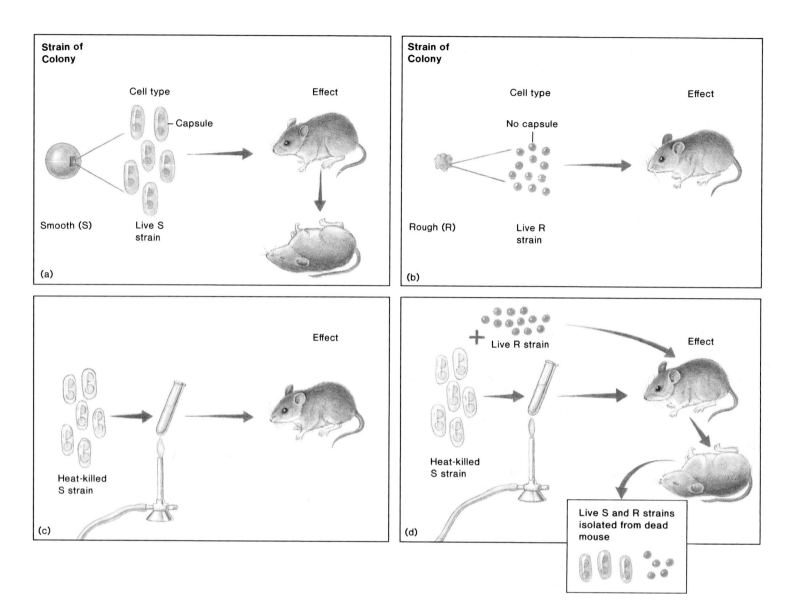

Strain of Colony

Cell type

Effect

Capsule

Smooth (S)

Live S strain

(a)

Strain of Colony

Cell type

Effect

No capsule

Rough (R)

Live R strain

(b)

Effect

Heat-killed S strain

(c)

Live R strain

Effect

Heat-killed S strain

Live S and R strains isolated from dead mouse

(d)

FIGURE 6.2   Griffith's transformation experiments.
(a) Virulent strain S *S. pneumoniae* kill their host; (b) avirulent
strain R bacteria cannot infect successfully, so the mouse survives;
(c) strain S bacteria that are heat-killed can no longer infect; (d) a
mixture of strain R and heat-killed strain S bacteria kills the mouse.
The killed virulent (S) bacteria have transformed the avirulent (R)
bacteria to virulent (S).

DNA: The Transforming Material

Oswald Avery, Colin MacLeod, and Maclyn McCarty supplied
the missing piece in 1944. They used a transformation test similar
to the one that Griffith had introduced, and they took pains to
define the chemical nature of the transforming substance from
virulent cells. First, they removed the protein from the extract
with organic solvents and found that it still transformed. Next,
they subjected it to digestion with various enzymes. Trypsin and
chymotrypsin, which destroy protein, had no effect on transformation. Neither did ribonuclease, which degrades RNA. These
experiments ruled out protein or RNA as the transforming material. On the other hand, Avery and his coworkers found that the
enzyme deoxyribonuclease (DNase), which breaks down DNA,
destroyed the transforming ability of the virulent cell extract.

Finally, direct physical–chemical analysis showed the purified transforming substance to be DNA. The analytical tools
Avery and his colleagues used were the following:

1. *Ultracentrifugation.* They spun the transforming substance
   in an ultracentrifuge to estimate its size. The material
   with transforming activity sedimented rapidly (moved
   rapidly toward the bottom of the centrifuge tube),
   suggesting a very high molecular weight, characteristic of
   DNA.

2. *Electrophoresis.* They placed the transforming substance in
   an electric field to see how rapidly it moved. The
   transforming activity had a relatively high mobility, also
   characteristic of DNA.

3. *Ultraviolet absorption spectrophotometry.* They placed a solution of the transforming substance in a spectrophotometer to see what kind of ultraviolet light it absorbed most strongly. Its absorption spectrum matched that of DNA. That is, the light it absorbed most strongly had a wavelength of about 260 nanometers (nm), in contrast to protein, which absorbs maximally at 280 nm.

4. *Elementary chemical analysis.* This yielded an average nitrogen/phosphorus ratio of 1.67, about what one would expect for DNA, which is rich in both elements, but vastly lower than the value expected for protein, which is rich in nitrogen but poor in phosphorus. Even a slight protein contamination would have raised the nitrogen/phosphorus ratio.

Further Confirmation

These findings should have settled the issue of the nature of the gene, but they had little immediate impact. The mistaken notion, from early chemical analyses, that DNA was a monotonous repeat of a four-nucleotide sequence, such as ACTG, persuaded many geneticists that it could not be the genetic material. Furthermore, controversy persisted about possible protein contamination in the transforming material, whether transformation could be accomplished with other genes besides those governing R and S, and even whether bacterial genes were like the genes of higher organisms.

Yet, by 1953, when James Watson and Francis Crick published the double-helical model of DNA structure, most geneticists agreed that genes were made of DNA. What had changed? For one thing, Erwin Chargaff had shown in 1950 that the bases were not really found in equal proportions in DNA as some geneticists believed, and that the base composition of DNA varied from one species to another. In fact, this is exactly what one would expect for genes, which also vary from one species to another. Furthermore, Rollin Hotchkiss had refined and extended Avery's findings. He purified the transforming substance to the point where it contained only 0.02% protein, and showed that it could still change the genetic characteristics of bacterial cells. He went on to show that such highly purified DNA could transfer genetic traits other than R and S.

Finally, in 1952, A. D. Hershey and Martha Chase performed another experiment that added to the weight of evidence that genes were made of DNA. This experiment involved a **bacteriophage** (bacterial virus) called T2 that infects the bacterium *Escherichia coli* (figure 6.3). (The term *bacteriophage* is usually shortened to *phage*.) During infection, the phage genes enter the host cell and direct the synthesis of new phage particles. The phage is composed of protein and DNA only. The question is: Do the genes reside in the protein or the DNA? The Hershey-Chase experiment answered this question by showing that, upon infection, most of the DNA entered the bacterium, along with only a little protein; the bulk of the protein stayed on the outside (figure 6.4). Since the DNA is the major component that gets into the host cells, it likely contains the genes. Of course,

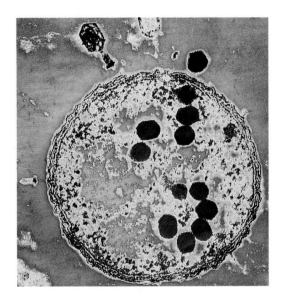

FIGURE 6.3   A false-color transmission electron micrograph of T2 phages infecting an *E. coli* cell. At top, a phage particle is about to inject its DNA into the host cell. However, another T2 phage has already infected the cell, and progeny phage particles are being assembled. The progeny phage heads are readily discernible as dark polygons inside the host cell.
From Lee D. Simon, Rutgers University, *Cell,* 4(6): Cover, June 1985.

this conclusion is not unequivocal; the small amount of protein that entered along with the DNA could conceivably carry the genes. But taken together with the work that had gone before, this study helped convince geneticists that genes really are made of DNA.

The Hershey-Chase experiment depended on radioactive labels on the DNA and protein—a different label for each. Such labeling became possible as a by-product of nuclear research before and during World War II. The labels used were phosphorus-32 (^{32}P) for DNA and sulfur-35 (^{35}S) for protein. These choices make sense, considering that DNA is rich in phosphorus while phage protein has none, and that protein contains sulfur but DNA does not.

Hershey and Chase allowed the labeled phages to attach by their tails to bacteria and inject their genes into their hosts. Then they removed the empty phage coats by homogenizing in a Waring blender. (Yes, they had those way back then.) Since we know that the genes must go into the cell, our question is: What went in, the ^{32}P-labeled DNA or the ^{35}S-labeled protein? As we have seen, it was the DNA. In general, then, genes are made of DNA. On the other hand, as we will see later in this chapter, other experiments showed that some viral genes consist of RNA.

Genes are made of nucleic acid, usually DNA. Some simple genetic systems such as viruses have RNA genes.

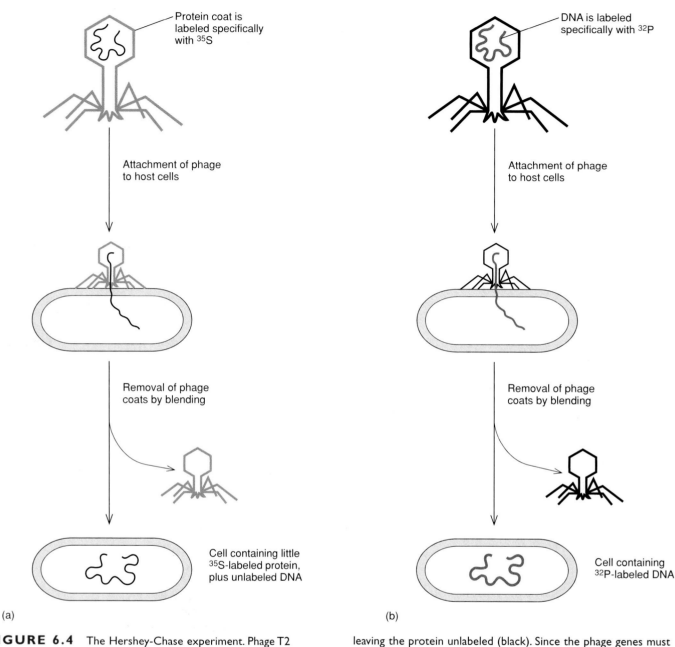

(a)   (b)

FIGURE 6.4   The Hershey-Chase experiment. Phage T2 contains genes that allow it to replicate in *E. coli.* Since the phage is composed of DNA and protein only, its genes must be made of one of these substances. To discover which, Hershey and Chase performed a two-part experiment. In the first part (*a*), they labeled the phage protein with ^{35}S (blue), leaving the DNA unlabeled (black). In the second part (*b*), they labeled the phage DNA with ^{32}P (red), leaving the protein unlabeled (black). Since the phage genes must enter the cell, the experimenters reasoned that the type of label found in the infected cells would indicate the nature of the genes. Most of the labeled protein remained on the outside and was stripped off the cells by blending (*a*), while most of the labeled DNA entered the infected cells (*b*). The conclusion was that the genes of this phage are made of DNA.

THE CHEMICAL NATURE OF POLYNUCLEOTIDES

By the mid-1940s, biochemists knew the fundamental chemical structures of DNA and RNA. When they broke DNA into its component parts, they found these constituents to be nitrogenous **bases, phosphoric acid,** and the sugar **deoxyribose** (hence the name deoxyribonucleic acid). Similarly, RNA yielded bases and phosphoric acid, plus a different sugar, **ribose.** The four bases found in DNA are **adenine (A), cytosine (C), guanine (G),** and **thymine (T).** RNA contains the same bases, except that **uracil (U)** replaces thymine. The structures of these bases, shown in figure 6.5, reveal that adenine and guanine are related to the parent molecule, purine. Therefore, we refer to these compounds as **purines.** The other bases resemble pyrimidine, so they

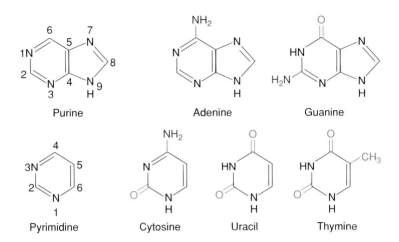

FIGURE 6.5 The bases of DNA and RNA. The parent bases, purine and pyrimidine, on the left, are not found in DNA and RNA. They are shown for comparison with the other five bases.

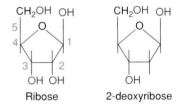

FIGURE 6.6 The sugars of nucleic acids. Note the OH in the 2-position of ribose, and its absence in deoxyribose.

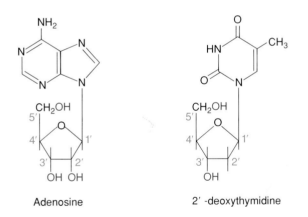

FIGURE 6.7 Two examples of nucleosides.

Base	Nucleoside (RNA)	Deoxynucleoside (DNA)
Adenine	Adenosine	Deoxyadenosine
Guanine	Guanosine	Deoxyguanosine
Cytosine	Cytidine	Deoxycytidine
Uracil	Uridine	Not usually found
Thymine	Not usually found	(Deoxy)thymidine

Because thymine is not usually found in RNA, the "deoxy" designation for its nucleoside is frequently assumed, and the deoxynucleoside is simply called **thymidine.** There does not seem to be much rhyme or reason to these names, but they are important. It is also worth learning the numbering of the carbon atoms in the sugars of the nucleosides (figure 6.7). Note that the ordinary numbers are used in the bases, so the carbons in the sugars are called by primed numbers. Thus, for example, the base is linked to the 1′-position of the sugar, and the 2′-position is deoxy in deoxynucleosides.

The structures in figure 6.5 were drawn using an organic chemistry shorthand that leaves out certain atoms for simplicity's sake. Figures 6.6 and 6.7 use a slightly different convention in which a straight line with a free end denotes a C-H bond. Figure 6.8 shows the structures of adenine and deoxyribose, first in shorthand, then with every atom included.

The subunits of DNA and RNA are **nucleotides,** which are nucleosides with a phosphate group attached through a phosphoester bond (figure 6.9). An ester is an organic compound formed from an alcohol (bearing a hydroxyl group) and an acid. In the case of a nucleotide, the alcohol group is the 5′-hydroxyl of the sugar, and the acid is phosphoric acid, which is why we call the ester a *phospho*ester. Figure 6.9 also shows the structure of one of the four precursors of DNA synthesis, deoxyadenosine-5′-triphosphate (dATP). Upon synthesis of DNA, two phosphate groups are removed from dATP, leaving deoxyadenosine-5′-monophosphate (dAMP). The other three nucleotides in DNA (dCMP, dGMP, and dTMP) have analogous structures and names.

are called **pyrimidines.** These structures constitute the alphabet of genetics.

Figure 6.6 depicts the structures of the sugars found in nucleic acids. Notice that they differ in only one place. Where ribose contains a hydroxyl (OH) group in the 2-position, deoxyribose lacks the oxygen and simply has a hydrogen (H) represented by the vertical line. Hence the name *deoxy*ribose. The bases and sugars in RNA and DNA are joined together into units called **nucleosides** (figure 6.7). The names of the nucleosides derive from the corresponding bases:

We will discuss the synthesis of DNA in detail in chapter 7. For now, notice the structure of the bonds that join nucleotides together in DNA and RNA (figure 6.10). These are called **phosphodiester bonds** because they involve phosphoric acid linked to *two* sugars—one through a sugar 5′-group, the other through a sugar 3′-group. You will notice that the bases have been rotated in this picture, relative to their positions in previous figures. This more closely resembles their geometry in DNA or RNA. Note also that this **trinucleotide** or string of three nucleotides, has polarity: The top of the molecule bears a free 5′-phosphate group, so it is called the **5′-end.** The bottom, with a free 3′-group, is called the **3′-end.**

Figure 6.11 introduces a shorthand way of representing a nucleotide or a DNA chain. This notation presents the deoxyribose sugar as a vertical line, with the base joined to the 1′-position at the top and the phosphodiester links to neighboring nucleotides through the 3′-(middle) and 5′-(bottom) positions.

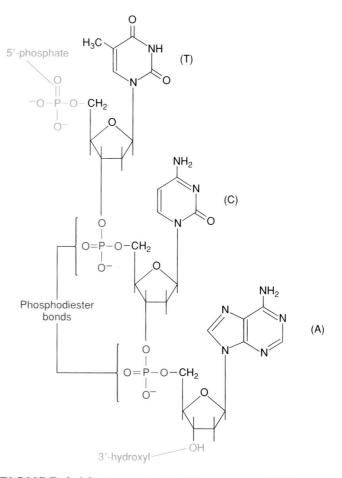

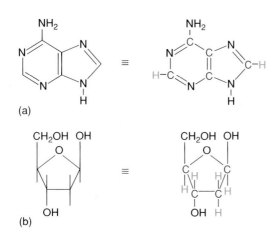

FIGURE 6.8   The structures of (*a*) adenine and (*b*) deoxyribose. Note that the structures on the left omit most or all of the carbons and some of the hydrogens. These are included in the structures on the right, in red and blue, respectively.

FIGURE 6.10   A trinucleotide. This little piece of DNA contains only three nucleotides linked together by phosphodiester bonds (red) between the 5′- and 3′-hydroxyl groups of the sugars. The 5′-end of this DNA is at the top, where there is a free 5′-phosphate group (blue); the 3′-end is at the bottom, where there is a free 3′-hydroxyl group (blue). The sequence of this DNA could be read as 5′pdTpdCpdA3′. This would usually be simplified to TCA.

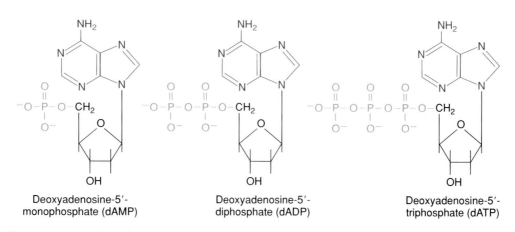

Deoxyadenosine-5′-
monophosphate (dAMP)

Deoxyadenosine-5′-
diphosphate (dADP)

Deoxyadenosine-5′-
triphosphate (dATP)

FIGURE 6.9   Three nucleotides. The 5′-nucleotides of deoxyadenosine are formed by phosphorylating the 5′-hydroxyl group. The addition of one phosphate results in deoxyadenosine-5′-monophosphate (dAMP). One more phosphate yields deoxyadenosine-5′-diphosphate (dADP). Three phosphates give deoxyadenosine-5′-triphosphate (dATP).

DNA and RNA are chainlike molecules composed of subunits called nucleotides. The nucleotides, in turn, contain a base linked to the 1'-position of a sugar (ribose in RNA or deoxyribose in DNA) and a phosphate group. The phosphate joins the nucleotides in a DNA or RNA chain through their 5'- and 3'-hydroxyl groups by phosphodiester bonds.

DNA STRUCTURE

All the facts about DNA and RNA just mentioned were known by the end of the 1940s. By that time it was also becoming clear that DNA was the genetic material and that it therefore stood at the very center of the study of life. Yet the three-dimensional structure of DNA was unknown. For these reasons, several researchers dedicated themselves to finding this structure.

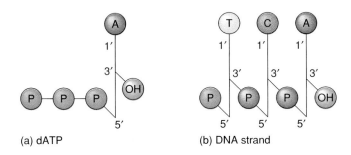

(a) dATP   (b) DNA strand

FIGURE 6.11 Shorthand DNA notation. (*a*) The nucleotide dATP. This illustration highlights four features of this DNA building block: (1) The deoxyribose sugar is represented by the vertical black line. (2) At the top, attached to the 1'-position of the sugar is the base, adenine (orange). (3) In the middle, at the 3'-position of the sugar is a hydroxyl group (OH, green). (4) At the bottom, attached to the 5'-position of the sugar is a triphosphate group (blue). (*b*) A short DNA strand. The same trinucleotide (TCA) illustrated in figure 6.10 is shown here in shorthand. Note the 5'-phosphate (blue), the phosphodiester bonds (red), and the 3'-hydroxyl group (green). According to convention, this little piece of DNA is written 5' to 3' left to right.

EXPERIMENTAL BACKGROUND

One of the scientists interested in DNA structure was Linus Pauling, a theoretical chemist at the California Institute of Technology. He was already famous for his studies on chemical bonding and for his elucidation of the α-helix, an important feature of protein structure. Indeed, the α-helix, held together by hydrogen bonds, laid the intellectual groundwork for the double helix model of DNA proposed by Watson and Crick. Another group trying to find the structure of DNA included Maurice Wilkins, Rosalind Franklin, and their colleagues at King's College in London. They were using X-ray diffraction to analyze the three-dimensional structure of DNA. Finally, there were James Watson and Francis Crick. Watson, still in his early twenties but already holding a Ph.D. degree from Indiana University, had come to the Cavendish Laboratories in Cambridge, England, to learn about DNA. There he met Crick, a physicist who at age thirty-five was retraining as a molecular biologist. Watson and Crick performed no experiments themselves. Their tactic was to use other groups' data to build a DNA model.

Erwin Chargaff was another very important contributor. We have already seen how his 1950 paper helped identify DNA as the genetic material, but the paper contained another piece of information that was even more significant. Chargaff's studies of the base compositions of DNAs from various sources revealed that the content of purines always equalled the content of pyridines. Furthermore, the amounts of adenine and thymine were always equal, as were the amounts of guanine and cytosine. These findings, known as Chargaff's rules, provided a valuable confirmation of Watson and Crick's model. Table 6.1 presents Chargaff's data. You will notice some deviation from the rules due to incomplete recovery of some of the bases, but the overall pattern is clear.

Perhaps the most crucial piece of the puzzle came from an X-ray diffraction picture of DNA taken by Franklin in 1952—a picture that Wilkins shared with James Watson over dinner in London on the night of January 30, 1953. The X-ray technique worked as follows: The experimenter made a very concentrated, viscous solution of DNA, then reached in with a needle and pulled out a fiber. This was not a single molecule, but a whole

Table 6.1 Composition of DNA in Moles of Base per Mole of Phosphate

	Human				Yeast		Avian Tubercle Bacilli	Bovine				
	Sperm		Thymus	Liver Carcinoma				Thymus			Spleen	
	#1	#2			#1	#2		#1	#2	#3	#1	#2
A:	0.29	0.27	0.28	0.27	0.24	0.30	0.12	0.26	0.28	0.30	0.25	0.26
T:	0.31	0.30	0.28	0.27	0.25	0.29	0.11	0.25	0.24	0.25	0.24	0.24
G:	0.18	0.17	0.19	0.18	0.14	0.18	0.28	0.21	0.24	0.22	0.20	0.21
C:	0.18	0.18	0.16	0.15	0.13	0.15	0.26	0.16	0.18	0.17	0.15	0.17
Recovery:	0.96	0.92	0.91	0.87	0.76	0.92	0.77	0.88	0.94	0.94	0.84	0.88

Source: E. Chargaff, *Experimentia* 6:206.

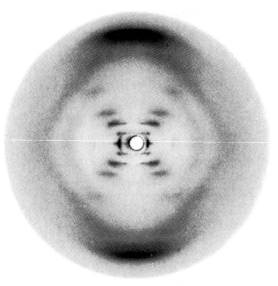

FIGURE 6.12   Franklin's X-ray picture of DNA. The regularity of this pattern indicated that DNA is a helix. The spacing between the strong bands at the top and bottom gave the spacing between elements of the helix (base pairs) as 3.4 angstroms. The spacing between neighboring lines in the pattern gave the overall repeat of the helix (the length of one helical turn) as 34 angstroms.

Courtesy of Professor M. H. F. Wilkins, Biophysics Department, King's College, London.

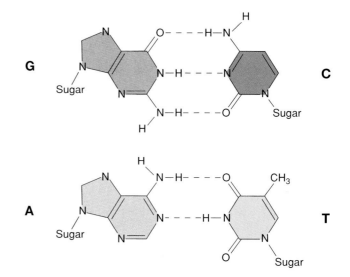

FIGURE 6.13   The base pairs of DNA. A guanine-cytosine pair (G–C), held together by three hydrogen bonds (dashed lines), has almost exactly the same shape as an adenine-thymine pair (A–T), held together by two hydrogen bonds.

batch of DNA molecules, forced into side-by-side alignment by the pulling action. Given the right relative humidity, this fiber was enough like a crystal that it diffracted X rays in an interpretable way. In fact, the X-ray diffraction pattern in Franklin's picture (figure 6.12) was so simple—a series of spots arranged in an **X** shape—that it indicated that the DNA structure itself must be very simple. By contrast, a complex, irregular molecule like a protein gives a complex X-ray diffraction pattern with many spots, rather like a surface peppered by a shotgun blast. Since DNA is very large, it can be simple only if it has a regular, repeating structure. And the simplest repeating shape that a long, thin molecule can assume is a corkscrew, or helix.

THE DOUBLE HELIX

Franklin's X-ray work strongly suggested that DNA was a helix. Not only that, it gave some important information about the size and shape of the helix. In particular, the spacing between adjacent spots in an arm of the **X** is inversely related to the overall repeat distance in the helix, 34 angstroms (34 Å), and the spacing from the top of the **X** to the bottom is inversely related to the spacing (3.4Å) between the repeated elements (**base pairs**) in the helix. However, even though the Franklin picture told much about DNA, it presented a paradox; DNA was a helix with a regular, repeating structure, but for DNA to serve its genetic function, it must have an *irregular* sequence of bases.

Watson and Crick saw a way to resolve this contradiction and satisfy Chargaff's rules at the same time: DNA must be a

double helix with its sugar-phosphate backbones on the outside and its bases on the inside. Moreover, the bases must be paired, with a purine always across from a pyrimidine. This way the helix would be uniform, it would not have bulges where two large purines are paired or constrictions where two small pyrimidines are paired. Watson has joked about the reason he seized on a double helix: "I had decided to build two-chain models. Francis would have to agree. Even though he was a physicist, he knew that important biological objects come in pairs."

But Chargaff's rules went further than this. They decreed that the amounts of adenine and thymine were equal and so were the amounts of guanine and cytosine. This fit very neatly with Watson and Crick's observation that an adenine-thymine base pair held together by hydrogen bonds has almost exactly the same shape as a guanine-cytosine base pair (figure 6.13). So Watson and Crick postulated that adenine must always pair with thymine, and guanine with cytosine. This way, the double-stranded DNA will be uniform, composed of very similarly shaped base pairs, regardless of the unpredictable sequence of either DNA strand by itself. This was their crucial insight, the key to the structure of DNA.

The double helix, often likened to a twisted ladder, is presented in several ways in figure 6.14. The curving sides of the ladder represent the sugar-phosphate backbones of the two DNA strands; the rungs are the base pairs. The spacing between base pairs is 3.4 Å, and the overall helix repeat distance is about 34 Å, meaning that there are about ten base pairs per turn of the helix. (One angstrom [Å] is one ten-billionth of a meter or one-tenth of a nanometer [nm].) The arrows indicate that the two strands are **antiparallel.** If one has $5' \rightarrow 3'$ polarity from top to bottom, the other must have $3' \rightarrow 5'$ polarity from top to bottom. In solution, DNA has a structure very similar to the one just described, but the helix contains about 10.5 base pairs per turn.

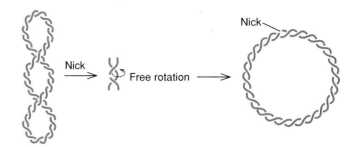

FIGURE 7.21 Nicking one strand relaxes supercoiled DNA. A nick in one strand of a supercoiled DNA (left) allows free rotation around a phosphodiester bond in the opposite strand (middle). This releases the strain that caused the supercoiling in the first place and so allows the DNA to relax to the open circular form (right).

FIGURE 7.20 Rubber band model of supercoiling in DNA. If you cut the rubber band and twist one free end through one complete turn while preventing the other end from rotating, the rubber band will relieve the strain by forming one supercoil.

about halfway down. With your other hand, twist one free end of the rubber band one full turn and hold it next to the other free end. You should notice that the rubber band resists the turning as strain is introduced, then relieves the strain by forming a supercoil. The more you twist, the more supercoiling you will observe: one superhelical turn for every full twist you introduce. Reverse the twist and you will see supercoiling of the opposite handedness or sign. DNA works the same way. Of course, the rubber band is not a perfect model. For one thing, you have to hold the severed ends together; in a circular DNA, chemical bonds join the two ends. For another, the rubber band is not a double helix.

If you release your grip on the free ends of the rubber band in figure 7.20, of course the superhelix will relax. In DNA, it is only necessary to cut one strand to relax a supercoil because the other strand can rotate freely, as demonstrated in figure 7.21.

Unwinding DNA at the replicating fork would form positive rather than negative supercoils if there were no way to relax the strain. That is because replication *permanently* unwinds one region of the DNA without nicking it; this means that the rest of the DNA would become *over*wound, and therefore positively supercoiled, to compensate. To visual-

ize this, look at the circular arrow ahead of the replicating fork in figure 7.19. Notice how twisting the DNA in the direction of the arrow causes unwinding behind the arrow but over-winding ahead of it. Imagine inserting your finger into the DNA just behind the fork and moving it in the direction of the moving fork to force the DNA strands apart. You can see how this would force the DNA to rotate in the direction of the circular arrow, which overwinds the DNA helix. This overwinding strain would resist your finger more and more as it moved around the circle. Therefore, unwinding the DNA at the replicating fork introduces *positive* superhelical strain that must be constantly relaxed so that replication will not be retarded. You can appreciate this when you think of how the rubber band increasingly resisted your twisting as it became more tightly wound. In principle, any enzyme that is able to relax this strain could serve as a swivel. In fact, of all the topo-isomerases in an *E. coli* cell, only one, DNA gyrase, appears to perform this function.

Topoisomerases are classified according to whether they operate by causing single- or double-stranded breaks in DNA. Those in the first class (type I topoisomerases, e.g., topo-isomerase I of *E. coli*) introduce temporary single-stranded breaks. Enzymes in the second class (type II topoisomerases, e.g., DNA gyrase of *E. coli*) break and reseal both DNA strands. Why is topoisomerase I incapable of providing the swivel function needed in DNA replication? Because it can relax only *negative* supercoils, not the positive ones that form in replicating DNA ahead of the fork. Obviously, the nicks created by these enzymes do not allow free rotation in either direction. But DNA gyrase pumps negative supercoils into closed circular DNA and therefore counteracts the tendency to form positive ones. Hence, it can operate as a swivel.

There is direct evidence that DNA gyrase is crucial to the DNA replication process. First of all, mutations in the genes for the two polypeptides of DNA gyrase are lethal and they block DNA replication. Second, antibiotics such as novobiocin, cou-mermycin, and nalidixic acid inhibit DNA gyrase and thereby prevent replication.

One or more enzymes called helicases use ATP energy to separate the two parental DNA strands at the replicating fork. As helicase unwinds the two parental strands of a closed circular DNA, it introduces a compensating positive supercoiling force into the DNA. The stress of this force must be overcome or it will resist progression of the replicating fork. The name given to this stress-release mechanism is the swivel. DNA gyrase, a bacterial topoisomerase, is the leading candidate for this role. Once the two strands are separated, single-strand-binding protein (SSB) keeps them that way.

INITIATION OF REPLICATION

We have already seen that a crucial aspect of initiation in DNA replication is synthesis of a short RNA primer. (There are a few exceptions to this rule, notably adenovirus, which uses a special protein to prime its DNA synthesis.) A rich variety of mechanisms exist to form these RNA primers, and each genetic system uses one of them. A large aggregate of about twenty polypeptides combine to make the primers in *E. coli* DNA replication. This **primosome** is so complex that the identification of its components would have taken much longer were it not for the existence of single-stranded DNA phages that use only parts of the primosome. These have been used as models for unraveling the mechanism of primer formation in *E. coli* itself.

Primer Synthesis

Two different experimental approaches have been used to identify the components of the DNA replication machinery, including the **primases,** or primer synthesizers, in these phage systems. The first is a combination genetic–biochemical approach, the strategy of which is to isolate mutants with defects in the ability to replicate phage DNA, then to complement extracts from these mutants with proteins from wild-type cells. The mutant extracts will be incapable of replicating the phage DNA in vitro unless the right wild-type protein is added. Using this system as an assay, the protein can be highly purified and then characterized. The second approach is the classical biochemical

one: purify all of the components needed and then add them all back together to reconstitute the replication system in vitro (see Box 7.1).

The first phage DNA replication system to be investigated was the simplest: the filamentous phage M13. As mentioned before, one of the striking early findings of this study was that the replication of this phage DNA is inhibited by rifampicin, an antibiotic that inhibits *E. coli* RNA polymerase. This means that the phage uses the host RNA polymerase as its primase. Another single-stranded DNA phage, G4, uses another host protein, the product of the *dnaG* gene, to make its primers. This is the same primase that *E. coli* uses. Still another well-studied phage, φX174, uses a group of *E. coli* proteins, including the *dnaG* protein (primase) for primer synthesis. These seem to be the three examples that all other single-stranded *E. coli* phages follow. The last has been the most help in understanding the host primosome because it resembles it most closely.

A combination of biochemical and genetic approaches has been used to identify the components of the primer-synthesizing systems of three classes of single-stranded DNA phages, all of which infect *E. coli*. The first class is represented by phage M13, which uses host RNA polymerase to make its primers; the second, represented by phage G4, uses the host primase (the *dnaG* protein); the third class, represented by the phage φX174, uses a complex primosome composed of the *dnaG* primase and other host polypeptides. This seems to be similar to the primosome the host employs.

Making Primers in *E. coli*

All the known elements of the DNA replicating machinery of *E. coli* make up a very complex system. For the moment, let us concentrate on some of the proteins involved in the synthesis of primers.

We have already introduced the primase, the enzyme that makes the primer, but we have not said much about how it operates. The difference between putting together a primer only

7.1 Biochemists Like to Grind Things Up

The tendency of biochemists to break things down to their component parts and then see how they go back together again has given rise to a joke about three scientists who were asked to figure out how a wristwatch worked. The physiologist stuck electrodes into it, the geneticist bombarded it with mutagenic radiation, and the biochemist ground it up in a Waring blender. In general, though, the separating and reconstituting strategy has proven very powerful, if not with wristwatches, then at least with

cells. Arthur Kornberg summed up his support for it with an aphorism "I remain faithful to the conviction that anything a cell can do, a biochemist should be able to do. He should do it even better. . . . Put another way, one can be creative more easily with a reconstituted system." In the investigation of DNA replication in *E. coli,* the marriage of this approach with genetics has been especially fruitful.

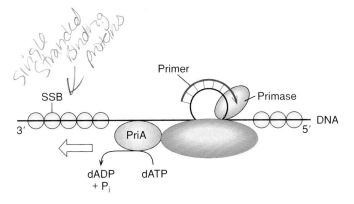

single stranded Binding proteins

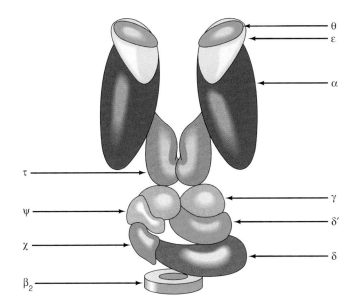

FIGURE 7.22 Primosome forming a primer on φX174 DNA. The complex primosome is moving in the 5′ → 3′ direction on the template strand (open arrow), while making a primer in the opposite direction (red arrow). The actual synthesis of the primer is performed by the primase (orange). The PriA protein (blue) provides energy for this process by cleaving energy-rich ATP or (as shown here) dATP. The rest of the primosome is represented by a green oval.

FIGURE 7.23 Schematic model of the *E. coli* DNA pol III holoenzyme. At top are the two core polymerase centers (subunits α, ε, and θ). These are joined through two τ subunits to the γ complex (subunits γ, δ, δ′, χ, and ψ). At bottom is the β clamp, made of two β subunits, not distinguished from each other in this drawing.

twelve nucleotides long and ordinary transcription of long genes is striking. It suggests that the primase works in a fundamentally different way than RNA polymerase involved in transcription.

This supposition is borne out by other characteristics of primase activity. During replication of double-stranded DNA, the primosome moves processively along the template for the lagging strand in the direction of motion of the replication fork. "Processively" means that it moves continuously along the DNA without dissociating from it. It is advantageous for the primosome to bind to the lagging strand template since that is where DNA is being made discontinuously and new primers are needed for each new Okazaki fragment. However, this presents a problem for the primase, because the direction the replicating fork is moving on the lagging strand template is 5′ → 3′, which means a primer synthesized complementary to that strand would be made in the 3′ → 5′ direction. RNAs, however, even primers, have to be made in the 5′ → 3′ direction. Actually, the primer *is* made in the 5′ → 3′ direction, even though the primosome moves in the opposite direction (figure 7.22). Just how this is accomplished is not known.

> The primosome is a complex structure containing many polypeptides that makes primers for DNA replication in *E. coli*. One of these proteins is the primase itself, which forms the primer. In *E. coli* DNA replication, the primosome moves along the template for the lagging strand and makes primers in the opposite direction.

ELONGATION

Once the primosome has made a primer, it is time for elongation to begin. The key enzyme involved in the elongation process is **DNA polymerase III holoenzyme,** a complex of ten different polypeptides (figure 7.23). The term "holoenzyme" simply means "the whole thing," including the active center (for DNA

synthesis in this case) as well as any attendant factors that participate in the reaction. We do not know for sure, but we suspect that the DNA polymerase III holoenzyme joins with the primosome to make a **replisome** that then carries out elongation of the DNA chain.

The *E. coli* Replisome

A major question immediately arises concerning the process of elongation: How does the replisome coordinate the synthesis of the leading and lagging strands? It appears at first that each replicating fork would need two replisomes, one for each strand, or that the replisome would have to be very flexible to carry out DNA synthesis in opposite directions on the two strands.

However, single DNA strands are considerably more flexible than proteins, so it is not unreasonable to expect that the lagging strand template loops through the replisome as shown in figure 7.24. If this happens, as Arthur Kornberg suggests, then the replisome could read both strands simultaneously because both would feed into the same replisome complex. The leading strand template would enter from the top, while the lagging strand template would enter from the bottom, causing a loop to form. This hypothesis implies that the replisome is an asymmetric dimer. One monomer synthesizes the leading strand; the other makes the lagging strand. This is consistent with the observation that the DNA polymerase III holoenzyme contains two DNA polymerase active centers. So the replisome really is double-headed.

This model has the great advantage that it permits one replisome to replicate both strands processively, without having to leave the replicating fork. This is important because once it

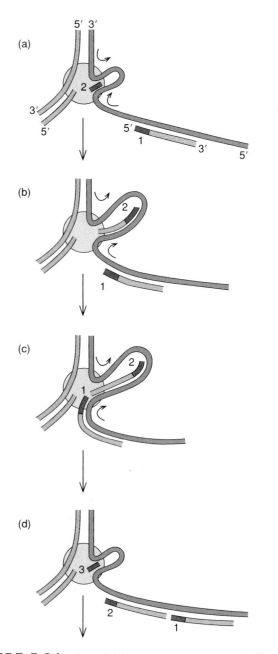

FIGURE 7.24 A model for simultaneous synthesis of both DNA strands. (a) The lagging template strand (blue) has formed a loop through the replisome (gold), and a new primer, labeled 2 (red), has been formed by the primase. A previously synthesized Okazaki fragment (green, with red primer labeled 1) is also visible. The leading strand template and its progeny strand are shown at left (gray), but the growth of the leading strand is not considered here. (b) The lagging strand template has formed a bigger loop by feeding through the replisome from the top and bottom, as shown by the arrows. The motion of the lower part of the loop (lower arrow) allows the second Okazaki fragment to be elongated. (c) Further elongation of the second Okazaki fragment brings its end to a position adjacent to the primer of the first Okazaki fragment. (d) The replisome releases the loop, including Okazaki fragments 1 and 2, and forms a new loop, which permits the primase to form a new primer (number 3). The process can now begin anew.

dissociates from the DNA, the replisome would waste much time rebinding. To appreciate this, consider that the rate of elongation in *E. coli* is about 1,000 nucleotides per second, and the rate of formation of a complex capable of initiation is one or two minutes. This means that while the replisome paused to rebind to the DNA, it could have been elongating the DNA chain by at least 60,000 nucleotides. Some time could be lost even in the looping and releasing process described here, but the intramolecular transfer involved must require only a split second.

In addition to DNA polymerases I and III, *E. coli* cells contain a third enzyme, DNA polymerase II. The role of this enzyme is not as clear as those of the other two, and cells with mutations in the polymerase II gene are quite viable.

Elongation during DNA synthesis in *E. coli* apparently requires a replisome made up of the primosome plus a DNA polymerase III holoenzyme composed of ten different polypeptides. Kornberg's model for simultaneous replication of both strands calls for a double-headed replisome with two DNA polymerase centers. One makes the leading strand continuously; the other synthesizes the lagging strand discontinuously, but without having to leave the replicating fork.

Multiple Eukaryotic DNA Polymerases

Much less is known about the proteins involved in eukaryotic DNA replication, but we do know that multiple DNA polymerases take part in the process, and we also have a good idea of the roles these enzymes play. Table 7.1 lists the known mammalian DNA polymerases and their probable roles.

Why do we think that polymerase α is responsible for lagging strand synthesis, while polymerase δ (and perhaps ϵ) is responsible for leading strand synthesis? At least two types of evidence point in this direction. First, polymerase α is the only one of the five enzymes to have primase activity. This capability is essential for the enzyme that makes the lagging strand, since each Okazaki fragment of this discontinuously synthesized

Table 7.1   Mammalian DNA Polymerases and Their Roles

Enzyme	Probable role
DNA polymerase α	DNA replication: synthesis of lagging strand
DNA polymerase δ	DNA replication: synthesis of leading strand
DNA polymerase ϵ	Unknown; structure similar to polymerase δ
DNA polymerase β	DNA repair
DNA polymerase γ	Replication of mitochondrial DNA

strand must be initiated anew with an RNA primer. On the other hand, only one primer is needed, at least in theory, for the continuously synthesized leading strand, so primase activity is much less crucial for the enzyme that makes that strand.

Furthermore, there is a big difference in processivity of polymerases α and δ. **Processivity** is the tendency of a polymerase to stick with the replicating job once it starts. The *E. coli* polymerase III holoenzyme is highly processive. Once it starts on a DNA chain, it remains bound to the template, making DNA for a long time. Since it does not fall off the template very often, which would require a pause as a new polymerase bound and took over, the overall speed of *E. coli* DNA replication is very rapid. Polymerase δ is much more processive than polymerase α. This makes sense if polymerase α is the enzyme that makes the lagging strand, since that enzyme has to dissociate and start anew at the beginning of each Okazaki fragment in any event. By contrast, the highly processive polymerase δ is a natural choice for the enzyme that makes the leading strand, since that strand grows continuously, with no dissociation required. Actually, much of the processivity of polymerase δ comes, not from the polymerase itself, but from an associated protein called **proliferating cell nuclear antigen,** or **PCNA.** This protein, which is enriched in proliferating cells that are actively replicating their DNA, enhances the processivity of polymerase δ by a factor of 40. That is, PCNA causes polymerase δ to travel 40 times farther elongating a DNA chain before falling off the template.

In marked contrast, polymerase β is not processive at all. It usually adds only one nucleotide to a growing DNA chain and then falls off, requiring a new polymerase to bind and add the next nucleotide. This fits with its postulated role as a repair enzyme that need only make short stretches of DNA to fill in gaps created when primers or mismatched bases are excised (see chapter 11). In addition, the level of polymerase β in a cell is not affected by the rate of division of the cell, which suggests that this enzyme is not involved in DNA replication. If it were, we would expect it to be more prevalent in rapidly dividing cells, as polymerases δ and α are.

Polymerase γ is found in mitochondria, not in the nucleus. It is therefore reasonable to infer that this enzyme is responsible for replicating mitochondrial DNA.

Mammalian cells contain five different DNA polymerases. Polymerases δ and α appear to participate in DNA replication, and to make the leading, and lagging strands, respectively. Polymerase β seems to have a function in DNA repair. Polymerase γ probably replicates mitochondrial DNA. The function of polymerase ε is not known.

Rolling Circle Replication

Certain circular DNAs replicate, not by the theta mode we have already examined, but by a mechanism called **rolling circle replication.** Lambda (λ) phage actually uses both means of

FIGURE 7.25 Rolling circle model for phage λ DNA replication. As the circle rolls to the right, the leading strand (red) elongates continuously. The lagging strand (blue) elongates discontinuously, using the unrolled leading strand as template. The progeny double-stranded DNA thus produced grows to many genomes in length (a concatemer) before one genome's worth is clipped off and packaged into a phage head.

replication. During the early phase of λ DNA replication, the phage follows the theta mode of replication to produce several copies of circular DNA. These circular DNAs are not packaged into phage particles; they serve as templates for rolling circle synthesis of linear λ DNA molecules that *are* packaged.

Figure 7.25 shows how this rolling circle operates. Here, the replicating fork resembles that in *E. coli* DNA replication, with continuous synthesis on the leading strand (the one going around the circle) and discontinuous synthesis on the lagging strand. It is not difficult to see how the name "rolling circle" arose. In fact, you can almost think of the replicating DNA as a roll of toilet paper unrolling as it speeds across the floor. The unrolled part represents the growing double-stranded progeny DNA. In λ, this progeny DNA reaches lengths that are several genomes long before it is packaged. These multiple-length DNAs are called **concatemers.** The packaging mechanism is designed to accept only one genome's worth of linear DNA into each phage head, so the concatemer must be cut during packaging. This rolling circle mechanism of DNA replication is also sometimes called **sigma mode replication** because the intermediate resembles the lowercase Greek letter sigma (σ).

Circular single-strand DNA phages such as φX174 also use a rolling circle mode of DNA replication (see chapter 15). First they produce a double-stranded replicative form of DNA. One strand of this DNA then serves as the template for making many copies of the other strand. This replication looks much like the model in figure 7.25 but no trailing strand is copied from the newly made leading strand.

Circular DNA can replicate by a rolling circle, or sigma mechanism. One strand of a double-stranded DNA is nicked and the 3′-end is extended, using the intact DNA strand as template. This displaces the 5′-end. In phage λ, the displaced strand serves as the template for discontinuous, lagging strand synthesis.

TERMINATION

To finish the job of DNA replication, a cell must fill in the gaps left where RNA primers were removed. With circular DNAs,

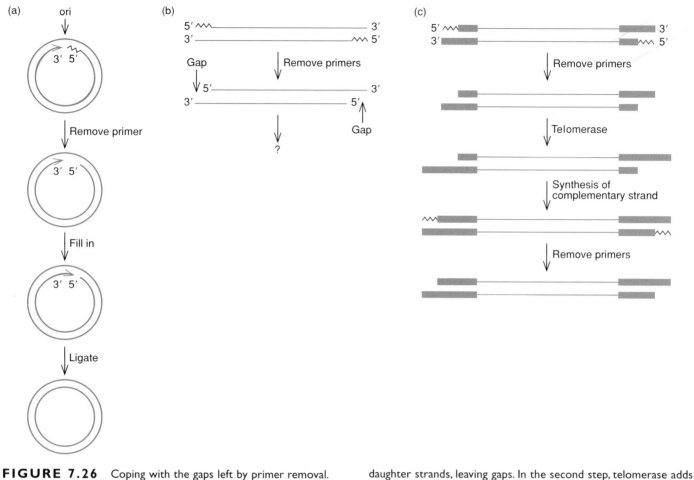

FIGURE 7.26 Coping with the gaps left by primer removal. (a) In prokaryotes, the 3′-end of a circular DNA strand can prime the synthesis of DNA to fill in the gap left by the first primer (red). For simplicity, only one replicating strand is shown. (b) Hypothetical model to show what would happen if primers were simply removed from the 5′-ends of linear DNA strands with no telomerase action. The gaps at the ends of chromosomes would grow longer each time the DNA replicated. (c) How telomerase can solve the problem. In the first step, the primers (red) are removed from the 5′-ends of the daughter strands, leaving gaps. In the second step, telomerase adds extra telomeric DNA (green boxes) to the 3′-ends of the other daughter strands. In the third step, DNA synthesis occurs, using the newly made telomeric DNA as a template. In the fourth step, the primers used in step three are removed. This leaves gaps, but the telomerase action has ensured that no net loss of DNA has occurred. The telomeres represented here are not drawn to scale with the primers. In reality, human telomeres are thousands of nucleotides long.

such as those in bacteria, there is no problem filling all the gaps because there is always another DNA 3′-end upstream to serve as a primer (figure 7.26a). But consider the problem faced by eukaryotes, with their linear chromosomes. Once the first primer on each strand is removed (figure 7.26b), it appears that there is no way to fill in the gaps, since DNA cannot be extended in the 3′ → 5′ direction, and there is no 3′-end upstream as there would be in a circle. If this were actually the situation, the DNA strands would get shorter every time they replicated, and genes would be lost forever. How do cells solve this problem?

Elizabeth Blackburn and her colleagues have provided the answer, which is summarized in figure 7.26c. The telomeres, or ends of eukaryotic chromosomes, contain no genes. Instead, they are composed of many repeats of short, GC-rich sequences. The exact sequence of the repeat in a telomere is species-specific. In *Tetrahymena,* a ciliated protozoan, it is TTGGGG/AACCCC;

in vertebrates, including humans, it is TTAGGG/AATCCC. These repeats are added to the very 3′-ends of DNA strands, not by semiconservative replication, but by an enzyme called **telomerase.** If telomerase does not follow a semiconservative replication mechanism, it cannot use one DNA strand as the template for the other. How, then, can it make telomeres with a specific sequence? Blackburn showed that this specificity resides in the telomerase itself, and is due to a small RNA in the enzyme. For example, the *Tetrahymena* telomerase has an RNA 159 nucleotides long that contains the sequence CAACCCCAA. Thus, it can serve as the template for the TTGGGG repeats in *Tetrahymena* telomeres, as shown in figure 7.27. Now the termination problem is solved; the telomerase adds many repeated copies of its characteristic sequence to the 3′-ends of chromosomes. Priming for synthesis of the opposite strand can then occur within these telomeres. There is no problem when the

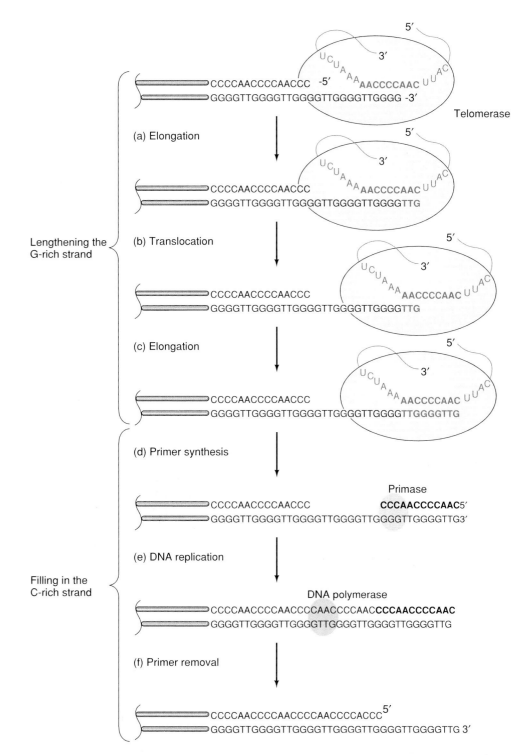

FIGURE 7.27   Forming telomeres in *Tetrahymena*.
(*a*) Telomerase (yellow) promotes hybridization between the 3'-end of the G-rich telomere strand and the template RNA (red) of the telomerase. The telomerase uses three bases (AAC) of its RNA as a template for the addition of three bases (TTG, boldface) to the 3'-end of the telomere. (*b*) The telomerase translocates to the new 3'-end of the telomere, pairing the left-hand AAC sequence of its template RNA with the newly incorporated TTG in the telomere. (*c*) The telomerase uses the template RNA to add six more nucleotides (GGGTTG) to the 3'-end of the telomere. Steps (*a*) through (*c*) can repeat indefinitely to lengthen the G-rich strand of the telomere. (*d*) When the G-rich strand is sufficiently long (probably longer than shown here), primase (orange) can make an RNA primer (boldface), complementary to the 3'-end of the telomere's G-rich, strand. (*e*) DNA polymerase (green) uses the newly made primer to prime synthesis of DNA to fill in the remaining gap on the C-rich telomere strand. (*f*) The primer is removed, leaving a 12- to 16-base overhang on the G-rich strand.

7.2   Telomeres, the Hayflick Limit, and Cancer

Everyone knows that organisms, including humans, are mortal. But biologists used to assume that cells cultured from humans were immortal. Each individual cell would ultimately die, of course, but the cell *line* would go on dividing indefinitely. Then in the 1960s Leonard Hayflick discovered that ordinary human cells are not immortal. They can be grown in culture for a finite period of time—about 50 generations. Then they enter a period of senescence, and finally they all die. This ceiling on the lifetime of normal cells is known as the Hayflick limit. But cancer cells do not obey any such limit. They *do* go on dividing generation after generation indefinitely.

Investigators have discovered a significant difference between normal cells and cancer cells that may explain why cancer cells are immortal while normal cells are not: Human cancer cells contain telomerase, whereas normal somatic cells lack this enzyme. (Germ cells must retain telomerase, of course, to safeguard the ends of the chromosomes handed down to the next generation.) Thus we see that cancer cells can repair their telomeres after every cell replication, but most normal cells cannot. Therefore, cancer cells can go on dividing without degrading their chromosomes, while normal cells' chromosomes grow shorter with each cell division. Sooner or later this loss of genetic material stops the replication of normal cells, so they die. But this does not happen to cancer cells; telomerase saves them from that fate.

One of the essential changes that must occur in a cell to make it cancerous is the reactivation of the telomerase gene. This is required for the immortality that is the hallmark of cancer cells. This discussion also suggests a potential treatment for cancer: turn off the telomerase gene in cancer cells, or, more simply, administer a drug that inhibits telomerase. Such a drug should not harm most normal cells, since they have no telomerase to begin with. Cancer researchers are already hard at work on this strategy.

---

primer on the very 5′-end is removed and not replaced, because only telomere sequences are lost, and these can always be replaced by telomerase and another round of replication.

Recently, biologists have discovered that normal mammalian somatic cells lack telomerase, whereas germ cells retain the enzyme. Furthermore, cancer cells also have telomerase. This has profound implications for the characteristics of cancer cells, and perhaps even for their control (see Box 7.2).

The ends of eukaryotic chromosomes are formed by an enzyme called telomerase, which adds DNA to the 3′-ends of the chromosomes according to the instructions of its own RNA template. This renewal of chromosome ends counteracts the tendency of chromosomes to lose telomeric DNA with every cell division.

FIDELITY OF REPLICATION

By this time, you may be asking yourself: "Why is DNA replication so complex? Why go to all the trouble of semidiscontinuous replication with RNA priming?" The answer seems to be that both of these complex mechanisms contribute to something very important to life: faithful DNA replication. Consider the alternative to this fidelity. What if every time our chromosomes replicated, the DNA synthesizing machinery made many mistakes? These mistakes are mutations, which would accumulate with every new cell division. We have already seen the devastating effects mutations can have; it is clear that life cannot tolerate too many of them. In other words, if our replicating machinery were much less faithful, our genome would have to be much smaller in order to minimize the number of mutations occurring each generation. Complex life-forms, relying on many genes, would be impossible.

How good are DNA polymerases at pairing up the right bases during replication? The *E. coli* DNA polymerase III holoenzyme, operating in vitro, makes about one pairing mistake in one hundred thousand—not a very good record, considering that even the *E. coli* genome contains about three million base pairs. At this rate, replication would introduce errors into a significant percentage of genes every generation. Fortunately, the DNA polymerase III holoenzyme includes its own **proofreading** activity—a 3′ → 5′ exonuclease—that allows it to synthesize DNA with only one error in up to ten billion nucleotides polymerized.

Figure 7.28 shows the proofreading process. The DNA polymerase III holoenzyme in the replisome can add a nucleotide to the growing strand only if the previously incorporated nucleotide is properly base-paired. This is a critical requirement. It means that the polymerase cannot add to a mismatched nucleotide. Therefore, if the replisome adds the wrong nucleotide, it will stall until the mismatch is repaired. The 3′ → 5′ exonuclease activity of DNA polymerase III holoenzyme can then remove the mismatched nucleotide, perhaps along with several others. With the proper base-paired primer restored, replication can continue.

Now consider the implications of this proofreading mechanism. It immediately explains the need for a primer. If DNA polymerase needs a base-paired nucleotide to add to, it clearly cannot start a new DNA chain unless a primer is already there. But why primers made of RNA? The reason seems to be the following: Primers are made without proofreading, and therefore accumulate more errors. Making primers out of RNA

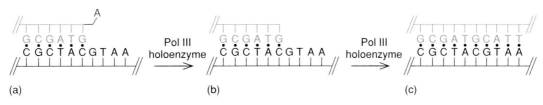

(a)   (b)   (c)

FIGURE 7.28 Proofreading in DNA replication. (*a*) An adenine nucleotide (red) has been mistakenly incorporated across from a guanine. This destroys the perfect base pairing required at the 3′-end of the primer, so the replicating machinery stalls. (*b*) This

then allows polymerase III holoenzyme to use its 3′ → 5′ exonuclease function to remove the mispaired nucleotide. (*c*) With the appropriate base pairing restored, polymerase III holoenzyme is free to continue replication.

guarantees that they will be recognized, removed, and replaced with DNA by extending the neighboring Okazaki fragments. The latter process is, of course, relatively error-free. Besides these theoretical arguments, there is the simple fact that RNA-synthesizing enzymes can start without primers, whereas DNA polymerase cannot.

The proofreading mechanism also helps rationalize semidiscontinuous replication. If there were a polymerase capable of synthesizing DNA in the 3′ → 5′ direction, thus making fully continuous replication possible, we would need a whole new 5′ → 3′ proofreading function, with the ability to restore the 5′-triphosphates that would be destroyed during this kind of proofreading. Semidiscontinuous replication, as cumbersome as it seems, is probably simpler.

Faithful DNA replication is essential to life. To help provide this fidelity, the DNA replicating machinery has a built-in proofreading system that requires priming. Only a base-paired nucleotide can serve as a primer for the replicating enzyme; therefore, if the wrong nucleotide is incorporated by accident, replication stalls until a nuclease erases the error by removing the most recently added nucleotide. The fact that the primers are made of RNA may mark them for degradation.

REPLICATION OF RNA GENOMES

The finely honed accuracy of the DNA replication systems we have been studying presents a stark contrast to the replication of genomes made of RNA. As we have seen, RNA-synthesizing enzymes have no need for primers, so there is no opportunity for proofreading. In addition, RNA is an inherently less stable molecule than DNA because its extra sugar hydroxyl group allows phosphodiester bond breakage (figure 7.29). The combination of lack of proofreading and instability imposes severe limitations on RNA genomes. They cannot grow too large, or the error rate (between one in a thousand and one in ten thousand) will doom them. It is therefore no surprise that the largest RNA genome known (in a coronavirus) is a single-stranded RNA with only about 25,000 bases. It is also not surprising that evolution has favored DNA as the repository of genetic information.

Even with their small sizes, the genomes of RNA viruses undergo rapid change because of their error-prone replication. This makes RNA viruses more dangerous because they can evolve rapidly to evade their hosts' immune systems. For example, HIV genomes are highly variable, which will make it difficult to design vaccines and drugs to combat all HIV variants.

Genomes made of RNA must replicate without benefit of a proofreading system. The errors that result from such imperfect replication severely restrict the size an RNA genome can attain, but they allow RNA viruses to evolve rapidly to evade host defenses.

THE MECHANISM OF RECOMBINATION

We have already seen many examples of progeny organisms with different combinations of genes than those found in either parent. Two mechanisms can explain how this happens. First, as we learned in chapter 2, genes on different chromosomes can simply assort independently. Second, as shown in chapter 4, **crossing over** can occur between homologous chromosomes. Indeed, DNA replication in meiotic cells is usually followed by such crossing over. This process scrambles the genes of maternal and paternal chromosomes, so that nonparental combinations occur in the progeny. This scrambling is valuable because the new combinations sometimes allow the progeny organisms a better chance of survival than their parents. Also, as we will see in chapter 11, recombination plays an important role in repair of DNA damage.

Both of these processes, independent assortment and crossing over, are central features of the science of genetics, and both can be called recombination, because of the new combinations of genes produced. However, recombination by crossing over is the process most geneticists associate with the term "recombination," and this is the way we will use the term for the remainder of this book.

Now that we have become well acquainted with DNA by studying its replication, we should pause to answer the question: How does recombination work?

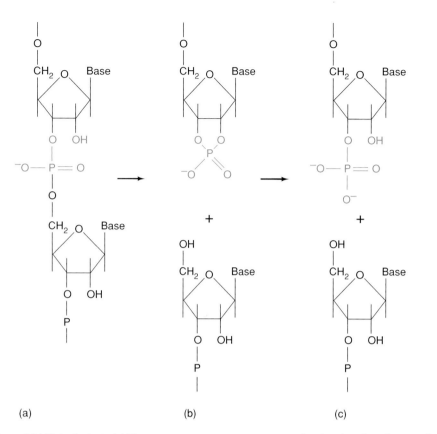

(a)   (b)   (c)

FIGURE 7.29   Mechanism of RNA hydrolysis. (*a*) The phosphodiester bond (blue) between the two ribonucleotides breaks, and (*b*) a new, cyclic phosphodiester bond forms between the 2′- and 3′-hydroxyl groups of the top sugar. (*c*) This cyclic structure opens, leaving the phosphate in the 3′-position. Since DNA has no 2′-hydroxyl group, it cannot participate in this kind of cleavage reaction; hence, DNA is a more stable molecule.

FORMS OF RECOMBINATION

The examples of recombination shown in figure 7.30 always involve a **strand exchange,** or crossing over event, that joins DNA segments that were previously separated. This does not mean the two segments must start out on separate DNA molecules. Recombination can be intramolecular, in which case crossing over between two sites on the same chromosome either removes or inverts the DNA segment in between. On the other hand, bimolecular recombination involves crossing over between two independent DNA molecules. Ordinarily, recombination is **reciprocal**—a "two-way street" in which the two participants trade DNA segments. DNA molecules can undergo one crossing over event, or two, or more, and the number of events strongly influences the nature of the final products.

In addition to the differences in recombination illustrated in figure 7.30, there are three fundamentally different kinds of recombination: (1) **Site-specific recombination** depends on limited sequence similarity between the recombining DNAs and always involves the same DNA regions, which explains why it is called site-specific. A prominent example of this kind of recombination is the insertion of λ phage DNA into the host DNA (chapter 8). (2) **Illegitimate recombination** also requires little if any sequence similarity between the recombining DNAs, but it is not site-specific. Transposable elements (chapter 12) usually use the illegitimate recombination pathway. (3) By far the most common form of recombination depends on extensive sequence similarity between the participating DNA molecules. This is the kind of recombination that occurs in meiosis. It is sometimes called **generalized recombination,** but is usually called **homologous recombination.**

A MODEL FOR HOMOLOGOUS RECOMBINATION

Our current models for the mechanism of homologous recombination share the following essential steps, although not necessarily in this order:(1) pairing of two **homologous** DNAs (DNAs having identical or very similar sequences); (2) breaking two of the homologous DNA strands; (3) re-forming phosphodiester bonds to join the two homologous strands (or regions of the same strand in intramolecular recombination); (4) breaking the other two strands and joining them. The basic mechanism, originally proposed by Robin Holliday, is presented in figure 7.31.

The Holliday model calls for pairing of two homologous DNA duplexes and creation of nicks in corresponding positions in two homologous DNA strands. Once both homologous strands are broken, the free ends can cross over and join with ends in the other duplex, forming intermolecular bonds, rather than rejoining to form the old intramolecular bonds.

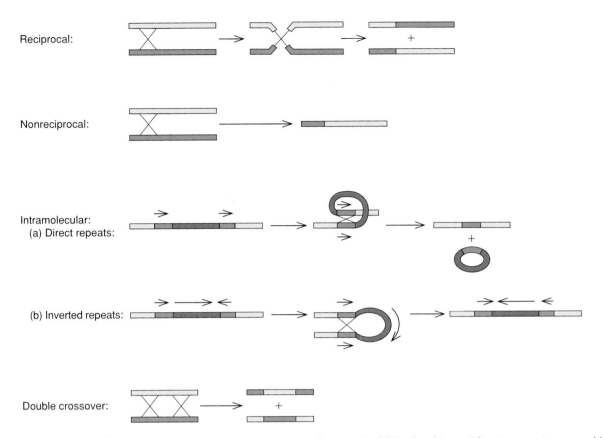

FIGURE 7.30 Examples of recombination. The ×'s represent crossing over events between the two chromosomes or parts of the same chromosome. To visualize how these work, look at the intermediate form of the reciprocal recombination on the top line. Imagine the DNAs breaking and forming new, interstrand bonds as indicated by the arms of the ×. This same principle applies to all the examples shown; in the nonreciprocal case, one of the recombinant DNAs is lost.

When the two strands have crossed over and DNA ligase has sealed the new intermolecular phosphodiester bonds, we have a **half chiasma,** or **chi structure,** a cross-shaped structure pictured in several ways in figure 7.31. The branch in the half chiasma can migrate in either direction simply by breaking old base pairs and forming new ones, in a process called **branch migration.** The chi structure is first shown with two of its strands crossed, but a 180-degree rotation of either the top or bottom half uncrosses these strands to give an intermediate with a diamond shape at the center where all the strands meet. If a few base pairs come apart at the junction point, the diamond, which is composed of single-stranded DNA, will expand. Figure 7.32 is an electron micrograph of recombining plasmid DNAs, which demonstrates beautifully that such structures really do form.

Of course, the two duplexes that are participating in the chi structure cannot remain linked indefinitely. The other two DNA strands must also break and rejoin to resolve the chi structure into two independent DNA duplexes. This can happen in two different ways. If the same strands that broke the first time break again and cross back, no true recombination results. What we get instead are two DNA duplexes with patches of **heteroduplex.** These patches contain one strand

from each of the DNAs involved in the chi structure. On the other hand, if the strands that did *not* break the first time are involved in the second break, the chi structure resolves into two recombinant molecules.

The RecBCD Pathway of Recombination

As far as we know, no natural recombination occurs exactly as depicted in figure 7.31. For example, no known mechanism includes the spontaneous appearance of two nicks conveniently located in the same positions in paired homologous DNA strands. In fact, it is the method of creation of the two nicks that serves to distinguish many of the major recombination pathways. Here we will describe the best studied of these, the RecBCD pathway of *E. coli* (figure 7.33).

This recombination process begins when the RecBCD protein, the product of the *recB,* -*C,* and -*D* genes, creates a nick at the 3'-end of a so-called chi site, which has the sequence 5'-GCTGGTGG-3. Chi sites are found on average every 5,000 base pairs in the *E. coli* genome. When the DNA near the nick "breathes," it forms a transient single-stranded tail, which can then be coated by RecA protein (the product of the *recA* gene) and SSB. SSB keeps the tail from re-forming a DNA duplex, and

FIGURE 7.31   Mechanism of recombination.
(a) Nicks occur in the same places in homologous chromosomes (blue and red). (b) Strands of the two chromosomes cross over. (c) Nicks are sealed, permanently joining the two chromosomes through two of their four strands and yielding a chi structure. (d) Branch migration (optional) occurs by breaking some base pairs and reforming others. (e) and (f) These are simply bends and twists in the chi structure to make the subsequent events easier to understand. (g) One kind of resolution of the chi structure. The same strands are nicked as were nicked in (a). (h) Sealing these nicks leaves two DNA duplexes with short stretches of "heteroduplex," containing one strand from each of the recombining partners. No true recombination has occurred. (i) The alternative resolution of the chi structure. The strands opposite those nicked originally are nicked. (j) When the nicks are sealed, two recombinant DNA duplexes emerge.

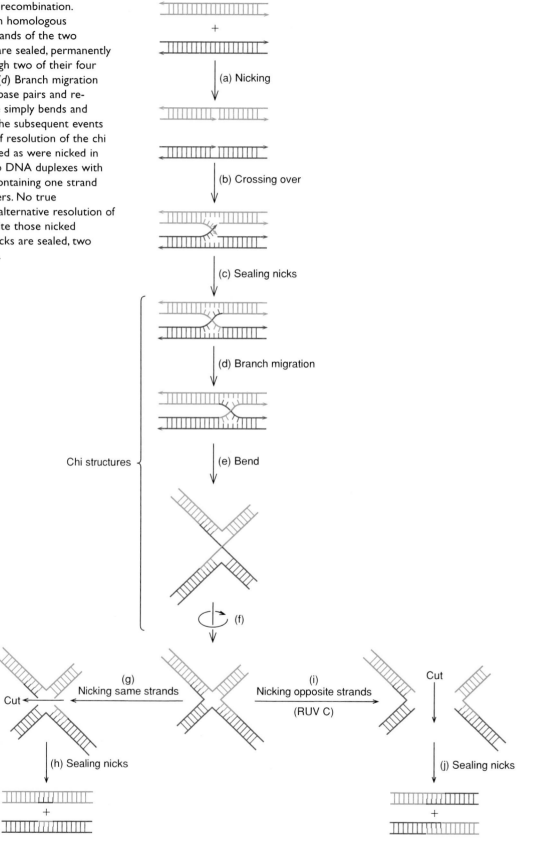

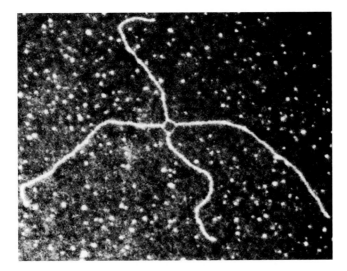

FIGURE 7.32 Chi structure generated during recombination. This corresponds to the intermediate generated in figure 7.31f, after some base pairs have opened up around the central "diamond" to produce short stretches of single-stranded DNA.

Courtesy of Dr. Huntington Potter and Dr. David Dressler.

RecA allows the tail to invade a double-stranded DNA, forming a **D-loop,** and beginning to scan the invaded DNA duplex for a region of homology. Once the tail finds a homologous region, a nick occurs in the D-looped DNA, possibly with the aid of RecBCD. This nick allows RecA and SSB to create a new tail that can pair with the gap in the other DNA. DNA ligase seals both nicks to generate a Holliday junction.

Branch migration does not occur spontaneously. Just as in DNA replication, DNA unwinding is required, and this in turn requires helicase activity and energy from ATP. Two proteins, **RuvA** and **RuvB,** collaborate in this function. Both have DNA helicase activity, and RuvB is an ATPase, so it can cleave ATP to harvest energy for the branch migration process. Finally, two DNA strands must be nicked to resolve the Holliday junction intermediate into heteroduplexes or recombinant products. The **RuvC** protein carries out this function.

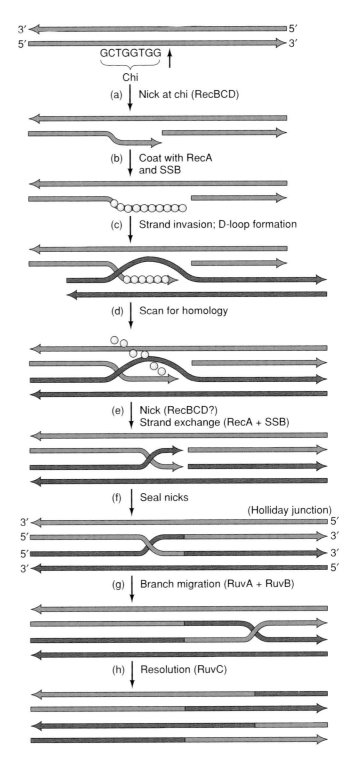

FIGURE 7.33 The RecBCD pathway of homologous recombination. (a) The RecBCD protein nicks one strand just to the 3′ side of the Chi sequence. This allows the DNA to "breathe," forming a single-stranded tail. (b) RecA and SSB (both represented as yellow spheres) coat the tail, preventing its reassociation with the complementary strand. (c) RecA promotes invasion of another DNA duplex, forming a D-loop. (d) RecA helps the invading strand scan for a region of homology in the recipient DNA duplex. Here, the invading strand has base-paired with a homologous region, releasing SSB and RecA. (e) Once a homologous region is found, a nick in the looped out DNA appears, perhaps caused by RecBCD. This permits the tail of the newly nicked DNA to base-pair with the single-stranded region in the other DNA, probably aided by RecA and SSB. (f) The remaining nicks are sealed by DNA ligase, yielding a four-stranded complex with a Holliday junction. (g) Branch migration occurs, sponsored by RuvA and RuvB. (h) Nicking by RuvC resolves the structure into two molecules, true recombinants in this example.

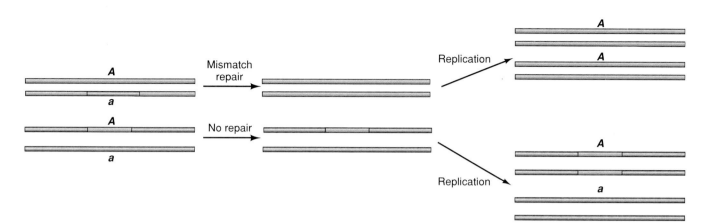

FIGURE 7.34   A model for gene conversion in sporulating *Neurospora*. A strand exchange event with branch migration during sporulation has resolved to yield two duplex DNAs with patches of heteroduplex in a region where there is a one-base difference between allele *A* (blue) and allele *a* (red). The other two daughter chromosomes are homoduplexes, one pure *A* and one pure *a* (not shown). The top heteroduplex undergoes mismatch repair to convert the *a* strand to *A*; the bottom heteroduplex is not repaired. Replication of the repaired DNA yields two *A* duplexes; replication of the unrepaired DNA yields one *A* and one *a*. Thus, the sum of the daughter duplexes pictured at right is three *A*'s and one *a*. Replication of the two DNAs not pictured yields two *A*'s and two *a*'s. The sum of all daughter duplexes is therefore five *A*'s and three *a*'s, instead of the normal four and four.

Homologous recombination begins with steps that cause nicks in two homologous strands on separate DNA duplexes. This gives rise to single-stranded regions with free ends that can cross over to the other DNA duplex to form new, intermolecular bonds. The resulting Holliday junction can undergo branch migration, in which the crossover point moves toward the 3′-ends of the crossed DNA strands. Finally, the Holliday junction intermediate is resolved by nicking two of its strands. This can yield two DNAs with patches of heteroduplex, or two true recombinant DNAs. In the RecBCD pathway, RecBCD performs the first nick, and probably the second; RecA and SSB promote strand invasion, which creates the crossover, or Holliday junction; RuvA and -B promote branch migration; and RuvC creates the nicks that resolve the structure into heteroduplexes or recombinant DNAs.

GENE CONVERSION

When fungi such as pink bread mold (*Neurospora crassa*) sporulate, two haploid nuclei fuse, producing a diploid nucleus that undergoes meiosis to give four haploid nuclei. These nuclei then experience mitosis to produce eight haploid nuclei, each appearing in a separate spore. In principle, if one of the original nuclei contained one allele (*A*) at a given locus, and the other contained another allele (*a*) at the same locus, then the mixture of alleles in the spores should be equal: four *A*'s and four *a*'s. It is difficult to imagine any other outcome (five *A*'s and three *a*'s, for example), because that would require conversion of one *a* to an *A*. In fact, aberrant ratios *are* observed about 0.1% to 1% of the time, depending on the fungal species. This phenomenon is called **gene conversion.**

The mechanism of recombination illustrated in figure 7.31, especially the branch migration step, suggests a mechanism for gene conversion during meiosis. Figure 7.34 depicts this hypoth-esis. We start with a nucleus in which DNA duplication has already occurred, so it contains four chromatids. In principle, there should be two chromosomes bearing the *A* allele and two bearing the *a* allele. But in this case, crossing over and branch migration have occurred, followed by resolution yielding two chromosomes with patches of heteroduplex, as illustrated in figure 7.31*g*. These heteroduplex regions just happen to be in the region where alleles *A* and *a* differ at one base, so each chromosome has one strand with one allele and the other strand with the other allele. If DNA replication occurred immediately, this situation would resolve itself simply, yielding two *A* duplexes and two *a* duplexes. However, before replication, one or both of the heteroduplexes may attract the enzymes that repair base mismatches (chapter 11). In the example shown here, only the top heteroduplex is repaired, with the *a* being converted to *A*. This leaves three strands with the *A* allele, and only one with the *a* allele. Now DNA replication will produce three *A* duplexes and only one *a* duplex. When we add the two *A* and two *a* duplexes resulting from the chromosomes that did not undergo heteroduplex formation, this gives a final ratio of five *A*'s and only three *a*'s.

Gene conversion is not confined to meiotic events. It is also the mechanism that switches the mating type of baker's yeast (*Saccharomyces cerevisiae*). This gene conversion event may involve the transient pairing of two similar, but nonidentical genes, followed by conversion of one gene sequence to that of the other.

When two similar but not identical DNA sequences line up beside each other, the possibility exists for gene conversion—the conversion of one DNA sequence to that of the other. The sequences participating in gene conversion can be alleles, as in meiosis, or nonallelic genes, such as the genes that determine mating type in yeast.

WORKED-OUT PROBLEMS

PROBLEM 1

(*a*) Draw a schematic diagram of a replicating bubble in *Drosophila* DNA, showing both parental and progeny strands after all Okazaki fragments have been ligated together.

(*b*) Assuming that replication began before labeled thymidine was added, show on the schematic where labeled and unlabeled DNA will be found.

(*c*) What would an autoradiograph of the bubble in (*b*) look like?

(*d*) What would this autoradiograph look like if the two daughter duplexes wound around each other (i.e., the bubble collapsed)?

SOLUTION

(*a*) Representing parental DNA with dark lines and progeny DNA with light lines, we can draw the replicating bubble as follows:

There are no breaks in the progeny strands because all Okazaki fragments have been joined.

(*b*) Representing labeled DNA with jagged lines, we have the following:

The progeny DNA in the middle is unlabeled because it replicated before the labeled thymidine was added.

(*c*)

Note: We only see labeled regions of the DNA (i.e., jagged lines) on an autoradiograph. Unlabeled DNA has no way of exposing the film and therefore remains invisible.

(*d*)

If the bubble collapses, the upper and lower labeled segments on each side will be superimposed and appear as one streak each on the film. Thus we will see two streaks (regions of developed grains) instead of four on the autoradiograph.

PROBLEM 2

How long would it take *E. coli* to replicate 100,000 base pairs of DNA?

SOLUTION

E. coli DNA elongates at the rate of about 1,000 nucleotides per second per fork. Since DNA replication is bidirectional and two forks are used, the effective rate of elongation is 2,000 nucleotides per second. If there were no pauses, 100,000 base pairs could replicate in 100,000/2,000, or 50 seconds.

PROBLEM 3

How long a streak on an autoradiograph would be produced by colE1 DNA replicating at a single fork for 50 seconds, assuming no pauses?

SOLUTION

A single fork moves at the rate of 1,000 nucleotides a second, so it will move 50,000 nucleotides in 50 seconds. Each nucleotide (or base pair) is 0.34 nm long; so 50,000 nucleotides would be 50,000 × 0.34 nm, or 17,000 nm, or 17 micrometers long. (1 micrometer = 1,000 nm.)

SUMMARY

DNA replicates in a semiconservative manner: when the parental strands separate, each serves as the template for making a new, complementary strand. The DNA that replicates under the control of a single origin of replication is called a replicon. The chromosomes of simple genetic systems like plasmids, viruses, and even bacteria contain just one replicon. DNA replication in most genetic systems is bidirectional. The replication machinery starts at a fixed point—the origin of replication—and two replicating forks move in opposite directions away from this point. DNA replication in prokaryotes and eukaryotes is semidiscontinuous. The leading strand grows continuously, $5' \rightarrow 3'$, in the direction of motion of the replicating fork. The other, lagging strand grows discontinuously, in pieces (Okazaki fragments) that are made, $5' \rightarrow 3'$, in a direction opposite to the motion of the replicating fork. DNA polymerase is incapable of initiating DNA synthesis by itself. It needs a primer to supply a free $3'$-end upon which it can build the nascent DNA. Very short pieces of RNA serve this priming function. After Okazaki fragments are made, their primers are removed, the resulting gaps are filled in, and the remaining nicks are sealed.

As helicase unwinds the two parental strands at the replicating fork of a closed circular DNA, it introduces superhelical strain into the DNA. DNA gyrase releases this strain, permitting replication to continue.

An enzyme called telomerase forms the ends of eukaryotic chromosomes. It adds DNA to the $3'$-ends of the DNA strands according to the instructions of its own RNA template.

Circular DNA can replicate by a rolling circle mechanism. One strand of a double-stranded DNA is nicked and the $3'$-end is extended, using the intact DNA strand as template. This displaces the $5'$-end. In phage λ, the displaced strand serves as the template for discontinuous, lagging strand synthesis.

DNA polymerase, like all enzymes, is not perfect. If left to itself, it would make about one mistake per one hundred thousand nucleotides of DNA. To improve on this intolerably poor performance, the DNA replicating machinery has a built-in proofreading system that depends on a base-paired nucleotide to serve as a primer for the replicating enzyme; therefore, if the wrong nucleotide is incorporated by accident, replication stalls until a nuclease erases the error by degrading the most recently made stretch of DNA. RNA genomes must replicate without benefit of a proofreading system. The errors that result from such imperfect replication severely restrict the size an RNA genome can attain.

DNA replication in meiotic cells is usually followed by recombination, using the homologous recombination mechanism:

Single-strand breaks occur in two homologous strands, creating free ends that can exchange, forming new, interstrand bonds, and yielding a chi structure. This cross-shaped structure is resolved into two recombinant DNA products when the other two strands break and exchange. Branch migration frequently takes place during recombination, creating heteroduplex regions. These can experience mismatch repair, causing deviations from the expected equal distribution of alleles after meiosis. This is one way that gene conversion can occur.

PROBLEMS AND QUESTIONS

1. Draw schematic diagrams of the DNA molecules in a Meselson-Stahl experiment after *three* generations in ^{14}N medium assuming: (*a*) dispersive, (*b*) conservative, and (*c*) semiconservative replication. Use solid lines for ^{15}N-labeled DNA and dotted lines for ^{14}N-labeled DNA.

2. Draw diagrams of the bands of DNA in CsCl gradients that would arise from the three types of replication in question 1. Show the positions of fully ^{15}N-labeled DNA and fully ^{14}N-labeled DNA.

3. Draw diagrams of the bands of DNA in CsCl gradients performed on heated (single-stranded) DNA that would arise after one generation from: (*a*) dispersive; (*b*) conservative; and (*c*) semiconservative replication. Show the positions of heavy and light DNA strands.

4. Assume the following sequence of events: Eukaryotic DNA replication begins; DNA is pulse-labeled with low specific activity thymidine; DNA is then pulsed with high specific activity thymidine. Show what the resulting autoradiograph will actually look like if the two daughter duplexes twist around each other (i.e., the "bubbles" collapse into straight lines, as in figure 7.16).

5. We pulse-label embryos with tritium-labeled thymidine, then autoradiograph the labeled DNA. What could we conclude from an autoradiographic pattern in which all pairs of silver streaks fall into two categories: _____ _____ or _____ _____ ? In both cases, these are *pairs* of streaks, not two independent ones.

6. Consider an autoradiography experiment in which you label colE1 DNA uniformly with low specific activity thymidine. Then, *after* a round of replication begins, you label with high specific activity thymidine for one-fourth of a replication cycle. Draw a picture of the expected autoradiogram. (Remember that colE1 replicates unidirectionally.)

7. An extraterrestrial being brings you a vial of bacteria from its planet and informs you that where it comes from all living things replicate their DNA only in a fully continuous manner. (*a*) Describe the results of a pulse labeling experiment you would perform that verifies this assertion. (*b*) What unearthly characteristic must the DNA polymerase of these bacteria have?

8. What role does a helicase play in DNA replication? Why is an ATPase a necessary part of this role?

9. Contrast the function of a helicase with that of a topoisomerase.

10. Why does the single-strand-binding protein (SSB) of *E. coli* bind so much better to single-stranded than to double-stranded DNA?

11. If there were no swivel, what kinds of supercoils would develop in a replicating closed circular DNA?

12. The *E. coli* primosome binds to the template for which strand? Why does it seem more useful for the primosome to bind to this strand?

13. What advantage does a double-headed replisome with two DNA polymerase centers offer?

14. Why is primer synthesis inherently more error-prone than DNA elongation?

15. What factors limit the size of RNA virus genomes?

16. What happens to the DNA between two direct repeats of DNA on the same chromosome when homologous recombination occurs between these two repeats? (Direct repeats are copies of a DNA segment in the same orientation.)

17. What happens to the DNA between two inverted repeats of DNA (one copy in one orientation, the other in the opposite orientation) when homologous recombination occurs between these two repeats?

18. (*a*) What would be the mixture of alleles in the gene conversion depicted in figure 7.34 if *both* heteroduplexes were converted to *A/A* by mismatch repair? (*b*) What if one heteroduplex were converted to *A/A* and the other to *a/a*?

19. How long would it take *E. coli* to replicate its entire genome (4.2×10^6 base pairs)?

20. How long a streak would *E. coli* DNA, replicating bidirectionally from a common origin for one minute, be able to produce on an autoradiograph?

21. Organisms of the planet Zorb carry out continuous replication of both strands of their DNA, which means that one strand must be made in the $3' \rightarrow 5'$ direction. Below is a diagram of the end of such a DNA strand that is growing from the 3'-end toward the 5'-end.

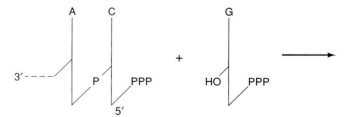

(*a*) What will this strand look like after one more nucleotide (the GTP pictured) is added? (*b*) If another nucleotide besides GTP is incorporated and then removed by the proofreading system, what would you expect the product to look like (before the correct nucleotide is inserted)? (*c*) Why would this pose a problem to an earthly organism?

Answers appear at end of book.

Transcription and Its Control in Prokaryotes

Learning Objectives

In this chapter you will learn:

1. How proteins bind to specific DNA sequences to regulate gene transcription.

2. How the enzyme RNA polymerase transcribes a gene, producing an RNA.

3. How bacterial genes of related function are arranged in functional units called operons, each controlled by a binding site for RNA polymerase called a promoter.

4. How expression of bacterial genes is controlled primarily by controlling which promoter the RNA polymerase recognizes.

5. How different programs of gene expression can be activated by altering the RNA polymerase.

The curiosity remains, though, to grasp more clearly how the same matter, which in physics and in chemistry displays orderly and reproducible and relatively simple properties, arranges itself in the most astounding fashions as soon as it is drawn into the orbit of the living organism. The closer one looks at these performances of matter in living organisms the more impressive the show becomes. The meanest living cell becomes a magic puzzle box full of elaborate and changing molecules, and far outstrips all chemical laboratories . . . in the skill of organic synthesis performed with ease, expedition, and good judgment of balance.

Max Delbrück

In chapter 6 we learned that most gene expression takes place in two phases—transcription (RNA synthesis) and translation (protein synthesis). Our job in this chapter is to explore the transcription process in bacteria, especially in *E. coli,* the best-studied organism in the world.

THE FUNDAMENTALS OF TRANSCRIPTION

Before we take up the detailed mechanism of transcription, let us look at the basics of the process, beginning with the story of the discovery of mRNA.

DISCOVERY OF MESSENGER RNA

The concept of messenger RNA carrying information from gene to ribosome developed in stages during the years following the publication of Watson and Crick's DNA model. In 1958, Crick himself proposed that RNA serves as an intermediate carrier of genetic information. He based his hypothesis in part on the fact that the DNA resides in the nucleus of eukaryotes, whereas proteins are made in the cytoplasm. This means that something must carry the information from one place to the other. Crick noted that ribosomes contain RNA and suggested that this ribosomal RNA (rRNA) is the information bearer. But rRNA is an integral part of ribosomes; it cannot escape. Therefore, Crick's hypothesis implied that each ribosome, with its own rRNA, would produce the same kind of protein over and over. This turned out to be wrong.

François Jacob and colleagues proposed an alternate hypothesis calling for nonspecialized ribosomes that translate RNAs called "messengers." The messengers are independent RNAs that bring genetic information from the genes to the ribosomes. They are independent in that they are not permanently bound to the ribosomes the way rRNA is. In 1961, Jacob, along with Sydney Brenner and Meselson, published their proof of the messenger hypothesis. This study used the same bacteriophage (T2) that Hershey and Chase had employed almost a decade earlier to show that genes were made of DNA. The premise of the experiments was this: When phage T2 infects *E. coli,* it subverts its host from making bacterial proteins to making phage proteins. If Crick's hypothesis were correct, this switch to phage protein synthesis should be accompanied by the production of new ribosomes equipped with phage-specific rRNAs.

To distinguish new ribosomes from old, these investigators labeled the ribosomes in uninfected cells with heavy isotopes of nitrogen (^{15}N) and carbon (^{13}C). This made "old" ribosomes heavy. Then they infected these cells with phage T2 (or the related phage T4) and simultaneously transferred them to medium containing light nitrogen (^{14}N) and carbon (^{12}C). Any "new" ribosomes made after phage infection would therefore be light and would separate from the old, heavy ribosomes upon density gradient centrifugation. Brenner and colleagues also labeled the infected cells with ^{32}P to tag any phage RNA as it was made. Then they asked this question: Was the radioactively labeled phage RNA associated with new or old ribosomes? Figure 8.1 shows that the phage RNA was found on old ribosomes whose rRNA was made before infection even began. Clearly, this old rRNA could not carry phage genetic information; by extension, it was very unlikely that it could carry host genetic information either. Thus, the ribosomes are unchanged. The nature of the proteins they make depends on the mRNA that associates with them. This relationship resembles that of a tape player and tape. The nature of the music (protein) depends on the tape (mRNA), not the player (ribosome).

Other workers had already identified a better candidate for the messenger: a class of unstable RNAs that associate transiently with ribosomes. Interestingly enough, in phage T2-infected cells, this RNA had a base composition (GC content) very similar to that of phage DNA—and quite different from that of bacterial DNA and RNA. This is exactly what we would expect of phage messenger RNA, and that is exactly what it is. On the other hand, host mRNA, unlike host rRNA, has a base composition similar to that of host DNA. This lends further weight to the hypothesis that the mRNA, not the rRNA, is the "instructional" molecule.

Most genes are expressed in a two-step process: transcription, or synthesis of an mRNA copy of the gene, followed by translation of this message to protein.

THE BASIC MECHANISM OF TRANSCRIPTION

As you might expect, transcription follows the same base-pairing rules as DNA replication: T, G, C, and A in the DNA pair with A, C, G, and U, respectively, in the RNA product. (Notice that uracil appears in RNA in place of thymine in DNA.) This base-pairing pattern ensures that an RNA transcript is a faithful copy of the gene (figure 8.2).

Of course, highly directed chemical reactions such as transcription do not happen by themselves—they are enzyme-catalyzed. The enzyme that directs transcription is called **RNA polymerase.** Figure 8.3 presents a schematic diagram of *E. coli* RNA polymerase at work. There are three phases of transcription: initiation, elongation, and termination. The initiation phase includes polymerase-DNA binding and formation of the first phosphodiester bond between nucleotides.

Initiation

First, the enzyme recognizes a region called a **promoter,** which lies just "upstream" from the gene. The polymerase binds tightly to the promoter and causes localized melting, or separation of the two DNA strands within the promoter. Approximately twelve base pairs are melted. Next, the polymerase starts building the RNA chain. The substrates, or building blocks, it uses for

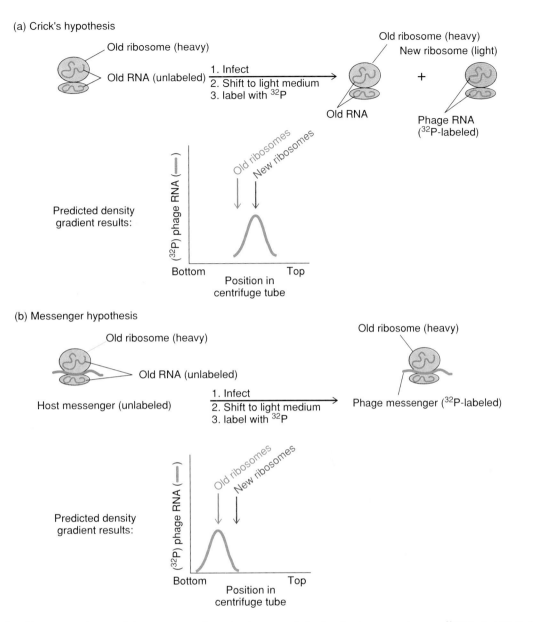

FIGURE 8.1 Experimental test of the messenger hypothesis. *E. coli* ribosomes were made heavy by labeling the bacterial cells with heavy isotopes of carbon and nitrogen. The bacteria were then infected with T2 and T4 phage and simultaneously shifted to "light" medium containing the normal isotopes of carbon and nitrogen, plus some ³²P to make the phage RNA radioactive. (*a*) Crick had proposed that ribosomal RNA carried the message for making proteins. If this were so, then whole new ribosomes with phage-specific ribosomal RNA would have been made after phage infection. In that case, the new, ³²P-labeled RNA (green) should have moved together with the new, light ribosomes (red). (*b*) Jacob and colleagues had proposed that a messenger RNA carried genetic information to the ribosomes. According to this hypothesis, phage infection would cause the synthesis of new, phage-specific messenger RNAs that would be ³²P-labeled (green). These would associate with old, heavy ribosomes (blue). Therefore, the radioactive label would move together with the old, heavy ribosomes in the density gradient. This was indeed what happened.

this job are the four **ribonucleoside triphosphates:** ATP, GTP, CTP, and UTP. The first, or initiating substrate is usually a purine nucleotide. After the first nucleotide is in place, the polymerase binds the second nucleotide and joins it to the first, forming the initial phosphodiester bond in the RNA chain. At this point, the initiation process is complete.

Elongation

During the elongation phase of transcription, RNA polymerase directs the sequential binding of ribonucleotides to the growing RNA chain in the 5′ → 3′ direction (from the 5′-end toward the 3′-end of the RNA). As it does so, it moves along the DNA **template,** and the "bubble" of melted DNA moves with it. This

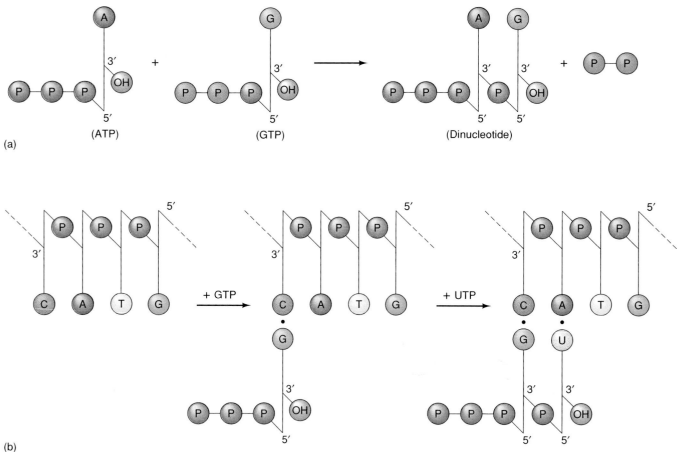

(a)

(b)

FIGURE 8.2   Making RNA. (*a*) Phosphodiester bond formation in RNA synthesis. ATP and GTP are joined together to form a dinucleotide. Note that the phosphorus atom (red) closest to the guanosine is retained in the phosphodiester bond. The other two phosphates (blue) are removed as a by-product called pyrophosphate. (*b*) Synthesis of RNA on a DNA template. The DNA template at top contains the sequence 3′-dC-dA-dT-dG-5′, and extends in both directions, as indicated by the dashed lines. To start

the RNA synthesis, GTP base-pairs with the dC nucleotide in the DNA template. Next, UTP provides a uridine nucleotide, which base-pairs with the dA nucleotide in the DNA template and forms a phosphodiester bond with the GTP. This produces the dinucleotide GU. In the same way, a new nucleotide will join the growing RNA chain at each step until transcription is complete. The pyrophosphate by-products are not shown.

melted region exposes the bases of the template DNA one by one so they can pair with the bases of the incoming ribonucleotides. As soon as the transcription machinery passes, the two DNA strands wind around each other again, re-forming the double helix.

Termination

Just as promoters serve as initiation signals for transcription, other regions at the ends of genes, called **terminators,** signal termination. These work in conjunction with RNA polymerase, sometimes aided by another protein, to loosen the association between the RNA product and DNA template. The result is that the RNA dissociates from the RNA polymerase and DNA; transcription is thereby terminated.

RNA sequences are usually written 5′ to 3′, left to right. This feels natural to a geneticist because RNA is made in a 5′ to 3′ direction, and as we will see, mRNA is also translated 5′ to 3′.

Thus, since ribosomes read the message 5′ to 3′, it is appropriate to write it 5′ to 3′ so *we* can read it like a sentence.

Genes are also usually written so that their transcription proceeds in a left to right direction. This "flow" of transcription from left to right gives rise to the term *upstream,* which refers to the leftward direction when the gene is written conventionally. Thus, most promoters lie just upstream from their respective genes. By the same token, we say that genes generally lie *downstream* from their promoters.

Having just learned the rudiments of transcription, we can appreciate two fundamental differences between transcription and DNA replication: (a) RNA polymerase makes only one RNA strand during transcription, which means that it copies only one DNA strand. Transcription is therefore said to be **asymmetric.** This contrasts with semiconservative DNA replication, in which both DNA strands are copied. (b) In transcription, DNA melting is limited and transient. Only enough strand separation occurs to allow the polymerase to "read" the DNA

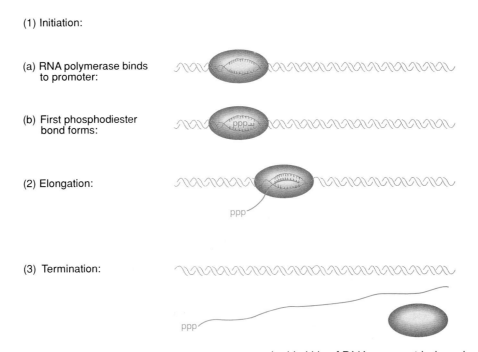

(1) Initiation:

(a) RNA polymerase binds
    to promoter:

(b) First phosphodiester
    bond forms:

(2) Elongation:

(3) Termination:

FIGURE 8.3   Transcription. (*a*) RNA polymerase (red) binds tightly to the promoter and "melts" a short stretch of DNA. (*b*) In the initiation step, the polymerase joins the first two nucleotides of the nascent RNA (blue) through a phosphodiester bond. The first nucleotide retains its triphosphate group. (*c*) During elongation, the melted bubble of DNA moves with the polymerase, allowing the enzyme to "read" the bases of the DNA template strand and make a complementary RNA. (*d*) Termination occurs when the polymerase reaches a termination signal, causing the RNA and the polymerase to fall off the DNA template.

template strand. However, during replication, the two parental DNA strands separate permanently.

Transcription takes place in three stages: initiation, elongation, and termination. Initiation involves binding RNA polymerase to the promoter, local melting, and forming the first phosphodiester bond; during elongation, the RNA polymerase links together ribonucleotides in the $5' \rightarrow 3'$ direction to make the rest of the RNA; finally, in termination, polymerase and RNA product dissociate from the DNA template.

THE DETAILED MECHANISM OF TRANSCRIPTION

The two most important participants in transcription are DNA, which is the template for the RNA product, and RNA polymerase, which does the transcribing. In order for transcription to begin, these two components must get together. But RNA polymerase doesn't start transcribing just anywhere on a bacterial DNA; it binds to special sites called promoters just before the "start points" of genes. Therefore, let us begin our discussion by considering these two elements: RNA polymerase and promoters.

Table 8.1   Ability of Core and Holoenzyme to Transcribe DNAs

	Transcription Activity	
DNA	*Core*	*Holoenzyme*
T4 (intact)	0.5	33.0
Calf thymus (nicked)	14.2	32.8

As early as 1960–1961, RNA polymerases were discovered in animals, plants, and bacteria, but the bacterial enzyme was the first to be studied in great detail. Richard Burgess, Andrew Travers, and their colleagues discovered in 1969 that they could physically separate the *E. coli* RNA polymerase holoenzyme into two components: a polypeptide called **sigma (σ),** and the remainder—a complex of four polypeptides collectively known as the **core polymerase,** or simply the **core.** This separation of σ from the core polymerase also caused a profound change in the enzyme's activity (table 8.1). Whereas the holoenzyme could transcribe intact phage T4 DNA in vitro quite actively, the core enzyme had little ability to do this. On the other hand, core polymerase retained its basic RNA polymerizing function, since it could still transcribe highly nicked templates very well. (As we will see, this transcription of nicked DNA has no biological significance.)

Adding σ back to the core reconstituted the enzyme's ability to transcribe un-nicked T4 DNA. (A control experiment showed that σ did not nick the T4 DNA.) Even more significantly, the holoenzyme transcribed only a certain class of T4 genes (called immediate early genes), but the core enzyme showed no such specificity; it transcribed the whole T4 genome, though weakly, in vitro. Furthermore, transcription of T4 DNA in vitro by the holoenzyme was asymmetric—just as it is in vivo—but the weak transcription by the core enzyme was symmetric. In other words, not only did the core enzyme read the whole genome, it read both strands. Clearly, depriving the holoenzyme of its σ subunit leaves a core enzyme with basic RNA synthesizing capability but lacking specificity. Adding σ back restores specificity. In fact, σ was named only after this characteristic came to light. The σ, or Greek letter *s*, was chosen to stand for "specificity."

> The key player in the transcription process is RNA polymerase. The *E. coli* enzyme is composed of a core, which contains the basic transcription machinery, and a σ factor, which directs the core to transcribe specific genes.

BINDING RNA POLYMERASE TO PROMOTERS

Why was core enzyme still capable of transcribing nicked DNA, but not intact DNA? Nicks and gaps in DNA provide initiation sites for RNA polymerase, even core polymerase, but this kind of initiation is necessarily nonspecific. There were few nicks or gaps on the intact T4 DNA, so the core polymerase encountered only a few such artificial initiation sites. On the other hand, when σ was present, the holoenzyme could recognize the authentic RNA polymerase binding sites—the promoters—on the intact T4 DNA and begin transcription there. This initiation is specific and resembles the initiation that occurs in vivo. Thus, σ operates by directing the polymerase to initiate at specific promoters.

How does σ change the way the core polymerase behaves toward promoters? Michael Chamberlin and his coworkers used binding studies to help answer this question. They measured the binding between phage DNA and RNA polymerase core or holoenzyme and found two different classes of binding sites. There were just a few sites that bound RNA polymerase holoenzyme extremely tightly. Once bound, it took 30–60 hours for half of the holoenzymes to dissociate from these tight binding sites, which are, of course, promoters. There were also many loose binding sites that were available to the core polymerase as well as to the holoenzyme.

The interpretation of these and other experiments is that the RNA polymerase can exist in one of two states: one appropriate for tight binding to promoters (the initiation state) and one appropriate for loose binding (the elongation state). The tight binding, or initiating, state can occur only in the presence of σ; after initiation is achieved, σ dissociates and leaves the core

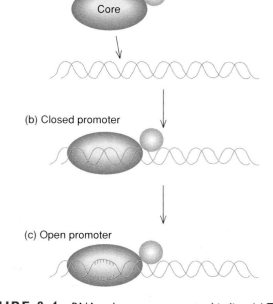

(a) RNA polymerase holoenzyme

(b) Closed promoter

(c) Open promoter

FIGURE 8.4 RNA polymerase-promoter binding. (*a*) The polymerase holoenzyme searches for a binding site on the DNA. (*b*) The holoenzyme has found a promoter and has bound loosely, forming a closed promoter complex. (*c*) The holoenzyme has bound tightly, melting a local region of DNA and forming an open promoter complex.

enzyme to carry out elongation. Furthermore, tight binding of the holoenzyme to promoters involves local melting of the DNA to form a so-called **open promoter complex.** This conversion of a loosely bound polymerase in a **closed promoter complex** to the tightly bound polymerase in the open complex requires σ, and this is what allows transcription to begin (figure 8.4). We can now appreciate how σ fulfills its role in determining specificity of transcription. It selects the promoters to which RNA polymerase will bind tightly. The genes adjacent to these promoters will then be transcribed.

> The σ factor allows initiation of transcription by causing the RNA polymerase holoenzyme to bind tightly to a promoter. This tight binding depends on local melting of the DNA to form an open promoter complex, and it is possible only in the presence of σ. The σ factor can therefore select which genes will be transcribed. After σ has participated in initiation, it dissociates from the core polymerase, leaving the core to carry on with the elongation process.

COMMON BASE SEQUENCES IN *E. COLI* PROMOTERS

What is the special nature of a prokaryotic promoter that attracts RNA polymerase? David Pribnow compared several *E. coli* and

FIGURE 8.5   A prokaryotic promoter. The positions of the −10 and −35 boxes and the unwound region are shown relative to the start of transcription for a typical *E. coli* promoter. Capital letters denote bases found in those positions in more than 50% of promoters examined; small letters denote bases found in those positions in 50% or fewer of promoters examined.

phage promoters and discerned a region they held in common: a sequence of six or seven bases centered approximately ten bases upstream from the start of transcription. This was originally dubbed the **Pribnow box,** but is now frequently referred to as the **−10 box.** Later, another short sequence, centered approximately thirty-five bases upstream from the transcription start, was found in a variety of *E. coli* and phage promoters; it is usually called the **−35 box.** More than one hundred promoters have now been examined, and a "typical" sequence for each of these boxes has emerged (figure 8.5).

These so-called **consensus sequences** represent probabilities. The capital letters in figure 8.5 denote bases that have a high probability of being found in the given position. The small letters correspond to bases that are usually found in the given position, but at a lower frequency than those denoted by capital letters. The probabilities are such that you rarely find −10 or −35 boxes that match the consensus sequences perfectly. However, when such perfect matches are found, they tend to occur in very strong promoters that initiate transcription unusually actively. As you might expect, strong promoters generally control genes whose products are needed in large quantities.

Mutations that destroy matches with the consensus sequences tend to be **down mutations.** That is, they make the promoter weaker, resulting in less transcription. For example, mutation of a −10 box sequence from TATAAT to TGTAAT would usually be a down mutation because it would weaken the promoter. By contrast, mutations that make promoter sequences more like the consensus sequences usually make the promoters stronger; these are called **up mutations.** For instance, mutation of a −10 box sequence from TATCTT to TATATT would probably be an up mutation. In chapter 9 we will see that eukaryotic promoters have their own consensus sequences, one of which closely resembles the −10 box.

> Prokaryotic promoters contain two regions centered at −10 and −35 bases upstream from the transcription start site. In *E. coli*, these bear a greater or lesser resemblance to two consensus sequences: TATAAT and TTGACA, respectively. In general, the more closely regions within a promoter resemble these consensus sequences, the stronger that promoter will be.

OPERONS: UNITS OF GENE EXPRESSION

Appendix C, Module 16.

In chapter 1 we discussed biochemical pathways that are catalyzed by groups of enzymes. Obviously, all of the enzymes in the pathway will be needed together, since one enzyme deficiency would stall the whole process. Therefore, if four enzymes are required for a pathway to operate, it would be wasteful to turn on the genes for three of them and leave the other turned off. In other words, for greatest efficiency, the genes for proteins that function together should be controlled together. One mechanism that allows coordinate control is to group functionally related genes together so they can be regulated together easily. Such a group of contiguous, coordinately controlled genes is called an **operon.**

THE *LAC* OPERON

The first operon to be discovered has become the paradigm of the operon concept. It contains three **structural genes**—genes that code for proteins. These three code for the enzymes that allow *E. coli* cells to use the sugar **lactose,** hence the name *lac* **operon.** Consider a flask of *E. coli* cells growing on the sugar **glucose.** We take away the glucose and replace it with lactose. Can the cells adjust to the new nutrient source? For a short time it appears that they cannot; they stop growing, but then, after this lag period, growth resumes. During the lag, the cells have been turning on the *lac* operon and beginning to accumulate the enzymes they need to metabolize lactose.

What are these enzymes? First, *E. coli* needs an enzyme to transport the lactose into the cells. The name of this enzyme is **galactoside permease.** Next, the cells need an enzyme to break the lactose down into its two component sugars, galactose and glucose. Figure 8.6 shows this reaction. Since lactose is composed of two simple sugars, we call it a *disaccharide.* As mentioned previously, these are the six-carbon sugars galactose and glucose; they are joined together by a linkage called a β-*galactosidic bond.* Lactose is therefore called a β-*galactoside,* and the enzyme that cuts it in half is called β-**galactosidase.** The genes for these two enzymes, galactoside permease and β-galactosidase, are found side-by-side in the *lac* operon, along with another structural gene—for

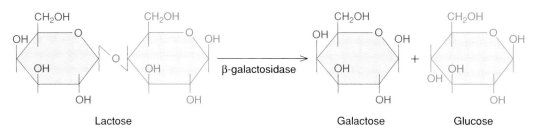

Lactose   Galactose   Glucose

FIGURE 8.6   The β-galactosidase reaction. The enzyme breaks the β-galactosidic bond (gray) between the two sugars, galactose (red) and glucose (blue), that compose lactose.

galactoside transacetylase—whose function in lactose metabolism is still not clear. These names give some insight into enzyme nomenclature. The *-ase* suffix is an almost universal feature of enzyme names; the rest of the name usually gives a clue to the reaction the enzyme catalyzes.

The three genes coding for enzymes that carry out lactose metabolism are grouped together in the following order: β-galactosidase (*lacZ*), galactoside permease (*lacY*), galactoside transacetylase (*lacA*). They are all transcribed together on one messenger RNA, called a **polycistronic message,** starting from a single promoter. Thus, they can all be controlled together simply by controlling that promoter. The term *polycistronic* comes from *cistron,* which is a synonym for *gene.* Therefore, a polycistronic message is simply a message with information from more than one gene.

> Lactose metabolism in *E. coli* is carried out by two enzymes, with possible involvement by a third. The genes for all three enzymes are clustered together and transcribed together from one promoter, yielding a polycistronic message. Therefore, these three genes, linked in function, are also linked in expression. They are turned off and on together.

Negative Control of the *lac* Operon

Figure 8.7 illustrates one aspect of *lac* operon regulation: negative control. The term **negative control** implies that the operon is turned on unless something intervenes to stop it. The "something" that can turn off the *lac* operon is the *lac* **repressor.** This repressor, the product of the **lacI gene** shown at the extreme left in figure 8.7, is a tetramer of four identical polypeptides; it binds to the **operator** just to the right of the promoter. When the repressor is bound to the operator, the operon is **repressed.** That is because the operator and promoter are contiguous, and when the repressor occupies the operator, it prevents RNA polymerase from moving through the operator to reach the three structural genes of the operon. Since its genes are not transcribed, the operon is off, or repressed.

The *lac* operon is repressed as long as no lactose is available. That is economical; it would be wasteful for the cell to produce enzymes needed to use an absent sugar. In fact, if all of an *E. coli*

cell's genes were turned on all the time, this would drain the cell of so much energy that it could not compete with more efficient bacteria. Thus, control of gene expression is essential to life.

On the other hand, when all the glucose is gone and lactose is present, there should be a mechanism for removing the repressor so the operon can turn on and take advantage of the new nutrient. How does this work? The repressor is a so-called **allosteric protein,** which means that it changes its conformation, and therefore its function, when it binds to a certain small molecule. (Greek: *allos* = other + *stereos* = solid, implying space, or shape.) The small molecule in this case is called the **inducer** of the *lac* operon, because it binds to the repressor, causing the protein to change to a conformation that favors dissociation from the operator, thus inducing the operon (figure 8.7*b*).

What is the nature of this inducer? It is actually an alternate form of lactose called **allolactose** (again, Greek: *allos* = other). When β-galactosidase cleaves lactose to galactose plus glucose, it rearranges a small fraction of the lactose to allolactose. Figure 8.8 shows that allolactose is just galactose linked to glucose in a different way than in lactose. (In lactose, the linkage is through a β-1, 4 bond; in allolactose, the linkage is β-1, 6.)

You may be asking yourself: How can lactose be metabolized to allolactose if there is no permease to get it into the cell and no β-galactosidase to perform the metabolizing, since the *lac* operon is repressed? The answer is that repression is somewhat leaky, and a low basal level of the *lac* operon products is always present. This is enough to get the ball rolling by producing a little inducer. It does not take much inducer to do the job, since there are only about ten tetramers of repressor per cell. Furthermore, the derepression of the operon will snowball as more and more operon products are available to produce more and more inducer.

Discovery of the Operon

The development of the operon concept by François Jacob and Jacques Monod and their colleagues was one of the classic triumphs of the combination of genetic and biochemical analysis. The story begins in 1940, when Monod began studying the inducibility of lactose metabolism in *E. coli.* Monod learned that an important feature of lactose metabolism was β-galactosidase, and that this enzyme was inducible by lactose and by other galactosides. Furthermore, he and Melvin Cohn had used an

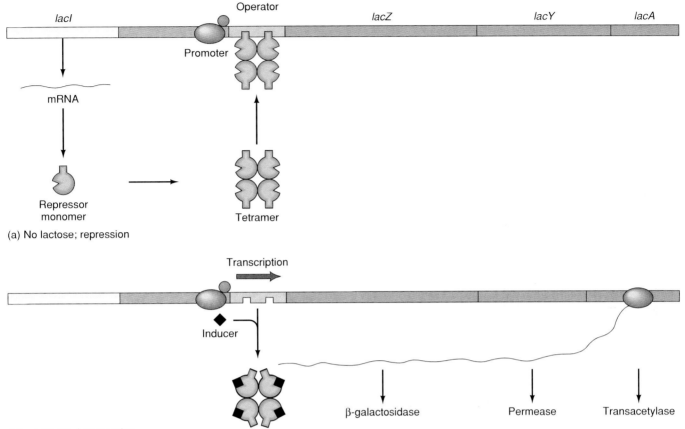

(a) No lactose; repression

(b) + lactose; derepression

FIGURE 8.7 Negative control of the *lac* operon. (*a*) No lactose; repression. The *lacI* gene produces repressor (green), which binds to the operator and blocks RNA polymerase from entering the *lac* structural genes. Instead, the polymerase spends its time making extremely short RNAs, no more than six nucleotides long. (*b*) Presence of lactose; derepression. The inducer (black) binds to the repressor, changing it to a form (bottom) that no longer binds well to the operator. This removes the repressor from the operator, allowing RNA polymerase to transcribe the structural genes. This produces a multicistronic mRNA that is translated to yield β-galactosidase, permease, and transacetylase.

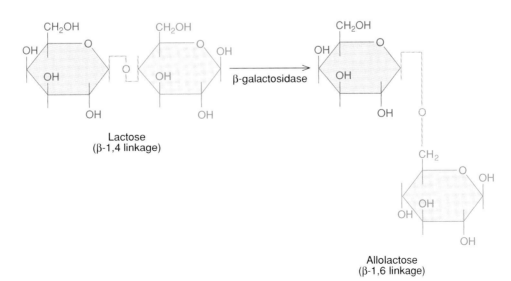

FIGURE 8.8 Conversion of lactose to allolactose. A side reaction carried out by β-galactosidase rearranges lactose to the inducer, allolactose. Note the change in the galactosidic bond from β-1,4 to β-1,6.

Table 8.2   Effect of Cryptic Mutant (*lacY⁻*) on Accumulation of Galactoside

Genotype	Inducer	Accumulation of Galactoside
Z^+Y^+	−	−
Z^+Y^+	+	+
Z^+Y^- (cryptic)	−	−
Z^+Y^- (cryptic)	+	−

anti-β-galactosidase antibody to detect β-galactosidase protein, and they showed that the amount of this protein increased upon induction. Therefore, since more gene product appeared in response to lactose, the β-galactosidase gene itself was apparently being induced.

To complicate matters, certain mutants ("cryptic mutants") were found that could make β-galactosidase but still could not grow on lactose. What was missing in these mutants? To answer this question, Monod and his coworkers added a radioactive galactoside to wild-type and cryptic bacteria. They found that uninduced wild-type cells did not take up the galactoside, and neither did the cryptic mutants, even if they were induced. Of course, induced wild-type cells did accumulate the galactoside. This revealed two things: First, a substance (galactoside permease) is induced along with β-galactosidase in wild-type cells and is responsible for transporting galactosides into the cells; second, the cryptic mutants seem to have a defective gene (*Y⁻*) for this substance (see table 8.2).

Monod named this substance galactoside permease, and then endured criticism from his colleagues for naming a protein before it had been isolated. He has since remarked, "This attitude reminded me of that of two traditional English gentlemen who, even if they know each other well by name and by reputation, will not speak to each other before having been formally introduced." In their efforts to purify galactoside permease, Monod and his colleagues identified another protein, galactoside transacetylase, which is induced along with β-galactosidase and galactoside permease.

Thus, by the late 1950s, Monod knew that three enzyme activities (and therefore at least three genes) were induced together by galactosides. He had also found some mutants, called **constitutive mutants,** that needed no induction. They produced the three gene products all the time. Monod realized that further progress would be greatly accelerated by genetic analysis, so he teamed up with François Jacob, who was working just down the hall at the Pasteur Institute.

In collaboration with Arthur Pardee, Jacob and Monod created merodiploids (partial diploid bacteria) carrying both the wild-type (inducible) and constitutive alleles. (We will learn how to generate such merodiploid bacteria in chapter 13.) The inducible allele proved to be dominant, demonstrating that wild-type cells produce some substance that keeps the *lac* genes turned

off unless they are induced. Because this substance turned off the genes from the constitutive as well as the inducible parent, it made the merodiploids inducible. Of course, this substance is the *lac* repressor. The constitutive mutants had a defect in the gene (*I*) for this repressor. These mutants are therefore *I⁻*.

The existence of a repressor required that there be some specific DNA sequence to which the repressor would bind. Jacob and Monod called this the operator. The specificity of this interaction suggested that it should be subject to genetic mutation; that is, some mutations in the operator should abolish its interaction with the repressor. These would also be constitutive mutations, so how are we to distinguish them from constitutive mutations in the repressor gene?

Jacob and Monod realized that they could do this by determining whether the mutation was dominant or recessive. Jacob explained this with an analogy. He likened the two operators in a merodiploid bacterium to two radio receivers, each of which controls a separate door to a house. The repressor genes in a merodiploid are like little radio transmitters, each sending out an identical signal to keep the doors closed. A mutation in one of the repressor genes would be like a breakdown in one of the transmitters; the other transmitter would still function and the doors would both remain closed. In other words, both *lac* operons in the merodiploid would still be repressible. Thus, such a mutation should be recessive (figure 8.9*a*), and we have already observed that it is.

On the other hand, a mutation in one of the operators would be like a breakdown in one of the receivers. Since it could no longer receive the radio signal, its door would swing open; the other, with a functioning receiver, would remain closed. In other words, the mutant *lac* operon in the merodiploid would be constantly derepressed, while the wild-type one would remain repressible. Therefore, this mutation would be dominant, but only with respect to the *lac* operon under the control of the mutant operator (figure 8.9*b*). We call such a mutation *cis*-**dominant** because it is dominant only with respect to genes on the same DNA (*in cis;* Latin: *cis* = here), not on the other DNA in the merodiploid. Jacob and Monod did indeed find such *cis*-dominant mutations, and they proved the existence of the operator. These mutations are called *Oᶜ*, for **operator constitutive.**

What about mutations in the repressor gene that render the repressor unable to respond to inducer? Such mutations should make the *lac* operon uninducible and should be dominant both *in cis* and *in trans* (on different DNA molecules; Latin: *trans* = across) because the mutant repressor will remain bound to both operators even in the presence of inducer or of wild-type repressor (figure 8.9*c*). We might liken this to a radio signal that cannot be turned off. Monod and his colleagues found two such mutants, and Suzanne Bourgeois later found many others. These are named *Iˢ* to distinguish them from constitutive repressor mutants (*I⁻*), which make a repressor that cannot recognize the operator.

Both kinds of constitutive mutants (*I⁻* and *Oᶜ*) affected all three of the *lac* genes (*Z, Y,* and *A*) in the same way. The genes had already been mapped and were found to be adjacent on the

Function   **Conclusion**

(a) Mutant repressor gene (I^-)

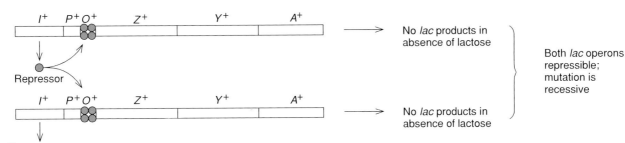

No *lac* products in absence of lactose

No *lac* products in absence of lactose

Both *lac* operons repressible; mutation is recessive

(b) Mutant operator (O^c)

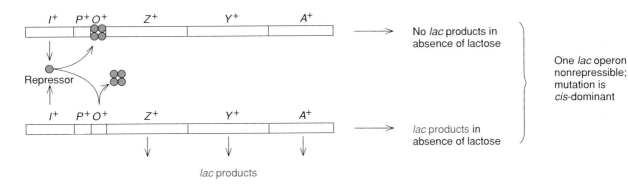

No *lac* products in absence of lactose

lac products in absence of lactose

One *lac* operon nonrepressible; mutation is *cis*-dominant

(c) Mutant repressor gene (I^s)

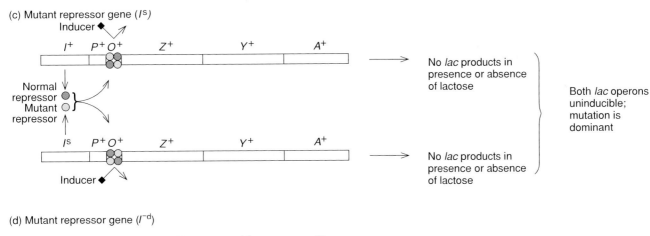

No *lac* products in presence or absence of lactose

No *lac* products in presence or absence of lactose

Both *lac* operons uninducible; mutation is dominant

(d) Mutant repressor gene (I^{-d})

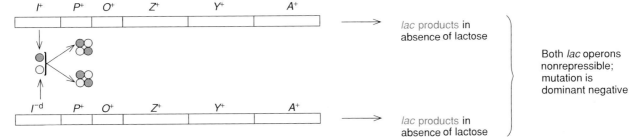

lac products in absence of lactose

lac products in absence of lactose

Both *lac* operons nonrepressible; mutation is dominant negative

FIGURE 8.9 Effects of regulatory mutations in the *lac* operon in merodiploids. (*a*) This merodiploid has one wild-type operon (top) and one operon (bottom) with a mutation in the repressor gene (I^-). The wild-type repressor gene (I^+) makes enough normal repressor (green) to repress both operons, so the I^- mutation is recessive. (*b*) This merodiploid has one wild-type operon (top) and one operon (bottom) with a mutation in the operator (O^c) that makes it incapable of binding repressor (green). The wild-type operon remains repressible, but the mutant operon is not; it makes *lac* products even in the absence of lactose. Since only the operon connected to the mutant operator is affected, this mutation is *cis*-dominant. (*c*) This merodiploid has one wild-type operon (top) and one operon (bottom) with a mutant repressor gene (I^s) whose product (yellow) cannot bind inducer. The mutant repressor therefore binds irreversibly to both operators and renders both operons uninducible. This mutation is therefore dominant. Notice that these repressor tetramers containing some mutant and some wild-type subunits behave as mutant proteins. That is, they remain bound to the operator even in the presence of inducer. (*d*) This merodiploid has one wild-type operon (top) and one operon (bottom) with a mutant repressor gene (I^{-d}) whose product (yellow) cannot bind to the *lac* operator. Moreover, mixtures (heterotetramers) composed of both wild-type and mutant repressor monomers still cannot bind to the operator. Thus, since the operon remains turned on even in the absence of lactose, this mutation is dominant. Furthermore, since the mutant protein poisons the activity of the wild-type protein, we call the mutation dominant negative.

chromosome. These findings strongly suggested that the operator lay near the three structural genes.

We now recognize yet another class of repressor mutants, those that are constitutive and dominant (I^{-d}). This kind of defective gene makes a defective product that can still form tetramers with wild-type repressor monomers. However, the defective monomers spoil the activity of the whole tetramer so it cannot bind to the operator. Hence the dominant nature of this mutation. These mutations are not *cis*-dominant, since the "spoiled" repressors cannot bind to either operator. This kind of "spoiler" mutation is widespread in nature, and it is called by the generic name **dominant negative.**

Notice that Jacob and Monod, by skillful genetic analysis, were able to develop the operon concept. They predicted the existence of two key control elements: the repressor gene and the operator. Deletion mutations revealed a third element (the promoter) that was necessary for expression of all three structural genes. Furthermore, they could conclude that all three structural genes (*lacZ, Y,* and *A*) were clustered into a single control unit: the *lac* operon. Subsequent biochemical studies have amply confirmed Jacob and Monod's beautiful hypothesis.

Negative control of the *lac* operon occurs as follows: The operon is turned off as long as repressor binds to the operator, because the repressor keeps RNA polymerase from moving through the operator to transcribe the three structural genes. When the supply of glucose is exhausted and lactose is available, the few molecules of *lac* operon enzymes produce a few molecules of allolactose from the lactose. The allolactose acts as an inducer by binding to the repressor and causing a conformational shift that encourages dissociation from the operator. With the repressor removed, RNA polymerase is free to transcribe the three structural genes. A combination of genetic and biochemical experiments revealed the two key elements of negative control of the *lac* operon: the operator and the repressor.

Positive Control of the *lac* Operon

As you might predict, positive control is the opposite of negative control; that is, the operon is turned off (actually, turned down to a basal level) unless something intervenes to turn it on. If negative control is like the brake of a car, positive control is the accelerator pedal. In the case of the *lac* operon, this means that removing the repressor from the operator (releasing the brake) is not enough to activate the operon. An additional, positive factor (stepping on the accelerator) is needed.

What is the selective advantage of this positive control system? Isn't negative control enough? Negative control alone can respond only to the presence of lactose. If there were no positive control, the presence of lactose alone would suffice to

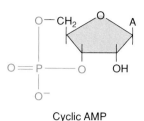

FIGURE 8.10   Cyclic AMP. Note the cyclic 5′–3′ phosphodiester bond (blue).

activate the operon. But that is inappropriate if glucose is still available, because *E. coli* cells metabolize glucose more easily than lactose; it would therefore be wasteful for them to activate the *lac* operon in the presence of glucose. In fact, *E. coli* cells keep the *lac* operon turned down as long as glucose is present. This selection in favor of glucose metabolism and against use of other energy sources has long been attributed to the influence of some breakdown product, or *catabolite,* of glucose. It is therefore known as **catabolite repression.**

The ideal positive controller of the *lac* operon would be a substance that sensed the lack of glucose and responded by activating the *lac* promoter so that RNA polymerase could bind and transcribe the structural genes (assuming, of course, that lactose is present and repressor is therefore gone). One substance that responds to glucose concentration is a nucleotide called **cyclic AMP (cAMP)** (figure 8.10). As the level of glucose drops, the concentration of cAMP rises.

Is cAMP the positive effector, then? Not exactly. The positive controller of the *lac* operon is a complex composed of two parts: cAMP and a binding protein called **CAP,** for **catabolite activator protein,** by its discoverers, Geoffrey Zubay and his colleagues. (CAP was also studied by Ira Pastan and coworkers, who named it CRP, for "cyclic AMP receptor protein.") The protein binds cAMP, and the resulting complex binds to the *lac* control region and helps RNA polymerase bind there. Figure 8.11 illustrates this mechanism. Notice that the *lac* promoter is divided into two parts, the CAP binding site on the left and the RNA polymerase binding site on the right.

How does binding CAP plus cAMP to one side of the promoter facilitate RNA polymerase binding to the other? X-ray diffraction studies of the CAP/cAMP/promoter complex have shown that when CAP/cAMP binds, it causes the DNA to bend noticeably (about 90 degrees). This bending apparently helps RNA polymerase bind, perhaps by making it easier for the RNA polymerase to separate the two DNA strands, forming an open promoter complex. CAP/cAMP may also stimulate transcription by direct protein–protein contact between CAP and RNA polymerase.

CAP and cAMP also stimulate transcription of other inducible operons, including the well-studied *ara* and *gal* operons. Just as the *lac* operon makes possible the metabolism of lactose, these other operons code for enzymes that break down the alternative

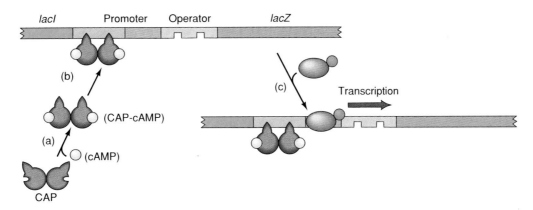

FIGURE 8.11   Positive control of the *lac* operon. (*a*) Without cyclic AMP (cAMP, yellow), CAP (purple) cannot bind to its target site in the *lac* control region. Cyclic AMP binds to CAP and renders it capable of binding to this DNA site. (*b*) The CAP-cAMP complex binds to its DNA target site. (*c*) Binding the CAP-cAMP complex to the DNA allows RNA polymerase to bind to the *lac* promoter. Transcription can now occur, assuming that lactose is present, and the operator is therefore unoccupied by repressor. We know that CAP-cAMP binding to its target site bends the DNA by about 90 degrees, but this bending is not illustrated here.

sugars arabinose and galactose, respectively. Thus, for greatest efficiency, all three operons should remain turned off as long as the cell has glucose available. Since cAMP responds to glucose concentration, it is not surprising that all three operons share a common regulatory mechanism involving cAMP. The CAP-cAMP complex binds at or near each operon's promoter and facilitates the binding of RNA polymerase.

> Positive control (catabolite repression) of the *lac* operon (or the *ara* or *gal* operon) works as follows: A complex composed of cAMP plus a protein known as catabolite activator protein (CAP) binds to the upstream part of the promoter and facilitates binding of RNA polymerase to the downstream part. This greatly enhances transcription of the operon. The physiological significance of this positive control mechanism is that it can operate only when cAMP concentration is elevated; this occurs when glucose concentration is low and there is a corresponding need to metabolize an alternate energy source.

THE *trp* OPERON

The **trp** (pronounced "trip") operon contains the genes for the enzymes that *E. coli* needs to make the amino acid tryptophan. Like the *lac* operon, it is subject to negative control by a repressor. However, there is a fundamental difference. The *lac* operon codes for **catabolic** enzymes—those that break down a substance. Such operons tend to be turned on by the presence of that substance, lactose in this case. The *trp* operon, on the other hand, codes for **anabolic** enzymes—those that build up a substance. Such operons are generally turned off by that substance. When the tryptophan concen-

tration is high, the products of the *trp* operon are not needed any longer, and we would expect the *trp* operon to be repressed. This is what happens. The *trp* operon also exhibits an extra level of control not seen in the *lac* operon. This is called attenuation.

Tryptophan's Role in Negative Control of the *trp* Operon

Figure 8.12 shows an outline of the structure of the *trp* operon. There are five structural genes, which code for the polypeptides in three enzymes that convert a tryptophan precursor, chorismic acid, to tryptophan. The first two genes, *trpE* and *trpD*, code for polypeptides that make up the first enzyme in the pathway. The third gene, *trpC*, codes for a single-polypeptide enzyme that catalyzes the second step. The last two genes, *trpB* and *trpA*, code for the two polypeptides of the enzyme that carries out the third and last step in the pathway. In the *lac* operon, the promoter and operator precede the structural genes, and the same is true in the *trp* operon. However, the *trp* operator lies wholly within the *trp* promoter, whereas the two loci are merely adjacent in the *lac* operon.

In the negative control of the *lac* operon, the cell senses the presence of lactose by the appearance of tiny amounts of its rearranged product, allolactose. In effect, this inducer pulls the repressor off the *lac* operator and derepresses the operon. In the case of the *trp* operon, a plentiful supply of tryptophan means that the cell does not need to spend any more energy making this amino acid. In other words, a high tryptophan concentration is a signal to turn off the operon.

How does the cell sense the presence of tryptophan? In essence, tryptophan helps the *trp* repressor bind to its operator. Here is how that occurs: In the absence of tryptophan, the *trp* repressor exists in an inactive form called the **aporepressor.** When the aporepressor binds tryptophan, it changes to a conformation with

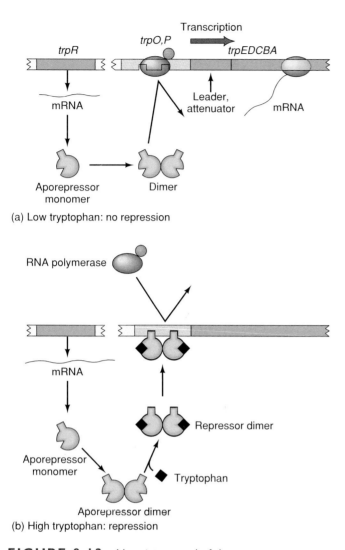

FIGURE 8.12   Negative control of the *trp* operon. (*a*) Derepression. RNA polymerase (red and blue) binds to the *trp* promoter and begins transcribing the structural genes (*trpE, D, C, B,* and *A*). Without tryptophan, the aporepressor (green) cannot bind to the operator. (*b*) Repression. Tryptophan, the corepressor (black), binds to the inactive aporepressor, changing it to repressor, with the proper shape for binding successfully to the *trp* operator. This prevents RNA polymerase from binding to the promoter, so no transcription occurs.

a much higher affinity for the *trp* operator (figure 8.12*b*). This is another allosteric transition like the one we encountered in our discussion of the *lac* repressor. The combination of aporepressor plus tryptophan is the *trp* repressor; therefore tryptophan is called a **corepressor.** When the cellular concentration of tryptophan is high, there is plenty of corepressor to bind and form the active *trp* repressor. Thus, the operon is repressed. When the tryptophan level in the cell falls, the amino acid dissociates from the aporepressor, causing it to shift back to the inactive conformation; the repressor-operator complex is thus broken, and the operon is derepressed. Later in this chapter, we will examine the nature of the conformational shift in the aporepressor that

occurs upon binding tryptophan, and see why this is so important in operator binding.

The negative control of the *trp* operon is, in a sense, the mirror image of the negative control of the *lac* operon. The *lac* operon responds to an inducer that pulls the repressor off the operator, derepressing the operon. The *trp* operon responds to a repressor that includes a corepressor, tryptophan, which signals the cell that it has made enough of this amino acid. The corepressor binds to the aporepressor, changing its conformation so it can bind better to the *trp* operator, thereby repressing the operon.

Control of the *trp* Operon by Attenuation

In addition to the standard, negative control scheme we have just described, the *trp* operon employs another mechanism of control called **attenuation.** Why is this extra control needed? The answer probably lies in the fact that repression of the *trp* operon is weak—much weaker, for example, than that of the *lac* operon. Thus, considerable transcription of the *trp* operon can occur even in the presence of repressor. In fact, in attenuator mutants where only repression can operate, the fully repressed level of transcription is only 70-fold lower than the fully depressed level. The attenuation system permits another 10-fold control over the operon's activity. This means that the combination of depression and attenuation controls the operon over a 700-fold range, from fully inactive to fully active; (70-fold [repression] × 10-fold [attenuation] = 700-fold). This is valuable because synthesis of tryptophan requires considerable energy.

Here is how attenuation works. Figure 8.12 points to the presence of two loci, the **trp leader** and the **trp attenuator,** in between the operator and the first structural gene, *trpE.* Figure 8.13 gives a closer view of the leader-attenuator, whose purpose is to attenuate, or weaken, transcription of the operon when tryptophan is relatively abundant. The attenuator operates by causing premature termination of transcription. In other words, transcription that gets started, even though the tryptophan concentration is high, stands a 90% chance of terminating in the attenuator region.

The reason for this premature termination is that the attenuator contains a transcription stop signal (terminator): An inverted repeat followed by a string of eight A–T pairs in a row. Because of the inverted repeat, the transcript of this region would tend to engage in intramolecular base-pairing, forming the "hairpin" shown in figure 8.14. Once the hairpin forms, the polymerase pauses and termination takes place. We will see below that this mimics the normal mechanism of transcription termination that takes place at the ends of mRNAs in *E. coli.*

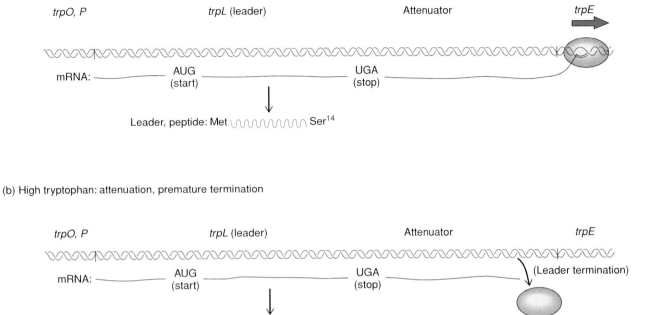

(a) Low tryptophan: transcription of *trp* structural genes

(b) High tryptophan: attenuation, premature termination

FIGURE 8.13 Attenuation in the *trp* operon. (*a*) In the presence of low tryptophan concentration, the RNA polymerase (red) reads right through the attenuator, so the structural genes are transcribed. (*b*) In the presence of high tryptophan, the attenuator causes premature termination of transcription, so the structural genes are not transcribed.

A ttenuation imposes an extra level of control on an operon, over and above the repressor-operator system. It operates by causing premature termination of transcription of the operon when the operon's products are abundant.

Defeating Attenuation

When tryptophan is scarce, the *trp* operon must be activated. Accordingly, there must be a means of overriding attenuation. Charles Yanofsky and his colleagues proposed such a mechanism, which goes like this: If something prevented the hairpin from forming, that would destroy the termination signal, attenuation would break down, and transcription would proceed. A look at figure 8.15*a* reveals not just one potential hairpin near the end of the leader transcript, but two. Furthermore, the two-hairpin arrangement is not the only one available; there is another, containing only one hairpin, shown in figure 8.15*c*. Note that this alternative hairpin contains elements from each of the two hairpins in the first structure. Figure 8.15*b* makes this concept clearer by labeling the sides of the original two hairpins 1, 2, 3, and 4. If the first of the original hairpins involves elements 1 and 2 and the second involves 3 and 4, then the alternative hairpin in the second structure involves 2 and 3. This means that the formation of the alternative hairpin

(figure 8.15*c*) precludes formation of the other two hairpins (figure 8.15*a*).

The two-hairpin structure involves more base pairs than the alternative, one-hairpin structure; therefore, it is more stable. So why should the less stable structure ever form? A clue comes from the base sequence of the leader region shown in figure 8.16. One very striking feature of this sequence is that there are two codons for tryptophan (UGG) in a row in element 1 of the first potential hairpin. This may not seem unusual, but tryptophan is a rare amino acid in most proteins; it is found on the average only once in every one hundred amino acids. So the chance of finding two tryptophan codons in a row *anywhere* is quite small, and the fact that they are found in the *trp* operon is very suspicious.

In bacteria, transcription and translation occur simultaneously. Thus, as soon as the *trp* leader region is transcribed, ribosomes begin translating this emerging mRNA. Think about what would happen to a ribosome trying to translate the *trp* leader under conditions of tryptophan starvation (figure 8.17*a*). Tryptophan is in short supply, and here are two demands in a row for that very amino acid. In all likelihood, the ribosome will not be able to satisfy those demands immediately, so it will pause at one of the tryptophan codons. And where does that put the stalled ribosome? Right on element 1, which should be participating in formation of the first hairpin. The bulky ribosome clinging to this RNA site effectively prevents its pairing with

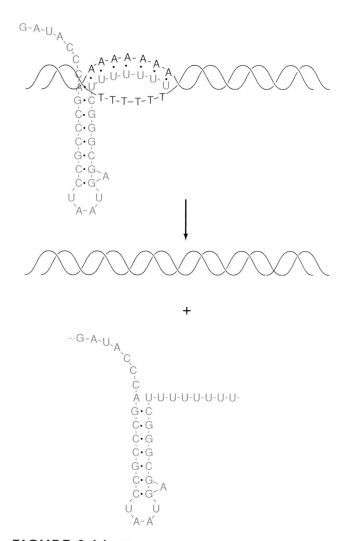

FIGURE 8.14 Mechanism of attenuation. The transcript of the leader-attenuator region (blue) contains an inverted repeat, so it can form a hairpin structure. When it does, the RNA polymerase pauses, then releases the transcript, so termination occurs before transcription can reach the structural genes.

element 2, which frees 2 to pair with 3, forming the one-hairpin alternative structure. Since the second hairpin cannot form, transcription does not terminate and attenuation has been defeated. This is desirable, of course, because when tryptophan is scarce, the *trp* operon should be transcribed.

Notice that this mechanism involves a coupling of transcription and translation, where the latter affects the former. It would not work in eukaryotes, where transcription and translation take place in separate compartments. It also depends on transcription and translation occurring at about the same rate. If RNA polymerase outran the ribosome, it might pass through the attenuator region before the ribosome had a chance to stall at the tryptophan codons.

You may be wondering how the polycistronic mRNA made from the *trp* operon can be translated if ribosomes are stalled in the leader at the very beginning. The answer is that each of the genes represented on the mRNA has its own translation start signal (AUG). So translation of *trpE*, for example, can occur, even if translation of the *trp* leader is stalled.

On the other hand, consider a ribosome translating the leader transcript under conditions of abundant tryptophan (figure 8.17b). Now the dual tryptophan codons present no barrier to translation, so the ribosome continues through element 1 until it reaches the stop signal (UGA) between elements 1 and 2 and falls off. With no ribosome to interfere, the two hairpins can form, completing the transcription termination signal that halts transcription before it reaches the *trp* structural genes. Thus, the attenuation system responds to the presence of adequate tryptophan and prevents wasteful synthesis of enzymes to make still more tryptophan.

Other operons besides *trp* use the attenuation mechanism. The most dramatic known use of consecutive codons to stall a ribosome occurs in the histidine (*his*) operon, where the leader region contains seven histidine codons in a row!

Attenuation operates in the *trp* operon as long as tryptophan is plentiful. When the supply of this amino acid is restricted, ribosomes stall at the tandem tryptophan codons in the *trp* leader. Since the *trp* leader is just being synthesized as this is taking place, the stalled ribosome will influence the way this RNA folds. In particular, it prevents the formation of a hairpin, which is part of the transcription termination signal that causes attenuation. Therefore, when tryptophan is scarce, attenuation is defeated and the operon remains active. This means that the control exerted by attenuation responds to tryptophan levels, just as repression does.

TERMINATION OF TRANSCRIPTION

The preceding discussion of attenuation is a fitting introduction to the topic of termination of transcription, because real termination at the ends of prokaryotic genes seems to work in much the same way as the premature termination in attenuation. In fact, attenuation has been used as an experimental probe of the mechanism of termination. Terry Platt and colleagues manipulated the *trp* attenuator region by introducing mutations and then looking to see what the effect on attenuation would be. The rationale is that if a certain sequence of the attenuator is important in causing termination, mutations in this sequence should block attenuation. Two features of the attenuator turned out to be important, and from what we know about the mechanism of attenuation, they are not surprising at all. These are (1) the inverted repeat that gives the transcript the ability to form a hairpin, and (2) a string of T's in the coding strand.

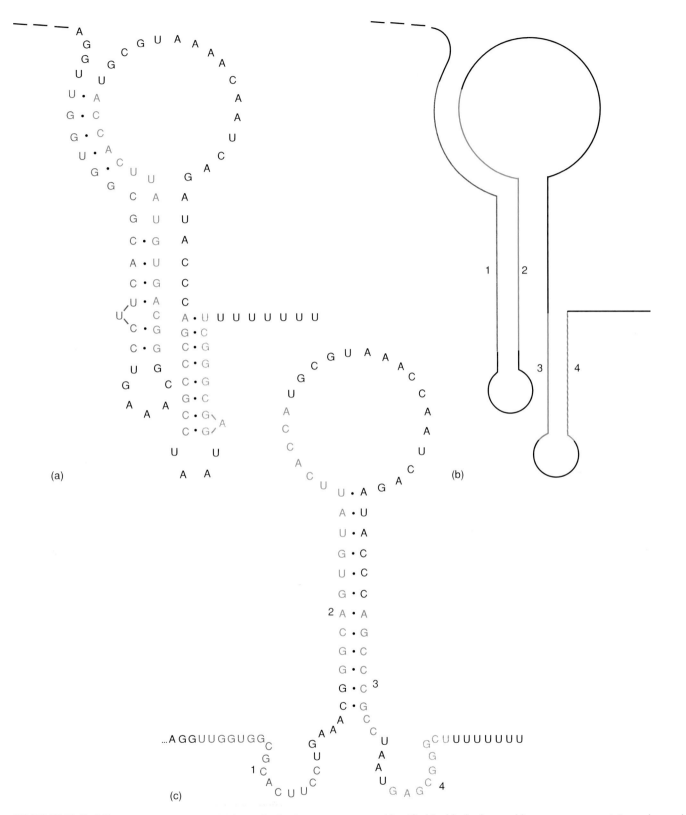

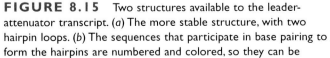

FIGURE 8.15 Two structures available to the leader-attenuator transcript. (*a*) The more stable structure, with two hairpin loops. (*b*) The sequences that participate in base pairing to form the hairpins are numbered and colored, so they can be identified in (*c*), the less stable structure, containing only one hairpin loop. The curved shape of the RNA at the bottom is not meant to suggest a shape for the molecule, it is drawn this way simply to save space.

Met Lys Ala Ile Phe Val Leu Lys Gly Trp Trp Arg Thr Ser Stop
pppA---AUGAAAGCAAUUUUCGUACUGAAAGGUUGGUGGCGCACUUCCUGA

FIGURE 8.16 Sequence of the leader. The sequence of part of the leader transcript is presented, along with the leader peptide it encodes. Note the two Trp codons in tandem (red).

(a) Tryptophan starvation

(b) Tryptophan abundance

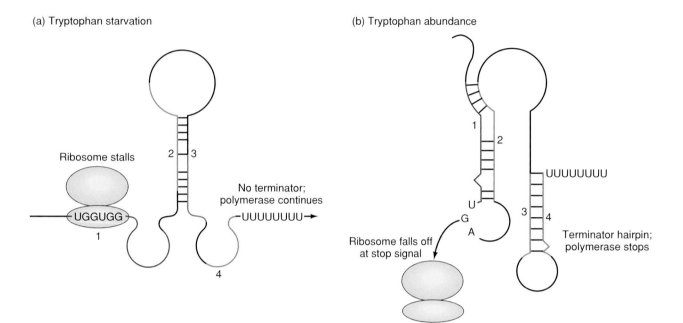

FIGURE 8.17 Overriding attenuation. (a) Under conditions of tryptophan starvation, the ribosome (yellow) stalls at the Trp codons and prevents element 1 (red) from pairing with element 2 (blue). This forces the one-hairpin structure, which lacks a terminator, to form, so no attenuation should take place. (b) Under conditions of tryptophan abundance, the ribosome reads through the two tryptophan codons and falls off at the translation stop signal (UGA), so it cannot interfere with base pairing in the leader transcript. The more stable, two-hairpin structure forms; this contains a terminator, so attenuation occurs.

INVERTED REPEATS AND HAIRPINS

Before we go on, we should understand how an inverted repeat predisposes a transcript to form a hairpin. Consider the inverted repeat:

5′-TACGAAGTTCGTA-3′
3′-ATGCTTCAAGCAT-5′

Such a sequence is symmetrical around its center, indicated by the dot; it would read the same if rotated 180 degrees in the plane of the paper, and we always read the strand that runs 5′ → 3′ left to right. Now observe that a transcript of this sequence

UACGAA**G**UUCGUA

is self-complementary around its center (the boldfaced G). That means that the self-complementary bases can pair to form a hairpin as follows:

U · A
A · U
C · G
G · C
A · U
A   U
G

The A and U at the apex of the hairpin cannot form a base pair because of the physical constraints of the turn in RNA.

The inverted repeat in the *trp* attenuator is not perfect, but eight base pairs are still possible between elements 3 and 4, and seven of these are strong G–C pairs, held together by three hydrogen bonds. The hairpin looks like this:

A · U
G · C
C · G
C · G
C · G
G · C
C · G‚A
C · G⟩
U   U
A   A

Notice that there is a small loop at the end of this hairpin because of the U–U and A–A combinations that cannot base-pair. Furthermore, one A has to be "looped out" to allow eight base pairs instead of just seven. Still, the hairpin should form and be relatively stable.

A MODEL FOR TERMINATION

The discovery that the inverted repeat and string of T's are key features of the transcription termination signal supports the termination hypothesis illustrated in figure 8.18. The RNA polymerase transcribes the inverted repeat, producing a transcript that tends to form a hairpin (figure 8.18*a*). When the hairpin forms (figure 8.18*b*), the RNA polymerase pauses for some unexplained reason. This leaves the base pairs between the dA's in the DNA and the rU's in the RNA as the only force holding the transcript and template together. Such rU–dA pairs are exceptionally weak; they have a melting temperature 20° lower than even rU–rA or rA–dT pairs. This leads to the hypothesis that the RNA can easily fall off the template, terminating transcription (figure 8.18*c*).

THE TERMINATION FACTOR ρ

The model of termination discussed above really applies only to termination that is carried out by the *E. coli* RNA polymerase itself. There are other termination events in *E. coli* that involve not only the polymerase but another factor called **rho (ρ)**. If σ

is thought of as the initiation, or specificity factor, ρ can be considered the termination, or release factor.

Rho-dependent termination follows somewhat different rules than ρ-independent termination. The most obvious difference is that ρ-dependent terminators have only one part—the inverted repeat. There is no string of T's or other second sequence motif. Therefore, the polymerase presumably pauses at the hairpin but cannot release the transcript by itself. Rho performs that function.

How does ρ do its job? Since we have seen that σ operates through RNA polymerase by binding tightly to the core enzyme, we might envision a similar role for ρ. But ρ seems not to have affinity for RNA polymerase. Whereas σ is a critical part of the polymerase holoenzyme, ρ is not. A more attractive possibility (figure 8.19) is that ρ binds to the transcript and follows the RNA polymerase. This chase would continue until the polymerase stalled in the terminator region just after making the hairpin. Then ρ could catch up and release the transcript. There is some evidence to support this idea. For example, it can be demonstrated that ρ probably does bind to RNA, at least under certain circumstances.

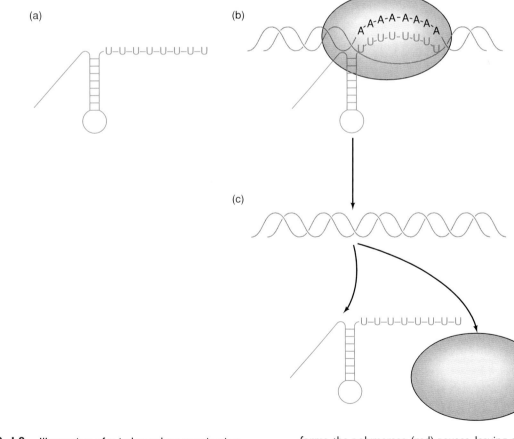

FIGURE 8.18 Illustration of ρ-independent termination. (*a*) The 3′-end of the transcript shows the two elements necessary for ρ-independent termination: a base-paired hairpin loop, followed by a second motif, in this case a string of U's. (*b*) When the hairpin forms, the polymerase (red) pauses, leaving only the A–U bonds holding the transcript to the template. (*c*) The bonds break, releasing the RNA (blue) and terminating transcription.

(a) ρ pursues polymerase

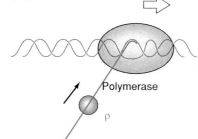

(b) Hairpin forms; polymerase pauses; ρ catches up

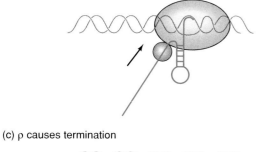

(c) ρ causes termination

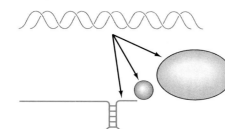

FIGURE 8.19   A model of ρ-dependent termination. (*a*) As polymerase (red) makes RNA, ρ (blue) binds to the transcript (green) and pursues the polymerase. (*b*) When the hairpin forms in the transcript, the polymerase pauses, giving ρ a chance to catch up. (*c*) ρ destabilizes the RNA–DNA bonds and releases the transcript.

Termination signals (terminators) recognized by RNA polymerase alone consist of an inverted repeat followed by another important base sequence, frequently a string of T's. The transcript of the inverted repeat can form a base-paired structure called a hairpin, which somehow stalls the polymerase. The second part of the terminator then ensures that the RNA will fall off, terminating transcription. Other terminators require an ancillary factor, ρ, in addition to RNA polymerase. These terminators contain only an inverted repeat. Rho may loosen transcript-template binding after RNA polymerase pauses at the hairpin.

TEMPORAL CONTROL OF TRANSCRIPTION

Up to this point we have been discussing the ways in which bacteria control the transcription of a very limited number of genes at a time. For example, when the *lac* operon switches on, only three structural genes are activated. There are other times in a bacterium's life when more radical shifts in gene expression take place. When a phage infects a bacterium, it usually subverts the host's transcription machinery to its own use. In the process, it establishes a time-dependent, or temporal, program of transcription. In other words, the early phage genes are transcribed first, then the later genes. By the time phage T4 infection of *E. coli* reaches its late phase, there is essentially no more host transcription—only phage transcription. This massive shift in specificity would be hard to explain by the operon mechanisms we have looked at so far. Instead, it is engineered by a fundamental change in the transcription machinery—a change in RNA polymerase itself. Another profound change in gene expression occurs during sporulation in bacteria such as *Bacillus subtilis*. Here, genes that are needed in the vegetative phase of growth switch off, and other, sporulation-specific genes switch on. Again, this switch is accomplished by changes in RNA polymerase. The remainder of this chapter is devoted to this important aspect of gene control.

MODIFICATION OF THE HOST RNA POLYMERASE

What part of RNA polymerase would be the logical candidate to change the specificity of the enzyme? We have already seen that σ is the key factor in determining specificity of T4 DNA transcription in vitro, so σ is the most reasonable answer to our question. Experimentation has shown it to be the correct answer. However, these experiments were not done first with the *E. coli*–T4 system, but with *B. subtilis* and its phages, especially phage SPO1.

SPO1, like T4, has a large DNA genome. It has a temporal program of transcription as follows: In the first five minutes or so of infection, the early genes are expressed; next, the middle genes turn on (about five to ten minutes post-infection); from about the ten-minute point until the end of infection, the late genes switch on. Since the phage has a large number of genes, it is not surprising that it uses a fairly elaborate mechanism to control this temporal program. Jan Pero and her colleagues have been the leaders in developing the model illustrated in figure 8.20.

The host RNA polymerase holoenzyme handles transcription of early SPO1 genes, which is analogous to the T4 model, where the earliest genes are transcribed by the host holoenzyme. This is necessary because the phage does not carry its own RNA polymerase. Therefore, when the phage first infects the cell, the host holoenzyme is the only RNA polymerase available. The

(a) Early transcription; specificity factor: host σ (◯)

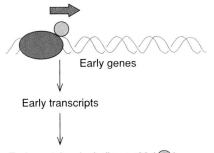

Early genes

↓

Early transcripts

↓

Early proteins, including gp28 (◯)

(b) Middle transcription; specificity factor: gp28 (◯)

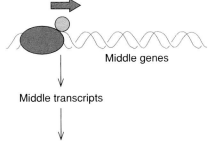

Middle genes

↓

Middle transcripts

↓

Middle proteins, including gp33 (⬤) and gp34 (◯)

(c) Late transcription; specificity factor: gp33 (⬤) + gp34 (◯)

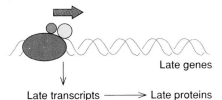

Late genes

↓

Late transcripts ⟶ Late proteins

FIGURE 8.20 Temporal control of transcription in phage SPO1-infected *B. subtilis*. (*a*) Early transcription is directed by the host RNA polymerase holoenzyme, including the host σ factor (blue); one of the early phage proteins is gp28 (green), a new σ factor. (*b*) Middle transcription is directed by gp28, in conjunction with the host core polymerase (red); two middle phage proteins are gp33 and gp34 (purple and yellow, respectively); together, these constitute yet another σ factor. (*c*) Late transcription depends on the host core polymerase plus gp33 and 34.

B. subtilis holoenzyme closely resembles the *E. coli* enzyme. Its core consists of two large (β and β′) and two small (α) polypeptides; its σ factor has a molecular weight of 43,000, somewhat smaller than *E. coli*'s σ (70,000). One of the genes transcribed in the early phase of SPO1 infection is called gene 28. Its product, **gp28,** associates with the host core polymerase, displacing the host σ (σ^{43}). With this new, phage-encoded polypeptide in place, the RNA polymerase changes specificity. It begins tran-

scribing the phage middle genes instead of the phage early genes and host genes. In other words, gp28 is a novel σ factor that accomplishes two things: It diverts the host's polymerase from transcribing host genes, and it switches from early to middle phage transcription.

The switch from middle to late transcription occurs in much the same way, except that two polypeptides team up to bind to the polymerase core and change its specificity. These are **gp33** and **gp34,** the products of two phage middle genes (genes 33 and 34, respectively). These proteins replace gp28 and direct the altered polymerase to transcribe the phage late genes in preference to the middle genes. Note that the polypeptides of the host core polymerase remain constant throughout this process; it is the progressive substitution of σ factors that changes the specificity of the enzyme and thereby directs the transcriptional program.

How do we know that the σ-switching model is valid? Two lines of evidence, genetic and biochemical, support it. First, genetic studies have shown that mutations in gene 28 prevent the early-to-middle switch, just as we would predict if the gene 28 product is the σ factor that turns on the middle genes. Similarly, mutations in either gene 33 or 34 prevent the middle-to-late switch, again in accord with the model. The biochemical studies have relied on purified RNA polymerases that contain one or another of the σ factors in question. When polymerase bearing host σ encounters phage DNA in vitro, it transcribes only the early genes; when the polymerase bears gp28, it transcribes the middle genes preferentially, and when it is bound to gp33 and gp34, it is specific for the late genes.

Transcription of phage SPO1 genes in infected *B. subtilis* cells proceeds according to a temporal program in which early genes are transcribed first, then middle genes, and finally late genes. This switching is directed by a set of phage-encoded σ factors that associate with the host core RNA polymerase and change its specificity from early to middle to late. The host σ is specific for the phage early genes; the phage gp28 switches the specificity to the middle genes; and the phage gp33 and gp34 switch to late specificity.

THE RNA POLYMERASE ENCODED IN PHAGE T7

Phage **T7** belongs to a class of relatively simple *E. coli* phages that also includes T3 and φII. These have a considerably smaller genome than SPO1 and, therefore, many fewer genes. In these phages we distinguish three phases of transcription, called classes I, II, and III. (They could just as easily be called early, middle, and late, to conform to SPO1 nomenclature.) One of the five class I genes (gene 1) is necessary for class II and class III gene

(a) Early transcription; specificity factor: host σ (⬤)

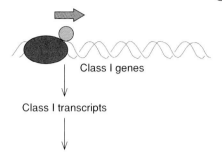

Class I genes

Class I transcripts

Class I proteins, including phage RNA polymerase (⬤)

(b) Late transcription; phage RNA polymerase (⬤)

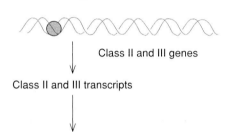

Class II and III genes

Class II and III transcripts

Class II and III proteins

FIGURE 8.21 Temporal control of transcription in phage T7-infected *E. coli*. (*a*) Early (class I) transcription depends on the host RNA polymerase holoenzyme, including the host σ factor (blue); one of the early phage proteins in the T7 RNA polymerase (green). (*b*) Late (class II and III) transcription depends on the T7 RNA polymerase.

expression. When it is mutated, only the class I genes are transcribed. Having just learned the SPO1 story, you are probably expecting to hear that gene 1 codes for a σ factor directing the host RNA polymerase to transcribe the later phage genes. In fact, this was the conclusion reached by the first workers on T7 transcription, but it was erroneous.

The gene 1 product is actually not a σ factor but a whole phage-specific RNA polymerase contained in one polypeptide. This polymerase, as you might expect, reads the T7 phage class II and III genes specifically, leaving the class I genes completely alone. Indeed, this enzyme is unusually specific; it will transcribe the class II and III genes of phage T7 and virtually no other natural template. The switching mechanism in this phage is thus quite simple (figure 8.21). When the phage DNA enters the host cell, the *E. coli* holoenzyme transcribes the five class I genes, including gene 1. The gene 1 product—the phage-specific RNA polymerase—then transcribes the phage class II and class III genes.

> **P**hage T7, instead of coding for a new σ factor to change the host polymerase's specificity from early to late, encodes a whole new RNA polymerase with absolute specificity for the later phage genes. This polymerase, composed of a single polypeptide, is a product of one of the earliest phage genes, gene 1. The temporal program in the infection by this phage is simple. The host polymerase reads the earliest (class I) genes, one of whose products is the phage polymerase, which then reads the later (class II and class III) genes.

CONTROL OF TRANSCRIPTION DURING SPORULATION

We have already seen how phage SPO1 changes the specificity of its host's RNA polymerase by replacing its σ factor. In the following section, we will show that the same kind of mechanism applies to changes in gene expression in the host itself during the process of sporulation. *E. coli*, the bacterium we have examined in greatest detail, lives its whole life in the **vegetative,** or growth state. If you starve a culture of *E. coli*, the cells stop growing and can remain for weeks or months in a stationary phase. But then they begin to die; they have relatively little defense against adverse conditions. But there are other bacteria that sense the onset of hard times and protect themselves by **sporulating.** *B. subtilis*, for example, forms **endospores**—tough, dormant bodies that can survive for years until favorable conditions return (figure 8.22).

Gene expression must change during sporulation; cells as different in morphology as vegetative and sporulating cells must contain at least some different gene products. In fact, when *B. subtilis* cells sporulate, they activate a whole new set of sporulation-specific genes. The switch from the vegetative to the sporulating state is accomplished by a complex σ-switching scheme that turns off transcription of some vegetative genes and turns on sporulation-specific transcription.

Sporulation is a fundamental change, involving large numbers of old genes turning off and new genes turning on. Furthermore, the change in transcription is not absolute; some genes that are active in the vegetative state remain active during sporulation. How is the transcription-switching mechanism able to cope with these complexities?

As you might anticipate, more than one new σ factor is involved in sporulation. In fact, there are at least three: σ^{29}, σ^{30}, and σ^{32} (now called σ^E, σ^H, and σ^C, respectively), in addition to the vegetative σ^{43} (now called σ^A). These polypeptides have molecular weights indicated in their names. Thus, σ^{43} has a molecular weight of 43,000. Each recognizes a different class of promoter. For example, the vegetative σ^{43} recognizes promoters that are very similar to the promoters recognized by the *E. coli* σ factor, with a −10 (Pribnow) box that looks something like TATAAT and a −35 box having the consensus sequence TTGACA. By contrast, the sporulation-specific factors σ^{32} and σ^{29} recognize quite different sequences.

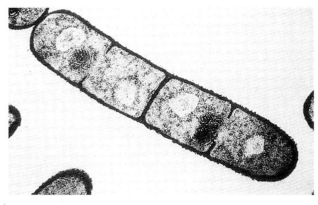

(a)

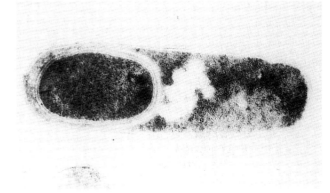

(b)

FIGURE 8.22 (*a*) *B. subtilis* vegetative cells and (*b*) a sporulating cell, with an endospore developing at the left end. Courtesy Dr. Kenneth Bott.

When the original σ was discovered in *E. coli* in 1969, Burgess and Travers speculated that many different σ's would be found, and that these would direct transcription of different classes of genes. This principle was first verified in *B. subtilis,* as we have seen, but it has also proven true in *E. coli.* For example, a distinct *E. coli* σ (σ^{32}) directs the transcription of a class of genes called *heat-shock genes* that switch on when the bacteria experience high temperature or other environmental insults. During nitrogen starvation, another σ factor (σ^{54}) directs transcription of genes that encode proteins responsible for nitrogen metabolism.

When the bacterium *B. subtilis* sporulates, a whole new set of sporulation-specific genes turns on, and many but not all vegetative genes turn off. This switch takes place largely at the transcription level. It is accomplished by several new σ factors that displace the vegetative σ factor from the core RNA polymerase and direct transcription of sporulation genes instead of vegetative genes. Each σ factor has its own preferred promoter sequence.

INFECTION OF *E. COLI* BY PHAGE λ

Many of the phages we have studied so far (T2, T4, T7, and SPO1, for example) are **virulent** phages. When they replicate, they kill their host by **lysing** it, or breaking it open. On the other hand, **lambda** (λ) is a **temperate** phage; when it infects an *E. coli* cell, it does not necessarily kill it. In this respect, λ is more versatile than many phages; it can follow two paths of reproduction (figure 8.23). The first is the **lytic** mode, in which infection progresses just as it would with a virulent phage. It begins with phage DNA entering the host cell and then serving as the template for transcription by host RNA polymerase. Phage mRNAs are translated to yield phage proteins, the phage DNA replicates, and progeny phages assemble from these DNA and protein components. The infection ends when the host cell lyses to release the progeny phages.

In the **lysogenic** mode, something quite different happens. The phage DNA enters the cell, and its early genes are transcribed and translated, just as in a lytic infection. But then a phage protein (the λ **repressor**) appears and binds to the two phage operator regions, shutting down transcription of all genes except for *cI* (pronounced "c-one," not "c-eye"), the gene for the λ repressor itself. Under these conditions, with only one phage gene active, it is easy to see why no progeny phages can be produced. Furthermore, when lysogeny is established, the phage DNA integrates into the host genome. A bacterium harboring this integrated phage DNA is called a **lysogen.** The integrated DNA is called a **prophage.** The lysogenic state can exist indefinitely and should not be considered a disadvantage for the phage, because the phage DNA in the lysogen replicates right along with the host DNA. In this way, the phage genome multiplies without the necessity of making phage particles; thus, it gets a "free ride." Under certain conditions, such as when the lysogen encounters mutagenic chemicals or radiation, lysogeny can be broken and the phage enters the lytic phase.

Phage λ can replicate in either of two ways: lytic or lysogenic. In the lytic mode, almost all of the phage genes are transcribed and translated, and the phage DNA is replicated, leading to production of progeny phages and lysis of the host cells. In the lysogenic mode, the λ DNA is incorporated into the host genome; after that occurs, only one gene is expressed. The product of this gene, the λ repressor, prevents transcription of all the rest of the phage genes. However, the incorporated phage DNA (the prophage) still replicates, since it has become part of the host DNA.

Lytic Reproduction of Phage λ

The lytic reproduction cycle of phage λ resembles that of the virulent phages we have studied in that it contains three phases of transcription. These phases are called **immediate early, delayed early,** and **late.** These genes are sequentially arranged on the phage DNA, which helps explain how they are regulated, as we will see. Figure 8.24 shows the λ genetic map in two forms:

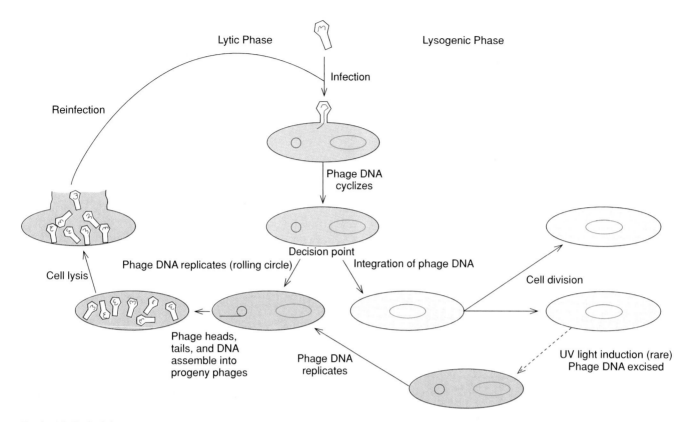

FIGURE 8.23 Lytic versus lysogenic infection by phage λ. Blue cells are in the lytic phase; yellow cells are in the lysogenic phase; green cells are uncommitted.

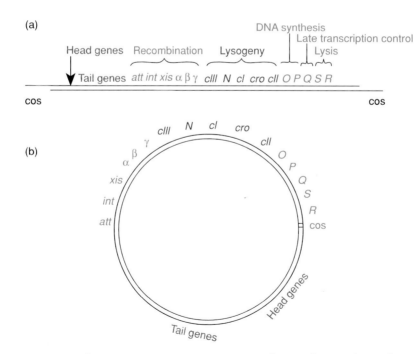

FIGURE 8.24 Genetic map of phage λ. (*a*) The map is shown in linear form, as the DNA exists in the phage particles; the cohesive ends (cos) are at the ends of the map. The genes are grouped primarily according to function. (*b*) The map is shown in circular form, as it exists in the host cell during a lytic infection after annealing of the cohesive ends.

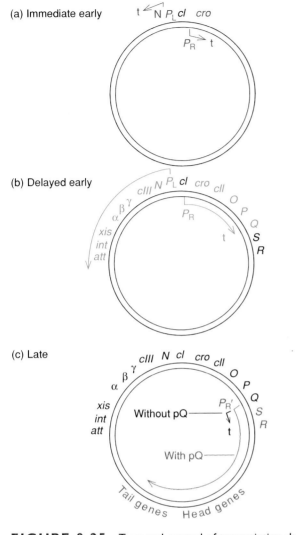

(a) Immediate early

(b) Delayed early

(c) Late

FIGURE 8.25 Temporal control of transcription during lytic infection by phage λ. (a) Immediate early transcription (red) starts at the rightward and leftward promoters (P_R and P_L, respectively) that flank the repressor gene (cI); transcription stops at the ρ-dependent terminators (t) after the N and cro genes. (b) Delayed early transcription (blue) begins at the same promoters, but bypasses the terminators by virtue of the N gene product, pN, which is an antiterminator. (c) Late transcription (gray) begins at a new promoter (P_R'); it would stop short at the terminator (t) without the Q gene product, pQ, another antiterminator. Note that O and P are protein-encoding delayed early genes, not operator and promoter.

linear, as the DNA exists in the phage particles, and circular, the shape the DNA assumes shortly after infection begins. The circularization is made possible by 12-base overhangs, or "sticky" ends, at either end of the linear genome. These cohesive ends go by the name **cos.** Note that cyclization brings together all the late genes, which had been separated at the two ends of the linear genome.

As usual, the program of gene expression in this phage is controlled by transcriptional switches, but λ uses a switch we have not seen before: **antitermination.** Figure 8.25 outlines this

scheme. Of course, the host RNA polymerase holoenzyme transcribes the immediate early genes first. There are only two of these genes, cro and N, which lie immediately downstream from the rightward and leftward promoters, P_R and P_L, respectively. At this stage in the lytic cycle, no repressor is bound to the operators that govern these promoters (O_R and O_L, respectively), so transcription proceeds unimpeded. When the polymerase reaches the ends of the immediate early genes, it encounters ρ-dependent terminators and stops short of the delayed early genes.

The products of both immediate early genes are crucial to further expression of the λ program. The **cro** gene product is an **antirepressor** that blocks transcription of the repressor gene, cI, and therefore prevents synthesis of repressor protein. This is, of course, necessary if any of the other phage genes are to be expressed. The **N** gene product, **pN,** is an **antiterminator** that permits RNA polymerase to ignore the terminators at the ends of the immediate early genes and continue transcribing right on into the delayed early genes. When this happens, the delayed early phase begins. Note that the same promoters (P_R and P_L) are used for both immediate early and delayed early transcription. The switch does not involve a new σ factor or RNA polymerase that recognizes new promoters and starts new transcripts, as we have seen with other phages; instead, it involves an extension of transcripts controlled by the same promoters.

The delayed early genes are important in continuing the lytic cycle, and, as we will see in the next section, in establishing lysogeny. Genes O and P code for proteins that are necessary for phage DNA replication, a key part of lytic growth. The **Q** gene product **(pQ)** is another antiterminator, which permits transcription of the late genes.

The late genes are all transcribed in the rightward direction, but not from P_R. The late promoter, P_R', lies just downstream from Q. Transcription from this promoter terminates after only 194 bases, unless pQ intervenes to prevent termination. The N gene product will not do; it is specific for antitermination after cro and N. The late genes code for the proteins that make up the phage head and tail, and for proteins that lyse the host cell so the progeny phages can escape.

The immediate early/delayed early/late transcriptional switching in the lytic cycle of phage λ is controlled by antiterminators. One of the two immediate early genes is cro, which codes for an antirepressor that allows the lytic cycle to continue. The other, N, codes for an antiterminator, pN, that overrides the terminators after the N and cro genes. Transcription then continues on into the delayed early genes. One of the delayed early genes, Q, codes for another antiterminator (pQ) that permits transcription of the late genes from the late promoter, P_R', to continue without premature termination.

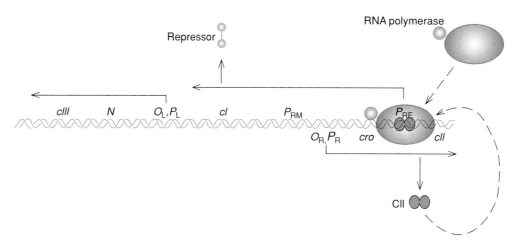

FIGURE 8.26   Establishing lysogeny. Delayed early transcription from P_R gives *cII* mRNA that is translated to CII (purple). CII allows RNA polymerase (blue and red) to bind to P_{RE} and transcribe the *cI* gene, yielding repressor (green).

Establishing Lysogeny

We have mentioned that the delayed early genes are required not only for the lytic cycle but for establishing lysogeny. The delayed early genes help establish lysogeny in two ways: (1) Some of the delayed early gene products are needed for integration of the phage DNA into the host genome, a prerequisite for lysogeny. (2) The products of the *cII* and *cIII* genes allow transcription of the *cI* gene and therefore production of the λ repressor, the central component in lysogeny.

There are two promoters for the *cI* gene (figure 8.26): P_{RM} and P_{RE}. The $_{RM}$ in P_{RM} stands for "repressor maintenance." This is the promoter that is used *during* lysogeny to ensure a continuing supply of repressor to maintain the lysogenic state. It has the peculiar property of requiring its own product—repressor—for activity. We will discuss the basis for this requirement; however, we can see immediately one important implication. This promoter cannot be used to establish lysogeny, because at the start of infection there is no repressor to activate it. Instead, the other promoter, P_{RE}, is used. The $_{RE}$ in P_{RE} stands for "repressor establishment." P_{RE} lies to the right of both P_R and *cro*. It directs transcription leftward through *cro* and then through *cI*. Thus, P_{RE} allows *cI* expression before any repressor is available.

Of course, the natural direction of transcription of *cro* is rightward from P_R, so the leftward transcription from P_{RE} gives an RNA product that is an "anti-sense" transcript of *cro,* as well as a "sense" transcript of *cI*. The *cI* part of the RNA can be translated to give repressor, but the anti-sense *cro* part cannot be translated. In fact, it probably hybridizes to sense *cro* RNA and prevents *its* translation.

P_{RE} has some interesting requirements of its own. It has −10 and −35 boxes with no clear resemblance to the consensus sequences ordinarily recognized by *E. coli* RNA polymerase. Indeed, it cannot be transcribed by this polymerase alone in vitro. However, the *cII* gene product, CII, helps RNA polymerase bind to this unusual promoter sequence. The product of the *cIII* gene, CIII, is also instrumental in this process,

but its effect is less direct. It retards destruction of CII by cellular proteases. In other words, the delayed early products CII and CIII make possible the establishment of lysogeny by activating P_{RE}.

> Phage λ establishes lysogeny by causing production of enough repressor to bind to the early operators and prevent further early RNA synthesis. The promoter used for establishment of lysogeny is P_{RE}, which lies to the right of P_R and *cro*. Transcription from this promoter goes leftward through the *cI* gene. The products of the delayed early genes *cII* and *cIII* also participate in this process: CII by directly stimulating polymerase binding to P_{RE}, CIII by slowing degradation of CII.

Autoregulation of the *cI* Gene during Lysogeny

Once the λ repressor appears, it binds to the λ operators, O_R and O_L. This has a double-barreled effect, both tending toward lysogeny. First, repressor turns off further early transcription, thus interrupting the lytic cycle. The turnoff of *cro* is especially important to lysogeny, because the *cro* protein (**Cro**) acts to counter repressor activity, as we will see. The second effect of repressor is that it encourages its own synthesis by activating P_{RM}.

Figure 8.27 illustrates how this self-activation works. The key to this phenomenon is the fact that both O_R and O_L are subdivided into three parts, each of which can bind repressor. The O_R region is more interesting, because it controls leftward transcription of *cI,* as well as rightward transcription of *cro*. The three binding sites in the O_R region are called O_R1, O_R2, and O_R3. Their affinities for repressor are quite different: Repressor binds tightest to O_R1, next tightest to O_R2, and least tightly to O_R3.

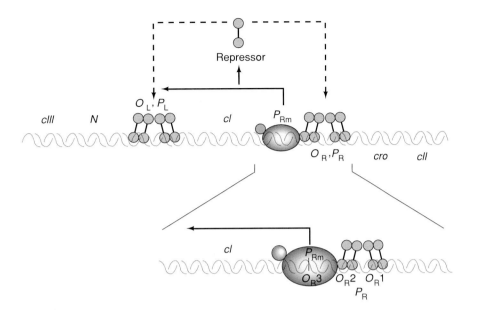

FIGURE 8.27   Maintaining lysogeny. (*bottom*) Repressor (green, made originally via transcription from P_{RE}) forms dimers and binds cooperatively to $O_R 1$ and 2. The protein-protein contact between repressor on $O_R 2$ and RNA polymerase (red and blue) allows polymerase to bind to P_{RM} and transcribe *cI*. (*top*) Transcription (from P_{RM}) and translation of the *cI* mRNA yields a continuous supply of repressor, which binds to O_R and O_L and prevents transcription of any genes aside from *cI*.

However, binding of repressor to $O_R 1$ and $O_R 2$ is cooperative. This means that as soon as repressor binds to its "favorite" site, $O_R 1$, it facilitates binding of another repressor to $O_R 2$. No cooperative binding to $O_R 3$ normally occurs.

Repressor protein is a dimer of two identical subunits, each of which is represented by a dumbbell shape in figure 8.27. This shape indicates that each subunit has two domains, one at each end of the molecule. The two domains have distinct roles: One is the DNA binding end of the molecule; the other is the site for the repressor–repressor interaction that makes dimerization and cooperative binding possible. Once repressor dimers have bound to both $O_R 1$ and $O_R 2$, the repressor occupying $O_R 2$ lies very close to the binding site for RNA polymerase at P_{RM}. So close, in fact, that the two proteins touch each other. Far from being a hindrance, as you might expect, this protein–protein contact is required for RNA polymerase to initiate transcription efficiently at this promoter.

With repressors bound to $O_R 1$ and $O_R 2$, there can be no more transcription from P_{RE}, because the repressors block *cII* and *cIII* transcription, and the products of these genes, needed for transcription from P_{RE}, break down very rapidly. However, this is not usually a problem, since lysogeny is already established and a small supply of repressor is all that is required to maintain it. That small supply of repressor can be provided as long as $O_R 3$ is left open, because RNA polymerase can transcribe *cI* freely from P_{RM}.

It is conceivable that the concentration of repressor could build up to such a level that it would even fill its weakest binding site, $O_R 3$. In that case, all *cI* transcription would cease, because even P_{RM} would be blocked. This would allow the repressor level to drop, at which time it would dissociate first from $O_R 3$, allowing *cI*

transcription to begin anew. This represents a means by which repressor can keep its own concentration from rising too high.

The promoter that is used to maintain lysogeny is P_{RM}. It comes into play after transcription from P_{RE} makes possible the burst of repressor synthesis that establishes lysogeny. This repressor binds to $O_R 1$ and $O_R 2$ cooperatively, but leaves $O_R 3$ open. RNA polymerase binds to P_{RM}, which overlaps $O_R 3$ in such a way that it just touches the repressor bound to $O_R 2$. This protein–protein interaction is required for this promoter to work.

Determining the Fate of a λ Infection: Lysis or Lysogeny

What determines whether a given cell infected by λ will enter the lytic cycle or lysogeny? There is a delicate balance between these two fates, and we usually cannot predict the actual path taken in a given cell. Support for this assertion comes from a study of the appearance of groups of *E. coli* cells infected by λ phage. When a few phage particles are sprinkled on a lawn of bacteria in a petri dish, they infect the cells. If a lytic infection takes place, the progeny phages spread to neighboring cells and infect them. After a few hours, we can see a circular hole in the bacterial lawn caused by the death of lytically infected cells. This hole is called a **plaque.** If the infection were 100% lytic, the plaque would be clear, because all the host cells would be killed. But λ plaques are not usually clear. Instead, they are turbid, indicating the presence of live lysogens. This means that even in the local environment of a plaque, some infected cells suffer the lytic cycle, while others are lysogenized.

Let us digress for a moment to ask this question: Why are the lysogens not infected lytically by one of the multitude of phages in the plaque? The answer is that if a new phage DNA enters a lysogen, there is plenty of repressor in the cell to bind to the new phage DNA and prevent its expression. We therefore can say that the lysogen is **immune** to **superinfection** by a phage with the same control region, or **immunity region,** as that of the prophage.

Now let us return to the main problem. We have seen that some cells in a plaque can be lytically infected, while others are lysogenized. The cells within a plaque are all genetically identical, and so are the phages, so the choice of fate is not genetic. Instead, it seems to represent a race between the products of two genes: cI and cro. This race is rerun in each infected cell, and the winner determines the pathway of the infection in that cell. If cI prevails, lysogeny will be established; if cro wins, the infection will be lytic. We can already appreciate the basics of this argument: If the cI gene manages to produce enough repressor, this protein will bind to O_R and O_L, prevent further transcription of the early genes, and thereby prevent expression of the late genes that would cause progeny phage production and lysis. On the other hand, if enough Cro is made, this protein can prevent cI transcription and thereby block lysogeny (figure 8.28).

The key to Cro's ability to block cI transcription is the nature of its affinity for the λ operators: Cro binds to both O_R and O_L, as does repressor, but its order of binding to the three-part operators is exactly opposite to that of repressor. Instead of binding in the order 1, 2, 3, as repressor does, Cro binds first to O_R3. As soon as that happens, cI transcription from P_{RM} cannot occur, because O_R3 overlaps P_{RM}. In other words, Cro acts as a repressor. Furthermore, when Cro fills up all the rightward and leftward operators, it prevents transcription of all the early genes from P_R and P_L, including cII and $cIII$. Without the products of these genes, P_{RE} cannot function; so all repressor synthesis ceases. Lytic infection is then assured.

Cro's turning off of early transcription is also required for lytic growth. Continued production of delayed early proteins late in infection somehow aborts the lytic cycle.

But what determines whether cI or cro wins the race? Surely it is more than just a flip of the coin. Actually, the most important factor seems to be the concentration of the cII gene product, CII. The higher the CII concentration within the cell, the more likely lysogeny becomes. This fits with what we have already learned about CII—it activates P_{RE} and thereby helps turn on the lysogenic program. We have also seen that the activation of P_{RE} by CII works against the lytic program by producing cro antisense RNA that can inhibit translation of cro sense RNA.

And what controls the concentration of CII? We have seen that CIII protects CII against cellular proteases, but high protease concentrations can overwhelm CIII, destroy CII, and ensure that the infection will be lytic. Such high protease concentrations occur under good environmental conditions—rich medium, for example. By contrast, protease levels are depressed under starvation conditions. Thus, starvation tends to favor lysogeny, while rich medium favors the lytic pathway. This is advantageous for

(a) *cI* wins, lysogeny results

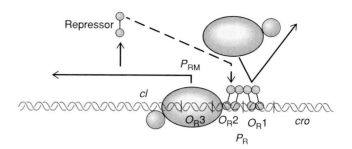
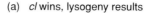

(b) *cro* wins, lytic cycle results

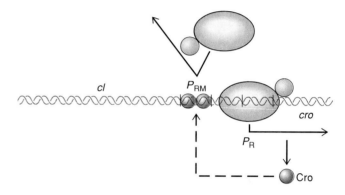

FIGURE 8.28   The battle between *cI* and *cro*. (a) *cI* wins. Enough repressor (green) is made by transcription of the *cI* gene from P_{RM} that it blocks polymerase (red and blue) from binding to P_R and therefore blocks *cro* transcription. Lysogeny results. (b) *cro* wins. Enough Cro (purple) is made by transcription from P_R that it blocks polymerase from binding to P_{RM} and therefore blocks *cI* transcription. The lytic cycle results.

the phage, since the lytic pathway requires considerable energy to make all the phage RNAs and proteins, and this much energy may not be available during starvation. In comparison, lysogeny is cheap; it requires only the synthesis of a little repressor.

Whether a given cell is lytically or lysogenically infected by phage λ depends on the outcome of a race between the products of the *cI* and *cro* genes. The *cI* gene codes for repressor, which blocks O_R1, O_R2, O_L1, and O_L2, turning off all early transcription, including transcription of the *cro* gene. This leads to lysogeny. On the other hand, the *cro* gene codes for Cro, which blocks O_R3 (and O_L3), turning off *cI* transcription. This leads to lytic infection. Whichever gene product appears first in high enough concentration to block its competitor's synthesis wins the race and determines the cell's fate. The winner of this race is determined by the cII concentration, which is determined by the cellular protease concentration, which is in turn determined by environmental factors such as the richness of the medium.

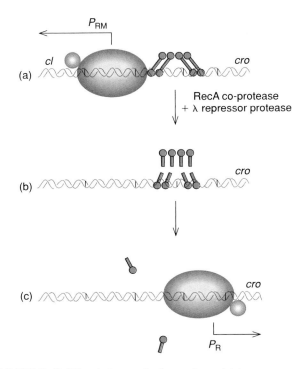

FIGURE 8.29   Inducing the λ prophage. (*a*) Lysogeny. Repressor (green) is bound to O_R (and O_L) and *cI* is being actively transcribed from the P_{RM} promoter. (*b*) The RecA co-protease (activated by ultraviolet light or other mutagenic influence) unmasks a protease activity in the repressor, so it can cleave itself. (*c*) The severed repressor falls off the operator, allowing polymerase (red and blue) to bind to P_R and transcribe *cro*. Lysogeny is broken.

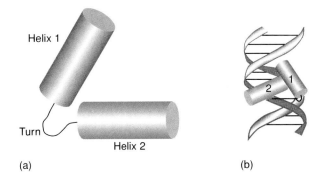

FIGURE 8.30   The helix-turn-helix motif as a DNA-binding element. (*a*) The helix-turn-helix motif of the λ repressor. (*b*) The fit of the helix-turn-helix motif of one repressor monomer with the λ operator. Helix 2 of the motif (blue) lies in the major groove of its DNA target; some of the amino acids on the back of this helix (away from the viewer) are available to make contacts with the DNA.

> When a lysogen suffers DNA damage, it induces the SOS response. The initial event in this response is the appearance of a co-protease activity in the RecA protein. This causes the repressors to cut themselves in half, removing them from the λ operators and inducing the lytic cycle. In this way, progeny λ phages can escape the potentially lethal damage that is occurring in their host.

Lysogen Induction

We mentioned that a lysogen can be induced by treatment with mutagenic chemicals or radiation. The mechanism of this induction is as follows: *E. coli* cells responds to environmental insults, such as mutagens or radiation, by inducing a set of genes whose collective activity is called the **SOS response;** one of these genes, in fact the most important, is ***recA***. The *recA* product (RecA) participates in recombination repair of DNA damage (which explains part of its usefulness to the SOS response), but environmental insults also induce a new activity in the RecA protein. It becomes a co-protease that stimulates a latent protease, or protein-cleaving activity, in the λ repressor. This protease then cleaves the repressor in half and releases it from the operators, as shown in figure 8.29. As soon as that happens, transcription begins from P_R and P_L. One of the first genes transcribed is *cro*, whose product shuts down any further transcription of the repressor gene. Lysogeny is broken and lytic phage replication begins.

Surely λ would not have evolved with a repressor that responds to RecA by chopping itself in half unless there were an advantage to the phage. That advantage seems to be this: The SOS response signals that the lysogen is under some kind of DNA-damaging attack. It is expedient under those circumstances for the prophage to get out by inducing the lytic cycle, rather like rats deserting a sinking ship.

SPECIFIC DNA–PROTEIN INTERACTIONS

We have seen several examples of proteins that bind to specific regions of DNA. These include the *lac, trp,* and λ repressors and Cro, which bind to their respective operators, and CAP, which binds to the CAP binding site in the *lac* control region. The binding of these proteins to their targets is very specific. They can locate and bind to one particular short DNA sequence among a vast excess of unrelated DNA.

How do these proteins accomplish such specific binding? All the proteins listed above have a similar structural motif: two α-helices connected by a short protein "turn." This **helix-turn-helix** motif allows the second helix (the **recognition helix**) to fit snugly into the major groove of the target DNA site, as illustrated in figure 8.30. Furthermore, the amino acids of the recognition helix of a DNA binding protein have chemical groups that fit hand-in-glove with the specific bases protruding into the major DNA groove, and with phosphate groups in the DNA backbone, at that protein's binding site. These contacts are usually made through hydrogen bonds. Other proteins with helix-turn-helix motifs will not be able to bind as well at that same site, because they do not have the correct amino acids in their recognition helices.

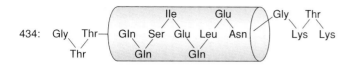

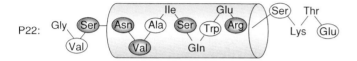

FIGURE 8.31 Key amino acids in the recognition helices of two repressors. The amino acid sequences of the recognition helices of the 434 and P22 repressors are shown, along with a few amino acids on either side. Amino acids that differ between these two proteins are circled; these are more likely to contribute to differences in specificity. Furthermore, the amino acids on the side of the helix that faces the DNA are most likely to be involved in DNA binding. These, along with one amino acid in the turn just before the helix (red), were changed to alter the binding specificity of the protein.

HELIX-TURN-HELIX MOTIFS IN λ-LIKE PHAGE REPRESSORS

Mark Ptashne and his colleagues have provided an excellent example of this kind of binding, using repressors from two λ-like phages, **434** and **P22,** and their respective operators. These two phages have very similar molecular genetics, but they have different immunity regions: They make different repressors that recognize different operators. Both repressors resemble the λ repressor in that they contain helix-turn-helix motifs. However, since they recognize operators with different base sequences, we would expect them to have different amino acids in their respective recognition helices, especially those amino acids that are strategically located to contact the bases in the DNA major groove.

Using X-ray diffraction analysis of operator-repressor complexes, Steve Harrison and Mark Ptashne identified the face of the recognition helix of the 434 phage repressor that contacts the bases in the major groove of its operator. By analogy, they could make a similar prediction for the P22 repressor. Figure 8.31 schematically illustrates the amino acids in each repressor that are most likely to be involved in operator binding.

If these are really the important amino acids, we ought to be able to change only these amino acids and thereby alter the specificity of the repressor. In particular, we should be able to employ such changes to alter the 434 repressor so that it recognizes the P22 operator instead of its own. This is exactly what Ptashne did. He started with a cloned gene for the 434 repressor and, using mutagenesis techniques (described in chapter 15), systematically altered the codons for five amino acids in the 434 recognition helix to codons for the five corresponding amino acids in the P22 recognition helix.

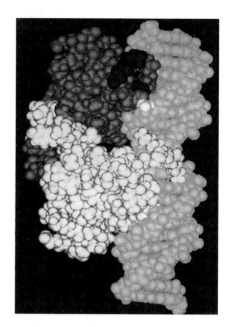

FIGURE 8.32 Computer model of λ repressor dimer binding to λ operator (O_R2). The DNA double helix (light blue) is at right. The two monomers of the repressor are in dark blue and yellow. The helix-turn-helix motif of the upper monomer (red and blue) is inserted into the major groove of the DNA. The arm of the lower monomer reaches around to embrace the DNA.

From Ann Hochschild, Nina Irwin, and Mark Ptashne, *Cell* 32:322, 1983, Cell Press. Photo by Richard Feldman.

Next, he expressed the altered gene in bacteria and tested the product for ability to bind to 434 and P22 operators. The five amino acid substitutions had completely changed the repressor's binding specificity. Instead of binding to the 434 operator, it bound to the P22 operator. This demonstrated that these five amino acids in the recognition helix, or perhaps only some of these five, actually do determine the specificity of the repressor.

THE HELIX-TURN-HELIX MOTIF OF CRO AND λ REPRESSOR

Cro also uses a helix-turn-helix DNA binding motif and binds to the same operators as the λ repressor, but it has the exact opposite affinity for the three different operators in a set. That is, it binds first to O_R3 and last to O_R1, rather than vice versa. Therefore, by changing amino acids in the recognition helices, we ought to be able to identify the amino acids that give Cro and repressor their different binding specificities. Again, Ptashne and his coworkers accomplished this task and found that amino acids 5 and 6 in the recognition helices are especially important. They also found that the arm at one end of the λ repressor, which embraces the operator, is important in the repressor's specificity.

Figure 8.32 shows a computer model of a dimer of λ repressor interacting with λ operator. In the repressor monomer at the top, the helix-turn-helix motif is visible projecting into the major groove of the DNA. At the bottom, we can see the arm of the other repressor monomer reaching around to embrace the DNA.

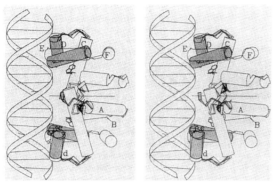

(a)

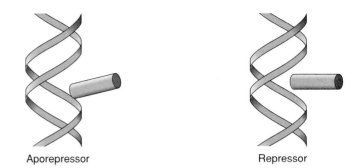

Aporepressor   Repressor

(b)

FIGURE 8.33 Comparison of the fit of *trp* repressor and aporepressor with *trp* operator. (*a*) Stereo diagram. The helix-turn-helix motifs of both monomers are shown in the positions they assume in the repressor (transparent) and aporepressor (gray). The position of tryptophan in the repressor is shown (black polygons). Note that the recognition helix (helix E) in the aporepressor falls back out of ideal position for inserting into the wide groove of the operator DNA. The two almost identical drawings constitute a stereo presentation that allows you to view this picture in three dimensions. To get this 3-D effect, use a stereo viewer, or alternatively, hold the picture one to two feet in front of you and let

your eyes relax as they would when you are staring into the distance or viewing a "magic eye" picture. After a few seconds, the two images should fuse into one in the center, which appears in three dimensions. This stereo view gives a better appreciation for the fit of the recognition helix and the wide groove of the DNA, but if you cannot get the 3-D effect, just look at one of the two pictures. (*b*) Simplified (nonstereo) diagram comparing the positions of the recognition helix (red) of the aporepressor (left) and the repressor (right) with respect to the DNA major groove. Notice that the recognition helix of the repressor points directly into the major groove, whereas that of the aporepressor points more downward.

THE DNA BINDING MOTIF OF THE *trp* REPRESSOR

The *trp* repressor is another DNA binding protein that uses a helix-turn-helix DNA binding motif. However, recall that the aporepressor (the protein without the tryptophan corepressor) is not active. Paul Sigler and colleagues have used X-ray crystallography of *trp* repressor and aporepressor to point out the subtle but important difference that tryptophan makes.

Figure 8.33 shows the results of this analysis. In panel (*a*), the *trp* repressor and aporepressor are drawn superimposed and making the best fit possible with the *trp* operator. The repressor is a dimer, and we can see the two monomers arranged vertically in the figure. Each one is making contact, through its recognition helix, with the major groove of the operator. The remaining parts of the two dimers are involved in binding to each other. The overall picture we have of the repressor dimer is a three-domain protein. The central domain is a kind of two-sided platform for the two helix-turn-helix motifs that protrude on either side. Panel (*b*) shows the positions of the helix-turn-helix domain in the aporepressor and repressor, respectively. The difference in the angle between recognition helix and DNA may appear subtle, but it is all-important in giving the correct fit of the protein with the DNA.

Sigler refers to these DNA-reading motifs as "reading heads," likening them to the heads in a tape recorder or the disk drive of a personal computer. In a computer, the reading heads can assume two positions: engaged and reading the disk, or disengaged and away from the disk. The *trp* repressor works the same way. When tryptophan is present, it inserts itself between the platform and

each reading head as illustrated in figure 8.33, and forces the reading heads into the best position for fitting snugly into the major groove of the operator. However, when tryptophan dissociates from the aporepressor, the gap it leaves allows the reading heads to fall back toward the central platform and out of position to fit with the operator.

> Many DNA binding proteins in prokaryotes employ a helix-turn-helix binding motif. The second (recognition) helix of each motif fits into the major groove of the target site on the DNA, and certain key amino acids in this helix make intimate contact with specific bases and phosphates in the DNA. The binding specificity of such a protein can be changed simply by changing these key amino acids. The *trp* repressor requires tryptophan to force the recognition helices of the repressor dimer into the proper position for interacting with the *trp* operator.

WORKED-OUT PROBLEMS

PROBLEM 1

The table below gives the genotypes (with respect to the *lac* operon) of several partial diploid *E. coli* strains. Fill in the phenotypes, using a "+" for β-galactosidase synthesis and "−" for no β-galactosidase synthesis. Glucose is absent in all cases. Give a brief explanation of your reasoning.

	Phenotype for β-galactosidase Production	
Genotype	No Inducer	Inducer
a. $I^+O^+Z^+/I^+O^+Z^+$		
b. $I^+O^+Z^-/I^+O^+Z^+$		
c. $I^-O^+Z^+/I^+O^+Z^+$		
d. $I^SO^+Z^+/I^+O^+Z^+$		
e. $I^+O^CZ^+/I^+O^+Z^+$		
f. $I^+O^CZ^-/I^+O^+Z^+$		
g. $I^SO^CZ^+/I^+O^+Z^+$		

SOLUTION

	Phenotype for β-galactosidase Production	
Genotype	No Inducer	Inducer
a. $I^+O^+Z^+/I^+O^+Z^+$	−	+
b. $I^+O^+Z^-/I^+O^+Z^+$	−	+
c. $I^-O^+Z^+/I^+O^+Z^+$	−	+
d. $I^SO^+Z^+/I^+O^+Z^+$	−	−
e. $I^+O^CZ^+/I^+O^+Z^+$	+	+
f. $I^+O^CZ^-/I^+O^+Z^+$	−	+
g. $I^SO^CZ^+/I^+O^+Z^+$	+	+

Reasoning:

a. The genotypes of both halves of the partial diploid are wild-type. Therefore, these cells will show the wild-type *lac* phenotype, which is repressed (turned off) in the absence of inducer and derepressed (turned on) in the presence of inducer. This means that these cells make no β-galactosidase without inducer, but they do make β-galactosidase in the presence of inducer.

b. This partial diploid has one mutant allele and one wild-type allele at the β-galactosidase (*Z*) locus. In this case the wild-type allele is dominant, since the mutant allele does not inhibit the function of the wild-type allele. Thus, these cells show the wild-type phenotype, as in (*a*), above.

c. This partial diploid has one constitutive (*I⁻*) and one wild-type allele at the *I* locus. Again, the wild-type allele is dominant, since the mutant allele cannot inhibit the function of the wild-type allele. Thus, these cells show the wild-type phenotype, as in (*a*).

d. This partial diploid has one wild-type and one (*Iˢ*) allele at the *I* locus. The mutant allele is dominant in this case, since it produces repressor that is incapable of binding to inducer. Thus, this repressor will remain bound to the operator and prevent induction of the *lac* operon, regardless of the fact that the wild-type allele still makes wild-type repressor. Therefore, the operon remains repressed even in the presence of inducer.

e. This partial diploid has one wild-type and one constitutive (*Oᶜ*) allele at the *O* locus. The constitutive allele is dominant because this operator cannot bind repressor even though the wild-type

operator can. Thus, these cells will produce β-galactosidase in the presence or absence of inducer.

f. One half of this partial diploid is double mutant (constitutive in the operator and defective in the β-galactosidase [*Z*] locus). The other half is wild-type at both loci. When these two mutations are found *in cis,* as they are here, they cancel each other. The *Oᶜ* mutation is dominant, but it is only *cis*-dominant. That is, it is only dominant with respect to the operon to which it is attached. In this case, that operon has a defective *Z* gene, so no β-galactosidase will be produced under any circumstances. On the other hand, there is nothing wrong with the other half of the partial diploid, so it will give these cells a wild-type phenotype.

g. One half of this partial diploid is double mutant (*Iˢ* at the *I* locus, and constitutive at the *O* locus). Since the *Oᶜ* operator cannot bind repressor, it will not bind the defective repressor that is the product of the *Iˢ* gene. Thus, this operon is turned on in the presence or absence of inducer, even though the wild-type operon is always turned off because of the mutant repressor provided *in trans.*

PROBLEM 2

(*a*) In the genotype listed in the table below, the letters *A, B,* and *C* correspond to the *lacI, lacO,* and *lacZ* loci, though not necessarily in that order. From the mutant phenotypes exhibited by the first three genotypes listed in the table below, deduce the identities of *A, B,* and *C* as they correspond to the three loci of the *lac* operon. The minus susperscripts (e.g., A⁻) can refer to the following aberrant functions: Z^-, O^C, or I^-.

(*b*) Determine the genotypes, in conventional *lac* operon genetic notation, of the partial diploid strains shown in lines 4 and 5 of the table below. Here, I^+, I^-, and I^S are all possible.

	Phenotype for β-galactosidase production	
Genotype	No Inducer	Inducer
1. $A^+B^+C^-$	+	+
2. $A^-B^+C^+$	+	+
3. $A^+B^+C^-/A^+B^+C^+$	+	+
4. $A^-B^+C^+/A^+B^+C^+$	−	−
5. $A^-B^+C^+/A^+B^+C^+$	−	+

SOLUTION

(*a*) Since the phenotype of Mutant strain #1 is constitutive (no inducer needed for activity), its genotype must carry a mutation in either the *I* or the *O* gene, i.e. I^- or O^C. Therefore *C* (the mutant gene in strain #1) is either *I* or *O*. Mutant #2 is also constitutive, so its genotype is also either I^- or O^C. Therefore, *A* is either *I* or *O.* So far we know that *A* and *C* are *I* and *O,* though not necessarily in that order. This means that *B* must correspond to gene *Z*, by elimination. Now we need to determine whether *A = I* and *C = O,* or vice-versa. The phenotype of Mutant #3 gives the answer. Because it is constitutive and the mutant form is dominant, C^- must be O^C. If it were I^-, the other constitutive possibility, the mutation would have been recessive, and no β-galactosidase would have been produced in the absence of inducer. Thus, $A = I$, $B = Z$, and $C = O$.

(b) The partial diploid in line 4 fails to make β-galactosidase even in the presence of inducer; that is, the mutation (A^-) is dominant, and A^- must, therefore, be I^s. On the other hand, the phenotype of the strain in line 5 is the same as that of a wild-type strain. Therefore, the mutation (A^-) in the genotype of the strain shown in line 5 must be recessive, i.e. $A^- = I^-$. Therefore, the two genotypes are

$$I^s O^+ Z^+ / I^+ O^+ Z^+$$
$$I^- O^+ Z^+ / I^+ O^+ Z^+$$

SUMMARY

RNA polymerase is the key player in the transcription process. The *E. coli* enzyme is composed of a core, which contains the basic transcription machinery, and a polypeptide called sigma (σ), which directs the core to transcribe specific genes. The σ factor allows initiation of transcription by causing the RNA polymerase holoenzyme to bind tightly to a promoter, melting a local region of DNA.

Bacterial genes that are linked in function are frequently clustered together in operons and are controlled together from a common promoter. For example, the genes for the three enzymes involved in lactose metabolism in *E. coli* are clustered together in the *lac* operon. This operon is controlled both positively (by cAMP plus catabolite activator protein) and negatively (by the *lac* repressor).

The *trp* operon codes for the enzymes that make tryptophan. It is controlled negatively by the *trp* repressor, including the corepressor, tryptophan. Attenuation imposes an extra level of control on the *trp* operon, over and above the repressor-operator system. It operates by causing premature termination of transcription of the operon when the operon's products are abundant.

Termination signals (terminators) recognized by RNA polymerase alone consist of an inverted repeat followed by a second sequence motif—usually a string of T's. The transcript of the inverted repeat can form a base-paired structure called a hairpin, which somehow stalls the polymerase. The polymerase then causes the RNA to fall off, terminating transcription. Other terminators require an ancillary factor, ρ, in addition to RNA polymerase.

Gene expression is controlled primarily at the transcription level in prokaryotes. The temporal control of transcription in *B. subtilis* cells infected by phage SPO1 is directed by a set of phage-encoded σ factors that associate with the host core RNA polymerase and change its specificity from early to middle to late. Phage T7, instead of coding for a new σ factor to change the host polymerase's specificity from early to late, encodes a whole new RNA polymerase with specificity for the later phage genes. Transcriptional switching in the lytic cycle of phage λ is controlled by antiterminators that override ρ-dependent terminators and allow transcription to continue into the next class of genes.

Phage λ can replicate in either of two ways: lytic or lysogenic. In the lytic mode, almost all of the phage genes are transcribed and translated, and the phage DNA is replicated, leading to production of progeny phage and lysis of the host cells. In the lysogenic mode, the λ DNA is incorporated into the host genome; when that occurs, only the λ repressor gene (*cI*) is expressed. The product of this gene, the λ repressor, binds to the two early phage operators and prevents transcription of all the rest of the phage genes. However, the incorporated phage DNA (the prophage) still replicates, since it has become part of the host DNA.

Whether a given cell is lytically or lysogenically infected by phage λ depends on the concentration of CII in the cell. This in turn depends on environmental conditions.

When a lysogen suffers DNA damage, it induces the SOS response. The initial event in this response is the appearance of a co-protease activity in the RecA protein. This unmasks a protease activity in the λ repressors, which cuts the repressors in half, removing them from the λ operators and inducing the lytic cycle. In this way, λ phages can escape the damage that is occurring in their host.

Several prokaryotic DNA binding proteins, such as the λ repressor, the *trp* repressor, and Cro, bind to their DNA targets via helix-turn-helix motifs that insert specifically into the DNA major groove. Amino acids in the second (recognition) helix in these motifs make specific contact with bases in the major groove and phosphates in the DNA backbone.

PROBLEMS AND QUESTIONS

1. Contrast the activities of RNA polymerase holoenzyme and core polymerase on un-nicked phage DNA in vitro.

2. Consider a −10 box with the following sequence: CATAGT. (a) Would a C → T mutation in the first position likely be an up or a down mutation? (b) Would a T → A mutation in the last position likely be an up or a down mutation?

3. What structural characteristic of an operon ensures that the genes contained in the operon will be coordinately controlled?

4. Consider *E. coli* cells, each having one of the following mutations:

   (a) a mutant *lac* operator (the O^c locus) that cannot bind repressor
   (b) a mutant *lac* repressor (the I^- gene product) that cannot bind to the *lac* operator
   (c) a mutant *lac* repressor (the I^s gene product) that cannot bind to allolactose
   (d) a mutant *lac* promoter that cannot bind CAP plus cAMP

   What effect would each mutation have on the function of the *lac* operon (assuming no glucose is present)?

5. The table below gives the genotypes (with respect to the *lac* operon) of several partial diploid *E. coli* strains. Fill in the phenotypes, using a "+" for β-galactosidase synthesis and a "−" for no β-galactosidase synthesis. Glucose is absent in all cases. Give a brief explanation of your reasoning.

**Phenotype for
β-galactosidase Production**

Genotype	No Inducer	Inducer
a. $I^+O^+Z^+/I^+O^+Z^+$		
b. $I^+O^+Z^-/I^+O^+Z^+$		
c. $I^-O^+Z^+/I^+O^+Z^+$		
d. $I^sO^+Z^+/I^+O^+Z^+$		
e. $I^+O^cZ^+/I^+O^+Z^+$		
f. $I^+O^cZ^-/I^+O^+Z^+$		
g. $I^sO^cZ^+/I^+O^+Z^+$		

6. The table below presents the genotypes and phenotypes of several *lac* mutants, but with one missing notation (represented by an X) in each case. From the information given, fill in the missing notation for each genotype. Give your reasoning. (Sometimes X can represent wild-type.) $Z = lacZ$, the gene for β-galactosidase.

**Phenotype for
β-galactosidase Production**

Genotype	No Inducer	Inducer
a. $I^xO^+Z^+$	+	+
b. $I^xO^+Z^+/I^+O^+Z^+$	−	−
c. $I^xO^+Z^+/I^+O^+Z^+$	−	+ (2 possible answers)
d. $I^xO^+Z^+/I^-O^+Z^+$	+	+
e. $I^+O^xZ^+/I^+O^+Z^+$	+	+
f. $I^+O^xZ^+/I^+O^+Z^+$	−	+
g. $I^+O^+Z^x/I^+O^+Z^-$	−	−

7. Consider mutations involving the promoter of the *lac* operon. These mutations prevent RNA polymerase binding and are designated P^-. Predict the phenotypes of the following partial diploids, assuming no glucose in the growth medium. $Z = lacZ$, the gene for β-galactosidase.

**Phenotype for
β-galactosidase Production**

Genotype	No Inducer	Inducer
a. $I^+O^+P^-Z^+/I^+O^+P^+Z^+$		
b. $I^+O^+P^-Z^+/I^+O^+P^-Z^+$		
c. $I^+O^+P^-Z^+/I^+O^+P^+Z^-$		
d. $I^-O^+P^-Z^+/I^-O^+P^-Z^+$		

8. There are five structural genes in the *trp* operon, yet the operon codes for only three enzymes. How can this be?

9. How much enhancement of *trp* transcription is possible in the following *E. coli* cells as the tryptophan concentration falls from abundant to very low? Why?

  (a) Wild-type
  (b) Mutants in which the two tryptophan codons in the leader are changed from UGGUGG to UGAUGG. (Hint: UGA is a stop codon.)
  (c) Mutants in which the two tryptophan codons in the leader are changed from UGGUGG to UCGUGC
  (d) Mutants in which the string of T's in the attenuator is replaced by a string of G's

10. Which phage genes would be transcribed in *B. subtilis* cells infected with SPO1 phages that carry mutations in the following genes? Why?

  (a) Gene 28
  (b) Gene 33
  (c) Gene 34

11. What would be the effect of a mutation in gene 1 of phage T7? Why?

12. What would be the effect of a mutation in the *B. subtilis* gene for σ^C or σ^E? Why?

13. Which phage gene would *not* be expressed after the λ lytic cycle is underway?

14. Which phage genes would *not* be expressed in a λ lysogen?

15. pN is an immediate early product that is required for delayed early transcription, yet it is very unstable. Why does this not pose a problem for continued delayed early expression in the lytic cycle?

16. RNA polymerase transcribing from P_{RE} traverses *cro*. Can this result in *cro* expression? Why, or why not?

17. Why can neither P_{RE} nor P_{RM} serve in initiating *cI* transcription before any other λ genes are expressed?

18. Would you classify the antiterminator activity of pN as positive or negative control? Why?

19. (a) Why are most plaques caused by λ phage turbid in appearance? (b) Would a *cI*− mutant give turbid plaques? Provide an explanation for your answer.

20. If a cell is infected by a λN^- mutant phage, will it grow lytically, form a lysogen, or neither? Why?

21. A mutant repressor is found that can bind to a mutated operator. What part of the mutant repressor is different from wild-type?

Answers appear at end of book.

Eukaryotic Gene Structure and Expression

Nature has a short menu, but the items on it are quite reliable.

Dagmar Ringe

American structural biologist

Learning Objectives

In this chapter you will learn:

1. How eukaryotic DNA is complexed with protein and condensed into chromosomes.

2. The roles of the three RNA polymerases found in eukaryotic cells.

3. The structures of the promoters recognized by the three eukaryotic RNA polymerases.

4. The roles of transcription factors in activating transcription from the three classes of promoters.

5. That many eukaryotic genes are interrupted.

6. The extra steps in eukaryotic gene expression that take place between transcription and translation.

The picture of prokaryotic gene structure and activity presented in chapter 8 seems rather complex. Nevertheless, compared to the situation in eukaryotes, it is simple. One obvious example of this difference between eukaryotes and prokaryotes is the structure of their chromosomes. The prokaryotic chromosome is a circle of DNA encumbered with relatively little protein, while the eukaryotic chromosome contains a large amount of protein in addition to the DNA. In this chapter we will examine the structure of eukaryotic chromosomes and the genes they contain. We will also learn that expression of these genes involves not only transcription and translation, but a variety of other events.

CHROMATIN STRUCTURE

Have you ever dealt with a length of kite string, or fishline, or thread, or even a garden hose? If so, you know that long, thin things tend to get hopelessly tangled. One look at figure 9.1 should convince you that the DNA in a chromosome is longer in relationship to its width than any kite string. In fact, you would have to fly a kite from sea level to an altitude far above Mt. Everest to make the length/width ratio of the string approach that of the DNA in an average human chromosome (more than ten million to one). How do those immensely long molecules stay untangled? As we will see, they are complexed with protein (forming a mixture called **chromatin**) and coiled up in a highly organized way.

When chromatin is lightly treated with a DNase called *micrococcal nuclease,* DNA-protein particles called **nucleosomes** are released (figure 9.2*a*). These nucleosomes, ubiquitous in the eukaryotic kingdom, are remarkably uniform in size, about 11 nanometers (nm) in diameter, and contain about 200 base pairs of DNA.

The abundance and uniformity of nucleosomes suggests that they are a repeating unit of chromatin structure, and direct evidence supports this idea. Consider, for example, the "minichromosome" of the mammalian tumor virus SV40. When this virus infects a host cell, its circular DNA attracts host proteins called histones to form a miniature chromosome. If we lower the salt concentration of a solution of such minichromosomes, we reveal a beaded structure (figure 9.2*b*). The beads are nucleosomes.

NUCLEOSOMES: THE FIRST ORDER OF CHROMATIN FOLDING

Nucleosomes contain, in addition to the 200 base pairs of DNA, a set of basic proteins called **histones.** Most eukaryotic cells contain five different kinds of histones called **H1, H2A, H2B, H3,** and **H4.** Figure 9.3 shows how these five species can be separated by gel electrophoresis. The "bead" of a nucleosome is actually a ball of histones with the DNA wound almost twice

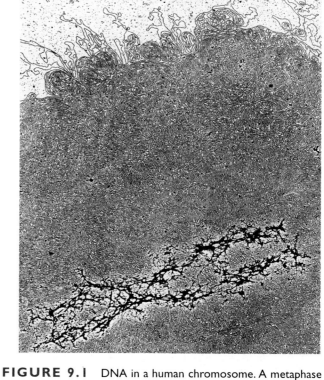

FIGURE 9.1 DNA in a human chromosome. A metaphase chromosome from HeLa (human) cells was extracted to remove most of its protein. This allowed the DNA to uncoil, leaving a scaffolding in the shape of the chromosome behind. The tangled web of DNA looks like a gray cloud throughout the lower part of the figure, but the individual DNA duplexes near the top are visible. Bear in mind that this is a single duplex molecule of DNA.

From James R. Paulson and U. K. Laemmli, *Cell* 12:820, 1977. © Cell Press.

around the outside. Each ball contains exactly eight histone molecules, a pair each of H2A, H2B, H3, and H4. A single molecule of histone H1 binds to "linker" DNA outside the ball, and can be removed by more stringent DNase treatment, which digests the "linker" DNA and yields a nucleosome *core particle* with about 150 bp of DNA. Figure 9.4 shows the arrangement of histones and DNA in the nucleosome core particle. It appears that the core histones are really not arranged in pairs. H3 and H4 form a tetramer at the center of the particle, and this tetramer is flanked by H2A–H2B dimers at top left and lower right.

Two properties of nucleosomes suggest that their function is fundamentally important. The first is the universality of nucleosomes in eukaryotes; the second is the extreme evolutionary conservation of most of the histones. The classic example of this conservation is that histone H4 from the cow differs from H4 from the pea plant by only two amino acids out of 102, and these are **conservative substitutions** that should not change the protein structure significantly. In other words, in all the eons since the cow and pea lines diverged from a common ancestor, only two amino acids in histone H4 have changed. Obviously,

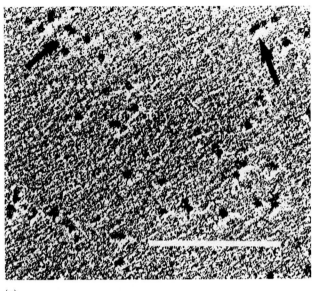

(a)

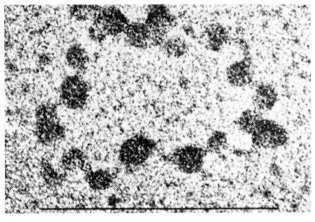

(b)

FIGURE 9.2   Nucleosomes. (*a*) Nucleosomes (arrow) released from chicken red blood cells by treatment of chromatin with micrococcal nuclease. The bar represents 250 nm. (*b*) Nucleosomes in the minichromosome of SV40 virus. The minichromosome has been relaxed by suspending it in low-salt buffer, which reveals the beaded structure. The bar represents 100 nm.

Photo (a) from P. Oudet, M. Gross-Bellard, and P. Chambon, *Cell* 4:287, 1975. © Cell Press. Photo (b) © Dr. Jack Griffith.

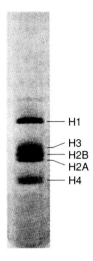

FIGURE 9.3   Separation of histones by electrophoresis. The histones of calf thymus were separated by polyacrylamide gel electrophoresis.

From S. Panyim and R. Chalkley, *Archives of Biochemistry and Biophysics* 130:343, 1969.

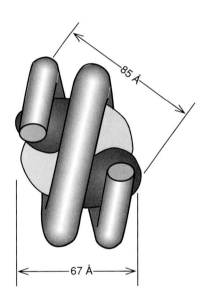

FIGURE 9.4   One interpretation of nucleosome structure, based on X-ray and neutron diffraction studies. The DNA (red) wraps around the core of histones, composed of the (H3–H4)$_2$ tetramer in the center (yellow) and an H2A–H2B dimer at each end (purple).

the histones, and the nucleosomes of which they are a part, play a vital role. Otherwise, their structures would have been allowed more evolutionary latitude.

The most likely role for nucleosomes is, of course, a structural one; nucleosomes provide the first order of condensing, or coiling, of the extremely long, thin chromosomal fiber. Figure 9.5 makes this point very clearly. SV40 DNA that has been stripped of its protein is shown next to an SV40 minichromosome at the same magnification. The apparent decrease in length of the DNA in the minichromosome is due to the formation of nucleosomes. In fact, the DNA achieves about a 7-fold condensation by coiling up into nucleosomes.

Eukaryotic DNA combines with basic protein molecules called histones to form structures known as nucleosomes. These structures contain four pairs of histones (H2A, H2B, H3, and H4) in a ball, around which is wrapped a stretch of about 200 base pairs of DNA. Histone H1 binds to DNA outside the ball. The first order of chromatin folding is represented by a string of nucleosomes.

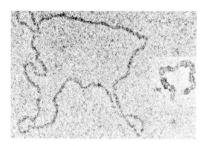

FIGURE 9.5   Condensation of DNA in nucleosomes. Deproteinized SV40 DNA is shown next to an SV40 minichromosome (inset) in electron micrographs enlarged to the same scale. The condensation of DNA afforded by nucleosome formation is apparent.

From Dr. Jack Griffith, "Chromatin Structure: Deduced from a Minichromosome," *Science*, 187:1202, 28 March 1975. © 1975 American Association for the Advancement of Science.

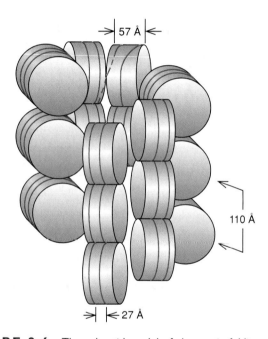

FIGURE 9.6   The solenoid model of chromatin folding. A string of nucleosomes coils into a hollow spiral to form the solenoid. Each nucleosome is represented by a blue cylinder with a wirelike DNA (red) coiled around it.

FURTHER FOLDING OF THE NUCLEOSOME STRING

Chromatin in interphase nuclei takes on various thicknesses. One of these, 11 nm (110 Å) in thickness, simply represents a string of nucleosomes, or **nucleosome fiber.** The next thickness, about 25 nm (250 Å), is due to further winding of the nucleosome to form a hollow coil called a **solenoid** (figure 9.6). Aaron Klug, who first described the solenoid, traced its formation by making electron micrographs of chromatin in solutions of increasing salt concentration (figure 9.7). At very low salt concentration, the chromatin appears as a string of nucleosomes,

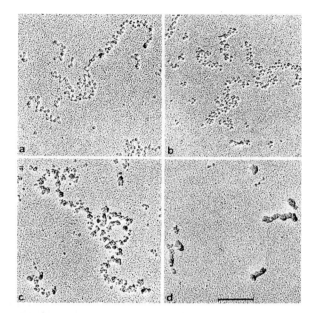

FIGURE 9.7   Formation of the solenoid. (*a*) A string of nucleosomes, extended due to very low ionic strength environment. (*b*) As the salt concentration is raised, the string begins to coil. (*c*) Still higher salt concentration, with more coiling. (*d*) At physiological salt concentration, the strings of nucleosomes have coiled all the way into tightly packed solenoids. The bar represents 200 nm.

From F. Thoma, T. Koller, A. Klug, *Journal of Cell Biology* 83:403–27, 1979. © Rockefeller University Press.

much like the SV40 minichromosome in figure 9.2. As the salt concentration rises, coiling takes place, until the typical solenoid structures appear. The DNA achieves another 6- to 7-fold condensation in coiling up into the solenoid. We know that histone H1 participates in this coiling, because coiling cannot occur in salt-treated chromatin that lacks H1.

Furthermore, it seems that H1-H1 interactions are important in coiling. We know that only one molecule of H1 is bound to each nucleosome. Therefore, since H1 interacts with other H1 molecules, these must be on *different* nucleosomes. This kind of interaction could easily *cause* chromatin folding; on the other hand, it could just as easily be a *result* of folding. However, other, more complex evidence suggests that the H1-H1 interactions are a cause, not an effect.

Further observations of chromatin suggest at least one more possible order of folding in the chromosome. First of all, it has been observed that eukaryotic chromosomes undergo supercoiling. We learned in chapter 8 that supercoiling can occur in bacterial and certain viral DNAs because the circular structure of these molecules prevents the release of strain. But eukaryotic chromosomes are linear, so how can the DNA in them be supercoiled? One possibility is that the chromatin fiber is looped and held fast at the base of each loop (figure 9.8). Each loop would be the functional equivalent of a circle, at least for supercoiling purposes, and the winding of DNA around the nucleosomes would provide the strain necessary for supercoiling.

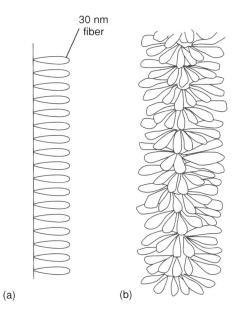

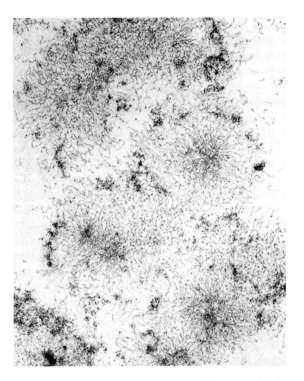

Radial loop models of chromatin folding. (*a*) This is only a partial model, showing some of the loops of chromatin attached to a central scaffold; of course, all the loops are part of the same continuous DNA molecule. (*b*) A more complete model, showing how the loops are arranged in three dimensions around the central scaffold.

Swollen chromosomes showing radial loops. The chromatids in the lower portion of the photograph are shown in cross section. The chromatid at the top is cut lengthwise.
From M. P. F. Marsden and U. K. Laemmli, *Cell* 17:851, 1979. © Cell Press.

Deproteinized chromosome showing looped structure. The scaffold is at the bottom, with loops of DNA extending upward. The bar represents 2 microns.
From James R. Paulson and U.K. Laemmli, *Cell* 12:823, 1977. © Cell Press.

several swollen chromosomes, further supports the notion of radial loops. Figure 9.11 summarizes our current thinking about chromatin folding, from the 2-nm bare DNA fiber to the postulated radial looped chromosome.

The second order of chromatin folding involves a coiling of the string of nucleosomes into a 25-nm-thick fiber called a solenoid. Histone H1 is thought to participate in this folding by interacting with other H1 molecules. The third order of folding is probably looping of the 25-nm fiber into a brushlike structure with the loops anchored to a central matrix.

CHROMATIN STRUCTURE AND GENE EXPRESSION

A strong relationship exists between the physical state of the chromatin in the vicinity of a particular gene and the activity of that gene. To begin with, there are two fundamental classes of chromatin structure: **heterochromatin** and **euchromatin.** Euchromatin is relatively extended and open. By contrast, heterochromatin is very condensed and its DNA is inaccessible. It even appears as clumps when viewed microscopically (figure 9.12). One example of heterochromatin is found at or near the centromeres of chromosomes. The DNA associated with the

The direct visualization of a human chromosome in figure 9.1 also supports the loop idea, although the DNA is so tightly bunched in that picture that the looping is difficult to see. Figure 9.9, also depicting a deproteinized chromosome, makes it very clear that loops begin and end at adjacent points on the chromosomal scaffold. Figure 9.10, showing cross sections through

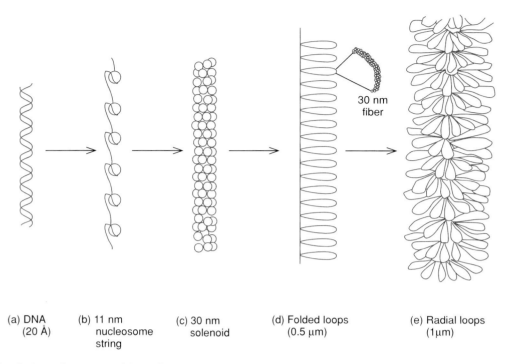

(a) DNA
(20 Å)

(b) 11 nm
nucleosome
string

(c) 30 nm
solenoid

(d) Folded loops
(0.5 μm)

(e) Radial loops
(1μm)

FIGURE 9.11   Orders of chromatin folding. Structures (a), (b), and (c) rest on solid experimental ground; (d) and (e), also shown in figure 9.8, are more speculative.

Reproduced, with permission, from the *Annual Review of Genetics*, Volume 12, © 1978 by Annual Reviews, Inc.

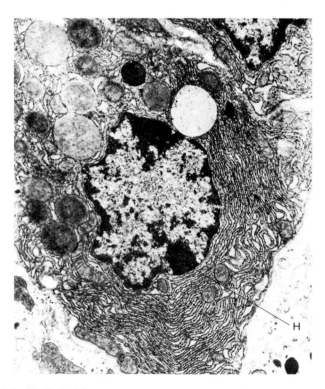

FIGURE 9.12   Interphase nucleus showing heterochromatin. Bat stomach lining cell with nucleus at center. Dark areas around periphery of nucleus are heterochromatin (H).

Courtesy Dr. Keith Porter.

centromeres belongs to a class of DNA that is repeated many times in the genome, in contrast to structural genes, which are usually found in only one or a few copies each. This **pericentromeric** DNA is complexed with protein and coiled tightly; obviously, DNA in such a state is not available to RNA polymerase and is not transcribed.

This does not imply that the relatively extended euchromatin is automatically transcribed, just because it does not physically exclude RNA polymerase. On the contrary, the majority of the euchromatin in a given cell is inactive. Why is this? That is an extremely important question; it is the same as asking what controls eukaryotic genes. In this chapter we will learn about the importance of general and gene-specific transcription factors. In chapter 14 we will examine this question in more detail.

Chromatin exists in eukaryotic cells in two forms: heterochromatin, which is condensed and transcriptionally inactive, and euchromatin, which is extended and at least potentially active.

RNA POLYMERASES AND THEIR ROLES

In chapter 8 we learned that prokaryotes have just one RNA polymerase (not counting primase), which makes all the different types of RNA the cell needs. True, the σ factor can change to meet the demands of a changing environment, but the core enzyme remains essentially the same. Furthermore, the normal holoenzyme can make all three major types of RNA: messenger RNA, ribosomal RNA, and transfer RNA. (The latter two classes of RNA are used in translation. We will discuss them in chapter 10.) Quite a different situation prevails in the eukaryotes. Robert Roeder and William Rutter showed in 1969 that eukaryotic cells have not one but three different RNA polymerases in their nuclei. Furthermore, these three enzymes have distinct roles in the cell; each makes a different kind of RNA.

It is intriguing that there are three different kinds of RNA polymerase and three different kinds of RNA to make: mRNA, rRNA, and tRNA. In the fourteenth century, William of Occam developed a maxim now called "Occam's razor," which states in effect that when you are confronted with several possible explanations for a phenomenon, you should choose the simplest. The simplest explanation for the existence of three kinds of polymerase and three kinds of RNA is that each polymerase makes one of the classes of RNA. In fact, there *is* such a one-to-one correspondence. In simplified form it goes like this: RNA polymerase I makes ribosomal RNA (the 18S, 28S, and 5.8S varieties); RNA polymerase II makes messenger RNA; RNA polymerase III makes transfer RNA.

Of course, there are some complicating factors. One of these is that the three kinds of RNA are not originally produced in mature form; instead, they are made as larger precursors that must later be cut down to size. So it is more correct to say, for example, that polymerase I makes an rRNA *precursor,* not rRNA itself. We have also glossed over the fact that there is a small 5S rRNA, in addition to the varieties made by RNA polymerase I. RNA polymerase III makes 5S rRNA, as well as a variety of other small RNAs with various cellular roles. Thus, RNA polymerase III specializes in synthesizing small RNAs. The roles of the three polymerases are summarized in table 9.1.

Eukaryotic cells contain three distinct RNA polymerases, named RNA polymerase I, RNA polymerase II, and RNA polymerase III. The roles of these three enzymes are as follows: Polymerase I makes a large rRNA precursor; polymerase II makes mRNA precursors; and polymerase III makes tRNA and 5S rRNA precursors.

Table 9.1   Roles of Eukaryotic RNA Polymerases

Polymerase	RNA Product
I	rRNA precursor (18S + 28S + 5.8S)
II	mRNA precursor
III	tRNA precursor; 5S rRNA precursor

PROMOTERS

Since there are three different RNA polymerases in eukaryotic nuclei, we would expect them to use three different kinds of promoters. We will see that this is almost true: Polymerases I and II recognize one kind of promoter each, but polymerase III uses two quite different classes of promoters. Moreover, eukaryotic RNA polymerases, unlike their prokaryotic counterparts, are incapable by themselves of binding to their respective promoters. Instead, they rely on proteins called **transcription factors** to show them the way. There are two classes of such factors: **general transcription factors** and **gene-specific transcription factors,** also called **gene activators.** The general transcription factors bind to DNA regions within promoters. By themselves, they can deliver the RNA polymerases to their respective promoters and thereby cause transcription. However, such transcription tends to occur at a low, or basal level. Furthermore, general transcription factors and the three polymerases alone allow for only minimal transcription control, whereas cells really exert exquisitely fine control over transcription. This fine control is especially pronounced in genes transcribed by polymerase II.

The gene-specific transcription factors have two roles: First, they usually boost the transcription rate beyond the basal level achieved with the general transcription factors, sometimes to spectacularly high levels; on the other hand, in certain circumstances they can actually inhibit transcription. Second, they allow for fine control of transcription. They perform these functions by binding to DNA sequences called **enhancers** that are usually distinct from the promoter itself. In this section we will survey the three kinds of promoters recognized by the three RNA polymerases. Then we will discuss the two classes of transcription factors that help RNA polymerases use these promoters. We will also examine some enhancers, the DNA targets of gene activators, and **silencers,** DNA elements that interact with transcription factors to inhibit transcription.

PROMOTERS RECOGNIZED BY POLYMERASE II

We start with the promoters recognized by polymerase II because they will be most familiar to you; they bear the closest resemblance to prokaryotic promoters. Moreover, polymerase II is responsible for transcribing the vast majority of genes—those that code for proteins. In general, these polymerase II promoters have more than one element.

The TATA Box

Closest to the transcription start site, approximately at position −30, is an A–T-rich sequence that is a very common feature of polymerase II promoters. Its consensus sequence is TATAAAA, although T's frequently replace the A's in the fifth and seventh positions. Its name, **TATA box,** is derived from its first four bases.

You may have noticed the close similarity between the eukaryotic TATA box and the prokaryotic −10 box. But notice that it is found farther upstream than its prokaryotic counterparts; −30 compared to −10. As is the case with all consensus sequences, there are exceptions to the rule. Sometimes G's and C's creep in, as in the "TATA" box of the rabbit β-globin gene, which starts with a C (CATA). Sometimes there is no recognizable TATA box at all.

Another conserved region in polymerase II promoters is not as prevalent as the TATA box. It is usually found about fifty base pairs farther upstream, approximately at position −75, and it has the consensus sequence GGCCAATCT. This sequence commonly goes by the delightful name **CCAAT box** (pronounced "cat box").

It is one thing to find conserved sequences upstream from the transcription start site; it is another to demonstrate that they actually function as parts of the promoter. The latter task has been accomplished for both the TATA box and the CCAAT box, and cloned genes have figured prominently in these studies. The experimental plan has been to clone (reproduce in large quantity) a particular gene transcribed by polymerase II, introduce mutations into its promoter, and then observe the effects of these mutations on transcription of the gene. These mutations differ from those that occur naturally at random in that we can choose just which bases we want to change and how. Thus, we call this process "site-directed mutagenesis." We will discuss this topic in greater detail in chapter 15.

To test for the effect of a mutation in the TATA box, we have to be able to add our cloned genes (mutant or wild-type) to some kind of transcription system so that we can observe the

difference in the way these genes are transcribed. The two basic types of transcription systems we use, in vitro and in vivo, are shown in figure 9.13. To make an in vitro system, we start by grinding up the cells, then slightly purifying their contents to yield a cell-free extract. This extract usually contains RNA polymerases and all the other factors needed to transcribe the DNAs that we add, as long as these DNAs contain promoters the extract can recognize. In vivo systems come in several types, but all are living cells to which we can add DNAs. We then check by one means or another to see if our added DNA is transcribed. (Both terms are derived from Latin: *vitro* = glass; *vivo* = life.)

When such tests were first applied to genes with mutations in their TATA boxes, something surprising happened. Instead of eliminating transcription, as one might have predicted, TATA box mutations in these genes caused an increase in the number of different transcription start sites. Figure 9.14 shows the results Pierre Chambon and his colleagues got when they tested TATA box mutants of the early promoter of the monkey tumor virus **SV40,** using an in vitro transcription system. Whereas the wild-type DNA gave transcription from three tightly clustered start sites, the TATA-less mutant showed starts at more than twenty different sites.

On the other hand, mutating the TATA box in certain other genes (the rabbit β-globin gene, for example) did not change the transcription initiation site. Instead, it caused a profound decrease in efficiency of transcription. Thus, the TATA box seems to play different roles in different promoters.

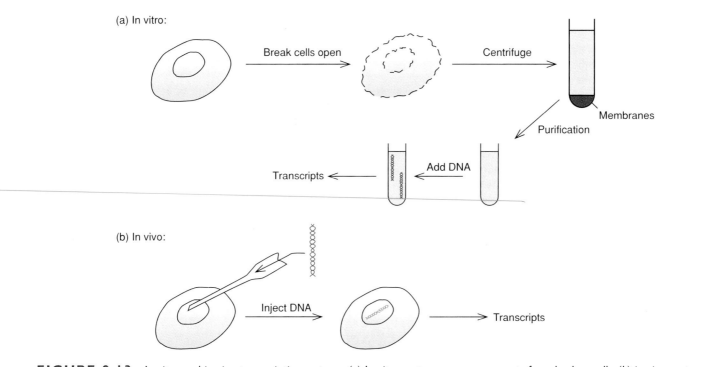

FIGURE 9.13   In vitro and in vivo transcription systems. (*a*) In vitro systems use components from broken cells; (*b*) in vivo systems use whole cells. In both cases, we introduce foreign DNA (red) into the system, and measure its transcription.

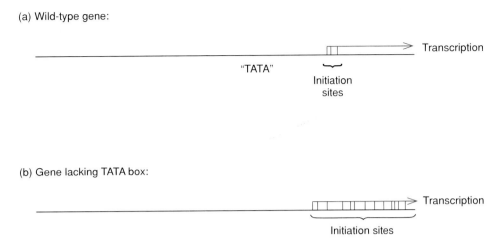

FIGURE 9.14   Effect of TATA box on transcription initiation. (*a*) In the wild-type SV40 early gene that contains a TATA box, transcription starts from three tightly clustered sites. (*b*) In the mutant that lacks a TATA box, we observe a multitude of transcription initiation sites.

Upstream Promoter Elements

Similar studies have been performed in which sequences upstream from the TATA boxes in a variety of genes have been deleted or changed. The CCAAT box, for example, seems to influence promoter activity. Thus, removing the CCAAT box from the rabbit β-globin promoter destroys most of that promoter's activity. The CCAAT box plays a similar role in the promoter for the thymidine kinase gene of herpes simplex virus (HSV). The SV40 early promoter also contains an important upstream element; instead of a CCAAT box, it is a 21-base-pair sequence that is repeated three times between positions −40 and −103 just upstream from the TATA box. Each of these 21-bp repeats contains two copies of a **GC box** with the sequence GGGCGG on the nontemplate strand. The HSV thymidine kinase promoter also contains GC boxes: one on each side of the CCAAT box. HSVs, by the way, cause cold sores and genital herpes infections in humans.

As a general rule then, promoters recognized by RNA polymerase II contain a TATA box near the transcription start site and at least one other important element farther upstream. We will see later in this chapter that these upstream elements serve as binding sites for transcription factors that activate the attached genes.

Many eukaryotic genes contain sequences farther upstream in the flanking regions (or even within the genes) that are crucial to active transcription. But these are not strictly part of the promoters. We call them enhancers, and we will discuss them in detail later in the chapter. Figure 9.15*a* and *b* show the arrangement of the promoter elements for the SV40 early and HSV thymidine kinase genes. Figure 9.15*c* presents a generalized view of the control region of a typical gene recognized by RNA polymerase II, including both promoter and enhancer. This is the arrangement of these elements in the SV40 early region—a favorite example of a polymerase II promoter-enhancer complex.

Most eukaryotic promoters that are recognized by RNA polymerase II have at least one feature in common: a consensus sequence found about thirty base pairs upstream from the transcription start site and containing the motif TATA. This TATA box usually acts to position the start of transcription about thirty base pairs downstream. Most polymerase II promoters also contain important sequences upstream from the TATA box. Examples are the CCAAT box and the GC box.

PROMOTERS RECOGNIZED BY POLYMERASE I

Several groups of investigators have searched for conserved sequences and have applied site-directed mutagenesis techniques to elucidate the nature of the promoter recognized by RNA polymerase I. Promoter is written here in the singular because each species should have only one kind of gene recognized by polymerase I: the rRNA precursor gene. It is true that this gene is present in hundreds of copies in each cell, but each copy is virtually the same as the others, and they all have the same promoter sequence. Robert Tjian (pronounced "Tee-jun" by his American colleagues) and his coworkers identified two important regions of the human rRNA promoter by mutagenesis (figure 9.16). By replacing 10 bp at a time with unrelated DNA and observing the effects of these replacements on transcription efficiency, they found a **core element** between positions −45 and +20, and an **upstream control element (UCE)** between positions −156 and −107 that were required for efficient transcription. These DNA elements resemble those of polymerase II promoters only in that they appear in the 5′-flanking region of the gene (that is, upstream from the start of transcription); their base sequences do not remind us at all of TATA, GC, or CCAAT boxes. Notice that the end of the gene where transcription

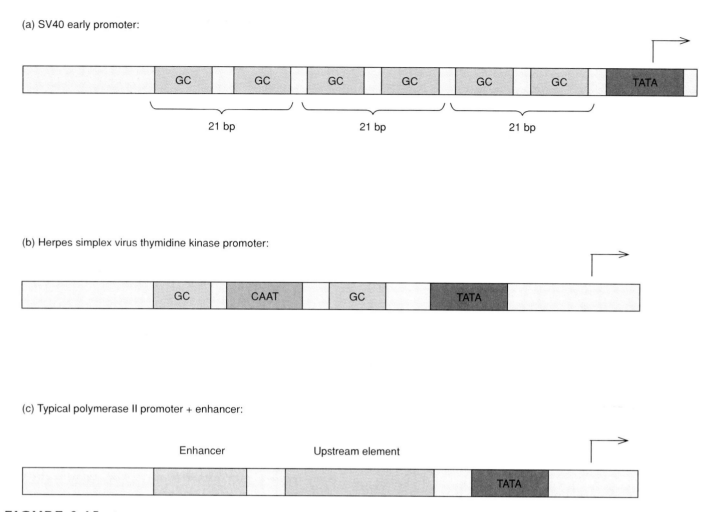

FIGURE 9.15 Structures of promoters recognized by RNA polymerase II. (*a*) The SV40 early promoter contains a TATA box (red) and an upstream element composed of three 21-bp repeats. Each of these repeats contains two 6-base GC boxes (green). (*b*) The HSV thymidine kinase promoter contains a TATA box (red) plus three upstream elements: two GC boxes (green) and a CCAAT box (purple). (*c*) A typical polymerase II promoter has a TATA box and at least one upstream element. It is also usually associated with an enhancer (blue), which generally lies upstream from the promoter.

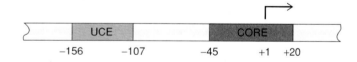

FIGURE 9.16 The human rRNA promoter. Two DNA elements, UCE (green) and core (red), are important for efficient transcription by RNA polymerase I.

begins is called the 5'-end. This refers to the 5'-end of the transcript, or of the nontemplate DNA strand.

> Promoters recognized by polymerase I, like polymerase II promoters, are found in the 5'-flanking region of the gene (the rRNA precursor gene). Two DNA blocks—a core sequence near the transcription start site and an upstream control element (UCE)—are required for efficient transcription.

PROMOTERS RECOGNIZED BY POLYMERASE III

The polymerase III promoters that govern the 5S rRNA and tRNA genes are radically different from those recognized by polymerases I and II and by bacterial polymerases. Instead of being located in the 5'-flanking region of the gene, as all these other promoters are, these polymerase III promoters have elements located *within* the genes they control.

Donald Brown and his coworkers performed the first experiments leading to this astonishing conclusion using a cloned 5S rRNA gene from *Xenopus*. The experimental plan was to mutate the 5S rRNA gene and measure the effects of these mutations on transcription of the gene in vitro. Correct transcription was scored by measuring the size of the transcripts by gel electrophoresis. An RNA of approximately 120 bases (the size of 5S RNA) was deemed an accurate transcript, even if it did not have the same sequence as real 5S rRNA. We have to allow for incorrect sequence in the transcript for the following reason: By changing the internal sequence of the gene in

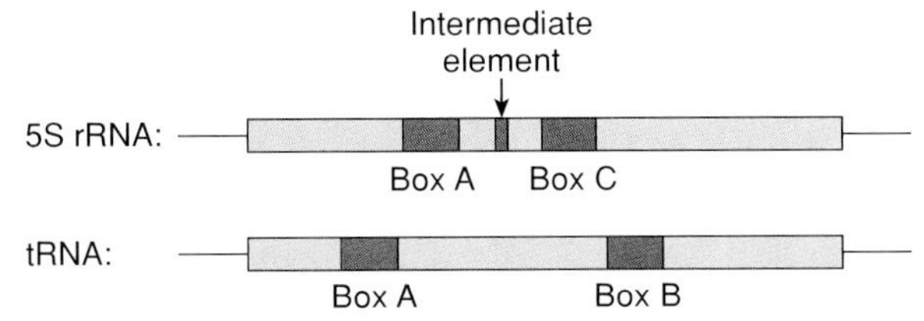

FIGURE 9.17 Promoters of two "classical" polymerase III genes. The promoters, or ICRs, are depicted as red boxes within the genes they control.

order to disrupt the promoter, we automatically change the sequence of the transcript.

The surprising result was that the entire 5′-flanking region of the gene could be removed without affecting transcription very much. Furthermore, big chunks of the gene could be removed, or extra DNA could be inserted into the gene, and a transcript of about 120 bases would still be made. However, there was one sensitive spot in the gene where one could not remove or insert DNA and still retain promoter function. This was a region between bases 50 and 83 of the transcribed sequence. This is the location of the promoter, or **internal control region (ICR)** of the *Xenopus* 5S gene.

More detailed analysis has shown that the 5S rRNA promoter is actually split into two major elements, called **box A** and **box C,** as well as a short sequence in the middle, called the **intermediate element.** Figure 9.17 summarizes the results of these experiments and similar ones performed on tRNA genes. The tRNA promoter is more obviously split than the 5S rRNA promoter. Like the 5S rRNA ICR, it has a box A, but it has no intermediate element or box C. Instead, it has a **box B.** The space in between these two elements can be altered without destroying promoter function, but there are limits to such alteration; if one inserts too much DNA between boxes A and B, the efficiency of transcription suffers.

The 5S rRNA and tRNA genes are "classical" polymerase III genes. Polymerase III also transcribes genes encoding other small RNAs, including one called 7SL RNA. Strangely enough, these "nonclassical" genes do not have ICRs. Instead, their promoters lie in the 5′-flanking regions of the genes, and at least one even has a TATA box.

RNA polymerase III transcribes a set of short genes, including those for 5S rRNA and tRNA. The promoters for these classical polymerase III genes consist of internal control regions (ICRs) that lie wholly within the genes. The 5S RNA gene has a three-part ICR lying between bases 50 and 83 of the transcribed region. The tRNA genes have two-part, divided ICRs. Nonclassical polymerase III genes, such as the 7SL RNA gene, have promoters in their 5′-flanking regions.

REGULATION OF TRANSCRIPTION

In chapter 8 we learned that control of gene expression in bacteria is a prerequisite for life; if all genes were switched on all the time, this would consume far more energy than the bacterial cell could afford. If anything, this statement is even more true of eukaryotes. To begin with, there are many more genes in a eukaryotic cell. Moreover, in higher eukaryotes with their complex organs, cells are much more specialized. As we will see in chapter 14 the cells of different tissues have different capabilities *because* different genes are expressed in each. Thus, complex organisms would not be possible if gene expression were uncontrolled.

As we have seen, bacteria control their gene expression primarily at the level of transcription. In general, if a cell does not need the product of a gene, it will not even transcribe that gene. Although there are exceptions, this generalization also holds true for eukaryotes. This means that the initiation of transcription is the most important control point in the expression of most eukaryotic genes. It is vital to know, therefore, what causes transcription initiation. This section will show that many promoters are associated with distinct *cis*-acting DNA regions called enhancers that can greatly stimulate transcription, and some are linked to negative control elements called silencers. Moreover, transcription of eukaryotic genes depends not only on RNA polymerase but on a collection of *trans*-acting proteins called transcription factors that interact with promoters and enhancers.

ENHANCERS AND SILENCERS

The early site-directed mutagenesis experiments on polymerase II promoters also pointed to very important areas of the 5′-flanking regions, aside from the TATA box and upstream promoter elements we have discussed. One of the best studied of these upstream sequences is a **72-base-pair repeat** in the early region of SV40 virus. The early region of SV40 codes for a protein called **T antigen** that is needed for viral DNA synthesis and also allows SV40 virus to transform normal cells to malignant ones. In other words, the early region is what makes SV40 a tumor virus. The 72-bp repeat is located just upstream from the three 21-bp repeats in the 5′-flanking region of the SV40 early gene.

The 72-bp repeat is just what its name implies: a side-by-side duplication of a sequence 72 base pairs long. When this element was deleted from the SV40 early region, transcription was greatly depressed. This is exactly the behavior we would expect from part of a promoter, so why did we not include the 72-bp repeat in our discussion of polymerase II promoters? Because this DNA element exhibits two characteristics that are very unpromoterlike: First, as figure 9.18 illustrates, it can be moved to a new location thousands of base pairs away from the SV40 early promoter (either in front of it or behind it) and still enhance early transcription. Second, it can be inverted without diminishing its effect. For this reason, we classify the 72-bp repeat and other elements with the same characteristics as

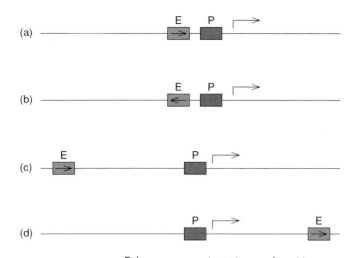

Enhancers are orientation- and position-independent. (*a*) The wild-type relationship between a typical enhancer (blue) and its associated promoter (red). (*b*) The enhancer has been inverted. (*c*) The enhancer has been moved a few thousand base pairs upstream from the promoter. (*d*) The enhancer has been moved a few thousand base pairs downstream from the promoter. In all four cases, the enhancer functions normally.

enhancers, which stimulate transcription from promoters but are not really part of the promoters themselves. The term "promoter" is reserved for elements that have a relatively fixed spatial relationship with the genes they control.

Enhancers have been found to be associated with a variety of genes—in eukaryotes as well as their viruses and in genes transcribed by all three RNA polymerases. They are usually tissue-specific; that is, a gene with a given enhancer can function only in certain kinds of cells. For example, a mouse immunoglobulin enhancer can function in cells that express the immunoglobulin genes (the cells of the immune system), but it does not function in mouse connective tissue cells. This tissue specificity derives from the fact that enhancers depend on transcription factors, and these factors are made in some tissues, but not others. Sometimes enhancers have sequences in common, but there are no real consensus sequences shared by all enhancers.

Sometimes transcription factors can repress, rather than stimulate, transcription. For example, the human thyroid hormone receptor stimulates transcription in the presence of thyroid hormone. However, in the absence of the hormone it binds to the same enhancer and represses transcription.

Enhancers are not the only DNA elements that can act at a distance to control transcription of a gene. **Silencers** also do this, and—as their name implies—they prohibit transcription. The mating system (*MAT*) of yeast provides a good example. The yeast chromosome III contains three loci of very similar sequence: *MAT, HML,* and *HMR.* Though *MAT* is expressed, the other two loci are not, and silencers located at least 1 kb away seem to be responsible for this genetic inactivity. We know that something besides the inactive genes themselves is at fault, because active yeast genes can be substituted for *HML* or *HMR* and the transplanted genes become inactive. Thus, they seem to be responding to some external negative influence. How do silencers work? It is

clear that they depend on **silencer proteins** that bind to the silencer DNA and somehow cause repression of surrounding genes. The mechanism of this repression is still unknown.

Enhancers are DNA elements that can greatly stimulate transcription from the eukaryotic promoters with which they associate. They differ from promoter elements in that they work in either orientation and can operate at some distance from the genes they control. Silencers also work at a distance, but they inhibit transcription.

RNA POLYMERASE II TRANSCRIPTION FACTORS

The great variety of promoters recognized by RNA polymerase II suggests that there should be a similar variety of polymerase II transcription factors, or "class II factors," and indeed there is. These can be separated into two groups. The first group includes the general transcription factors that are involved in transcription of most, and probably all, polymerase II genes. Because they seem to have such broad specificity, we would predict that they interact with something found in most polymerase II promoters (the TATA box), or with other factors, or with polymerase II itself. Indeed, we will soon see that this prediction has proven true. The class II general transcription factors that cooperate with polymerase II are **TFIIA, TFIIB, TFIID, TFIIE, TFIIF,** and **TFIIH.** (The "TFII" in these names stands for "**T**ranscription **F**actor for polymerase **II.**")

The best studied of these factors is TFIID. In higher eukaryotes, including mammals, this factor contains many polypeptide subunits, but the most important is the **TATA box binding protein,** or **TBP.** The other subunits in TFIID are known as **TBP-associated factors,** or **TAFs** (pronounced "taffs"). TBP does what its name suggests: it binds to the TATA box, bringing the rest of TFIID (the TAFs) along with it. Once bound to the promoter, TFIID serves as a nucleus around which all the other factors organize at the promoter. The factors bind in this order: D, B, F + polymerase II, E, and H. TFIIA can apparently bind anytime. Polymerase II and TFIIF bind simultaneously; in fact, it is the job of TFIIF to deliver polymerase II to the growing complex. The promoter with all factors plus RNA polymerase II bound is called a **preinitiation complex** because it is primed to initiate transcription. Figure 9.19 shows a hypothetical scheme of assembly of the class II preinitiation complex.

What about promoters that lack a TATA box? TFIID, including TBP, is still essential for transcription from such promoters, and apparently still performs the same organization function, but it requires extra factors to help it bind to DNA without a TATA box.

In addition to these general transcription factors, we find a diverse group of gene-specific factors. Some of these bind to sequences such as GC boxes and CCAAT boxes that are present in many mammalian promoters. Others bind to more specific elements, usually enhancers, that are associated with smaller

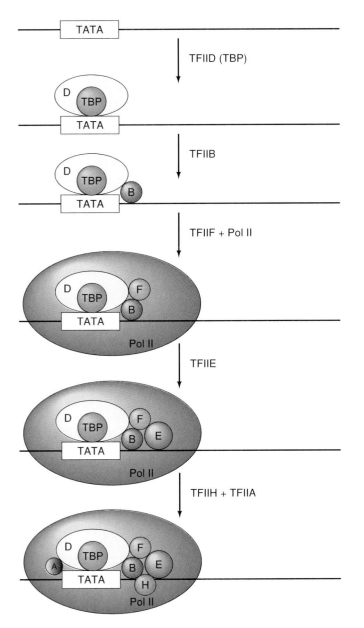

Assembly of a class II preinitiation complex. The sizes of the factors are not necessarily to scale, and their arrangement is hypothetical. TFIIA may bind earlier in the assembly process.

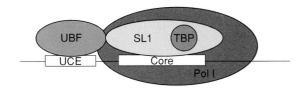

The class I preinitiation complex. The two parts of the rRNA promoter (UCE and Core) are represented by white boxes and are labeled at the bottom.

RNA POLYMERASE I TRANSCRIPTION FACTORS

The preinitiation complex that forms at rRNA promoters is much simpler than the polymerase II preinitiation complex we have just discussed. It involves polymerase I, of course, in addition to just two transcription factors, called **SL1** and **upstream binding factor (UBF).** Robert Tjian discovered SL1 in 1985, and showed that it does not bind by itself to the rRNA promoter. Later experiments demonstrated that purified polymerase I also cannot bind by itself to the promoter, even with the help of SL1. Finally, in 1988 Tjian and coworkers showed that another factor, UBF, binds to the upstream control element (UCE) of the rRNA promoter and causes polymerase I to bind near the transcription start site. Once UBF binds, SL1 can join the complex, strengthening polymerase I binding, and therefore promoting formation of the preinitiation complex. Interestingly enough, SL1 is similar to TFIID in that it contains TATA box binding protein (TBP), even though rRNA promoters do not have TATA boxes. In addition to TBP, SL1 has three TAFs, which are different from the TAFs associated with TBP in TFIID. Figure 9.20 shows the class I preinitiation complex.

> The class I preinitiation complex contains RNA polymerase I plus two transcription factors. The upstream binding factor (UBF) binds to the UCE and provokes binding of polymerase I to the core element near the start of transcription. SL1, a TBP-containing factor, strengthens polymerase I binding to the promoter.

groups of class II genes. We will return to these specific factors later in this chapter.

> RNA polymerase II is assisted by at least six general transcription factors, TFIIA, B, D, E, F, and H, as well as by gene-specific transcription factors, most of which bind to enhancers associated with certain genes. TFIID contains the TATA box binding protein (TBP), which binds to the promoter's TATA box and serves as an organizing center for the formation of the preinitiation complex.

RNA Polymerase III Transcription Factors

Donald Brown and Robert Roeder and their colleagues, working independently, have supplied most of our knowledge of the factors that interact with RNA polymerase III to effect the transcription of the 5S rRNA and tRNA genes. Three factors are involved in 5S rRNA transcription: TFIIIA, TFIIIB, and TFIIIC. Two factors—TFIIIB and TFIIIC—participate in tRNA gene transcription. It now appears that TFIIIB is the true transcription initiation factor for RNA polymerase III; the other two are assembly factors that allow TFIIIB to bind to its DNA site, which actually lies upstream from the transcription initiation site. TFIIIB then helps RNA polymerase III bind just

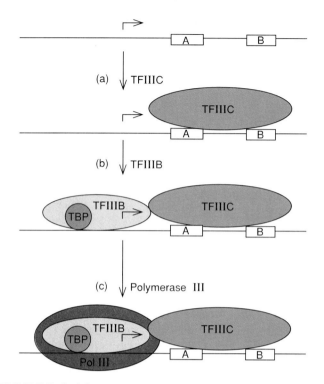

FIGURE 9.21 Hypothetical scheme for assembly of the preinitiation complex on a classical polymerase III promoter (tRNA). (*a*) TFIIIC (green) binds to the internal promoter's A and B boxes. (*b*) TFIIIC promotes binding of TFIIIB (yellow), with its TBP (blue) to the region upstream of the transcription start site. (*c*) TFIIIB promotes polymerase III (red) binding at the start site, ready to begin transcribing.

downstream, in a position to start transcribing the gene. Thus, these three factors participate in a binding cascade.

First, TFIIIA and C (in the case of the 5S rRNA genes) or TFIIIC alone (in the case of the tRNA genes) bind to the internal promoter and, by protein–protein interaction, allow TFIIIB to bind near the transcription start site. TFIIIB then positions RNA polymerase III at the transcription start site. Figure 9.21 illustrates this sequence of events with a tRNA promoter. This looks similar to the class I picture we saw in figure 9.20. Here, however, TFIIIB plays the same role that SL1 seems to play in RNA polymerase I binding, and TFIIIA and C, or C alone, play the UBF role. It is not surprising that TFIIIB and SL1 act in similar ways, because TFIIIB, like SL1, contains TBP. Like class I promoters, however, classical class III promoters lack TATA boxes.

Now we have seen examples of TATA-less promoters recognized by all three classes of RNA polymerase. In every case, a TBP-containing protein is involved: SL1, TFIID, and TFIIIB, respectively. Since all three of these factors contain the same TBP, yet recognize three very different classes of promoters, the other proteins in these factors—the TAFs—must provide the specificity. Also, in every case, assembly factors are required. These are UBF (polymerase I) and TFIIIA and C, or just TFIIIC (polymerase III). There appear to be several assembly factors for TATA-less class II promoters. We will discuss one of these, Sp1, later in this chapter.

Classical class III promoters rely on two kinds of general transcription factors: assembly factors, TFIIIA and C (5S rRNA gene) or TFIIIC alone (tRNA genes); and an initiation factor, TFIIIB, that actually positions polymerase III for transcription.

GENE-SPECIFIC FACTORS FOR POLYMERASE II

In addition to the general transcription factors, we find a diverse group of gene-specific factors. Some of these bind to sequences such as GC boxes and CCAAT boxes that are present in many mammalian promoters. Therefore, these factors are found in a wide variety of cells. Other factors interact with enhancers associated with smaller sets of genes.

TRANSCRIPTION FACTORS THAT INTERACT WITH PROMOTER ELEMENTS

Robert Tjian has purified a protein factor called **Sp1** from human tumor cells (HeLa cells, Box 9.1) that greatly stimulates transcription from the SV40 early promoter and from the HSV thymidine kinase promoter. At the same time, this factor actually seems to decrease transcription from the major late promoter of another human virus called **adenovirus.** Thus, this factor has the ability to select certain promoters for transcription while ignoring others.

The key to Sp1's activity is its ability to bind to the GC boxes in the 5′-region of some promoters and to stimulate transcription from those promoters. As we discussed earlier, the GC boxes in the SV40 early promoter are arranged in the 21-bp repeats between the gene's TATA box and enhancer (see figure 9.15). The spacing between GC boxes is almost exactly two full turns of the DNA double helix, which puts the bound Sp1 molecules all on the same face of the DNA molecule. Sp1 can bind to each of the GC boxes individually, but because some of the sites are a little closer together than others, only five of them are effectively occupied.

The CCAAT boxes in the HSV thymidine kinase promoter and in other promoters interact with their own transcription factor, called the **CCAAT box transcription factor (CTF).** Figure 9.22 depicts the interactions between two well-studied polymerase II promoters (SV40 early and herpes simplex virus [HSV1] thymidine kinase) and their respective transcription factors.

TRANSCRIPTION FACTORS THAT INTERACT WITH ENHANCERS

Enhancer-binding transcription factors are more specific than Sp1 or CTF, since the number of genes with a given enhancer is smaller than those with GC or CCAAT boxes. Nevertheless, many enhancers are found near a wide variety of genes, and so

When people see the term *HeLa cells* for the first time, they may think it is a misprint. Shouldn't the *L* be lowercase? No, HeLa is correct; the reason for capitalizing the *H* and the *L* is that the term stands for the name of the woman who donated the cells: Henrietta Lacks. In 1952, George Gey was trying to grow human tumor cells in culture at Johns Hopkins University. No one had succeeded in this enterprise before, but Gey managed to get human cervical cancer cells to grow in a serum clot. One of these specimens came from Henrietta Lacks, so Gey called the culture HeLa (pronounced hee'la) cells. Lacks died of her cancer not long afterward, but she has achieved a kind of immortality in that her cells are still growing in laboratories all over the world. They are one of the favorite sources of human cells for molecular genetics investigations. In fact, they grow so easily—almost like weeds— that many researchers have cultured them inadvertently. That is, their cultures of other cells became contaminated with HeLa cells, which then outgrew their neighbors until only HeLa cells remained. This change sometimes went undetected, and several studies, ostensibly on the behavior of other kinds of cells, have actually been investigations of already well-studied HeLa cells.

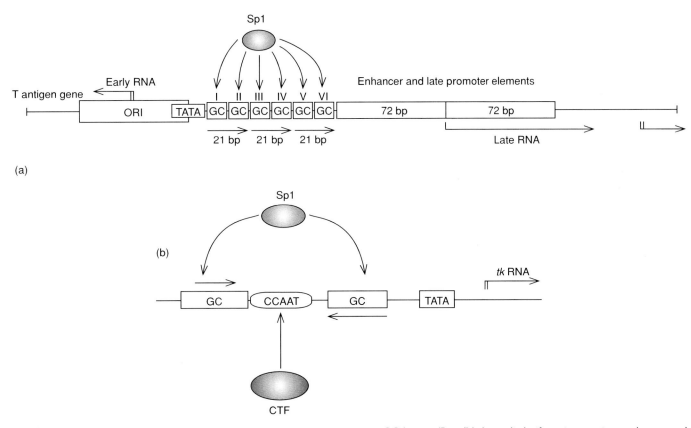

FIGURE 9.22 Interactions of transcription factors with two promoters. (*a*) The whole SV40 control region, including both late and early promoters. Note that the early promoter is in the opposite orientation from that in figure 9.14. The drawing also depicts the interaction between the transcription factor Sp1 and the six GC boxes. (Box IV shows little if any interaction under normal conditions.) (*b*) The HSV thymidine kinase promoter elements and their interaction with the transcription factors Sp1 and CTF (CCAAT box transcription factor).

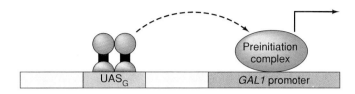

FIGURE 9.23 Action of GAL4. The transcription activator GAL4 binds to its target site, UAS$_G$, upstream from the promoter of one of the *GAL* genes, *GAL1* in this example. This stimulates assembly of the preinitiation complex (RNA polymerase plus general transcription factors) and therefore stimulates transcription.

the gene-specific transcription factors that bind to them are capable of controlling many genes. Let us look at a few examples.

The **GAL4** protein is one of the prototypes of a gene-specific transcription factor, or enhancer-binding protein. It is a yeast activator that controls a set of genes responsible for metabolism of galactose. Each of these GAL4-responsive genes contains a GAL4 target site upstream from the transcription start site. Such a target site is called a **UAS$_G$**, which stands for "**U**pstream **A**ctivating **S**equence for *GAL* genes." GAL4 binds to a UAS$_G$ as a dimer and activates transcription from the promoter downstream, as illustrated in figure 9.23.

The **nuclear receptors** form another important group of enhancer-binding proteins. These proteins interact with a variety of steroid and related hormones that diffuse through the cell membrane. They form hormone-receptor complexes that function as activators by binding to enhancers called **hormone response elements,** and stimulating transcription of their associated genes. Some of the hormones that work this way are the sex hormones (androgens and estrogens); progesterone, the hormone of pregnancy and the principal ingredient of common birth control pills; the glucocorticoids, such as cortisol; the mineralocorticoids, which regulate kidney function; vitamin D, which regulates calcium metabolism; and thyroid hormone and retinoic acid, which regulate gene expression during development. Each hormone binds to its specific receptor and together they activate their own set of genes. We will discuss this topic in more detail in chapter 14. Some nuclear receptors (e.g., the glucocorticoid receptor) exist in the cytoplasm complexed with a protein called heat shock protein 90 (hsp90). When they meet their ligands, or hormones, in the cytoplasm, they dissociate from hsp90, change conformation, and bind to the hormone. This hormone-receptor complex can then move into the nucleus to perform its function (figure 9.24). Other receptors (e.g., the thyroid hormone receptor) spend all their time in the nucleus. In the absence of hormone, they bind to their respective enhancers and repress transcription. In the presence of hormone, they form hormone-receptor complexes in the nucleus, then bind to the same enhancers and function as gene activators.

STRUCTURE OF GENE-SPECIFIC TRANSCRIPTION FACTORS

Many transcription factors, including enhancer binding proteins, share a structure composed of three functional domains: a DNA-binding domain, a transcription activation domain, and a dimerization (or oligomerization) domain. We can discuss the DNA-binding and transcription activating domains separately, but we will combine our discussion of DNA-binding and dimerization domains.

1. *DNA-binding domains.* Most of these structural motifs fall into the following classes.

   a. *Zinc fingers.* Zinc fingers were discovered first in TFIIIA and later in Sp1. By now, many other transcription factors have been found to have zinc fingers. Structural studies of Sp1 revealed a repeating domain that was predicted to be finger-shaped (figure 9.25*a*). Because each of these domains binds a zinc ion, they were called **zinc fingers.** Figure 9.25*b* shows the actual three-dimensional structure of one of the zinc fingers from the *Xenopus* protein *Xfin,* an activator of polymerase II promoters. Note that this structure really is finger-shaped, as the fingers in TFIIIA and Sp1 were predicted to be. It is also worth noting that this finger shape by itself does not confer any binding specificity, since there are many different finger proteins, all with the same shape fingers but each binding to its own unique DNA target sequence. Thus, it is the precise amino acid sequence of one or more of the fingers, or of neighboring parts of the protein, that determines the DNA sequence to which the protein will bind.

   How do the fingers interact with their DNA targets? Carl Pabo and his colleagues have used X-ray crystallography to obtain the structure of the complex between DNA and the mouse zinc finger protein Zif268. This is the product of a so-called **immediate early gene,** which is one of the first genes to be activated when resting mammalian cells are stimulated to divide. The Zif268 protein has three adjacent zinc fingers that fit into the major groove of the DNA double helix. Figure 9.26 presents the structure of finger 1 as an example. This presentation is perhaps not so obviously finger-shaped as in figure 9.25. Still, upon close inspection we can see the finger contour, which is indicated by the dashed line. Notice that, in contrast to figure 9.25, this finger is represented pointing down. As in the Xfin zinc finger, the right side of each Zif268 finger is an α-helix. This is connected by a short loop at the bottom to the left side of the finger, an antiparallel β-ribbon. Do not confuse this β-ribbon itself with the whole finger; it is only one-half of it. The zinc ion is in the middle, coordinated by two histidines in the α-helix and by two cysteines in the β-ribbon.

   All three fingers of Zif268 have almost exactly the same shape. Figure 9.27 shows the three fingers lining up in the major groove of the target DNA. In fact, the three fingers are arranged in a curve, or C shape, which matches the curve of the DNA double helix. All the fingers approach the DNA from essentially the same

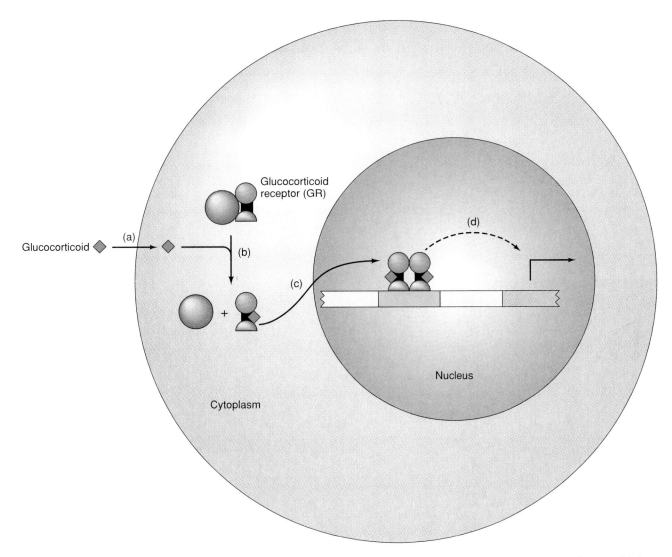

FIGURE 9.24   Glucocorticoid action. The glucocorticoid receptor (GR) is typical of a group of nuclear receptors that exist in an inactive form in the cytoplasm complexed with heat shock protein 90 (hsp90). (*a*) The glucocorticoid (blue diamond) diffuses across the cell membrane and enters the cytoplasm. (*b*) The glucocorticoid binds to its receptor (GR, red and green), which changes conformation and dissociates from hsp90 (orange). (*c*) The hormone-receptor complex (HR) enters the nucleus, dimerizes with another HR, and binds to a hormone-response element, or enhancer (pink) upstream from a hormone-activated gene (brown). (*d*) Binding of the HR dimer to the enhancer activates (dashed arrow) the associated gene, so transcription occurs (bent arrow).

angle, so the geometry of protein–DNA contact is very similar in each case. Most of these contacts occur between amino acids in the α-helix of the finger and bases in the major groove of the DNA. This is another example of the utility of an α-helix as a DNA-recognition element (see chapter 8).

Steroid hormone receptors also have zinc-containing domains, but they have four cysteines (and no histidines) coordinated to the zinc atom. This causes these zinc domains to have a somewhat different shape than that of the classic zinc fingers. The estrogen and progesterone binding proteins bind to slightly different target sequences. These different specificities result from different amino acids at the bases of the zinc domains in the two proteins.

b. *Homeodomains (HDs).* **Homeodomains** containing about 60 amino acids were discovered in DNA-binding proteins that regulate development in the fruit fly (chapter 14). HDs are found in a variety of transcription factors, including Pit-1, OCT-1 (and -2), and Unc-86. In these three classes of factors, the HD is paired with another motif, called a *POU box,* and the two together are known as the *POU domain.* The acronym comes from the first letters of Pit-1, OCT-2, and Unc-86 and is pronounced "pow." The HDs and POU boxes strongly resemble the helix-turn-helix domains of prokaryotic DNA-binding proteins such as the λ phage repressor (chapter 8). This similarity extends to the mode of DNA binding, in which a recognition helix of the protein contacts the major groove of the DNA target.

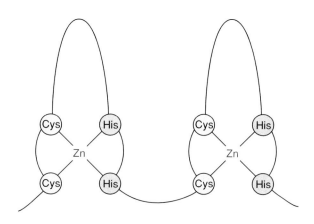

(a)

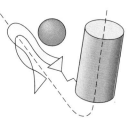

FIGURE 9.26 Schematic diagram of zinc finger 1 of the Zif 268 protein. The left-hand side of the finger is an antiparallel β-ribbon (yellow), and the right-hand side is an α-helix (red). Two cysteines in the β-ribbon and two histidines in the α-helix coordinate the zinc ion in the middle (blue). The dashed line traces the outline of the "finger" shape.

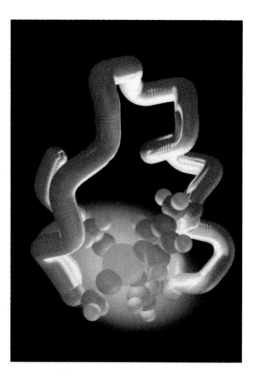

(b)

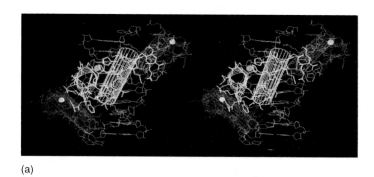

(a)

(b)

FIGURE 9.25 Zinc finger structure. (*a*) Schematic diagram of two adjacent zinc fingers. The zinc (magenta) in each finger is bound to four amino acids: two cysteines (yellow) and two histidines (blue), holding the finger in the proper shape for DNA binding. These four key amino acids are found in the same place in each finger. (*b*) Three-dimensional structure of one of the zinc fingers of the *Xenopus* protein *Xfin*. The zinc is represented by the turquoise sphere at bottom center. The sulfurs of the two cysteines are represented by the yellow spheres at lower left. The two histidines are represented by the blue and green structures at lower right. The backbone of the finger is represented by the purple tube.

(*b*) From Peter Wright and N. S. Lee, "Three Dimensional Solution Structure of a Single Zinc Finger DNA-Binding Domain," *Science,* 245:639, August 1989. © 1989, The American Association for the Advancement of Science. Photograph provided by Michael Pique and Peter E. Wright, Dept. of Molecular Biology, Research Institute of Scripps Clinic, La Jolla, California.

FIGURE 9.27 Arrangement of the three zinc fingers of Zif 268 in a curved shape to fit into the major groove of the DNA. (*a*) Stereo diagram of the protein-DNA complex, with finger 1 in orange, finger 2 in yellow, finger 3 in purple, and the DNA in blue. The zinc ions are rendered as light blue spheres. As usual, the cylinders and ribbons stand for α-helices and β-ribbons. (*b*) Schematic diagram of the same protein DNA complex. The zinc ions are not included.

(*b*) From Pavletich and Pabo, "Zinc finger—DNA Recognition: Crystal Structure of a Zif 268-DNA Complex at 2.1Å, *Science,* 252:749–888. © 1991, American Association for the Advancement of Science.

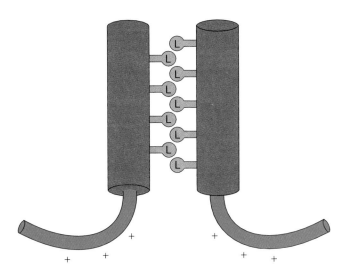

FIGURE 9.28 Model of leucine zipper. Two identical protein molecules have α-helices, represented by the red cylinders. The leucines (circled L's, blue) on the surfaces of these α-helices are thought to interact like two halves of a zipper. The basic regions (green) outside the leucine zipper (+ charges) participate in DNA binding, but only when the zipper is "zipped shut," that is, when the protein is in dimer form. This model is only hypothetical. It is not based on three-dimensional structural studies.

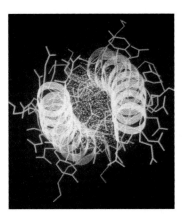

(a)

(b)

FIGURE 9.29 Structure of a leucine zipper. (*a*) Kim and Alber and colleagues crystallized a 33-amino-acid peptide containing the leucine zipper motif of the transcription factor GCN4. X-ray crystallography on this peptide yielded this view along the axis of the zipper with the coiled coil pointing out of the plane of the paper. (*b*) A side view of the coiled coil with the two α-helices colored red and blue. Notice that the amino ends of both peptides are on the left. Thus, this is a parallel coiled coil.

(a) From O'Shea, Klemm, Kim and Alber, "X-ray Structure of the GCN4 Leucine Zipper, a Two Stranded, Parallel Coiled Coil," *Science,* 254:541, 25 October, 1991. © 1991 American Association for the Advancement of Science.

c. *bZIP domains.* Most transcription factors bind to DNA as dimers, or occasionally as tetramers. This allows for cooperative binding, which increases the affinity between protein and DNA and permits transcription factors to bind effectively even though they are present in cells in tiny amounts. Thus, most transcription factors have dimerization (or oligomerization) domains. The bZIP domain is a perfect example of a case where it is difficult to separate the discussion of DNA-binding and dimerization domains: The DNA-binding domain is actually created by two monomers coming together, as we will see. Without dimerization, in this case, there is no DNA-binding domain.

The ZIP part of a bZIP domain is a "zipper," usually a **leucine zipper,** which allows dimerization. Each monomer of a bZIP protein has an α-helix with leucines, or other hydrophobic (fatty) amino acids spaced seven amino acids apart. This puts all the leucines on one face of the α-helix, which puts them in position to interact with an identical set of leucines on the corresponding domain of another bZIP protein. In this way, the two helices act like the two halves of a zipper. Figure 9.28 shows a schematic representation of a leucine zipper.

Figure 9.28 is only schematic, and must be qualified in at least two ways. First, the α-helices of the zipper actually coil around each other, forming a so-called **coiled coil,** as illustrated in figure 9.29. Second, the green tails connected to the helices in figure 9.28 have plus signs, representing basic protein domains. These are

the DNA-binding parts of the bZIP domain, and they do not really flap in the breeze as these seem to be doing here. Instead, they are extensions of the zipper α-helices. Figure 9.30 illustrates the binding between a bZIP domain and its DNA target. Notice that the leucine zipper not only brings the two monomers together, it also places the two basic parts of the domain in position to grasp the DNA like a pair of forceps, or fireplace tongs, with the basic domains fitting into the DNA major groove.

This list is certainly not exhaustive. Several transcription factors have now been identified that do not fall into any of these categories.

2. *Transcription activating domains.* Most transcription factors have one of these domains, and some have more than one. So far most of these domains fall into three classes, as follows:

a. *Acidic.* The yeast transcription factor GAL4 typifies this group. It has a 49 amino acid domain with 11 acidic amino acids.

b. *Glutamine-rich.* The transcription factor Sp1 has two such domains, which are about 25% glutamine (an amino acid). One of these has 39 glutamines in a span of 143 amino acids. In addition, Sp1 has two other activating domains that do not fit into any of these three main categories.

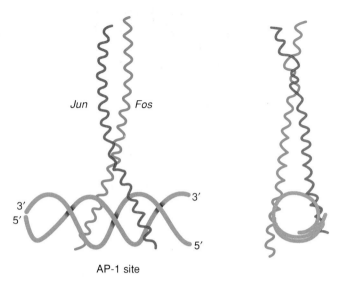

Interaction between a bZIP domain and its DNA target. These are two views of one of the two structures determined by X-ray crystallography of a bZIP domain-DNA complex. The bZIP domain is from a human gene-specific transcription factor called AP-1, which is a heterodimer composed of one molecule each of polypeptides called Jun (red) and Fos (blue). The DNA (green) used to make the crystals contains a target for AP-1, called an AP-1 site. Note that this diagram also shows the coiled coil nature of the bZIP domain.

c. *Proline-rich.* The transcription factor CTF, for instance, has a domain of 84 amino acids, 19 of which are proline.

How do these transcription activating domains activate? They appear to interact with other proteins, particularly general transcription factors like TFIIB and TFIID, or perhaps RNA polymerase. By this interaction, they help assemble the preinitiation complex and thereby stimulate transcription.

Distinct DNA-Binding and Transcription Activating Domains

The transcription activating domains we have discussed are structurally and functionally distinct from the DNA-binding domains on the same proteins. Mark Ptashne and his colleagues performed a clever experiment to prove this, using the yeast transcription activator GAL4. Recall that this transcription factor activates the yeast *GAL* genes by binding to an upstream control element called UAS_G (figure 9.31*a*). By cloning and manipulating the *GAL4* gene, Ptashne produced a half-protein containing only the DNA-binding domain. As figure 9.31*b* shows, this protein can still bind to UAS_G, but cannot activate transcription. Next, Ptashne made a hybrid protein containing the DNA-binding domain of a bacterial repressor called LexA and the transcription activating domain of GAL4. We call such hybrids **chimeras,** after the mythological beast that combined the features of a lion, a goat, and a snake. Obviously, this chimeric protein would not bind to UAS_G, so it would not be able to activate *GAL* gene transcription. But then Ptashne placed a *lexA* operator near the transcription start site of a *GAL* gene (figure 9.31*c*). Now, by virtue of its LexA DNA-binding domain, the chimeric protein binds to the *lexA*

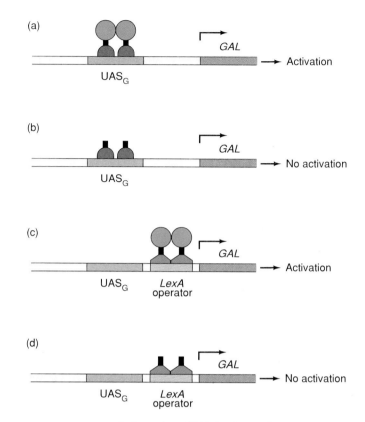

Independent DNA-binding and transcription activating regions of a transcription factor. (*a*) The GAL4 protein has two domains: one for DNA binding (red) and one for transcription activation (green). The protein is shown here interacting with its binding site, UAS_G (pink), upstream from a *GAL* gene. Transcription is activated. (*b*) A fragment of GAL4 containing only the DNA-binding domain can still bind to UAS_G, but no transcription activation occurs. (*c*) A chimeric protein containing the transcription activation domain of GAL4 (green) and the DNA-binding domain of the bacterial repressor LexA can bind to a *lexA* operator (light blue) placed upstream from the *GAL* gene and activate transcription. (*d*) A control experiment using the LexA DNA-binding domain and no transcription activating domain gave no activation. Together these experiments demonstrate the independence of the two domains of the GAL4 protein. The chimeric protein is shown here interacting with the DNA as a dimer, although that is not proven.

operator upstream from the *GAL* gene. And, using its GAL4 transcription activating domain, it activates *GAL* transcription. Thus, the DNA-binding and transcription activating parts of GAL4 are clearly separable, and even interchangeable with other proteins.

Gene-specific factors that stimulate transcription by RNA polymerase II can bind either to promoter elements such as GC boxes or CCAAT boxes, or to enhancers associated with genes. Most of these factors contain three different domains that allow them to do their job: a DNA-binding domain; a transcription activation domain; and a dimerization domain. Sometimes, as in bZIP proteins, the DNA-binding and dimerization domains are inseparable.

THE RELATIONSHIP BETWEEN GENE STRUCTURE AND EXPRESSION

In our discussion of transcription of eukaryotic genes, we have thus far been ignoring a central fact: Many eukaryotic genes are fundamentally different from their prokaryotic counterparts. A typical prokaryotic structural gene is simple. Once it begins, it continues uninterrupted until it ends. In fact, this seems such an obvious way of doing things that nobody imagined eukaryotic genes would be organized any differently, but they are. Eukaryotic genes are frequently arranged in pieces; in other words, they are interrupted. For example, if we expressed the sequence of the human β-globin gene as a sentence, this is how it might read:

This is *bhg* the human β-globin *qwtzptlrbn* gene.

INTERRUPTED GENES

There are obviously two regions (italicized) within the β-globin gene that make no sense; they contain sequences totally unrelated to the globin coding sequences surrounding them. These are sometimes called **intervening sequences,** or **IVSs,** but they usually go by the name Walter Gilbert gave them: **introns.** Similarly, the parts of the gene that make sense are sometimes called coding regions, or expressed regions, but Gilbert's name for them is more popular: **exons.** Some genes have no introns at all; most genes in higher eukaryotes have an abundance. The current record, sixty introns, is held by the human dystrophin gene, whose failure causes Duchenne muscular dystrophy.

The usefulness of introns, if any, remains something of a mystery; fortunately, we can do a lot better at describing them—how we know they are there and what their implications are. Consider the major late locus of adenovirus, a favorite system for studying gene expression and the first place introns were found—by Phillip Sharp and Richard Roberts and their colleagues in 1977. Several lines of evidence converged at that time to show that the genes of the adenovirus major late locus are interrupted, but perhaps the easiest to understand comes from studies using a technique called *R-looping,* which allows direct visualization of introns by electron microscopy.

In R-looping experiments, we hybridize RNA to its DNA template. In other words, we separate the DNA template strands and form a double-stranded hybrid between one of these strands and the RNA product. Such a hybrid double-stranded polynucleotide is actually a bit more stable than a double-stranded DNA under the conditions of the experiment. After we form the hybrid, we examine it by electron microscopy. There are two basic ways to do these experiments: (1) using DNA whose two strands are separated only enough to let the RNA hybridize, or (2) completely separating the two DNA strands before hybridization. Sharp used the latter method, hybridizing single-stranded adenovirus DNA to mature mRNA for one of the viral coat proteins: the hexon protein. Figure 9.32 shows the results. (Do not be confused by the similarity between the terms "exon" and "hexon." There is no relationship; "hexon" is just the name of a viral coat protein.)

If there were no introns in the hexon gene, there should be a smooth, linear hybrid where the mRNA lined up with its

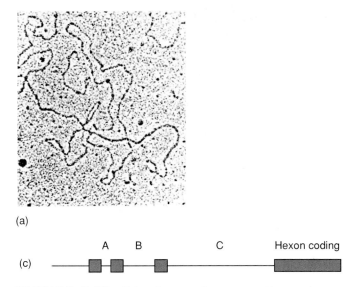

(a)

A   B   C   Hexon coding

(c) ──────■──■────■─────────────────■■■■■──

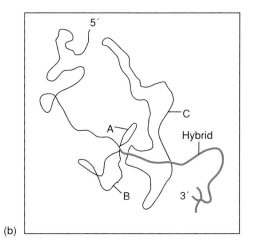

(b)

FIGURE 9.32 R-looping experiments reveal introns in adenovirus. (*a*) Electron micrograph of a cloned fragment of adenovirus DNA containing the 5′-part of the late hexon gene, hybridized to mature hexon mRNA. The loops represent introns in the gene that cannot hybridize to mRNA. (*b*) Interpretation of the electron micrograph, showing the three intron loops (labeled A, B, and C), the hybrid (heavy red line), and the unhybridized region of DNA upstream from the gene (upper left). The fork at the lower right is due to the 3′-end of the mRNA, which cannot hybridize

because the 3′-end of the gene is not included. Therefore, the mRNA forms intramolecular double-stranded structures that have a forked appearance. (*c*) Linear arrangement of the hexon gene, showing the three short leader exons, the two introns separating them (A and B), and the long intron (C) separating the leaders from the coding exon of the hexon gene. All exons are represented by red boxes.

Photo (a) From Susan M. Berget, Claire Moore, and Phillip Sharp. *Proceedings of the National Academy of Sciences,* 74:3173. 1977.

DNA template. But what if there *are* introns in this gene? Clearly, there can be no introns in the mature mRNA, or they would code for nonsense that would appear in the protein product. Therefore, introns are sequences that occur in the DNA but are missing from mRNA. That means the hexon DNA and hexon mRNA will not be able to form a smooth hybrid. Instead, the intron regions of the DNA will not find counterparts in the mRNA and so will form unhybridized loops. Of course, that is exactly what happened in the experiment shown in figure 9.32. The loops here are made of DNA, but we still call them R-loops, since hybridization with RNA caused them to form.

The electron micrograph shows an RNA-DNA hybrid interrupted by three single-stranded DNA loops (labeled A, B, and C). These loops represent the introns in the hexon gene. Each loop is preceded by a short hybrid region, and the last loop is followed by a long hybrid region. This means there are four exons in the gene: three short ones near the beginning, followed by one large one. The three short exons are transcribed into **leader** regions that appear at the 5′-end of the hexon mRNA before the coding region; the long exon contains the coding region of the gene.

Frequently throughout this chapter, we have referred to experiments involving eukaryotic viruses. That is because many of the experiments that have unraveled the process of gene expression in eukaryotes were first done in viruses. Why is that? Viruses are much simpler genetically than their hosts, so they are much easier to study. Moreover, DNA viruses like adenovirus and SV40 replicate in the host cell nuclei, so they use the same gene expression machinery as their hosts. This means that the things we learn about eukaryotic viral gene expression are very likely to tell us about cellular gene expression.

Nevertheless, when we discover something as surprising as introns in a virus, we wonder whether it is just a bizarre viral phenomenon that has no relationship to eukaryotic cellular processes. Thus, it was important to determine whether cellular genes also have introns. One of the first such demonstrations was an R-looping experiment done by Pierre Chambon and colleagues, using the chicken ovalbumin gene. They observed six DNA loops of various sizes that could not hybridize to the mRNA, so this gene contains six introns spaced among seven exons. It is also interesting that most of the introns were considerably longer than most of the exons. This preponderance of introns is typical of higher eukaryotic genes. Introns in lower eukaryotes such as yeast are much rarer.

So far we have discussed introns only in mRNA genes, but some tRNA genes also have introns, and even rRNA genes sometimes do. The introns in both these latter types of genes are a bit different from those in mRNA genes. For example, tRNA introns are relatively small, ranging in size from four to about fifty bases long. Not all tRNA genes have introns; those that do have only one, and it is adjacent to the DNA bases corresponding to the anticodon of the tRNA. Genes in mitochondria and chloroplasts can also have introns. Indeed, as we will see, these introns are some of the most interesting.

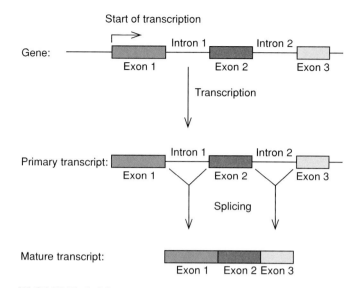

FIGURE 9.33   Outline of splicing. The introns in a gene are transcribed along with the exons (colored boxes) in the primary transcript. Then they are removed as the exons are spliced together.

Most higher eukaryotic genes coding for mRNA and tRNA, and a few coding for rRNA, are interrupted by unrelated regions called introns. The other parts of the gene, surrounding the introns, are called exons; the exons contain the sequences that finally appear in the mature RNA product. Genes for mRNAs have been found with anywhere from zero to sixty introns. Transfer RNA genes have either zero or one.

RNA Splicing

Consider the problem introns pose. They are present in genes but not in mature RNA. How is it that the information in introns does not find its way into the mature RNA products of the genes? There are two main possibilities: (1) The introns are never transcribed; the polymerase somehow jumps from one exon to the next and ignores the introns in between. (2) The introns are transcribed, yielding a **primary transcript,** an overlarge gene product that is cut down to size by removing the introns. As wasteful as it seems, the latter possibility is the correct one. The process of cutting introns out of immature RNAs and stitching together the exons to form the final product is called **RNA splicing.** The splicing process is outlined in figure 9.33, although, as we will see later in the chapter, this picture is considerably oversimplified.

How do we know splicing takes place? Actually, at the time introns were discovered, circumstantial evidence to support splicing already existed. A class of large nuclear RNAs called **heterogeneous nuclear RNA (hnRNA),** widely believed to be precursors to mRNA, had been found. These hnRNAs are the right size (larger than mRNAs) and have the right location (nuclear) to be unspliced mRNA precursors. Furthermore, hnRNA turns over very rapidly, which means it is made and

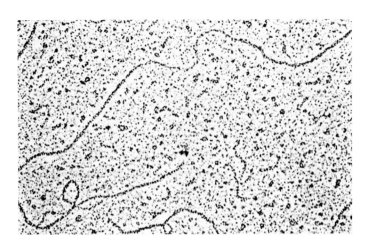

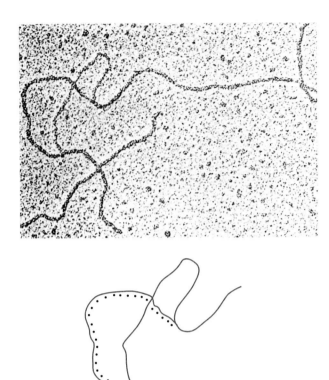

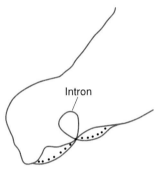

(a)

(b)

FIGURE 9.34 Introns are transcribed. (a) R-looping experiment in which mouse globin mRNA precursor was hybridized to a cloned mouse β-globin gene. A smooth hybrid formed, demonstrating that the introns are represented in the mRNA precursor. (b) Similar R-looping experiment in which mature mouse globin mRNA was used. Here, the large intron in the gene looped

out, showing that this intron was no longer present in the mRNA. The small intron was not detected in this experiment. In the interpretive drawings below, the dotted black lines represent RNA and the solid red lines represent DNA.

Photos (a and b) from Shirley Tilgham, Peter Curtis, David Tiemeier, Philip Leder, and Charles Weissmann, *Proceedings of the National Academy of Science* 75:1312, 1978.

converted to smaller RNAs quickly. This too suggested that these RNAs are merely intermediates in the formation of more stable RNAs. However, no direct evidence existed to show that hnRNA could be spliced to yield mRNA.

The mouse β-globin mRNA and its precursor provided an ideal place to look for such evidence. The mouse globin mRNA precursor is a member of the hnRNA population. It is found only in the nucleus, turns over very rapidly, and is about twice as large (1,500 bases) as mature globin mRNA (750 bases). Also, immature red blood cells make so much globin (about 90% of their protein) that α- and β-globin mRNAs are abundant and can be purified relatively easily; even their precursors exist in appreciable quantities. This made experiments feasible. Furthermore, the β-globin precursor is the right size to contain both exons and introns. Charles Weissmann and Philip Leder and their coworkers used R-looping to test the hypothesis that the precursor still contained the introns.

The experimental plan was to hybridize mature globin mRNA, or its precursor, to the cloned globin gene, then observe the resulting R-loops (figure 9.34). We know what the results with the mature mRNA should be. Since this RNA has no intron sequences, the introns in the gene will loop out. On the other hand, if the precursor RNA still has all the intron sequences, no such loops will form. Of course, that is what happened. You may have a little difficulty recognizing the structures in figure 9.34 because this R-looping was done with double-stranded DNA instead of single-stranded. Thus, the RNA simply hybridizes to one of the DNA strands, displacing the other. The precursor RNA gave a smooth, uninterrupted R-loop; the mature mRNA gave an R-loop interrupted by an obvious loop of double-stranded DNA, which represents the large intron. The small intron was not visible in this experiment. Notice that the term *intron* can be used for intervening sequences in either DNA or RNA.

Messenger RNA synthesis in eukaryotes occurs in stages. The first stage is synthesis of the primary transcription product, an mRNA precursor that still contains introns copied from the gene, if any were there. This precursor is part of a pool of large nuclear RNAs called hnRNAs. The second stage is mRNA maturation. Part of the maturation of an mRNA precursor is the removal of its introns in a process called splicing. This yields the mature-sized mRNA.

Splicing Signals

Pause to consider the importance of accurate splicing. If too little RNA is removed from an mRNA precursor, the mature RNA will be interrupted by "garbage." If too much is removed, important sequences may be left out.

Given the importance of accurate splicing, signals must occur in the mRNA precursor that tell the splicing machinery exactly where to "cut and paste." What are these signals? One way to find out is to look at the base sequences of a number of different genes, locate the intron boundaries, and see what sequences they have in common. In principle, these common sequences could be part of the signal for splicing. The most striking observation, first made by Chambon, is that almost all introns in nuclear mRNA precursors begin and end the same way:

exon/GU-intron-AG/exon

In other words, the first two bases in the intron of a transcript are GU and the last two are AG. This kind of conservation does not occur by accident; surely the GU—AG motif is part of the signal that says: "Splice here." However, GU—AG motifs occur too frequently. A typical intron will contain several GU's and AG's within it. Why are these not used as splice sites? The answer is that GU—AG is not the whole story; splicing signals are more complex than that. The sequences of many intron/exon boundaries have now been examined and a consensus sequence has emerged:

$$5'-^{C}_{A}AG/GU^{A}_{G}AGU-intron-Y_NNAG/G-3'$$

where Y_N denotes a string of about nine pyrimidines (U's and C's) and N is any base.

Again, finding consensus sequences is one thing; showing that they are really important is another. Several research groups have found ample evidence supporting the importance of these splice junction consensus sequences. Their experiments were of two basic types. In one, they mutated the consensus sequences at the splice junctions in cloned genes, then checked whether proper splicing still occurred. In the other, they collected defective genes from human patients with presumed splicing problems and examined the genes for mutations near the splice junctions. Both approaches gave the same answer: Disturbing the consensus sequences usually destroys normal splicing.

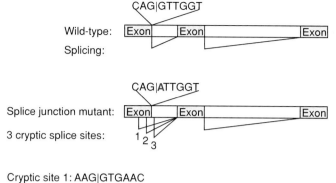

Cryptic site 1: AAG|GTGAAC
Cryptic site 2: GTG|GTGAGG
Cryptic site 3: AAG|GTTACA

Wild-type: CAG|GTTGGT

FIGURE 9.35   Importance of consensus sequences at splice junctions. With the wild-type splice junctions, splicing proceeds normally; when the first G of the intron changes to A (red), splicing no longer occurs at that location, but at three alternate, cryptic sites. The DNA sequences of the three cryptic splice junctions are shown, with the wild-type sequence below for comparison. Since these are DNA sequences, the introns begin with GT instead of GU.

For example, consider the β-globin gene in several cases of human **β-thalassemia,** an often fatal disease in which no functional β-globin mRNA appears in the cytoplasm. This can be for a number of different reasons; in the cases cited here, it is because the patients carry a β-globin gene with defective splice junctions. Figure 9.35 shows the sequence of the mutated β-globin genes of some thalassemia patients and the consequences of these mutations. In these three cases, a single base change in the crucial GT destroyed a normal splice site at the beginning of the first intron, so the splicing system used a series of alternate sites (*cryptic sites*) where splicing does not normally occur. Two of these are upstream from the normal site, within the first exon; the other is downstream, in the intron.

The sequences of the cryptic sites resemble that of the normally used site. In this way, the splicing machinery is like a child who always buys the same candy bar until one day that particular brand is sold out. For the first time, the child notices other brands that don't look too bad, and buys not one but three different kinds. These alternate brands had been cryptic, or hidden, to the child before, even though they were in plain sight, because the favorite was there.

Unfortunately these cryptic splice sites are in the wrong places, so the wrong sequences are removed from the mRNA precursors. In the first two cases, parts of exon 1 are spliced out; in the third case, part of intron 1 is left in. In none of these cases is functional mRNA produced. Thus, even though the globin coding sequence may have been perfectly normal in these mutated genes, the fact that they could not be spliced properly rendered them nonfunctional, with life-threatening consequences.

The splicing signals in nuclear mRNA precursors are remarkably uniform. The first two bases of the intron are almost always GU, and the last two are almost always AG. The 5'- and 3'-splice sites have consensus sequences that extend beyond the GU and AG motifs. The whole consensus sequences are important to proper splicing; when they are mutated, abnormal splicing can occur.

PARTICIPATION OF SMALL NUCLEAR RNAS IN SPLICING

Joan Steitz and her coworkers first focused attention on a class of **small nuclear RNAs (snRNAs)** as potential agents in splicing (figure 9.36). These snRNAs exist in the cell coupled with proteins in complexes called **small nuclear ribonucleoproteins (snRNPs),** sometimes referred to informally as "snurps." Steitz noticed that one snRNA, called U1, has a region whose sequence is almost perfectly complementary to both 5'- and 3'-splice site consensus sequences. In principle, this would allow base-pairing between U1 RNA and an mRNA precursor. In practice, such base-pairing does occur, but probably only at the 5'-splice site.

How do we know that snRNPs are involved in splicing? One line of evidence comes from in vitro splicing studies using antibodies directed against snRNPs. Antibodies react with a specific substance and frequently interrupt its normal function. Therefore, if snRNPs are needed for splicing, one would expect that antibodies directed against snRNPs would prevent splicing. This is what the experiments showed.

We also know that splicing of nuclear mRNA precursors occurs on a complex particle called a **spliceosome.** The mammalian spliceosome is about the size of a large ribosomal subunit (50–60S). It includes the mRNA precursor as well as a set of snRNPs: U1, U2, U4, U5, and U6, although U1 is relatively easily lost from the spliceosome during purification.

A LARIAT-SHAPED INTERMEDIATE IN SPLICING

The detailed mechanism by which cells splice mRNA precursors seems at first so illogical that it would be best to present the conclusion first, then the evidence for it. Figure 9.37 illustrates the two-step **lariat model** of mRNA splicing. The first step is the formation of a looped intermediate that looks like a lariat or lasso. This occurs when a nucleotide in the middle of the intron attacks the phosphodiester bond between the first exon and the intron, forming the loop and simultaneously breaking the phosphodiester bond. The splicing is completed when the 3'-end of the first exon attacks the 5'-end of the second exon, forming the exon-exon phosphodiester bond and releasing the intron, in lariat form, at the same time.

This mechanism seemed unlikely enough that rigorous proof had to be presented in order for it to be accepted. There

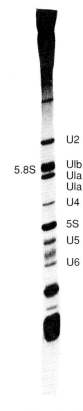

FIGURE 9.36 Electrophoresis of snRNAs. Small nuclear RNAs from mouse tumor cells are separated according to size by polyacrylamide gel electrophoresis; 5S and 5.8S rRNAs are also present and serve as size markers. The origin of electrophoresis, by convention, is at top.
From Michael R. Lerner, John A. Boyle, Stephen M. Mount, Sandra L. Wolin, and Joan A. Steitz. *Nature* 283:221, 1980 © Macmillan Magazines Limited.

is, in fact, very good evidence for the existence of all the intermediates shown in figure 9.37, most of it collected by Sharp and his research group. Only one piece of this evidence will be presented here. The key to the whole scheme is the branched adenine nucleotide at the "knot" of the lariat. This nucleotide will therefore be unique in participating in three phosphodiester bonds instead of the usual two. (See parentheses in figure 9.37.) The extra bond joins the 2'-hydroxyl group of the adenosine to the 5'-phosphate of the guanine nucleotide at the 5'-end of the intron. Therefore, we ought to be able to digest the lariat with an appropriate RNase and leave only the branch:

$$pA\begin{array}{c} \nearrow pG \\ \searrow pG \end{array}$$

Sharp found this structure, just as predicted.

Figure 9.38 shows the lariat model again, including our current understanding of the involvement of snRNPs in splicing. The first complex to form, containing the unspliced RNA plus U1 base-paired at the upstream (5') splice site, is called the **commitment complex (CC).** As its name

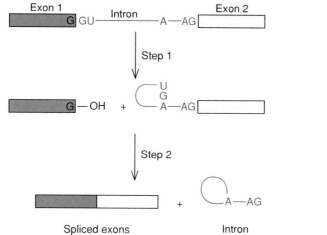

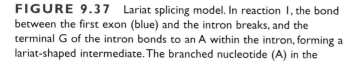

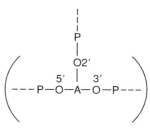

FIGURE 9.37   Lariat splicing model. In reaction 1, the bond between the first exon (blue) and the intron breaks, and the terminal G of the intron bonds to an A within the intron, forming a lariat-shaped intermediate. The branched nucleotide (A) in the intron is shown in parentheses at right. In reaction 2, the first exon (blue) is spliced to the second (yellow), liberating the lariat-shaped intron (red).

implies, the commitment complex is committed to splicing out the intron at which it assembles. Next, with help from ATP, U2 base-pairs to the intron at the site that will be the branch of the lariat, to form the **A complex.** Next, U5 and a complex of U4 and U6 join to form the **B1 complex** (not pictured). U4 then dissociates from U6 to allow base-pairing between U2 and U6, and formation of the **B2 complex.** In this complex, U5 binds to both exons and positions them to be spliced together at a later step. Also, U6 displaces U1 from its binding site at the 5′ splice site. ATP then provides the energy for the first splicing step, which separates the two exons and forms the lariat-shaped splicing intermediate; all these components are held in the **C1 complex.** With energy from a second molecule of ATP, the second splicing step occurs, joining the two exons and removing the lariat-shaped intron, with all components retained in the **C2 complex.** In the next step, the spliced, mature mRNA exits the complex, leaving the intron bound to the **I complex.** Finally, the I complex dissociates into its component snRNPs, which can be recycled into another splicing complex, and the lariat intermediate, which is debranched and degraded.

The lariat model raises intriguing questions. For example: What is to be gained by splicing RNA this way, instead of by a simpler path? The answer seems to be that this mechanism is an evolutionary relic. We will see that some eukaryotic RNA precursors can splice themselves and that one class of RNAs carries out such self-splicing via a lariat intermediate. Eukaryotes probably adapted this established mechanism, adding extra proteins and RNAs that ultimately took over the job of splicing nuclear pre-mRNAs.

A question can also be raised about splitting the first exon away from the rest of the precursor RNA before a new bond can form with the second exon. This seems to create a danger of losing exon 1 before it can bind to exon 2. However, it is prob-able that exon 1 is held tightly to the spliceosome, so it does not miss the opportunity to join exon 2.

A class of small nuclear RNAs (snRNAs) appears to be important in splicing nuclear mRNA precursors. In particular, U1 snRNA has a region that is complementary to the consensus splice site sequences in mRNA precursors. This RNA has also been shown to base-pair with mRNA precursors at the 5′-splice site in vivo. The mechanism of mRNA precursor splicing involves intramolecular attack by a nucleotide within the intron on the phosphodiester bond between the first exon and the G at the 5′-end of the intron, breaking the 5′-exon/intron bond and creating a lariat-shaped intermediate. Then the two exons are joined, releasing the intron, still in its lariat shape.

ALTERNATIVE SPLICING

We have seen that genes can have alternative splice sites (cryptic sites) that are used only when the normal sites are mutated. However, some genes have true alternative splice sites that can be used to control gene expression. If the cell splices the pre-mRNA one way, one protein product will be made; if it splices the alternative way, a different protein, or perhaps no protein at all, will be made.

A good example is the regulation of sex determination in *Drosophila.* Sex in the fruit fly is determined by a pathway that includes several alternative RNA processing steps. One of the first steps involves the product of the *Sex lethal* (*Sxl*) gene, which governs the splicing of its own pre-mRNA, and that of the next gene in the pathway, *transformer* (*tra*). Figure 9.39 shows how this works.

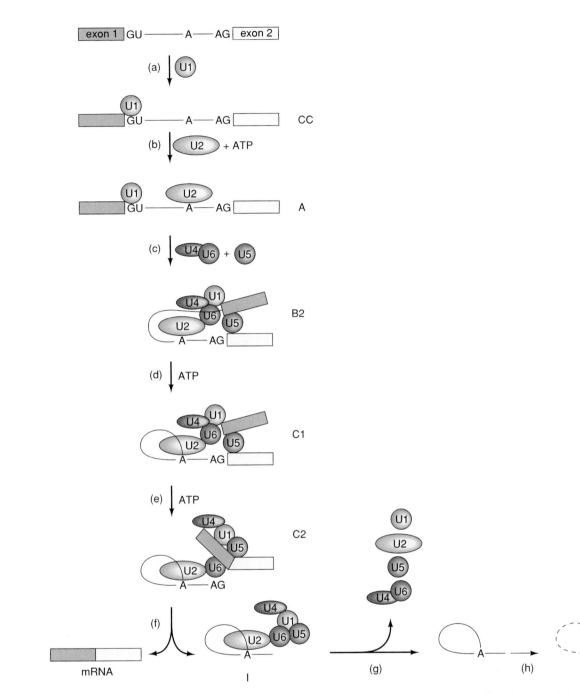

FIGURE 9.38 Involvement of snRNPs in splicing. (*a*) U1 (orange) binds to the pre-mRNA by base-pairing to the 5′-splice site, forming the committed complex (CC). (*b*) U2 (green), with the help of ATP, base-pairs to a region of the intron containing the adenosine that will be the branch point of the lariat. This produces the A complex. (*c*) U4 (brown) and U6 (purple), base-paired with each other, join the complex. In the process, U4 dissociates from U6, which then base-pairs with the 5′-splice site, displacing U1. At the same time, U5 (red) binds to the ends of both exons and positions them for splicing. The result is the B2 complex. (*d*) The lariat intermediate forms, with breakage of the bond between exon 1 (blue) and the intron. ATP is required for this step, which forms the C1 complex. (*e*) The two exons are ligated together, releasing the lariat-shaped intron. All the participants are still part of the same complex, called C2. Again, ATP is required. (*f*) The mature mRNA with its exons ligated together is released, leaving the lariat intron associated with the snRNPs in complex I. (*g*) The snRNPs leave the lariat intron, which then becomes debranched and degraded (*h*).

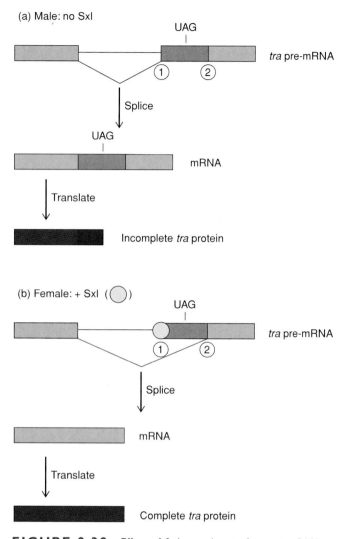

FIGURE 9.39 Effect of *Sxl* on splicing of *tra* pre-mRNA. (*a*) In males, there is no *Sxl* protein; both 3′-splice sites (1 and 2) in the *tra* pre-mRNA are available, so the nearest one (#1) is used. This produces a nonfunctional *tra* mRNA with a premature translation termination signal (UAG) that prevents translation of the second exon. Thus, only incomplete protein is made and *tra* is not expressed. (*b*) In females, *Sxl* protein is made. It binds to 3′-splice site #1 and prevents splicing from occurring there. Splicing therefore takes place only at site #2, which yields a functional mRNA and production of the active *tra* protein. Thus we can see that *Sxl* controls expression of *tra* by controlling the way its pre-mRNA is spliced.

The *tra* gene can be spliced in two different ways. One way leads to an mRNA with a long coding region that encodes a useful protein product. The other splicing product has a premature stop codon (UAG) that prevents synthesis of a useful protein. The *Sxl* product appears to influence the splicing of *tra* pre-mRNA by binding to the nonproductive 3′-splice site and suppressing splicing there. This forces splicing to occur at the productive 3′-splice site, so active *tra* product is made. This is what happens in females. On the other hand,

males make no *Sxl* product. Therefore, the nonproductive 3′-splice site remains available, and splicing of *tra* pre-mRNA occurs there. Thus, *tra* is not effectively expressed in males. Maleness is therefore the "default" condition in *Drosophila*. In order to override this condition, females must express *tra*, which in turn requires *Sxl*.

Self-Splicing rRNA Precursors

As we will see, the large rRNAs from all eukaryotes are encoded in a common gene. In other words, the 5.8S, 18S, and 28S rRNAs, or their equivalents, are transcribed together on a precursor RNA, which is cut, or **processed,** to liberate the mature RNAs. Part of this processing involves removing a spacer RNA between the 18S and 28S species, but this spacer is not an intron, and its removal is not splicing. The difference is that the two RNA species remain independent after processing; they do not join together the way exons do. However, the 26S part of the protozoan *Tetrahymena* rRNA precursor does contain an intron, and it can be spliced in a surprising way: The RNA splices itself, without any help from outside enzymes! (Note that the *Tetrahymena* 26S rRNA is the equivalent of the mammalian 28S rRNA.)

Figure 9.40*a* shows the *Tetrahymena* rRNA precursor. It includes the 17S, 5.8S, and 26S rRNA sequences, as well as regions in between that must be removed. It also includes the intron within the 26S region. To examine the processing of this molecule, Thomas Cech (pronounced "Check") and his collaborators cloned a piece of its gene (indicated with a bracket in figure 9.40*a*) and transcribed this gene fragment with *E. coli* RNA polymerase. This gave a piece of RNA containing the intron. And, much to everyone's surprise, this little piece of RNA could splice itself.

Figure 9.40*b* presents the splicing scheme. In the first step, a guanosine nucleotide (GMP) attacks the 5′-end of the intron, breaking the exon/intron bond. Then the 3′-end of the first exon attacks the 5′-end of the second exon, joining the two exons (the splicing step) and removing the intron in linear form. The boundaries of this intron do not look at all like those of a nuclear mRNA intron.

The introns in the *Tetrahymena* nuclear rRNA precursor and the yeast mitochondrial rRNA precursor belong to a class called **group I introns.** These all self-splice with the help of a guanosine nucleotide, which initiates the first step. Another class of introns, called **group II introns,** also self-splice, but they do not need assistance from guanosine to start the reaction. Instead, the initiating entity is an adenosine nucleotide within the intron of the RNA itself. When this attacks the bond between the first exon and the intron, it forms a lariat, just as we have observed for nuclear mRNA splicing.

It is always easier to believe something if we can see it with our own eyes. In this case, Arnberg and colleagues have provided us with electron micrographs (figure 9.41) that actually show the lariat-shaped group II splicing intermediate (intron plus exon 2) and product (intron). These RNAs come from splicing an rRNA precursor in yeast. This is not the rRNA precursor of the yeast nuclei; instead, it comes from the mitochondria, the energy-producing organelles of the cell.

(a)

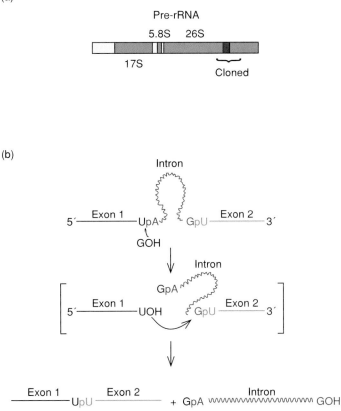

(b)

FIGURE 9.40   Self-splicing of *Tetrahymena* rRNA precursor. (*a*) Structure of the rRNA precursor, containing the 17S, 5.8S, and 26S sequences. Note the intron within the 26S region (red). The cloned segment used in subsequent experiments is indicated by a bracket. (*b*) Self-splicing scheme. In the first step (top), a guanine nucleotide attacks the adenine nucleotide at the 5′-end of the intron, releasing exon 1 (black) from the rest of the molecule and generating the hypothetical intermediates shown in brackets. In the second step, exon 1 (black) attacks exon 2 (blue), performing the splice reaction that releases a linear intron (red) and joins the two exons together. Finally, in a series of reactions not shown here, the linear intron loses 19 bases from its 5′-end.

The main difference between nuclear mRNA precursors and RNAs bearing group II introns is that the latter can handle the splicing all by themselves, without spliceosomes. What kinds of genes have group II introns? Some mitochondrial genes that encode proteins have such introns. Figure 9.42 summarizes the splicing of all three classes of introns.

Self-splicing RNAs, besides tickling the fancy of geneticists, have a number of interesting implications. One of these is the fact that an RNA can act like an enzyme, at least in that it has catalytic activity. Before the 1980s, it had been widely assumed that protein was the only biological molecule that had this ability. Catalytic RNA may help explain how life originated. If we assume that only proteins can act as enzymes, the origin of life is plagued by a "catch 22": The first organism needed proteins to catalyze the replication of its genes, but it could not make proteins without catalysts. Catalytic nucleic acids help explain away

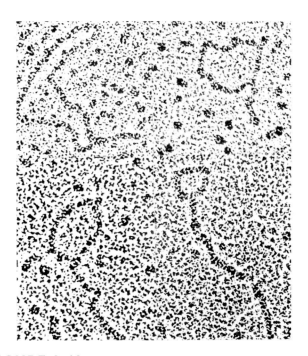

FIGURE 9.41   Lariat-shaped splicing intermediate and product. Electron micrographs of intermediate and product of yeast mitochondrial rRNA splicing provide direct visual evidence for lariats. (Top) Two lariat-shaped introns. (Bottom) Two lariat-shaped intermediates containing an intron plus the second exon.
From Annika Arnberg, Gerda Van der Horst, and Henk F. Tabak, *Cell,* 144(2): 31 January 1986. © Cell Press.

this paradox: RNA could have served as both genome and catalyst in the earliest forms of life.

The rRNA precursor of the protozoan *Tetrahymena* contains an intron in the 26S part of the molecule. This group I intron can be spliced out in vitro by the RNA precursor itself. The mechanism involves a series of phosphodiester bond breakings and joinings that results in exon-exon bonding and splitting out the intron as a linear molecule. Group II intron splicing is also autocatalytic. It begins with an attack by a nucleotide within the intron on the 5′-exon/intron boundary. This leads to a lariat-shaped intermediate and intron product.

OTHER RNA PROCESSING EVENTS

Splicing is one thing that happens to RNA between the time it is synthesized and the time it is exported to the cytoplasm for use in protein synthesis. But several other things occur during this **posttranscriptional** period. Some of them are peculiar to mRNA; others pertain to rRNA and tRNA only; and some involve all three types of RNA. All these events can be lumped together under the heading **RNA processing.** Among these processing events are: (1) trimming RNA precursors outside the coding regions; (2) methylating (adding

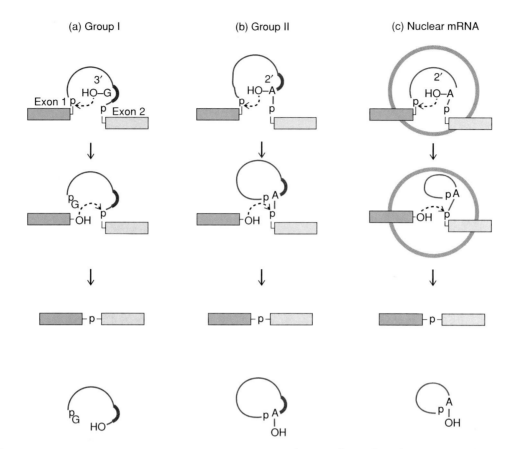

(a) Group I   (b) Group II   (c) Nuclear mRNA

FIGURE 9.42   Summary of three splicing schemes. The major differences in these mechanisms lie in the first step. The self-splicing of group I introns (*a*) is initiated by a guanosine nucleotide that presumably resides in a pocket in the intron (represented by a thickened semicircle). This guanosine attacks the phosphate linking exon I (blue) and the intron (red). In group II (*b*), an adenosine nucleotide that is part of the intron itself plays this initiation role, resulting in a lariat-shaped intermediate. This adenosine is represented as adjacent to a pocket similar to the one in group I introns that harbors the initiating guanosine. Nuclear mRNA precursors (*c*) follow a splicing scheme remarkably similar to that used by group II introns. The major difference is that nuclear mRNA splicing requires a spliceosome (purple).

methyl groups to) RNA; (3) adding nucleotides to the 3′-ends of RNAs; and (4) blocking the 5′-ends of mRNAs with special structures called caps.

TRIMMING RIBOSOMAL RNA PRECURSORS

The ribosomal RNA genes in eukaryotes are repeated several hundred times and clustered together in the nucleolus of the cell. Their arrangement in amphibians has been especially well studied, and as figure 9.43*a* shows, they are separated by regions called **nontranscribed spacers.** This is to distinguish them from **transcribed spacers,** regions of the gene that are transcribed as part of the rRNA precursor and then removed in the processing of the precursor to mature rRNA species.

This clustering of the reiterated rRNA genes in the nucleolus made them easy to find and thereby provided Oscar Miller and his colleagues with an excellent opportunity to observe genes in action. These workers examined amphibian nuclei with the electron microscope and uncovered a visually appealing phe-

nomenon, shown in figure 9.43*b*. The DNA containing the rRNA genes can be seen winding through the picture, but the most obvious feature of the micrograph is a series of "Christmas tree" structures. These include the rRNA genes (the trunk of the tree) and growing rRNA transcripts (the branches of the tree). We must remember that these transcripts are actually rRNA precursors, not mature rRNA molecules (more about this later). The spaces between "Christmas trees" are of course the nontranscribed spacers. You can even tell the direction of transcription from the lengths of the transcripts within a given gene; the shorter RNAs are at the beginning of the gene and the longer ones are at the end.

Ribosomal and transfer RNAs are sometimes called *stable RNAs* because they persist so much longer than mRNA in the cell. It is sometimes useful to lump rRNA and tRNA together this way, because they behave differently than mRNA in several respects. One of these is in processing. Ribosomal RNAs and tRNAs first appear as precursors that sometimes need splicing, but they also have excess nucleotides at their ends, or even between regions that will become separate mature RNA

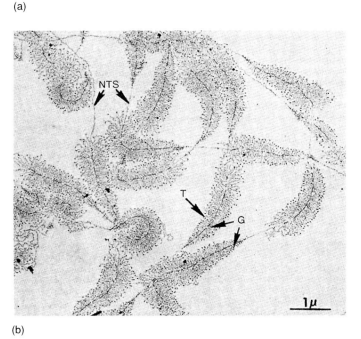

(b)

FIGURE 9.43 Transcription of rRNA precursor genes. (*a*) A portion of the newt (amphibian) rRNA precursor gene cluster, showing the alternating rRNA genes (blue) and nontranscribed spacers (NTS, red). (*b*) Electron micrograph of part of a newt nucleolus, showing rRNA precursor transcripts (T) being synthesized in a "Christmas tree" pattern on the tandemly duplicated rRNA precursor genes (G). At the base of each transcript is a polymerase I, not visible in this picture. The genes are separated by nontranscribed spacer DNA (NTS).

(b) From O. L. Miller, et al. *Symposia on Quantitative Biology,* 35:506, 1970. © Cold Spring Harbor.

sequences (e.g., 5.8S, 18S, and 28S rRNA sequences). These excess regions must also be removed.

The rRNA precursors of vertebrates are more typical than the one found in *Tetrahymena* because they contain no introns. Figure 9.44 depicts the structure and scheme of processing of the 45S human rRNA precursor. It contains the 28S, 18S, and 5.8S sequences, imbedded between transcribed spacer RNA regions. The processing of this precursor occurs in the nucleolus. The first step is to cut off the spacer at the 5′-end, leaving a 41S intermediate. The next step involves cleaving the 41S RNA into two pieces, 32S and 20S, that contain the 28S and 18S sequences, respectively. The 32S precursor also retains the 5.8S sequence. Finally, the 32S precursor is split to yield the mature 28S and 5.8S RNAs, which base-pair with each other, and the 20S precursor is trimmed to mature 18S size.

The details of this processing scheme are not universal; even the mouse does things a little differently, and the frog precursor is only 40S, which is quite a bit smaller than 45S. Still, the basic mechanism of rRNA processing, including the order of mature sequences in the precursor, is preserved throughout the eukaryotic kingdom.

POLY(A) TAILS

We have already seen that hnRNA is a precursor of mRNA. One finding that suggested such a relationship between these two types of RNA was that they shared a unique structure at their 3′-ends: a long chain of AMP residues called **poly(A).** Neither rRNA nor tRNA has a poly(A) tail. James Darnell and his coworkers performed much of the early work on poly(A)

and the process of adding poly(A) to RNA (called **polyadenylation**). They released the poly(A) intact from the rest of the mRNA by treating with two enzymes: ribonuclease A, which cuts RNA after the pyrimidine nucleotides C and U, and ribonuclease T1, which cuts after G nucleotides. In other words, the RNA was cut after every nucleotide except the A's so only runs of A's were preserved. The poly(A) prepared this way was electrophoresed to determine its size, which turned out to be an average of 150 to 200 nucleotides. This also implies that the poly(A) is not interrupted by other kinds of nucleotides, or it would have been broken up by the RNases.

It is apparent that the poly(A) goes on the 3′-end of the mRNA or hnRNA because it can be released very quickly with an enzyme that degrades RNAs from the 3′-end inward. We also know that poly(A) is added posttranscriptionally, because there are no runs of T's in the DNA long enough to encode it. There is an enzyme in nuclei called **poly(A) polymerase** that adds AMP residues one at a time to mRNA precursors once they are complete. We know that poly(A) is added to mRNA precursors because it is found on hnRNA. When we look closely, we can even see that specific unspliced mRNA precursors (the 15S mouse globin mRNA precursor, for example) contain poly(A). Once an mRNA enters the cytoplasm, its poly(A) "turns over"; in other words, it is constantly being shortened by RNases and rebuilt by a cytoplasmic poly(A) polymerase.

Most mRNAs contain poly(A). One famous exception is histone mRNA, which somehow manages to perform its function without a detectable poly(A) tail. The near universality of poly(A) suggests that it has an important function. What is it? One function of poly(A) is to help protect the vulnerable mRNA from

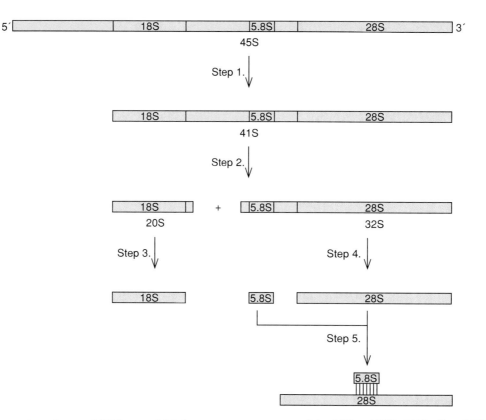

FIGURE 9.44   Processing scheme of 45S human (HeLa) rRNA precursor. Step 1: The 5′-end of the 45S precursor RNA is removed, yielding the 41S precursor. Step 2: The 41S precursor is cut into two parts, the 20S precursor of the 18S rRNA, and the 32S precursor of the 5.8S and 28S rRNAs. Step 3: The 3′-end of the 20S precursor is removed, yielding the mature 18S rRNA. Step 4: The 32S precursor is cut to liberate the 5.8S and 28S rRNAs. Step 5: The 5.8S and 28S rRNAs associate by base-pairing.

degradation by ribonucleases lurking about in the cell. Another function is to stimulate translation of mRNAs. In fact, some mRNAs cannot be translated at all until they are polyadenylated.

> Most eukaryotic mRNAs carry a poly(A) tail about 150–200 bases long on their 3′-ends. Poly(A) polymerase adds this tail to mRNA precursors in the nucleus and the poly(A) apparently remains with the message through splicing and transport to the cytoplasm. Poly(A) seems to protect mRNA from degradation; in many cases it also stimulates translation.

POLYADENYLATION

It would be logical to assume that poly(A) polymerase simply waits for a transcript to be finished, then adds poly(A) to its 3′-end. However, this is not what ordinarily happens. Instead, the mechanism of polyadenylation usually involves clipping an mRNA precursor, sometimes even before transcription has terminated, and then adding poly(A) to the newly exposed 3′-end (figure 9.45). This means that, contrary to expectations, polyadenylation and transcription termination are not necessarily linked. RNA polymerase II can still be elongating an RNA chain as it approaches a termination signal, while the polyadenylating

apparatus has already located a polyadenylation signal somewhere upstream, cut the growing RNA, and polyadenylated it.

What constitutes a polyadenylation signal? One consensus sequence appears in the transcript about twenty bases upstream from the polyadenylation site. This is usually AAUAAA in higher eukaryotes, although deviations, especially AUUAAA, are also found.

When the base sequences of 269 mRNAs were compared, the "correct" bases were found in the consensus region with the following frequencies:

$$A_{98}A_{86}U_{98}A_{98}A_{95}A_{96}$$
$$U_{12}$$

In other words, some "incorrect" sequences are tolerated, the most common of which, AUUAAA, is found in about 12% of polyadenylation signals. Another part of most polyadenylation signals is a GU-rich sequence just downstream from the polyadenylation site. Of course, this sequence participates only in the clipping phase, since it is removed before polyadenylation occurs.

What about the termination signals, or terminators, as opposed to the polyadenylation signals, in eukaryotes? Unfortunately, our knowledge of these sequences is still meager. The trouble seems to be that termination of most polymerase II transcripts is sloppy; the terminations occur in several different places spread out over hundreds or even thousands of bases.

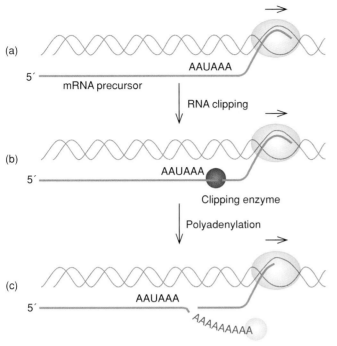

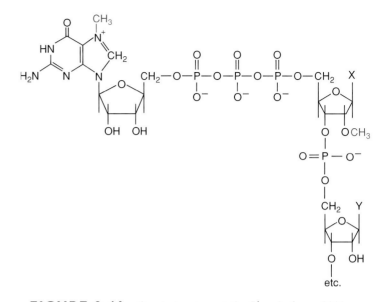

FIGURE 9.46 Cap I structure at the 5'-end of an mRNA. The methyl groups of the cap are shown in red. X and Y represent the bases on the second and third nucleotides, respectively. Cap 0 would have no methyl group on the penultimate ribose. Cap 2 (rare) would have both methyl groups of cap I, plus another on the ribose linked to base Y.

FIGURE 9.45 Polyadenylation. (*a*) The mRNA precursor (blue) is being synthesized by RNA polymerase II (orange). (*b*) A clipping enzyme (purple) recognizes the polyadenylation signal (AAUAAA) and cuts the growing RNA about twenty bases downstream. (*c*) The poly(A) polymerase (yellow) adds AMP residues (red) stepwise to the new 3'-end, as transcription continues. The transcript beyond the polyadenylation site is presumably discarded.

This is apparently not a problem, since eukaryotes have the clipping and polyadenylation system to tidy up the 3'-ends of their mRNAs.

Polyadenylation occurs by clipping a transcript upstream from its 3'-end, sometimes while transcription is still in progress, then adding poly(A) to the newly created 3'-end. The signal for such clipping and polyadenylation, at least in higher eukaryotes, includes the consensus sequence AAUAAA upstream from the clipping site, plus a GU-rich sequence downstream from the clipping site. The termination signals in genes transcribed by polymerase II have not been identified.

EUKARYOTIC mRNA CAP STRUCTURES

The 3'-end of a eukaryotic messenger is not the only place that differs from its prokaryotic counterpart. The 5'-end also has something different—a **cap.** The structure of a cap can vary somewhat, but its typical architecture is shown in figure 9.46. The cap contains a methylated guanine nucleotide linked by a triphosphate to the penultimate (next-to-last) nucleotide. The penultimate nucleotide is frequently methylated also, but the

methyl group is on the 2'-hydroxyl group of the ribose instead of on the base.

The capping process is illustrated in figure 9.47. The growing RNA starts with a triphosphate group at its 5'-end, just as a prokaryotic mRNA does, but a phospho transferase removes the terminal phosphate, leaving a diphosphate group. Then, before the RNA has grown to fifty bases long, a capping enzyme adds a GTP to its 5'-end. This reaction is unlike the normal RNA polymerase reaction in several respects: (1) The incoming nucleotide adds to the 5'-end of the RNA chain instead of the 3'-end. (2) Instead of a simple phosphodiester bond forming between the two end nucleotides, a triphosphate linkage forms. (3) The triphosphate does not link the 3'-site of one nucleotide with the 5'-site of the next; instead, the two end nucleotides are linked through their 5'-sites. Thus, the capping nucleotide "backs in" to join the growing RNA chain. Two enzymes, called **methyl transferases,** perform the methylation that finishes capping. They transfer methyl groups from the methyl donor, **S-adenosyl methionine (SAM),** to the appropriate positions on the cap. Note that, strictly speaking, capping cannot be considered a posttranscriptional process, because it occurs before transcription is complete. It might be better to call it a "cotranscriptional" process.

Yasuhiro Furuichi and Aaron Shatkin and their colleagues first discovered caps in viral mRNA in 1974. As frequently happens when such a surprising discovery occurs, geneticists wondered whether it would be a general phenomenon or an isolated curiosity. It turns out that caps are ubiquitous among eukaryotes. What, then, is the function of these widespread structures? Furuichi showed that caps have two important roles. First, they

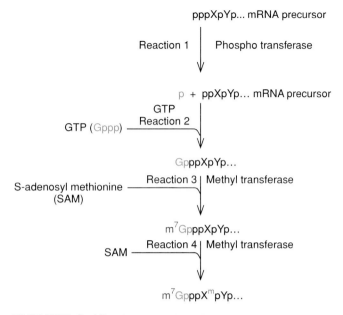

FIGURE 9.47   Capping and methylating mechanism. Reaction 1: A nucleotide phospho transferase removes the 5′-terminal phosphate from the mRNA precursor. Reaction 2: GTP caps the mRNA precursor; the blue p indicates that only one of GTP's phosphate groups is preserved in the cap. Reaction 3: A methyl transferase transfers a methyl group (red) from SAM to the 7-position of the capping guanine. Reaction 4: Another methyl transferase methylates (gray) the 2′-hydroxyl group of the penultimate nucleotide.

protect mRNA from degradation, just as poly(A) does. We can envision the cap and poly(A) as tough blocking structures at each end of the mRNA, protecting the tender midsection from attack by ribonucleases. Second, caps are necessary for binding mRNA to the ribosome. Little if any translation of decapped messages occurs.

The 5′-ends of eukaryotic mRNAs are blocked with structures called caps. A cap is added by the capping enzyme, which joins a GTP through a 5′-5′-triphosphate linkage to the penultimate nucleotide. Methyl transferases then add methyl groups to the 7-nitrogen of the terminal guanine and to the 2′-hydroxyl of the penultimate ribose. Caps serve a dual purpose: They protect messages from degradation, and they allow them entry to ribosomes for translation.

WORKED-OUT PROBLEMS

PROBLEM 1

Here is the sequence of the 5′-end of the coding strand of a hypothetical mammalian gene recognized by RNA polymerase II:

5′GAATGTATAAATGCGCAATCGAACGTGACATCGTCCTTG CCTGA

Identify the most likely transcription start site (cap site).

SOLUTION

First find the TATA box. There is a sequence (TATAAAT) near the 5′-end of the sequence that matches closely the TATA box consensus sequence (TATAAAA). Next, since the cap site of a gene transcribed by mammalian RNA polymerase II usually lies about 30 bases downstream (to the 3′-side) of the middle of the TATA box, count 30 bases to the 3′-side of the middle of this sequence (the second A). This brings us to a region (CTT**G**CCT) in which the boldfaced **G** is exactly 30 bases from the middle of the TATA box. Since transcription usually begins with a purine, this is the most likely start site, although the G and A just downstream are also candidates, and could be alternate start sites in a real gene.

PROBLEM 2

Here is the structure of a simple gene with one intron.

The boxes represent exons, and the line, the intron. It is possible to hybridize the DNA from such a gene (including DNA sequences surrounding the gene) to the mRNA product of the gene, then visualize the hybrid by electron microscopy. In the following sections (a)–(d), show schematically what the hybrids would look like. Label the RNA and DNA parts of the hybrid, and show the locations of the exons and introns. Ignore the poly(A) in parts (a)–(c). (a) If this hybridization were done with single-stranded DNA and mature, spliced mRNA, what would the hybrid look like? (b) If the hybridization were done with double-stranded DNA and mature, spliced mRNA, what would the hybrid look like? (c) If the hybridization were done with double-stranded DNA and the unspliced mRNA precursor, what would the hybrid look like? (d) What would the hybrid in part (c) look like if the poly(A) on the mRNA precursor is considered?

SOLUTION

(a)   Since the intron has been spliced out of the mRNA, the intron in the single-stranded DNA has nothing to hybridize with, and it will loop out as follows:

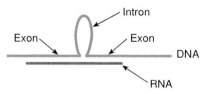

(b)   Since the intron has been spliced out of the mRNA, the intron in the double-stranded DNA has nothing to hybridize with, and it will loop out. On either side of this double-stranded intron loop, we can see RNA hybridizing to one strand of the DNA and displacing the other DNA strand in what we call an R-loop.

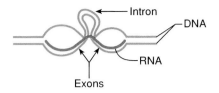

(c)   Since the mRNA precursor has not been spliced, the intron still remains in the RNA, and it will hybridize to the intron in the DNA. Of course, the exons will also hybridize, so the hybrid will show one long R-loop.

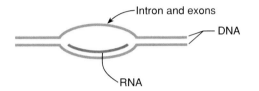

(d)   The hybrid will be the same as in part (c), but the mRNA will have a poly(A) tail that will not hybridize to the DNA, because it is not encoded in the DNA. Thus, the hybrid will look like this:

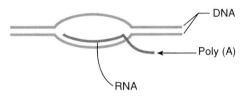

P R O B L E M   3

Here is the structure of a gene you are studying:

The boxes represent exons (e_1, e_2, and e_3) and the single lines represent introns (i_1 and i_2). Draw the results of an R-looping experiment performed on this gene, using mature mRNA and single-stranded DNA, encompassing the whole gene as well as its 5′- and 3′-flanking regions. Provide two diagrams: (a) a simple version of how the actual experimental result would appear, with single-stranded DNA represented as a thin line and RNA-DNA hybrid as a thick line; and (b) an interpretive diagram showing the DNA as a solid line and the mRNA as a dashed line. Ignore the poly(A). Label the parts of your diagrams (e_1, i_1, etc.).

S O L U T I O N

Messenger RNA is already processed to remove the introns, so it will hybridize only to the exons in the single-stranded DNA. The introns in the DNA will therefore form loops whose sizes correspond to the sizes of the respective introns.

(a)   The electron micrograph of the RNA-DNA hybrid will show the loops, but it will not show the difference between the DNA and RNA strands. Thus, the R-loops will look like this:

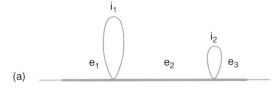

Notice that the exons are represented by thick lines, which denote RNA-DNA hybrids, whereas the introns are thin lines, denoting single-stranded DNA that had no RNA to which they could hybridize. Notice also that the relative sizes of the hybrid regions and loops correspond to the relative sizes of the exons and introns.

(b)   The interpretive diagram shows the mRNA (dashed line) hybridizing to the exons, but not the introns:

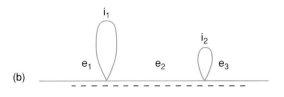

SUMMARY

Eukaryotic chromosomes consist of chromatin: DNA combined with five basic proteins called histones and a large group of nonhistone proteins. This chromatin is organized by several types of folding. The first order of chromatin folding is represented by a string of nucleosomes. These structures contain four pairs of histones (H2A, H2B, H3, and H4) in a ball, around which are wrapped about 200 base pairs of DNA. The fifth histone, H1, binds to DNA outside the ball. The second order of chromatin folding involves a coiling of the string of nucleosomes into a 25-nm-thick fiber called a solenoid. This folding is apparently mediated by H1-H1 interactions. The third order of folding is probably looping of the 25-nm fiber into a brushlike structure with the loops anchored to a central matrix.

Chromatin exists in eukaryotic cells in two forms: heterochromatin, which is condensed and transcriptionally inactive, and euchromatin, which is extended and at least potentially active.

Eukaryotic cells contain three distinct RNA polymerases, named RNA polymerase I, RNA polymerase II, and RNA polymerase III. The roles of these three enzymes are as follows: Polymerase I makes a large rRNA precursor; polymerase II makes mRNA precursors; and polymerase III makes tRNA and 5S rRNA precursors. Eukaryotic promoters that are recognized by RNA polymerase II have at least one feature in common: a consensus sequence found about 30 base pairs upstream from the transcription start site and containing the motif TATA. This TATA box acts either to position the start of transcription about 30 base pairs downstream or to increase efficiency of transcription, or both. The TATA box is recognized by the TATA box binding protein (TBP), which is part of a general transcription factor called TFIID. Once this protein binds, it organizes the binding of five other general transcription factors (TFIIA, -B, -E, -F, and -H), as well as RNA polymerase II, to form a preinitiation complex.

Polymerase II promoters generally contain other important sequences upstream from the TATA box. These sequences are typically recognition sites for specific activating proteins called transcription factors. Polymerase I promoters also lie in the 5′-flanking regions of the genes they control, but classical polymerase III promoters are internal control regions (ICRs) that lie wholly within their genes.

Polymerase I promoters interact with two general transcription factors, SL1 and UBF. Polymerase III promoter–transcription factor interaction is also relatively simple. Assembly factors TFIIIA and/or TFIIIC bind to the internal promoter and help the transcription initiation factor, TFIIIB, bind upstream from the transcription start site.

Many eukaryotic genes also contain, or reside close to, position- and orientation-independent elements called enhancers. These are not promoter elements, but they strongly stimulate transcription of nearby genes. Some genes also reside near DNA elements called silencers, which inhibit transcription. The various domains of eukaryotic enhancers and silencers interact with different transcription factors that control promoter activity.

Most higher eukaryotic genes coding for mRNA and tRNA, and a few coding for rRNA, are interrupted by unrelated regions called intervening sequences, or introns. The other parts of the gene, surrounding the introns, are called exons. Part of the maturation of an mRNA precursor is the removal of its introns in a process called splicing.

The splicing signals in eukaryotic mRNA precursors are remarkably uniform. The first two bases of the intron are almost always GU and the last two are almost always AG. In addition, there are consensus sequences at the 5'- and 3'-splice sites. These splice sites are recognized by a large structure called a spliceosome, which contains several small nuclear RNPs plus other proteins that participate in splicing.

The rRNA precursor of the protozoan *Tetrahymena* contains an intron in the 26S part of the molecule. This intron can be spliced out in vitro by the RNA precursor itself. Some mitochondrial mRNA precursors are also self-splicing.

Most eukaryotic mRNAs carry a poly(A) tail about 150–200 bases long on their 3'-ends. Poly(A) polymerase adds this tail to mRNA precursors in the nucleus, and the poly(A) apparently remains with the message through splicing and transport to the cytoplasm. Poly(A) seems to protect mRNA from degradation and stimulates translation of many mRNAs. The 5'-ends of eukaryotic mRNAs are blocked with structures called caps. Caps also serve a dual purpose: they protect messages from degradation and they allow them entry to ribosomes for translation.

PROBLEMS AND QUESTIONS

1. Why do changes in classical polymerase III promoters change the sequence of the gene product?

2. Cite two features of nucleosomes that suggest that these structures are important to life.

3. Without histone H1, would the 11-nm chromatin fiber still form? The 25-nm fiber? What is the rationale for your predictions?

4. In each somatic cell of a female mammal, one X chromosome is euchromatic and the other is heterochromatic. Which chromosome carries the active genes?

5. What important characteristic do eukaryotic transcription factors have in common with prokaryotic σ factors? In what important respect do they differ?

6. Here is the structure of a gene you want to examine:

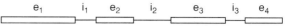

The boxes represent exons, and the single lines introns. Draw a picture of the results of an R-looping experiment on this gene, using mRNA and single-stranded DNA. Label the parts of your diagram (e_1, i_1, etc.)

7. What would the experimental results look like if the gene in problem 7 were left in double-stranded form and R-looped with mRNA? Assume that all the introns are big enough to produce a visible loop.

8. What would the results be if the experiment in problem 7 were run with an unspliced mRNA precursor?

9. Here is the sequence of the 5'-end of the nontemplate strand of a hypothetical eukaryotic gene transcribed by RNA polymerase II.

5'AATTGCGTATAAATTCGAGACTACTTACGGCTACCCGTT AGCT

Identify the most likely start (cap) site(s).

10. Draw a diagram of the "Christmas tree" pattern of transcription of the rRNA gene cluster in amphibian nucleoli. Point out: (*a*) an rRNA precursor gene; (*b*) a nontranscribed spacer; (*c*) an rRNA precursor being synthesized; (*d*) an RNA polymerase; (*e*) the direction of transcription.

11. What would a cap look like in the absence of methylation?

12. You have discovered a new form of life that shares characteristics with both eukaryotes and prokaryotes. You are curious to know whether its mRNAs have poly(A) at their 3'-ends. What experiments would you perform to answer this question?

13. You have discovered a new class of eukaryotic algae growing near thermal vents deep in the ocean. These organisms lack TATA boxes, but have a conserved GAGA motif in about the same position. Outline an experiment to test the importance of this GAGA box. What effect would you expect if the GAGA box behaves like the TATA box in the SV40 early promoter?

14. If you radioactively labeled the phosphate between the last nucleoside of an exon and the guanosine at the 5'-end of the adjoining intron, as shown by the boldface in exon-G**p**/GpUp-intron, would that radioactivity wind up with the intron or the exon, assuming the lariat model of splicing?

15. If you radioactively labeled the phosphate between the last nucleoside of an intron and the guanosine at the 5'-end of the adjoining exon, as shown by the boldface in intron-ApG**p**/Gp-exon, would that radioactivity wind up with the intron or the exon, assuming the lariat model of splicing?

Answers appear at end of book.

CHAPTER 10

Translation

Learning Objectives

In this chapter you will learn:

1. The fundamentals of the nature of proteins.
2. The structures of two important elements of translation: the ribosome and tRNA.
3. The nature of the genetic code.
4. The mechanism of translation.
5. How some antibiotics interfere with translation.

W hat is the secret of life?" I asked. "Protein," the bartender declared.

Kurt Vonnegut

Cat's Cradle

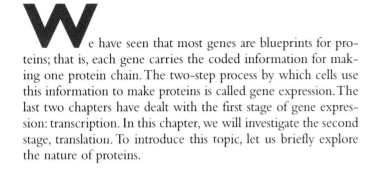

e have seen that most genes are blueprints for proteins; that is, each gene carries the coded information for making one protein chain. The two-step process by which cells use this information to make proteins is called gene expression. The last two chapters have dealt with the first stage of gene expression: transcription. In this chapter, we will investigate the second stage, translation. To introduce this topic, let us briefly explore the nature of proteins.

PROTEIN STRUCTURE

Proteins, like nucleic acids, are chainlike polymers of small subunits. In the case of DNA and RNA, the links in the chain are nucleotides. The chain links of proteins are **amino acids.** Whereas DNA contains only four different nucleotides, proteins contain twenty different amino acids. The structures of these compounds are shown in figure 10.1. The amino acid subunits join together in proteins via **peptide bonds,** as shown in figure 10.2. Another name for a chain of amino acids is **polypeptide.**

The linear order of amino acids constitutes a protein's **primary structure.** The way these amino acids interact with their neighbors gives a protein its **secondary structure.** The α-*helix* is a common form of secondary structure. It results from hydrogen bonding among near-neighbor amino acids, as shown in figure 10.3. Another common secondary structure found in proteins is the β-pleated sheet (figure 10.4). This involves extended protein chains, packed side by side, that interact by hydrogen bonding. The packing of the chains next to each other creates the sheet appearance. Silk is a protein very rich in β-pleated sheets. A third example of secondary structure is simply a turn. Such turns connect the α-helices and β-pleated sheet elements in a protein.

The total three-dimensional shape of a polypeptide is its **tertiary structure.** Figure 10.5 illustrates how the protein myoglobin folds up into its tertiary structure. Elements of secondary structure are apparent, especially the several α-helices of the molecule and the turns that connect them. Note the overall roughly spherical shape of myoglobin. Most polypeptides take this form, which we call *globular.*

Many proteins are large enough that they contain more than one compact structural region. Each of these regions is called a **domain.** Antibodies (the proteins our white blood cells make to repel invaders) provide a good example of domains. Each of the four polypeptides in the IgG-type antibody contains globular domains, as shown in figure 10.6. This figure also illustrates the highest level of protein structure: **quaternary structure.** This is the way two or more individual polypeptides fit together in a complex protein.

Figure 10.7 shows another convention for depicting protein structure. The β-sheets are represented by side-by-side flat arrows, which point toward the carboxyl terminus. The α-

helices are represented by the corkscrew-shaped ribbons at top. The α-helices are also frequently represented by cylinders (see figures 9.26–9.27, for example). There are two amino termini (N) and two carboxyl termini (C) in this protein, which tells us it is a dimer composed of two polypeptides, the α-chain (yellow, pink, and aqua) and the β-chain (blue). This drawing illustrates three levels of protein structure: (1) secondary structure (α-helices, β-sheets, and turns); (2) tertiary structure (the globular domains labeled α_1, α_2, α_3, and β_2m); and (3) quaternary structure (the two polypeptides (α and β). This is the structure of a human histocompatibility antigen, HLA-AZ (see chapter 3).

It was long assumed that a protein's amino acid sequence determined all of its higher levels of structure, much as the linear sequence of letters in this book determines word, sentence, and paragraph structure. However, this is an oversimplification. Many proteins cannot fold properly by themselves outside their normal cellular environment. Some cellular factors besides the protein itself are required in these cases.

> Proteins are polymers of amino acids linked through peptide bonds. The sequence of amino acids in a polypeptide (primary structure) gives rise to that molecule's local shape (secondary structure), overall shape (tertiary structure), and interaction with other polypeptides (quaternary structure).

PROTEIN FUNCTION

Why are proteins so important that most genes are dedicated to their creation? Some proteins provide the structure that helps give cells integrity and shape. Others serve as hormones to carry signals from one cell to another. For example, the pancreas secretes the hormone insulin that signals liver and muscle cells to take up the sugar glucose from the blood. Proteins can also bind and carry substances. The protein hemoglobin carries oxygen from the lungs to remote areas of the body; myoglobin stores oxygen in muscle tissue until it is used. Proteins also bind to genes and control their activity, as we have seen in chapters 8 and 9. Perhaps most important of all, proteins serve as enzymes that catalyze the thousands of chemical reactions necessary for life. Left to themselves, these reactions would take place far too slowly at ordinary body temperatures; without enzymes, life as we know it would be impossible.

THE RELATIONSHIP BETWEEN GENES AND PROTEINS

Our knowledge of the gene-protein link dates back as far as 1902, when Archibald Garrod, a physician, noticed that a human disease, alcaptonuria, behaved as if it were caused by a single recessive gene. Fortunately, Mendel's work had been rediscovered

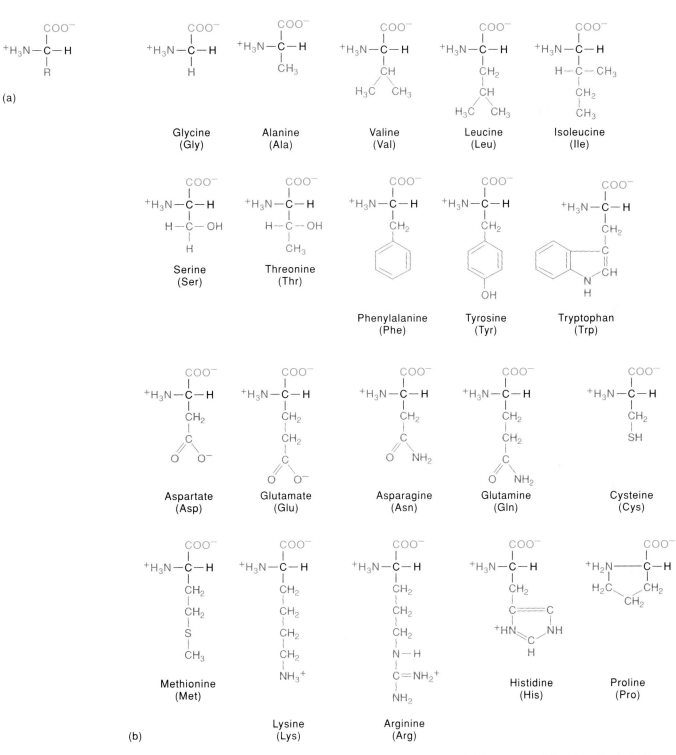

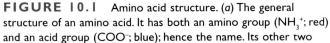

FIGURE 10.1 Amino acid structure. (*a*) The general structure of an amino acid. It has both an amino group (NH_3^+; red) and an acid group (COO^-; blue); hence the name. Its other two positions are occupied by a proton (H) and a side chain (R, gray). (*b*) There are twenty different side chains on the twenty different amino acids, all of which are illustrated here.

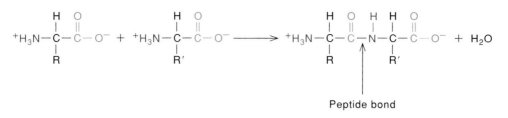

FIGURE 10.2   Formation of a peptide bond. Two amino acids with side chains R and R′ combine through the acid group of the first and the amino group of the second to form a dipeptide, two amino acids linked by a peptide bond. One molecule of water also forms as a by-product.

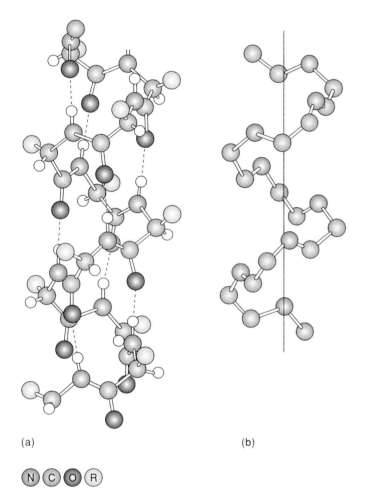

(a)   (b)

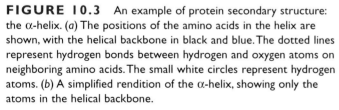

FIGURE 10.3   An example of protein secondary structure: the α-helix. (*a*) The positions of the amino acids in the helix are shown, with the helical backbone in black and blue. The dotted lines represent hydrogen bonds between hydrogen and oxygen atoms on neighboring amino acids. The small white circles represent hydrogen atoms. (*b*) A simplified rendition of the α-helix, showing only the atoms in the helical backbone.

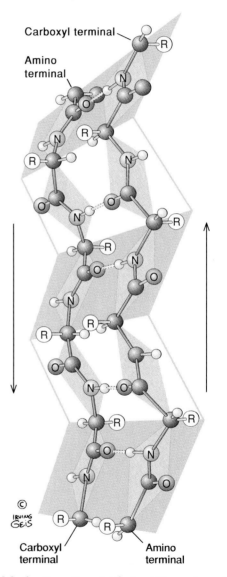

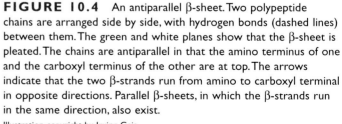

FIGURE 10.4   An antiparallel β-sheet. Two polypeptide chains are arranged side by side, with hydrogen bonds (dashed lines) between them. The green and white planes show that the β-sheet is pleated. The chains are antiparallel in that the amino terminus of one and the carboxyl terminus of the other are at top. The arrows indicate that the two β-strands run from amino to carboxyl terminal in opposite directions. Parallel β-sheets, in which the β-strands run in the same direction, also exist.

Illustration copyright by Irving Geis.

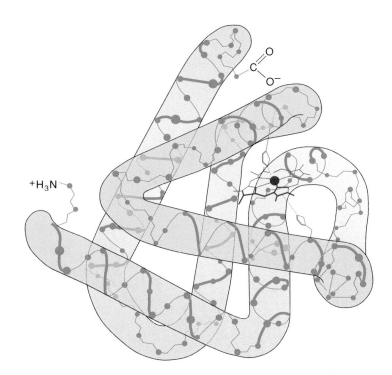

FIGURE 10.5 Tertiary structure of myoglobin. The several α-helical regions of this protein are represented by blue-green corkscrews. The overall molecule seems to resemble a sausage, twisted into a roughly spherical or globular shape. The heme group is shown in red, bound to two histidines (green polygons) in the protein.

two years earlier and provided the theoretical background for Garrod's observation. Patients with alcaptonuria excrete copious amounts of homogentisic acid, which has the startling effect of coloring their urine black. Garrod reasoned that the abnormal buildup of this compound resulted from a defective metabolic pathway. Somehow, a blockage somewhere in the pathway was causing the intermediate, homogentisic acid, to accumulate to abnormally high levels, much as a dam causes water to accumulate behind it. Several years later, Garrod postulated that the protein came from a defect in the pathway that degrades the amino acid phenylalanine (figure 10.8).

By that time, metabolic pathways had been studied for years and were known to be controlled by enzymes—one enzyme catalyzing each step. Thus, it seemed that alcaptonuria patients carried a defective enzyme. And since the disease was inherited in a simple Mendelian fashion, Garrod concluded that a gene must control the enzyme's production. When that gene is defective, it gives rise to a defective enzyme. This established the crucial link between genes and proteins.

George Beadle and E. L. Tatum carried this argument a step further with their studies of a common bread mold, *Neurospora*, in the 1940s. They performed their experiments as follows: First, they bombarded the peritheca (spore-forming parts) of *Neurospora* with X rays or ultraviolet radiation to cause mutations. Then, they collected the spores from the irradiated mold and germinated them separately to give pure strains of mold. They screened many thousands of strains to find a few mutants.

The mutants revealed themselves by their inability to grow on minimal medium composed only of sugar, salts, inorganic nitrogen, and the vitamin biotin. Wild-type *Neurospora* grows readily on such a medium, but the mutants had to be fed something extra—a vitamin, for example—in order to survive.

Next, Beadle and Tatum performed biochemical and genetic analyses on their mutants. By carefully adding substances, one at a time, to the mutant cultures, they pinpointed the biochemical defect. For example, the last step in the synthesis of the vitamin pantothenate involves putting together the two halves of the molecule: pantoic acid and β-alanine. One "pantothenateless" mutant would grow on pantothenate, but not on the two halves of the vitamin. This demonstrated that the last step in the biochemical pathway leading to pantothenate was blocked, so the enzyme (enzyme 3 in figure 10.9) that carries out that step must have been defective.

The genetic analysis was just as straightforward. *Neurospora* is an *ascomycete,* in which nuclei of two different mating types fuse and undergo meiosis to give eight haploid *ascospores,* borne in a fruiting body called an *ascus* (see chapter 5). Therefore a mutant can be crossed with a wild-type strain of the opposite mating type to give eight spores (figure 10.10). If the mutant phenotype is due to a mutation in a single gene, then four of the eight spores should be mutant and four should be wild-type. Beadle and Tatum collected the spores, germinated them separately, and checked the phenotypes of the resulting molds. Sure enough, they found that four of the eight spores gave rise to

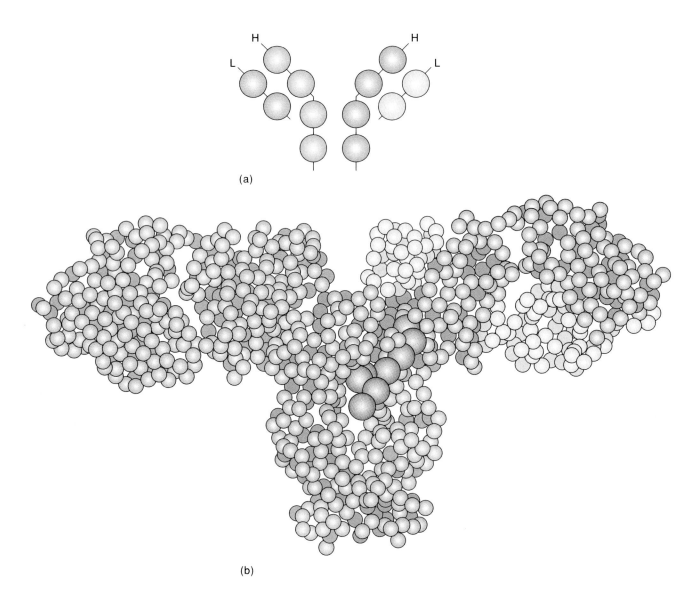

(a)

(b)

FIGURE 10.6   The globular domains of immunoglobulin. (a) Schematic diagram, showing the four polypeptides that comprise the immunoglobulin: two light chains (L) and two heavy chains (H). The light chains each contain two globular regions, while the heavy chains have four globular domains apiece. (b) Space-filling model of immunoglobulin. The colors correspond to those in (a). Thus, the two H-chains are in red and blue, while the L-chains are in green and yellow. A complex sugar attached to the protein is shown in gray. Note the globular domains in each of the polypeptides. Also note how the four polypeptides fit together to form the quaternary structure of the protein.

mutant molds, demonstrating that the mutant phenotype was controlled by a single gene. This happened over and over again, leading these investigators to the conclusion that each enzyme in a biochemical pathway is controlled by one gene. Subsequent work has shown that many enzymes contain more than one polypeptide chain and that each polypeptide is usually encoded in one gene. This is the **one gene–one polypeptide hypothesis.**

Most genes contain the information for making one polypeptide.

RIBOSOMES: PROTEIN SYNTHESIZING MACHINES

Figure 10.11 shows the apparent shapes of the *E. coli* **ribosome** and its two subunits, the *50S* and *30S* particles. The numbers 50S and 30S refer to the **sedimentation coefficients** of the two subunits. These coefficients are a measure of the speed with which the particles move through a solution when spun in an ultracentrifuge. The 50S particle, with a larger sedimentation coefficient, migrates more rapidly to the bottom of the centrifuge tube under the influence of a centrifugal force. The coefficients are functions of the mass and shape of the particles. Heavy particles sediment more rapidly than light ones; spherical

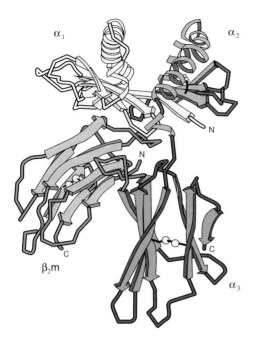

FIGURE 10.7   Schematic diagram of human histocompatibility antigen HLA-A2. Note the three levels of protein structure. (1) Three kinds of secondary structure are readily apparent: α-helices, represented by helical ribbons at top; β-sheets, represented by flat arrows; and turns, which are most obvious between β-strands at bottom. (2) The overall shape, dominated by the four globular domains (α₁, yellow; α₂, pink; α₃, aqua; and β₂m, light blue) is the protein's tertiary structure. (3) The subunit composition (one α-chain and one β-chain) is the quaternary structure. The α-chain is colored yellow, pink, and aqua; the β-chain is colored light blue. Yellow spheres represent the side chains of the amino acid cysteine, which link together through disulfide bonds. These bonds help hold the protein in the proper shape.

particles migrate faster than extended or flattened ones—just as a skydiver falls more rapidly in a tuck position than with arms and legs extended. The 50S particle is actually about twice as massive as the 30S. Together, the 50S and 30S subunits compose a 70S ribosome. Notice that the numbers do not add up. This is because the sedimentation coefficients are not proportional to the particle weight; actually, they are roughly proportional to the 2/3-power of the particle weight.

Each ribosomal subunit contains RNA and protein. The 30S particle includes one molecule of **ribosomal RNA (rRNA)** with a sedimentation coefficient of 16S, plus twenty-one ribosomal proteins. The 50S particle is composed of two rRNAs (23S + 5S) and thirty-four proteins (figure 10.12). All these ribosomal proteins are of course gene products themselves. Thus, a ribosome is produced by dozens of different genes.

Note that rRNAs participate in protein synthesis but do not code for proteins. Transcription is the only step in expression of the genes for rRNAs, aside from some trimming of the transcripts (chapter 9). No translation of these RNAs occurs.

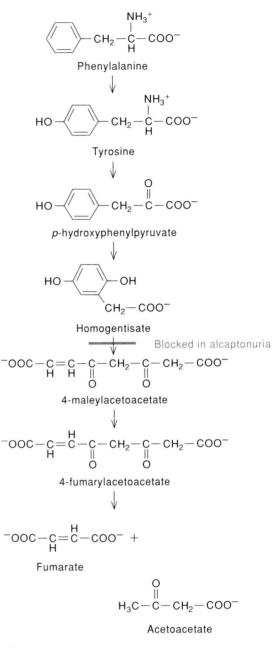

FIGURE 10.8   Pathway of phenylalanine breakdown. Alcaptonuria patients are defective in the enzyme that converts homogentisate to 4-maleylacetoacetate.

Ribosomes are the cell's protein factories. Bacteria contain 70S ribosomes with two subunits, called 50S and 30S. Each of these contains ribosomal RNA and many proteins.

SELF-ASSEMBLY OF RIBOSOMES

We know exactly how many proteins ribosomes contain, because several research groups, primarily Masayasu Nomura's, have taken ribosomes apart and put them back together again.

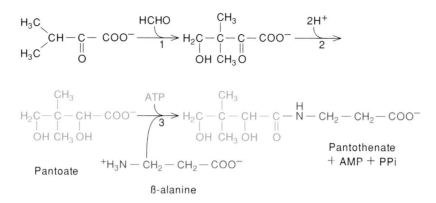

FIGURE 10.9 Pathway of pantothenate synthesis. The last step (step 3), formation of pantothenate from the two half-molecules, pantoate acid (blue) and β-alanine (red), was blocked in one of Beadle and Tatum's mutants. The enzyme that carries out this step must have been defective.

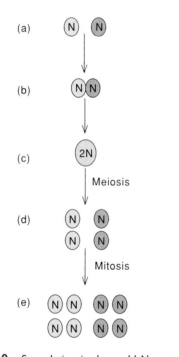

FIGURE 10.10 Sporulation in the mold *Neurospora crassa.* (a) Two haploid nuclei, one wild-type (yellow) and one mutant (blue), have come together in the immature fruiting body of the mold. (b) The two nuclei begin to fuse. (c) Fusion is complete, and a diploid nucleus (green) has formed. One haploid set of chromosomes is from the wild-type nucleus and one set is from the mutant nucleus. (d) Meiosis occurs, producing four haploid nuclei. If the mutant phenotype is controlled by one gene, two of these nuclei (blue) should have the mutant allele and two (yellow) should have the wild-type allele. (e) Finally, mitosis occurs, producing eight haploid nuclei, each of which will go to one ascospore. Four of these nuclei (blue) should have the mutant allele and four (yellow) should have the wild-type allele. If the mutant phenotype is controlled by more than one gene, the results will be more complex.

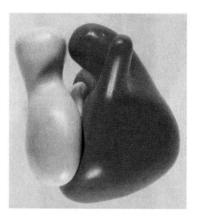

(a)

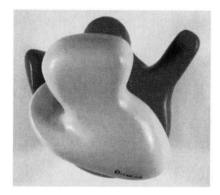

(b)

FIGURE 10.11 *E. coli* ribosome structure. (a) The 70S ribosome is shown from the "side" with the 30S particle (yellow) and the 50S particle (red) fitting together. (b) The 70S ribosome is shown rotated 90° relative to the view in (a). The 30S particle (yellow) is in front, with the 50S particle (red) behind.

From James A. Lake, *Journal of Molecular Biology* 105:131–59, 1976. © Academic Press.

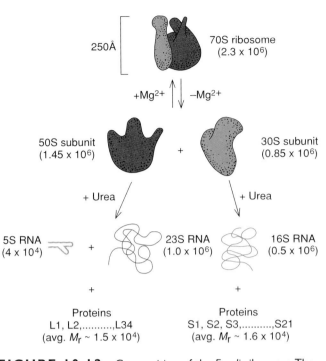

FIGURE 10.12   Composition of the *E. coli* ribosome. The arrows at the top denote the dissociation of the 70S ribosome into its two subunits when magnesium ions are withdrawn. The lower arrows show the dissociation of each subunit into RNA and protein components in response to the protein denaturant urea. The masses (M_r in daltons) of the ribosome and its components are given in parentheses.

FIGURE 10.13   Resolution of *E. coli* 30S ribosomal proteins. The proteins were electrophoresed on a polyacrylamide gel, which gave only partial resolution (fewer than twenty bands appear, so some contain more than one protein). The proteins that were incompletely resolved here were separated completely by chromatography.

From P. Traub and M. Nomura, Laboratory of Genetics, University of Wisconsin, Madison. *Proceedings of the National Academy of Science*, 59(3):777–784. March 1968.

This work, in the best biochemical tradition of separating and reconstituting, allows us to identify the components that are necessary to yield an active ribosome.

Figure 10.13 demonstrates the resolution of the proteins from the 30S particle by polyacrylamide gel electrophoresis. The proteins are named S1–S21, the S standing for "small," since they derive from the small ribosomal subunit. The proteins of the large subunit are therefore named L1–L31, where the L, of course, stands for "large." In some cases, two or more proteins co-migrated, but further purification separated these proteins completely. Once the proteins were completely resolved, they could be added back to 16S rRNA to produce fully active 30S particles. The test for activity was to combine these reconstituted 30S particles with 50S particles to give 70S ribosomes that could carry out translation in vitro.

What about the spatial relationship of the ribosomal proteins to each other and to the rRNA? Two techniques, neutron scattering and immune electron microscopy, have been most valuable in providing answers to these questions, at least for the small ribosomal particle of *E. coli*. Neutron scattering works according to the same principle as X-ray crystallography. The pattern of scattering of neutrons passing through a sample of ribosomes is related to the structure of the ribosome. Immune electron microscopy relies on antibodies that bind specifically to

one type of ribosomal protein. We can observe these antibodies on the surface of a ribosome by electron microscopy; in effect, the antibody is an arrow that points out the corresponding protein's location. Figure 10.14*a* shows the arrangement of the 30S ribosomal proteins relative to each other, and figure 10.14*b* superimposes the location of part of the 16S rRNA on this collection of proteins.

The 50S particle proved much more difficult to reconstitute than the small particle. In fact, the first experiments failed because the temperatures needed for reassociation of the 50S particle were so high that some of the proteins were inactivated. Later, tricks were discovered that could allow 50S ribosomes from *E. coli* to reassemble at a lower temperature. This means that the whole ribosome is a **self-assembly system;** it can be reconstituted from its component parts, with no intervention from outside enzymes or factors.

We assume that a similar assembly process occurs in vivo in prokaryotes. But what about assembly of eukaryotic ribosomes? Do they also self-assemble, and if so, where? We cannot answer the first part of that question because the complex eukaryotic ribosome has not been self-assembled in vitro. However, we do know the site of assembly of the eukaryotic ribosome: the nucleolus. The large ribosomal RNA genes are located in the nucleolus, and therefore the 5.8S, 18S, and 28S rRNAs are made there. Proteins are not synthesized in the nucleus, so the ribosomal proteins are made on cytoplasmic ribosomes. Then they must be imported into the nucleolus, where they join with rRNAs and build new ribosomes, which are exported to the cytoplasm to begin their task of making proteins.

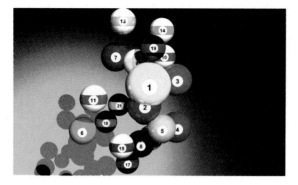

(a)

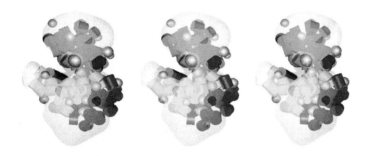

(b)

FIGURE 10.14   Arrangement of components in the *E. coli* 30S ribosomal particle. (*a*) Positions of proteins. The relative positions of the small subunit ribosomal proteins (numbered 1–21) are shown, with white contour lines indicating the overall shape of the 30S particle. The orientation of the particle is close to that presented in figure 10.11*a*. The different colors have no meaning other than to help distinguish the different proteins. (*b*) Spatial relationship of 16S rRNA and proteins. Silver spheres represent the positions of the proteins as determined by neutron diffraction. Protein diameters have been scaled to 50% of their actual values, to permit visualization of the double helical regions of the 16S RNA (cylinders). These are color coded, with the 5′-domain shown in red, the central domain in yellow, and the 3′-domain in blue. The surface of the subunit, as determined by electron microscopy, is shown as a transluscent gray surface. This view is from the cytoplasmic side of the 30S subunit, so the interface with the 50S subunit is on the backside of this structure. This is the same orientation as in panel (*a*). The image is a stereo triple: the left pair of images can be viewed with a stereo viewer, or by relaxing the eyes, while the pair on the right is for cross-eyed viewing.

(*a*) Courtesy of Peter Moore, Yale University, and Malcolm Capel, Brookhaven National Lab. (*b*) Image courtesy of Dr. Stephen Harvey, University of Alabama at Birmingham. © 1994, Academic Press, Ltd.

Bacterial ribosomal subunits can be dissociated into their component parts—RNAs and proteins—and then reassociated. Thus, the bacterial ribosome is a self-assembly system that needs no help from the outside. Eukaryotic ribosomes are assembled in the nucleolus, then exported to the cytoplasm, where they play their role in protein synthesis.

FUNCTIONS OF RIBOSOMAL PROTEINS

The in vitro reassociation of ribosomal particles makes it possible to examine the role of each ribosomal protein in protein synthesis. The idea of such an experiment is to create a mutant with a suspected defect in its ribosomes, then reassemble the particles in several different reactions. In each reaction there is one ribosomal protein from the mutant bacterium, plus all the other proteins from a wild-type bacterium. The protein from the mutant is changed in each reaction until a defective ribosome results. The mutant protein in that reassembly reaction must be the one that caused the defect.

The classic example of this approach is the identification of the protein responsible for resistance to the antibiotic **streptomycin.** Streptomycin kills bacterial cells by causing their ribosomes to misread mRNAs. For example, ribosomes given poly(U) as mRNA would ordinarily incorporate only phenylalanine into polyphenylalanine. In the presence of streptomycin, however, ribosomes also incorporate isoleucine, serine, and leucine. Apparently, streptomycin distorts the ribosome in such a way that it partially destroys fidelity of translation. Where does streptomycin act? To answer that question, we can take components from wild-type (streptomycin-sensitive) ribosomes and from streptomycin-resistant ribosomes, mix them, and see whether the reconstituted ribosomes are streptomycin-sensitive or -resistant.

For example, when the 30S particle came from a streptomycin-resistant bacterium and the 50S particle came from a streptomycin-sensitive bacterium, the reconstituted ribosome was streptomycin-resistant. Therefore, streptomycin apparently acts on the 30S particle. Does the streptomycin act on the ribosomal RNA or on protein? Reconstitution experiments showed that if the proteins came from a resistant bacterium, the ribosome was resistant, regardless of the origin of the rRNA. So streptomycin involves the protein, not the RNA. But which protein(s) are involved? The same kind of experiment showed that when the protein S12 came from a resistant bacterium, the reconstituted ribosome was resistant, regardless of the origin of the other proteins. Thus, S12 is the apparent site of action of streptomycin.

This does not necessarily mean that S12 is the binding site for streptomycin. In fact, the antibiotic will still bind tightly to 30S particles in the absence of S12, but only if two other proteins, S3 and S5, are present. Furthermore, an antibody against S12 only faintly inhibits streptomycin binding to ribosomes. Apparently, therefore, even though S12 seems to be the locus of action of streptomycin, other ribosomal proteins define the antibiotic binding site.

In vitro reconstitution experiments with mutant and wild-type ribosomal proteins allow us to pinpoint the defective component by mixing one mutant protein at a time with all other wild-type components and observing the effects on the reconstituted ribosomes. This strategy showed that protein S12 is the site of action of the antibiotic streptomycin.

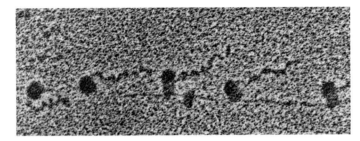

(a)

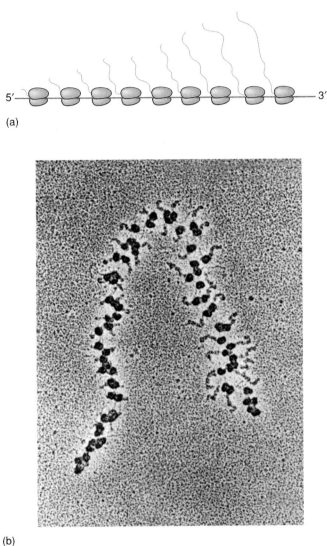

(b)

(c)

FIGURE 10.15 Polysomes. (a) Schematic diagram of a polysome. The ribosomes (green) are translating the mRNA (red), producing nascent polypeptide chains (blue). Since the polypeptides grow longer toward the right, translation is moving left-to-right. The 5′-end of the mRNA is therefore on the left. (b) Electron micrograph of a polysome from the midge *Chironomus*. The 5′-end on the mRNA is at lower left, and the mRNA bends up and then down to the 3′-end at lower right. The dark blobs attached to the mRNA are ribosomes. The fact that there are many (74) of them is the reason for the name *polysome*. Nascent polypeptides extend away from each ribosome, and grow longer as the ribosomes approach the end of the mRNA. The faint blobs on the nascent polypeptides are not individual amino acids but domains containing groups of amino acids. (c) High resolution image of a *Chironomus* polysome that was stretched during preparation. Individual ribosomal subunits are visible, and the nascent protein chains seem to emerge from a point between the two subunits.

(b and c) From Francke, et al., "Electron Microscopic Vizualization of a Discrete Class of Giant Translation Units in Salivary Gland Cells of *Chirohomus tetans*." *EMBO* I:(1–6)59–62, 1982. European Molecular Biology Organization.

POLYSOMES

We have seen in chapters 8 and 9 that more than one RNA polymerase can transcribe a gene at one time. The same is true of ribosomes and mRNA. In fact, it is common for many ribosomes to be traversing the same mRNA in tandem at any given time. The result is a polyribosome, or **polysome,** such as the partial one pictured in figure 10.15b. Even in this part of the polysome we can count at least 40 ribosomes translating the mRNA simultaneously. We can also tell which end of the polysome is which, by looking at the nascent polypeptide chains. These grow longer and longer as we go from the 5′-end (where translation begins) to the 3′-end (where translation ends). Therefore the 5′-end is at left, while the 3′-end is at right.

The polysome in figure 10.15 is from a eukaryote (the midge). Since transcription and translation occur in different compartments in eukaryotes (transcription in the nucleus, translation in the cytoplasm), polysomes will always occur in the cytoplasm. Prokaryotes also have polysomes, but the picture in these organisms is complicated by the fact that transcription and translation of a given gene and its mRNA occur simultaneously, and in the same location. Thus, we can see nascent mRNAs being synthesized by RNA polymerase and being translated by ribosomes at the same time. Figure 10.16 shows just such a situation in *E. coli*. We can see two segments of the bacterial chromosome running parallel from left to right. Only the segment on top is being transcribed. We can tell that transcription is occurring from left to right in this picture because the polysomes are getting longer as we move in that direction; as they get longer, they have room for more and more ribosomes. Do not be misled by the difference in scale between figures 10.15 and 10.16; the ribosomes are smaller, and the nascent protein chains are not visible in the latter picture. Remember also that the strands running across figure 10.16 are DNA, whereas the one in figure 10.15 is mRNA. The mRNAs are more or less vertical in figure 10.16.

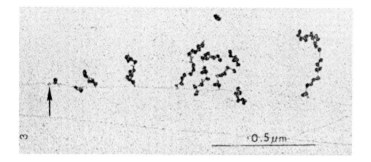

FIGURE 10.16 Simultaneous transcription and translation in *E. coli*. Two DNA segments stretch across the picture from left to right. The top segment is being transcribed from left to right. As the mRNAs grow, more and more ribosomes attach and carry out translation. This gives rise to polysomes, which are arrayed perpendicular to the DNA. The nascent polypeptides are not visible in this picture. The arrow at left points to a faint spot, which may be an RNA polymerase just starting to transcribe the gene. Other such spots denoting RNA polymerase appear at the bases of some of the other polysomes, where the mRNAs join the DNA.

From O. L. Miller, "Visualization of Bacterial Genes in Action," *Science*, 169:394, July 1970. © 1970 American Association for the Advancement of Science.

Most mRNAs are translated by more than one ribosome at a time; the result, a structure in which many ribosomes translate an mRNA in tandem, is called a polysome. In eukaryotes, polysomes are found in the cytoplasm. In prokaryotes, transcription of a gene and translation of the resulting mRNA occur simultaneously. Therefore, active genes are associated with many polysomes.

Transfer RNA: The Adapter Molecule

The transcription mechanism was easy to predict. RNA resembles DNA so closely that it follows the same base-pairing rules. By following these rules, RNA polymerase produces replicas of the genes it transcribes. But what rules govern the ribosome's translation of mRNA to protein? This is a true translation problem. A nucleic acid language must be translated to a protein language. Francis Crick suggested the answer to this problem in a 1958 paper before much experimental evidence was available to back it up. What is needed, Crick reasoned, is some kind of adapter molecule that can recognize the nucleotides in the RNA language as well as the amino acids in the protein language. He was right. He even noted that there was a type of small RNA of unknown function that might play the adapter role. Again, he guessed right. Of course there were some bad guesses in this paper as well, but even they were important. By their very creativity, Crick's ideas stimulated the research (some from Crick's own laboratory) that led to solutions to the puzzle of translation.

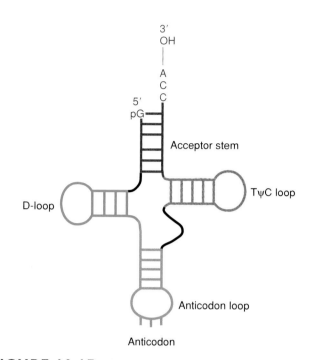

FIGURE 10.17 Cloverleaf structure of yeast tRNA^{Phe}. At top is the acceptor stem (red), where amino acid binds to 3'-terminal adenosine. At left is the dihydro U loop (D-loop, blue), which contains at least one dihydrouracil base. At bottom is the anticodon loop (green), containing the anticodon. The TψC loop (right, gray) contains the virtually invariant sequence TψC. Each loop is defined by a base-paired stem of the same color.

The Adapter Function of tRNA

The adapter molecule in translation is indeed a small RNA that recognizes both RNA and amino acids; it is called **transfer RNA (tRNA).** Figure 10.17 shows a schematic diagram of a tRNA that recognizes the amino acid phenylalanine (Phe). Later in this chapter we will discuss the structure and function of tRNA in detail. For the present, the *cloverleaf* model, though it bears scant resemblance to the real shape of tRNA, will serve to point out the fact that the molecule has two "business ends." One end (the top of the model) attaches to an amino acid. Since this is a tRNA specific for phenylalanine (tRNA^{Phe}), only phenylalanine will bind. An enzyme called phenylalanine-tRNA synthetase catalyzes this binding. The generic name for such enzymes is **aminoacyl-tRNA synthetase.**

The other end (the bottom of the model) contains a three-base sequence that pairs with a complementary three-base sequence in an mRNA. Such a triplet in mRNA is called a **codon;** naturally enough, its complement in a tRNA is called an **anticodon.** The codon in question here has attracted the anticodon of a tRNA bearing a phenylalanine. That means that this codon says: "Please insert one phenylalanine into the protein at this point." A schematic of codon-anticodon recognition appears in figure 10.18. The recognition between codon and anticodon, mediated by the ribosome, obeys the same Watson-Crick rules as any other double-stranded polynucleotide, at least

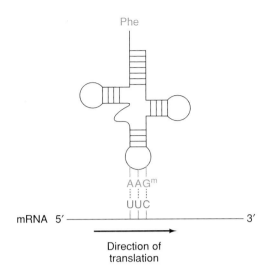

FIGURE 10.18 Codon-anticodon recognition. The recognition between a codon in an mRNA and a corresponding anticodon in a tRNA obeys essentially the same Watson-Crick rules as apply to other polynucleotides. Here, a 3'AAG^m5' anticodon (green) on a tRNA^Phe is recognizing a 5'UUC3' codon (red) for phenylalanine in an mRNA. The G^m denotes a methylated G, which base-pairs like an ordinary G. Notice that the tRNA is pictured backwards (3' → 5') relative to normal convention, which is 5' → 3', left to right. That was done to put its anticodon in the proper orientation (3' → 5', left to right) to base-pair with the codon, shown conventionally reading 5' → 3', left to right. Remember that the two strands of DNA are antiparallel; this applies to any double-stranded polynucleotide, including one as small as the three-base codon-anticodon pair.

in the case of the first two base pairs. The third pair is allowed somewhat more freedom, as we will see.

> Two important sites on tRNAs allow them to recognize both amino acids and nucleic acids. They are therefore capable of serving the adapter role postulated by Crick and are the key to the mechanism of translation.

Secondary Structure of tRNA

Look again at figure 10.17, which shows the familiar cloverleaf structure of tRNA. Remember that the molecule does not really look like this in three dimensions; the cloverleaf model simply points out the secondary structure of tRNA—that is, the base-paired regions of the molecule. At the top of the diagram lie the two terminal regions of the RNA: the 5'-end on the left and the 3'-end on the right. These two regions base-pair with one another to form the so-called **acceptor stem.** Three other base-paired stems occur, defining three loops: the **dihydrouracil loop (D-loop),** named for the modified uracil bases this region always contains; the **anticodon loop,** named for the all-important anticodon near its apex; and the **TψC**

loop, which takes its name from a nearly invariant sequence of three bases, TψC. The psi (ψ) stands for a modified nucleoside, **pseudouridine.** It is the same as normal uridine, except that the base is linked to the ribose through the 5-carbon of the base instead of the 1-nitrogen. The region between the anticodon loop and the TψC loop in figure 10.17 is called the **variable loop** because it varies in length from four to thirteen nucleotides in different tRNAs; some of the longer variable loops contain base-paired stems.

Since tRNAs contain many modified bases in addition to dihydrouracil, a reasonable question would be: Is the tRNA made with modified bases, or are the bases modified after synthesis of the tRNA is finished? The answer is that tRNAs are made with the same four ordinary bases as any other RNA. Then, once transcription is complete, multiple enzyme systems modify the bases. What effects, if any, do these modifications have on tRNA function? We know they are important, because at least two tRNAs have been made in vitro with the four normal, unmodified bases, and they were unable to bind amino acids.

> The cloverleaf model of tRNA shows its secondary (base-pairing) structure and reveals four base-paired stems that define three loops of RNA: the D-loop, the anticodon loop, and the TψC loop. Transfer RNAs contain many modified bases, at least some of which are vital to tRNA activity.

Three-Dimensional Structure of tRNA

Alexander Rich and his colleagues used X-ray diffraction techniques to reveal the three-dimensional structure, or tertiary structure, of several tRNAs. Since all tRNAs have essentially the same secondary structure, represented by the cloverleaf model, it is perhaps not too surprising that they all have essentially the same tertiary structure as well. This L-shaped structure is shown in figure 10.19.

The key to understanding the three-dimensional structure of tRNA is to realize that the longer the base-paired regions in a tRNA are, the more stable they will be. The centipede metaphor we used in chapter 6 helps illuminate this point. Each base pair is like a pair of legs on two embracing centipedes; the more legs, the tighter the grip. However, this imagery does not convey the idea that ten legs in a row are much better than two separated sets of five, which is certainly true of base pairs. Transfer RNA maximizes the length of its base-paired stems by lining them up in two extended regions. One of these lies horizontally at the top of the molecule and encompasses the acceptor stem and the TψC stem; the other forms the vertical axis of the molecule and includes the D stem and the anticodon stem. Even though the two parts of each stem are not aligned perfectly and the stems therefore bend slightly, the alignment confers stability.

Figure 10.19 also demonstrates that the bases of the anticodon do not extend inward to base-pair with other parts of the

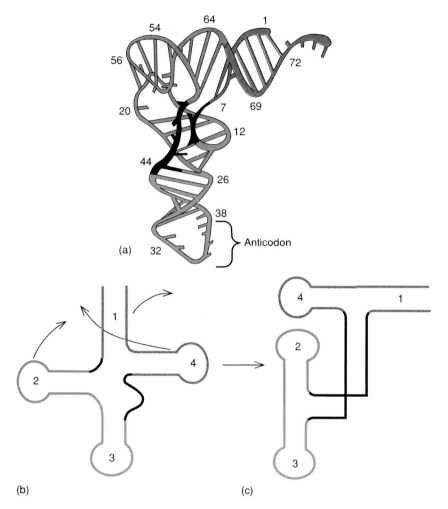

FIGURE 10.19   Three-dimensional structure of tRNA. (*a*) A planar projection of the three-dimensional structure of yeast tRNA^Phe. The various parts of the molecule are color-coded to correspond to (*b*) and (*c*). (*b*) Familiar cloverleaf structure of tRNA with same color scheme as (*a*). Arrows indicate the contortions this cloverleaf would have to go through to achieve the approximate shape of a real tRNA, shown in (*c*).

tRNA molecule. Rather, they protrude outward to the right so they can base-pair with a codon in an mRNA. The three bases of the anticodon are stacked parallel to one another, suggesting a helical arrangement. In fact, the structure of the tRNA twists the anticodon into a helical shape that greatly facilitates its base-pairing with the codon. The ribosome presumably makes this binding even easier.

> The three-dimensional structure of tRNA consists of an L-shaped framework in which the four stems stack upon one another to produce two long, double-helical regions that help stabilize the molecule. The bases of the anticodon protrude to one side and are arranged in a helical shape so they can more easily form a partial double helix by pairing with the three-base codon.

AMINO ACID BINDING TO tRNA

When an amino acid binds to the 3′-end of a tRNA, we say the tRNA is **charged** with that amino acid. All tRNAs have the same three bases (CCA) at their 3′-ends, and the terminal adenosine is the target for charging. Figure 10.20 gives the chemical structure of this terminal CCA and shows an amino acid attached through its acid group to the 3′-hydroxyl group of the adenosine's ribose. As we learned in chapter 6, this kind of linkage between an acid group and a hydroxyl group is called an ester bond.

Charging takes place in two steps, both catalyzed by the enzyme aminoacyl-tRNA synthetase. In the first step (figure 10.21), the amino acid is activated, using energy from ATP; the product of the reaction is aminoacyl-AMP. The pyrophosphate by-product is simply the two end phosphate groups from the ATP, which were lost in forming AMP. The bonds between phosphate groups in ATP (and the other nucleoside triphosphates) are "high energy bonds." When they are broken, this

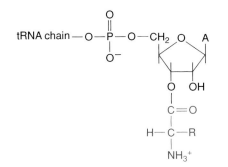

FIGURE 10.20 Linkage between tRNA and an amino acid. Some amino acids are bound initially by an ester linkage to the 3'-hydroxyl group of the terminal adenosine of the tRNA as shown, but some bind initially to the 2'-hydroxyl group. In any event, the amino acid is transferred to the 3'-hydroxyl group before it is incorporated into a protein.

energy is released. In this case, the energy is trapped in the **aminoacyl-AMP,** which is why we call this an **activated amino acid.** The principle is the same as burning wood to release energy, which is then used to boil water. The boiling water is activated in much the same way as the aminoacyl-AMP. In the second reaction of charging, the energy in the aminoacyl-AMP is used to transfer the amino acid to a tRNA, forming **aminoacyl-tRNA.** This is comparable to using the boiling water to cook an egg.

Just like other enzymes, an aminoacyl-tRNA synthetase plays a dual role. Not only does it catalyze the reaction leading to an aminoacyl-tRNA, but it determines the specificity of this reaction. A given organism has only twenty synthetases, one for each amino acid. And the synthetases are very specific; each will almost always place an amino acid on the right kind of tRNA.

Consider what would happen if these enzymes made many mistakes. For example, if serine-tRNA synthetase, which attaches the amino acid serine to its cognate tRNA (tRNA^{Ser}), suddenly started attaching lysine instead, the result would be catastrophic (figure 10.22). The lysine-tRNA^{Ser} would still recognize serine codons, because its anticodon would be unchanged and the amino acid plays no role in codon-anticodon recognition. But this aminoacyl-tRNA contains lysine instead of serine, so it would insert lysines into proteins where serines ought to go. Since a protein's structure and function depend absolutely on its primary structure, or amino acid sequence, these changes would in all likelihood destroy the protein's usefulness.

SPECIFICITY OF AMINOACYL-tRNA SYNTHETASES

Fortunately, the aminoacyl-tRNA synthetases are *very* specific, so mistakes occur only rarely. What determines the great specificity of these enzymes? Surely each one must interact in a very

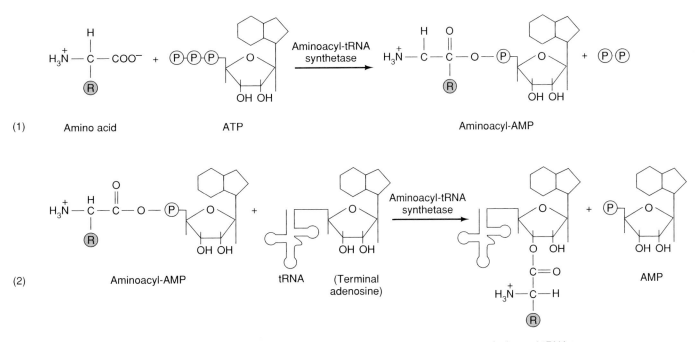

FIGURE 10.21 Aminoacyl-tRNA synthetase activity. Reaction 1: The aminoacyl-tRNA synthetase couples an amino acid to AMP, derived from ATP, to form an aminoacyl-AMP, with pyrophosphate (P-P) as a by-product. Reaction 2: The synthetase replaces the AMP in the aminoacyl-AMP with tRNA, to form an aminoacyl-tRNA, with AMP as a by-product. The amino acid is joined to the 3'-hydroxyl group of the terminal adenosine of the tRNA. For simplicity, the adenine base is merely indicated in outline.

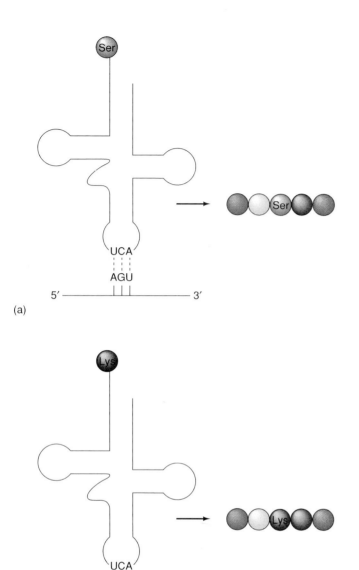

(a)

(b)

FIGURE 10.22 Effects of charging mistakes made by an aminoacyl-tRNA synthetase. (a) Normal situation. This tRNA^{Ser} has a 3′ UCA 5′ anticodon, so it recognizes the serine codon AGU. It is charged with serine (green), as it should be, so a serine is inserted into the growing polypeptide chain (right). (b) Abnormal situation. This is the same tRNA^{Ser}, with the same anticodon, so it still recognizes the serine codon AGU. But it has been mischarged with a lysine (red), so it inserts lysine into the growing polypeptide chain, where a serine belongs. This could drastically change the characteristics of the finished protein. Note that these tRNAs are rendered backwards (3′ → 5′ left to right) so they base-pair with the mRNAs presented conventionally (5′ → 3′ left to right).

specific way with a particular tRNA, but what does it recognize in this tRNA that is present in no other? One thing that distinguishes one tRNA from all others is its anticodon, so we would predict that this contributes to recognition. In fact, that is part of

the answer, but the whole picture is considerably more complex. It now appears that the synthetases recognize several tRNA regions, dubbed **recognition elements** by Thomas Steitz and his colleagues.

The first of these elements is the acceptor stem of the tRNA. In 1972, Dieter Söll and his coworkers pointed out that the identity of the base next to the CCA in the acceptor stem correlates with the nature of the cognate amino acid of that tRNA. Thus, for example, tRNAs with an A in that position tend to be charged with nonpolar amino acids, while tRNAs with a G in that position tend to be charged with polar ones. Therefore, these workers called this the **discriminator base.** Clearly, however, this can be only a rough discrimination since there are only four possible bases in this position, and twenty different amino acids. Over twenty-five years later, Paul Schimmel and his colleagues demonstrated the importance of one base pair in the acceptor stem. They showed that changing a single base pair (a G–U pair) in the acceptor stem of tRNA^{Ala} destroys its ability to be charged with alanine. Moreover, introducing a G–U pair into the same position of the acceptor stem of a cysteine or phenylalanine tRNA converts them to alanine tRNAs. Thus, the G–U pair in that position is a key recognition element of an alanine tRNA. The G–U base pair is a so-called wobble pair. We will consider such unorthodox combinations later in this chapter.

More recently, Söll and Steitz and their colleagues have used X-ray crystallography to determine the three-dimensional structure of glutaminyl-tRNA synthetase bound to tRNA^{Gln}. Figure 10.23 presents this structure. Near the top, we see a deep cleft in the enzyme that enfolds the acceptor stem, including the discriminator base and the counterpart of the important G–U pair in tRNA^{Ala}. This base pair actually appears to be broken in the complex with the synthetase. At the lower left, we observe a smaller cleft in the enzyme into which the anticodon of the tRNA protrudes. This would allow for specific recognition of the anticodon by the synthetase. In addition, we see that most of the left side of the enzyme is in intimate contact with the inside of the L of the tRNA. Thus, bases in this region may also be recognition elements.

Some investigators once assumed that there is a simple recognition code between synthetase and tRNA, rather like the genetic code that determines the recognition between codon and anticodon. This simple hypothesis gave rise to the phrase "second genetic code" to describe the synthetase-tRNA recognition. We now see that this term is much too simple to do justice to the complex interactions between a synthetase and the multiple recognition elements on its corresponding tRNA.

Each aminoacyl-tRNA synthetase couples its specific amino acid to one of its cognate tRNAs. The reaction, which is very specific for both amino acid and tRNA, requires activation of the amino acid by ATP to form aminoacyl-AMP, then transfer of the activated amino acid to the appropriate tRNA. Each synthetase recognizes multiple recognition elements on its tRNA.

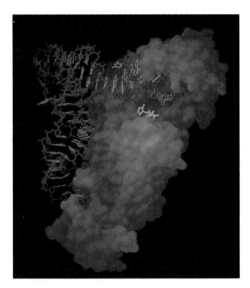

FIGURE 10.23 Three-dimensional structure of glutaminyl-tRNA synthetase complexed with tRNA and ATP. The synthetase is shown in blue, the tRNA in red and yellow, and the ATP in green. Note the three areas of contact between enzyme and tRNA: (1) the deep cleft at top that holds the acceptor stem of the tRNA, and the ATP; (2) the smaller pocket at lower left into which the tRNA's anticodon protrudes; and (3) the area in between these two clefts, which contacts much of the inside of the L of the tRNA.

From Dr. M. A. Rould, *Science*, 236:cover, 7 December 1989. © 1989 American Association for the Advancement of Science.

THE GENETIC CODE

The term **genetic code** refers to the set of three-base code words (codons) in mRNAs that stand for the twenty amino acids in proteins. Like any code, this one had to be broken before we knew what the codons stood for. Indeed, before 1960, other more basic questions about the code were still unanswered. These included: Do the codons overlap? Are there gaps, or "commas," in the code? How many bases make up a codon? These questions were answered in the 1960s by a series of imaginative experiments, which we will describe here.

NONOVERLAPPING CODONS

In a nonoverlapping code, each base is part of at most one codon. In an overlapping code, one base may be part of two or even three codons. Consider the following micromessage:

AUGUUC

Assuming that the code is triplet (three bases per codon) and this message is read from the beginning, the codons will be AUG and UUC if the code is nonoverlapping. On the other hand, an overlapping code might yield four codons: AUG, UGU, GUU, and UUC. As early as 1957, Sydney Brenner realized that an overlapping triplet code like this would be impossible. He reasoned as follows: consider the UUC codon above. If the code is fully overlapping, the next codon upstream

would be GUU. We could change the G to any of the other three bases to create three other codons, AUU, CUU, and UUU. These four codons could code for at most four different amino acids upstream of the amino acid encoded by UUC. But looking at almost any given amino acid in all the proteins that had been sequenced even in 1957, Brenner found more than four different neighbors on the amino-terminal side. For example, lysine had 17 different N-neighbors. This analysis was complicated by the fact that lysine might have more than one codon (it actually has two); nevertheless, even without knowing the number of codons for each amino acid, Brenner was able to conclude that a fully overlapping triplet code is impossible.

Given the data available in 1957, a partially overlapping code remained possible, but A. Tsugita and H. Frankel-Conrat laid it to rest with the following line of reasoning: If the code is nonoverlapping, a change of one base in an mRNA (a **missense mutation**) would change no more than one amino acid in the resulting protein. For example, consider another micromessage:

AUGCUA

Assuming that the code is triplet (three bases per codon) and this message is read from the beginning, the codons will be AUG and CUA if the code is nonoverlapping. A change in the third base (G) would change only one codon (AUG) and therefore at most only one amino acid. On the other hand, if the code were overlapping, base G could be part of three adjacent codons (AUG, UGC, and GCU). Therefore, if the G were changed, up to three adjacent amino acids could be changed in the resulting protein. But when the investigators introduced one-base alterations into mRNA from tobacco mosaic virus (TMV), they found that these caused changes in only one amino acid. Hence, the code must be nonoverlapping.

NO GAPS IN THE CODE

If the code contained untranslated gaps, or commas, mutations that add or subtract a base from a message might change a few codons, but we would expect the ribosome to get back on track after the next comma. In other words, these mutations might frequently be lethal, but there should be many cases where the mutation occurs just before a comma in the message and therefore has little if any effect. If there were no commas to get the ribosome back on track, these mutations would be lethal except when they occur right at the end of a message.

Such mutations do occur, and they are called **frame-shift mutations;** they work as follows. Consider another tiny message:

AUGCAGCCAACG

If translation starts at the beginning, the codons will be AUG, CAG, CCA, and ACG. If we insert an extra base (X) right after base U, we get:

AUXGCAGCCAACG

Now this would be translated from the beginning as AUX, GCA, GCC, AAC. Notice that the extra base changes not only the codon (AUX) in which it appears, but every codon from that point on. The **reading frame** has shifted one base to the left; whereas C was originally the first base of the second codon, G is now in that position. This could be compared to a filmstrip viewing machine that slips a cog, thus showing the last one-third of one picture together with the first two-thirds of the next.

On the other hand, a code with commas would be one in which each codon is flanked by one or more untranslated bases, represented by Z's in the following message. The commas would serve to set off each codon so the ribosome could recognize it:

AUGZCAGZCCAZACGZ

Deletion or insertion of a base anywhere in this message would change only a single codon. The comma (Z) at the end of the damaged codon would then put the ribosome back on the right track. Thus, addition of an extra base (X) to the first codon would give the message:

AUXGZCAGZCCAZACGZ

The first codon (AUXG) is now wrong, but all the others, still neatly set off by Z's, would be translated normally.

When Francis Crick and his colleagues treated bacteria with acridine dyes that usually cause single-base insertions or deletions, they found that such mutations were very severe; the mutant genes gave no functional product. This is what we would expect of a "comma-less" code with no gaps; base insertions or deletions cause a shift in the reading frame of the message that persists until the end of the message.

Moreover, Crick found that adding a base could cancel the effect of deleting a base, and vice versa. This phenomenon is illustrated in figure 10.24, where we start with an artificial gene composed of the same codon, CAT, repeated over and over. When we add a base, G, in the third position, we change the reading frame so that all codons thereafter read TCA. When we start with the wild-type gene and delete the fifth base, A, we change the reading frame in the other direction, so that all subsequent codons read ATC. Crossing these two mutants sometimes gives a recombined "pseudo-wild-type" gene like the one on line 4 of the figure. Its first two codons, CAG and TCT, are wrong, but thereafter the insertion and deletion cancel, and the original reading frame is restored; all codons from that point on read CAT.

THE TRIPLET CODE

Crick and Leslie Barnett discovered that a set of three adjacent insertions or deletions could produce a wild-type gene (figure 10.24, line 5). This of course demands that a codon consist of three bases. As Crick remarked to Barnett when he saw the experimental result: "We're the only two [who] know it's a triplet code!"

1. Wild-type:	CAT	CAT	CAT	CAT	CAT
2. Add a base:	CAG	TCA	TCA	TCA	TCA
3. Delete a base:	CAT	CTC	ATC	ATC	ATC
4. Cross #2 and #3:	CAG	TCT	CAT	CAT	CAT
5. Add 3 bases:	CAG	GGT	CAT	CAT	CAT

FIGURE 10.24 Frameshift mutations. Line 1: An imaginary gene has the same codon, CAT, repeated over and over. The vertical dotted lines show the reading frame, starting from the beginning. Line 2: Adding a base, G (red), in the third position changes the first codon to CAG and shifts the reading frame one base to the left so that every subsequent codon reads TCA. Line 3: Deleting the fifth base, A (marked by the triangle), from the wild-type gene changes the second codon to CTC and shifts the reading frame one base to the right so that every subsequent codon reads ATC. Line 4: Crossing the mutants in lines 2 and 3 sometimes gives pseudo-wild-type revertants with an insertion and a deletion close together. The end result is a DNA with its first two codons altered, but all the other ones put back into the correct reading frame. Line 5: Adding three bases, GGG (red), after the first two bases disrupts the first two codons, but leaves the reading frame unchanged. The same would be true of deleting three bases.

In 1961, Marshall Nirenberg and Johann Heinrich Matthaei performed a groundbreaking experiment that laid the groundwork for confirming the triplet nature of the code, and for breaking the genetic code itself. The experiment was deceptively simple; it showed that synthetic RNA could be translated in vitro. In particular, when Nirenberg and Matthaei translated poly(U), a synthetic RNA composed only of U's, they made polyphenylalanine. Of course, that told them that there is a codon for phenylalanine that contains only U's. This finding by itself was important, but the long-range implication was that one could design synthetic mRNAs of defined sequence and analyze the protein products to shed light on the nature of the code. Gobind Khorana and his colleagues were the chief practitioners of this strategy.

Here is how Khorana's synthetic messenger experiments confirmed that the codons contain three bases: First, if the codons contain an odd number of bases, then a repeating dinucleotide [poly(UC) or UCUCUCUC . . .] should contain two alternating codons (UCU and CUC, in this case), no matter where translation starts. The resulting protein would be a repeating dipeptide—two amino acids alternating with each other. If codons have an even number of bases, only one codon (UCUC, for example) should be repeated over and over. Of course, if translation started at the second base, the single repeated codon would be different (CUCU). In either case, the resulting protein would be a homopolypeptide, containing only one amino acid repeated over and over. Khorana found that poly(UC) translated to a repeating dipeptide, poly(serine-leucine) (figure 10.25a), proving that the codons contained an odd number of bases.

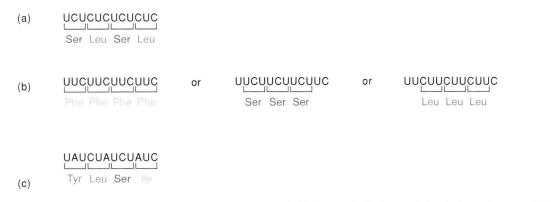

(a)   UCUCUCUCUCUC
      Ser Leu Ser Leu

(b)   UUCUUCUUCUUC   or   UUCUUCUUCUUC   or   UUCUUCUUCUUC
      Phe Phe Phe Phe   Ser Ser Ser   Leu Leu Leu

      UAUCUAUCUAUC
(c)   Tyr Leu Ser Ile

FIGURE 10.25   Coding properties of several synthetic mRNAs. (*a*) Poly(UC) contains two alternating codons, UCU and CUC, which code for serine (Ser) and leucine (Leu), respectively. Thus, the product is poly(Ser-Leu). (*b*) Poly(UUC) contains three codons, UUC, UCU, and CUU, which code for phenylalanine (Phe), serine (Ser), and leucine (Leu), respectively. Therefore, the product is poly(Phe), or poly(Ser), or poly(Leu), depending on which of the three reading frames the ribosome uses. (*c*) Poly(UAUC) contains four codons in a repeating sequence: UAU, CUA, UCU, and AUC, which code for tyrosine (Tyr), leucine (Leu), serine (Ser), and isoleucine (Ile), respectively. The product is therefore poly(Tyr-Leu-Ser-Ile).

Repeating triplets translated to homopolypeptides, as had been expected if the number of bases in a codon was three or a multiple of three. For example, poly(UUC) translated to polyphenylalanine plus polyserine plus polyleucine (figure 10.25*b*). The reason for three different products is that translation can start at any point in the synthetic message. Therefore, poly(UUC) can be read as UUC, UUC etc., or UCU, UCU etc., or CUU, CUU etc., depending on where translation starts. In all cases, once translation begins, only one codon is encountered, as long as the number of bases in a codon is divisible by three.

Repeating tetranucleotides translated to repeating tetrapeptides. For example, poly(UAUC) yielded poly(tyrosine-leucine-serine-isoleucine) (figure 10.25*c*). As an exercise, you can write out the sequence of such a message and satisfy yourself that it is compatible with codons having three bases, or nine, or even more, but not six. (We already know six cannot be right because it is not an odd number.) Since codons are not likely to be as cumbersome as nine bases long, three is the best choice. Look at the problem another way: Three is the lowest number that gives enough different codons to specify all twenty amino acids. (The number of permutations of four different bases taken three at a time is 4^3, or 64.) There would be only sixteen two-base codons ($4^2 = 16$), not quite enough. But there would be over 200,000 ($4^9 = 262,144$) nine-base codons. Nature is usually more economical than that.

The genetic code is a set of three-base code words, or codons, in mRNA that instruct the ribosome to incorporate specific amino acids into a polypeptide. The code is nonoverlapping: that is, each base is part of only one codon. It is also devoid of gaps, or commas; that is, each base in the coding region of an mRNA is part of a codon.

BREAKING THE CODE

Obviously, Khorana's synthetic mRNAs gave strong hints about some of the codons. For example, since poly(UC) yields poly(serine-leucine), we know that one of the codons (UCU or CUC) codes for serine and other codes for leucine. The question remains: Which is which? Nirenberg developed a powerful assay to answer this question. He found that a trinucleotide was usually enough like an mRNA to cause a specific aminoacyl-tRNA to bind to ribosomes. For example, the triplet UUU will cause phenylalanyl-tRNA to bind, but not lysyl-tRNA or any other aminoacyl-tRNA. Therefore, UUU is a codon for phenylalanine. This method was not perfect; some codons did not cause any aminoacyl-tRNA to bind, even though they were authentic codons for amino acids. But it provided a nice complement to Khorana's method, which by itself would not have given all the answers either, at least not easily.

Here is an example of how the two methods could be used together: Translation of the polynucleotide poly(AAG) yielded polylysine plus polyglutamate plus polyarginine. There are three different codons in that synthetic message: AAG, AGA, and GAA. Which one codes for lysine? All three were tested by Nirenberg's assay, yielding the results shown in figure 10.26. Clearly, AGA and GAA cause no binding of (^{14}C)lysyl-tRNA to ribosomes, but AAG does. Therefore, AAG is the lysine codon in poly(AAG). Something else to notice about this experiment is that the triplet AAA also causes lysyl-tRNA to bind. Therefore, AAA is another lysine codon. This illustrates a general feature of the code: In most cases, more than one triplet codes for a given amino acid. In other words, the code is **degenerate**.

Figure 10.27 shows the entire genetic code. As predicted, there are sixty-four different codons and only twenty different amino acids, yet all of the codons are used. Three are "stop" codons found at the ends of messages, but all the others specify amino acids, which means that the code is highly

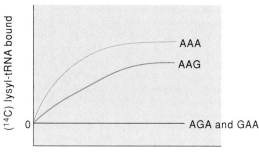

FIGURE 10.26 Binding of lysyl-tRNA to ribosomes in response to various codons. Lysyl-tRNA was labeled with radioactive carbon (^{14}C) and mixed with *E. coli* ribosomes in the presence of the following trinucleotides: AAA, AAG, AGA, and GAA. Lysyl-tRNA-ribosome complex formation was measured by binding to nitrocellulose filters. (Unbound lysyl-tRNA does not stick to these filters, but a lysyl-tRNA-ribosome complex does.) AAA was a known lysine codon, so binding was expected with this trinucleotide.

Second position

First position (5'-end)	U	C	A	G	Third position (3'-end)
U	UUU ⎱ Phe UUC ⎰ UUA ⎱ Leu UUG ⎰	UCU ⎱ UCC ⎰ Ser UCA ⎱ UCG ⎰	UAU ⎱ Tyr UAC ⎰ UAA ⎱ STOP UAG ⎰	UGU ⎱ Cys UGC ⎰ UGA STOP UGG Trp	U C A G
C	CUU ⎱ CUC ⎰ Leu CUA ⎱ CUG ⎰	CCU ⎱ CCC ⎰ Pro CCA ⎱ CCG ⎰	CAU ⎱ His CAC ⎰ CAA ⎱ Gln CAG ⎰	CGU ⎱ CGC ⎰ Arg CGA ⎱ CGG ⎰	U C A G
A	AUU ⎱ AUC ⎰ Ile AUA ⎰ AUG Met	ACU ⎱ ACC ⎰ Thr ACA ⎱ ACG ⎰	AAU ⎱ Asn AAC ⎰ AAA ⎱ Lys AAG ⎰	AGU ⎱ Ser AGC ⎰ AGA ⎱ Arg AGG ⎰	U C A G
G	GUU ⎱ GUC ⎰ Val GUA ⎱ GUG ⎰	GCU ⎱ GCC ⎰ Ala GCA ⎱ GCG ⎰	GAU ⎱ Asp GAC ⎰ GAA ⎱ Glu GAG ⎰	GGU ⎱ GGC ⎰ Gly GGA ⎱ GGG ⎰	U C A G

FIGURE 10.27 The genetic code. All sixty-four codons are listed, along with the amino acid for which each codes. To find a given codon—ACU, for example—we start with the wide horizontal row labeled with the name of the first base of the codon (A) on the left border. Then we move across to the vertical column corresponding to the second base (C). This brings us to a box containing all four codons beginning with AC. It is now a simple matter to find the one among these four we are seeking, ACU. We see that this triplet codes for threonine (Thr), as do all the other codons in the box: ACC, ACA, and ACG. This is an example of the degeneracy of the code. Notice that three codons (red) do not code for amino acids; instead, they are stop signals.

degenerate. Leucine, serine, and arginine have six different codons; several others, including proline, threonine, and alanine, have four; isoleucine has three; and many others have two. Just two amino acids, methionine and tryptophan, have only one codon.

The genetic code was broken by using either synthetic messengers or synthetic trinucleotides and observing the polypeptides synthesized or aminoacyl-tRNAs bound to ribosomes, respectively. There are sixty-four codons in all. Three are stop signals, and the rest code for amino acids. This means that the code is highly degenerate.

UNUSUAL BASE PAIRS BETWEEN CODON AND ANTICODON

How does an organism cope with multiple codons for the same amino acid? One way would be to have multiple tRNAs **(isoaccepting species)** for the same amino acid, each one specific for a different codon. In fact, this is part of the answer, but there is more to it than that; we can get along with considerably fewer tRNAs than that simple hypothesis would predict. Again Francis Crick anticipated experimental results with insightful theory. In this case, Crick hypothesized that the first two bases of a codon must pair correctly with the anticodon according to Watson-Crick base-pairing rules, but the last base of the codon can "wobble" from its normal position to form unusual base pairs with the anticodon. This proposal was called the **wobble hypothesis.** In particular, Crick proposed that a G in an anticodon can pair not only with a C in the third position of a codon (the **wobble position**), but also with a U. This would give the **wobble base pair** shown in figure 10.28*b*. Notice how the U has wobbled from its normal position to form this base pair.

Furthermore, Crick noted that one of the unusual nucleosides found in tRNA is **inosine(I),** which has a structure similar to that of guanosine. This nucleoside can ordinarily pair like G, so we would expect it to pair with C (Watson-Crick base pair) or U (wobble base pair) in the third position of a codon. But Crick proposed that inosine could form still another kind of wobble pair, this time with A in the third position of a codon (figure 10.28*c*). That means an anticodon with I in the first position can potentially pair with three different codons ending with C, U, or A.

The wobble phenomenon obviously reduces the number of tRNAs required to translate the genetic code. For example, consider the two codons for phenylalanine, UUU and UUC, listed at the top left of figure 10.27. According to the wobble hypothesis, they can both be recognized by an anticodon that reads 3'AAG5' (figure 10.29*a*). The G in the 5' position of the anticodon could form a Watson-Crick G-C base pair with the C in the UUC, or a G-U wobble base pair with the U in UUU. Similarly, the two leucine codons in the same box, UUA and

Anticodon (first base)   Codon (third base)

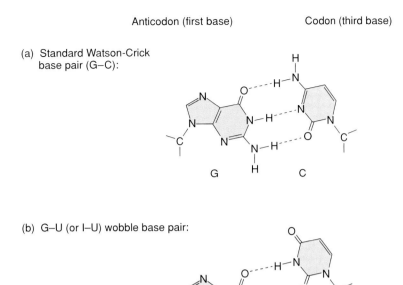

(a) Standard Watson-Crick base pair (G–C):

G   C

(b) G–U (or I–U) wobble base pair:

G   U

(c) I–A wobble base pair:

I   A

FIGURE 10.28 Wobble base pairs. (*a*) Relative positions of bases in a standard (G–C) base pair. The base on the left here and in the wobble base pairs (*b*) and (*c*) is the first base in the anticodon. The base on the right is the third base in the codon. (*b*) Relative positions of bases in a G–U (or I–U) wobble base pair. Notice that the U has to "wobble" upward to pair with the G (or I). (*c*) Relative positions of bases in an I–A wobble base pair. The A has to rotate (wobble) counterclockwise in order to form this pair.

UUG can both be recognized by the anticodon 3′AAU5′ (figure 10.29*b*). The U can form a Watson-Crick pair with the A in UUA, or a wobble pair with the G in UUG.

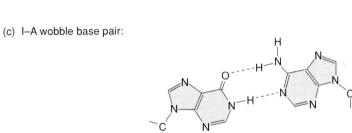

Part of the degeneracy of the genetic code is accommodated by isoaccepting species of tRNA that bind the same amino acid but recognize different codons. The rest is handled by wobble, in which the third base of a codon is allowed to move slightly from its normal position to form a non-Watson-Crick base pair with the anticodon. This allows the same aminoacyl-tRNA to pair with more than one codon. The wobble pairs are G-U (or I-U) and I-A.

THE (ALMOST) UNIVERSAL CODE

In the years after the genetic code was broken, all organisms examined, from bacteria to humans, were shown to share the same code. Therefore it was generally assumed (incorrectly, as we will see) that the code was universal, with no deviations whatsoever. This apparent universality led in turn to the notion of a single origin of present life on earth.

The reasoning for this idea goes like this: There is nothing inherently advantageous about each specific codon assignment we see. There is no obvious reason, for example, why UUC should make a good codon for phenylalanine, whereas AAG is a good one for lysine. Rather, the genetic code may be an "accident"; it just happened to evolve that way. However, once these codons were established, there was a very good reason why they did not change: A change that fundamental would almost certainly be lethal.

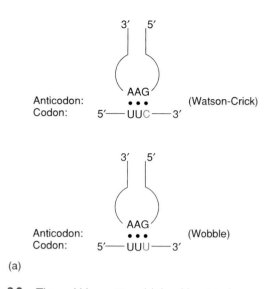

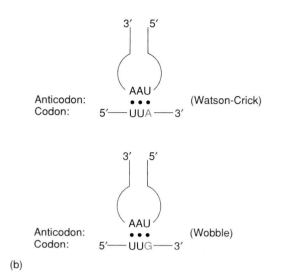

(a)

(b)

FIGURE 10.29   The wobble position. (*a*) An abbreviated tRNA with anticodon 3'-AAG-5' is shown base-pairing with two different codons for phenylalanine, UUC and UUU. The wobble position (the third base of the codon) is highlighted in red. The base-pairing with the UUC codon (top) uses only Watson-Crick pairs; the base-pairing with the UUU codon (bottom) uses two Watson-Crick pairs in the first two positions of the codon, but requires a wobble pair (G–U) in the wobble position. (*b*) A similar situation, in which a tRNA with anticodon AAU base-pairs with two different codons for leucine, UUA and UUG. Pairing with the UUG codon requires a G–U wobble pair in the wobble position.

Consider, for instance, a tRNA for the amino acid cysteine and the codon it recognizes, UGU. In order for that relationship to change, the anticodon of the cysteinyl-tRNA would have to change so it can recognize a different codon, say UCU, which is a serine codon. At the same time, all the UCU codons in that organism's genome that code for important serines would have to change to alternate serine codons so they would not be recognized as cysteine codons. The chances of all these things happening together, even over vast evolutionary time, are negligible. That is why the genetic code is sometimes called a "frozen accident"; once it was established, for whatever reasons, it had to stay that way. So a universal code would be powerful evidence for a single origin of life. After all, if life started independently in two places, we would hardly expect the two lines to evolve the same genetic code by accident!

In light of all this, it is remarkable that the genetic code is *not* absolutely universal; there are some exceptions to the rule. The first of these to be discovered were in the genomes of mitochondria. In mitochondria of the fruit fly *D. melanogaster,* UGA is a codon for tryptophan rather than for "stop." Even more remarkably, AGA in these mitochondria codes for serine, whereas it is an arginine codon in the universal code. Mammalian mitochondria show some deviations, too. Both AGA and AGG, though they are arginine codons in the universal code, have a different meaning in human and bovine mitochondria; there they code for "stop." Furthermore, AUA, ordinarily an isoleucine codon, codes for methionine in these mitochondria.

These aberrations might be dismissed as relatively unimportant, occurring as they do in mitochondria, which have very small genomes coding for only a few proteins and therefore more latitude to change than nuclear genomes. But there are also exceptional codons in nuclear genomes and prokaryotic genomes. In at least three ciliated protozoa, including *Paramecium,* UAA and UAG, which are normally stop codons, code for glutamine. In the prokaryote *Mycoplasma capricolum,* UGA, normally a stop codon, codes for tryptophan. Deviations from the standard genetic code are summarized in table 10.1.

Clearly, the so-called universal code is not really universal. Does this mean that the evidence now favors more than one origin of present life on earth? If the deviant codes were radically different from the standard one, this might be an attractive possibility, but they are not. In many cases, the novel codons are stop codons that have been recruited to code for an amino acid: glutamine or tryptophan. There is a well-established mechanism for this sort of occurrence, as we will see later in this chapter. The only examples of codons that have switched their meaning from one amino acid to another occur in mitochondria. Again, mitochondrial genomes, because they code for far fewer proteins than nuclear genomes or even prokaryotic genomes, might be expected to change a codon safely every now and then. In summary, even if there is no universal code, there is a standard code from which the deviant ones almost certainly evolved. Therefore, the evidence still strongly favors a single origin of life.

Table 10.1   Deviations from the "Universal" Genetic Code

Source	Codon	Usual Meaning	New Meaning
Fruit fly mitochondria	UGA	Stop	Tryptophan
	AGA & AGG	Arginine	Serine
	AUA	Isoleucine	Methionine
Mammalian mitochondria	AGA & AGG	Arginine	Stop
	AUA	Isoleucine	Methionine
	UGA	Stop	Tryptophan
Yeast mitochondria	CUN★	Leucine	Threonine
	AUA	Isoleucine	Methionine
	UGA	Stop	Tryptophan
Higher plant mitochondria	UGA	Stop	Tryptophan
	CGG	Arginine	Tryptophan
Protozoa nuclei	UAA & UAG	Stop	Glutamine
Mycoplasma	UGA	Stop	Tryptophan

★N = Any base

The genetic code is not strictly universal. In certain eukaryotic nuclei and mitochondria and in at least one bacterium, codons that cause termination in the standard genetic code can code for amino acids such as tryptophan and glutamine. In several mitochondrial genomes, the sense of a codon is changed from one amino acid to another. These deviant codes are still closely related to the standard one from which they probably evolved.

MECHANISM OF TRANSLATION

Translation, like transcription, is divided into three phases: initiation, elongation, and termination. Let us look at each of these in turn.

INITIATION

Initiation of translation in *E. coli* takes place in several steps (figure 10.30a). First, a ribosome that has finished translating an mRNA must be separated into its component subunits. An **initiation factor, IF-3,** influences this step, binding to the 30S subunit and preventing its reassociation with free 50S subunits. **IF-1** is also thought to participate by stimulating this dissociation process. Next, the 30S subunit binds a new mRNA in a reaction directed by IF-3. **Initiation factor 2 (IF-2)** then joins the growing collection of molecules and attaches the first aminoacyl-tRNA to the 30S subunit. GTP also binds to this **initiation complex.** Next, the 50S subunit joins the complex, and IF-3 dissociates. Finally, GTP is broken down to GDP and phosphate, and IF-2 departs, leaving the complete ribosome

bound to a fresh mRNA and the first aminoacyl-tRNA. Initiation is now complete.

In prokaryotes, the first amino acid, the one that initiates protein synthesis, is always the same: **N-formyl methionine (fMet).** The structure of fMet is given in figure 10.30b. N-formyl methionyl-tRNA (fMet-tRNA) forms in two steps. First, ordinary methionine binds to a special tRNA; then the methionine on the tRNA receives the formyl group. This unique amino acid participates only in initiation; it never goes into the interior of a polypeptide.

A Special tRNA for Initiation

The codon for fMet is AUG (or sometimes GUG). But these codons cause initiation only when they occur near the beginning (the 5′-end) of the coding region of an mRNA. If they appear in the middle of a message, they simply cause elongation, coding for methionine and valine, respectively. Thus, something special about the context of the codons AUG and GUG at the beginning of a message identifies them as initiation codons. These initiation codons are recognized by a special tRNA called **tRNA$_f^{Met}$**, which bears fMet. AUG and GUG in the interior of a message attract tRNA$_m^{Met}$ and tRNA^{Val}, bearing methionine and valine, respectively.

Eukaryotes perform the initiation step by a mechanism similar to that used by prokaryotes, but ordinary methionine, instead of fMet, plays the initiating role. The methionine may be ordinary, but it is set apart by being bound to a special tRNA, analogous to tRNA$_f^{Met}$, that participates only in initiation. This eukaryotic initiator tRNA is called **tRNA$_i^{Met}$** or **tRNA$_i$**.

Having served as the initiating amino acid, the methionine is usually, if not always, removed from eukaryotic proteins. In prokaryotes, the formyl group on the initiating fMet is always removed, and the whole fMet is removed from most proteins.

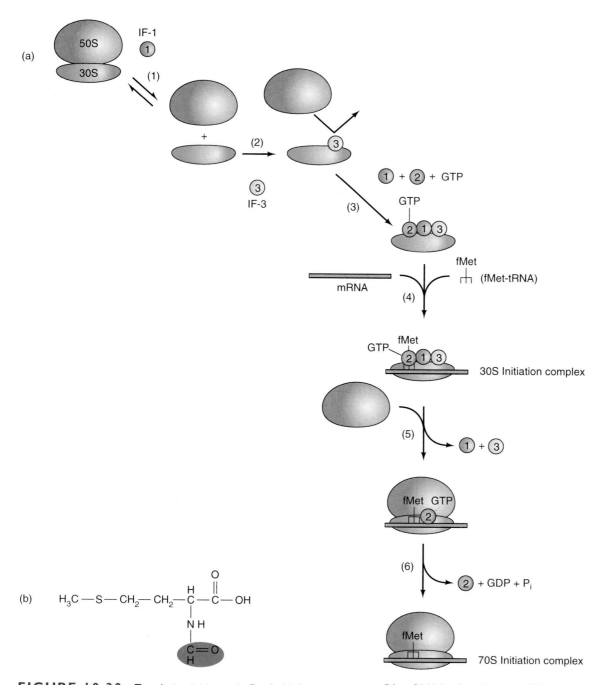

FIGURE 10.30 Translation initiation in *E. coli.* (*a*) Summary of steps in initiation. Step 1: A 70S ribosome has just completed a round of translation and has dissociated from an mRNA. IF-1 stimulates the dissociation of the 70S ribosome into 50S and 30S subunits. Step 2: IF-3 binds to the 30S subunit and prevents its reassociation with the 50S subunit. Step 3: IF-1 and IF-2 (the latter complexed with GTP) bind to the 30S subunit. Step 4: mRNA and fMet-tRNA bind, yielding the 30S initiation complex. Step 5: IF-1 and IF-3 dissociate from the complex, and the 50S subunit joins the 30S subunit, reconstituting a 70S ribosome. Step 6: GTP is hydrolyzed to GDP and phosphate, allowing IF-2 to dissociate from the complex. The result is a 70S initiation complex that is ready to begin elongation. (*b*) Structure of N-formyl methionine, with formyl group in red.

Eukaryotic Initiation Factors

The eukaryotic initiation process is similar to that in prokaryotes in that the same steps occur: The ribosome dissociates into 40S and 60S particles, Met-tRNA and mRNA join with the 40S particle, and finally the 60S particle joins to form an initiation complex. One difference is that the Met-tRNA binds to the 40S particle before the mRNA binds, rather than simultaneously or after as in prokaryotes. Of course, these initiation steps require initiation factors, which eukaryotes have in abundance.

Figure 10.31 summarizes our current understanding of the initiation process in eukaryotes, and the roles of the initiation

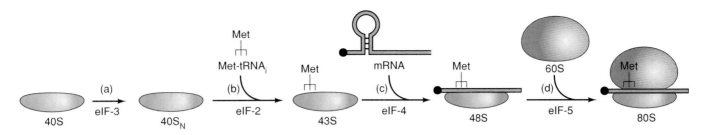

FIGURE 10.31 Summary of translation initiation in eukaryotes. (*a*) The eIF-3 factor converts the 40S ribosomal subunit to 40S$_N$, which is ready to accept the initiator aminoacyl-tRNA. (*b*) With the help of eIF-2, Met-tRNA$_i$ binds to the 40S$_N$ particle, forming the 43S complex. (*c*) Aided by eIF-4, the mRNA binds to the

43S complex, forming the 48S complex. (*d*) The eIF-5 factor helps the 60S ribosomal particle bind to the 48S complex, yielding the 80S complex that is ready to begin translating the mRNA.

Source: R. E. Rhoads, "Regulation of Eukaryotic Protein Synthesis of Initiation Factors" *J. Biol. Chem.* 268:3108, 1993.

factors in each step. Note that eukaryotic initiation factors are denoted by an *e*, to distinguish them from prokaryotic ones. Notice also that eIF-2 behaves like *E. coli* IF-2 in binding GTP and Met-tRNA and attaching them to the small ribosomal subunit. A complex, labeled eIF-4 in figure 10.31, helps in binding eukaryotic mRNA to the small ribosomal subunit, just as IF-3 does in prokaryotes. eIF-4 includes eIF-4F and eIF-4B. The eIF-4F factor contains eIF-4E, which is a cap-binding protein that recognizes caps on eukaryotic mRNAs. It also contains eIF-4A and another protein called p220. eIF-4B is a helicase that opens up RNA hairpins at the 5′-ends of mRNAs. We will see later in this chapter how important this is. Finally, eIF-5 helps the 60S ribosomal subunit bind to the initiation complex.

Initiation consists of binding an mRNA and first aminoacyl-tRNA to the ribosome to form an initiation complex. In prokaryotes this complex is built in stages by three initiation factors (IF-1, IF-2, and IF-3). Eukaryotes have their own set of initiation factors, most of which operate similarly to their bacterial counterparts. Prokaryotes use a special amino acid (N-formyl methionine, fMet) and tRNA (tRNA$_f^{Met}$) for initiation. Eukaryotes use unmodified methionine and a unique tRNA, tRNA$_i^{Met}$. AUG is ordinarily the initiating codon in both prokaryotes and eukaryotes.

mRNA Ribosome Binding Sites

We have mentioned that there must be something special about the context of AUG codons at the beginning of a message—something that identifies them as initiation codons rather than ordinary methionine codons. Presumably, that "something" could be either a special primary structure (sequence of bases) or a characteristic secondary structure of the mRNA, such as a hairpin loop. Whatever the signal, it could be expected to be part of the ribosome binding site, the place on the mRNA where ribosomes first bind to initiate translation.

Using the RNA phage R17 as a model, Joan Steitz and her colleagues showed at the end of the 1960s that there is no sec-

ondary structure common to all initiation codons, even in this one phage. Furthermore, relaxing the secondary structure of the phage RNA increases, rather than decreases, its overall efficiency of translation. Therefore, it appears that bacterial ribosomes do not recognize a common secondary structure as a binding signal.

The alternative was to look for a common primary structure. J. Shine and L. Dalgarno pointed out that all the initiation regions recognized by *E. coli* ribosomes contain at least part of the sequence 5′-AGGAGGU-3′ a few bases upstream from the initiation codon. This sequence is complementary to a sequence contained at the very 3′-end of *E. coli* 16S rRNA: 5′-CACCUCCUUA-3′. Therefore, base-pairing could occur as follows:

16S rRNA: 3′-AUUCCUCCAC-5′
Initiation site: 5′-AGGAGGU-3′

This would provide a convenient mechanism for binding an mRNA to a ribosome. The 16S rRNA in the ribosome would simply base-pair with the **Shine-Dalgarno sequence (SD sequence)** in the initiation region of the mRNA. Most mRNAs do not have such a good fit with the 16S rRNA sequence. The R17 coat gene, for instance, has only three bases that can base-pair, yet it is still translated actively. But genes that can form fewer than three base pairs are not translated well.

Eukaryotic mRNAs do not contain a sequence analogous to the Shine-Dalgarno sequence of prokaryotes. Instead, the cap at the 5′-end of a eukaryotic mRNA (chapter 9) seems to provide a "docking site" for the initiation factor eIF-4F, which contains a cap-binding protein, eIF-4E. This is part of a sequence of events that lead to formation of an initiation complex.

Most prokaryotic mRNAs have a sequence of bases (the Shine-Dalgarno sequence) just to the 5′-side of the initiation codon that is more or less complementary to a sequence at the very 3′-end of the 16S rRNA in the ribosome. Base-pairing between these two regions of RNA presumably helps the ribosome recognize and bind to the initiation regions of mRNAs. Eukaryotic mRNAs have no Shine-Dalgarno sequences. Instead, the cap ensures binding of mRNA to ribosomes.

Finding the Initiator AUG in a Eukaryotic mRNA: The Scanning Model

Once the 40S complex (actually a 43S complex, including the protein factors and Met-tRNA$_i$) has bound to the mRNA, it scans toward the 3′-end of the mRNA, searching for the initiation codon. This codon is almost always an AUG, and it is usually the first AUG, but there are exceptions. Sometimes the first AUG is not in the right context. That is, the surrounding bases do not tend to promote initiation at that AUG, so the 43S complex scans on to the next one. Marilyn Kozak has made mutations in each of the bases surrounding an initiation codon and found that the following sequence is optimal:

$$GCC\frac{A}{G}CC\underline{AUG}G$$

where the underlined AUG is the initiation codon. The most important bases are the purine (A or G, boldface) at position −3 and the G (boldface) at position +4. (We designate the A of the AUG as position +1.) Kozak also compared dozens of real initiation codons in vertebrate genes and found that the consensus sequence exactly matches the optimal sequence given above.

The strength of an initiation codon is important, since many eukaryotic genes are regulated at the translation level. Kozak and others have found five factors that play a major role. All these factors are consistent with the **scanning model.** They are:

1. Context, which we have already discussed.

2. The presence of a cap. Almost all eukaryotic mRNAs have caps, which is a good thing because most mRNAs without caps cannot be translated. The cap has a central role in the scanning model: It is the cap that attracts the initiation factors that help assemble the 43S complex, which carries out the scanning process.

3. Position. The first AUG in good context is used in most cases, probably about 90% of the time. That makes sense in terms of the scanning model, since the first AUG is the first one the scanning ribosome encounters. At least some ribosomes will probably use this first AUG, even if it is in a poor context.

4. Leader length. The **leader** is the RNA between the cap and the initiation codon. Sometimes it is very short, only a few nucleotides, but sometimes it is hundreds of nucleotides long. You might predict that the shorter the leader, the better the chance of initiation, but experiment has shown the opposite to be true. Extremely short leaders can actually hinder initiation: Approximately half the ribosomes translating an mRNA with an AUG within 12 nucleotides of the cap missed the first AUG and scanned on to a second. This fits with the scanning model, in that the initiation factors must have room to bind and assemble the 43S complex. Twelve nucleotides may not allow enough room for efficient recognition of the first AUG codon. This great inhibitory effect diminishes when the leader is lengthened to 20 nucleotides, but initiation strength remains proportional to leader length over a range of 17 to 80 nucleotides. The reason for this effect seems to be that long leaders allow many ribosomes to line up to translate the mRNA.

5. Secondary structure. The secondary structure (presence or absence of hairpins) can have either positive or negative effects on initiation (figure 10.32). If a hairpin occurs just downstream from an AUG in a weak context, it usually has a strong stimulatory effect, because it makes a scanning ribosome pause long enough to recognize the AUG and initiate. This obviously fits right in with the scanning model. But hairpins in the leader *before* an AUG can inhibit initiation. This is especially so if the hairpin occurs in the first 12 nucleotides of the RNA. Again, this fits with the notion that the initiation factors and 40S ribosomal subunits must bind in that region to begin their scan. Hairpins more remote from the 5′-end of the mRNA can also inhibit scanning, but they must be quite stable (composed of many base pairs, especially G–C pairs) to do so. We know that the ribosome plows through (unwinds) relatively unstable hairpins, rather than jumping over them, because it can actually initiate at an AUG codon that is involved in base-pairing within a hairpin. All of these effects are consistent with the scanning model, as opposed to a model in which a ribosome would first bind to the interior of an mRNA, at an AUG codon.

The 40S ribosomal particle binds first near the cap of a eukaryotic mRNA and scans toward the 3′-end of the message until it finds a favorable AUG at which to initiate translation. The factors that influence the favorability of a given AUG as an initiation codon are: its context (surrounding nucleotides); its position (whether or not it is the closest AUG to the 5′-end); the secondary structure (presence of hairpins) nearby; and the length of the leader. Almost all mRNAs require a cap for initiation to occur. All of these effects make sense in terms of the scanning model.

ELONGATION: ADDING AMINO ACIDS TO THE GROWING CHAIN

As suggested in the preceding discussion, ribosomes translate an mRNA in the 5′ → 3′ direction, the same direction in which the RNA was made. We will see that the protein grows in the amino (N) to carboxyl (C) direction. In other words, the N-terminal amino acid is added first, and the C-terminal amino acid is added last. Therefore, the 5′-end of the mRNA encodes the amino end of the protein.

The elongation step in protein synthesis is better visualized than described verbally. Figure 10.33 schematically depicts two rounds of elongation (adding two amino acids to a growing polypeptide chain) in *E. coli*. We start with fMet-tRNA bound

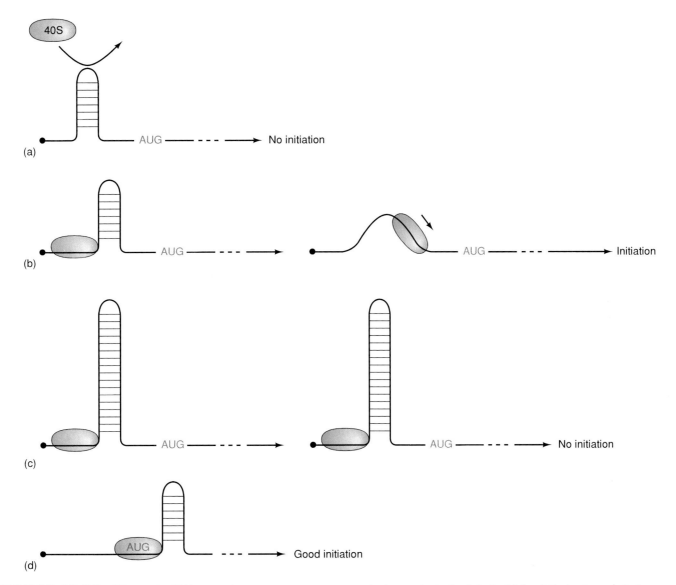

FIGURE 10.32 Effects of mRNA secondary structure on translation initiation. (*a*) A hairpin, even a relatively weak one, very close to the cap prevents ribosome binding and therefore translation. (*b*) A relatively weak hairpin in the leader region does not impede the scanning of the ribosome, which apparently disrupts the base pairs in the hairpin. (*c*) A relatively strong hairpin cannot be disrupted by the scanning ribosome, which stalls before it can reach the initiation codon. (*d*) A hairpin just downstream from the initiation codon stalls the ribosome there long enough to recognize the codon. This stimulates initiation at an AUG in a poor context.

to a ribosome. There are two binding sites for aminoacyl-tRNAs on the ribosome, called the **P (peptidyl) site** and the **A (aminoacyl) site.** In our schematic diagram, the P site is on the left and the A site is on the right. The fMet-tRNA is in the P site. Here are the elongation events:

1. To begin elongation, we need another amino acid to join with the first. This second amino acid arrives bound to a tRNA, and the nature of this aminoacyl-tRNA is dictated by the second codon in the message. The second codon is in the A site, which is otherwise empty, so our second aminoacyl-tRNA will bind to this site. Such binding requires a catalyst and energy. The catalyst is a protein **elongation factor (EF)** known as **EF-Tu.** GTP provides the energy. The eukaryotic homolog of EF-Tu is eEF-1α.

2. Next, the first peptide bond forms. An enzyme called **peptidyl transferase**—an integral part of the large ribosomal subunit—transfers the fMet from its tRNA in the P site to the aminoacyl-tRNA in the A site. This forms a two-amino acid unit called a *dipeptide* linked to the tRNA in the A site. The whole assembly is called a *dipeptidyl-tRNA*. The tRNA in the P site, minus its amino acid, is said to be *deacylated*.

   The peptidyl transferase step in prokaryotes is inhibited by an important antibiotic called **chloramphenicol,** whose structure is given in figure 10.34, along with those of

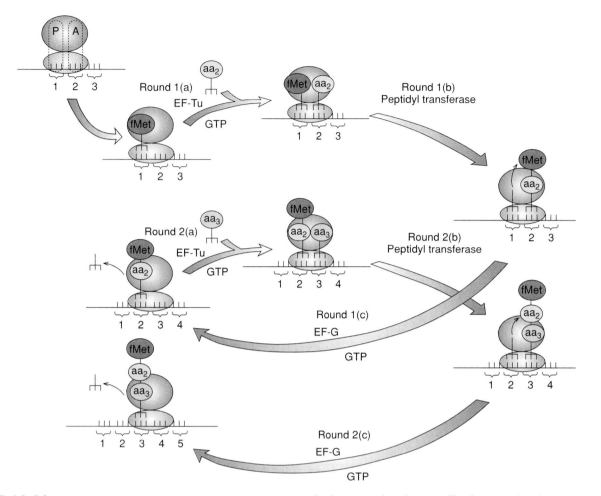

FIGURE 10.33 Elongation in translation. Note first of all that this is a highly schematic view of protein synthesis. For example, tRNAs are represented by forklike structures that show only the two business ends of the molecule. Upper left: A ribosome with an mRNA attached is shown to illustrate the two sites, P and A, roughly delineated with dotted lines. The diagram at upper left shows the locations of the P and A sites in the schematic ribosome. Round 1: (*a*) EF-Tu brings in the second aminoacyl-tRNA (yellow) to the A site on the ribosome. The P site is already occupied by fMet-tRNA (red). (*b*) Peptidyl transferase forms a peptide bond between fMet-tRNA and the second aminoacyl-tRNA. (*c*) In the translocation step, EF-G shifts the message and tRNAs one codon's width to the left. This moves the dipeptidyl-tRNA into the P site, bumps the deacylated tRNA aside, and opens up the A site for a new aminoacyl-tRNA. In round 2, these steps are repeated to add one more amino acid (green) to the growing polypeptide.

some other common antibiotics. This drug has no effect on most eukaryotic ribosomes, which makes it selective for bacterial invaders in higher organisms. However, the mitochondria of eukaryotes have their own ribosomes, and chloramphenicol does inhibit their peptidyl transferase. Thus, chloramphenicol's selectivity for bacteria is not absolute. Table 10.2 summarizes the modes of action of several important antibiotics, including chloramphenicol, that block translation by interacting with ribosomes.

3. In the next step, called **translocation,** the mRNA with its peptidyl-tRNA attached in the A site moves one codon's length to the left. This has the following results: (a) The deacylated tRNA in the P site (the one that lost its amino acid during the peptidyl transferase step when

the peptide bond formed) leaves the ribosome. (b) The dipeptidyl-tRNA in the A site, along with its corresponding codon, moves into the P site. (c) The codon that was "waiting in the wings" to the right moves into the A site, ready to interact with an aminoacyl-tRNA. Translocation requires an elongation factor called **EF-G** plus energy from GTP. The eukaryotic equivalent of EF-G is eEF-2.

The process then repeats itself to add another amino acid: (1) EF-Tu, in conjunction with GTP, brings the appropriate aminoacyl-tRNA to match the new codon in the A site. (2) Peptidyl transferase brings the dipeptide from the P site and joins it to the aminoacyl-tRNA in the A site, forming a tripeptidyl-tRNA. (3) EF-G, using energy from GTP, translocates the tripeptidyl-tRNA, together with its mRNA codon, to the P site.

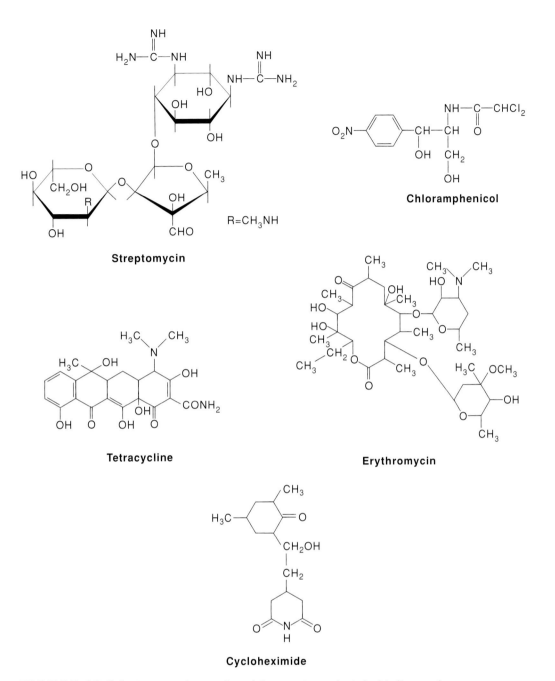

Streptomycin

Chloramphenicol

Tetracycline

Erythromycin

Cycloheximide

FIGURE 10.34 Some antibiotics that inhibit protein synthesis by binding to ribosomes.

We have now completed two rounds of peptide chain elongation. We started with an aminoacyl-tRNA (fMet-tRNA) in the P site, and we have lengthened the chain by two amino acids to a tripeptidyl-tRNA. This process continues over and over until the ribosome reaches the last codon in the message. The protein is now complete; it is time for chain termination.

Some controversy continues about the two-site model presented here, but there is general agreement about the existence of at least the A and P sites, based on experiments with the antibiotic **puromycin** (figure 10.35). This drug has a structure that resembles an aminoacyl-tRNA; as such, it can be accepted by the A site of the ribosome. Then, since puromycin has an amino group in the same position as an aminoacyl-tRNA would, peptidyl transferase forms a peptide bond between the peptide in the P site and puromycin in the A site, yielding a *peptidyl-puromycin*. At this point, the ruse is over. Puromycin has no association with the mRNA and cannot be translocated to the P site. Therefore, the peptidyl-puromycin dissociates from the ribosome, aborting translation prematurely. This is why puromycin kills bacteria and other cells.

The link between puromycin and the two-site model is this: Before translocation, because the A site is occupied by a peptidyl-tRNA, puromycin cannot bind and release the peptide; after translocation, the peptidyl-tRNA has moved to the P site,

Table 10.2 Some Antibiotics that Interact with Ribosomes

Antibiotic	Action
Tetracycline	Blocks binding of aminoacyl-tRNAs to small subunit of prokaryotic ribosome.
Chloramphenicol	Inhibits prokaryotic peptidyl transferase.
Cycloheximide	Inhibits eukaryotic peptidyl transferase.
Fusidic acid	Blocks dissociation of EF-G-GDP from ribosome and thereby inhibits translation elongation.
Erythromycin	Inhibits prokaryotic translocation.
Puromycin	Forms covalent bond with growing polypeptide and releases it from ribosome, causing premature termination.
Streptomycin	Causes misreading of codons, so reduces fidelity of translation in prokaryotes. Also inhibits prokaryotic translation initiation.

and the A site is open. At this point, puromycin can bind and release the peptide. We therefore see two states the ribosome can assume: puromycin-reactive and puromycin-unreactive. Those two states require at least two binding sites on the ribosome for the peptidyl-tRNA.

We have known for years that the large ribosomal particle carries out the peptidyl transferase reaction, and for most of that time we assumed that one or more of the proteins in this particle had the peptidyl transferase activity. However, since the discovery of catalytic RNA (chapter 9), many geneticists and biochemists have been intrigued by the notion that one of the ribosomal RNAs may actually catalyze peptide bond formation. We now have good evidence that this is so.

Harry Noller and colleagues treated ribosomes with a harsh detergent and phenol, both of which denature proteins so they can no longer function. To further degrade the ribosomal proteins, they added a proteinase, which chops proteins into small pieces. The next step was to see if the remaining rRNA had peptidyl transferase activity. This presents a problem, because true peptidyl transferase requires the whole ribosome, including

(a)

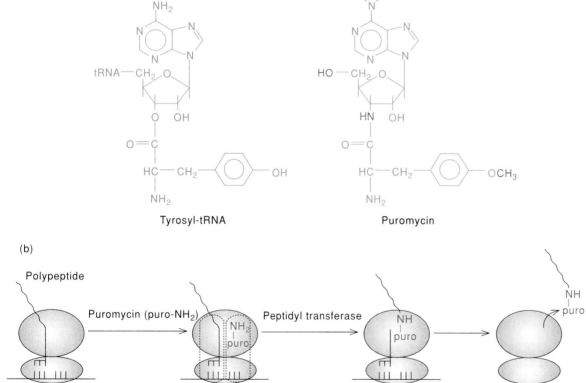

Tyrosyl-tRNA   Puromycin

(b)

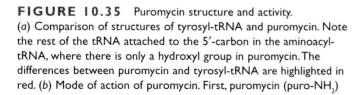

FIGURE 10.35 Puromycin structure and activity.
(a) Comparison of structures of tyrosyl-tRNA and puromycin. Note the rest of the tRNA attached to the 5′-carbon in the aminoacyl-tRNA, where there is only a hydroxyl group in puromycin. The differences between puromycin and tyrosyl-tRNA are highlighted in red. (b) Mode of action of puromycin. First, puromycin (puro-NH$_2$) binds to the open A site on the ribosome. (The A site must be open in order for puromycin to bind.) Next, peptidyl transferase joins the peptide in the P site to the amino group of puromycin in the A site. Finally, the peptidyl-puromycin dissociates from the ribosome, terminating translation prematurely.

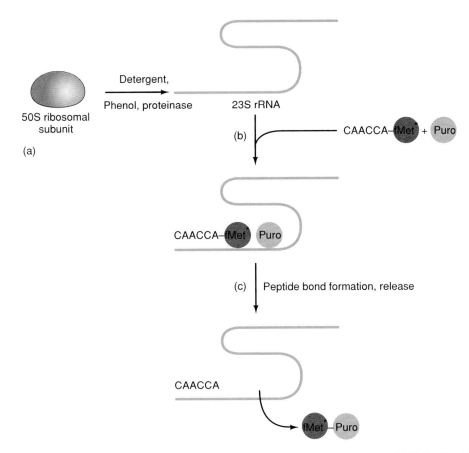

FIGURE 10.36 23S ribosomal RNA carries out the fragment reaction, a model for peptidyl transferase. (*a*) Noller and coworkers treated *E. coli* 50S ribosomal particles (pink) with detergent, phenol, and proteinase to destroy the protein, leaving only the rRNAs (5S and 23S). The 23S rRNA is pictured here, in an imaginary conformation. (*b*) Next, the investigators added puromycin (puro, green) plus the terminal six-nucleotide fragment of the fMet-tRNA (red), with the amino acid attached. The fMet is radioactively labeled (asterisk). These two substances apparently bind to the rRNA. (*c*) Peptide bond formation. The rRNA catalyzes the formation of a peptide bond between the fMet and puromycin and releases the product. This indicates that purified 23S rRNA has peptidyl transferase activity.

peptidyl- and aminoacyl-tRNAs. It would be too much to expect a mere rRNA to be able to duplicate the whole natural peptidyl transferase reaction. But it can do something almost as good—catalyze an analogous process called the "fragment reaction."

Figure 10.36 illustrates the plan of the fragment reaction. Noller and his group mixed together three components: 1) ribosomal RNA, purified by detergent, phenol, and proteinase extraction from *E. coli* ribosomes; 2) a fragment of fMet-tRNA containing the last six nucleotides of the RNA, plus radioactively labeled fMet; and 3) puromycin. The fMet-tRNA fragment is enough like the whole initiator aminoacyl-tRNA that it can bind to the P site of an intact ribosome. And, as we have seen, puromycin is enough like an ordinary aminoacyl-tRNA that it can bind to the A site of a ribosome. Thus, if we were dealing with a whole ribosome, we could add the fMet-tRNA fragment plus puromycin, and the peptidyl transferase activity of the ribosome would couple the radioactive fMet and the puromycin together through a peptide bond. The question is, can purified rRNA do the same thing? The answer, apparently,

is yes, because we can detect radioactive fMet-puromycin after the reaction. Furthermore, chloramphenicol blocks this reaction, just as it blocks authentic peptidyl transferase. Finally, ribonuclease, which destroys RNA, blocks the reaction. Thus, it looks as if ribosomal RNA, not ribosomal protein, has the essential peptidyl transferase activity. Of course, the proteins might still augment and modulate this activity.

Elongation takes place in three steps: (1) EF-Tu, with energy from GTP, binds an aminoacyl-tRNA to the ribosomal A site. (2) Peptidyl transferase, an activity that appears to reside in rRNA, forms a peptide bond between the peptide in the P site and the newly arrived aminoacyl-tRNA in the A site. This lengthens the peptide by one amino acid and shifts it to the A site. (3) EF-G, with energy from GTP, translocates the growing peptidyl-tRNA, with its mRNA codon, to the P site.

10.1   How the Amber Mutation Got Its Name

Molecular geneticists frequently invent colorful names for the materials and phenomena they discover, and some of these names actually make it into the normally dry scientific literature. The CCAAT box is one example we have encountered; another, which we have not discussed, is "magic spot," an unusual nucleotide that accumulates when bacteria are starved for amino acids. The amber codon is another well-known example, but the origin of its name is somewhat mysterious.

Here is the story, related to us by the man who did the naming, Dr. Harris Bernstein. One evening, when Bernstein was a student at Cal Tech, he went down to Seymour Benzer's lab to try to persuade his friends, Charles Steinberg and Richard Epstein, to go to the movies with him. He was unsuccessful, because Steinberg and Epstein were busy making mutant phages and could not leave. Instead, they persuaded Bernstein to make himself

useful and help them. As they worked, they explained the experiments to Bernstein, who became intrigued with the project and made a prediction about its outcome. The others disagreed, so they made a bet. Whoever guessed correctly about the characteristics of the mutant phage would get to name the mutant. Bernstein guessed that it would grow on a mutant strain of *E. coli,* but not on wild-type. He proposed to name the mutant phage for his mother, Mrs. Bernstein.

Sure enough, the mutant behaved as Bernstein predicted. He chuckles now about the fact that his reasons were faulty and that he managed to kill all the bacteria he plated that night because he heated his wire loop too much. Nevertheless, the honor of naming was his. As it turned out, the mutant was not exactly called *Bernstein* but *amber,* which is the English equivalent of the German word *Bernstein.* The name has stuck.

TERMINATION: RELEASE OF THE FINISHED POLYPEPTIDE

In our discussion of the genetic code, we saw that three codons (UAG, UAA, and UGA), instead of coding for amino acids, cause termination of protein synthesis. In the early 1960s, Seymour Benzer and his colleagues discovered the first stop codon (later determined to be UAG) as a mutation in phage T4 that rendered the phage incapable of growing, except in special **suppressor** strains of *E. coli* that counteract the effect of the mutation. These workers called the mutation **amber,** and that name has stuck to the codon **UAG** ever since (see box 10.1). The amber mutation is lethal because it causes premature termination in the middle of messages. This was first demonstrated with amber mutations in the alkaline phosphatase gene of *E. coli.* Such mutants produce no full-length alkaline phosphatase; instead, they make fragments that begin at the amino terminus of the protein and stop somewhere in the middle.

Later, a new mutation occurred that could not grow in **amber suppressor** strains, but could grow in certain other suppressor strains. In keeping with the colorful name amber, these new mutants were called **ochre;** they involve the codon **UAA,** which also causes premature termination. The third stop codon, **UGA,** has been named in kind: **opal.**

The sequences of the amber and ochre codons were discovered in 1965, not by sequencing mRNA, but by sequencing protein products. The rationale was as follows: Amber mutants regularly **reverted** to a phenotype that no longer required the amber suppressor. But these **revertants** did not usually give a product exactly like that in wild-type cells. Instead, one amino acid, the one corresponding to the position of the amber mutation, was different. Martin Weigert and Alan Garen studied such revertants of an amber mutation at one position in the alkaline phosphatase gene of *E. coli.* The amino

acid at this position in wild-type cells was tryptophan, whose sole codon is UGG. Since the amber mutation originated with a one-base change, we already know that the amber codon is related to UGG by a one-base change. To find out what this change was, Weigert and Garen determined the amino acids inserted in this position by several different revertants. The amino acids found there were serine, tyrosine, leucine, glutamate, glutamine, and lysine. Even without a computer, it was possible to deduce that UAG is the only codon that can yield at least one codon for each of those amino acids, including tryptophan, by a one-base change (figure 10.37).

By the same logic, including the fact that amber mutants can mutate by single-base changes to ochre mutants, Sydney Brenner and his colleagues reasoned that the ochre codon must be UAA. Severo Ochoa and his group verified that UAA is a stop signal when they showed that the synthetic message AUGUUUUAAA—directed the synthesis and release of the dipeptide fMet-Phe. (AUG codes for fMet; UUU codes for Phe; and UAA codes for stop.) With UAG and UAA assigned to the amber and ochre codons, respectively, UGA must be the opal codon. These three codons, because they code for no amino acid, are frequently referred to as **nonsense codons.** This is really a misnomer. They make perfect sense; they say, "Stop."

How do suppressors overcome the lethal effects of premature termination signals? A clue came from an experiment showing that purified tRNA from an amber suppressor strain could suppress an amber mutation in vitro. Therefore, tRNA is the suppressor molecule. Furthermore, suppressor strains insert amino acids into polypeptides in the position corresponding to the stop codon. Together, these data suggest that a tRNA from the suppressor strain recognizes the stop codon and inserts an amino acid, thus preventing termination. Proof of this mechanism came from experiments on an amber suppressor strain

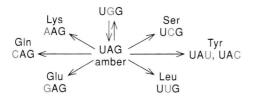

FIGURE 10.37 The amber codon is UAG. The amber codon (middle) came via a one-base change from the tryptophan codon (UGG), and the gene reverts to a functional condition in which one of the following amino acids replaces tryptophan: serine, tyrosine, leucine, glutamic acid, glutamine, or lysine. The red represents the single base that is changed in all these revertants, including the wild-type revertant that codes for tryptophan.

that inserts a tyrosine in the position corresponding to an amber codon. A tRNA^Tyr from this strain has an altered anticodon. Instead of GUA, which would respond to the tyrosine codon UAC, it has the anticodon CUA, which responds to the amber codon UAG and inserts a tyrosine (figure 10.38). Thus, the suppressor is itself a mutant. Its mutation is in a gene for a tRNA—to be more precise, in the part of the gene that specifies the anticodon. (This mutation is not lethal because there is another isoaccepting species of tRNA^Tyr with a normal anticodon that can still recognize tyrosine codons.)

We have seen that three codons, UAG, UAA and UGA, can cause premature termination when they occur in the middle of a message. Are these actually the termination signals at the ends of messages? Every gene for an mRNA sequenced to date

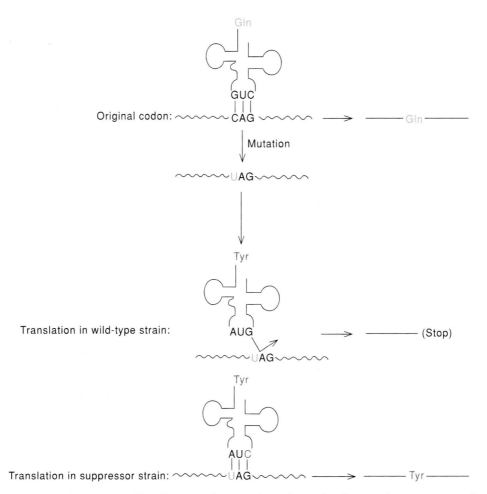

FIGURE 10.38 Mechanism of suppression. Top: The original codon in the wild-type *E. coli* gene was CAG, which was recognized by a glutamine tRNA. Middle: This codon mutated to UAG, which was translated as a stop codon by a wild-type strain of *E. coli*. Notice the tyrosine tRNA, whose anticodon (AUG) cannot translate the amber codon. Bottom: A suppressor strain contains a mutant tyrosine tRNA with the anticodon AUC instead of AUG. This altered anticodon recognizes the amber codon and causes the insertion of tyrosine (gray) instead of allowing termination.

(except in certain mitochondria) has at least one of these codons at the end, just after the codon for the last amino acid in the protein. Sometimes there are two stop codons in a row, such as the UAAUAG sequence at the end of the coat gene in R17 phage.

By now we can see that a prokaryotic mRNA contains at least one **open reading frame (ORF),** a sequence of nucleotides with the following characteristics:

1. It begins with an initiation codon (usually AUG).

2. It contains no stop codons in the same reading frame as the AUG until the end of the coding region (hence the term "open").

3. At the end, it has at least one stop codon in the same reading frame.

Because a special tRNA participates in initiation of translation, it would seem logical to propose a similar mechanism for termination, especially since the termination signals are three-base codons just like all other codons. Indeed, it was suggested that special tRNAs recognize the termination signals. However, when Mario Capecchi devised an assay for termination in 1967, he discovered that when he primed the ribosome with a polynucleotide containing an early stop codon, termination was triggered, not by RNA, but by protein factors. These factors, termed **release factor–1** and **release factor–2** (**RF-1** and **RF-2**), recognize the stop codons as follows: RF-1 reacts to UAA and UAG; RF-2 reacts to UAA and UGA. Both cause breakage of the bond between the polypeptide and its tRNA, releasing the product from the ribosome. Another factor, **RF-3,** binds GTP and helps the other two factors bind to the ribosome. In eukaryotes, a single release factor, **eRF,** causes termination. This reaction, like the termination reactions in prokaryotes, requires GTP, but the GTP-binding protein has not been found.

Translation release factors RF-1 and RF-2 recognize the stop signals UAA, UAG, and UGA, and along with RF-3 and GTP cause termination in prokaryotes. With the help of GTP, and presumably one other protein, eRF performs the same function in eukaryotes. When a mutation creates one of these stop signals in the middle of a message, premature termination occurs. Such mutations can be overridden by suppressors—tRNAs that respond to termination codons by inserting an amino acid, thus preventing termination.

TRANSLATIONAL CONTROL

We learned in chapter 8 that most of the control of prokaryotic gene expression occurs at the transcription level. The very short lifetime (only 1–3 minutes) of the great majority of prokaryotic mRNAs is consistent with this scheme, since it allows bacteria to

respond quickly to changing circumstances. It is true that different cistrons on a polycistronic transcript can be translated better than others. For example, the *lacZ, Y,* and *A* cistrons yield protein products in a molar ratio of 10:5:2. However, this ratio is constant under a variety of conditions, so it seems to reflect nothing more than the relative efficiencies of the Shine-Dalgarno sequences of the three cistrons. However, there are examples of real control of prokaryotic translation. Let us consider two.

PROKARYOTIC TRANSLATIONAL CONTROL

Messenger RNA secondary structure can play a role in translation efficiency, at least in bacteriophage gene expression. The RNA phage MS2 and its relatives provide a good example. These are **positive strand phages,** which means that their genomes are also their mRNAs. Thus, these phages provide a convenient source of pure mRNA. These phages are also very simple; each has only three genes, which encode the A protein (or maturation protein), the coat protein, and the replicase. The sequences surrounding the translation start site of all three cistrons suggest considerable base-pairing, or secondary structure. In vitro translation experiments have shown that the intramolecular base-pairing at all three start sites is inhibitory; relaxing these secondary structures actually enhances initiation. This is particularly true of the replicase gene, where the initiation codon is buried in a double-stranded structure that also involves part of the coat gene, as illustrated for MS2 phage RNA in figure 10.39a. This explains why the replicase gene of these phages cannot be translated until the coat protein is translated: The ribosomes moving through the coat gene open up the secondary structure that hides the initiation codon of the replicase gene (figure 10.39b).

Some prokaryotic translation is subject to feedback repression. The best-studied example is translation of the cistrons on the polycistronic mRNAs encoding the *E. coli* ribosomal protein genes. One such mRNA includes cistrons encoding both the L11 and L1 ribosomal proteins, as shown in figure 10.40. The L1 protein, when made in moderate quantities, binds tightly to a hairpin-loop structure in 23S rRNA. However, when L1 (and L11) are more abundant than 23S rRNA, L1 binds to a similar stem-loop structure near the translation start site of the L11 cistron. This represses translation of both the L11 and L1 cistrons, because their translation is coupled. This saves energy because ribosomal proteins in excess of ribosomal RNAs go to waste.

Prokaryotic mRNAs are very short-lived, so control of translation is not common in these organisms. However, some translational control does occur. Messenger RNA secondary structure can govern translation initiation, as in the replicase gene of the MS2 class of phages, or ribosomal proteins can feedback-inhibit the translation of their own mRNAs. L1 does this by binding to a hairpin loop at the beginning of the mRNA that encodes L11 and L1.

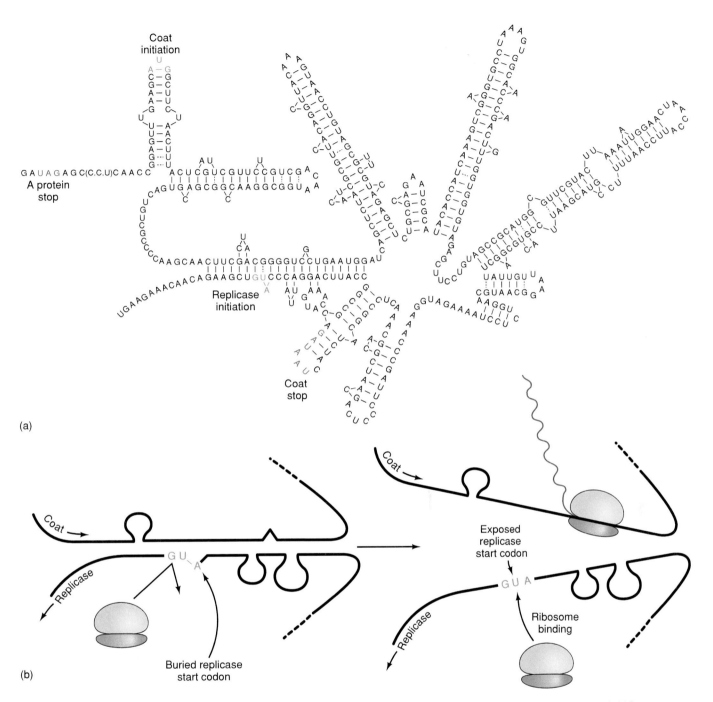

(a)

(b)

FIGURE 10.39 Potential secondary structure in MS2 phage RNA and its effect on translation. (*a*) The sequence of the coat gene and surrounding regions in the MS2 RNA, and potential secondary structure. Independent evidence suggests that some of the "petals" in this "flower model" really do form. Initiation and termination codons are boxed and labeled. (*b*) Effect of translation of coat gene on replicase translation. At left, the coat gene is not being translated, and the replicase initiation codon (AUG, green, written right to left here) is buried in a stem that is base-paired to part of the coat gene. Thus, the replicase gene cannot be translated. At right, a ribosome is translating the coat gene. This disrupts the base-pairing around the replicase initiation codon and opens it up to ribosomes that can now translate the replicase gene.

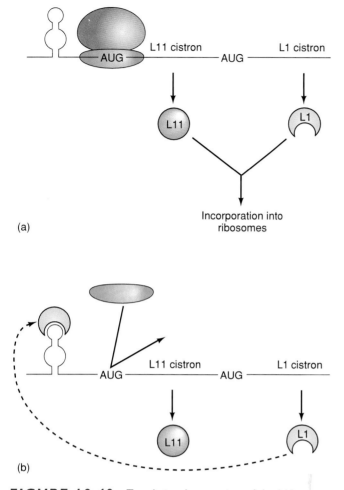

(a)

(b)

FIGURE 10.40 Translational repression of the L11 operon. (a) Moderate amounts of the operon's products, the L11 and L1 ribosomal proteins, are made. All the products can be incorporated into ribosomes, where L1 binds to a stem-loop structure on 23S rRNA similar to the one pictured on the mRNA near the initiation codon for the L11 cistron. (b) Too much L11 and L1 proteins are made. The excess L1 finds no 23S rRNA to which to bind, so it binds to the stem-loop near the initiation codon for the L11 cistron. This blocks ribosomes' access to the L11 cistron, and since translation of the two cistrons is linked, it blocks access to the L1 cistron as well. Thus, L11 and L1 protein synthesis is blocked until the excess L1 disappears.

EUKARYOTIC TRANSLATIONAL CONTROL

Eukaryotic mRNAs are much longer lived than prokaryotic ones, so there is more opportunity for translational control. The rate limiting step in translation is usually initiation, so we would expect to find most control exerted at this level. In fact, the best-studied mechanism of such control is phosphorylation of initiation factors, and we know of cases where such phosphorylation can be inhibitory, and others where it can be stimulatory.

The best-known example of inhibitory phosphorylation occurs in reticulocytes, which make one protein, hemoglobin, to the exclusion of almost everything else. But sometimes reticulocytes are starved for heme, the iron-containing part of hemoglobin, so it would be wasteful to go on producing α- and β-globins, the protein parts. Instead of stopping the production of the globin mRNAs, reticulocytes block their translation (figure 10.41). The absence of heme unmasks the activity of a protein kinase called the **heme-controlled repressor,** or **HCR** (also known as heme-regulated inhibitor, HRI). This enzyme phosphorylates one of the subunits of eIF-2, known as **eIF-2α.** The phosphorylated form of eIF-2 binds more tightly than usual to eIF-2B, which is an initiation factor whose job it is to exchange GTP for GDP on eIF-2. When eIF-2B is stuck fast to phosphorylated eIF-2, it cannot get free to exchange GTP for GDP on other molecules of eIF-2, so eIF-2 remains in the inactive GDP-bound form and cannot attach Met-tRNA$_i$ to 40S ribosomes. Thus, translation initiation grinds to a halt.

The antiviral proteins known as **interferons** follow this same pathway. In the presence of interferon and double-stranded RNA, which appears in many viral infections but not in normal cellular life, another eIF-2α kinase is activated. This one is called **DAI,** for **double-stranded RNA-activated inhibitor of protein synthesis.** The effect of DAI is the same as that of HCR: blocking translation initiation. This is useful in a virus-infected cell, because the virus has taken over the cell, and blocking translation will block production of progeny viruses, thus short-circuiting the infection.

The rate-limiting step in translation initiation is cap binding by the cap-binding factor eIF-4E, so it is intriguing that eIF-4E is also subject to phosphorylation, and that this seems to stimulate, rather than repress, translation initiation. We saw that the conditions that favor eIF-2α phosphorylation and translation repression (e.g., heme starvation and virus infection) are unfavorable for cell growth. This suggests that the conditions that favor eIF-4E phosphorylation and translation stimulation should be favorable for cell growth, and this is generally true. Indeed, stimulation of cell division with insulin or mitogens leads to an increase in eIF-4E phosphorylation.

Insulin and various growth factors, such as epidermal growth factor (EGF), also stimulate translation by an alternate pathway that involves eIF-4E. Figure 10.42 shows that these factors all stimulate cellular activity by a pathway involving a protein called Ras, which we will investigate in more detail in chapter 17. This pathway causes phosphorylation of a protein called **MAP kinase** (mitogen-activated protein kinase). One of the targets of MAP kinase is a protein called PHAS-I, which binds to eIF-4E and inhibits its activity. But once phosphorylated by MAP kinase, PHAS-I dissociates from eIF-4E, which is then free to engage in unhindered cap binding. Thus, translation is stimulated.

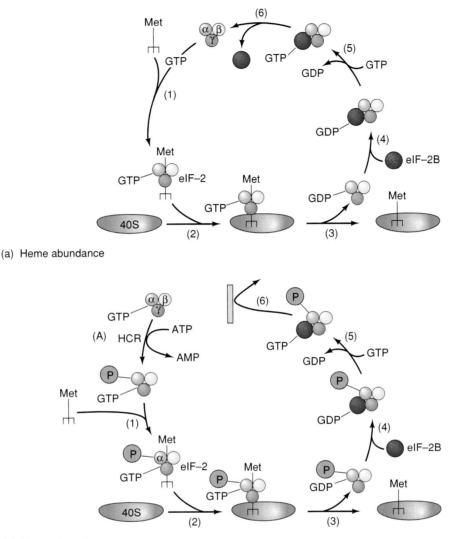

(a) Heme abundance

(b) Heme starvation

FIGURE 10.41 Repression of translation by phosphorylation of eIF-2α. (a) Heme abundance, no repression. Step 1, Met-tRNA$_i$ binds to the eIF-2-GTP complex, forming the ternary Met-tRNA$_i$-GTP-eIF-2 complex. The eIF-2 factor is a trimer of nonidentical subunits (α, green; β, yellow; and γ, purple). Step 2, the ternary complex binds to the 40S ribosome (blue). Step 3, GTP is hydrolyzed to GDP and phosphate, allowing the GDP-eIF-2 complex to dissociate from the 40S ribosome, leaving Met-tRNA$_i$ attached. Step 4, eIF-2B (red) binds to the eIF-2-GDP complex. Step 5, eIF-2B exchanges GTP for GDP on the complex. Step 6, eIF-2B dissociates from the complex. Now eIF-2-GTP and Met-tRNA$_i$ can get together to form a new complex to start a new round of initiation. (b) Heme starvation leads to translational repression. Step A, HCR (activated by heme starvation) attaches a phosphate group (orange) to the α-subunit of eIF-2. Then, steps 1–5 are identical to those in panel (a), but step 6 is blocked because the high affinity of eIF-2B for the phosphorylated eIF-2α prevents its dissociation. Now eIF-2B will be tied up in such complexes, and translation initiation will be repressed.

Eukaryotic mRNA lifetimes are relatively long, so there is more opportunity for translation control than in prokaryotes. The α subunit of eIF-2 is a favorite target for translation control. In heme-starved reticulocytes, HCR is activated, so it can phosphorylate eIF-2α and inhibit initiation. In virus-infected cells, another kinase, DAI, is activated; it also phosphorylates eIF-2α, and inhibits translation initiation. The cap-binding protein eIF-4E is also a target for phosphorylation, but this generally happens when conditions are favorable for growth, and it leads to stimulation of translation. Insulin and a number of growth factors stimulate a pathway involving Ras and MAP kinase. One of the targets for MAP kinase is a protein called PHAS-I. Upon phosphorylation by MAP kinase, this protein dissociates from eIF-4E and releases it to participate in more active translation initiation.

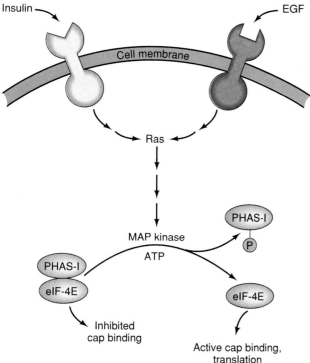

FIGURE 10.42 Stimulation of translation by phosphorylation of PHAS-I. Insulin, or a growth factor such as EGF, binds to its receptor at the cell surface. Through a series of steps, this activates the signal transducer Ras, which, through another series of steps, activates MAP kinase. This kinase has many targets, one of which is PHAS-I. When PHAS-I is phosphorylated by MAP kinase, it dissociates from eIF-4E, releasing it to participate in translation initiation.

WORKED-OUT PROBLEMS

PROBLEM 1

Below is a double-stranded bacterial DNA. The bottom strand is the template strand. Find and write the sequence of the open reading frame in the mRNA that would be transcribed from this gene.

5'ATCCGATGAAACCGTGGACACCCAGATAAATCG3'
3'TAGGCTACTTTGGCACCTGTGGGTCTATTTAGC5'

SOLUTION

Remembering that T in DNA replaces U in RNA, we need to look for an ATG initiation codon near the 5'-end of the top strand. (This is the nontemplate strand, so it is the one with the same sense as the mRNA.) Such a sequence begins with the A in the sixth position from the 5'-end of the top strand. Now write the sequence of bases in groups of three until you encounter a stop codon (TAG, TAA, or TGA, corresponding to UAG, UAA, or UGA). There is a TAA beginning with the T that is the seventh base from the 3'-end. Writing the sequence of the mRNA open reading frame in groups of three (as codons) as follows will help you verify that there are no intervening stop codons:

5' AUG AAA CCG UGG ACA CCC AGA UAA 3'

PROBLEM 2

(a) Use the genetic code to predict the amino acid sequence of the small polypeptide encoded in this prokaryotic mini-message, assuming that only physiologically significant initiation signals are used.

5'AUGAGAUACCAUGGGCUAAUGUGAAAA3'

What amino acid sequences would result if the following changes occurred in the message?
(b) The first C is changed to a G.
(c) The first U is changed to a G.
(d) The first C is changed to a U.
(e) The second G is changed to an A.
(f) The first C is deleted.
(g) An extra G is added after the first G.

SOLUTION

(a) Since physiologically significant initiation signals (AUGs) are the only ones used, translation of this message will begin with the first codon (AUG). In order to simplify the decoding of the message, it helps to write down the codons in order, separated by spaces, as follows:

AUG AGA UAC CAU GGG CUA AUG UGA AAA

Next, we consult the genetic code for the meaning of each of these codons, i.e. the amino acids specified by these codons.

fMet Arg Tyr His Gly Gln Met Stop Lys
AUG AGA UAC CAU GGG CUA AUG UGA AAA

The next-to-last codon (UGA) is a stop signal, so translation will stop after the methionine (Met) codon (AUG). Thus, the last codon (AAA, lysine) will remain untranslated, and the amino acid sequence of the peptide encoded in this little message is: fMet-Arg-Tyr-His-Gly-Gln-Met.

(b) The first C is changed to a G. The new message will read AUG AGA UA**G** etc., where the boldface indicates the altered base. Since UAG is a stop codon, only the first two codons will be translated, and the product will be fMet-Arg.

(c) The first C is changed to a U. The new message will read AUG AGA UA**U** etc., where the boldface indicates the altered base. UAU codes for tyrosine (Tyr), just as UAC does, so there will be no change in the amino acid sequence.

(d) The second G is changed to an A. The new message will read AUG A**A**A etc., where the boldface indicates the altered base. AAA codes for lysine (Lys), so the product will have Lys in the second position, and the rest will remain the same: fMet-Lys-Tyr-His-Gly-Gln-Met.

(e) The first G is changed to a U. This changes the first codon from AUG to AUU, which is no longer an initiation codon. Thus, initiation will occur at the next AUG to the right. Note that this does not have to be in the original reading frame, and in fact it is not in this case. The first AUG in this new message begins at the 11th base: AUUAGAUACC AUG GGC UAA UGU GAA AA. The coding sequence and the corresponding amino acid sequence of this new open reading frame will be:

fMet Gly Stop
AUG GGC UAA UGU GAA AA

Thus, this message codes for fMet-Gly.

(f) The first C is deleted. The new message will read:

AUG AGA U**AC** AUG GGC UAA UGU GAA AA

where the boldfaced bases lie on either side of the deleted base. The coding sequence and corresponding amino acid sequence are as follows:

fMet Arg Tyr Met Gly Stop Cys Glu
AUG AGA UAC AUG GGC UAA UGU GAA AA

Thus, this message codes for fMet-Arg-Tyr-Met-Gly.

(g) An extra G is added after the first G. The new message will read:

AUG **G**AG AUA CCA UGG GCU AAU GUG AAA A

where the boldface represents the inserted base. This message will have the following coding properties:

fMet Glu Ile Pro Trp Ala Asn Val Lys
AUG GAG AUA CCA UGG GCU AAU GUG AAA A

Thus, this message codes for fMet-Glu-Ile-Pro-Trp-Ala-Asn-Val-Lys.

P R O B L E M 3

A certain amber suppressor inserts tryptophan (Trp) in response to the amber codon. What change in the anticodon of tRNA^{Trp} probably created this suppressor strain?

S O L U T I O N

The tryptophan codon is 5'UGG3', which is recognized by the anticodon 3'ACC5' of the wild-type tRNA^{Trp}. The amber codon is 5'UAG3', which is most likely recognized by the anticodon 3'AUC5' of the amber suppressor tRNA. Thus, comparing this anticodon (3'AUC5') with the wild-type one (3'ACC5') shows that the change in the tRNA^{Trp} anticodon that created the amber suppressor was 3'ACC5' → 3'AUC5'. In other words, a C was changed to a U.

SUMMARY

Most genes contain the information for making one polypeptide. Polypeptides (or proteins) are polymers of amino acids linked together through peptide bonds. Ribosomes are the cell's protein factories.

Ribosomal subunits can be dissociated into their component parts, RNAs and proteins, and then reassociated. Thus, the ribosome is a self-assembly system that needs no help from the outside. Most mRNAs are translated by several ribosomes in tandem. The resulting structure is called a polysome.

The cloverleaf model of tRNA shows its secondary (base-pairing) structure and reveals four base-paired stems that define three loops of RNA: the D-loop, the anticodon loop, and the TψC loop. Transfer RNAs contain modified bases, at least some of which are vital to tRNA activity. The three-dimensional structure of tRNA consists of an L-shaped framework, with the bases of the anticodon protruding to one side and arranged in a helical shape for easy pairing with the three-base codon. Each aminoacyl-tRNA synthetase couples a specific amino acid to one of its cognate tRNAs.

The genetic code is a set of three-base code words, or codons, in mRNA that instruct the ribosome to incorporate specific amino acids into a polypeptide. The code is nonoverlapping; that is, each base is part of only one codon. It is devoid of gaps, or commas; each base in the coding region of an mRNA is part of a codon.

The code is also degenerate. More than one codon can code for one amino acid. Part of the degeneracy of the code is accommodated by isoaccepting species of tRNA that bind the same amino acid but recognize different codons. The rest is handled by wobble, in which

the third base of a codon is allowed to move slightly from its normal position to form a non-Watson-Crick base pair with the first base of the anticodon.

Initiation of translation consists of binding the first aminoacyl-tRNA to the ribosome-mRNA complex. Prokaryotes use a special amino acid (N-formyl methionine, fMet) and tRNA (tRNA$_f^{Met}$) for initiation. A sequence of bases (the Shine-Delgarno sequence) just to the 5'-side of the initiation codon (usually AUG) in prokaryotic mRNAs is more or less complementary to a sequence at the very 3'-end of the 16S rRNA in the ribosome. Base pairing between these two regions of RNA presumably helps the ribosome recognize and bind to the initiation regions of the mRNAs.

Eukaryotic mRNAs have no Shine-Dalgarno sequences. Instead, the cap serves as the "ticket" to the ribosome. eIF-4F binds first to the cap; then the 40S ribosomal particle, with the Met-tRNA; and other factors, scan along the mRNA until they find the initiation codon (AUG). Initiation codons can be bypassed if they are not in the right context (A or G in position −3 and G in position +4).

Elongation takes place in three steps: (1) EF-Tu, with GTP, binds an aminoacyl-tRNA to the ribosomal A site. (2) Peptidyl transferase forms a peptide bond between the peptide in the P site and the newly arrived aminoacyl-tRNA in the A site. This lengthens the peptide by one amino acid and shifts it to the A site. (3) EF-G, using energy from GTP, translocates the growing peptidyl-tRNA, with its mRNA codon, to the P site. Peptidyl transferase activity apparently resides on the large rRNA.

Translation release factors RF-1 and RF-2 recognize the stop signals UAA, UAG, and UGA, and along with RF-3 and GTP, cause termination in prokaryotes. With GTP and probably another protein, eRF performs the same function in eukaryotes. When a mutation creates one of these stop signals in the middle of a message, premature termination occurs. Such mutations can be overridden by suppressors—tRNAs that respond to termination codons by inserting an amino acid, thus preventing termination.

Both prokaryotes and eukaryotes have mechanisms for controlling the rate of translation.

PROBLEMS AND QUESTIONS

1. Would two tRNAs with complementary anticodons bind to each other less tightly, more tightly, or to the same degree as two complementary trinucleotides? Why? (*Hint*: Free trinucleotides have no special shape, but tRNAs do.)

2. If you charged a methionine tRNA with methionine, then chemically modified the amino acid to alanine and used this alanyl-tRNA^{Met} in an in vitro translation reaction, what do you predict would be the result? In other words, would the alanine go into the protein where an alanine is supposed to go or where a methionine is supposed to go? Why?

3. What does the answer to question 2 say about the importance of accurate aminoacyl-tRNA synthetases in faithful translation?

4. What are the products of the reaction catalyzed by aminoacyl-tRNA synthetase?

5. Consider this short message: 5'-AUGGCAGUGCCA-3'. Answer the following questions, assuming first that the code is fully overlapping and then that it is nonoverlapping. (*a*) How many codons would be represented in this oligonucleotide? (*b*) If the

second G were changed to a C, how many codons would be changed?

6. What would be the effect on reading frame and gene function if: (*a*) two bases were inserted into the middle of a message? (*b*) three bases were inserted into the middle of a message? (*c*) one base were inserted into one codon and one subtracted from the next?

7. If codons were six bases long, what kind of product would you expect from a repeating tetranucleotide such as poly(UUCG)?

8. If a mutant *E. coli* cell had an aminoacyl-tRNA synthetase that should attach the amino acid phenylalanine to its tRNA (tRNA^Phe), but at the elevated temperature of 42°C attached the amino acid arginine instead, what consequence would this have for the proteins that the cell made at elevated temperature? Would this affect the way these proteins function? Why, or why not?

9. What would be the effect of a mutation in the initiating codon of an mRNA?

10. (*a*) Use the genetic code to predict the amino acid sequence of the small polypeptide encoded in this prokaryotic mini-message:

    5'-AUGUUCAAGAUGGUGACUUGGUAAAUC-3'

    (*b*) What amino acid sequence would result if the first G in this message were changed to C, assuming that only physiologically meaningful start codons are used? (*c*) What amino acid sequence would result if the first C were changed to a U? (*d*) If the first C were changed to a G? (*e*) If the last G were changed to an A?

11. Find and write the sequence of the longest open reading frame in the following mRNA.

    5'AAUGUUCUGAUGAACCACGACUUAUAGC3'

12. (*a*) Which three amino acids have codons with 6-fold degeneracy? (*b*) Which two amino acids have only one codon each?

13. How many different codons would exist in a genetic code that had codons that were four bases long?

14. Assuming wobble can occur, what two codons could be recognized by the anticodon 3'-GAG-5'? Write the codons in the 5' → 3' direction, and remember that the pairing between a codon and an anticodon is antiparallel.

15. What anticodon could recognize all three codons for isoleucine, assuming wobble? Write the anticodon in the 3' → 5' direction.

16. A ribonuclease called cloacin DF13 specifically cleaves off 49 bases from the 3'-end of 16S rRNA in the intact *E. coli* ribosome. What effect would you expect this RNase to have on translation? Why?

17. When the λ phage *cI* (repressor) gene is transcribed from the P$_{RM}$ promoter, the resulting mRNA has no leader. That is, the initiation codon (AUG) lies right at the 5'-end of the mRNA. Would you predict this to be an actively translated mRNA? Why, or why not?

18. Which of the following eukaryotic mRNAs is likely to be translated most actively from the given initiation codon, which is in a weak context? Why would the others be translated less actively?

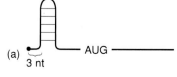

(a)
3 nt

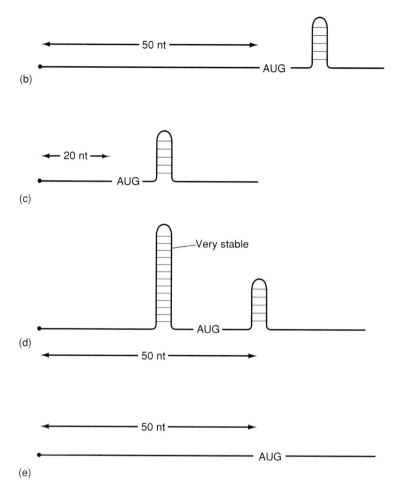

(b)
50 nt — AUG

(c)
20 nt — AUG

(d)
Very stable — AUG — 50 nt

(e)
50 nt — AUG

19. Where would chain elongation be stalled if mutations were to occur in the genes for the following proteins: (*a*) EF-G; (*b*) EF-Tu; (*c*) peptidyl transferase.

20. Of all the aminoacyl-tRNAs, only one enters the P site on the ribosome. Which one is it? Why is it important that this aminoacyl-tRNA not stay in the A site?

21. A bacterium makes a protein that contains 403 amino acids, counting the terminal methionine that is retained. How many peptidyl transferase reactions occur during synthesis of this protein?

22. If every stop signal in phage QT were UAGUGA, what would happen to all the phage QT proteins made in an amber suppressor strain in which the suppressor tRNA is tRNA^Tyr? Why?

23. A certain ochre suppressor inserts glutamine (Gln) in response to the ochre codon. What is the likeliest change in the anticodon of a tRNA^Gln that created this suppressor strain?

24. You have isolated a *Neurospora* mutant that cannot make amino acid *z*. Normal *Neurospora* cells make *z* from a common cellular substance *x* by a pathway involving three intermediates, *a, b,* and *c*: *x* → *a* → *b* → *c* → *z*. How would you establish that your mutant contains a defective gene for the enzyme that catalyzes the *b* → *c* reaction?

Answers appear at end of book.

Gene Mutation

And nothin's quite as sure as change.

John Phillips

The Mamas and the Papas

Learning Objectives

In this chapter you will learn:

1. The different phenotypic effects of mutations.
2. The different ways in which chemicals and radiation can damage DNA and thereby cause mutations.
3. How DNA damage can be repaired.
4. How DNA-damaging chemicals can be identified.

Chapters 7–10 have treated in detail two fundamental characteristics of genes: They replicate faithfully, and they store information for the production of RNAs and proteins. A third characteristic of genes is that they accumulate alterations, or mutations. We have already mentioned mutations several times in this book; indeed, there could hardly be a science of genetics without mutations, since we usually do not notice a gene until a mutation changes it. So far, however, we have not said much about the mechanisms by which genes are changed. That is the main purpose of this chapter.

Sometimes a mutation goes beyond the alteration of one or a few bases and involves a larger part of a chromosome. Such changes are called chromosome mutations. Genomic mutations, or aneuploidies, are even more noticeable and involve loss or gain of whole chromosomes. We have already discussed chromosome mutations and aneuploidies in chapter 5 and will not consider them further here. Instead, we will focus on **gene mutations,** alterations that are confined to a single gene. We will see that many mutations are harmful, but many others are innocuous, and a few are actually helpful.

TYPES OF GENE MUTATIONS

We can look at gene mutations in several different ways. For example, we can classify them according to their effects on the phenotype of the mutant organism or according to their effects on the genetic material itself. Let us first examine the various ways that mutations affect phenotype.

SOMATIC VERSUS GERM-LINE MUTATIONS

Multicellular organisms that reproduce sexually can experience mutations in their sex cells, in which case the gametes can be altered and the mutation passed on to the progeny. Such mutations are called **germ-line,** or **germinal, mutations.** A germ-line mutation probably happened to Queen Victoria of Great Britain (chapter 3), because some of her male descendants were afflicted with hemophilia, a sex-linked blood clotting disorder, yet none of her ancestors had this disease.

On the other hand, mutations in the nonsex cells, or **somatic cells,** may change the phenotype of the individual that suffers the mutation, but the mutant trait will stop with that individual. Since the mutation does not affect the gametes, it cannot be passed on.

Examples of such **somatic mutations** abound. Consider, for instance, a kernel of Indian corn such as the one pictured in figure 11.1. A wild-type kernel would be solid purple, but a mutation has inactivated the color gene, so this kernel is mostly white. However, you can easily see that the kernel is not solid white, but speckled. Each spot of color represents a back-mutation, or reversion, that reactivated the color gene in one cell, which then grew into a patch of colored cells. If the muta-

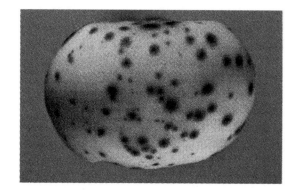

FIGURE 11.1   Somatic mutation. Each colored spot on this speckled kernel of corn represents a group of cells descended from one that experienced a back-mutation in a color gene. Without the back-mutations, the kernel would be all white.
From Nina V. Fedoroff, "Transposable Genetic Elements in Maize," *Scientific American,* June 1984. © Scientific American, Inc. Photo by F. W. Goro.

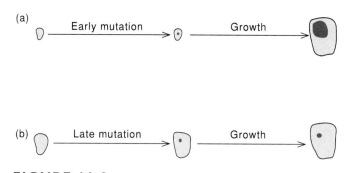

FIGURE 11.2   Influence of timing on the effect of somatic mutation. (*a*) The mutation occurs early in development of the seed, so many colored cells descend from the one with the back-mutation, producing a large colored spot. (*b*) The mutation occurs late in seed development, so not many more cell divisions take place. As a result, the colored spot is small.

tion takes place early in kernel development (figure 11.2*a*), when there are few cells, the mutant cell will give rise to a large group of colored progeny cells, so the colored patch will be large and will sometimes encompass the entire kernel. On the other hand, if the mutation takes place late in seed development (figure 11.2*b*), when there are many cells and few cell divisions left to go, the colored spot will be tiny. It is obvious that these mutations must have occurred during the lifetime of this corn plant and in the color-determining endosperm tissue, which is made up of somatic cells. Therefore, they must be somatic mutations; we will see in chapter 12 that these particular corn mutations are caused by mobile pieces of DNA called **transposable genetic elements.**

It is also likely that somatic mutations play a major role in cancer. Many cancers are derived from a single cell that has

FIGURE 11.3   An albino mouse (top) with a wild-type littermate.
From R. L. Brinster, *Cell*, 27: Cover, November 1981. © Cell Press.

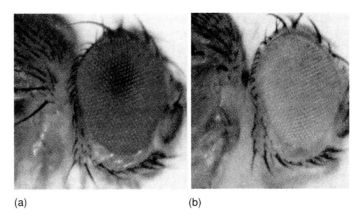

(a)   (b)

FIGURE 11.4   Morphological mutation in *Drosophila*. (a) A wild-type fly with red eyes is contrasted with (b) a white-eyed mutant fly.
From Gerald M. Rubin, *Cell*, 40: Cover, April 1985. © Cell Press.

changed its behavior from normal to malignant. It appears that somatic mutations cause such behavior changes. We will explore this topic further in chapter 17.

> **G**erm-line mutations affect the gametes, so the mutation can be transmitted to the next generation. Somatic mutations affect the nonsex, or somatic cells. Their effects are felt in the organism that experiences the mutation, but are passed no further.

MORPHOLOGICAL MUTATIONS

Complex multicellular organisms carrying **morphological mutations,** or **visible mutations,** can usually be distinguished readily from wild-type organisms because of their altered appearance. For example, **albino** mammals, including humans, have a mutation in a gene that is responsible for dark coat (or skin) pigment. This is ordinarily the gene for tyrosinase, the key enzyme in the pathway that leads to melanin, the black pigment in hair, eyes, and skin. A mutated tyrosinase gene may produce no active enzyme, so no melanin can be made. As a result, albino humans have very fair skin and hair, and light blue eyes; albino mice have white fur and pink eyes (figure 11.3).

Many morphological mutants of the fruit fly *Drosophila* have been observed. Among these are *white* flies, with white eyes instead of the usual red (figure 11.4); *miniature,* with tiny wings; *Bar,* with bar-shaped eyes instead of round; and *Curly,* with curly wings instead of straight.

Microorganisms and even viruses also show a few morphological mutations. The mutant bacterium *Streptococcus pneumoniae,* used by Avery and his colleagues to demonstrate that DNA

is the genetic material (chapter 6), had a different morphology (rough) than the wild-type (smooth). A mutant yeast is called *petite* because it forms colonies much smaller than normal. The bacterial virus T4 can be mutated to the *rapid lysis,* or *r,* form, which clears out a bigger and clearer hole, or plaque, in a lawn of infected bacteria than does the wild-type phage (chapter 13). Thus, the altered morphology we observe in microorganisms is usually not in a single organism but in colonies of organisms—or, in the case of a virus, in plaques.

> **M**orphological mutations cause a visible change in the affected organism.

NUTRITIONAL MUTATIONS

Although some morphological mutations can be distinguished in microorganisms, these are necessarily few in number. That is because these simple organisms have a limited number of observable morphological features. Thus, we must usually rely on more subtle mutant phenotypes. The most common of these is the inability to grow on simple growth media. Certain wild-type organisms can grow on "minimal medium," which contains just salts and an energy source such as glucose. These organisms, called **prototrophs** (Greek: *proto* = first + *trophe* = nourishment), can make all the amino acids, nucleotides, vitamins, and other substances they need to sustain life. Nutritional mutants, also called **auxotrophs** (Latin: *auxilium* = help), need one or more extra substances in order to survive, so they die on minimal medium.

For example, a *bio* mutant of *E. coli* will die unless it is supplied with the vitamin biotin, while a *leu* mutant will die without the amino acid leucine. Similarly, as we saw in chapter 10, a pantothenate-deficient mutant of *Neurospora* will die without the vitamin pantothenate. The reason for these deficiencies, of course, is that genes for enzymes in the biotin, leucine, and pantothenate synthesis pathways, respectively, are defective. Since these genes do not make active enzymes, no biotin, leucine, or pantothenate can be made and the cells die.

Nutritional mutations render an organism dependent on added nutrients that a wild-type organism does not require.

LETHAL MUTATIONS

Some mutations are so severe that an organism carrying them cannot survive at all; these mutations are called **lethals** (chapter 3). An example is an inactivating mutation in the gene for one of the subunits of RNA polymerase. If the mutation rendered the RNA polymerase inactive, the mutant organism could make no RNA and therefore could not live. Haploid organisms with lethal mutations die immediately because they have no wild-type gene to compensate. But diploid organisms can carry a lethal mutation, masked by a wild-type allele, for generations with no ill effects. It is only when two heterozygotes mate that we may discover the lethal consequences of the mutation. In chapter 3 we saw one such example in yellow mice. The heterozygotes are viable, but homozygous yellow mice are never seen—they die in the mother's uterus. Note that the yellow coat color in these mice is dominant, but the *lethality* is recessive. Thus, most lethal mutations in diploid organisms are recessive. If one parent contributes a defective gene for an essential protein and the other contributes a wild-type gene, the latter will usually allow the cell to make enough protein to compensate. It is only when two defective genes come together in a homozygote that lethality ordinarily results.

Huntington's disease in humans is an example of a dominant lethal. You may be wondering how a dominant lethal mutation allows a person to live long enough to have a disease, let alone pass the mutant gene along to his or her children. We do not know exactly how the disease operates, but we do know that heterozygotes typically live normal lives well past the childbearing years, then begin to show signs of the degenerative disease that ultimately kills them. We will discuss genetic mapping techniques used with Huntington's disease and other serious human genetic disorders in chapter 16.

Lethal mutations occur in indispensable genes. Haploid organisms with lethal mutations die immediately, but lethal mutations are usually recessive, so diploid organisms can tolerate them in heterozygotes.

CONDITIONAL MUTATIONS

As we have seen, lethal mutations in haploid organisms usually cause death immediately. Hence, they are of little use to geneticists

who are interested in studying the altered gene product. With no organism, there can be no gene product. One way around this problem is to look for **conditional lethals.** These are mutations that are lethal only under certain circumstances.

Consider, for example, a **temperature-sensitive (ts) mutation.** This allows growth at low temperature (the **permissive** temperature) but not at normal growth temperature (the **restrictive,** or **nonpermissive,** temperature), so lethality in this case is conditional on temperature. It is important to realize that it is the protein product that is temperature-sensitive, not the gene itself. The mutation creates an altered protein that is more easily unfolded, or **denatured,** than the wild-type protein. Since heat is a powerful protein denaturant, the mutant protein will be heat-sensitive. One has only to consider what happens to a boiled egg to appreciate heat denaturation of protein: The transparent solution of protein (mostly ovalbumin) in the egg white becomes a rubbery, opaque mass upon heating. In effect, this is what happens to the protein product of a temperature-sensitive gene at the nonpermissive temperature. It can happen to the wild-type protein too, but at a considerably higher temperature. Temperature-sensitive mutations also occur in humans. One kind of cystic fibrosis (Δ508) is caused by a mutant gene whose protein product cannot fold properly at body temperature, so it remains inactive. At lower temperatures, however, it functions normally.

Another example of a conditional mutation is the amber mutation, introduced in chapter 10. This mutation creates a premature translation stop signal—a "nonsense" mutation—in the middle of a gene, an event that would be lethal in a vital gene of an ordinary, nonpermissive strain. It is only conditionally lethal because the gene will usually still work in an amber suppressor strain (a permissive strain). Recall from chapter 10 that such a strain contains tRNA with the ability to recognize an amber codon and insert an amino acid, thus preventing termination. Therefore, when a phage bearing an amber mutation infects an amber suppressor strain, it will usually survive.

Mutations can be conditional without being conditional lethal. An example is the Siamese cat (figure 11.5). These animals have a mutation in the gene for dark coat color. But Siamese cats are not pure white like albino mice. Instead, they have dark

FIGURE 11.5 A Siamese cat.
© Carl W. May/Biological Photo Service.

patches on their feet, faces, and ears—areas of the body where the temperature is somewhat lower than the rest. This indicates that the mutation in the color gene is temperature-sensitive. Most of the cat's coat is warm enough to inactivate the color-producing enzyme, so it is white, but the extremities are cool enough that the enzyme operates, producing the dark color.

> Conditional mutations affect the mutant organism only under certain conditions. Temperature-sensitive mutations, for example, cause a protein to be inactive at high temperature but active at lower temperature. Amber mutations do not produce active product under ordinary circumstances but do so in organisms with an amber suppressor tRNA. If a conditional mutation causes death under nonpermissive conditions, it is a conditional lethal.

How Mutations Affect the Genetic Material

A second way of classifying mutations is to consider their effects on DNA. The mutations we are considering in this chapter involve the alteration, insertion, or deletion of one or a few bases at a time, and so are called **point mutations.** Let us consider the mechanisms of such point mutations, and their effects on DNA.

Missense and Nonsense Mutations

Many point mutations are **missense mutations,** in which a base change alters the sense of a codon from one amino acid to another. This causes an improper amino acid to be inserted into the protein product of the mutated gene. For example, a missense mutation might change the proline codon CCG to the arginine codon CGG.

An excellent example of such a defect is **sickle-cell disease,** a true genetic disease. People who are homozygous for this condition have normal looking red blood cells when their blood is rich in oxygen. The shape of normal cells is a *biconcave disc;* that is, the disc is concave viewed from both the top and bottom. However, when these people exercise, or otherwise deplete the oxygen in their blood, their red blood cells change dramatically to a sickle, or crescent, shape (figure 11.6). This has dire consequences. The sickle cells cannot fit through tiny capillaries, so they clog and rupture them, starving parts of the body for blood and causing internal bleeding and pain. Furthermore, the sickle cells are so fragile that they burst, leaving the patient anemic. Without medical attention, patients undergoing a sickling crisis are in mortal danger.

What causes this sickling of red blood cells? The problem is in **hemoglobin,** the red, oxygen-carrying protein in the red blood cells. Normal hemoglobin remains soluble under ordinary physiological conditions, but the hemoglobin in sickle

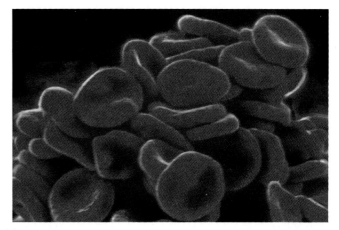
(a)

(b)

FIGURE 11.6 Normal red blood cells and sickle cells. These scanning electron micrographs contrast (*a*) the regular, biconcave shape of normal cells with (*b*) the distorted shape of red blood cells from a person with sickle-cell disease. Magnifications: (*a*) ×3,333; (*b*) ×5,555.

(a) © Jeroboam/Photo Researchers, Inc.; (b) © Omikron/Photo Researchers, Inc.

cells precipitates when the blood oxygen level falls, forming long, fibrous aggregates that distort the blood cells into the sickle shape.

What is the difference between normal hemoglobin (HbA) and sickle-cell hemoglobin (HbS)? Vernon Ingram answered this question in 1957 by determining the amino acid sequences of parts of the two proteins using a process that was invented by Frederick Sanger and is known as **protein sequencing.** Ingram focused on the β-globins of the two proteins. β-globin is one of the two different polypeptide chains found in the tetrameric (four-chain) hemoglobin protein. First, Ingram cut the two polypeptides into pieces with an enzyme that breaks selected peptide bonds. Then he separated these pieces, called **peptides,** by paper chromatography run in two dimensions (figure 11.7). He used one solvent to cause partial resolution of the peptides in the first dimension, then turned the paper 90 degrees and repeated the chromatography with another solvent to separate

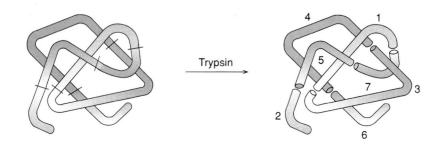

Trypsin

(a) Cutting protein to peptides

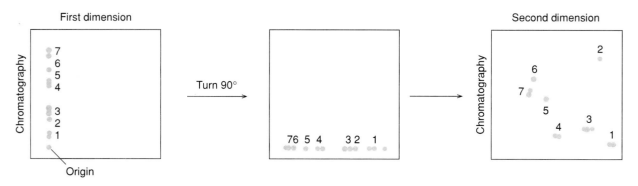

First dimension

Chromatography

Origin

Turn 90°

76 5 4   3 2 1

Second dimension

Chromatography

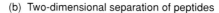

(b) Two-dimensional separation of peptides

FIGURE 11.7   Fingerprinting a protein. (*a*) A hypothetical protein, with its six trypsin-sensitive sites indicated by slashes. After digestion with trypsin, seven peptides are released. (*b*) These tryptic peptides separate partially during chromatography in the first dimension, then fully after the paper is turned 90 degrees and rechromatographed in the second dimension with another solvent.

the peptides still further. The peptides usually appear as spots on the paper; different proteins, because of their different amino acid compositions, give different patterns of spots. These patterns are aptly named **fingerprints.**

When Ingram compared the fingerprints of HbA and HbS, he found that all the spots matched except for one (figure 11.8). This spot had a different mobility in the HbS fingerprint than in the normal HbA fingerprint, which indicated that it had an altered amino acid composition. Ingram checked the amino acid sequences of the peptides in these two spots. He found that they were the amino-terminal peptides located at the very beginning of both proteins and that they differed in only one amino acid. The glutamate (glutamic acid) in the sixth position of HbA becomes a valine in HbS (figure 11.9). This is the only difference in the two proteins, yet it is enough to cause a profound change in the protein's behavior.

Knowing the genetic code, we can ask: What change in the β-globin gene caused the change Ingram detected in its protein product? The two codons for glutamate (Glu) are GAA and GAG; two of the four codons for valine (Val) are GUA and GUG. If the glutamate codon in the HbA gene is GAG, a single-base change to GTG would alter the mRNA to GUG, and the amino acid inserted into HbS would be valine instead of glutamate. A similar argument can be made for a GAA → GTA change. Figure 11.10 presents a summary of the mutation

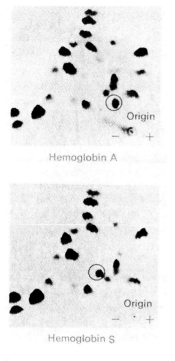

FIGURE 11.8   Fingerprints of hemoglobin A and hemoglobin S. The fingerprints are identical except for one peptide (circled), which shifts up and to the left in hemoglobin S.
Courtesy of Dr. Corrado Baglioni.

and its consequences. We can see how changing the blueprint does indeed change the product.

Sometimes, as we saw in chapter 10, a codon can be converted to a nonsense codon, which causes termination of translation. For example, the tryptophan codon UGG can be converted by a one-base change to either UGA or UAG, both of which are stop (or nonsense) codons. Of course, the mutation occurs at the DNA level, so TGG would be converted to TGA or TAG in the example above; we present the RNA codons here because they are the familiar way of writing codons. In chapter 10 we also learned that single-base insertions of deletions can alter the translation reading frame in the middle of a gene; these are called frameshift mutations.

We can also consider the chemistry of a single base change. Viewed in the simplest terms, such changes fall into two classes: (1) **transitions,** in which a pyrimidine replaces a pyrimidine (C → T or T → C) or a purine replaces a purine (A → G or G → A), or (2) **transversions,** a more drastic kind of change, in which a purine replaces a pyrimidine or vice versa.

Why do we say that a transversion is "more drastic" than a transition? This makes sense chemically, because one purine resembles another purine (or one pyrimidine another pyrimidine) far more than a purine resembles a pyrimidine. It also makes sense genetically, because of the degeneracy of the genetic code. Recall from chapter 10 that two related codons ending in a pyrimidine, such as UUU and UUC, are more likely to code for the same amino acid (phenylalanine, in this case) than are two related codons, one of which ends in a pyrimidine and the other of which ends in a purine (e.g., UUU and UUG, which code for phenylalanine and leucine, respectively). Similarly, two

related codons ending in a purine are likely to code for the same amino acid (e.g., GAA and GAG both code for glutamate, while GAU and GAC both code for aspartate). Thus, changing one pyrimidine to another or one purine to another, at least in the third position, is less likely to change the sense of a codon than is changing a purine to a pyrimidine, or vice versa.

Missense mutations alter the sense of a codon from one amino acid to another. A classic example is sickle-cell disease, a human genetic disorder. It results from a single-base change in the gene for β-globin. The altered base causes insertion of the wrong amino acid into one position of the β-globin. This altered protein results in distortion of red blood cells under low-oxygen conditions. This example illustrates a fundamental genetic concept: A change in a gene can cause a corresponding change in the protein product of that gene. A single-base change can also create a "nonsense" mutation—a stop codon in the middle of a gene—which causes premature termination of translation. Single-base changes are called transitions if they involve a change of a purine to another purine or a pyrimidine to another pyrimidine. A change of a purine to a pyrimidine or a pyrimidine to a purine is called a transversion.

SPONTANEOUS MUTATIONS

All mutations have a cause, of course, but sometimes mutations occur in the absence of **mutagens,** or mutation-causing agents. Such spontaneous mutations have several causes.

Mutations Caused by the DNA Replication Machinery

We saw in chapter 8 that replication is very faithful, but it is not perfect, so a certain number of mutations will occur simply due to fallible DNA synthesis.

We can see the effects of unfaithful replication especially clearly in mutant strains of *E. coli* called **mutators.** These bacteria make more than the usual number of mistakes during DNA replication, so their mutation rates are higher than normal. Mutator mutations have been mapped to several different

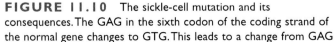

FIGURE 11.9 Sequences of amino-terminal peptides from normal and sickle-cell β-globin. The numbers indicate the positions of the corresponding amino acids in the mature protein. The only difference is in position 6, where a valine (Val) in HbS replaces a glutamate (Glu) in HbA.

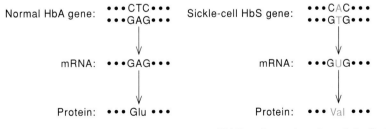

FIGURE 11.10 The sickle-cell mutation and its consequences. The GAG in the sixth codon of the coding strand of the normal gene changes to GTG. This leads to a change from GAG

to GUG in the sixth codon of the β-globin mRNA of sickle cells. This, in turn, results in insertion of a valine in the sixth amino acid position of sickle-cell β-globin, where a glutamate ought to go.

genes, some of which have been identified. For example, *mutD* encodes the epsilon (ε) subunit of the DNA polymerase III holoenzyme. This is the polypeptide that gives the holoenzyme its $3' \rightarrow 5'$ exonuclease activity. Without this activity, proofreading cannot occur, so the newly replicated DNA is left with an excess of mutations (chapter 7). Mutations in *mutH, mutL,* and *mutS* are also mutator mutations. They impair mismatch repair, a mechanism for repairing mismatches that the proofreading system missed. We will discuss this mechanism later in the chapter.

As unlikely as it may seem, there are also mutations that have the opposite effect—they make DNA replication even more faithful than normal. This raises a question: If a more faithful DNA replication system is available, why have *E. coli* evolved with a less effective one? The answer is probably that mutants with these extra-faithful systems have too slow a rate of evolution; they are not flexible enough to compete with more changeable organisms in adapting to a shifting environment.

Mistakes in Replication Caused by DNA Bases

The bases in DNA ordinarily exist in one of two possible forms, or **tautomers.** For example, the base thymine usually assumes the **keto** form, as illustrated in figure 11.11*a*. This is the structure presented in chapter 6 that base-pairs naturally with adenine. Occasionally, the keto form of thymine switches to the **enol** form, also shown in figure 11.11. Note that both tautomers have the same atoms but they are arranged slightly differently. The enol form pairs naturally with guanine instead of adenine, so if a thymine happens to be in the enol form at the moment it takes a partner during replication, a guanine will be inserted in place of an adenine. If this error goes uncorrected, DNA replication will perpetuate it and a mutation will result.

Figure 11.11*b* shows the two tautomeric forms of adenine. The **amino** form is by far the more common one; as we have seen, it base-pairs with thymine. Occasionally, adenine can assume the alternate, **imino** form, which base-pairs with cytosine. If this happens to an adenine in DNA just as DNA polymerase is supplying it with a partner, the polymerase will insert a C instead of a T. Again, if this mistake persists until the next round of replication, it will cause a mutation that will last: a GC pair instead of an AT pair. Figure 11.11*c* shows the **tautomerization** of cytosine to its rare imino form and of guanine to its rare enol form. By inspecting these rare tautomers, you can satisfy yourself that they form abnormal CA and GT base pairs, respectively.

So far, we have considered what would happen only if the bases in the replicating DNA strand form the rare tautomer during base-pairing. It is also possible that a newly inserted base would exist in the rare tautomeric form just at the moment of base-pairing. If so, it would result in the same kind of faulty base-pairing we have just observed.

Spontaneous Frameshift Mutations during Replication

Sometimes DNA replication causes the insertion or deletion of one or more bases in the middle of a coding region, which changes the translational reading frame from that point on (chapter 10). These **frameshift mutations** are very severe because they change every codon from the point of the mutation to the end of the mRNA. Since one of these new codons may be a stop codon, premature termination is often the result of a frameshift mutation. Figure 11.12 shows how this works: (*a*) is a hypothetical gene fragment, showing the translation of each codon; (*b*) is the same gene fragment, showing how insertion of a single A in the fourth position of the gene shifts the reading frame one base to the left. All the codons after that point are different. In fact, one of these new codons, at the fair right, is a stop codon, so protein synthesis will stop prematurely.

If small insertions and deletions cause frameshift mutations, the next question is: How do these insertions and deletions occur? We do not understand exactly how this works, but figure 11.13 presents one hypothesis that calls for the DNA replication machinery to "slip a cog" every now and then. The idea is that a base in one strand sometimes fails to pair with its partner in the complementary strand. This looping out of a base seems especially likely in stretches of DNA where one base is repeated over and over. Note that if a base in the template strand loops out, one too few bases will be incorporated into the progeny strand, resulting in a one-base deletion. On the other hand, if a base in the progeny strand loops out, one too many bases will be incorporated, producing a one-base insertion.

Spontaneous Mutations Caused by Deamination

Spontaneous mutations can occur by mechanisms other than mistakes in DNA replication. For example, bases, especially cytosine, have a slight tendency to lose their amino groups in a process called **deamination.** When cytosine is deaminated, it receives a carbonyl oxygen in place of its amino group; this converts it to uracil, which base-pairs with adenine instead of guanine. Adenine can also be deaminated, yielding the base hypoxanthine, which base-pairs with cytosine instead of thymine. In both of these cases, deamination can potentially cause a mutation because a new base, with new pairing properties, is created. Figure 11.14 shows the changes and new base-pairing caused by deamination of cytosine and adenine.

The most common kind of deamination, the conversion of cytosine to uracil, does not usually lead to a mutation, because cells have a mechanism for removing uracils that find their way into DNA by mistake. It involves an enzyme called uracil-DNA glycosylase (figure 11.15), which cuts the bond between the uracil and its deoxyribose, thus removing the uracil and leaving behind a DNA strand containing one sugar without a base. Another enzyme soon supplies a cytosine to pair with the guanine on the opposite strand.

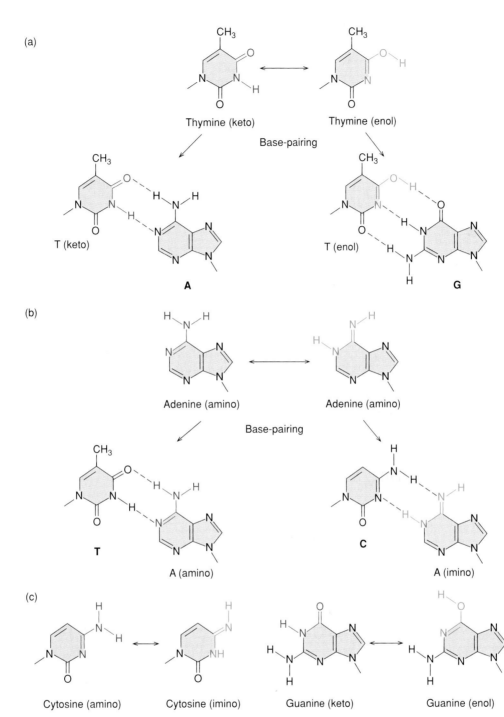

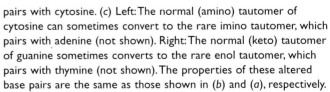

FIGURE 11.11 Spontaneous mutation induced by tautomerization. (*a*) The normal (keto) tautomer of thymine (left) base-pairs with adenine. After a rare conversion to the enol tautomer (right), the thymine base-pairs with guanine. The dotted lines represent hydrogen bonds between bases. (*b*) The normal (amino) tautomer of adenine (left) base-pairs with thymine. After a rare conversion to the imino tautomer (right), the adenine base-pairs with cytosine. (*c*) Left: The normal (amino) tautomer of cytosine can sometimes convert to the rare imino tautomer, which pairs with adenine (not shown). Right: The normal (keto) tautomer of guanine sometimes converts to the rare enol tautomer, which pairs with thymine (not shown). The properties of these altered base pairs are the same as those shown in (*b*) and (*a*), respectively.

The DNA of some organisms contains a small number of modified bases in addition to the usual four. The most common of these is 5-methylcytosine (figure 11.16a), which base-pairs with guanine exactly as ordinary cytosine does. However, 5-methylcytosine does not behave in every respect like cytosine. In particular, the sites in DNA that contain 5-methylcytosine can become "hot spots" for spontaneous mutation via deamination.

If deamination of cytosine is usually not mutagenic, why is deamination of 5-methylcytosine so likely to cause mutations? The answer is simple: Although the deaminated product of cytosine is uracil, which is easily recognized and removed, the deaminated product of 5-methylcytosine is thymine, one of the

```
Met Ala Leu Trp Ile Arg Phe Ile Arg
ATGGCCCTGTGGATCCGCTTCATTAGG---
```
(a)

```
Met Ser Pro Val Asp Pro Leu His Stop
┌─┐┌─┐┌─┐┌─┐┌─┐┌─┐┌─┐┌─┐┌─┐ ── New reading frame
ATGAGCCCTGTGGATCCGCTTCATTAGG---
└─┘└─┘└─┘└─┘└─┘└─┘└─┘└─┘└─┘ ── Old reading frame
Met   Ala  Leu Trp Ile Arg Phe Ile Arg
```
(b)

FIGURE 11.12 Frameshift mutation. (*a*) Normal reading frame of part of a gene. (*b*) The adenine nucleotide inserted in the fourth position of the coding region (gray) shifts the reading frame one base to the left. The new reading frame with the corresponding codon meanings is shown by the brackets above the base sequence (red); the old reading frame, which corresponds to the unmutated base sequence, is denoted by brackets below the sequence (blue). A deletion of a base, instead of an insertion, would have shifted the reading frame one base to the right.

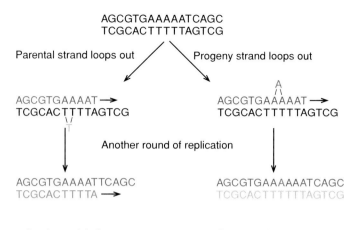

One-base deletion   One-base insertion

FIGURE 11.13 Hypothetical mechanism for frameshift mutation. A parental DNA duplex (top) contains a string of five A–T pairs. During replication of this DNA, transient looping out of bases within this A–T stretch can occur. On the left, one of the T's (pink) in the parental strand loops out. This leaves only four T's to pair with A's in the growing progeny strand (red). After one more round of replication, these four A's serve as the template for the incorporation of four T's in the new progeny strand (blue). Now we have four A–T pairs instead of five; a one-base deletion has occurred. On the right, one of the newly incorporated A's (gray) in the progeny strand loops out, allowing six A's to be inserted instead of five. When this DNA replicates again, the strand with six A's will dictate the incorporation of six T's in the new progeny strand (light blue). A one-base insertion results.

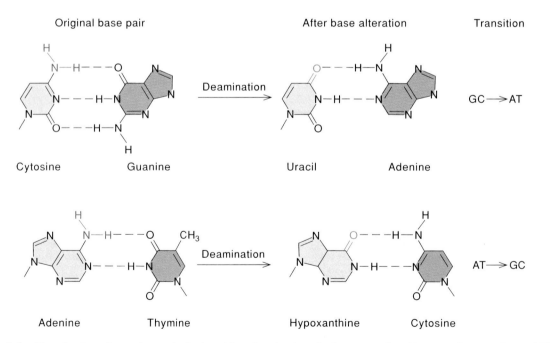

FIGURE 11.14 Deamination of cytosine and adenine. After deamination, the bases uracil and hypoxanthine are formed. These pair with adenine and cytosine, respectively, so transitions occur in each case.

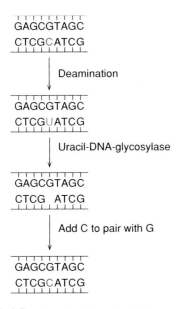

FIGURE 11.15 Repair of deaminated cytosine. A deamination event (top) has converted a C (red) to a U (blue). The enzyme uracil-DNA glycosylase recognizes the U as foreign and removes it (middle), leaving a sugar without a base in the bottom DNA strand. In the second step of the repair process (bottom), a C (red) is added to pair with the G in the top strand.

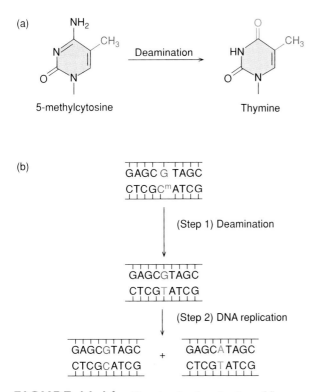

FIGURE 11.16 Mutation by deamination of 5-methylcytosine. (*a*) The deamination step converts 5-methylcytosine to thymine. (*b*) Making the change permanent. In step 1, deamination converts the 5-methylcytosine (red) to thymine (blue). Since thymine cannot be recognized as foreign, it is not removed. In step 2, DNA replication yields one duplex with a wild-type GC pair (red) and one with a mutant AT pair (blue).

ordinary DNA bases and therefore not recognizable as foreign. This means that once 5-methylcytosine deaminates, a G–T mismatch appears. After this DNA replicates, one duplex will have the wild-type G–C pair, but the other will have a mutant A–T pair (figure 11.16*b*).

The pattern of methylation of cytosines in bacteria and in higher eukaryotes is specific. That is, a certain few cytosines are selected for methylation, while all the others remain unmodified. In mammals, for example, cytosines in certain CG sequences are targets for methylation. This specificity has predictable consequences. The sequence CG has become much rarer than any other dinucleotide sequence, presumably because deamination has led to conversion of most of the CG's to TG's during evolution. In bacteria, too, methylated C's become hot spots for mutation.

Triplet Repeats

Since 1991, an important and totally unexpected type of human mutation has been discovered. In a variety of different human genetic diseases—including spinal and bulbar muscular atrophy, fragile X syndrome, myotonic dystrophy (DM), and Huntington's disease—the defective gene has experienced an expansion of a repeated base triplet. The triplet is repeated in the normal allele of the gene in each case, as shown in figure 11.17, but is repeated even more in the mutant state.

Fragile X syndrome is one of the most common types of mental retardation, affecting about 1 in 1,250 males. It is a sex-linked recessive disorder with characteristic physical deformities

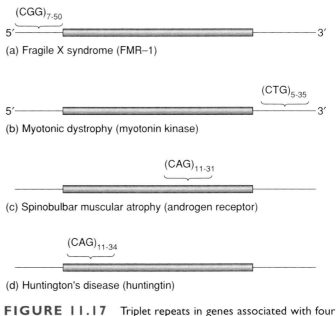

FIGURE 11.17 Triplet repeats in genes associated with four human genetic diseases. The positions of the triplet repeats with respect to the coding region (red) of each gene is shown, along with the approximate range of number of triplet repeats in normal individuals. The number of repeats in individuals affected by each disease is higher. (*a–d*) The name of the genetic disorder is given first, followed by the name of the gene product in parentheses.

as well as moderate to severe retardation. The target gene, *FMR-1,* normally has about 6 to 54 repeats of the triplet CGG in its 5′ untranslated region. In the disease state, the number of repeats expands to hundreds or even thousands. Huntington's disease is a progressive, fatal nervous deterioration that usually strikes in middle age but sometimes begins in childhood. In this disease, the repeated triplet, which apparently occurs in the target gene's coding region, is CAG. Its expansion is less dramatic than that of the CGG triplet in fragile X syndrome; the normal range of 11 to 34 repeats expands to 42 to 86 repeats in most Huntington's patients. Nevertheless, this alteration is devastating, and the number of repeats seems to correlate with the age of onset of the disease. That is, the more repeats, the earlier the onset.

We do not understand how triplet repeats occur, but they are common in the human genome. They are also exceptionally unstable, which means that the number of repeats at a given locus is likely to vary from one individual to another. We will discuss the practical implications of this phenomenon in chapter 16. We do not know for sure how the repeats expand, giving rise to disease. One possibility is that the repeat regions sometimes cause slippage in the replicating machinery, so too many (or occasionally too few) repeats are inserted as the DNA replicates. This process also seems to have a threshold effect. That is, once the number of repeats has passed a certain threshold, the instability of the repeat region increases dramatically, and the number of repeats increases rapidly into the danger zone. This instability leads to the generic term **dynamic mutations.**

Spontaneous mutations can occur in several ways: (1) The DNA replicating machinery can simply make mistakes that go uncorrected. (2) The bases in the DNA template strand or in the newly inserted nucleotide can shift to an alternate tautomeric form that base-pairs incorrectly. (3) Too many or too few bases can be inserted, causing frameshift mutations. (4) Certain bases can deaminate, altering their base-pairing properties. (5) The number of triplet repeats in a gene can expand, altering the properties of the corresponding gene product. These dynamic mutations may be a variation on (1), since the cause of the increase in number of triplet repeats may be slippage during DNA replication.

Measuring Spontaneous Mutation Rates

How frequently do spontaneous mutations occur? This is a difficult determination to make. To begin with, many mutations are undetectable by ordinary means, so they are not counted. We will consider such silent mutations later. Furthermore, mutations are rare, so we must examine many organisms to find a significant number of mutants. Once we have detected a given number of mutants, we must determine the rate at which they are produced—that is, the number of mutations per unit of time. But geneticists measure time in different ways than physicists do. Instead of mutations per minute, day, or year, we speak of

mutations per generation or per cell division. This is the **mutation rate.** In higher eukaryotes, including humans, the number of generations is usually taken as the number of gametes in the population under study. For example, if we are considering a population of 50,000 people, the number of gametes that produced this population is 100,000.

Table 11.1 presents mutation rates for several organisms. We do not fully understand why these numbers vary so much from one organism to another, but one contributing factor is likely to be the efficiency of proofreading during DNA synthesis (chapter 7). Another consideration may be the efficiency of mismatch repair, which we will consider later in this chapter.

Spontaneous mutation rates—the number of mutations arising per gamete or per generation—vary considerably from one organism to another. The spontaneous mutation rate for a given human mutation can be calculated by dividing the number of individuals in a population having that mutation by twice the total number of individuals in the population.

CHEMICAL MUTAGENESIS

We have just seen that mutations can occur spontaneously, but they occur much more frequently in response to environmental agents: chemicals and radiation. We will consider chemical agents first.

Some chemicals act by accelerating the rate of mutations that occur spontaneously. We have already discussed several mechanisms of spontaneous mutation, including deamination, frameshifting, and tautomerization. Deamination can be greatly enhanced by agents such as nitrous acid or bisulfite, or even by heat. Frameshift mutations can be accelerated by planar molecules like the acridine dyes (figure 11.18), which **intercalate,** or insert themselves between the flat base pairs of DNA. This disruption may stabilize the looping out process that seems to be important in frameshift mutations (see figure 11.8).

Mutations Caused by Nucleoside Analogues

Some synthetic compounds can enhance the frequency of tautomerization and thereby induce mutations. The classic example of such a mutagen is **5-bromodeoxyuridine (BrdU),** which resembles thymidine except for the substitution of a bromine atom for a methyl group in the 5-position (figure 11.19). However, once BrdU is incorporated in place of thymidine, it can cause trouble. The trouble derives from the enhanced tendency of BrdU to switch occasionally to the enol tautomer that base-pairs as C instead of T.

Mutations Caused by Alkylation of Bases

Some substances in our environment, both natural and synthetic, are **electrophilic.** This name means electron- (or negative charge-)loving; thus, electrophiles seek centers of negative charge

Table 11.1 A Sample of Spontaneous Mutation Rates in Different Organisms

Organism	Character	Rate	Units
Bacteriophage T2	Lysis inhibition, $r \to r^+$	1×10^{-8}	Per gene per replication
	Host range, $h^+ \to h$	3×10^{-9}	
Bacteria: Escherichia coli	Lactose fermentation, $lac^- \to lac^+$	2×10^{-7}	
	Phage T1 sensitivity, T1-$s \to$ T1-r	2×10^{-8}	
	Histidine requirement, $his^- \to his^+$	4×10^{-8}	
	$his^+ \to his^-$	2×10^{-6}	Per cell per division
Algae: Chlamydomonas reinhardi	Streptomycin sensitivity, str-$s \to$ str-d	1×10^{-9}	
	str-$d \to$ str-s	1×10^{-8}	
	Streptomycin sensitivity, str-$s \to$ str-r	1×10^{-6}	
Fungi: Neurospora crassa	Inositol requirement, $inos^- \to inos^+$	8×10^{-8}	Mutant frequency among asexual spores
	Adenine requirement, $ade^- \to ade^+$	4×10^{-8}	
Corn: Zea mays	Shrunken seeds, $Sh \to sh$	1×10^{-5}	
	Purple, $P \to p$	1×10^{-6}	
Fruit fly: Drosophila melanogaster	Yellow body, $y^+ \to y$, in males	1×10^{-4}	Mutant frequency per gamete per sexual generation
	$y^+ \to y$, in females	1×10^{-5}	
	White eye, $w^+ \to w$	4×10^{-5}	
	Brown eye, $bw^+ \to bw$	3×10^{-5}	
Mouse: Mus musculus	Piebald coat color, $S \to s$	3×10^{-5}	
	Dilute coat color, $D \to d$	3×10^{-5}	
Human: Homo sapiens	Normal $\to$ hemophilic	3×10^{-5}	
	Normal $\to$ albino	3×10^{-5}	
Human bone marrow cells in tissue culture	Normal $\to$ 8-azaguanine resistant	7×10^{-4}	Per cell per division
	Normal $\to$ 8-azaguanosine resistant	1×10^{-6}	

Source R. Sager and F. J. Ryan, *Cell Heredity.* John Wiley & Sons, Inc., New York, 1961.

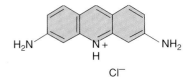

FIGURE 11.18 Proflavin (hydrochloride), one of the acridine dyes.

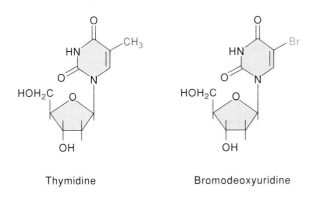

Thymidine   Bromodeoxyuridine

FIGURE 11.19 Comparison of thymidine and bromodeoxyuridine.

in other molecules and bind to them. Many other environmental substances are metabolized in the body to electrophilic compounds. One of the most obvious centers of negative charge in biology is the DNA molecule. Every nucleotide contains one full negative charge on the phosphate and partial negative charges on the bases. When **electrophiles** encounter these negative centers, they attack them, usually adding carbon-containing groups called **alkyl groups.** Thus, we refer to this process as **alkylation.**

Aside from the phosphodiester bonds, the favorite sites of attack by alkylating agents are the N^7 of guanine and the N^3 of adenine (figure 11.20). Since neither of these positions is involved in base-pairing, such alkylations do not lead immediately to mispairing. However, they do make the bond between sugar and base more labile, or more apt to break. When this break occurs, it leaves an apurinic site, a sugar without its purine. This obviously cannot be replicated properly unless it is first

repaired, but cells sometimes attempt to replicate apurinic DNA anyway. If they do, they frequently insert the wrong base across from an apurinic site, and this generates a mutation. Alkylation can also enhance the tendency of a base to form the rare (wrong) tautomer. As we have seen, this can lead to mutations. Moreover, all of the nitrogen and oxygen atoms involved in base-pairing are also subject to alkylation, which can directly disrupt base-pairing and lead to mutation (figure 11.21).

Many environmental **carcinogens,** or cancer-causing agents, are electrophiles that seem to act by attacking DNA and alkylating it, as we will see in chapter 17. Many of the favorite mutagens used in the laboratory for the express purpose of creating mutations are also alkylating agents. One example is ethylmethane sulfonate (EMS), which transfers ethyl (CH_3–CH_2) groups to DNA (figure 11.21).

> **D**ifferent chemicals induce different kinds of DNA damage. Nitrous acid and bisulfite cause deamination of bases, especially cytosine, causing changes in their base-pairing properties. Alkylating agents like ethylmethane sulfonate add bulky alkyl groups to bases, either disrupting base-pairing directly or causing loss of bases, either of which can lead to faulty DNA replication or repair. Base analogues such as BrdU are incorporated into DNA and then can lead to abnormal base-pairing. Certain planar molecules, such as the acridine dyes, cause addition of extra bases or too few bases during DNA replication, which can alter the reading frame of a gene.

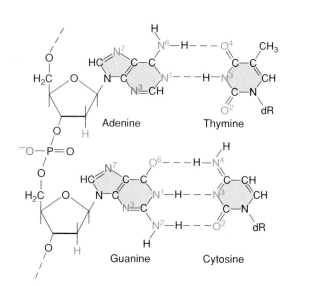

FIGURE 11.20   Electron-rich centers in DNA. The targets most commonly attacked by electrophiles are the phosphate groups, N^7 of guanine, and N^3 of adenine (red); other targets are in blue.

RADIATION-INDUCED MUTATIONS

Ultraviolet, gamma-, and X-radiation are the common types of mutagenic radiation found in nature and used in mutagenesis experiments. These kinds of radiation differ greatly in energy; therefore, they differ greatly in the kinds of DNA damage they cause.

Ultraviolet Radiation

Ultraviolet radiation is relatively weak, so the damage it causes is relatively modest; it cross-links adjacent pyrimidines on the same DNA strand, forming dimers, usually **thymine dimers.** Figure 11.22 shows the structure of a thymine dimer and illustrates how it interrupts base-pairing between the two DNA strands. This blocks DNA replication because the replication machinery cannot tell which bases to insert opposite the dimer. As we will see, replication sometimes proceeds anyway, and bases are inserted at random. If these are the wrong bases (and they usually are), a mutation results.

The kind of ultraviolet radiation that is most damaging to DNA has a wavelength of about 260 nm, which is not surprising, since this is the wavelength of radiation that is absorbed most strongly by DNA. Such radiation has great biological significance; it is abundant in sunlight, so most forms of life are exposed to it to some extent. The mutagenicity of ultraviolet light explains why sunlight can cause skin cancer: Its ultraviolet component damages the DNA in skin cells, which sometimes causes those cells to lose control over their division.

Given the dangers of ultraviolet radiation, we are fortunate to have a shield—the ozone layer—in the earth's upper atmosphere to absorb the bulk of such radiation. However, scientists have recently noticed alarming holes in this protective shield—the most prominent one located over Antarctica. The causes of this ozone depletion are still somewhat controversial, but they probably include compounds traditionally used in air conditioners and in plastics. Unless we can arrest the destruction of the ozone layer, we are destined to suffer more of the effects of ultraviolet radiation, including skin cancer.

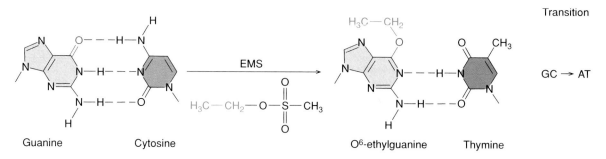

FIGURE 11.21   Alkylation of guanine by EMS. At the left is a normal guanine-cytosine base pair. Note the free O^6 oxygen (red) on the guanine. Ethylmethane sulfonate (EMS) donates an ethyl group (blue) to the O^6 oxygen, creating O^6-ethylguanine (right), which base-pairs with thymine instead of cytosine.

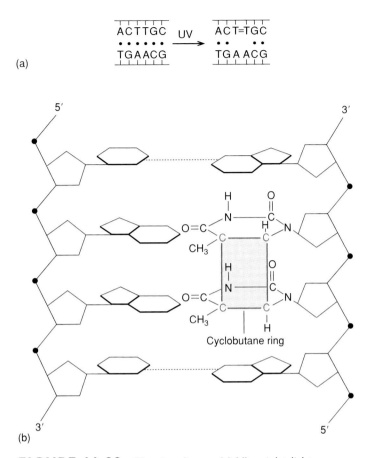

(a)

(b)

FIGURE 11.22 Thymine dimers. (*a*) Ultraviolet light cross-links the two thymine bases on the top strand. This distorts the DNA so that these two bases no longer pair with their adenine partners. (*b*) The two bonds joining the two thymines form a four-membered cyclobutane ring (red).

Gamma and X Rays

The much more energetic **gamma rays** and **X rays,** like ultraviolet rays, can interact directly with the DNA molecule. However, they cause most of their damage by ionizing the molecules, especially water, surrounding the DNA. This forms **free radicals,** chemical substances with an unpaired electron. These free radicals, especially those containing oxygen, are extremely reactive, and they immediately attack neighboring molecules. When such a free radical attacks a DNA molecule, it can change a base, but it frequently causes a single- or double-stranded break. Single-stranded breaks are ordinarily not serious because they are easily repaired, but double-stranded breaks are very difficult to repair properly, so they frequently cause a lasting mutation. Because ionizing radiation can break chromosomes, it is referred to not only as a muta-gen, or mutation-causing substance, but also as a **clastogen,** which means "breaker."

> Different kinds of radiation cause different kinds of mutations. Ultraviolet rays have comparatively low energy, and they cause a moderate type of damage: thymine dimers. Gamma and X rays are much more energetic. They ionize the molecules around DNA and form highly reactive free radicals that can attack DNA, altering bases or breaking strands.

SILENT MUTATIONS

Most noticeable mutations are harmful, leading to the production of faulty proteins or RNAs. Still, since life is not perfect and there is always room for improvement, mutations can sometimes actually improve genes. They can change their products so they work better or allow their hosts to survive and reproduce better.

Many mutations are not detectable by ordinary genetic means. We call them **silent mutations.** For example, consider the change of a codon from UCA to UCG. This is a base change, so it is a real mutation. But both triplets code for the same amino acid (serine), so no change occurs in the protein product of the gene. The mutation is silent. The only way we would ever know for sure that it had happened would be to determine the mutant gene's exact base sequence. Such mutations in the third base of a codon (the **wobble position**) are more likely to be silent since they frequently do not change the sense of the codon. Therefore, these mutations are tolerated and we find them much more frequently than mutations in the first two positions of a codon. Other types of silent mutations occur within introns, where they usually have no effect on the gene's function, or entirely outside of genes—in intergenic regions—where they are also usually without phenotypic effect. On the other hand, some mutations outside a gene's coding region may have profound effects. For example, a mutation in an intron may prevent splicing, or a mutation in a promoter may prevent transcription of the gene. These are certainly not silent mutations.

> Mutations with no measurable phenotypic effect, i.e., those that do not change a gene's product or activity, are called silent mutations.

REVERSION

Point mutations, especially frameshift mutations, may have drastic effects, but they are reversible. **Reversion,** or **back-mutation,** to the original phenotype can occur in two ways: a **true reversion** is the alteration of the mutated base back to its original identity; alternatively, a change can occur elsewhere in the same gene to compensate for the original mutation. For example, consider a gene that codes for a protein having a pos-itively charged amino acid (arginine) that interacts with a nega-tively charged amino acid glutamate, as in figure 11.23. This

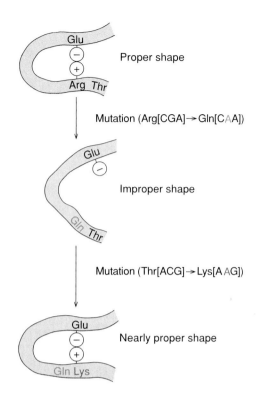

FIGURE 11.23 Second-site reversion. The original G → A mutation (blue) changed an arginine, which bears a positive charge, to a glutamine (blue), which is uncharged. This destroyed an important interaction with a negatively charged glutamate and resulted in an inactive protein. The C → A reversion (red) changed a neighboring threonine to a positively charged lysine (red), restoring the vital interaction with the glutamate and therefore restoring the protein's function.

interaction helps hold part of the protein in the proper shape—a fold that is vital to the protein's function.

Suppose that we cause a mutation (a **forward mutation**—away from wild-type) in the codon for the arginine. This will prevent proper folding of the protein, so the protein will not function. Now suppose we cause another mutation that creates a new positively charged amino acid (lysine) near the position of the lost arginine. If this lysine can substitute for the missing arginine in holding the protein in the right shape, it will restore its function. The mutation that created the new lysine codon will therefore be a reversion. It is not a true reversion, because the reverted gene is different from the original. Instead, we call it a **second-site reversion.**

Second-site reversion can also be considered **suppression,** in that it is a compensation by one mutation for the effects of another. In particular, it is an **intragenic suppression,** since both mutations occur in the same gene. **Intergenic suppression**—suppression of a mutation in one gene by a mutation in another—also happens. Because two genes are involved, intergenic suppression is not the same as reversion. A familiar example of intergenic suppression is suppression of nonsense codons, such as amber codons, by mutant tRNAs (e.g., amber suppressors) that recognize these codons (chapter 10).

Frameshift mutations can revert in a number of ways (figure 11.24): adding a base to compensate for a deleted base; losing a base to balance an added one; losing two more bases if one has already been lost, bringing the total lost to three and eliminating the frameshift; or gaining extra bases to bring the total number gained to three. However, such gains and losses are not likely to

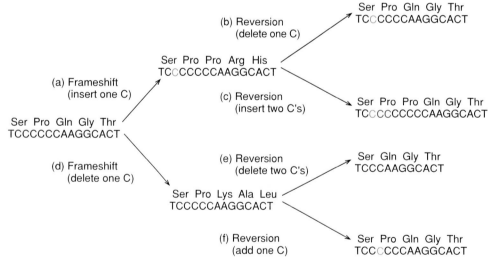

FIGURE 11.24 Revision of frameshift mutation. The original coding strand is presented at left. It contains a string of six C's. (*a*) A frameshift occurs, involving the insertion of an extra C (blue). This shifts the reading frame at the point of the insertion and alters the coding, starting with the third codon. (*b*) This frameshift can revert simply by deleting one of the C's, returning the reading frame and coding to their exact original state. (*c*) The frameshift can also revert by inserting two more C's (red). Now the total number of inserted bases is three, so one whole new codon (CCC) has been created and the original reading frame has been restored. This

results in a slightly altered protein with one extra proline. (*d*) A frameshift occurs, involving the deletion of one C. This shifts the reading frame at the point of the deletion and alters the coding, starting with the third codon. (*e*) This frameshift can revert by deleting two more C's, restoring the original reading frame but resulting in the loss of one codon (CCC). The corresponding protein will lack one proline. (*f*) The frameshift can also revert simply by inserting one C (pink) to compensate for the one that was lost. This restores the exact original reading frame and coding.

happen naturally, so further mutagenesis with frameshifting agents is generally required to obtain reversion of frameshift mutations.

A reversion is a mutation back to the original phenotype. This can happen in two ways: (1) a true reversion, or change of the altered base to the original one; or (2) a second-site reversion, or a new mutation that compensates for the original one.

DNA Repair

We have discussed many different ways that DNA can be damaged. Since much of this damage is serious enough to threaten life, it is not surprising that living things have evolved mechanisms for dealing with it. One way to cope with a mutation is to repair it, or restore it to its original, undamaged state. There are two basic ways to do this: (1) Directly undo the damage, or (2) remove the damaged section of DNA and fill it in with new, undamaged DNA.

Directly Undoing DNA Damage

In the late 1940s, Albert Kelner was trying to measure the effect of temperature on repair of ultraviolet damage to DNA in the bacterium *Streptomyces.* However, he noticed that damage was repaired much faster in some bacterial spores than in others kept at the same temperature. Obviously, some factor other than temperature was operating. Finally, Kelner noticed that the spores whose damage was repaired fastest were the ones kept most directly exposed to light from a laboratory window. When he performed control experiments with spores kept in the dark, he could detect no repair at all. Renato Dulbecco soon observed the same effect in bacteria infected with ultraviolet-damaged phages. It now appears that most forms of life share this important mechanism of repair, which is termed **photoreactivation, or light repair.**

It was discovered in the late 1950s that photoreactivation is catalyzed by an enzyme called **photoreactivating enzyme** or **DNA photolyase.** This enzyme operates by the mechanism sketched in figure 11.25. First, the enzyme detects and binds to the ultraviolet-damaged DNA site (a pyrimidine dimer). Then the enzyme absorbs visible light, which activates it so it can break the bonds holding the pyrimidine dimer together. This restores

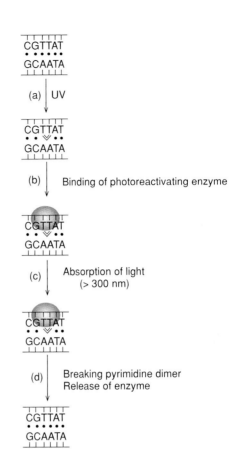

FIGURE 11.25   Model for photoreactivation. (*a*) Ultraviolet radiation causes a thymine dimer to form. (*b*) The photoreactivating enzyme (red) binds to this region of the DNA. (*c*) The enzyme absorbs visible light. (*d*) The enzyme breaks the dimer, and finally dissociates from the repaired DNA.

the pyrimidines to their original independent state. Finally, the enzyme dissociates from the DNA; the damage is repaired.

Organisms ranging from *E. coli* to human beings can directly reverse another kind of mutation, alkylation of the O^6 of guanine. After DNA is methylated or ethylated, an enzyme called **O^6 methylguanine methyl transferase** comes on the scene to repair the damage. It does this by accepting the methyl or ethyl group itself, as outlined in figure 11.26. The acceptor site on the enzyme for the alkyl group is the sulfur atom of an amino acid called cysteine. Strictly speaking, this means that the methyl transferase does not fulfill one part of the definition of an enzyme—that it be regenerated unchanged after the reaction. Instead, this protein seems to be

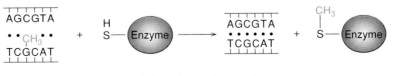

O^6-methylguanine methyl transferase

FIGURE 11.26   Mechanism of O^6-methylguanine methyl transferase. A sulfhydryl group of the enzyme accepts the methyl group (blue) from a guanine on the DNA, thus inactivating the enzyme.

irreversibly inactivated, so we call it a "suicide enzyme" to denote the fact that it "dies" in performing its function. The repair process is therefore expensive; each repair event costs one protein molecule.

One more property of the O⁶-methylguanine methyl transferase is worth noting. The enzyme, at least in *E. coli,* is induced by DNA alkylation. This means bacterial cells that have already been exposed to alkylating agents are more resistant to DNA damage than cells that have just been exposed to such mutagens for the first time.

Ultraviolet damage to DNA (pyrimidine dimers) can be directly repaired by a DNA photolyase that uses energy from visible light to break the bonds holding the two pyrimidines together. O⁶-alkylations on guanine residues can be directly reversed by the suicide enzyme O⁶-methylguanine methyl transferase, which accepts the alkyl group onto one of its amino acids.

EXCISION REPAIR

The percentage of mutations that can be handled by direct reversal is necessarily small. Most mutations involve neither pyrimidine dimers nor O⁶-alkylguanine, so they must be handled by a different mechanism. Most are removed by a process called **excision repair.** The damaged DNA is first removed, then replaced with fresh DNA. This occurs by one of two mechanisms: **base excision repair** or **nucleotide excision repair.**

Base Excision Repair

Certain mutations are recognized by an enzyme called **DNA glycosylase,** which breaks the **glycosidic bond** between the damaged base and its sugar (figure 11.27). This leaves an **apurinic** or **apyrimidinic site (AP site),** which is a sugar without its purine or pyrimidine base. Once the AP site is created, it is recognized by **AP endonucleases** that cut, or **nick,** the DNA strand on either side of the AP site; removing the AP sugar-phosphate. (The "endo" in endonuclease means the enzyme cuts *inside* a DNA strand, not at a free end; Greek: *endo* = within.) DNA polymerase fills the resulting gap by inserting a nucleotide to pair with the one in the opposite strand. DNA ligase seals the remaining nick to complete the job.

Nucleotide Excision Repair

Bulky base damage, including thymine dimers, can also be removed directly, without help from a DNA glycosylase. In this pathway (figure 11.28), the incising enzyme system makes cuts on either side of the mutation, removing an oligonucleotide with the damage. The key enzyme *E. coli* cells use in this process is called the *uvrABC* **endonuclease** because it contains three polypeptides, the products of the *uvrA, uvrB,* and *uvrC* genes. This enzyme generates an oligonucleotide that is twelve to thirteen bases long,

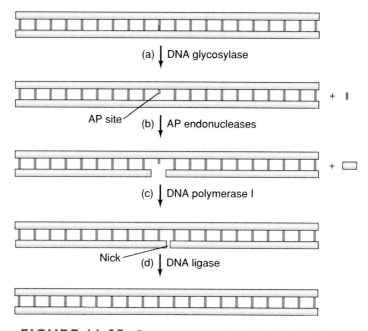

FIGURE 11.27   Base excision repair in *E. coli.* (*a*) DNA glycosylase removes the damaged base (red), leaving an apurinic or apyrimidinic (AP) site on the bottom DNA strand. (*b*) AP endonucleases cut the DNA on either side of the AP site and remove the AP sugar phosphate (yellow block). (*c*) DNA polymerase I fills in the gap with the correct nucleotide, which pairs with the nucleotide in the top strand. (*d*) DNA ligase seals the nick left by the DNA polymerase.

depending on whether the damage affects one nucleotide (alkylations) or two (thymine dimers). A more general term for the enzyme system that catalyzes nucleotide excision repair is excision nuclease, or **excinuclease.** The excinuclease in eukaryotic cells is considerably more complex than that in *E. coli* cells, and removes an oligonucleotide 27 to 30 nucleotides long, rather than a 12- to 13-mer. In any case, DNA polymerase fills in the gap left by the excised oligonucleotide and DNA ligase seals the final nick.

Most DNA damage is corrected by excision repair, in which the mutated DNA is removed and replaced with normal DNA. This can occur by two different mechanisms. (1) The damaged base can be clipped out by a DNA glycosylase, leaving an apurinic or apyrimidinic site that attracts the DNA cutting enzymes that remove the damaged region. (2) The damaged DNA can be clipped out directly by cutting on both sides with an endonuclease to remove the damaged DNA as part of an oligonucleotide. DNA polymerase I fills in the gap and DNA ligase seals the final nick.

MISMATCH REPAIR

So far, we have been discussing repair of DNA damaged by external agents. What about DNA that simply has a mismatch due to

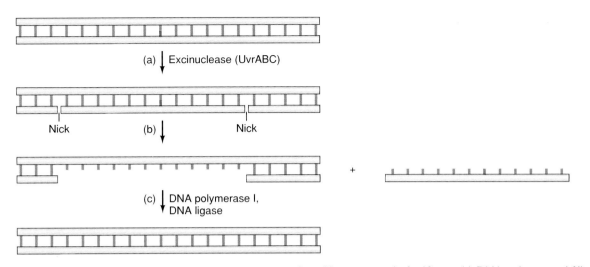

FIGURE 11.28 Nucleotide excision repair in *E. coli*. (*a*) The UvrABC excinuclease cuts on either side of a damaged base (red). This cause removal (*b*) of an oligonucleotide 12 nucleotides long. If the damage were a thymine dimer, then the oligonucleotide would be a 13-mer instead of a 12-mer. (*c*) DNA polymerase I fills in the missing nucleotides, using the top strand as template, and then DNA ligase seals the nick to complete the task, as in base excision repair (figure 11.27*d*).

incorporation of the wrong base and failure of the proofreading system (chapter 7)? At first, it would seem tricky to repair such a mistake, because of the apparent difficulty in determining which strand is the newly synthesized one that has the mistake and which is the parental one that should be left alone. At least in *E. coli* this is not a problem, because the parental strand has identification tags that distinguish it from the progeny strand. These tags are methylated adenines, created by a methylating enzyme that recognizes the sequence GATC and places a methyl group on the A. Since this four-base sequence occurs approximately every 250 base pairs, there is usually one not far from a newly created mismatch.

Moreover, GATC is a palindrome, so the opposite strand also reads GATC in the $5' \rightarrow 3'$ direction. This means that a newly synthesized strand across from a methylated GATC is also destined to become methylated, but a little time elapses before that can happen. The **mismatch repair** system (figure 11.29) takes advantage of this delay; it uses the methylation on the parental strand as a signal to leave that strand alone and correct the nearby mismatch in the unmethylated progeny strand. This process must occur fairly soon after the mismatch is created, or both strands will be methylated and no distinction between them will be possible. Eukaryotic mismatch repair is not as well understood as that in *E. coli,* but the genes encoding the repair enzymes are very well conserved, so the mechanisms are likely to be similar.

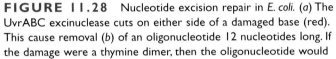

The *E. coli* mismatch repair system recognizes the parental strand by its methylated adenines in GATC sequences. Then it corrects the mismatch in the complementary (progeny) strand.

COPING WITH DNA DAMAGE WITHOUT REPAIRING IT

The direct reversal and excision repair mechanisms described so far are all true repair processes. They eliminate the defective DNA entirely. However, cells have other means of coping with mutations that do not remove the damage but simply skirt around it. These are usually called repair mechanisms, even though they really are not.

Recombination Repair

Recombination repair is the most important of these mechanisms. Because it requires DNA replication before it can operate, it is also sometimes called **postreplication repair.**

Figure 11.30 shows how recombination repair works. First, the DNA must be replicated. This creates a problem for DNA with pyrimidine dimers because the dimers stop the replication machinery. Nevertheless, after a pause, replication continues, leaving a gap across from the dimer. (A new primer is presumably required to restart DNA synthesis.) Next, recombination occurs between the gapped strand and its homologue on the other daughter DNA duplex. This recombination depends on the *recA* gene product, which exchanges the homologous DNA strands (chapter 7). The net effect of this recombination is to fill in the gap across from the pyrimidine dimer and to create a new gap in the other DNA duplex. However, since the other duplex has no dimer, the gap can easily be filled in by DNA polymerase and ligase. Note that the DNA damage still exists, but the cell has at least managed to replicate its DNA. Sooner or later, true DNA repair could presumably occur.

Error-Prone Repair

So-called **error-prone repair** is another way of dealing with damage without really repairing it. This pathway is induced by

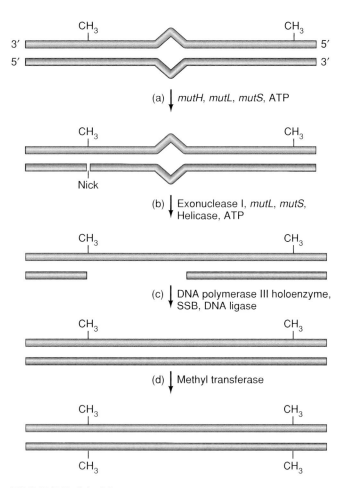

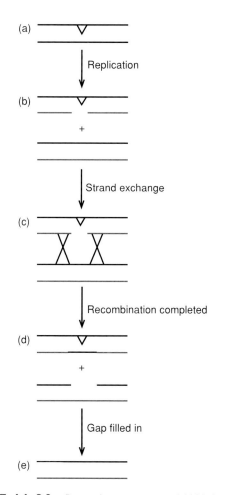

FIGURE 11.29 Mismatch repair in *E. coli.* (*a*) The products of the *mutH, L,* and *S* genes, along with ATP, recognize a base mismatch (center), identify the newly synthesized strand by the absence of methyl groups on GATC sequences, and introduce a nick into that new strand, across from a methylated GATC and upstream from the incorrect nucleotide. (*b*) Exonuclease I, along with MutL, MutS, DNA helicase, and ATP, removes DNA downstream from the nick, including the incorrect nucleotide. (*c*) DNA polymerase III holoenzyme, with help from single-stranded binding protein (SSB), fills in the gap left by the exonuclease, and DNA ligase seals the remaining nick. (*d*) A methyl transferase methylates GATC sequences in the progeny strand across from methylated GATC sequences in the parental strand. Once this happens, mismatch repair cannot occur because the progeny and parental strands are indistinguishable.

FIGURE 11.30 Recombination repair. (*a*) We begin with DNA with a pyrimidine dimer, represented by a V shape. (*b*) During replication, the replication machinery skips over the region with the dimer, leaving a gap; the complementary strand is replicated normally. The two newly synthesized strands are shown in red. (*c*) Strand exchange between homologous strands occurs. (*d*) Recombination is completed, filling in the gap opposite the pyrimidine dimer, but leaving a gap in the other daughter duplex. (*e*) This last gap is easily filled, using the normal complementary strand as template.

DNA damage, including ultraviolet damage, and depends on the product of the *recA* gene. We have encountered *recA* before in our discussion of the induction of a λ prophage during the SOS response (chapter 8), as well as in our consideration of recombination in chapter 7 and in the preceding paragraph. Indeed, error-prone repair is also part of the SOS response.

The chain of events seems to be as follows (figure 11.31): Ultraviolet light or another mutagenic treatment somehow activates the RecA co-protease activity. This co-protease has several targets. One we have studied already is the λ repressor, but its main target is the product of the *lexA* gene. This product, **LexA,**

is a repressor for many genes, including repair genes; when it is stimulated by RecA co-protease to cleave itself, all these genes are induced.

Two of the newly induced genes are *umuC* and *umuD,* which make up a single operon (*umuDC*). Somehow the products of these genes promote error-prone repair. We know this is so because mutations in either *umuC* or *umuD* prevent error-prone repair, but we do not know how these genes work. One thing seems clear: Error-prone repair requires DNA replication. When a gene for the replicating enzyme DNA polymerase III is mutated, no error-prone repair occurs. Our best guess at this time is that error-prone repair involves replication of DNA across from the pyrimidine dimer even though correct "reading" of the defective strand is impossible. This avoids leaving a gap, but it usually puts the wrong bases into the new DNA strand

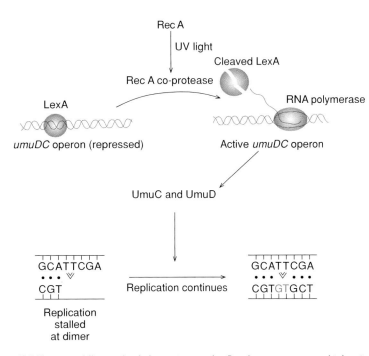

FIGURE 11.31   Error-prone (SOS) repair. Ultraviolet light activates the RecA co-protease, which stimulates the LexA protein (purple) to cleave itself, releasing it from the *umuDC* operon. This results in synthesis of UmuC and UmuD proteins, which somehow allow DNA synthesis across from a thymine dimer, even though mistakes (blue) will usually be made.

(hence the name "error-prone"). When the DNA replicates again, these errors will be perpetuated.

> Cells can employ nonrepair methods to circumvent DNA damage. One of these is recombination repair, in which the gapped DNA strand across from a damaged strand recombines with a normal strand in the other daughter DNA duplex after replication. This solves the gap problem but leaves the original mutation unrepaired. Another mechanism to deal with DNA damage is to induce the SOS response, which causes the DNA to replicate even though the damaged region cannot be read correctly. This results in errors in the newly made DNA, so the process is called error-prone repair.

SEVERE CONSEQUENCES OF DEFECTS IN DNA REPAIR MECHANISMS

Wild-type *E. coli* cells can tolerate as many as fifty pyrimidine dimers in their genome without ill effect because of their active repair mechanisms. Bacteria lacking one of the *uvr* genes cannot carry out excision repair, so their susceptibility to ultraviolet damage is greater. However, they are still somewhat resistant to DNA damage. On the other hand, double mutants in *uvr* and *recA* can perform neither excision repair nor recombination repair, and they are very sensitive to ultraviolet damage. Under these conditions, only one to two pyrimidine dimers per genome is a lethal dose.

A similar situation prevails in humans. Congenital defects in DNA repair cause a group of human diseases, including xeroderma pigmentosum, Fanconi's anemia, and Bloom's syndrome. The latter two conditions are characterized by increased incidence of chromosomal abnormalities, especially gaps and breaks. This is understandable in the case of Bloom's syndrome, because that condition involves a deficiency in DNA ligase, the enzyme that repairs DNA breaks. Whether this defect explains the failure to repair damaged chromosomes is still unclear, but the result of this failure is that patients with any of these diseases are more susceptible to cancer than is the general population.

Xeroderma pigmentosum is better understood. Most patients with this disease are thousands of times more likely to develop skin cancer than normal people if they are exposed to the sun (figure 11.32). In fact, their skin can become literally freckled with skin cancers. This sun-sensitivity reflects the fact that xeroderma pigmentosum cells are defective in excision repair and therefore cannot repair pyrimidine dimers effectively. In addition, patients with xeroderma pigmentosum have a somewhat higher than average incidence of internal cancers. This probably means that these people are also defective in repairing DNA damage caused by chemical mutagens. Notice the underlying assumption here that unrepaired genetic damage can lead to cancer. We will expand on this theme in chapter 17.

What repair steps are defective in xeroderma pigmentosum cells? There are at least seven answers to this question. The problem has been investigated by fusing cells from different patients to see if the fused cells still show the defect. Frequently they do not; instead, the genes from two different patients complement

FIGURE 11.32 Inhabitants of a darkened world. These two girls have inherited the genetic disorder xeroderma pigmentosum. They must avoid sunlight and other ultraviolet light sources since their cells are defective in repairing DNA damage caused by ultraviolet light.

People Weekly, © Kim Komenich.

each other (chapter 3). This probably means that a different gene was defective in each patient. So far, seven different complementation groups have been identified this way, suggesting that the defect can lie in any of seven different genes (XPA–XPG). Most often, the first step in excision repair, incision, or cutting the effective DNA strand, seems to be defective. Besides these seven genes, at least three others are required for excision repair in humans. Defects in these genes may not lead to xeroderma pigmentosum, but they can cause other human disorders, such as Cockayne's syndrome, a neurodevelopmental disease.

Two of the xeroderma pigmentosum genes, *XPB* and *XPD,* code for two subunits of the general transcription factor TFIIH. This factor aids in transcription of genes in two ways: First, it phosphorylates one of the subunits of RNA polymerase II and thereby permits it to start transcribing. Second, TFIIH is also a DNA helicase that can open up a double helix and help initiate transcription. The involvement of a transcription factor in excision repair suggests a tie-in between transcription and DNA repair, and so does the finding that the transcribed strand is preferentially repaired. The current hypothesis to explain this relationship is that RNA polymerase stalls at a damaged base and then so-called coupling factors recruit TFIIH and other components necessary for repair. The role of TFIIH seems to be to use its helicase activity to open up the double helix around the damaged DNA prior to nicking one strand on either side of the damage. This allows for the oligonucleotide that contains the damage to be excised. If the DNA were still in double-stranded form, it would be difficult for such a long oligonucleotide to be removed.

Failure of mismatch repair also has serious consequences, including cancer. One of the most common forms of hereditary cancer is **hereditary nonpolyposis colon cancer (HNPCC).** Approximately one American in 200 is affected by this disease, and it accounts for about 15% of all colon cancers—20,000 cases per year in the United States. One of the characteristics of HNPCC patients is **microsatellite instability,** which means that DNA microsatellites, tandem repeats of 1- to 4-bp sequences, including the triplet repeats we discussed earlier in this chapter, change in size (number of repeats) during the patient's lifetime. This is unusual; the number of repeats in a given microsatellite may differ from one normal individual to another, but it should be the same in all tissues and remain constant throughout the individual's lifetime. The relationship between microsatellite instability and mismatch repair is that the mismatch repair system is responsible for recognizing and repairing the "bubble" created by the inaccurate insertion of too many or too few copies of a short repeat because of "slippage" during DNA replication. When this system breaks down, such slippage goes unrepaired, leading to mutations in many genes whenever DNA replicates in preparation for cell division. This kind of genetic instability presumably leads to cancer, by mechanisms involving mutated oncogenes and tumor suppressors, which we will discuss in chapter 17.

Failure to correct genetic damage is very harmful to an organism. *E. coli* cells that are blocked in both excision repair and recombination repair can tolerate only a small fraction of the ultraviolet radiation that would be innocuous to wild-type bacteria. Humans defective in excision repair suffer from a variety of conditions, the best understood of which is xeroderma pigmentosum. Most patients with this disease cannot carry out excision repair. As a result, ultraviolet damage persists in their skin cells and leads to multiple skin cancers. The failure of human mismatch repair leads to microsatellite instability, and ultimately to cancer.

DETECTING MUTAGENS

We have seen that many chemicals are now recognized as mutagens, and that most mutations are deleterious. In fact, as we will see in chapter 17, environmental mutagens are thought to be responsible for the majority of human cancers. It is important, therefore, to identify mutagenic chemicals so we can limit our exposure to them. But because there are so many potential mutagens to test and because massive numbers of organisms must be examined to find significant numbers of mutants, using test animals such as mice is extremely expensive and terribly time-consuming

Fortunately, excellent mutagen testing systems have been developed using bacteria. As a test organism, bacteria have tremendous advantages: They are cheap and easy to culture, and billions of cells can be tested in a day or two. The best known of the bacterial assays was created by Bruce Ames and is known as the **Ames test.** Ames started with auxotrophic histidine mutants—strains of *Salmonella typhimurium* with mutations in the *his* operon that rendered them incapable of growing in the absence of the amino acid histidine. The basic idea of the assay is to observe the frequency of reversions to wild-type in the *his* operon (figure 11.33). This is very simple to do. One just looks for

colonies that can grow on plates lacking histidine. Without any mutagen, these reversions occur only rarely, so few colonies can grow. On the other hand, a powerful mutagen can greatly accelerate the mutation rate—including the rate of back-mutations, or reversions—and many colonies can be observed.

This concept sometimes gives students trouble because the mutation we are looking for is a beneficial back-mutation to wild-type, rather than a deleterious forward mutation. Nevertheless, it is a mutation, and it is much easier to observe than the forward mutation. To understand this, consider the ease of counting the colonies produced by a few prototrophs that can grow, among a vast excess of auxotrophs that cannot; we usually see just a few colonies on an otherwise clear plate. Then

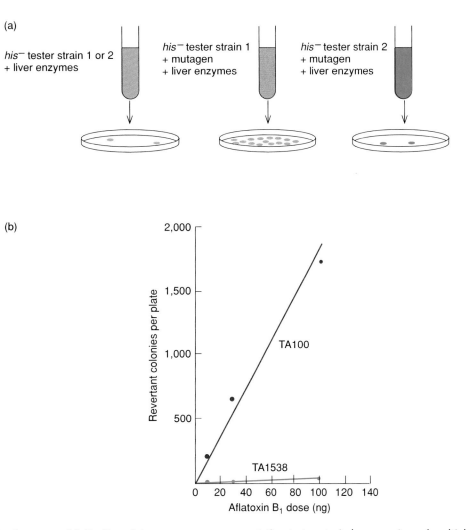

FIGURE 11.33 The Ames test. (*a*) Outline of the procedure. In this example, two *his* tester strains are used. Strain 1 contains a missense mutation in the *his* operon and so will detect mutagens that can cause missense back-mutations, or reversions. Strain 2 contains a frameshift mutation in the *his* operon and so is sensitive to frameshift reversions. These strains are mixed with liver enzymes, which can metabolize compounds to their mutagenic forms. In the control on the left, no mutagen is added, thus only a few spontaneous reversions occur. In the middle, the missense

mutation tester strain has experienced multiple reversions, showing that the compound under study can cause missense mutations. On the right, the frameshift mutation tester strain shows no significant difference from the control, so this compound does not cause frameshift mutations. (*b*) Data from a test of the mutagen aflatoxin B_1. The tester strain TA100, like the hypothetical strain 1 above, tests for missense mutations. Clearly, aflatoxin is a powerful mutagen in this strain. By contrast, the frameshift tester strain, TA1538, detects little if any tendency of aflatoxin to cause frameshift mutations.

compare this with the hopelessness of finding a few auxotrophs that cannot grow, among countless prototrophs that can. We would simply see a uniform lawn of bacteria; the few absent auxotrophs would never be missed.

Ames had to make some additions and modifications to his bacteria in order to make them work well in detecting mutagens. First, he took note of the fact that most chemical mutagens must be metabolized by the body before they become mutagenic. This metabolism, carried out by liver enzymes, is part of the normal method by which we excrete toxic materials. Since bacteria do not perform such metabolism, Ames added a rat liver extract to his bacteria to mimic the metabolism a potential mutagen would experience in the body. In addition, Ames mutated his bacteria to make them leaky to exogenous chemicals and deficient in excision repair. He also added a plasmid bearing an aberrant *umuDC* operon, which produces enzymes that are extra efficient in error-prone SOS repair. This converts most kinds of DNA damage to mutations.

How valid are the results of this kind of mutagenicity testing? Are these strains too sensitive to mutation? Are compounds that cause mutations in bacteria also mutagens in humans? It is true that we have much better defenses against mutagens than these severely weakened bacteria possess, but the Ames test does give a good idea about the relative mutagenicities of different chemical agents. And since bacteria and humans have the same kind of genetic material, agents that mutate their genes are likely to do the same to ours. On the other hand, despite high hopes, the Ames test is a relatively poor predictor of carcinogenicity in humans. In other words, compounds that cause mutations in bacteria do not necessarily cause human cancer. We will return to this topic in chapter 17.

The Ames test uses a set of bacterial tester strains to screen for chemical mutagens. These *Salmonella* cells are histidine auxotrophs and are exquisitely sensitive to mutation by chemicals. To perform the Ames test, the researcher adds a suspected mutagen to these cells (plus liver enzymes to metabolize the chemical to its mutagenic form) and observes the appearance of histidine prototrophic colonies. The number of revertant colonies is a measure of the potency of the mutagen.

WORKED-OUT PROBLEMS

PROBLEM 1

Here is the double-stranded DNA sequence of a portion of a bacterial gene:

5′CACTATGCTTGCGTGGACGCATTAAC3′
3′GTGATACGAACGCACCTGCGTAATTG5′

The bottom strand is the template strand.

(a) Assuming that transcription starts with the first A (in the RNA) and continues to the end, what would be the sequence of the mRNA transcribed from this gene fragment?

(b) What is the amino acid sequence encoded by this mRNA?

What effects would the following events have on this DNA sequence, and on its protein product, assuming that any damage goes unrepaired and finally affects *both* DNA strands?

(c) A tautomerization to the enol form of T in the first T of the nontemplate strand during DNA replication.

(d) A tautomerization to the imino form of A in the second A of the template strand during DNA replication.

(e) Deamination of the third C of the nontemplate strand prior to DNA replication.

(f) O⁶-alkylation of the fourth G of the template strand prior to DNA replication.

(g) Deletion of the third C in the nontemplate strand.

SOLUTION

(a) Since the top strand is the nontemplate strand, it is the one that has the same polarity and sense as the mRNA. Therefore, this is the strand you should use as your model for the mRNA transcript of this gene. The only changes you need to bear in mind are that T's in the DNA strand will be replaced by U's in the mRNA, and that since transcription starts with the first A, you should ignore the first C. Thus, the mRNA will read:

ACUAUGCUUGCGUGGACGCAUUAAC

(b) using the genetic code in chapter 10, and the strategy in the second worked-out problem in that same chapter, we can find the first AUG, beginning with the fourth base in the mRNA. The reading frame and coding from there on is:

fMet	Leu	Ala	Trp	Thr	His	Stop
AUG	CUU	GCG	UGG	ACG	CAU	UAA

Thus, the protein product will be fMet-Leu-Ala-Trp-Thr-His.

(c) A tautomerization to the enol form of the first T of the nontemplate strand will make it pair like C, resulting in a T → C transition at that point. This will change the mRNA sequence from ACU etc., to ACC etc. However, this will have no effect on the coding properties of the mRNA, since the mutation occurred outside the coding region of the gene.

(d) A tautomerization to the imino form of A in the second A in the template strand will make it pair like G, resulting in an A → G transition at that point. This will in turn cause a T → C transition in the nontemplate strand. The change in the gene can be summarized as follows:

$$\text{CACTATG---} \rightarrow \text{CACTACG---}$$
$$\text{GTGATAC---} \quad\;\; \text{GTGATGC---}$$

where the boldface denotes the altered base pair. This will change the mRNA sequence from ACUAUG etc., to ACUACG etc. The change of the initiation codon from AUG to ACG will destroy its function, so the ribosome will search for another AUG (corresponding to ATG in DNA). Since there is none, no protein product will be made.

(e) Deamination of the third C of the nontemplate strand will convert it to a U. If this change goes unrepaired, it will result in a C → T transition, and the new mRNA sequence and coding properties will be:

fMet   Phe
ACU AUG **UUU** etc.,

where the boldface indicates the altered base. Thus, the protein product will now be fMet-Phe-Ala-Trp-Thr-His, with a phenylalanine (Phe) in place of the original leucine (Leu).

(f) O[6]-alkylation of the fourth G of the template strand will make it pair like A, resulting in a C → T transition in the nontemplate strand. The change in the gene can be summarized as follows:

5′CACTATGCTTGCG---   CACTATGCTTG**T**G---
3′GTGATACGAACGC--- → GTGATACGAAC**A**C---

where the boldface denotes the altered base pair. The new mRNA sequence and coding properties will be:

fMet   Leu   Val
ACU AUG CUU GUG etc.,

Thus, the protein product will now be fMet-Leu-Val-Trp-Thr-His, with a valine (Val) in place of the original alanine (Ala).

(g) Deletion of the third C in the nontemplate strand will result in the following DNA sequence:

CACTAT**GT**TGCGTGGACGCATTAAC
GTGATA**CA**ACGCACCTGCGTAATTG

where the boldfaced bases flank the deleted base. The new mRNA sequence and coding properties will be:

fMet   Leu   Arg   Gly   Arg   Ile   Asn
ACU AUG UUG CGU GGA CGC AUU AAC

Thus, the protein product will be fMet-Leu-Arg-Gly-Arg-Ile-Asn. Note that the frameshift has caused a drastic difference in the gene product.

PROBLEM 2

You have isolated a protein from a wild-type organism and the corresponding, defective protein from a mutant. You find that the wild-type and mutant proteins differ in only one amino acid, as follows:

Wild-type: Arg   Met   Ser
Mutant: Arg   Ile   Ser

Three different missense mutations could have occurred. Using the genetic code, show the nontemplate strand of the wild-type gene and the three different missense mutant genes. Show only the codon that is involved in the mutation and present your answer in a table as follows:

Organism	Codon	Transition or Transversion?
Wild-type		
Mutant #1		
Mutant #2		
Mutant #3		

SOLUTION

We need to account for the change from Met to Ile in the mutant protein. The genetic code shows that the only codon for Met (methionine) is AUG (ATG in DNA), so that is the codon we need

to use for the wild-type gene. Ile (isoleucine) has three codons: AUU, AUC, and AUA (ATT, ATC, and ATA in DNA). Thus, the mutant gene has one of these latter three Ile codons. If the Ile codon is ATA, then a G → A transition has occurred. A purine (G) has been replaced by another purine (A). If it is one of the others, then a transversion has occurred, because a purine (G) has been replaced by a pyrimidine (C or T). Thus, the table can be filled in as follows:

Organism	Codon	Transition or Transversion?
Wild-type	ATG	—
Mutant #1	ATA	Transition
Mutant #2	ATC	Transversion
Mutant #3	ATT	Transversion

PROBLEM 3

Here is the sequence of a small piece of an mRNA and the protein it encodes.

Ser   Thr   Ala
AGU ACG GCU

A point mutation within its DNA coding sequence can change the sequence of amino acids in the protein to:

Arg   Tyr   Gly

What was the mutation? For the purposes of this problem, a point mutation is defined as changing, inserting, or deleting one base.

SOLUTION

Since three codons in a row are changed, the mutation must be a frameshift rather than a missense mutation. The original amino acid sequence Ser Thr Ala, was changed to Arg Tyr Gly after the mutation. Since the first codon was changed, and a frameshift mutation affects only the codon in which it occurs and codons downstream (toward the 3′-end), the mutation must have occurred in the first codon.

Next we need to determine the direction of the frameshift, by looking at the genetic code. Since arginine (Arg) has six codons, it would be a bit time consuming to examine all its codons; furthermore, the mutation occurred in the first codon, so that confuses the issue somewhat. It is easier to look downstream and see how changing the reading frame one base to the left (−1) or right (+1) could have given codons for tyrosine (Tyr) and glycine (Gly). Tyrosine has only two codons (UAU and UAC). You should look at the −1 and +1 reading frames for one of these codons. There is a UAC codon in the −1 reading frame, beginning with the third base of the wild-type mRNA. In order to shift to the −1 reading frame, the gene must have acquired a one-base insertion in the first codon, before the U. Notice that this will make the third codon GGC, which codes for glycine (Gly), as expected.

Now where was the base inserted, after the A, or after the G in the AGU codon? By consulting the genetic code, we discover that the only way inserting a base can convert the serine codon AGU to an arginine codon is by placing either an A or a G between the second and third bases to give AGA or AGG, both of which are arginine codons. This must have been the mutation in the DNA. Therefore, the sequence of the mutant mRNA is: AGA UAC GGC U or AGG UAC GGC U.

PROBLEM 4

The records of 94,075 children at Lying-In Hospital in Copenhagen show that ten were afflicted with achondroplasia. Two of the ten came from families with a history of such dwarfism, so only eight represent new mutations. What is the mutation rate for the achondroplasia gene?

SOLUTION

Since the number of new mutations is 8, and the total number of gametes involved is twice the total number of individuals, the mutation rate is $8/(2)(94,075) = 4.2 \times 10^{-5}$.

SUMMARY

Mutations are heritable changes in the genetic material. They can be categorized according to their effects on the mutant organism's phenotype. Germ-line mutations affect the gametes, so the mutation can be transmitted to the next generation. Somatic mutations affect the nonsex, somatic cells, and thus are not passed on. Morphological mutations cause a visible change in the affected organism. Nutritional mutations render an organism dependent on added nutrients that a wild-type organism does not require. Lethal mutations occur in indispensable genes. Haploid organisms with lethal mutations die immediately, but lethal mutations are usually recessive, so diploid organisms can tolerate them in heterozygotes. Conditional mutations affect the mutant organism only under certain conditions, such as high temperature. If a conditional mutation causes death under nonpermissive conditions, it is a conditional lethal.

A number of different kinds of agents can cause genetic mutations. Radiation can either damage DNA directly or produce free radicals that do the damage. Ultraviolet light causes pyrimidine dimers, while the more energetic X rays and gamma rays usually cause breaks in DNA. Chemicals can injure DNA in a variety of ways, including deamination and alkylation of bases or by forcing the incorporation of the wrong base, or too many or too few bases, into a replicating DNA. Chemicals usually cause small lesions called point mutations that can revert readily to wild-type. One common kind of point mutation is a missense mutation, in which an altered DNA base leads to an altered mRNA codon, which leads in turn to an altered amino acid in the gene's protein product. The classic example of a human missense mutation is one that causes sickle-cell disease.

All living things have means of reversing genetic damage. Some mutations can be directly undone. For example, the photoreactivating enzyme can break pyrimidine dimers, and the suicide enzyme O^6-methylguanine methyl transferase can remove methyl and ethyl groups from guanine. However, most DNA damage is corrected by excision repair, in which the mutated DNA is removed and replaced with normal DNA. Cells can also employ nonrepair methods to replicate their DNA even though it is damaged. Two examples are recombination repair and error-prone (SOS) repair.

Mutagens can be readily detected by the Ames test, which measures the ability of a compound to cause reversions in auxotrophic strains of bacteria.

PROBLEMS AND QUESTIONS

1. You notice a rose that is all red except for a small patch of white. Is this likely to be due to a somatic or a germ-line mutation? Why?

2. Two mice, heterozygous for the yellow trait, were mated and produced progeny that were ⅔ yellow and ⅓ wild-type in color. The yellow mice are heterozygotes; the homozygous mutants died in utero. Is the yellow color dominant or recessive? Is the lethality of the yellow gene dominant or recessive?

3. You are mapping a DNA polymerase III gene in a new strain of bacteria. Can you use nonconditional mutations that inactivate the gene? Why, or why not?

4. What kinds of mutations would be useful in the mapping studies in question 3?

5. Is a temperature-sensitive gene more easily melted (denatured) than a wild-type gene?

6. A Siamese cat was wounded and had to have a patch of white fur shaved. The fur grew in dark instead of white. With time, this dark fur was gradually replaced with white. Present a plausible explanation for these observations.

7. Does tautomerization of thymine cause transitions or transversions?

8. (a) What kind of DNA damage does ultraviolet light cause? (b) Why is this damage harmful to a cell?

9. (a) What kind of DNA damage do X rays cause? (b) Why is this damage harmful to a cell?

10. A hypothetical mRNA, AUGCGCCUAAAGAGG, codes for fMet-Arg-Leu-Lys-Arg. What happens to the coding sequence and the encoded protein if you delete the first C?

11. What would be the simplest way to cause a second-site reversion of the mutation in question 10?

12. Which two enzymes catalyze direct reversal of DNA damage?

13. What is the difference between an exonuclease and an endonuclease?

14. In excision repair in *E. coli,* two mechanisms could be used for removing damaged DNA and filling in the gap: (1) The old DNA could be removed and replaced in a single pass of the DNA polymerase I by nick translation, or (2) the old DNA could be removed and then, in a separate step, replaced. What DNA feature would occur at an intermediate stage in the second case but not in the first?

15. Why are recombination repair and error-prone repair not real repair processes?

16. A certain wild-type protein and its mutant counterpart have the same sequence except in one region, where one amino acid is changed:

Wild-type:	Ala	Trp	Leu
Mutant:	Ala	Cys	Leu

    (a) Present the wild-type and mutant codons (DNA nontemplate strand) at the point of the mutation.
    (b) Is this a missense (transition or transversion), nonsense, or frameshift (+1 or −1) mutation?

17. Here is a small piece of an mRNA and the protein it encodes:

Glu	His	Lys	
GAG	CAU	AAG	C

    Through two different frameshift mutations in the DNA sequence, the encoded protein can be altered to: (a) Asp Ala Stop, or (b) Glu Ile Ser. What were these mutations?

18. Draw the chemical structure of the base pair between the imino form of cytosine (figure 11.11) and adenine.

19. Draw the chemical structure of the base pair between the enol form of guanine (figure 11.11) and thymine.

20. Here is the sequence of a portion of a bacterial gene:

    5′GTATCGTATGCATGCACGTGAC3′
    3′CATAGCATACGTACGTAGCACTG5′

    The template strand is on the bottom.

    (a) Assuming that transcription starts with the first A (in the RNA) and continues to the end, what would be the sequence of the mRNA derived from this gene fragment?
    (b) What is the amino acid sequence of the peptide encoded by this mRNA?

    What effects would the following events have on this DNA sequence, the corresponding mRNA sequence, and protein product, assuming that any damage goes unrepaired and finally affects both DNA strands:

    (c) Deamination of the third C in the template strand.
    (d) Tautomerization to the imino form of the second A in the template strand.
    (e) Tautomerization to the enol form of the next-to-last T in the nontemplate strand.
    (f) Deamination of the next-to-last C in the template strand.
    (g) Deletion of the third A of the nontemplate strand.

Answers appear at end of book.

Transposable Elements

All is flux, nothing stays still.

Heraclitus

Greek philosopher

Learning Objectives

In this chapter you will learn:

1. How transposable genetic elements can move from one place to another in the genome.

2. How transposable elements can foster mutation.

3. How prokaryotic transposable elements can switch gene expression.

4. The characteristics of a class of eukaryotic transposable elements that replicate in much the same way that retroviruses do.

We have already learned that an organism's genome is not absolutely fixed from the beginning to the end of its life. In addition to the random mutations that occur and the occasional insertion of viral genes, rearrangements of genetic material take place. This can be useful, as in the rearrangement of immunoglobulin gene segments to form a functioning gene, or in the activation of a trypanosome surface protein gene by moving it to an expression site (chapter 14). But pieces of DNA whose primary function is their own replication also move around the genomes of both prokaryotes and eukaryotes. These **transposable elements,** or **transposons,** are mobile in the sense that they can insert copies of themselves throughout the genome. They can also cause rearrangements of host DNA, giving rise to chromosome mutations (chapter 4). We will discuss the characteristics of bacterial and eukaryotic transposons in turn, showing how they can cause mutations, including chromosomal mutations.

BACTERIAL TRANSPOSONS

Bacterial transposons were originally noticed by James Shapiro and others in the late 1960s as mutations that did not behave normally. For example, they did not revert readily the way point mutations do, and the mutant genes contained long stretches of extra DNA. Shapiro demonstrated this by taking advantage of the fact that a λ phage will sometimes pick up a piece of host DNA during lytic infection of E. coli cells, incorporating this "passenger" DNA into its own genome. He allowed λ phages to pick up either a wild-type E. coli galactose utilization gene (gal⁺) or its mutant counterpart (gal⁻), then measured the sizes of the **recombinant DNAs,** which contained λ DNA plus host DNA (figure 12.1). He measured the DNA sizes by measuring the densities of the two types of phage using cesium chloride gradient centrifugation (chapter 7). Since the phage coat is made of protein and always has the same volume, and since DNA is much denser than protein, the more DNA the phage contains, the denser it will be. It turned out that the phages harboring the gal⁻ gene were denser than the phages with the wild-type gene, and therefore held more DNA. The simplest explanation is that foreign DNA had inserted into the gal gene and thereby mutated it. Indeed, later experiments revealed 800–1,400 bp inserts in the mutant gal gene, which were not found in the wild-type gene. In the rare cases when such mutants did revert, they lost the extra DNA. These extra DNA's that could inactivate a gene by inserting into it were the first transposons discovered in bacteria. They are called **insertion sequences (ISs).**

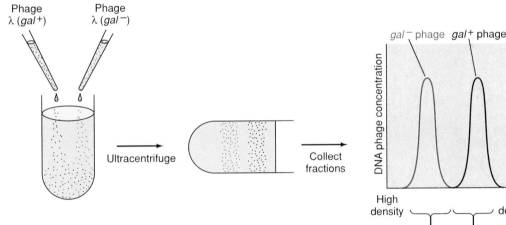

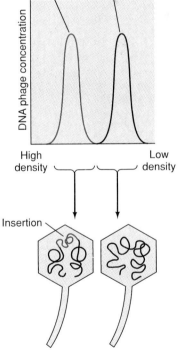

FIGURE 12.1   Demonstration of mutation by insertion. λ phage bearing a wild-type (gal⁺) or mutant (gal⁻) gene from E. coli was subjected to CsCl gradient centrifugation. Two bands of phage separated clearly, the denser (red) bearing the mutant phage. Since the denser phage must contain more DNA, this means that the mutation in the gal gene was caused by an insertion of extra DNA.

Adapted from "Transposable Genetic Elements" by Stanley N. Cohen and James A. Shapiro. Copyright © 1980 by Scientific American, Inc. All rights reserved. Reprinted by permission.

INSERTION SEQUENCES: THE SIMPLEST TRANSPOSONS

Bacterial insertion sequences contain only the elements necessary for transposition. The first of these elements is a set of special sequences at a transposon's ends, one of which is the inverted repeat of the other. The second element is the set of genes that code for the enzymes that catalyze transposition.

Since the ends of an insertion sequence are inverted repeats, if one end of an insertion sequence is 5′-ACCGTAG, the other end of that strand will be the reverse complement: CTACGGT-3′. This is akin to the inverted repeats we discussed in chapter 8, but the two parts of the inverted repeat are interrupted by hundreds of base pairs, corresponding to the body of the insertion sequence. The seven-nucleotide inverted repeats given here are hypothetical and are presented to illustrate the point. Typical insertion sequences have somewhat longer inverted repeats, from fifteen to twenty-five base pairs long. IS1, for example, has inverted repeats twenty-three base pairs long. Larger transposons can have inverted repeats hundreds of base pairs long.

Stanley Cohen graphically demonstrated the presence of inverted repeats at the ends of a transposon with the experiment illustrated in figure 12.2. He started with a **plasmid,** an independently replicating little circle of DNA from a bacterial cell. This plasmid contained a transposon, and had the structure shown at the upper left in figure 12.2a. The original plasmid was linked to the ends of the transposon, which were inverted repeats. Cohen reasoned that if the transposon really had inverted repeats at its ends, he could separate the two strands of the recombinant plasmid, and get the inverted repeats on one strand to base-pair with each other, forming a stem-loop structure as shown on the right in figure 12.2a. The stems would be double-stranded DNA composed of the two inverted repeats; the loops would be the rest of the DNA in single-stranded form. The electron micrograph in figure 12.2b shows the expected stem-loop structure.

The main body of an insertion sequence codes for at least two proteins that catalyze transposition. These proteins are collectively known as **transposase;** we will discuss their mechanism

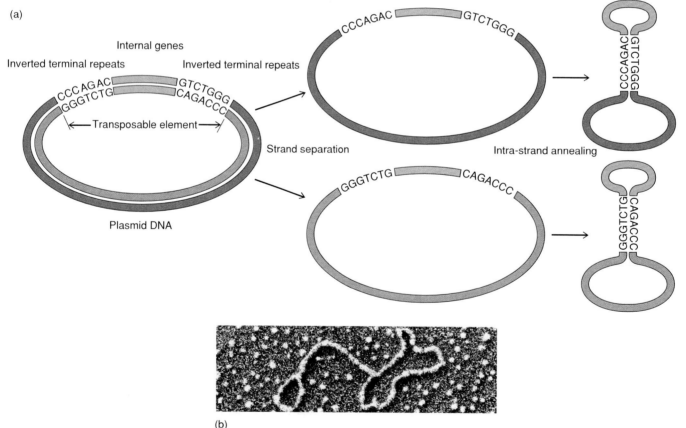

FIGURE 12.2 Transposons contain inverted terminal repeats. (a) Schematic diagram of experiment. The two strands of a transposon-bearing plasmid were separated and allowed to anneal with themselves separately. The inverted terminal repeats will form a base-paired stem between two single-stranded loops corresponding to the internal genes of the transposon (small loop, green) and host plasmid (large loop, purple and red). (b) Experimental results. The

DNA was shadowed with heavy metal and subjected to electron microscopy. The loop-stem-loop structure is obvious. The stem is hundreds of base pairs long, demonstrating that the inverted terminal repeats in this transposon are much longer than the seven base pairs shown for convenience in (a).

Photo (b) courtesy of Stanley N. Cohen, Stanford University.

of action later in this chapter. We know that these proteins are necessary for transposition, because mutations in the body of an insertion sequence can render that transposon immobile.

One other feature of an insertion sequence, shared with more complex transposons, is found just outside the transposon itself. This is a pair of short direct repeats in the DNA immediately surrounding the transposon. These repeats did not exist before the transposon inserted; they result from the insertion process itself and tell us that the transposase cuts the target DNA in a staggered fashion rather than with two cuts right across from each other. Figure 12.3 shows how staggered cuts in the two strands of the target DNA at the site of insertion lead automatically to direct repeats. The length of these direct repeats depends on the distance between the two cuts in the target DNA strands. This distance depends in turn on the nature of the insertion sequence. The transposase of IS1 makes cuts nine base pairs apart and therefore generates direct repeats that are nine base pairs long.

> Insertion sequences are the simplest of the transposons. They contain only the elements necessary for their own transposition; short inverted repeats at their ends and at least two genes coding for an enzyme called transposase that carries out transposition. Transposition causes duplication of a short sequence in the target DNA; one copy of this short sequence flanks the insertion sequence on each side after transposition.

MORE COMPLEX TRANSPOSONS

Insertion sequences are sometimes called "selfish DNA." You should not interpret this term to mean that ISs think about what they do; of course they cannot. Rather, it means that an IS replicates at the expense of its bacterial host and apparently provides nothing useful in return. Other transposons do carry genes that

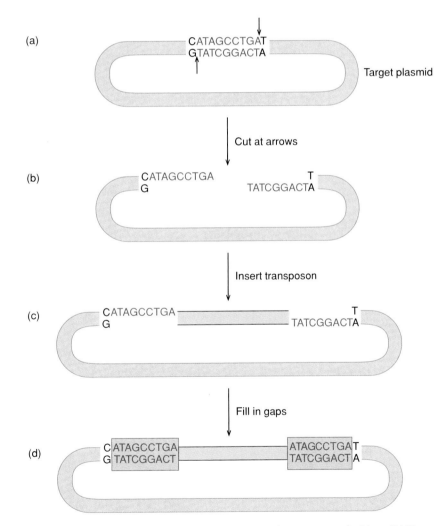

FIGURE 12.3   Generation of direct repeats in host DNA flanking a transposon. (*a*) The arrows indicate where the two strands of host DNA will be cut in a staggered fashion, nine base pairs apart. (*b*) After cutting. (*c*) The transposon (yellow) has been ligated to one strand of host DNA at each end, leaving two nine-base gaps. (*d*) After the gaps are filled in, there are nine base-pair repeats of host DNA (red boxes) at each end of the transposon.

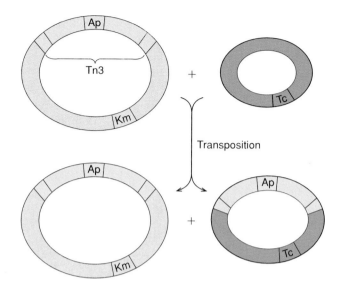

FIGURE 12.4   Tracking transposition with antibiotic resistance genes. We begin with two plasmids: The larger (blue) encodes kanamycin resistance (Km) and bears the transposon Tn3 (yellow), which codes for ampicillin resistance (Ap); the smaller (green) encodes tetracycline resistance (Tc). After transposition, the smaller plasmid bears both the Tc and Ap genes.

are valuable to their hosts, the most familiar being genes for antibiotic resistance. Not only is this a clear benefit to the bacterial host, it is also valuable to geneticists, because it makes the transposon much easier to track.

For example, consider the situation in figure 12.4, in which we start with a donor plasmid containing a gene for kanamycin resistance (Km) and harboring a transposon **(Tn3)** with a gene for ampicillin resistance (Ap); in addition, we have a target plasmid with a gene for tetracycline resistance (Tc). After transposition, Tn3 has replicated and a copy has moved to the target plasmid. Now the target plasmid confers both tetracycline and ampicillin resistance, properties that we can easily monitor by transforming antibiotic-sensitive bacteria with the target plasmid and growing these host bacteria in medium containing both antibiotics. If the bacteria survive, they must have taken up both antibiotic resistance genes; therefore, Tn3 must have transposed to the target plasmid.

MECHANISMS OF TRANSPOSITION

Because of their ability to move from one place to another, transposons are sometimes called "jumping genes." However, the term is a little misleading, since it implies that the DNA always leaves one place and jumps to the other. This mode of transposition does occur and is called **conservative transposition,** because both strands of the original DNA are conserved as they move together from one place to the other. However, transposition frequently involves DNA replication, so one copy of the transposon remains at its original site as another copy inserts at the new site. This is called **replicative transposition,** because when a transposon moves by this

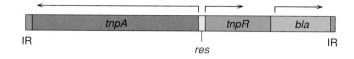

FIGURE 12.5   Structure of Tn3. The *tnpA* and *tnpR* genes are necessary for transposition; *res* is the site of the recombination that occurs during the resolution step in transposition; the *bla* gene encodes β-lactamase, which protects bacteria against the antibiotic ampicillin. This gene is also called Ap and Amp^r. Inverted repeats (IR) are found on each end. The arrows indicate the direction of transcription of each gene.

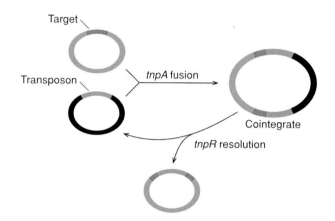

FIGURE 12.6   Simplified scheme of the two-step Tn3 transposition. In the first step, catalyzed by the *tnpA* gene product, the plasmid (black) bearing the transposon (blue) fuses with the target plasmid (green, target in red) to form a cointegrate. During cointegrate formation, the transposon replicates. In the second step, catalyzed by the *tnpR* gene product, the cointegrate resolves into the target plasmid, with the transposon inserted, plus the original transposon-bearing plasmid.

route, it also replicates itself. Let us discuss how both kinds of transposition take place.

Replicative Transposition of Tn3

Tn3, whose structure is shown in figure 12.5, illustrates one well-studied mechanism of transposition. In addition to the *bla* gene, which encodes ampicillin-inactivating β-**lactamase,** Tn3 contains two genes that are instrumental in transposition. Tn3 transposes by a two-step process, each step of which requires one of the Tn3 gene products. Figure 12.6 shows a simplified version of the sequence of events. We begin with two plasmids; the donor, which harbors Tn3, and the target. In the first step, the two plasmids fuse, with Tn3 replication, to form a **cointegrate** in which they are coupled through a pair of Tn3 copies. This step requires recombination between the two plasmids, which is catalyzed by the product of the Tn3 transposase gene *tnpA*. Figure 12.7 shows a detailed picture of how all four DNA strands involved in transposition might interact to form the cointegrate.

Most of the recombinations you have encountered before (chapters 4, 5, and 7) required homology between the two

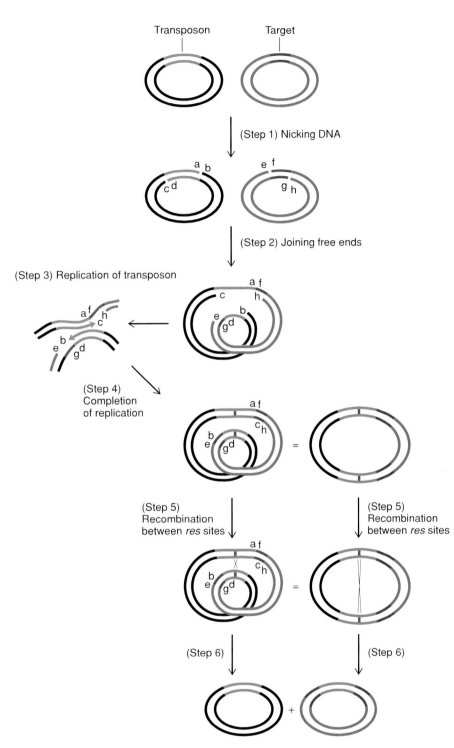

FIGURE 12.7   Detailed scheme of Tn3 transposition. Step 1: The two plasmids are nicked to form the free ends labeled a–h. Step 2: Ends a and f are joined, as are g and d. This leaves b, c, e, and h free. Step 3: Two of these remaining free ends (b and c) serve as primers for DNA replication, which is shown in a blowup of the replicating region. Step 4: Replication continues until end b reaches e and end c reaches h. These ends are ligated to complete the cointegrate. Notice that the whole transposon (blue) has been replicated. The paired *res* sites (purple) are shown for the first time here, even though one *res* site existed in the previous steps. The cointegrate is drawn with a loop in it, so its derivation from the previous drawing is clearer; however, if the loop were opened up, the cointegrate would look just like the one in figure 12.6 (shown here at right). Steps 5 and 6: A crossover occurs between the two *res* sites in the two copies of the transposon, leaving two independent plasmids, each bearing a copy of the transposon. This process is shown for the two equivalent forms of the cointegrate, at right and at left.

recombining DNAs, but the recombination catalyzed by the Tn3 transposase requires little if any homology between donor and target DNAs. This means that transposition is relatively independent of DNA sequence, so Tn3 can transpose to a wide variety of target loci. This kind of **illegitimate recombination,** which shows little dependence on DNA sequence, is a common feature of transposons. Figures 12.6 and 12.7 illustrate transposition between two plasmids, but the donor and target DNAs can be other kinds of DNA, including phage DNAs or the bacterial chromosome itself.

The second step in Tn3 transposition is a **resolution** of the cointegrate, in which the cointegrate breaks down into two independent plasmids, each bearing one copy of Tn3. This step, catalyzed by the product of the **resolvase** gene *tnpR,* is a recombination between homologous sites on Tn3 itself, called *res* sites. Several lines of evidence show that Tn3 transposition is a two-step process. First, mutants in the *tnpR* gene cannot resolve cointegrates, so they cause formation of cointegrates as the final product of transposition. This demonstrates that the cointegrate is normally an intermediate in the reaction. Second, even if the *tnpR* gene is defective, cointegrates can be resolved if a functional *tnpR* gene is provided by another DNA molecule—the host chromosome or another plasmid, for example.

Conservative Transposition

Figures 12.6 and 12.7 illustrate the replicative transposition mechanism, but transposition does not always work this way. Sometimes the transposon moves conservatively, as an intact entity, leaving the donor DNA and appearing in the target DNA. How does this occur? It may be that conservative transposition starts out in the same way as replicative transposition, by nicking and joining strands of the donor and target DNAs, but then something different happens (figure 12.8). Instead of replication occurring through the transposon, new nicks appear in the donor DNA on either side of the transposon. This releases the donor DNA but leaves the transposon still bound to the target DNA. Once the remaining nicks are sealed, we have a donor DNA that has lost its transposon and a target DNA to which the transposon has moved.

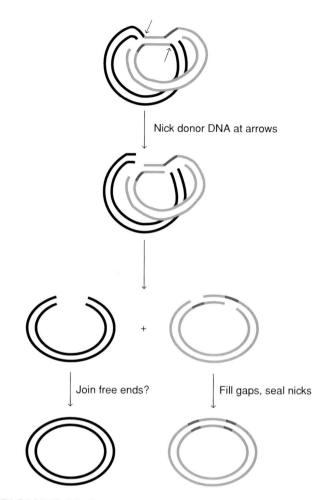

FIGURE 12.8   Conservative transposition. The first two steps are just like those in replicative transposition, and the structure at the top is the same as that between steps 2 and 3 in figure 12.7. Next, however, new nicks occur at the positions indicated by the arrows. This liberates the donor plasmid minus the transposon, which remains attached to the target DNA. Filling gaps and sealing nicks completes the target plasmid with its new transposon. The free ends of the donor plasmid may or may not join. In any event, this plasmid has lost its transposon.

> Many transposons contain genes aside from the ones necessary for transposition. These are commonly antibiotic-resistance genes. For example, Tn3 contains a gene that confers ampicillin resistance. Tn3 and its relatives transpose by a two-step process (replicative transposition). First the transposon replicates and the donor DNA fuses to the target DNA, forming a cointegrate. In the second step, the cointegrate is resolved into two DNA circles, each of which bears a copy of the transposon. An alternate pathway, not used by Tn3, is conservative transposition, in which no replication of the transposon occurs.

TRANSPOSONS AS MUTAGENIC AGENTS

A bacteriophage called **Mu** that infects *E.coli* also behaves as a transposon. In fact, its replication depends on transposition. Phage Mu particles contain a linear DNA genome 37 kb long that can replicate either lytically or lysogenically (figure 12.9). (Recall the lytic and lysogenic replication of phage λ described in chapter 8.) In the lysogenic phase, the Mu genome inserts into the host chromosome, and a phage repressor prevents most phage gene expression, just as in the λ lysogenic phase. However, the lytic phase of Mu is very different from that of λ. Instead of remaining free, the phage DNA integrates into the host chromosome. More strikingly, the phage DNA remains integrated throughout the lytic phase and replicates by transposing. The Mu prophage exists in two forms: In one, it is flanked by direct

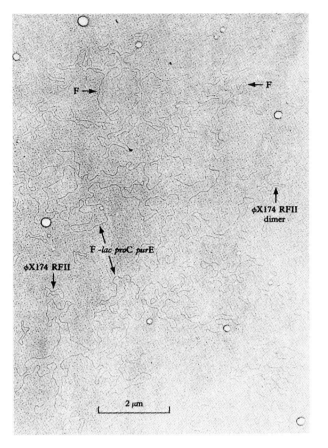

FIGURE 13.7 The F plasmid. This electron micrograph shows two circular molecules of F plasmid DNA among other circular DNAs of different sizes.
© Dr. Richard Deonier.

DNA appeared in the recipient cells, and its density matched that of *E. coli,* not that of the recipient bacterium. This new DNA must be the F factor, since it is the component that is transferred readily during conjugation. As we will see, this is not true conjugation because only a small amount of DNA, the F factor, is transferred from the *E. coli* donor to the *S. marcescens* recipient. Subsequent work has shown that the F factor is a **plasmid,** a circular extra-chromosomal DNA (figure 13.7) that replicates independently of the host chromosome. It is about 2% as large as the host chromosome, or 94,500 base pairs (94.5 kb). Now that we know what it is, we refer to the F factor as the **F plasmid,** and transfer of fertility can be seen as simply transferring the F plasmid from an F⁺ cell to an F⁻ cell.

The fertility of F⁺ *E. coli* is due to an F plasmid of 94.5 kb that can transfer at high frequency into F⁻ bacteria.

HFR STRAINS

In the 1950s, several investigators discovered a new type of *E. coli* in addition to the F⁺ and F⁻ already under study. The new type

was called **Hfr,** which stands for "high-frequency recombinant." Hfr bacteria are mutant F⁺ strains, but they have lost the ability to transfer fertility at high frequency. Instead, they transfer host chromosomal genes at high frequency.

For example, consider an Hfr strain isolated by Hayes in 1953. When Hayes conjugated this Hfr with auxotrophic recipient cells, it generated prototrophic recombinants a thousand times more efficiently than did its F⁺ parent. Hayes also found that this transfer of host genes closely resembled the transfer by F⁺ cells in two respects: It was not inhibited by treating the Hfr donor with streptomycin, and the recombinant genotype came mostly from the recipient, not from the donor. We will see that an F⁺ cell converts to an Hfr when its F plasmid inserts into the host chromosome.

Hfr Cell Transfer of Host Genes

There are two possible explanations for the fact that F⁺ strains contain a few Hfr cells. First, there may be an equilibrium between F⁺ and Hfr cells such that each F⁺ spends a small fraction of its life as an Hfr. Second, F⁺ cells may occasionally mutate to Hfr cells. The difference is that a mutation is permanent (if we neglect reversion, which is rare). By the mid 1950s, Elie Wollman and François Jacob had resolved this question in favor of mutation using the **fluctuation test** devised by Max Delbrück and Salvador Luria. Here is how this test works: The investigators took a very dilute F⁺ culture (only 250 cells per milliliter), split it in half, and then split the first half into fifty equal parts. The second half was not divided. After the bacteria in the cultures had multiplied to a million times their original concentration, each was tested for the ability to transfer host genes, thereby forming recombinant prototrophs.

If the nonmutational hypothesis is correct, then each F⁺ cell has a small but equal probability of changing briefly to an Hfr and transferring host genes. Thus, each of the fifty subcultures should show an equal ability to do this. On the other hand, if the mutational hypothesis is correct and the occasional Hfr cells in the F⁺ population are the only ones that can transfer genes, the fifty cultures should vary widely in this ability. This is because the original dilute cultures contained so few cells that most of them included no Hfr cells at all. Of course, as the cells multiplied, there was a good chance of Hfr cells appearing in the cultures by mutation, and once an Hfr appears, it divides to yield Hfr progeny. Thus, a culture in which an F⁺ → Hfr conversion occurred early in its growth would wind up with the most cells competent to transfer host genes. This is represented on a small scale in figure 13.8*b,* where culture 2 experiences an F⁺ → Hfr conversion in generation 2 and therefore accumulates eight Hfr cells after four generations, while culture 4, with no F⁺ → Hfr conversions, finishes with no Hfr cells.

When the experiment was performed, the small cultures did indeed show great variation—from one prototrophic recombinant in one culture to 116 in another. As a control, the half of the culture that was not split before the cells multiplied was finally divided into fifty equal parts and each of these was tested immediately in the same way. They varied much less: from

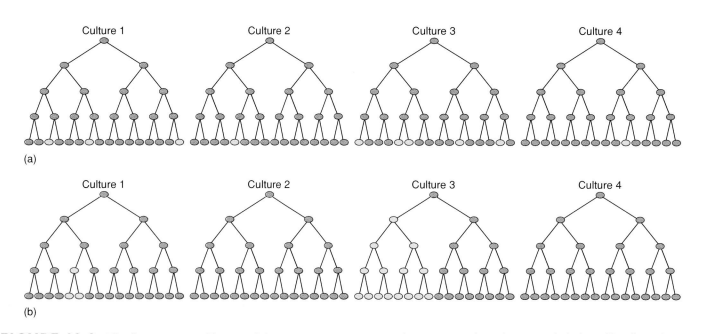

FIGURE 13.8   The fluctuation test. Two possibilities are illustrated: (*a*) Each culture has the same number of F⁺ cells, and each cell has a small but equal chance of transferring host genes during conjugation (represented by the yellow cells). Therefore, each culture should give roughly the same amount of host gene transfer. (*b*) Only Hfr cells (mutant F⁺ cells, yellow) are competent to transfer host genes. Therefore, cultures in which an F⁺ → Hfr conversion occurs early in their growth (culture 3) will wind up with many competent cells, whereas other cultures can easily wind up with few (if the F⁺ → Hfr conversion occurs late in the growth phase; culture 1) or none (if no conversion occurs; culture 4). In other words, the competence of the cells in host gene transfer will fluctuate widely. That is what Wollman and Jacob demonstrated.

a low of ten prototrophic recombinants to a high of twenty-three. Taken together, these results demonstrate that a genetic change in a small proportion of the total F⁺ population renders them capable of transferring genes to auxotrophic F⁻ cells and converting them to prototrophs. This small subpopulation is composed of Hfr cells.

F Plasmid Insertion into the Host Chromosome

How can Hfr cells transfer host genes so efficiently? A clue comes from the similarity between transfer of the F plasmid and transfer of the host genes. It is the F plasmid itself that causes both. An F⁺ cell becomes an Hfr cell when the F plasmid ceases its solitary existence and inserts into the host chromosome. When this happens, the F plasmid loses the ability to transfer itself directly into a recipient cell, but confers this ability on the host chromosome. This leads to the hypothesis of F⁺ → Hfr conversion shown in figure 13.9. The F plasmid recombines with the host chromosome at any of a number of different locations. This mobilizes the chromosome, which is presumably nicked within the F plasmid part, initiating replication by the rolling circle mechanism (see chapter 7). During replication, part of the integrated F plasmid acts as a locomotive and leads the newly synthesized chromosomal DNA into the F⁻ cell.

The bridge between the two cells established by the F pilus is fragile and usually breaks before DNA transfer is complete. The part of the Hfr chromosome that has entered the recipient cell then recombines with the F⁻ chromosome, changing its genotype. Occasionally, the whole Hfr chromosome makes it across before the bridge breaks. When this happens, the recipient cell is converted to Hfr, because the entire integrated F plasmid has been transferred. Part of it led the chromosome across the bridge; the rest of it came across last, like a caboose.

> Occasionally, the F plasmid recombines with its host's chromosome, converting the cell to an Hfr. This mobilizes the Hfr cell's chromosome to transfer to recipient cells, presumably by the same mechanism the independent F plasmid would normally use. The favorite hypothesis for this mechanism involves a rolling circle-style replication of the Hfr chromosome, with one of the daughter DNA molecules entering the recipient cell and recombining with its chromosome to change its genotype.

MAPPING BY INTERRUPTED CONJUGATION

The natural interruption of conjugation comes from random movements of the conjugating partners that can rupture the fragile link connecting them. Elie Wollman and François Jacob learned that they could stop conjugation at a defined time simply by placing the bacteria in a blender. This vigorous treatment

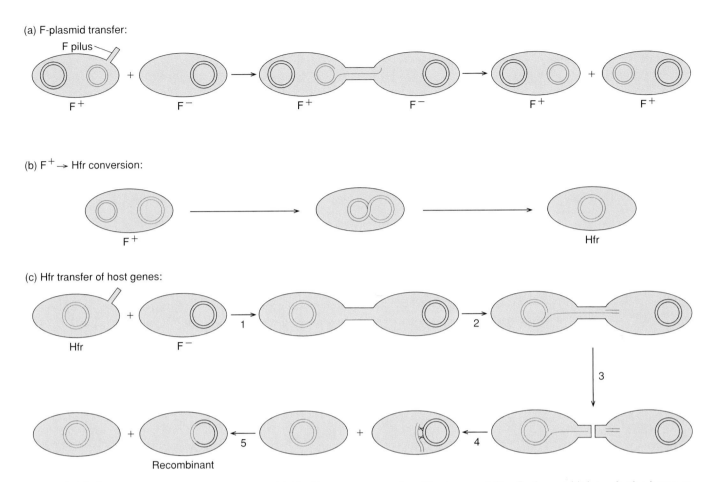

FIGURE 13.9   Model for F⁺ → Hfr conversion. (*a*) The F plasmid (blue-green) normally passes from cell to cell, leaving a copy in the donor cell. This presumably involves rolling circle-style replication of the plasmid, as pictured. The F plasmid and host chromosome are not drawn to scale; the host chromosome is actually about fifty times larger than the F plasmid. (*b*) In F⁺ → Hfr conversion, the F plasmid (blue-green) recombines with the host chromosome (red), mobilizing the whole chromosome. (*c*) Hfr transfer of host genes: 1. The F pilus establishes a bridge between the conjugating cells. 2. The Hfr chromosome begins replicating, with one daughter DNA molecule passing across the bridge to the recipient cell. 3. The bridge breaks during conjugation, leaving a piece of donor DNA in the recipient cell. 4. The donor DNA recombines with the recipient chromosome. 5. After the recipient cell replicates, it contains a recombinant DNA, with part (red) derived from the donor and part (black) from the recipient.

guaranteed that all the conjugating partners would be separated. Furthermore, Wollman and Jacob took advantage of this procedure to build genetic maps.

Here is the basic strategy used by Wollman and Jacob: They conjugated two auxotrophs for a defined length of time, interrupted their conjugation by placing them in a Waring blender, then checked to see whether a given gene had been transferred. For example, consider an F⁻ Leu⁻ Lac⁻ Gal⁻ Strʳ auxotroph (an F⁻ auxotroph incapable of growing on media lacking the amino acid leucine or on media containing the sugars lactose or galactose, but resistant to streptomycin). Imagine that we conjugate it with an Hfr Leu⁺ Lac⁺ Gal⁺ Strˢ (streptomycin-sensitive) auxotroph for two minutes. Then we select Strʳ Leu⁺ recombinants by growing them in medium containing streptomycin and lacking leucine. But after only two minutes of conjugation, no Strʳ recombinants that are Leu⁺ are formed. This tells us that in two minutes, transfer of the *leu* genes from the streptomycin-sensitive Hfr cell to the

streptomycin-resistant F⁻ cell has not yet occurred. On the other hand, if we allow conjugation for four minutes, a significant number of these recombinants will arise. Therefore, transfer of the *leu* genes begins about four minutes after conjugation begins.

Taking the procedure a step further, if conjugation proceeds for ten minutes, some Leu⁺ Lac⁺ Strʳ recombinants appear. This means that the Hfr cell begins to transfer its *lac* genes about ten minutes after the onset of conjugation. Furthermore, if conjugation proceeds for nineteen minutes, some Leu⁺ Lac⁺ Gal⁺ Strʳ recombinants appear. Therefore, the Hfr cell has begun to transfer its *gal* genes by this time. Figure 13.10 presents an idealized picture of the results of such an experiment. Since we are following the transfer of genes from the Hfr donor to the F⁻ recipient, it is important to select only recipient cells after conjugation. The streptomycin sensitivity of the Hfr cells allows us to do just that: We can kill them with streptomycin, leaving only the antibiotic-resistant recipient cells.

These time values tell us something very important about the *E. coli* genetic map. If we assume that the rate of transfer is constant along the whole length of the bacterial chromosome, then the difference in the times it takes to transfer a given pair of genes is proportional to their separation on the genetic map. For example, the difference in the transfer times of the *leu* and *lac* genes is ten minutes minus four minutes, or six minutes; we

therefore say that the *leu* and *lac* loci lie about 6 **minutes** apart on the genetic map. Similarly, the differences in times of transfer of the *gal* and *lac* genes is 19 − 10, or 9 minutes. Thus, these two loci lie 9 minutes apart, and the *leu* and *gal* loci are 9 + 6, or 15 minutes apart. It takes just one hundred minutes for the entire *E. coli* chromosome to be transferred, including its F "caboose." Therefore, the *E. coli* map is composed of 100 minutes.

In his Nobel prize address, François Jacob made the following tongue-in-cheek observation about this long conjugation time: "Marvelous organism, in which conjugal bliss can last for nearly three times the life span of the individual." However, Jacob forgot to note that this "conjugal bliss" is almost always interrupted.

The Circular *E. coli* Genetic Map

Not all Hfr strains transfer the *leu* genes first; each strain has its own characteristic starting point. Moreover, some strains transfer genes in an order opposite to what we have just described. Thus, *gal* would be transferred before *lac,* and *leu* would be transferred last of all. Table 13.2 lists eleven different Hfr strains and the order in which each transfers host genes. Even though the orders vary, you can see that they can all be made the same by considering them not as open-ended lines, but as closed circles. For example, *thr* is the first and *thi* is the last gene transferred by Hfr Hayes, whereas these two genes are transferred one after the other (*thr,* then *thi*) in Hfr 1. All we must do to make these two orders identical is to form a circle of each one. Then the only difference will be the location of O, the origin of replication, which determines which gene will be transferred first. Now look at the order of Hfr 5. You will see that the order of transfer of these two genes is reversed: *thi,* then *thr.* Again, the order of gene transfer by these two strains can be reconciled by forming a circle of each one and turning one of them over as you would flip a coin.

What does all this mean in physical terms? First, it tells us that the *E. coli* chromosome is circular and that the F plasmid inserts in different places in different Hfr strains. This accounts for the fact that the *order* of transfer is the same in different

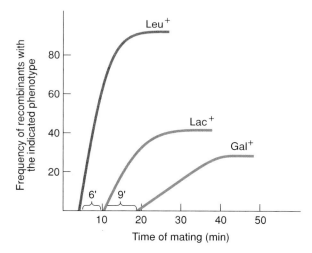

FIGURE 13.10   Timing of gene transfer in a bacterial conjugation experiment. An Hfr Leu⁺ Lac⁺ Gal⁺ was mated with an F⁻ Leu⁻ Lac⁻ Gal⁻ recipient and the conjugation was interrupted at various times by blending. The phenotypes of the recipients were tested by growing in medium lacking leucine or containing lactose or galactose, instead of glucose, as the sole carbon source. Some Leu⁺ recombinants (red line) began to appear at four minutes after the onset of conjugation, whereas Lac⁺ and Gal⁺ recombinants began to appear at ten and nineteen minutes, respectively (blue and green lines). This means that the time between transfer of the *leu*⁺ and *lac*⁺ markers was six minutes, while that between *lac*⁺ and *gal*⁺ was nine minutes. This tells us the relative positions and spacing of these three markers. The lower and lower plateau levels are due to increased incidence of interrupted mating.

Table 13.2   Order of Gene Transfer during Conjugation with a Variety of Hfr Strains

Hfr Strain	Order of Gene Transfer
Hayes	O–thr–leu–azi–ton–pro–lac–pur–gal–trp–his–gly–str–mal–xyl–mtl–ile–met–thi
Hfr 1	O–leu–thr–thi–met–ile–mtl–xyl–mal–str–gly–his–trp–gal–pur–lac–pro–ton–azi
Hfr 2	O–pro–ton–azi–leu–thr–thi–met–ile–mtl–xyl–mal–str–gly–his–trp–gal–pur–lac
Hfr 3	O–pur–lac–pro–ton–azi–leu–thr–thi–met–ile–mtl–xyl–mal–str–gly–his–trp–gal
Hfr 4	O–thi–met–ile–mtl–xyl–mal–str–gly–his–trp–gal–pur–lac–pro–ton–azi–leu–thr
Hfr 5	O–met–thi–thr–leu–azi–ton–pro–lac–pur–gal–trp–his–gly–str–mal–xyl–mtl–ile
Hfr 6	O–ile–met–thi–thr–leu–azi–ton–pro–lac–pur–gal–trp–his–gly–str–mal–xyl–mtl
Hfr 7	O–ton–azi–leu–thr–thi–met–ile–mtl–xyl–mal–str–gly–his–trp–gal–pur–lac–pro
AB311	O–his–trp–gal–pur–lac–pro–ton–azi–leu–thr–thi–met–ile–mtl–xyl–mal–str–gly
AB312	O–str–mal–xyl–mtl–ile–met–thi–thr–leu–azi–ton–pro–lac–pur–gal–trp–his–gly
AB313	O–mtl–xyl–mal–str–gly–his–trp–gal–pur–lac–pro–ton–azi–leu–thr–thi–met–ile

strains, but that the starting points (defined by the site of F plasmid insertion) vary. Second, the F plasmid can insert in either orientation, thus inverting the order of gene transfer. These phenomena are illustrated in figure 13.11. In the 1960s, John Cairns used electron micrographs (chapter 7) to show that *E. coli* DNA is circular, so we accept the proposition of a circular map naturally. Nevertheless, in 1957, when Jacob and Wollman proposed the idea, it was a radical one. A partial circular map of the *E. coli* chromosome is presented in figure 13.12.

B y interrupting conjugation between an Hfr donor and an F⁻ recipient, we can determine the length of time it takes to transfer a given gene. The differences in the transfer times of different genes tell us their relative positions on the circular genetic map of *E. coli*.

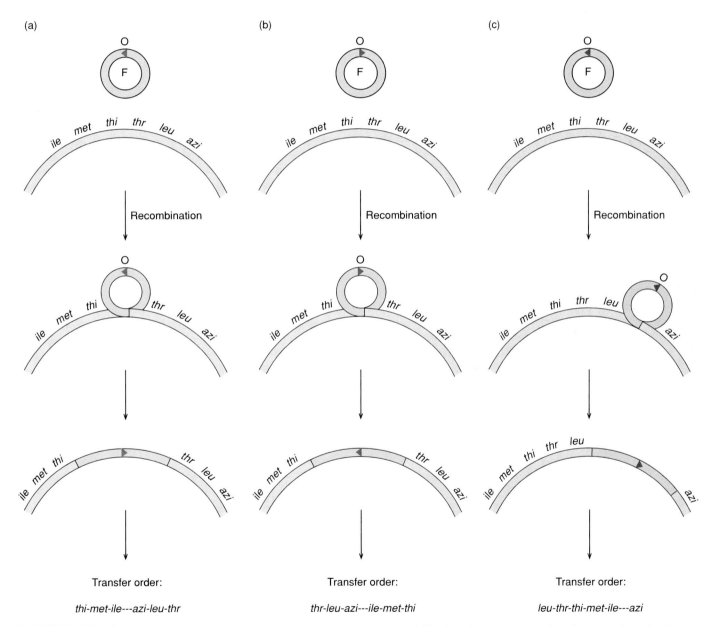

(a)

Transfer order:

thi-met-ile---azi-leu-thr

(b)

Transfer order:

thr-leu-azi---ile-met-thi

(c)

Transfer order:

leu-thr-thi-met-ile---azi

FIGURE 13.11 The F plasmid portion of an Hfr chromosome determines the direction and order of host gene transfer. (*a*) The F plasmid (yellow circle with red arrowhead showing orientation) inserts between *thi* and *thr* in a rightward orientation; thus, the host genes will be transferred in the order *thi-met-ile---azi-leu-thr,* where the three dashes in the middle represent the entire remainder of the Hfr chromosome (not shown in this picture). The last three genes are therefore transferred only in those rare conjugations that last throughout the entire one hundred minutes. (*b*) The F plasmid inserts between *thi* and *thr* in a leftward orientation; thus, the host genes will be transferred in the order *thr-leu-azi---ile-met-thi.* (*c*) The F plasmid inserts between *leu* and *azi* in a rightward orientation; thus, the order of transfer is *leu-thr-thi-met-ile---azi.*

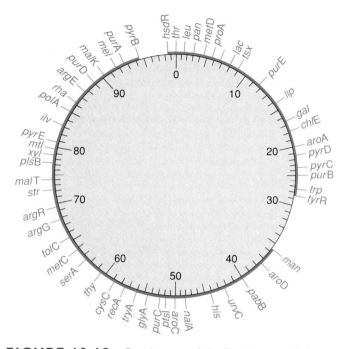

FIGURE 13.12   Circular map of the *E. coli* genome. This map was prepared in 1976 and contains only a fraction of the genes that have now been mapped.

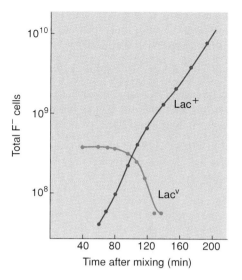

FIGURE 13.13   Recombination is the final step in conjugation. The Lac^V phenotype, indicative of unstable association between donor and recipient chromosomes, decays, beginning about 50 minutes after the onset of conjugation. At the same time, the Lac$^+$ phenotype, indicative of a stable association that results from recombination between the two chromosomes, begins a steady rise.

COMPLETION OF CONJUGATION

The fact that a given gene enters an F$^-$ cell during conjugation does not mean that the gene will be expressed indefinitely in the recipient. In order to become a stable part of the recipient cell's genetic makeup, the newly arrived gene must recombine with the recipient cell's chromosome. If it does not, it has no way of replicating and will be lost. J. Tomizawa demonstrated this phenomenon clearly in a conjugation experiment with an Hfr Lac$^+$ donor and an F$^-$ Lac$^-$ recipient. After various times, he interrupted conjugation and screened the merodiploids for the Lac$^+$ phenotype by growing them on indicator plates containing EMB-lactose agar. Lac$^+$ cells metabolize the lactose, causing chemical changes that turn the indicators in the medium red; Lac$^-$ cells cannot metabolize the lactose and so remain white.

At the earliest times that Lac$^+$ cells appear, they produce colonies that are not pure red, but red and white **variegated,** a phenotype called **Lac^V.** This is because the *lac* genes in the merodiploids are not yet stably integrated into their new hosts' chromosomes. Thus, they can be lost, giving rise to Lac$^-$ cells that will be white. On the other hand, when Tomizawa tested the merodiploids for longer periods of time, he observed a drop in the Lac^V phenotype, beginning about fifty minutes after the onset of conjugation, and a corresponding rise in the number of pure Lac$^+$ colonies (figure 13.13). This tells us that integration of the *lac* genes from the Hfr cell into the F$^-$ chromosome begins about fifty minutes after conjugation begins, or about forty minutes after the *lac* genes entered the recipient cell. This integration of the *lac* genes into the *lac$^-$* genome stabilizes the Lac$^+$ phenotype.

This integration process is a recombination event, and requires the products of the cell's *rec* genes. This same set of genes is responsible for other recombination events between homologous DNA regions, including recombination repair, which we discussed in chapter 11. We have also seen that the product of one of the *rec* genes, recA, figures prominently in a process unrelated to recombination: the SOS response. This allows for error-prone repair of DNA damage (chapter 11) and for the induction of the lytic cycle so λ phage can escape from a threatened cell (chapter 8).

The completion of conjugation in *E. coli* requires recombination between the donated piece of Hfr chromosome and the F$^-$ chromosome. Otherwise, the donated DNA cannot replicate and will be lost. This recombination requires extensive sequence similarity between the recombining DNAs and therefore follows the homologous recombination pathway.

GENETIC MAPPING WITH *E. COLI*

We have seen that conjugation occurs in *E. coli* and that this transfer of DNA allows us to construct genetic maps by following the progressive entry of one gene after another into the recipient F$^-$ cell. This technique works fairly well when the distance separating the genes is great, but it is too crude a method for mapping closely spaced genes. For such fine mapping, more

FIGURE 13.14 Formation of F-*lac*. The integrated F plasmid (red) in an Hfr chromosome is clipped out along with some host DNA, including the nearby *lac* operon (blue). This produces a recombinant DNA called F-*lac*, part F plasmid and part host DNA, including *lac*.

delicate genetic tools have been developed. We will examine the most important of these.

F′ Plasmids

In 1959, E. H. Adelberg discovered that one of his Hfr strains was contaminated with an F⁺ revertant. At first this seemed like an impediment, but it turned out that the F⁺ revertant had very interesting properties. In fact, this was one of those occasions when an annoyance turned out to be a boon.

Adelberg's F⁺ revertant did not behave normally. Whereas an F plasmid usually inserts randomly into the host chromosome to generate a new Hfr, the F plasmid in this revertant always went back to the same locus at which it had resided in the original Hfr. And it generated Hfrs with a much higher frequency than normal. Adelberg recognized that the F factor in his F⁺ revertant had suffered a genetic change, so he gave it a new name: **F′.** He also guessed the reason for the unusual behavior of the F′ plasmid: It had picked up a piece of the host chromosome during its sojourn there in the Hfr (figure 13.14). This host DNA gave the F′ plasmid a "memory" of its former location, since it provided homologous sequences on the plasmid and the host chromosome that could serve as sites for homologous recombination. This ensured that recombination between F′ plasmid and host chromosome would be rapid and would always occur at the same locus on the chromosome.

Apart from its theoretical interest, the F′ plasmid also provides a handy genetic tool. We can use it to introduce a defined small segment of the *E. coli* genome into an F⁻ cell very efficiently. This process, sometimes called **F-duction,** or even **sex-duction,** is so efficient because the segment of host DNA is now part of the F plasmid and transfers with the same high frequency as the F plasmid itself. Therefore, conjugation between an F′ cell and an F⁻ produces a merodiploid for genetic analysis.

For example, consider an F′ called **F-*lac*** that carries part of the *lac* operon. When an F⁺ cell bearing F-*lac* is conjugated with an F⁻ Lac⁻ cell, it converts the recipient to F⁺ Lac⁺ (figure 13.15). Thus, even though the recipient retains a mutant *lac* gene in its chromosome, the wild-type gene on the F-*lac* plasmid provides the missing enzyme. Note that no recombination is necessary; the F-*lac* plasmid reproduces along with the host cell. This cross tells us that the wild-type *lac* gene is dominant over the *lac*⁻ gene in the F⁻ chromosome. Of course, merodiploids make possible many other, more sophisticated genetic experiments, which we

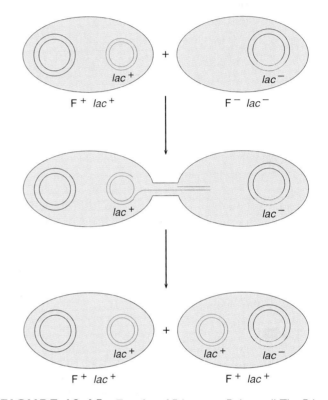

FIGURE 13.15 Transfer of F-*lac* to an F⁻ *lac*⁻ cell. The F-*lac* plasmid (blue, with *lac*⁺ genes in red) crosses the connecting bridge, just as an F⁺ normally would. This converts the F⁻ *lac*⁻ recipient to F⁺ *lac*⁺. The host chromosomes are black, with *lac*⁻ genes in gray. Again, the host chromosome and F-*lac* plasmid are not drawn to scale.

will also investigate. These experiments have been aided considerably by a collection of F′ plasmids that represent the entire *E. coli* genome.

We have seen that plasmids can be efficient vehicles for passing genes from one cell to another. Thus, it is not surprising that many other plasmids besides the F plasmid participate in bacterial genetics. We have already discussed at least two examples: In chapter 7 we encountered the colE1 plasmid as an example of unidirectional DNA replication. This plasmid confers on its host the ability to kill competing bacteria. In chapter 12 we mentioned the antibiotic resistance plasmids that allow bacteria to evade the lethal effects of antibiotics. In chapter 15 we will see how these plasmids have been engineered so they can be used as agents to carry foreign genes into bacteria.

Sometimes a recombination event within an Hfr chromosome clips out the F plasmid along with some of the host chromosome. This produces an F′ plasmid that includes the clipped out piece of host DNA. The host genes thus become part of a highly mobile element that can deliver them to another cell. We can take advantage of this efficient sharing of limited pieces of DNA to perform genetic experiments.

THE CIS-TRANS COMPLEMENTATION TEST

Suppose we have isolated two strains of *E. coli* that cannot grow on the sugar lactose. Two genes (*lacZ* and *lacY*) code for the two enzymes a cell needs to metabolize this sugar. Therefore, we can pose the following important question about these two Lac⁻ mutants: Do both strains have mutations in the same gene, or are two different genes affected? We can answer this by performing F-duction to form a merodiploid, thus placing both mutant *lac* operons together in the same cell. If the two mutations are in different genes, they will **complement** each other in the merodiploid to give a Lac⁺ cell. In other words, the donor provides the gene product that the recipient is missing, and vice versa. In the case illustrated in figure 13.16*a*, the donor (F-*lac*) cell contained a wild-type *lacZ* gene, but a mutant *lacY* gene. Thus, the wild-type *lacZ* gene product from the donor cooperated with the wild-type *lacY* gene product from the recipient to allow the merodiploid to use lactose.

On the other hand, if the mutations are in the same gene in the two cells, they will usually not complement each other, so the merodiploid will be Lac⁻. Figure 13.16*b* illustrates what happens when both mutations are in the same gene—the *lacY* gene, in this case. Both the donor plasmid and the recipient chromosome provide wild-type *lacZ* products, but mutant *lacY* products. They cannot complement because neither can provide a functional *lacY* product. Notice the emphasis here on gene *products*. In order for complementation to work, the gene products have to be **diffusible.** That is, they must be capable of moving within the cell so they can cooperate.

As an aid to understanding the complementation concept, consider the following analogy: Two factories in the same town make the same kind of automobile. Each has two assembly lines, one for the engines and one for the bodies. One day the engine line of one factory and the body line of the other factory break down. Nevertheless, the two factories are able to continue producing cars by cooperating—workers put the engines from the second factory together with the bodies from the first. This is complementation, and you can see how it depends on the portable nature of the product. Of course, if the engine assembly lines of both factories break down at the same time, the two defective factories cannot complement each other and there will be no way to continue making cars.

We should make several qualifications about the complementation tests we have just discussed. First, as we have seen, the merodiploid will be Lac⁻ if the two mutations are in the same

gene (*lacY,* for example). But how do we know that the cells we are testing are actually merodiploids? The Lac⁻ phenotype may simply result from a failure of conjugation. To show that we really have transferred the F′ plasmid, we need a selectable marker. So, for example, we could place a wild-type *leuA* gene on the F′ plasmid and make the recipient cell *leuA⁻*. We could also make the donor *str^s*, and the recipient *str^r*. After conjugation we would select for cells that are *leu⁺* and *str^r* by growing them in medium lacking leucine and containing streptomycin. The donor cells could not grow because they are *str^s*, and any recipient cells that have not received the F′ plasmid would not grow because they are *leu⁻*. Only recipients that have received the F′ plasmid could grow. If these are still Lac⁻, we know that the two *lac* mutations were in the same gene.

Another factor to bear in mind is that the complementation test should be performed in a situation where recombination does not occur to a significant extent. If recombination is common, it can take place between two homologous mutant genes, creating one gene with no mutations and one with two (figure 13.16*c*). Thus, the two mutations would seem to complement each other and would be scored as existing in different genes even though they actually started out in homologous genes in the two chromosomes. One way to eliminate this problem is to use a RecA⁻ recipient that cannot carry out homologous recombination.

Another kind of problem can arise if one of the mutations affects the expression of several other genes. Such a **pleiotropic** mutation behaves as if it were mutations in several genes instead of just one, which complicates the complementation test. A final problem is that mutations in homologous genes in two different chromosomes can sometimes complement each other. This occurs when the gene product is a polypeptide that aggregates with one or more identical polypeptides to form an **oligomeric protein.** β-galactosidase, a **tetramer** (composed of four identical polypeptides, or monomers), is a good example. Each *lacZ* gene in figure 13.16*d* is nonfunctional; each gives rise to an inactive tetramer of four identical defective β-galactosidase polypeptides. However, when the two different *lacZ* genes are present in the same cell, they produce two different defective polypeptides that somehow cooperate to form an active tetramer. Therefore, these two *lacZ* mutations complement each other; if we did not know better, we would say they were on different genes. Instead, they are on the same gene, or **cistron,** and they therefore engage in **intracistronic complementation.**

Note that this intracistronic complementation is very different from intercistronic complementation, or complementation by the products of different genes. The automobile factory analogy works well to illustrate the latter, because we can readily visualize the body and engine as separate gene products that can complement each other. But the same analogy does not work well to illustrate intracistronic complementation, because it would require envisioning a car normally made with two (or more) identical, cooperating engines. Further, we would have to envision a car made with two defective engines, each of which somehow compensated for

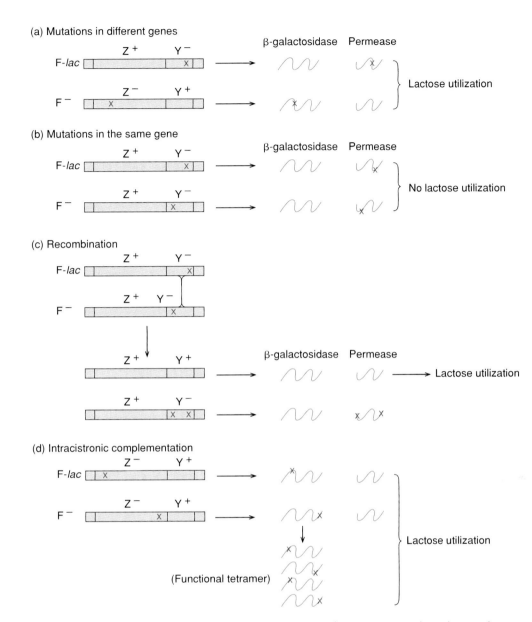

FIGURE 13.16 The *cis-trans* complementation test.
(a) Mutations, represented by small red x's, are in different genes—
lacZ and *lacY*, respectively. Therefore, their products can complement
each other, and the cell will be *lac*⁺. (b) The mutations are in the
same gene (*lacY*). Now both chromosomes will produce active *lacZ*
product, but inactive *lacY* product, so the cell will be *lac*⁻. (c) and
(d) are complications that must be considered when running the *cis-
trans* test. (c) Sometimes recombination can occur between markers

in the same gene on homologous chromosomes. This produces a
wild-type gene and a double-mutant gene. The cell is therefore *lac*⁺,
which makes it look as if the two mutations were in different genes.
(d) Sometimes the products of two homologous genes with different
mutations (two different *lacZ* mutants, in our example) can get
together to make an active, multi-subunit protein, in this case an
active β-galactosidase. This makes it look as if the two mutations
were in different genes.

the defect in the other to produce a functioning car. This does
not work for cars, but it sometimes does for proteins.

We can use the complementation examples just given to
illustrate some important genetic terms. The mutant genes in
these cases were supplied **in trans**—on separate chromosomes
(Latin: *trans* = across). Alternatively, we could supply the mutant
genes **in cis**—on the same chromosome (Latin: *cis* = here).
These two possible configurations of the mutant genes—*cis* and

trans—have given the complementation test another name: the
cis-trans test. That in turn has provided the term "cistron,"
which Seymour Benzer coined to describe a genetic unit that
could be distinguished from another genetic unit by the *cis-trans*
test. Of course, this is a unit of genetic function—a piece of
chromosome that gives rise to a protein (or RNA) product. As
such, it is synonymous with our current understanding of the
term "gene."

We also use the terms *cis* and *trans* to describe the way two genetic loci interact. Promoters, the binding sites for RNA polymerase (chapters 8 and 9), are **cis-acting** elements: They must be on the same DNA molecule as the gene in question in order to influence that gene's activity. To return to our factory analogy, they are like the main power switch. If it breaks down, the factory cannot operate and the defect cannot be complemented by a product of the other factory. By contrast, genes that encode proteins are **trans-acting.** Each protein can interact with other genes or their products. The mobility of proteins allows them to act at a distance.

Two different mutations that affect the same function can be subjected to the *cis-trans* test, or complementation test, to determine whether they occur in the same or in different genes. In this test, the two mutant genes are supplied to the same cell *in trans* (on separate chromosomes). If the mutations complement each other to give wild-type function, they are probably in separate genes. If the mutations fail to complement, they affect the same gene.

TRANSDUCTION

The term **transduction** refers to the use of phages to carry bacterial genes into new host cells. Two different transduction mechanisms are widely used: **specialized transduction,** in which λ or a similar phage DNA carries the bacterial genes, and **generalized transduction,** in which **P1** or other phage heads carry the bacterial genes instead of phage genes.

λ Phage as a Carrier of Host Genes

Lambda (λ) is a temperate phage that infects *E. coli* cells. Temperate phages, unlike T2, T4, and other virulent phages, do not necessarily kill their hosts. Instead, as we saw in chapter 8, λ DNA can incorporate itself into the host genome. This gives rise to a lysogenic infection in which the phage DNA becomes a passenger and replicates along with the host DNA. On the other hand, the λ phage can initiate a lytic infection in which the phage DNA remains independent of the host chromosome. Under these circumstances, host cells fill up with progeny phage particles and finally burst, or lyse, releasing the new phages to infect other cells. The spreading infection started by a single phage replicating lytically on a lawn of bacteria can produce a hole, or **plaque,** devoid of living cells.

During lysogenic infection, λ DNA inserts into a site called *att* **B,** which lies between the *gal* and *bio* genes in the *E. coli* DNA. We also saw that mutagens can induce the λ lytic cycle, causing the phage DNA to be excised from the host chromosome. Usually, the λ DNA excises perfectly, but sometimes it comes out carrying some excess baggage: a piece of host DNA.

Since the λ DNA usually inserts between the host *gal* and *bio* genes, these are the genes that λ DNA picks up when it

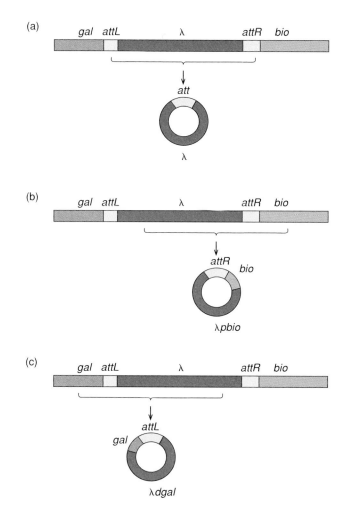

FIGURE 13.17   Creation of λ transducing phages. (a) Normal λ excision: The λ DNA (red) is correctly cut out of the host genome (colored purple on the left and green on the right) by clipping in the *att* sites (yellow) on either side of the λ DNA. (b) The cutting occurs too far to the right, thus including the host *bio* operon (green), but missing the left end of the λ genome. The resulting phage will still be able to replicate and can transduce the *bio* genes; it is called λ*pbio*. (c) The cutting occurs too far to the left, thus including the host *gal* genes (purple), but missing the right end of the λ genome, which is essential for replication. The resulting transducing phage is called λ*dgal*.

excises improperly (figure 13.17). And since the room available inside a λ phage head is limited, the amount of extra DNA that can be accommodated is only about 10% of the size of the λ DNA itself, or about 5 kb. In practice, more host DNA than this is usually included, which means that some phage DNA must be excluded. If a large person sits down on one end of a crowded bench, someone on the other end gets pushed off. By the same token, if the λ DNA picks up the *E. coli bio* locus at one end, it loses some of its regulatory region from the other. The result is a **transducing phage,** a particle that contains a recombinant DNA: part *E. coli bio* and part λ. The λ part is missing a regulatory region required for lysogeny but not for the λ lytic cycle, so the resulting transducing phage can replicate lytically to form plaques. It is therefore called λ*pbio,* where the *p* stands for "plaque-forming."

On the other hand, if the λ DNA picks up the *gal* locus on one end, it loses genes on the other end that are essential for the lytic cycle, and the resulting transducing phage (λ*dgal*) is defective. Indeed, the *d* stands for "defective." Such a defective transducing phage needs a nondefective **helper phage** to supply the functions it requires to replicate.

What can we do with specialized transduction? We can use it instead of conjugation to form merodiploids for complementation tests. It is somewhat more limited than conjugation in that the *gal* and *bio* loci are the only ones that usually participate. However, λ DNA sometimes goes to alternate sites in the host genome, so it can pick up and transfer a variety of genes. This makes specialized transduction more versatile than it seems at first glance.

There is a significant difference between the physical relationship of the two genes in merodiploids resulting from F-duction and specialized transduction. After F-duction, the two complementing genes are on different DNA molecules: one on the host chromosome and the other on the F′ plasmid. After specialized transduction, both complementing genes are part of the host chromosome, because the λ DNA, with its piece of host DNA attached, inserts into the recipient genome. Nevertheless, both genes are presumably active in each case, so complementation can occur equally well in both.

> **W**hen λ DNA escapes from the host chromosome after lysogenic infection, it sometimes carries a piece of host DNA along with it. A phage carrying such a recombinant DNA is called a transducing phage because it can be used to "transduce" a piece of host DNA into a new host cell to form a merodiploid. Geneticists use such merodiploids for complementation tests. λ transduction is called specialized transduction because it usually carries only the *gal* or *bio* genes that flank the λ integration site.

P1 Phage Mediation of Generalized Transduction

In specialized transduction, a temperate phage such as λ carries host genes as passengers on its genome. In generalized transduction, little if any phage DNA is involved. Instead, the phage head forms around a piece of host DNA, producing a transducing particle that can deliver this DNA to a new host. Of course, the resulting infection cannot produce new phages in the absence of phage genes, but the donated host DNA can recombine with the recipient's chromosome and change its genotype. In fact, recombination is necessary for successful generalized transduction, just as it is for mating between Hfr

and F⁻ strains. In both cases, linear pieces of DNA enter new host cells and must recombine with the host chromosome or be lost.

The favorite phage for generalized transduction is P1, whose genome normally contains 91.5 kb of DNA. This means that a P1 phage head can hold up to 91.5 kb of donor DNA if it does not have to accommodate any phage DNA. This is about 2.4% of the whole *E. coli* genome, enough room for about seventy-five average-size genes—many times more DNA than can be transferred by specialized transduction.

About 0.3% of the phages from a given P1 infection are transducing phages; the rest are normal. Since a given transducing phage contains at most 2.4% of the host genome, picked up more or less at random, the chance of finding a given host locus in a given phage particle is at most 0.024×0.003, or 7.2×10^{-5}. This is less than one in ten thousand; in practice, it is only about one in one hundred thousand, so the experimenter must perform an efficient selection to find this "needle in a haystack."

Consider an experiment in which we want to transduce a *leu⁺* gene into a *leu⁻* recipient (figure 13.18). This is relatively straightforward. We simply infect the *leu⁺* donor with P1, collect the progeny phage, including the transducing phage, then infect the *leu⁻* recipient and transfer the cells to medium lacking the amino acid leucine. Cells infected by normal P1 phage will lyse, or will perhaps fail to grow or even to replicate the phage because of the lack of leucine. Cells infected by transducing phages carrying genes other than *leu⁺*, and cells not infected at all, will fail to grow in the absence of leucine. Only those rare cells infected by transducing phages carrying the *leu⁺* gene will survive. These will be easy to spot as opaque colonies on petri dishes, surrounded by transparent regions of nongrowing cells. This is another example of a selection. It removes all undesirable cells—in this case, by preventing them from reproducing. Only the desired cells (*leu⁺* in this example) can grow.

Now suppose we want to transduce DNA from *leu⁻* strain (allele *1*) into a *leu⁻* (allele *2*) recipient. (We will see the usefulness of such an experiment shortly.) If the mutations in the two alleles are in the same gene of the *leu* operon, this poses a problem. We cannot select for the Leu⁻ phenotype by leaving out leucine, because the Leu⁻ cells fail to grow under those conditions. Instead, we can pick a selectable marker near the nonselectable one. Figure 13.19 shows how this works in the *leu⁻* case we are trying to solve. Since the *leu* locus is very close to the *ara* locus, which codes for enzymes that metabolize the sugar arabinose, chances are that a transducing phage containing *leu* will also contain *ara*. Therefore, we can infect an Ara⁺ Leu*1*⁻ donor strain with P1, collect the transducing phages, infect an Ara⁻ Leu*2*⁻ recipient strain with them, and select for Ara⁺ transductants (cells converted to Ara⁺ by transduction) by plating on medium containing arabinose, but no glucose. Some of the Ara⁺ transductants will also be carrying the *leu⁻* gene from the donor cell, since *leu* and *ara* are closely linked. This process, called **cotransduction,** narrows our search considerably.

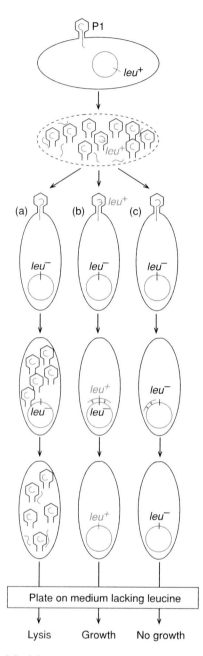

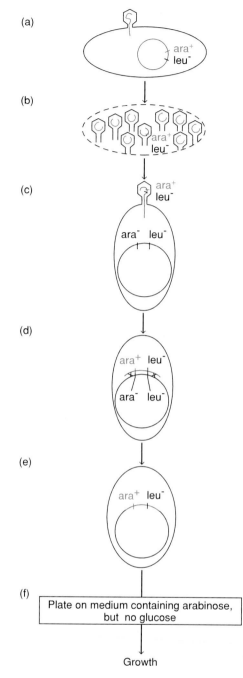

FIGURE 13.18 Generalized transduction. Infection of a *leu*⁺ bacterium by the phage P1 generates three kinds of particles: (*a*) a normal P1 particle containing only phage DNA (gray); (*b*) a transducing particle containing host DNA (red) that includes the *leu* operon (blue); (*c*) a transducing particle containing non-*leu* host genes only (red). These three kinds of particles can be tested by using them to infect *leu*⁻ hosts, then plating on medium lacking leucine. The normal P1 particles (left) will infect and destroy these cells. The transducing particles with the *leu* genes (middle) will render the host cells *leu*⁺, so they will grow in the leucine-free medium. The transducing particles lacking the *leu* genes (right) cannot confer the *leu*⁺ genotype, so the infected cells will not grow on the leucine-free medium.

FIGURE 13.19 Cotransduction screen. The goal is to transduce a *leu*⁻ locus, which is nonselectable. Therefore, we select for the nearby *ara* locus, which confers the ability to grow on the sugar arabinose. (*a*) Cells with the phenotype Ara⁺ Leu⁻ are infected with phage P1. (*b*) Some of the particles resulting from this infection are transducing phages, containing host DNA (red), including the *ara*⁺ (blue) and *leu*⁻ loci. (*c*) These particles are used to infect Ara⁻ Leu⁻ hosts; most particles will be normal P1 phage or transducing particles with irrelevant genes, but a few, like the one pictured, will contain the host *ara* and *leu* genes. (*d*) Recombination occurs between the transduced DNA and the new host chromosome. (*e*) The result is an Ara⁺ Leu⁻ cell that can grow on medium containing only arabinose as energy source (*f*). Those cells that received the *ara* genes from the transducing particle are likely to have received the *leu* genes as well, since these loci are closely linked.

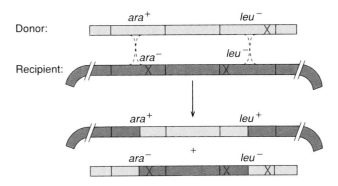

FIGURE 13.20 Crossing over between two mutations. The donor, transducing DNA, contains the *ara⁺* and *leu⁻* alleles; the chromosome of the recipient cell (purple) contains the *ara⁻* and *leu⁻* alleles. Recombination occurs between these two DNAs (dashed lines), resulting in an Ara⁺ Leu⁺ cell. The cell becomes Leu⁺ because the recombination crossover occurred between the two mutations (red) in the two *leu* genes, placing both of them on the short, linear piece of DNA (bottom), which is lost. This leaves the chromosomal DNA free of mutations in both *ara* and *leu*.

P1 phage can mediate generalized transduction by incorporating host DNA instead of phage DNA. Such particles can deliver up to 91.5 kb of host DNA to a new cell, allowing recombination between the donor DNA and the recipient chromosome. Since only a small fraction of the phages from a P1 infection will carry a given gene, a selection must be performed to locate the transductant of interest. This is easiest if transduction creates a prototrophic recombinant.

Mapping Mutations

Why would we want to transduce a DNA from a *leu⁻* cell into yet another *leu⁻* cell? One good reason would be to see if the two *leu⁻* mutations giving rise to alleles *1* and *2* are at identical nucleotide sites, and if not, how far apart they are: in other words, to map the mutations. This can be done by trying to grow many of our *ara⁺* cotransductants on plates lacking leucine. If all of these are still *leu⁻*, chances are the *leu⁻* mutations (alleles *1* and *2*) in the donor and recipient cell, respectively, are in precisely the same place. But this is unlikely; it is much more probable that some of the *ara⁺* transductants will also be *leu⁺*. This would mean that a recombination occurred between the two *leu* mutations, as shown in figure 13.20.

We learned in chapter 5 that the farther apart two genes are, the better the chance that a recombination event will occur between them. This applies equally well to the two *leu* mutations in figure 13.20. They are very far apart in the figure, so the chance of a crossover between them is relatively high. As the figure shows, the crossover results in a *leu⁺* recombinant. Generalized transduction makes this type of fine mapping within a gene relatively convenient. The linkage between two or more different genes can also be examined by this same kind of recombination analysis. Again, the farther apart the genes are, the greater the frequency of recombination between them.

A Direct Mapping Scheme Based on Cotransduction

We have seen that cotransduction is a useful way to "piggyback" a gene we want to study with another that is selectable. Cotransduction also gives us another mapping method. First, if two genes cotransduce, we immediately know that they lie within 91.5 kb of each other and are therefore fairly closely linked. Furthermore, the closer the linkage between two genes, the more likely they are to cotransduce. Notice that this is just the opposite of the recombination test. There, the tighter the linkage between two genes, the less likely they are to recombine. In mapping studies, we frequently refer to genes as **markers** because they mark positions on the genome, much as milestones mark positions on a highway. T. T. Wu has proposed a formula for calculating the frequency (x) of cotransduction of two markers:

$$x = (1 - d/L)^3$$

where d is the distance between the two markers, and L is the length of the DNA that can be accommodated by the transducing particle. Using this expression, we find that markers separated by 60 kb (kilobase pairs), 30 kb, and 10 kb are expected to cotransduce with frequencies of about 4%, 30%, and 70%, respectively. An average bacterial gene is only about 1 kb long, which means that two markers within a gene cotransduce with a frequency higher than 96%.

We can also use transduction to deduce the order of genes. For example, consider three genes: *A, B,* and *C.* We know from previous experiments that genes *B* and *C* are close together and *A* is farther away. The question is this: Is the order of these genes *A B C,* as shown in figure 13.21*a,* or *A C B,* as shown in figure 13.21*b?* We can answer this by doing a three-factor cross as illustrated in figure 13.21.

We start with a recipient cell with the phenotype A⁺ B⁺ C⁻ and transduce DNA from a cell with the phenotype A⁻B⁻ C⁺. Then we select for transductants that are B⁺ C⁺. In order to produce this phenotype, crossovers 1 and 2, at least, must occur. Among these B⁺ C⁺ transductants, we then screen for the *A* alleles. We ask: How many transductants are A⁺ B⁺ C⁺, and how many are A⁻ B⁺ C⁺? Figure 13.21*a* shows us what to expect if the gene order is A B C: Crossovers 1 and 2 alone will produce A⁺ B⁺ C⁺, but two additional crossovers (3 and 4) must occur to bring the A⁻, B⁺, and C⁺ alleles together. Clearly, two crossovers are much more likely to occur than four, so if the gene order is A B C, we would expect to find many more A⁺ B⁺ C⁺ transductants than A⁻ B⁺ C⁺.

On the other hand, if the gene order is *A C B,* figure 13.21*b* shows us what to expect. Here, crossovers 1 and 2 or 1 and 3 will produce the B⁺ C⁺ phenotype we first screen for. The difference is that crossovers 1 and 2 will produce A⁺ B⁺ C⁺, whereas 1 and 3 will give A⁻ B⁺ C⁺. And crossover 3 is much more likely to

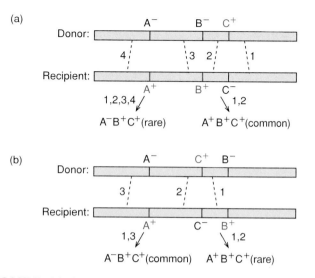

FIGURE 13.21 Three-factor cross to determine the order of genes A, B, and C. We know that genes B and C are closer together than either is to gene A. We want to know if the order is A B C as shown in (a), or A C B as shown in (b). We perform a transduction with an A⁻ B⁻ C⁺ donor and an A⁺ B⁺ C⁻ recipient, and screen for B⁺ C⁺ transductants. Then we check to see whether among these transductants we get more A⁺ or A⁻. In (a) we know that crossovers 1 and 2 have occurred to bring together the B⁺ and C⁺ alleles on the recipient chromosome. If those are the only crossovers, the transductant will also be A⁺; if crossovers 3 and 4 also occur, the transductant will be A⁻. But two crossovers will occur much more frequently than four, so A⁺ will be much more common if (a) is correct. In (b) we again know that crossovers 1 and 2 or 1 and 3 have occurred to bring together B⁺ and C⁺. But 1 and 3 are more likely than 1 and 2, because there is much more space for 3 to the left (not fully shown) than for 2, which must occur between A and C. And 1 and 3 give A⁻, while 1 and 2 give A⁺. Therefore, A⁻ will be more common if (b) is correct.

occur than 2 because there is much more room at the left end of both DNAs (not fully shown in the figure) than between genes *A* and *C*. Therefore, if the gene order is *A C B*, we would expect to find more A⁻ B⁺ C⁺ transductants than A⁺ B⁺ C⁺. This is the opposite of what we expect for the *A B C* gene order.

Transduction allows us to do several types of genetic mapping. We can transduce two genes, or two loci within the same gene, and observe the frequency of recombination between them. The closer together they are, the lower their frequency of recombination. Alternatively, we can observe the frequency of cotransduction of two markers. The closer together the markers, the higher their frequency of cotransduction. We can also deduce the order of the markers on the genome by performing three-factor crosses in which we transduce three markers.

GENETICS OF T-EVEN PHAGES

Scientists began to work on bacterial viruses as early as 1915, when F. W. Twort reported that very small, invisible agents could destroy bacteria. He proposed that these agents were viruses that parasitized their bacterial hosts. Twort's work, like other genetic studies we have encountered, escaped recognition, allowing F. d'Herelle to rediscover the same phenomenon two years later. D'Herelle popularized the notion of a bacterial virus and gave this agent the name **bacteriophage** (Greek: *phagein* = to devour; hence, *bacteriophage* = bacteria-devourer). Before the antibiotic era, it seemed that these phages might help in the fight against bacterial infections. Delbrück and his colleagues initiated the modern era of phage investigation in 1938 when they began to focus on the genetics of a small group of phages called **T1–T7,** where the T stands for "type." This focusing of attention was important, especially since it was directed mostly on the closely related **T-even** phages (T2, T4, and T6). Thus, new findings could be compared with one another and organized into a coherent picture, rather than remaining isolated fragments of information about a bewildering variety of very different phages. Figure 13.22 illustrates some of the various forms of *E. coli* phages.

STRUCTURE AND GROWTH OF T-EVEN PHAGES

Figure 13.23 is a schematic diagram of the structure of T4, one of the T-even phages. The most obvious features are a polyhedral **head,** which contains the phage genome—175 kb of double-stranded DNA tightly coiled inside; a **tail,** which the phage uses to inject its DNA into the host bacterium *E. coli;* and **tail fibers** and a **baseplate,** by which the phage attaches to the surface of the host cell. Figure 13.24 dramatically illustrates the tight coiling of phage DNA inside the head by showing that when the head bursts upon osmotic shock, the DNA spews out in all directions. Indeed, it is difficult to believe all that DNA could fit into the phage head until you remember that the DNA is heavily shadowed with metal to make it show up; it is really far thinner than it appears here.

Even though phages are invisible to all but the electron microscopist, they can be easily counted by observing their effects on their hosts. In this procedure, called a **plaque assay,** an experimenter mixes phage particles with bacterial cells, suspends the mixture in a small volume of molten agar containing nutrients for the bacteria, then pours the mixture on top of solid agar medium in a petri dish. The agar "overlay" quickly solidifies, and the bacteria grow overnight to form a uniform, turbid "lawn." However, as figure 13.25 shows, phages make their presence known by creating clear holes, or **plaques,** in the bacterial lawn.

Here is how phages create plaques: Each phage that infects a bacterial cell in the original mixture takes over the host machinery for DNA, RNA, and protein synthesis, forcing the

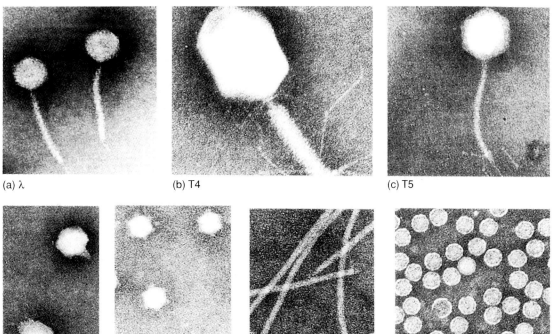

FIGURE 13.22 (a–g) Structures of some phages of *E. coli*. All are shown at approximately equal magnification.

From Robley C. Williams and Harold W. Fisher, *An Electron Micrographic Atlas of Viruses*, 1974. Courtesy of Charles C. Thomas, Publisher, Springfield, Illinois.

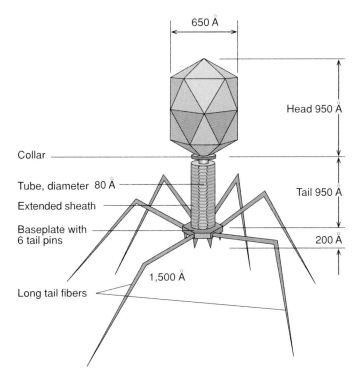

FIGURE 13.23 Schematic diagram of a T-even phage particle.

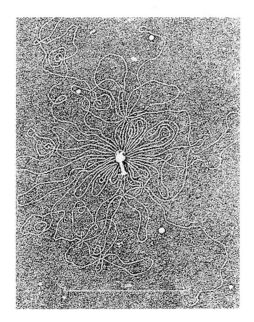

FIGURE 13.24 Lysed T4 phage. The phage ghost is at center, with escaped phage DNA all around.

From A. K. Kleinschmidt, et al. "Darstellung und Langenmessungen des gesamtem Desoxyribosenucleinsaure-Inhaltes Von T2-Bacteriophagen." *Biochemica et Biophysica Acta* 61(1962):857–64, 1962. Reprinted with permission of Elsevier Science. NL, Sara Burgerhartstraat 25, 1055 KV Amsterdam, The Netherlands.

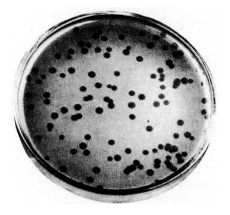

FIGURE 13.25   Plaque assay for T2 phage. The phages were mixed with *E. coli* cells and plated on nutrient agar in petri dishes. Each hole, or plaque, in the lawn of bacteria started from a single phage infecting a single cell. The infection spread to surrounding cells to form the plaque.

From Gunther S. Stent, *Molecular Biology of Bacterial Viruses.* © 1963 W. H. Freeman and Company.

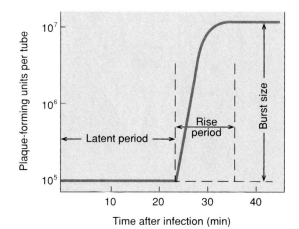

FIGURE 13.26   The one-step growth experiment. Plaque-forming units per tube were recorded as a function of time after infection. No change occurred for almost twenty-five minutes (the latent period). At the end of this time, a rapid rise in phage numbers began (the rise period). The difference in phage number at the beginning and end of the rise period is the burst size.

cell to make phage products instead of its own. Ultimately, the phage DNAs and proteins assemble to form complete progeny phages, and about twenty-five minutes after the start of infection, the cell lyses, releasing hundreds of progeny phages. By that time, however, the agar has hardened, so the progeny phages cannot float away; instead, they are restricted to infecting adjacent cells. This process repeats itself over and over, until the phages, by lysing one host cell after another, succeed in clearing out a visible plaque in the bacterial lawn.

Each infectious phage in the original suspension will, in principle, cause one plaque to form. Therefore, by counting plaques, we get an idea of the number of infectious phages, or **plaque-forming units (pfu),** in a given volume. For example, suppose we start with a phage suspension, dilute it 1:100,000 (a 10^5-fold dilution), and spread 1/4 ml on a plate of bacteria. If we obtain one hundred plaques, that means we had $100 \times 10^5 \times 4 = 4 \times 10^7$ pfu, or infectious phages, per milliliter of the original suspension. Why did we do the dilution? If we had plated the concentrated suspension, there would have been so many plaques that they all would have blended into one, and there would have been no way to count them. In practice, we perform several different dilutions so that some of the plates in every experiment have a reasonable number of plaques.

Our description of phage replication implies a discontinuity. One phage infects a cell and, after a period in which no new phages appear, gives rise to a burst of hundreds of progeny phages. This idea comes from the "single-step growth experiment" performed by Delbrück and E. Ellis at the beginning of their investigation into phage genetics. In this study, Delbrück and Ellis infected *E. coli* cells with T4 phages and then measured the number of phages (pfu) per milliliter at various times thereafter. Figure 13.26 shows that there was no increase in the phage concentration for almost twenty-five minutes. This is the **latent period** (or **lag period**). Then, very rapidly, the concentration

rose from about 10^5 per milliliter to more than 10^7, an increase of more than 100-fold. This is the **rise period.** The difference between the final and initial phage concentrations (more than 100) is the **burst size.**

The actual production of infectious particles within the host cell may be somewhat more gradual than suggested by the rise period, but each infected cell will be counted as only one pfu in the plaque assay until it lyses. In other words, the rise period really reflects **lysis,** or rupture (Greek: *lysis* = a loosening), of infected cells, which allows each progeny phage to produce a separate infection. Before lysis, all the infectious particles are contained in one cell; thus that cell and all the infectious particles within still behave as only one infectious unit.

T-even phages have polyhedral heads, containing the double-stranded DNA genome of about 175 kb, and tails through which they inject the DNA into their host, *E. coli.* The phages take over their hosts' DNA, RNA, and protein synthesizing machinery, reproduce themselves, lyse their hosts, and go on to infect other cells. Since T-even phages lyse their hosts, they can be conveniently counted by observing the number of plaques, or areas devoid of living cells, they produce on a lawn of bacteria.

PHAGE PHENOTYPES AND PHAGE GENETICS

Since phages are detected by the plaques they form, it is perhaps not surprising that some of the first T-even phage mutants discovered were those that formed abnormal plaques. Among these was the *r* mutant, which, for reasons we still do not fully understand, forms extra-large plaques. A. D. Hershey (the same

13.1   Inadvertent Sharing of a Phage

Many stories circulate in the scientific world, some apocryphal, some true. Here is one concerning phages that shows how cleverness can triumph over selfishness. It is probably true, but no names will be used.

It is traditional for biologists to share materials with each other; when one scientist reads about a new mutant, plasmid, or virus that another has isolated or made, all he or she usually has to do is call or write the first scientist requesting a sample, and it will be delivered.

This system has obvious advantages: It greatly accelerates the pace of research because it puts new materials into many hands, instead of bottling them up in one lab. However, nowadays these biological materials, especially clones of bacteria harboring useful eukaryotic genes, can be exploited commercially. This makes them potentially very valuable, and regrettably, leads to less sharing than there used to be.

Our story predates the era of genetic engineering, so no immediate profit motive was involved. Instead, the story simply demonstrates the quite understandable desire to avoid the intense competition that characterizes much genetic research.

It seems that a certain phage geneticist (let us call him A) discovered a novel phage, and a second scientist (B) soon got wind of the discovery. B wrote to A and asked for a sample of the new phage. A wrote back and declined to cooperate, claiming the right to exploit the phage himself, without competition from B.

Nevertheless, A had already unwittingly cooperated, simply by writing back, because B had a clever idea. He cut the letter into pieces and dropped it into a culture of *E. coli*. Soon, the bacteria were infected and producing the new phage. How can a letter be infectious? Since A was already actively growing the phage in his lab, the air in the lab contained significant numbers of phage particles. Some of these settled on the letter A sent to B, so without intending to, A actually did send B the phage!

Hershey who later participated in the famous Hershey-Chase experiment, chapter 6) discovered this mutant in 1946 and called it *r*, for "rapid lysis," because it seemed that rapid lysis of the host cells was what allowed the plaques to grow so large. The wild-type allele of the *r* gene is *r⁺*. Hershey also demonstrated that the *r*-mutant phenotype was indeed due to a mutated gene, since phages from an *r*-mutant plaque gave rise only to *r* plaques, not to wild-type plaques. In other words, the *r* phenotype was heritable.

Another early T-even mutant, also discovered in 1946, was T2*h,* where the *h* stands for **host-range.** This is a T2 phage that can infect *E. coli* cells (Tto^r) that cannot be infected by wild-type T2 phage. Tto is simply a genetic way of writing T2, so Tto^r is T2-resistant. (Similarly, Ton^r means T1-resistant.) Since wild-type T2 phages (*h⁺*) cannot infect Tto^r cells, they will not produce plaques on this strain of cells. Nevertheless, we can get them to form plaques by plating them on a mixture of Tto^r and Tto^s cells. The T2 phages will infect the Tto^s cells in the lawn and therefore form turbid plaques that still contain unlysed Tto^r cells.

The existence of two different T2 mutants allowed phage geneticists to do crosses and look for recombinants. Here is how the experiments were performed: A mixture of Tto^r and Tto^s cells was infected with a high concentration of two different phages, T2*h r⁺* and T2*h⁺ r.* The high concentration of phages gave a high ratio of phages to bacteria (a high **multiplicity of infection, or m.o.i.**) and ensured that most bacteria would be infected by more than one phage. This allowed the two different phage DNAs to recombine to produce the nonparental types, T2*h⁺ r⁺* and T2*h r,* in the progeny.

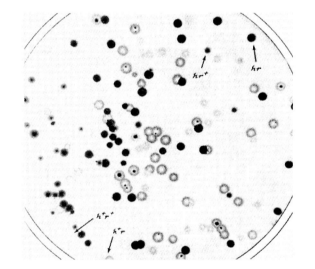

FIGURE 13.27 Distinguishing four phenotypes of phage T2. The phages were plated on a mixture of Tto^s and Tto^r *E. coli*. Four different kinds of plaques are visible, corresponding to *h⁺ r⁺, h⁺ r, h r⁺,* and *h r* phages. The *h⁺r⁺* plaque is small, faint, and difficult to see. The *h⁺r* plaque is not as obviously large as some of the other *r* plaques.
From Gunther S. Stent, *Molecular Biology of Bacterial Viruses,* © 1963 W. H. Freeman and Company.

The mixed lawn of Tto^r and Tto^s cells offers a neat way to distinguish all four possible combinations of the two pairs of genotypes in the progeny phages (figure 13.27): T2*h⁺ r⁺* phages will produce small, turbid plaques (small because of the *r⁺* gene, turbid because of the *h⁺* gene); T2*h⁺ r* phages will produce large, turbid plaques; T2*h r⁺* phages will produce small, clear plaques; and T2*h r* phages will produce large, clear plaques. In fact, all four

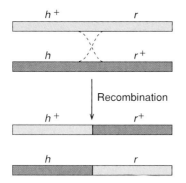

FIGURE 13.28 Recombination between *h* and *r* genes of phage T2. Bacteria were coinfected with phages of genotypes *h⁺ r* and *h r⁺*. Recombination between these two markers produced progeny phage of genotype *h⁺ r⁺* and the double-mutant *h r*.

(a)

(b)

FIGURE 13.29 (*a* and *b*) Two possible arrangements of the markers: *h, r*13, and *r*7.

combinations were observed. Does this prove that recombination has occurred? Since the progeny phages went on producing nonparental plaques, this new plaque morphology bred true. Thus, recombination, not just complementation, has indeed occurred (figure 13.28).

Recombination between two different strains of a T-even phage makes it possible to map phage genes by methods analogous to those used for bacterial genes. For example, Hershey and his colleagues isolated several different *r* mutants of phage T2, numbered *r*1, *r*2, and so on, and studied the recombination frequency of each one with an *h* mutant. Let us consider two *r* mutants whose recombination frequencies with *h* differ significantly: *r*7 and *r*13. In fact, recombinants between *r*7 and a particular *h* mutation are found in 12.3% of the progeny phages, but recombinants between *r*13 and this same *h* mutation are found in only 1.7% of the progeny. This means that the *r*13 and *h* mutations lie much closer on the T2 genetic map than do the *r*7 and *h* mutations. This can be explained by two possible arrangements of *r*7, *r*13, and *h,* as illustrated in figure 13.29.

To distinguish between these two possibilities, we need to examine the recombination frequency between *r*13 and *r*7 and see if it is greater or less than that between *h* and *r*7. If it is greater, *r*13 and *r*7 are farther apart than *h* and *r*7, and figure 13.29*b* is correct. If it is less, *r*13 and *r*7 are closer than *h* and *r*7, and figure 13.29*a* is correct. In fact, it is less, so diagram (*a*) is the right one. You may be thinking that this was a foregone conclusion, since the two *r* mutations surely lie in the same gene and that gene is not likely to be interrupted by another gene, *h.* However, it turns out that *r*7 and *r*13 lie in *different* genes, but mutations in either gene lead to the same mutant plaque phenotype. So the experiment was not so trivial after all.

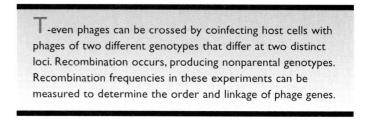

T-even phages can be crossed by coinfecting host cells with phages of two different genotypes that differ at two distinct loci. Recombination occurs, producing nonparental genotypes. Recombination frequencies in these experiments can be measured to determine the order and linkage of phage genes.

FINE STRUCTURE MAPPING

We have already examined some mapping of T-even phage *r* mutants. Now we are going to focus on a class of T4 *r* mutants that Hershey mapped to a region he named the **rII** locus. These *rII* mutants were especially interesting because they exhibited a phenotype in addition to the *r*-mutant plaque morphology. They could grow on *E. coli* strain B, but not on strain K(λ), which harbors a λ phage DNA in its chromosome. Thus, *rII* mutants are a kind of conditional lethal—the lethality in this case depends on the host strain. Benzer exploited this trait to perform an exhaustive study of the *rII* locus, mapping a large number of very closely linked *rII* mutants.

Benzer's rII Mapping

This kind of detailed mapping, or **fine structure mapping,** required that Benzer isolate many mutants in a single locus and map them relative to one another. Obviously, if all these mutations are in one gene, they will all be relatively close together. In fact, some will be only one or a few nucleotides apart, and therefore will recombine with a very low frequency. Thus, Benzer had to hunt through thousands of unrecombined, parental (T4*rII*) phages to find one recombinant phage. This is where the conditional lethality of *rII* mutants came in handy. The unrecombined parental phages could not grow on *E. coli* strain K(λ); only the rare, recombinant, wild-type (T4*r⁺*) phages could do so. This simplified the search tremendously. It was much easier to see a wild-type plaque against a smooth lawn of undisturbed bacteria than to pick one out of a sea of mutant plaques. This is another example of the power of selection in bacterial and phage genetics.

Benzer's conditional lethal screening method was so powerful that he could have detected a recombination frequency of one in a million. In fact, the lowest real frequency of *r⁺* recombinants he observed was one in ten thousand, or 0.01%. This seems to represent the closest that two mutations can be and still recombine.

How close are two mutations that give *r⁺* recombinants with a frequency of 0.01%? We can make an estimate as follows: First, we must realize that if *r⁺* recombinants appeared with a frequency of 0.01%, another 0.01% of double-mutant recombinants went undetected. So the real rate of recombination was 0.02%. We know from other mapping data that the circular phage map consists of about 1,500 **map units,** where a map unit is the distance separating two loci that recombine with a frequency of 1% (chapter 5). Thus, two loci that recombine with a frequency of 0.02% are separated by a distance equal to 0.02/1,500, or 1.3×10^{-5} of the total phage genome, assuming that the recombination frequency is constant throughout the genome. Since the

T4 phage genome contains about 1.75×10^5 base pairs, the distance between recombinable sites is about two base pairs.

The mechanisms we envision for recombination (chapter 7) predict that it should be able to occur between mutations on adjacent nucleotides. Benzer's experiments were not quite precise enough to demonstrate that, but later experiments nailed down the point: Recombination can indeed occur between mutations on adjacent nucleotides.

The finding that recombination could occur *within* a gene was an important one. It destroyed the classical notion of the gene as an indivisible unit. Instead of the gene itself being the unit that recombines, a subgenic unit that Benzer called the "recon" was recognized as the real indivisible unit of recombination. When it became obvious in later years that a recon is really just a base pair, Benzer's term fell into disuse.

Deletion Mapping

So far, we have pretended that Benzer dealt only with point mutations. If that had actually been the case, his task would have been vastly more difficult. The reason: Mapping point mutations would have required crossing every mutant with every other mutant; with thousands of mutants, that means millions of crosses—a daunting prospect indeed.

Wisely, Benzer employed an alternative, very efficient method for initial screening of his mutants that pinned them down to forty-seven different groups, depending on their location within the *rII* locus. Once he had assigned a mutant to one of these groups, he had only to cross it with other members of the same group to define its location precisely.

The initial screening method is called **deletion mapping,** and it deals with mutations called **deletions.** A point mutation involves alteration of one or a very few adjacent base pairs. By contrast, a deletion involves loss of a significant part of a gene, or a whole gene, or even more than one gene. Consider a simple case in which we have identified two different deletions (deletions 1 and 2) within a gene, as depicted in figure 13.30. We can use these two deletions to define three different areas of the gene. Point mutants are crossed with mutants containing deletions 1 and 2 and scored for wild-type recombinants. If a point mutant can give a wild-type recombinant with deletion 2, but not 1, we know that its defective base pair lies in region a. Similarly, we can see that a DNA with a defect in region b or c cannot recombine with deletion 2 DNA to give a wild-type DNA. On the other hand, if a point mutant can give a wild-type recombinant with deletion 1, but not 2, its defect must lie in region c. By the same reasoning, if a point mutant cannot give a wild-type recombinant with either deletion 1 or 2, its defect must lie in the region of overlap between the two deletions, region b.

Using this type of deletion mapping, Benzer identified forty-seven different regions of the *rII* locus, each defined by the ends of a set of deletions. These regions, and the deletions that define them, are shown in figure 13.31. Here is how Benzer's mapping worked: First, he crossed a new point mutant with each of the seven deletion mutants (1272–638) listed at the top of the figure. This allowed a rough assignment of the site of the mutation. For

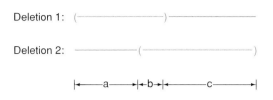

FIGURE 13.30 Deletion mapping. Two deletion mutants (deletion 1 and deletion 2), with deleted DNA denoted by blue lines within parentheses, are crossed with various point mutants. Deletion 2 can give wild-type recombinants only with mutants in region a; deletion 1 can give wild-type recombinants only with mutants in region c; neither deletion can give wild-type recombinants with mutants in region b.

example, if the mutant could give a wild-type recombinant with deletion 1241, but not 1272, the mutation lay at the left end of the *rII* locus, in region A1a, A1b1, or A1b2. To determine which of these three subregions contained the mutation, Benzer could cross them with deletions 1364 and EM66. Suppose that the mutant could recombine with deletion EM66, but not with 1364. In that case, the mutation must lie in subregion A1b1. Benzer could then complete his fine mapping of this mutant by crossing it only with other mutants that mapped to this small area of the *rII* locus. Obviously, his job was simplified enormously.

When Benzer plotted the locations of hundreds of spontaneous mutations in the *rII* region, he found that the mutations were not evenly distributed throughout the region. Instead, some sites were not mutated at all, whereas others were mutated repeatedly. One site was affected in more than five hundred of the 1,612 mutants screened. Benzer called these highly mutable sites **hot spots.** The same phenomenon has been observed with mutagen-induced mutations.

Seymour Benzer collected thousands of mutants in the *rII* region of phage T4 and mapped their mutations exhaustively. This and other fine structure mapping revealed that recombination can occur between mutations as close as one nucleotide pair apart. Benzer also showed that his mutations were not evenly distributed throughout the *rII* region. Instead, many were clustered in hot spots.

APPLYING THE *CIS-TRANS* TEST TO THE *RII* REGION

We have already observed, in the context of bacterial genetics, the usefulness of the *cis-trans* complementation test in determining whether two mutations affect the same or different genes. The idea is to bring the two mutant genomes together in one organism and see whether a wild-type phenotype results. If so, the two mutant genomes have complemented each other, demonstrating that their mutations lie in different genes. On the other hand, if they cannot complement each other, the mutations must be in the same gene. You may already have recognized that this is really the same thing that geneticists do when they mate two mutant

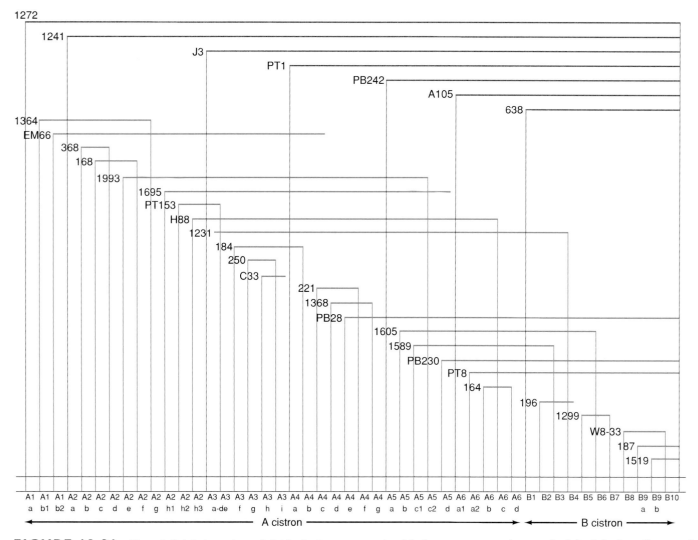

FIGURE 13.31 Map of *rII* deletions that subdivide the locus into forty-seven segments. The two genes (*rIIA* and *rIIB*) are divided into forty-seven segments (A1a through B10) according to the ends of the deletions denoted by the heavy horizontal bars. The names of the deletion mutants are given to the left of the bars. Some ends do not define segments; these are not connected to a vertical line. Major divisions are denoted by red lines; subdivisions within these are represented by blue lines.

eukaryotic organisms. They place two mutant genomes in the same offspring and observe the phenotype of that offspring.

In 1955, Benzer applied this approach to phage genetics—in particular to the *rII* region of phage T4. We have already seen that recombination between some *rII* mutants is quite frequent, which tells us that the mutated loci are relatively far apart, and therefore suggests that more than one gene is involved in the *rII* phenotype. Benzer proved this point as follows: He coinfected bacteria with pairs of *rII* mutants at a high ratio of phages to bacteria to ensure that each host cell would receive both mutant phage genomes. (Notice that this is the equivalent of producing a diploid by crossing bacteria or higher organisms.) The question was: Would the two *rII* mutants complement each other? The answer was that many of them would. Thus, more than one gene was indeed involved. How many genes? Benzer found that all of his *rII* mutants fell into only two **complementation groups,** or cistrons. That is, members of one group would complement

members of the other, but not members of their own group. These two cistrons represent the two genes of the *rII* region, which Benzer called *rIIA* and *rIIB*.

Figure 13.32 shows the arrangement of these two genes and some of the mutations that lie in each one. At the time Benzer carried out these studies his term *cistron* became popular as a substitute for the word *gene,* because the latter term was tied to the outdated notion of an indivisible unit. However, *gene* has proven to be a very durable word and is again widely used.

Benzer applied the *cis-trans* test to the *rII* region of phage T4 and identified two complementation groups, or cistrons, that are responsible for the *rII*-mutant phenotype. Thus, the *rII* region consists of two genes.

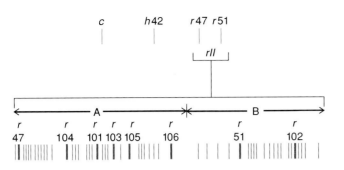

FIGURE 13.32 The two cistrons of the *rII* locus. The *rIIA* and *B* genes and a few of their mutations are depicted.

Source: From S. Benzer, "Fine Structure of a Genetic Region in Bacteriophage," *Proceedings of the National Academy of Science* 41:344, 1955.

WORKED-OUT PROBLEMS

PROBLEM 1

Consider an experiment in which we do not know anything about the order of three genes, X, Y, and Z, but we do know that they are close enough together to cotransduce. We use an $X^+ Y^+ Z^+$ donor and an $X^- Y^- Z^-$ recipient. We select for X^+ transductants and find the following frequencies of transduction:

103 are $X^+ Y^- Z^-$
315 are $X^+ Y^+ Z^-$
57 are $X^+ Y^+ Z^+$
0 are $X^+ Y^- Z^+$

We can ask the following questions:

1. What is the order of these three markers?

2. What is the cotransduction frequency of X and Y? Of X and Z?

3. What are the DNA distances between X and Y? Between X and Z?

SOLUTION

First, what is the marker order? We will see that all of the combinations but one can be achieved by two crossovers between donor and recipient DNA. The fourth can only be produced by four crossovers and will therefore be the rarest. The rarest combination is $X^+ Y^- Z^+$, which appeared in none of the transductants. How can four crossovers between donor and recipient produce this combination? There is only one way, as shown in the figure below, where Y is in the middle. Therefore, the order of markers is XYZ.

Here is another way of looking at this: If there are four crossovers, there must be one before and after each of the three

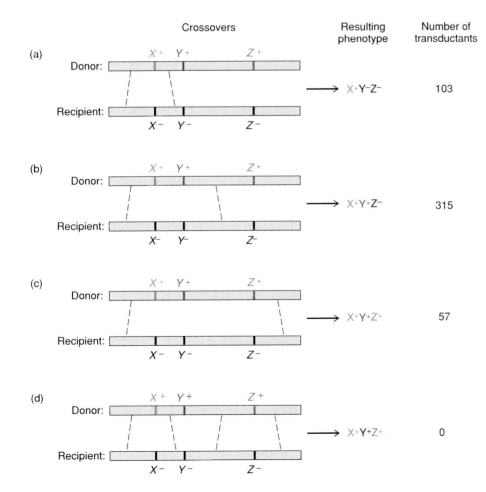

	Crossovers	Resulting phenotype	Number of transductants
(a)	Donor: X^+ Y^+ Z^+ / Recipient: X^- Y^- Z^-	$X^+Y^-Z^-$	103
(b)	Donor: X^+ Y^+ Z^+ / Recipient: X^- Y^- Z^-	$X^+Y^+Z^-$	315
(c)	Donor: X^+ Y^+ Z^+ / Recipient: X^- Y^- Z^-	$X^+Y^+Z^+$	57
(d)	Donor: X^+ Y^+ Z^+ / Recipient: X^- Y^- Z^-	$X^+Y^+Z^+$	0

Three-factor cross to determine gene order and cotransduction frequency. At least two crossover events must occur to transfer the X^+ marker from the donor DNA to the recipient. Two crossovers can transfer just X as in (*a*), X + Y (*b*), or X + Y + Z (*c*). In order to transfer the donor markers on the ends (X^+ and Z^+), and leave the middle recipient marker (Y^-), four crossovers must occur (*d*). Since four crossovers will be much more infrequent than two, the rarest transductant ($X^+ Y^- Z^+$) will have the marker with the recipient genotype in the middle. This is Y^-, telling us that the marker order is X-Y-Z (or Z-Y-X, which is indistinguishable from X-Y-Z in this experiment). The cotransduction frequency is obtained by summing the number of times two donor markers are transduced together (e.g., 315 + 57 = 372 for X^+ and Y^+) and dividing by the total number of transductants (103 + 315 + 57 = 475).

markers. This will automatically give the transductant the donor alleles of the two end markers, and the recipient allele of the middle marker. Thus, a shortcut is to find the rarest combination ($X^+Y^-Z^+$ in this case) and determine which of the three loci has the recipient phenotype (Y^- in this case). Therefore, Y must be in the middle.

What are the cotransduction frequencies? For X and Y: $(315 + 57)/475 = 0.78$; for X and Z: $57/475 = 0.12$. The figure shows the crossovers that create each of the four marker combinations. Notice again that the first three involve only two crossovers, while the fourth involves four. Also note that a crossover before X is required in all cases, since only X^+ transductants were selected.

Finally, what are the DNA distances among the markers? Rearrangement of the expression $x = (1 - d/L)^3$ for cotransduction frequency (x) gives:

$$d = L(1 - \sqrt[3]{x})$$

Given a length (L) of 91.5 kb, and a cotransduction frequency of 0.78 for X and Y, the distance between X and Y would be 91.5 kb $(1 - \sqrt[3]{.78})$, or 7.3 kb. Similarly, the distance between X and Z would be 91.5 kb $(1 - \sqrt[3]{.12})$, or 46.7 kb.

PROBLEM 2

An exotic bacterium takes just 60 minutes to conjugate. You perform a conjugation (mating) experiment with this bacterium and find that the following markers are transferred at the respective times: *pud,* 10 minutes; *spa,* 15 minutes; *ara,* 25 minutes; *dip,* 40 minutes; and *lip,* 50 minutes. Draw a genetic map of this bacterium, assuming a linear rate of DNA transfer during conjugation, and ignoring any lag at the beginning of conjugation.

SOLUTION

Since the total conjugation time is 60 minutes, the genetic map of this organism looks just like a clock face. Each gene goes on the clock at the position corresponding to the time of transfer. For example, *pud* took 10 minutes to be transferred, so it goes at the 10 minute position (two o'clock). Similarly, *spa* took 15 minutes to be transferred, so it goes at the 15 minute position (three o'clock), and so forth. If the conjugation time were not 60 minutes, the process would not be quite as simple, but the principle would be the same. For example, if the conjugation time were 150 minutes, a gene transferred in 50 minutes would map at a position 50/150, or 1/3 of

the way around the circle. This would correspond to 20 minutes (or four o'clock) on a 60-minute map.

PROBLEM 3

You want to determine the titre (plaque-forming units [pfu] per milliliter) of phage particles in a phage suspension. You dilute the suspension by a factor of 10^7 and plate 0.2 ml. You observe 120 plaques. What is the titre of the original phage suspension?

SOLUTION

Simply multiply the number of plaques observed by the dilution factor and divide by the volume in milliliters:

$$\frac{120 \text{ pfu} \times 10^7}{0.2 \text{ ml}} = 6 \times 10^9 \text{ pfu/ml}$$

SUMMARY

Two mutant (auxotrophic) bacterial strains can transfer DNA by direct contact, producing wild-type (prototrophic) recombinants. This process, called conjugation, is mediated in F^+ *E. coli* by an F plasmid of 94.5 kb that can transfer at high frequency into F^- bacteria. Occasionally, the F plasmid recombines with its host's chromosome, converting the cell to an Hfr and mobilizing the Hfr chromosome to transfer to recipient cells. By interrupting mating between an Hfr donor and an F^- recipient, we can determine the length of time it takes to transfer a given gene. The differences in the transfer times of different genes tell us their relative positions on the circular genetic map of *E. coli*. The completion of mating in *E. coli* requires recombination between the donated piece of Hfr chromosome and the F^- chromosome.

Sometimes a recombination event brings together a host DNA and an F plasmid, creating an F′ plasmid. The host genes thus become part of a highly mobile element that can deliver them to another cell. We can take advantage of this efficient transfer of limited pieces of DNA to perform genetic experiments. For example, two different bacterial mutations that affect the same function can be subjected to the *cis-trans* test, or complementation test. If they complement each other to restore wild-type function, the mutations are usually in separate genes.

Phages can also be used to carry, or transduce, DNA from one bacterium to another. Transduction by phage λ is called specialized transduction because it usually carries only the *gal* or *bio* genes that flank the λ integration site. P1 phage heads can mediate generalized transduction by incorporating up to 91.5 kb of host DNA instead of phage DNA. Transduction allows us to do several types of genetic mapping. We can transduce two genes, or two markers within a gene, and observe the frequency of recombination between them. The closer together they are, the lower their frequency of recombination. Alternatively, we can observe the frequency of cotransduction of two markers. The closer together the markers, the higher their frequency of cotransduction. We can also perform three-factor crosses, transducing three markers and then deducing the order of the markers on the genome.

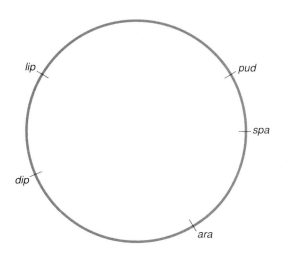

T-even phages can be crossed by coinfecting host cells with phages of two different genotypes differing at two different loci. Recombination occurs, producing nonparental genotypes. Recombination frequencies in these experiments can be measured to determine the order and linkage of phage genes. Conditional lethal mutations are used in essential genes to avoid destroying the phage's viability.

Fine structure mapping of the *rII* region of phage T4 revealed that recombination can occur between mutations as close as one nucleotide pair apart, and that the mutations were clustered in hot spot regions. The *cis-trans* test, applied to the *rII* region of phage T4, identified two complementation groups, or cistrons, that are responsible for the *rII* phenotype. Thus, the *rII* region consists of two genes.

PROBLEMS AND QUESTIONS

1. Using standard bacterial genetic notation, write the genotypes and phenotypes of bacterial strains that have the following characteristics. The three-letter abbreviations for each locus are given in parentheses.

    (*a*) Unable to metabolize the sugar lactose (lac).
    (*b*) Able to make the amino acid leucine (leu).
    (*c*) Unable to make the vitamin thiamine (thi).
    (*d*) Resistant to the antibiotic streptomycin (str).
    (*e*) Sensitive to the antibiotic puromycin (pur).
    (*f*) Able to make the vitamin biotin (bio).

2. Which of the strains in problem 1 are prototrophs, and which are auxotrophs? Assume that the strains are wild-type for all genes but the one in question.

3. Given a reversion rate of one in a million and a recombination rate of one in ten million in Lederberg and Tatum's first experiment, was Lederberg justified in his use of double mutants? Why, or why not?

4. Why is a prototrophic recombinant in a bacterial conjugation experiment rarely a true zygote? What do we call it instead?

5. You have discovered that certain *E. coli* cells are resistant to infection by phage T1 (*ton*r). You would like to know whether this characteristic is caused by a mutation in the bacterial genome or by an adaptation of the cells to the presence of the phage. To answer this question, you perform a fluctuation test and find that twenty individual cultures vary quite a bit in their resistance to T1—from a low of zero to a high of 107 *ton*r colonies per culture. Is the characteristic caused by mutation or adaptation? Why?

6. What event converts an F+ cell to an Hfr? How does this work?

7. Why does an Hfr cell rarely transfer its whole genome in conjugation experiments?

8. The giant sandworms of the planet Arrakis harbor a gut bacterium (*Xenobacterium gigantii*) that undergoes conjugation. However, like everything in these worms, the circular bacterial chromosome is large, and conjugation takes 200 minutes. You perform a conjugation experiment with this organism and find that the following markers are transferred at the following respective times: *ple*, 25 minutes; *bar*, 75 minutes; *gap*, 100 minutes; *zap*, 150 minutes; *chr*, 200 minutes. Draw a genetic map of this exotic bacterium, assuming a linear rate of transfer of DNA during conjugation, and ignoring any lag at the beginning of conjugation.

9. In the experiment in problem 8, you find that the efficiency of transfer of the *ple* marker is very much greater than that of the *chr* marker. What is the probable explanation for this phenomenon?

10. In the experiment in problem 8, if you used an Hfr strain that begins its transfer just after *zap*, what difference in transfer times would you expect between *chr* and *ple*?

11. In another Hfr strain of *X. gigantii*, the fertility plasmid inserts between *gap* and *zap* in the orientation opposite to that in problem 8. What would be the order of transfer of markers by this Hfr?

12. One strain of *X. gigantii* harbors an F′ plasmid. F-*zap*, that contains the *zap* operon and is composed of two genes: *zapA* and *zapB*. You have two Zap⁻ mutants and you want to determine whether their mutations lie in the same or different genes. Assuming (a) that you can readily place one of these mutant genes on the F-*zap* plasmid, and (b) that the F-zap plasmid carries the selectable marker *hisB*+ (for synthesis of the amino acid histidine), and (c) that you can arrange for *X. gigantii* cells to be either sensitive or resistant to the antibiotic lethalmycin, outline an experiment to answer your question.

13. You perform the experiment described in the answer to problem 12 (see answer section at end of book) and obtain only a few Zap+ colonies. What are two possibilities for the cause of these Zap+ cells, assuming back mutation cannot occur?

14. Why should we not expect to find both *bio* and *gal* markers in a λ transducing phage?

15. In *E. coli*, the markers *thr* and *leu* are separated by about 2% of the total bacterial genome. Would you expect these genes to cotransduce in P1 phage? Why, or why not? If so, at what frequency?

16. The genes *zip* and *zap* lie less than 10 kb apart on the *X. gigantii* genome. A generalized transducing phage called Q2 infects this organism. You have isolated two different *zip* mutants and want to map the locations of the corresponding mutations (*zip*₁ and *zip*₂) within the *zip* gene. Outline the transduction experiment you would perform.

17. You know that a gene called *bop* is tightly linked to both *zip* and *zap* in the *X. gigantii* genome and readily cotransduces with these other markers in Q2 phage. You want to know whether the gene order is *zip bop zap* or *zip zap bop*. Therefore, you perform a three-factor cross by transduction as follows: You transduce DNA from a Zip+ Zap⁻ Bop+ strain into a Zip⁻ Zap+ Bop⁻ recipient and select for Bop+ transductants. The lowest frequency transductant has the phenotype Zip+ Zap+ Bop+. What is the order of these three genes? What results would you expect if the other gene order were correct? (Drawing diagrams of the two possibilities should help you figure out the answer.)

18. You perform a three-factor cross in which you use Q2 phage to transduce DNA with the three markers Awk⁺ Kat⁺ and Nrd⁺ into an Awk⁻ Kat⁻ Nrd⁻ recipient. You select for Awk⁺ transductants and observe the following:

Awk⁺ Kat⁻ Nrd⁻:	638
Awk⁺ Kat⁻ Nrd⁺:	309
Awk⁺ Kat⁺ Nrd⁺:	115
Awk⁺ Kat⁺ Nrd⁻:	1

    What is the order of these markers? What are the cotransduction frequencies between *awk* and *kat* and between *awk* and *nrd*?

19. You have a suspension of phages and you want to know its titer (plaque-forming units per milliliter). You dilute the suspension by a factor of 10^6 and plate 0.05 ml. You observe 250 plaques. What was the titer of the original suspension?

20. You perform a phage growth experiment with phage T2. Instead of waiting for the phage to lyse the host cells, you break the cells open at various times after infection and assay for plaque-forming units. How will your growth curve differ from the one-step growth curve? Why?

21. You infect *X. gigantii* cells with different strains of Q2 phage and observe the following frequencies of recombination among the three markers *K*, *L*, and *M*: between *K* and *L*, 18%; between *K* and *M*, 2.4%; between *L* and *M*, 15%. What is the order of these three markers?

Answers appear at end of book.

Developmental Genetics

I do not want to be a fly!
I want to be a worm!

Charlotte Perkins Stetson Gilman

British poet

Learning Objectives

In this chapter you will learn:

1. How gene expression and cell differentiation are related.
2. How eukaryotic gene expression is controlled.
3. The difference between cell differentiation and determination.
4. Theories to explain determination.
5. How a hierarchical group of genes cooperates to control development in the fruit fly.
6. How genes themselves can sometimes be altered during development.

One of the greatest mysteries, if not the central mystery, of modern biology is this: How does a fertilized egg develop into a human being? The question is phrased in human terms because we tend to be most curious about ourselves, but the development of any other complex organism is just as mystifying. Through the ages, scientists have offered explanations of increasing sophistication, but we are still a long way from solving the mystery.

In the seventeenth century, biologists believed that a tiny, preformed person, or *homunculus* (figure 14.1), resided in each sperm cell. They assumed that the homunculus simply grew in the fertilized egg until it reached the proportions of a newborn baby. Of course there were serious theoretical problems with this idea—one being that the homunculus itself would also have to contain miniature sperm cells enclosing even tinier homunculi, and so on, to include all possible future generations.

Scientists studying the embryology of chickens found no support for the homunculus theory. In fact, as they looked farther and farther back in the development of the chick, they noticed that the embryo not only became smaller, it also became simpler. This gave rise to the idea of **epigenesis,** which holds that tissues do not exist in the earliest embryo, but arise from simpler structures by developmental change. Taken to its logical conclusion, this means that a single fertilized egg cell develops into an adult human composed of trillions of cells that are organized into dozens of different organs, each of which performs a different, highly specialized function.

THE GENETIC BASIS OF CELL DIFFERENTIATION

We have already seen that specialized cellular functions depend on proteins: structural proteins to give cells their shapes; enzymes to catalyze specific chemical reactions; hormones to carry signals from cell to cell; and so forth. For example, consider two different specialized cells in the human body: a red blood cell and a β cell from the pancreas (figure 14.2). The blood cell is highly specialized to carry oxygen through the bloodstream. Its smooth disk shape allows it to glide easily through tiny capillaries, and it contains a large concentration of a single protein, hemoglobin. As blood flows through the lungs, this protein binds oxygen and turns bright red. Later, it releases the oxygen to peripheral tissues, sustaining their lives.

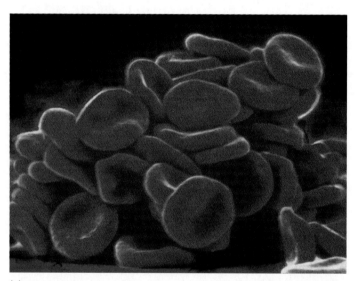

(a)

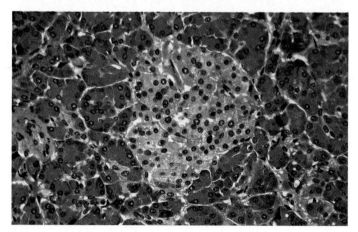

(b)

FIGURE 14.2   Contrasting cell morphologies. (*a*) Mature red blood cells; (*b*) an islet of Langerhans (light area at center) composed of insulin-producing β cells.

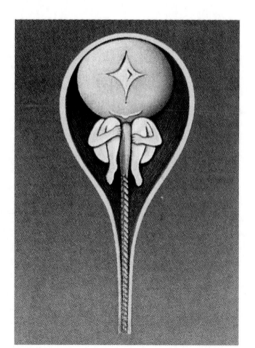

FIGURE 14.1   Homunculus.

By contrast, β cells do not move throughout the body; they are stuck in one place—in small conglomerations of cells called islets of Langerhans in the pancreas. The main job of the β cells is to produce a hormone called **insulin.** The cells secrete this small protein into the bloodstream, after which it travels to remote areas of the body, signaling cells to alter their metabolism. For example, insulin responds to a high level of sugar (glucose) in the blood by promoting the uptake of glucose by cells and depressing the breakdown of fats in adipose (fat) tissue that would normally produce more sugar. The result of these and other actions of insulin is a lowering of the blood glucose level.

Obviously, the functions of red blood cells and pancreatic β cells are vastly different. These different functions are clearly reflected in the proteins the two types of cells make. Immature red blood cells make huge amounts of hemoglobin and very little else; in particular, they do not make insulin. On the other hand, β cells make lots of insulin, but no hemoglobin. Yet both of these cell types descended directly from the same fertilized egg cell. The early embryo contained many cells, all of which looked more or less alike, and none of which made either hemoglobin or insulin. However, at some point during development, the cells underwent **differentiation** so that they began to manufacture these specialized products. Some became blood cells and made hemoglobin; others became β cells and made insulin.

All cells, even though they begin to synthesize different specialized products, continue to make the key proteins essential for life. For example, they all make RNA polymerase and ribosomal proteins. Because of their maintenance function, the genes for these fundamental proteins are sometimes called **housekeeping genes.** By contrast, genes that code for specialized cell products like hemoglobin or insulin are sometimes called **luxury genes.** They are a "luxury" only in that their products are not immediately vital to the survival of the cells that make them; they are certainly not a luxury to the whole organism.

In short, different specialized cells make different specialized proteins, and those proteins determine what the cells can do. Moreover, as we have already learned, proteins are products of genes. This means that the genes for the globin proteins in hemoglobin, are turned on in immature red blood cells, and the gene for insulin is turned on in β cells of the pancreas. Cell differentiation is therefore a reflection of differential gene activity, and those who seek to understand differentiation and development must ultimately ask: What allows expression of certain genes in one kind of cell and other genes in a second kind of cell? We will examine two theories that attempt to answer this question.

MOSAIC DEVELOPMENT AND REGULATIVE DEVELOPMENT

One simple explanation of cell differentiation would be that a differentiating cell loses all its genes, except the ones it needs for housekeeping and for its specialized function. This would be a true genetic change, since the genetic material itself would be altered or lost. A version of this theory, presented by Wilhelm Roux, held that embryonic cells begin losing genetic potential from the very

beginning, so that each cell division during development involves the unequal partitioning of genetic material between the daughter cells. This would produce an embryo whose genetically dissimilar cells resemble a mosaic of many different pieces. Hence the term **mosaic embryo.**

Mosaic Development

Roux's theory found support in the fact that several invertebrate embryos, even at a very early stage, are composed of cells that behave quite differently; in fact, they seem to be irreversibly programmed to develop along preset lines. A favorite example of such a mosaic embryo is the ascidian *Styela*. Even at the two-cell stage, the two embryonic cells **(blastomeres)** in this organism are different. These differences continue at the four-cell stage, as shown by an experiment performed by E. G. Conklin in 1905, which is depicted in figure 14.3. Conklin forced four-cell *Styela* embryos through a pipette to destroy some of the blastomeres, then looked to see what would happen to the remaining live cells.

In figure 14.3*a,* the two cells on the left (anterior cells) were killed, but the two on the right (posterior cells) went on to differentiate as they would have in a complete embryo, giving rise to chorda cells and muscle cells. In figure 14.3*b,* the bottom left (anterior-vegetal) and top right (posterior-animal) cells were killed, but again the two remaining cells carried out their own developmental programs. The top left cell made neural plate cells, and the bottom right cell again made muscle cells. In theory, this could be caused by the embryo's dividing up its genes with each cell division, so that the top left cell received the genes for making neural plate cells and the bottom right cell got the genes for muscle. No evidence exists for such a partitioning of genes among embryonic cells, and it is increasingly unlikely that any will be found.

Regulative Development

Even if we someday find an unusual organism that does divide its genes among embryonic cells, this cannot be a general phenomenon. Embryos from higher eukaryotes, such as amphibians,

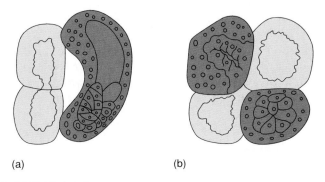

(a)   (b)

FIGURE 14.3   Demonstration of mosaic development. *Styela* embryos were forced through a pipette at the four-cell stage to kill some of the blastomeres, then allowed to develop. (*a*) Both cells on the far left (yellow) were killed, but the two remaining cells (red) continued to divide and differentiate. (*b*) The lower left and upper right cells (yellow) were killed, but again the remaining cells (red) developed into partial embryos.

can be separated into individual cells as late as the eight-cell stage, and all of the resulting cells are **totipotent,** which means they have the potential to become any cell type in the organism. Clearly, these individual embryonic cells have not lost any of their genetic potential. Even in humans, we know that embryonic cells retain all of their genes at least through the first three cell divisions, because identical quintuplets have occurred. Humans and frogs are examples of organisms that have **regulative embryos.** The cells in these embryos look and behave identically much longer than those of mosaic embryos. Of course, other organisms, such as sea urchins, fall in between these extremes, and sometimes it is difficult to label an embryo either mosaic or regulative.

We have established the fact that early regulative embryos contain totipotent cells. But what about differentiated cells? Are they still totipotent, or have they lost some of their genes? A number of experiments have shown that they are sometimes totipotent. Figure 14.4 illustrates John Gurdon's classic nuclear transplantation experiment, which worked as follows: First, Gurdon used ultraviolet light to kill the nuclei in several eggs from the South African clawed frog (*Xenopus laevis*), producing **enucleate eggs** (eggs without nuclei). Then he extracted nuclei from the intestinal epithelium of a *Xenopus* tadpole and transplanted these nuclei into the enucleate eggs by **microinjection** with a very fine glass pipette.

The transplanted nuclei were diploid, just like **zygote** (fertilized egg) nuclei, and many of the eggs began dividing, just like zygotes, to produce early "hollow ball" embryos called **blastulas.** Gordon took nuclei from these blastulas and transplanted them to new enucleate eggs. Some of the resulting eggs actually developed into tadpoles, and a few made it all the way to the adult frog stage. This demonstrated that even a differentiated intestinal cell nucleus retained the genetic potential to produce a whole adult frog. In other words, the nucleus was still totipotent. It could not have lost all the superfluous genes it did not need for housekeeping and intestinal function. In fact, since it gave rise to every cell type in a whole frog, it seems not to have lost any genes at all.

A potential objection to this experiment could be that the ultraviolet light failed to kill all of the original egg nuclei, and these nuclei provided the genes that caused **embryogenesis** (embryonic development). To counter this argument, Gurdon used transplanted nuclei that were genetically different from the original nuclei in that they had only one nucleolus, whereas the original nuclei had two. The resulting tadpoles and frogs consisted of cells with only one nucleolus, proving that the transplanted nuclei really had directed their development.

The nuclear transplantation experiment has been remarkably difficult to duplicate in other animal species. Even before Gurdon's experiment on *Xenopus,* Robert Briggs and Thomas King showed that another frog species, *Rana pipiens,* does not contain totipotent nuclei after gastrulation, which is a relatively early phase of embryogenesis. However, totipotent adult cells

are abundant in plants. F. C. Steward provided the first such example in the early 1950s by regenerating a whole carrot plant from a single carrot root cell. Since then, several different kinds of plants, including tobacco, petunias, and peas, have been grown from single cells. Notice that this furnishes us with a way to **clone** (produce many identical copies of) a plant. In fact, the word "clone" comes from the Greek *klon,* meaning "twig," and refers to a centuries-old method of propagating plants by taking cuttings.

We are left with the somewhat messy conclusion that whereas some animals and plants do contain totipotent differentiated cells, many seem not to. We can at least state that loss of genetic material is not necessary for cell differentiation. Specialized cells that are not totipotent may have suffered genetic changes, and we will see some examples of these later in this chapter, but there is certainly no evidence for wholesale loss of genes. Instead, most cell differentiation seems to be caused by **epigenetic** changes. These are changes in gene expression, rather than changes in the genes themselves.

Multicellular organisms with specialized tissues develop from a single cell, or zygote, by a process called embryogenesis. During embryogenesis, cells differentiate to produce tissues with specialized functions. Differentiation seems to require little if any change in a cell's genetic makeup, since differentiated cells or nuclei from some plants and animals are still totipotent. Instead, the pattern of expression of a common set of genes varies from one kind of differentiated cell to another.

DETERMINANTS

Even before a cell becomes differentiated and starts expressing its specialized functions, it undergoes a process called **determination.** This is the stage at which the cell passes the point of no return in its developmental program. It may not yet look or behave any differently than its neighbors, because it has not differentiated, but it has become irreversibly committed to its fate. Indeed, determination is a prerequisite for differentiation. We will look in detail at possible mechanisms for determination later in the chapter; for now, let us consider some examples of determination that imply the existence of cytoplasmic effector substances called **determinants.**

The very existence of mosaic embryos, in which the position of a blastomere in the embryo determines absolutely the fate of that cell, suggests that determinants are present in the egg and that they are unequally distributed. The simplest form of unequal distribution of determinants would be a gradient from the top (**animal pole**) to the bottom (**vegetal pole**) of the egg. But a second gradient, from one side to the other, could coexist with the vertical one (figure 14.5).

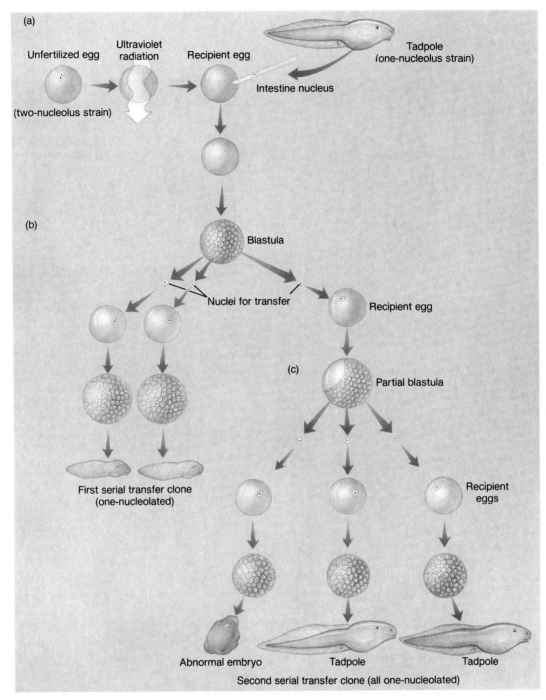

FIGURE 14.4 Demonstration of regulative development.
(a) An unfertilized frog egg from a two-nucleolus strain was enucleated by ultraviolet radiation, then given a replacement nucleus from the intestine of a one-nucleolus strain. The artificially fertilized egg subsequently developed into a blastula. (b) Nuclei from this blastula were transplanted to new enucleate eggs from the two-nucleolus strain. These developed into partial embryos or tadpoles.

(c) Nuclei from a partial embryo (a blastula) were transplanted to new enucleate eggs, some of which developed into normal, one-nucleolated tadpoles. The first transfer rarely produces normal tadpoles because the differentiated nuclei cannot divide fast enough to support normal development. After the first transfer, the nuclei seem to be geared up to divide more rapidly.

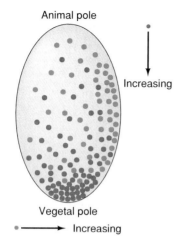

Animal pole

Increasing

Vegetal pole

Increasing

FIGURE 14.5 Gradients of determinants in eggs. A hypothetical gradient of two determinants in an egg. One type of determinant (red dots) is arranged in a gradient from top to bottom, where the highest concentration is at the bottom. The other determinant (blue dots) is arranged in a horizontal gradient, where the highest concentration is on the right.

Determinants are cytoplasmic effector substances that cause cells to become determined, or irreversibly committed, to a given fate. They are present and unequally distributed in the egg.

INDUCERS

Substances that one cell makes to influence the development of another cell are called **inducers.** This property of signaling from one cell to another sets the inducers apart from determinants, whose effects are limited to the cells that contain them. An inducer can produce rather startling effects if its signal goes to the wrong address. For example, consider the development of feathers in an embryonic chicken. Feathers are products of one of the three classical embryonic tissue layers, the **ectoderm.** But the underlying **mesoderm** plays an inductive role; that is, it produces substances (inducers) that dictate what kind of feathers the ectoderm will make. We know this because we can transplant embryonic thigh mesoderm to the wing of a developing chick and produce thigh feathers on that wing. The reverse experiment yields wing feathers on the thigh.

Another example of induction sheds additional light on determination. We can transplant embryonic cells of the amphibian *Triturus* to new locations in the embryo and follow their fates. Up to the gastrula stage, the transplanted cells are still totipotent; they will differentiate just like their neighbors in their new location, not like the cells in their old location. Clearly, the transplanted cells are responding to inducers they would not have experienced in their original home. After gastrulation, something quite different happens. The transplanted cells have become determined in their original locations, so they

are no longer responsive to external influences; they follow their predetermined program in spite of the inducers they encounter after transplantation.

Embryonic inducers are substances that diffuse from one embryonic site to another and cause differentiation of the target cells.

MORPHOGENS

Morphogens (Greek: *morphe* = form + *genes* = born) are a subclass of determinants and inducers that act to determine shape in a developing embryo. For example, John Saunders and Mary Gasseling provided an example of a vertebrate morphogen two decades ago when they showed that a piece of developing chick limb bud grafted onto another limb bud would cause the host tissue to develop an extra set of digits that grew as a mirror image of the normal set. Thus, the transplanted tissue apparently contained a so-called zone of polarizing activity (ZPA) whose cells produced a morphogen that was inducing morphological change in the host embryo. What is this morphogen? At first it appeared that it was retinoic acid, a compound similar to vitamin A.

The case for retinoic acid as the morphogen in chick limb buds depended on experiments of the following type: A bead impregnated with retinoic acid caused the same duplication of digits that the ZPA did. However, retinoic acid induced some other cellular changes that were not compatible with its role as a morphogen, so the search for the real morphogen continued. It now appears that we have a good candidate. The vertebrate limb bud morphogen seems to be the product of a gene that is homologous to the *hedgehog* gene of *Drosophila*. The *Drosophila hedgehog* gene is responsible for determining the polarity of segments in embryonic flies, so its homologue's role in establishing the polarity of limb buds in vertebrates is not surprising. The vertebrate gene is named *Sonic hedgehog* for a computer game cartoon character.

One line of evidence supporting the role of the *Sonic hedgehog* product as a morphogen is that the gene is expressed in the same place and time as the ZPA. Figure 14.6 shows an experiment performed by Cliff Tabin and colleagues that leads to this conclusion. These workers made a labeled *Sonic hedgehog* probe and hybridized it to whole chick limb bud mounts to detect the production of *Sonic hedgehog* mRNA. Then they compared the location of this mRNA to the ZPA activity at various stages of embryonic development. By comparing panels (*a*) and (*b*) in figure 14.6, we can see the correspondence in space and time between *Sonic hedgehog* mRNA and ZPA activity. The only time the two diverge is very early, at stage 17. At that time, the posterior region of the prelimb bud has ZPA activity, but very little *Sonic hedgehog* mRNA. This lack of correlation may be explained as follows: Although the posterior prelimb bud tissue is not yet expressing *Sonic hedgehog* to a great extent, it is determined to do so. Thus, after this tissue is transplanted in an assay for ZPA, the *Sonic hedgehog* gene is switched

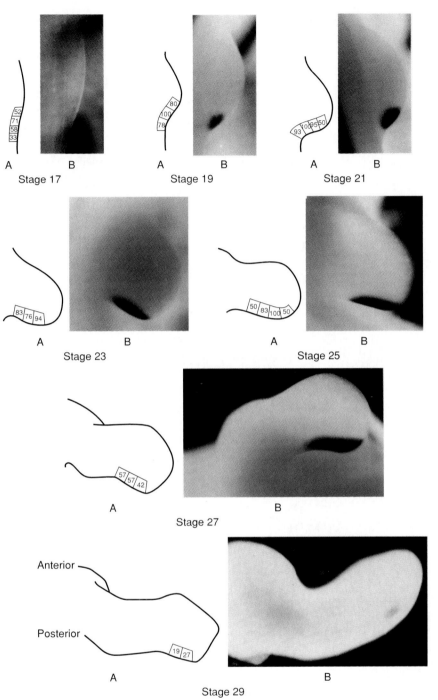

Stage 17

Stage 19

Stage 21

Stage 23

Stage 25

Stage 27

Anterior

Posterior

Stage 29

FIGURE 14.6 Coincidence of *Sonic hedgehog* expression and zone of polarizing activity (ZPA) in chick limb bud. Tabin and colleagues assayed chick embryos at various stages of development (stage 17–29, as indicated) for expression of *Sonic hedgehog* by making whole limb bud mounts and hybridizing a labeled *Sonic hedgehog* probe to them. The dark parts of the limb buds in panel B of each stage indicate label, and therefore extent of *Sonic hedgehog* expression. Other workers assayed for ZPA by taking sections of the limb buds and transplanting them to other embryos and measuring the ability to induce production of an extra set of digits. The degree of ZPA of each limb bud section is indicated by numbers in boxes in the limb buds diagrammed in panel A of each stage. There is a good space and time correlation between ZPA and *Sonic hedgehog* expression.

From Riddle, et al., "Sonic hedgehog Mediates the Polarizing Activity of the ZPA," *Cell*, 75:1401–1416, Dec. 31, 1993. © Cell Press.

on to a greater degree, and that is why this tissue ultimately shows strong ZPA activity. These and other data suggest that the *Sonic hedgehog* product is really the morphogen.

Morphogens are a subset of determinants and inducers that determine shape in a developing embryo. The *Sonic hedgehog* product is a good candidate for a vertebrate morphogen.

LEVELS OF GENE EXPRESSION CONTROL

Prokaryotes undergo radical changes in gene expression and cellular activity that resemble cell differentiation in higher organisms. A prime example, discussed in chapter 8, is sporulation in bacteria. In that case, changes in RNA polymerase σ factors allow recognition of new promoters, shifting transcription from vegetative genes to sporulation genes. Other examples of massive gene control are found in the replication cycles of bacteriophages, which effect complex, time-dependent shifts in gene expression. Again, these changes in gene expression take place at the transcription level.

Given that cell differentiation occurs via the turn-on of selected genes, differentiation is largely a question of how gene expression is controlled. Chapters 8, 9, and 10 demonstrated that gene expression involves three main levels: transcription, translation, and, especially in eukaryotes, an intervening, **posttranscriptional** phase during which transcripts are processed (by splicing, clipping, and modification) into mature form. In addition, many proteins experience posttranslational modifications of various sorts. This extra complexity gives a eukaryotic cell more opportunities to control gene expression. In principle, control could occur at the transcriptional, posttranscriptional, or translational level.

We will see examples of eukaryotic gene control at all three levels (figure 14.7), but the following general rules seem to apply: (1) Specialized genes, ones that are expressed in great amounts in certain cells, are controlled primarily at the transcriptional level. For example, the α- and β-globin genes are expressed thousands of times more actively in red blood cells than in any other vertebrate cell. This reflects the much greater rate of synthesis of globin mRNA in red blood cells than in other cells. (2) Housekeeping genes, which are essential for the life of every cell (RNA polymerase, for example), tend to be controlled at the post-transcriptional level—mainly through control of mRNA stability.

TRANSCRIPTIONAL CONTROL

What is the evidence that transcription is the primary level of control of specialized genes? The best evidence comes from direct examination of transcription products in differentiated cells. To begin with, we can look at the levels of mature mRNA in the cytoplasms of various differentiated cells. These RNA levels can be measured in several ways, but most are variations on the theme of **hybridization.** Recall from chapter 6 that a strand of RNA and a complementary strand of DNA can form a double-stranded helix, much like a DNA double helix. Therefore, if we obtain a piece of DNA complementary to an mRNA whose concentration we want to measure, we can hybridize it to radioactive mRNA and measure the amount of radioactivity in the RNA-DNA hybrid (figure 14.8).

One way to produce a piece of DNA complementary to our mRNA of interest is to use reverse transcriptase from a tumor virus. As we learned in chapter 12, this enzyme does just what its name implies—the reverse of transcription; it uses RNA as a template and makes a DNA copy. If we supply the enzyme with a little bit of the mRNA we want to detect, it makes a complementary DNA, or copy DNA **(cDNA).** This cDNA can

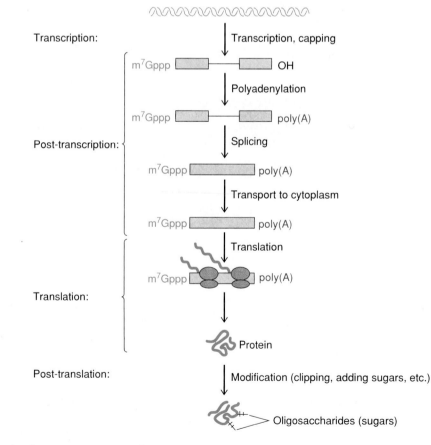

FIGURE 14.7   Levels of gene expression in eukaryotes.

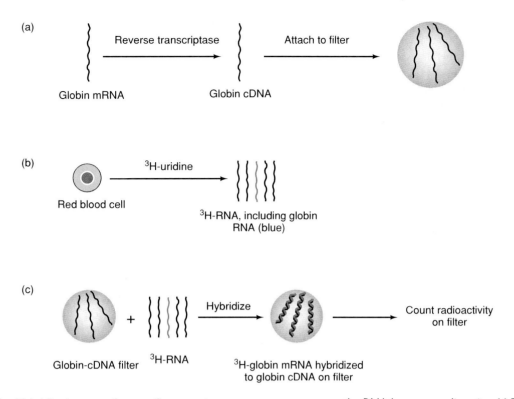

FIGURE 14.8 Hybridization assay for specific transcripts. (a) Reverse transcriptase makes a globin cDNA from purified globin mRNA. The cDNA is attached to a filter. (b) Red blood cells make RNA, including globin mRNA (blue), in the presence of a radioactive precursor, so the RNA becomes radioactive. (c) This radioactive RNA is hybridized to the globin cDNA on the filter; the more globin mRNA is present, the more radioactivity will stick to the filter.

then be cloned so it can be reproduced in very large quantity (see chapter 15 for a fuller explanation).

When we hybridize an α- or β-globin cDNA to mRNA from the cytoplasms of several different kinds of cells, including red blood cells, we find that the cDNA hybridizes successfully only to mRNA from red blood cells. In fact, we can show that red blood cells contain about 50,000 molecules of α- and β-globin mRNAs, compared to less than one molecule in other types of cells. Does this mean that these globin genes are not turned on in the other types of cells? It would be tempting to say so, but there are other possible explanations. For example, perhaps these globin genes are active in all types of cells, but the globin RNAs are degraded before they reach the cytoplasm in all cells except red blood cells.

In order to prove our thesis, we need to examine newly synthesized RNA before it has a chance to be degraded. That means looking in the nucleus, not in the cytoplasm. Furthermore, we must observe periods of RNA synthesis that are short enough that the RNA has not suffered significant degradation. In other words, we need to look at pulse-labeled RNA (chapter 7). In case after case where this has been done, we find evidence for transcriptional control. For example, red blood cells make vast quantities of hemoglobin (about ten million molecules of α- and β-globins per cell) but no ovalbumin (the major egg white protein). Accordingly, their nuclei contain many transcripts of these globin genes, but almost no transcripts

Table 14.1 Synthesis of Ovalbumin mRNA in Various Tissues[1]

Nuclei From	Ovalbumin mRNA (% of Total)
Oviduct + DES	0.238
Oviduct − DES	0.002
Spleen	<0.001
Liver	<0.001

[1]The limit of detectability was 0.001%.

Source: Data from Swaneck et al., *Proceedings of the National Academy of Science,* 76:1049, 1979.

of the ovalbumin gene. Conversely, oviduct cells make great amounts of ovalbumin, but no hemoglobin; as expected, their nuclei contain abundant ovalbumin transcripts, but almost no globin transcripts.

Table 14.1 shows the data that led to these conclusions. Bert O'Malley and his colleagues isolated nuclei from a variety of chicken cells, then allowed these nuclei to continue RNA synthesis for several minutes in the presence of a radioactive RNA precursor in order to label the RNA. Isolated nuclei usually cannot initiate RNA chains, so it is assumed that the RNA made in these nuclei represents RNA chains that were started in vivo. Next, these investigators hybridized the labeled RNA to filters containing a cloned ovalbumin gene. Only ovalbumin mRNA will hybridize, so the radioactivity on the filters is a direct measure of ovalbumin mRNA synthesis in the isolated nuclei, and therefore of ovalbumin gene activity in vivo. In table 14.1 we see that spleen and liver nuclei support no detectable ovalbumin mRNA synthesis. On the other hand, oviduct nuclei from chickens exposed to the hormone diethylstilbestrol (DES) make a considerable amount of ovalbumin mRNA. Nuclei from chickens that have been withdrawn from DES for three days drop back to a barely detectable level of ovalbumin mRNA synthesis. These results are certainly not compatible with a model in which all genes are transcribed equally in all cells. On the contrary, it appears that DES stimulates transcription of the ovalbumin gene in oviduct cells by at least a factor of 100. In other words, a great deal of transcriptional control occurs.

Most specialized genes of eukaryotes are controlled, at least primarily, at the transcriptional level. They are significantly active only in those differentiated cells where their protein products are made.

POSTTRANSCRIPTIONAL CONTROL

Theoretically, cells could control their gene expression at the posttranscriptional level in several different ways. For example, they could add caps or poly(A) tails to certain mRNAs, but not to others; or they could transport certain mRNAs, but not others, to the cytoplasm; or they could selectively degrade certain mRNAs or their precursors. In practice, cells do not seem to use all of these potential control mechanisms: (1) Cells do not withhold capping from certain types of messages. If they did, we would expect to find some uncapped messages, or at least message precursors, but virtually all mRNA precursors are capped at a very early time after transcription initiation. (2) With one exception, cells seem not to withhold polyadenylation from certain types of messages. The lone exceptions are the histone mRNAs, which do not seem to be polyadenylated in higher eukaryotes.

One way that posttranscriptional control does appear to occur is by selective degradation of mRNAs in the cytoplasm. The response of mammary gland tissue to the hormone prolactin provides a good example of this mechanism. When cultured mammary gland tissue is stimulated with prolactin, it responds by producing the milk protein casein. One would expect an increase in casein mRNA concentration to accompany this casein buildup, and it does. The number of casein mRNA molecules increases about 20-fold in twenty-four hours following the hormone treatment. But this does not mean the rate of casein mRNA synthesis has increased 20-fold. In fact, it increases only about 2- to 3-fold. The rest of the increase in casein mRNA level depends on an approximately 20-fold increase in stability of the casein mRNA.

Table 14.2   Effect of Prolactin on Half-life of Casein mRNA[1]

Species	RNA Half-life (H$_r$)	
	Without Prolactin	*With Prolactin*
rRNA	>90	>90
Poly(A)$^+$ RNA (short-lived)	3.3	12.8
Poly(A)$^+$ RNA (long-lived)	29	39
Casein mRNA	1.1	28.5

[1]Prolactin has relatively little effect on rRNA and total mRNA (poly[A]$^+$ RNA), but causes a great increase in the half-life of casein mRNA.

Source: Data from Guyette, et al., *Cell,* 17:1013, 1979.

A **pulse-chase** experiment was performed to measure the **half-life** of casein mRNA. The half-life is the time it takes for half the RNA molecules to be degraded. Casein mRNA was radioactively labeled in cultured mammary gland cells for a short time in the presence or absence of prolactin. In other words, cells were given a pulse of radioactive nucleotides, which they incorporated into mRNA. Then the cells were transferred to "cold" medium lacking radioactivity. This chases the radioactivity out of the RNA, as "hot" RNAs break down and are replaced by "cold" ones. After various chase times, the level of labeled casein mRNA remaining was measured by hybridizing it to a cloned casein gene. The faster the labeled casein mRNA disappeared, the shorter its half-life. The surprising conclusion, shown in table 14.2, was that the half-life of casein mRNA increased dramatically, from 1.1 hours to 28.5 hours, in the presence of prolactin. At the same time, the half-life of total polyadenylated mRNA increased only 1.3- to 4-fold in response to the hormone. It appears prolactin causes a selective stabilization of casein mRNA that is largely responsible for the enhanced expression of the casein gene.

Of the possible mechanisms of posttranscriptional control, there is good evidence for altering mRNA stability. Differential capping and polyadenylation appear not to be significant means of controlling gene expression.

TRANSLATIONAL CONTROL

In chapter 10 we looked at some examples of how translation can be controlled in eukaryotic cells. Translational control also occurs during development. **Maternal messages** in eggs provide good examples of how genes can be controlled at the translational level. Maternal messages are mRNAs made in great quantity during **oogenesis** (the development of an egg), then stored (complexed with protein) in the egg cytoplasm. These stored mRNAs are abundant enough that they can support the early development of the embryo even in the absence of transcription.

Maternal messages in sea urchin eggs are translated only poorly until fertilization, at which time their translation increases at least 50-fold. We know that this represents enhanced translation of maternal messages, because the increase occurs even when new RNA synthesis is blocked, or even in cells that lack nuclei.

It is also possible to control the location within an embryo where translation of a given mRNA occurs. This is frequently done by controlling the location of the corresponding maternal message. For example, the *Drosophila bicoid* mRNA is a maternal message located in the very anterior end of the oocyte. Considering this, it is no wonder that the *bicoid* protein is most concentrated at the anterior end of the developing embryo, although it diffuses more easily than the mRNA, so its localization is not as pronounced (figure 14.9).

> The expression of some genes, at least in early embryogenesis, can be regulated at the level of translation. The location of translation of a given mRNA within the embryo can also be governed by controlling the location of that mRNA.

POSTTRANSLATIONAL CONTROL

In theory, at least, there is one more level at which gene expression can be controlled. This is the **posttranslational** level, or what happens to a protein after it is released from the ribosome. Cells modify and decorate proteins in a variety of ways. Proteins can be clipped by proteases and thus converted from immature to mature form, or they can be coupled to sugars, phosphate groups, or other compounds.

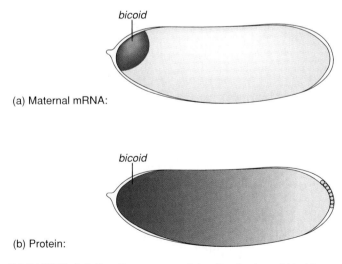

(a) Maternal mRNA:

(b) Protein:

FIGURE 14.9 Comparison of the distribution of *bicoid* mRNA and protein in the *Drosophila* embryo. (*a*) Distribution of *bicoid* mRNA. Note extreme anterior localization of the mRNA. (*b*) Distribution of *bicoid* protein. The protein diffuses more easily than the mRNA, so the gradient of its concentration extends farther toward the posterior end of the embryo.

One type of clipping actually occurs *during* synthesis of certain proteins. It involves the removal of a stretch of about twenty amino acids at the protein's **amino terminus**—the end with the free amino group, where protein synthesis began. This portion of the protein, called the **signal peptide,** anchors the nascent protein and its polysome to membranes called **endoplasmic reticulum** (**ER,** chapter 1), which simply means "network in the cell." As figure 14.10 demonstrates, the signal peptide also serves as a "needle" to pull the protein "thread" into the interior, or lumen, of the ER. Once inside, the signal peptide is removed. The shortened protein moves through the ER to the **Golgi apparatus,** where it is packaged for export. Therefore, this type of processing is characteristic of proteins that are destined for export from a cell.

The classic example of **proteolytic** (protein-cleaving) processing is provided by the hormone insulin, which appears first as a precursor called **preproinsulin.** After the signal peptide of preproinsulin is removed, we call it **proinsulin,** because there is still clipping to be done. During this latter process, an enzyme clips out a part of the proinsulin called the C peptide, leaving two pieces, the A peptide and B peptide, that compose mature insulin (figure 14.11). If these clipping steps did not occur, active insulin would not be produced, and expression of the insulin gene would be blocked as surely as if it were never transcribed. Expression of the insulin gene is apparently not controlled this way, but the expression of some other genes may be.

> Many proteins experience posttranslational alterations such as proteolytic clipping or addition of sugars. Because these alterations are frequently important for the formation of an active protein, they are steps at which gene expression might be controlled.

MECHANISMS OF TRANSCRIPTION CONTROL

We have seen that most actively expressed eukaryotic luxury genes are regulated primarily at the transcription level. Let us examine a few examples, bearing in mind that these genes may also be subject to other levels of control.

ACTIVATION OF GENES BY HORMONES

In chapter 9 we discussed the nuclear receptors, which bind to their respective ligands (hormones) and then activate sets of genes by binding to nearby enhancers known as hormone response elements (HREs). One interesting nuclear receptor in the context of development is the **retinoic acid receptor (RAR).** Let us now consider the interaction between the RAR and its DNA target, the **retinoic acid response element (RARE).**

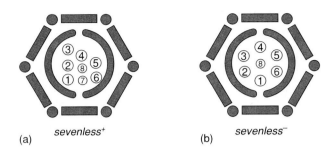

(a) *sevenless⁺* (b) *sevenless⁻*

FIGURE 14.40 A genetic experiment to demonstrate cell-autonomy of the sevenless gene. Tomlinson made a mosaic fly containing cells that were either *white⁻ sevenless⁻* or *white⁺ sevenless⁺*. Clones of the former in the eye were white, while clones of the latter were red, so they were easy to distinguish. (*a*) Arrangement of photoreceptors (numbered) in a *sevenless⁺* omatidium.
(*b*) Arrangement of photoreceptors in a *sevenless⁻* omatidium, which is clearly distinguishable from the arrangement in (*a*). (*c*) Results of the experiment. Cells in each omatidium are colored red to denote *sevenless⁺* (and *white⁺*), or white to denote *sevenless⁻* (and *white⁻*). Note that every R7 cell is *sevenless⁺*, even in the omatidium (arrow) where all the other cells are *sevenless⁻*.

(c)

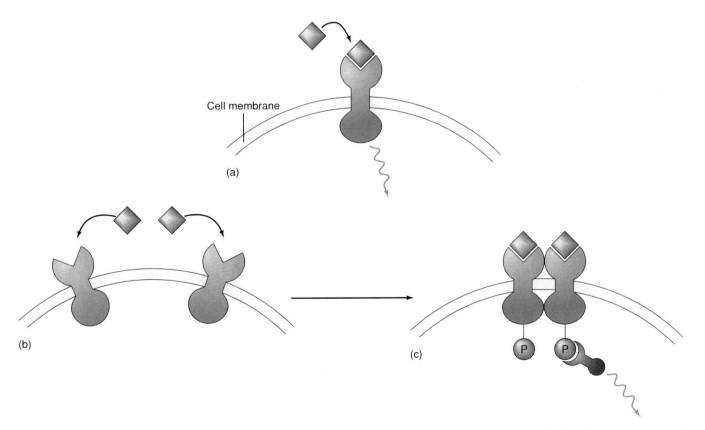

FIGURE 14.41 Transmembrane receptors. (*a*) General principle. The ligand (red) binds to the extracellular domain of the receptor (blue), causing a change in the intracellular domain, so it passes a signal (green) to the interior of the cell. (*b*) Receptor tyrosine kinase action. Binding the ligand (red) to the extracellular domains of two receptors causes the receptors to come together in the membrane, forming a dimer (*c*). This allows the intracellular domains, which have tyrosine kinase activity, to autophosphorylate, adding phosphate groups (orange) to specific tyrosine residues. An adaptor protein (purple) binds to a phosphotyrosine and, through a series of steps, passes the signal to the cell nucleus.

14.1   Two Methods in *Drosophila* Developmental Genetics

I. MITOTIC RECOMBINATION

We can create mosaic flies in several ways, but perhaps the best method uses mitotic recombination. Recombination does not ordinarily occur during mitosis, but we can force it to happen by irradiating embryos with X-rays or gamma-rays. As we learned in chapter 11, such ionizing radiation breaks chromosomes. The free ends can sometimes re-join properly, but sometimes the chromosome break induces a recombination event such that one

part of a maternal chromosome joins with the other part of a paternal chromosome, as illustrated in figure B14.1.

If the embryo we irradiate is heterozygous for the two markers *white* and *sevenless*, and the chromosome break occurs between the *sevenless* locus and the centromere, then the two daughter cells after a mitotic recombination will be either homozygous wild-type (*white⁺ sevenless⁺*) or homozygous mutant

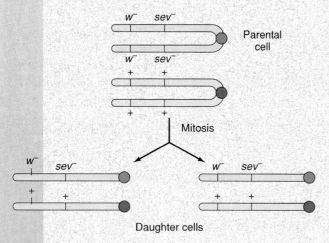

(a) Normal mitosis

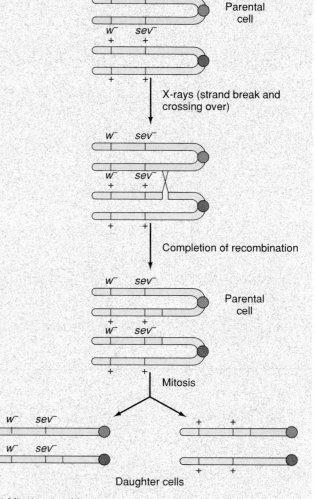

(b) Mitotic recombination

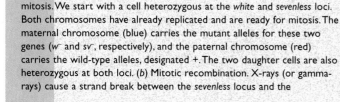

FIGURE B14.1   Mechanism of mitotic recombination. (*a*) Normal mitosis. We start with a cell heterozygous at the *white* and *sevenless* loci. Both chromosomes have already replicated and are ready for mitosis. The maternal chromosome (blue) carries the mutant alleles for these two genes (*w⁻* and *sv⁻*, respectively), and the paternal chromosome (red) carries the wild-type alleles, designated +. The two daughter cells are also heterozygous at both loci. (*b*) Mitotic recombination. X-rays (or gamma-rays) cause a strand break between the *sevenless* locus and the

centromere in the paternal chromosome. This induces a recombination event between the paternal and maternal chromosomes as denoted by the X. This yields two recombinant chromosomes, as shown. Mitosis yields two daughter cells, each bearing one recombinant chromosome. The daughter cell on the left is homozygous mutant at both loci, and the daughter cell on the right is homozygous wild-type. The homozygous mutant cell can now establish a mutant clone of cells.

14.1 (continued)

(**white⁻** *sevenless⁻*) (figure B14.1b). If such a homozygous mutant cell establishes a clone of cells in the embryonic eye, these cells will be white and will be easy to detect in a background of heterozygous (red) cells. Also, if the recombination event occurs early in the development of the embryo, the clone of mutant cells will be large, but if the event occurs late in embryonic life, the clone will be small.

2. P ELEMENT-MEDIATED TRANSFORMATION WITH HEAT-SHOCK PROMOTER-DRIVEN GENES

To express a cloned gene in the wrong place or at the wrong time, we can place the cloned wild-type gene under the control of the *hsp-70* heat-shock promoter and insert the whole construct into an embryo, using a transposable P element vector. (See chapter 12 for a discussion of P elements.) The heat-shock genes are normally inactive, but their promoters can be activated by heating to 37 °C. Thus, at various times during development, we can heat the embryo to activate the heat-shock promoter and express the gene we have placed next to it. This expression will occur throughout the embryo, mostly in inappropriate locations. The effects (often lethal) can shed light on the gene's normal function.

To test this approach, John Lis and colleagues placed the *E. coli lacZ* gene under the control of the *hsp-70* promoter in a P element vector that carried the selectable *rosy⁺* gene. They injected this construct into early embryos, along with a plasmid bearing a P element, then selected for *rosy⁺* flies. To verify that these flies carried the *lacZ* gene driven by the *hsp-70* promoter, they dissected the flies' abdomens and suspended the dissected flies in buffer containing X-gal to detect the *lacz* product, β-galactosidase. Figure B14.2 shows that the heat-shocked flies' abdomens turned dark blue, indicating abundant expression of the *lacZ* gene. Unshocked flies showed little color. Expression of the *lacZ* gene made no difference in development of heat-shocked flies, but ectopic expression of a *Drosophila* gene can make a huge difference.

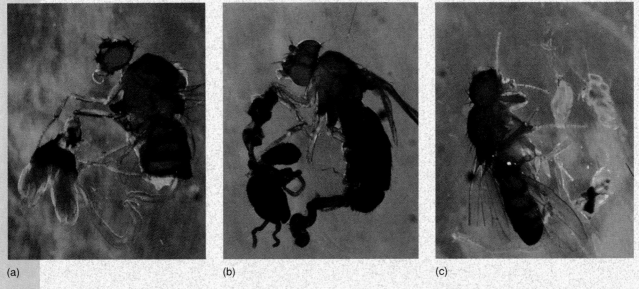

(a) (b) (c)

FIGURE B14.2 Test of *hsp-70*-driven gene expression in transformed flies. Lis and colleagues used a P element vector to introduce the *lacZ* gene, under control of the *hsp-70* promoter, into fly embryos, then allowed the embryos to develop into adults. They placed these flies in medium containing X-gal to detect *lacZ* activity by the appearance of blue color, then dissected the animals' abdomens to allow X-gal to contact the internal organs. (*a*) A transformant that was not heat-shocked. Very little blue color is apparent, indicating very little *lacZ* gene activity. (*b*) A heat-shocked transformant. Abundant blue color appeared, indicating high level *lacZ* activity. (*c*) A heat-shocked nontransformed fly, which is a negative control.

From J. T. Lis, J. A. Simon, and C. A. Sutton, "New Heat Shock Puffs and B-Galactosidase Activity Resulting from Transformation of *Drosophila* with an HSP-*lacZ* Hybrid Gene," *Cell,* 35:407, Dec. 1983. © Cell Press.

signal on, usually to the nucleus, where it causes a change in gene expression.

We can label the cloned gene and use it as a probe to reveal where *sevenless* is expressed. Figure 14.42 shows that R7 expresses *sevenless* at its border with R8. (R3 and R4 also express *sevenless* in this way, but that seems to be irrelevant, since *sevenless* mutations do not cause detectable effects in these two cells.) The location of *sevenless* expression at a site adjacent to R8 suggests that R8 produces the molecule (ligand) that activates the *sevenless* protein. In fact, a candidate for this ligand has been found. It is another cell-surface protein, made by R8, and is the product of the *bride of sevenless*

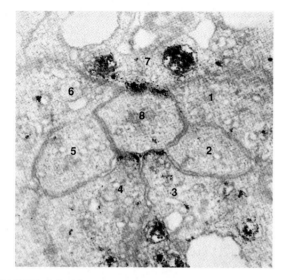

FIGURE 14.42 Immunostaining to locate *seventless* protein in photoreceptors. Tomlinson reacted a tagged anti-*seventless* protein antibody with a thin section through the photoreceptor cells of an omatidium. Then he examined the image for staining, which located the *seventless* protein. The staining occurred in R7, R3, and R4, at their junctions with R8. This suggested that R8 has the ligand that binds to *seventless*.

Courtesy of Dr. Andrew Tomlinson.

(*boss*) gene. The *boss* gene is not required for R8 development, but it is required for R7 development: R7 fails to develop in *boss* mutants. Furthermore, mosaic experiments show that R7 development is normal if *boss* is wild-type in R8 of the same omatidium. The state of *boss* in surrounding omatidia is irrelevant.

Similar lines of genetic analysis have led to the hypothetical signal transduction pathway illustrated in figure 14.43. Stimulation of the *seventless* protein by the *boss* protein leads to a cascade of protein phosphorylations that culminates in phosphorylation of two transcription regulators: the products of the *pointed* and *yan* genes. This phosphorylation activates the *pointed* protein, which is a transcription activator, and inactivates the *yan* gene, which is a transcription repressor. The net effect is activation of the set of genes controlled by these two proteins. One of these is *Rh4,* which, appropriately enough, encodes a photoreceptor pigment, rhodopsin Rh4, that confers a unique light-sensitivity to R7.

In chapter 17 we will see that this same pathway is important in controlling cell division in humans, and therefore is important in the cancer process. When one or more of the members of the human pathway malfunction, the affected cell can become cancerous. The pathway is named the *ras* pathway, for a vital gene, *ras.* Mutations in human *ras* can lead to cancer, so this gene is called an oncogene.

Some gene products activate genes in other cells via signal transduction pathways. A recurring theme in such pathways is protein phosphorylation. In the *seventless* pathway, the *boss* protein on the surface of photoreceptor R8 binds to the *seventless* protein on the surface of R7 and activates it. This initiates a series of protein phosphorylations that results in the activation of genes necessary for R7 development.

EPISTASIS AND DEVELOPMENT

This chapter has introduced several examples of developmental pathways. Some involve direct interaction between genes: A transcription factor encoded by one gene directly influences the activity of another. Some are less direct, and may involve signal transduction pathways. Geneticists use the phenomenon of **epistasis** to discover whether or not two or more genes participate in the same pathway.

In chapter 2 we learned that epistasis is the alteration or masking of the phenotype of one mutation by a mutation in another gene. Developmental geneticists exploit this phenomenon to show whether two genes operate in the same pathway or not. Sometimes, this sort of analysis can also tell which gene is downstream and which is upstream in the pathway, but we have to be careful in making that kind of determination.

For example, consider a hypothetical pathway (figure 14.44*a*) that leads to purple flower pigment. This pathway is catalyzed by two enzymes that are the products of two genes, *A* and *B*. The *A* gene product converts a colorless (or white) compound to a red pigment. The *B* gene product, in turn, converts the red pigment to a purple pigment. Thus, a *B⁻* mutation (figure 14.44*b*) would give a red flower phenotype because there would be no B enzyme to convert the red pigment to purple. But an *A⁻ B⁻* double mutant (figure 14.44*c*) would give a white flower phenotype because there would be no A enzyme to convert the colorless starting material to the red pigment. With no red pigment to convert to purple, the presence or absence of the B enzyme is irrelevant. Thus, the *B* mutation does not matter; it has been masked by the *A* mutation. In other words, *A* is epistatic to *B*. In this case, the epistatic gene is the one that is upstream in the pathway.

But in developmental genetics we frequently find that the epistatic gene is downstream. This can happen in several ways, and signaling pathways are a rich source of examples. For instance, consider the hypothetical pathway pictured in figure 14.45*a*. Three genes, (*A, B,* and *C*) code for the three proteins (A, B, and C) that participate in this pathway. A is a ligand (hormone or growth factor) that binds to a receptor (B) and induces it to accept the third protein (C) and phosphorylate it. This phosphorylated protein then passes the signal to other members of the pathway, which finally causes the cell harboring these genes to become a neuron. In panel (*b*), gene *A* is mutated such that it makes a defective A protein that cannot

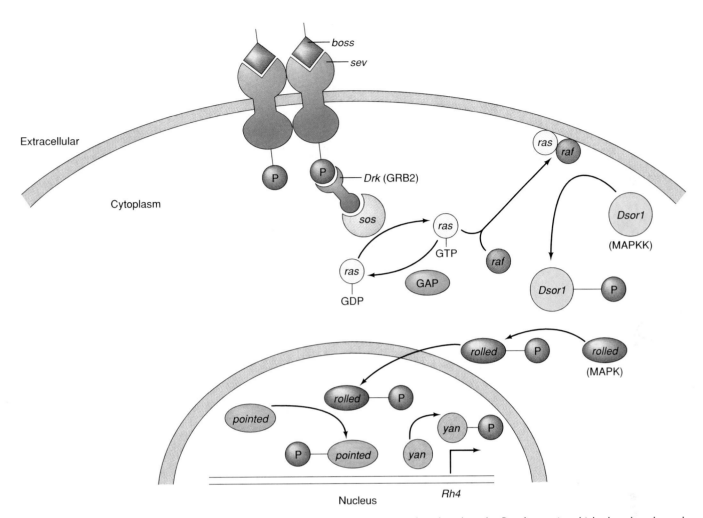

FIGURE 14.43 The *sevenless* signal transduction pathway. At top, the *sevenless* (*sev*) protein dimer is interacting with its ligand, the *boss* protein. (We refer to proteins in this figure by their genes' names.) This results in dimerization of the *sev* protein and autophosphorylation by the intracellular tyrosine kinase domain. The tyrosine phosphate is recognized by an adaptor protein (the *Drk* product), which in turn associates with the *Son of sevenless* (*Sos*) protein. This protein exchanges GTP for GDP on the *ras* protein, activating it, and allowing it to carry the *raf* protein to the inner surface of the cell membrane. There, the *raf* protein is activated, so it

can phosphorylate the *Dsor1* protein, which phosphorylates the *rolled* protein. This protein, also called MAPK, for "mitogen-activated-protein kinase," enters the nucleus and phosphorylates two transcription factors, the *pointed* and *yan* proteins. The former is a transcription activator, and phosphorylation causes it to stimulate transcription of target genes, such as *Rh4*, which encodes a visual pigment protein, rhodopsin 4. The *yan* protein is a transcription repressor, and its phosphorylation causes it to cease repressing transcription of *Rh4*. Thus, both effects of *rolled* activity stimulate *Rh4* transcription, which stimulates development of the R7 photoreceptor.

interact with B. This means that B is not activated to phosphorylate C. As a result, the cell becomes an epidermal cell (the default fate in this example) instead of a neuron. Panel (*c*) depicts a double mutant with defective *A,* as in (*b*), plus a constitutive *B* gene that codes for a B receptor protein that is locked in the "on" position, so it can phosphorylate C without first interacting with A. Since C is phosphorylated, the cell becomes a neuron in spite of the mutation in *A*. Thus, the mutation in *A* is no longer relevant; the mutation in *B* has masked it, and in this case, *B* is epistatic to *A*. Finally, panel (*d*) depicts a triple mutant in which *A* produces a nonfunctional protein as in (*c*), B is constitutive, also as in (*c*), but *C* is mutated

so it produces a product that cannot be phosphorylated by B. Now the signal cannot get through, even though B is independent of A, so the cell becomes an epidermal cell. Here, *C* is epistatic to *B*. In a pathway like this, it is usually the downstream gene—the one closest to the final outcome—that determines the outcome, or phenotype. In these situations the downstream gene is epistatic to the others.

This hypothetical pathway can be considered a genetic switch. If C is phosphorylated, the switch is thrown one way, and the cell differentiates into a neuron. If C fails to be phosphorylated, the switch is thrown the other way, and the cell assumes the fate of an epidermal cell. Consider an actual

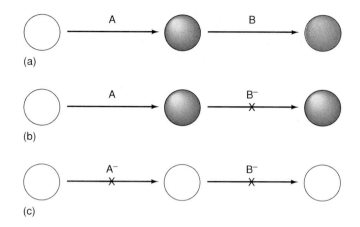

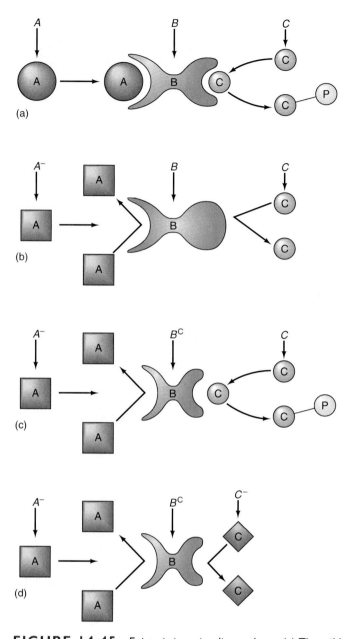

FIGURE 14.44 Epistasis in a biosynthetic pathway for a purple flower pigment. (*a*) The wild-type pathway. Genes *A* and *B* encode two proteins, A and B, that catalyze the two reactions in the pathway. Reaction 1, catalyzed by A, converts a white compound to a red one. Reaction 2, catalyzed by B, converts the red compound to a purple one, yielding purple flowers. (*b*) A *B* mutant. When protein B does not function, it cannot convert the red compound to purple, so the flowers will be red. (*c*) An *A B* double mutant. When neither *A* nor *B* function, the flowers will remain white. In fact, they will be white whether or not protein B is active because there will be no red intermediate for B to act on. Thus, *A* is epistatic to *B*.

example of a genetic switch that involves two genes, *gurken* and *K10*. These maternal effect genes are involved in establishing the dorsal-ventral embryonic axis, in the same way that *bicoid* is involved in establishing the anterior-posterior axis. We find that *gurken* (*grk*) is required to induce cells to become dorsalized (adopt a dorsal fate). In *grk⁻* embryos, these cells will adopt a ventral fate, which is the default condition. *K10* mutants have the opposite effect: They cause cells to become dorsalized. What happens in *grk⁻ K10* double mutants? The cells are ventralized, showing that *grk* is epistatic to *K10*; therefore, these two genes are likely to be members of the same developmental pathway. But which gene is upstream in the pathway? If this were like the flower pigment pathway, the epistatic gene, *grk*, would be upstream. But *grk* is actually downstream from *K10*. The *K10* product works *through* the *grk* product in this way: The *K10* protein causes the *grk* mRNA to localize to the anterior dorsal corner of the embryo. Thus, a *grk⁻* embryo, which lacks *grk* mRNA, has nothing for the *K10* product to act on. That is why *K10* mutations are irrelevant in *grk* mutants.

If one gene is epistatic to another, this suggests strongly that the two genes participate in the same developmental pathway. But to determine which gene is upstream of the other in the pathway, we need to have independent information about the nature of the pathway. In many developmental pathways, downstream genes are epistatic to upstream genes.

FIGURE 14.45 Epistasis in a signaling pathway. (*a*) The wild-type pathway. Genes *A*, *B*, and *C* encode three proteins that interact as shown. A (red) binds to a receptor (B, blue), which responds by binding C (green) long enough to phosphorylate it. (*b*) A single mutation in *A*. The defective gene produces a defective product (A, orange) that cannot bind B, so B cannot phosphorylate C. (*c*) An *A B* double mutant. The defective A still cannot bind to B, but the *B* gene in this mutant is constitutive: It produces a mutant B protein that can phosphorylate C even without binding to A. In this case, *B* is epistatic to *A* because the *B* mutation overrides the effect of the *A* mutation. (*d*) A triple mutant. The defective A still cannot bind to B, and the constitutive B still can phosphorylate a normal C, but the *C* gene is now defective, and its product (C, blue-green) cannot be phosphorylated by B. Here, the *C* mutation overrides the effect of the *B* mutation, so *C* is epistatic to *B*.

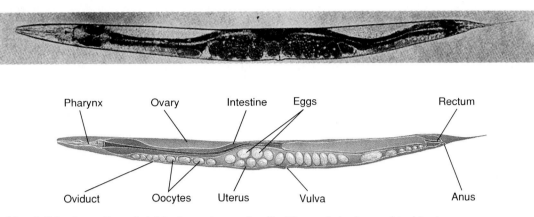

FIGURE 14.46 Self-fertilizing *Caenorhabditis elegans* hermaphrodite. The testis is obscured in this picture.

Photo: from J. E. Sulston and H. R. Horvitz, *Developmental Biology,* 56:111. © 1977 Academic Press.

DETERMINATION IN THE ROUNDWORM

The roundworm *Caenorhabditis elegans* is an especially attractive model for studying development for several reasons. First, although the adult worm has well-defined organ systems, it contains a total of only 959 somatic cells, a third of which are nerve cells. This is approximately the square root of the number of cells in *Drosophila,* which is in turn roughly the square root of the number of cells in a human (several trillion). Obviously, tracing the development of fewer than one thousand cells is incalculably simpler than doing the same with trillions of cells. The job is simplified further by the fact that *C. elegans* is transparent (figure 14.46), so the cells can be seen in the intact animal. Moreover, *C. elegans* is *hermaphroditic,* or self-fertilizing. (In Greek mythology, Hermaphroditos was the son of Hermes and Aphrodite. One day while he was bathing he and a nymph became fused into one body. Thus, a **hermaphrodite** combines the sexual organs of both male and female in one body.) This greatly simplifies genetic experiments because a mutant hermaphroditic worm can be "selfed," or mated with itself, like a pea plant. Nevertheless, mapping all those cells was an overwhelming task, requiring years of attention by several research groups, led by Sydney Brenner's group. The result is that we now know the fate of every cell at every stage in this animal's development. Conversely, we can trace the development of every one of the adult's 959 somatic cells back to the zygote that started it all. We also know about hundreds of *C. elegans* mutations, some of which affect the animal's development.

One of the interesting findings to come out of the *C. elegans* work so far is the identification of homeotic genes. Certain mutations in one of these, called **lin-12,** cause failure of vulva development in the hermaphroditic form of the worm. Robert Horvitz and his colleagues have traced the reason for this failure to the lack of a gonadal anchor cell that would normally induce the development of the vulva. By consulting the developmental fate map of *C. elegans,* we can see what went wrong. One cell

(Z1.ppp) would normally develop into the anchor cell, and another (Z1.aaa) would become a uterine precursor cell. But the *lin-12* mutants develop either two anchor cells or two uterine precursor cells, not one of each.

Mutations that cause hyperactivity of the *lin-12* gene give rise to two uterine precursor cells; mutations that cause depression of *lin-12* activity result in two anchor cells. This suggests that the *lin-12* gene is another example of a genetic switch. If it is on, a cell goes one way; if it is off, the cell follows the other path. We would therefore predict *lin-12* to be expressed strongly in uterine precursor cells but weakly in anchor cells at the time of determination. Furthermore, *lin-12* affects the development of other pairs of cells that follow quite different fates from those we have just considered. Thus, the switching activity of *lin-12* is not confined to just one branch point, but is more general.

This barely scratches the surface of the developmental genetics of *C. elegans.* Many of the same phenomena we have discussed in the section on *Drosophila* apply to *C. elegans,* but we do not have the space to describe them here.

The entire fate map of the roundworm *C. elegans* has been worked out so that the lineage of every somatic cell in the adult is known. This has revealed the existence of homeotic genes, such as *lin-12,* whose malfunction can change the fate of one cell to that of another.

GENE REARRANGEMENT DURING DEVELOPMENT

We have been assuming that an organism's genome remains constant during development as gene expression changes. This is generally true, but there are conspicuous exceptions.

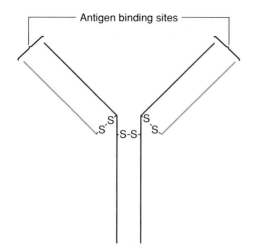

Antigen binding sites

FIGURE 14.47   The antibody is composed of two light chains (blue) bound through disulfide bridges to two heavy chains (red), which are themselves held together by a disulfide bridge. The antigen binding sites are at the ends of the protein chains, where the variable regions lie.

REARRANGEMENT OF IMMUNOGLOBULIN GENES

One example of rearrangement involves the mammalian genes that produce **antibodies,** or **immunoglobulins,** the proteins that help us fight infections. As mentioned in chapter 7, an antibody is composed of four polypeptides: two heavy chains and two light chains. Figure 14.47 illustrates an antibody schematically and shows the sites that combine with an invading antigen. These sites, called **variable regions,** vary from one antibody to the next and give these proteins their specificities; the rest of the molecule (the **constant region**) does not vary from one antibody to another within an antibody class, though there is some variation between the few classes of antibodies. Any given immune cell can make antibody with only one kind of specificity. Remarkably enough, humans have immune cells capable of producing antibody to react with virtually any foreign substance we would ever encounter. That means we can make literally millions, perhaps even billions, of different antibodies.

Does this imply that we have millions or billions of different antibody genes? That is an untenable hypothesis; it would place an impossible burden on our genomes to carry all the necessary genes. So how do we solve the antibody diversity problem? As unlikely as it may seem, a cell that is destined to make an antibody rearranges its genome to bring together separate parts of its antibody genes. The machinery that puts together the gene selects these parts at random from heterogeneous groups of parts, rather like ordering from a Chinese menu ("Choose one from column A and one from column B"). This arrangement greatly increases the variability of the genes. For instance, if there are 200 possibilities in "column A" and 5 in "column B," the total number of combinations of

A + B is 200 × 5, or 1,000. Thus, from 205 gene fragments, we can assemble 1,000 genes. And this is just for one of the antibody polypeptides. If a similar situation exists for the other, the total number of antibodies will be 1,000 × 1,000, or one million. This description, though correct in principle, is actually an oversimplification of the situation in the antibody genes; as we will see, they have somewhat more complex mechanisms for introducing diversity, which lead to an even greater number of possible antibody products.

Antibody genes have been studied most thoroughly in the mouse because of very handy mouse tumors called myelomas. A myeloma is a clone of malignant immune cells that overproduce one kind of antibody. Studies on the antibodies made by these myelomas have revealed two families of mouse antibody light chains called kappa (κ) and lambda (λ). Figure 14.48a illustrates the arrangement of the gene parts for a κ light chain. "Column A" of this Chinese menu contains 90–300 variable region parts (V); "Column B" contains 5 **joining region** parts (J). The J segments actually encode the last thirteen amino acids of the variable region, but they are located far away from the rest of the V region and close to a single constant region part. This is the situation in the animal's germ cells, before the antibody-producing cells differentiate and before rearrangement brings the two unlinked regions together. The rearrangement and expression events have been worked out by several groups, including those led by Susumo Tonegawa, Philip Leder, and Leroy Hood; these events are depicted in figure 14.48b.

First, a recombination event brings one of the V regions together with one of the J regions. In this case, V_3 and J_3 fuse together, but it could just as easily have been V_1 and J_4; the selection is random. After the two parts of the gene assemble, transcription occurs, starting at the beginning of V_3 and continuing until the end of C. Next, the splicing machinery joins the J_3 region of the transcript to C, removing the extra J regions and the intervening sequence between the J regions and C. It is important to remember that the rearrangement step takes place at the DNA level, but this splicing step occurs at the RNA level by mechanisms we studied in chapter 9. The messenger RNA thus assembled moves into the cytoplasm to be translated into an antibody light chain with a variable region (encoded in both V and J) and a constant region (encoded in C).

Why does transcription begin at the beginning of V_3 and not farther upstream? The answer seems to be that an enhancer in the intron between the J regions and the C region activates the promoter closest to it: the V_3 promoter in this case. This also provides a convenient way of activating the gene after it rearranges; only then is the enhancer close enough to turn on the promoter.

The rearrangement of the heavy chain gene is even more complex, because there is an extra set of gene parts in between the V's and J's. These gene fragments are called D, for "diversity," and they represent a third column on our Chinese menu. Figure 14.49 shows that the heavy chain is assembled from 100 to 200 V regions, approximately 20 D regions, and 4 J regions. On this

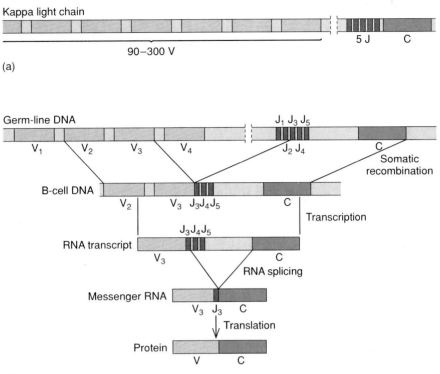

FIGURE 14.48 Rearrangement of an antibody light chain gene. (*a*) The kappa antibody light chain is encoded in 90–300 variable gene segments (V; light green), five joining segments (J; red), and one constant segment (C; blue). (*b*) During maturation of an antibody-producing cell, a DNA segment is deleted, bringing a V segment (V$_3$, in this case) together with a J segment (J$_3$ in this case).

The gene can now be transcribed to produce the mRNA precursor shown here, with extra J segments and intervening sequences. The material between J$_3$ and C is then spliced out, yielding the mature mRNA, which is translated to the antibody protein shown at the bottom. The J segment of the mRNA is translated into part of the variable region of the antibody.

basis alone, we could put together up to 200 × 20 × 4, or 16,000 different heavy chain genes. Furthermore, 16,000 heavy chains combined with 1,000 light chains yield more than ten million different antibodies.

But there are even more sources of diversity. The first derives from the fact that the mechanism joining V, D, and J segments is not precise. It can delete a few bases on either side of the joining site, or even add bases that are not found in the original segments. This leads to extra differences in antibodies' amino acid sequences.

Another source of antibody diversity is somatic mutation, or mutation in an organism's somatic (nonsex) cells. In this case, the mutations occur in antibody genes, probably at the time that a clone of antibody-producing cells proliferates to meet the challenge of an invader. This offers an even wider variety of antibodies and, therefore, a better response to the challenge. Assuming that the extra diversity introduced by imprecise joining of antibody gene segments and somatic mutation is a factor of about 100, this brings the total diversity to one hundred times ten million, or about one billion, surely enough different antibodies to match any attacker.

The immune systems of vertebrates can produce millions, if not billions, of different antibodies to react with virtually any foreign substance. These immune systems generate such enormous diversity by three basic mechanisms: (1) assembling genes for antibody light chains and heavy chains from two or three component parts, respectively, each part selected from heterogeneous pools of parts; (2) joining the gene parts by an imprecise mechanism that can delete bases or even add extra bases, thus changing the gene; and (3) causing a high rate of somatic mutations, probably during proliferation of a clone of immune cells, thus creating slightly different genes.

REARRANGEMENT OF TRYPANOSOME COAT GENES

Do the mechanisms for introducing diversity during development of the immune system apply to development in general? Probably not. In most cases, a gene in the germ cells is the same as the corresponding one in somatic cells in the adult

Heavy chain

100–200 V   20 D   4 J   C

FIGURE 14.49   Structure of an antibody heavy chain gene. The heavy chain is encoded in 100–200 variable segments (V; light green), 20 diversity segments (D; purple), four joining segments (J; red), and one constant segment (C; blue).

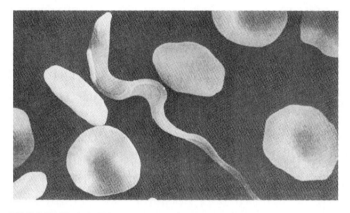

FIGURE 14.50   A trypanosome amid red blood cells. This scanning electron micrograph shows these cells enlarged 5,500 times. The undulations in the trypanosome are caused by its single flagellum, which extends down one side of the cell.

organism. However, changes during an animal's lifetime are not unique to antibody genes. Another well-known example is the rearrangement of surface protein genes during the life of the **trypanosomes,** the protozoa that cause African sleeping sickness (figure 14.50). These parasites have protein coats that are easily recognized as foreign, yet they manage to evade human and other mammalian immune defenses by changing the nature of their coat proteins just as our bodies are mounting an effective response to the previous coat. The change in coat comes from switching on new coat protein genes, but in an unorthodox way: The genome rearranges periodically by homologous recombination to bring new coat protein genes into **expression sites** at the ends of chromosomes where they can be transcribed efficiently. This is rather like taking a lightbulb out of a dead socket and screwing it into a live one.

> Trypanosomes change their coats to elude mammalian immune systems. They do this by rearranging their genomes to bring one coat protein gene after another into an active expression site.

GENOMIC IMPRINTING

In our discussion of transmission genetics, we have assumed that it made no difference which allele came from the female parent and which came from the male. Thus, in garden peas, crossing a homozygous green-seeded and a homozygous yellow-seeded parent will produce all yellow-seeded hybrid offspring, regardless of which parent contributed the gamete that specified yellow seeds. This is indeed generally true, but there are some important exceptions. These fall into three categories, the first two of which can be explained easily. First, as we saw in chapters 3 and 4, some genes are sex-linked. For example, the outcome of a cross between white-eyed and red-eyed fruit flies is different if the female parent is white-eyed than if the male parent is white-eyed. However, since the eye color gene resides on the X chromosome in fruit flies, and since females have two X chromosomes and males have one X and one Y chromosome, these results still conform to Mendelian principles. The second category of exceptions includes traits that are carried on genes that lie outside the nucleus—in the mitochondria or chloroplasts of a cell. For instance, the gene for a human neuromuscular disease called encephalomyopathy resides on a mitochondrial gene. Since mitochondria come exclusively from the mother, in the egg cytoplasm, this disease can be inherited only from the mother. As in the sex-linked case, this exception turns out to be readily explainable.

On the other hand, we now recognize a third class of exceptions that are not so easily explained. The genes in this case are not necessarily sex-linked, and they inhabit the nucleus, not a cytoplasmic organelle like the mitochondrion. Still, traits governed by these genes are inherited in a non-Mendelian fashion: that is, it matters whether the gene involved comes from the mother or from the father.

The first evidence for the existence of this phenomenon came from studies with mouse eggs just after fertilization, in which the maternal and paternal nuclei had not yet fused. At this stage, the maternal nucleus can be removed and replaced with a second paternal nucleus. Similarly, the paternal nucleus can be removed and replaced with a second maternal nucleus. In either case, the embryo will have chromosomes contributed by only one parent. In principle, that should not have made a big difference, because the parental mice were from an inbred strain in which all the individuals are genetically identical (except, of course, for the XY versus XX difference between males and females). In fact, however, it made a tremendous difference. All of these embryos died during development, most at a very early stage. Those that made it the longest before dying showed an interesting difference, depending on whether their genes came from the mother or the father. Those with genes derived only from the mother had few abnormalities in the embryo itself but had abnormal and stunted placentas and yolk sacs. Embryos with genes derived only from the father were small and poorly formed, but had relatively normal placentas and yolk sacs. How can we account for this difference if the genes contributed by the mother and father are identical? One explanation for this phenomenon is that the genes—that is, the base sequences of the genes—are identical, but they are somehow modified, or **imprinted,** differently in males and females.

Bruce Cattanach provided more evidence for **imprinting** with his studies on mice with fused chromosomes. For example, in some mice, chromosome 11 is fused, so it cannot separate during mitosis. This means that some gametes produced by such a mouse will have two copies of chromosome 11, while others will have none. These mice made it possible for Cattanach to produce offspring with both chromosomes 11 from the father, or both from the mother (using sperm with a double dose of chromosome 11 and eggs with no chromosome 11 in the former case, and sperm with no chromosome 11 and eggs with a double dose in the latter). Again, if inheritance were strictly Mendelian, it should not matter which parent contributes the chromosomes 11, as long as the embryo gets two copies. But it did matter. In cases where both chromosomes came from the mother, the pups were abnormally small; if both chromosomes came from the father, the pups were giants.

Furthermore, these experiments demonstrated that the imprint is erased at each generation. That is, a runty male mouse whose chromosomes 11 came from his mother generally would produce normal size offspring himself. The production of male gametes somehow erased the maternal imprint.

Genomic imprinting also occurs in humans, sometimes with tragic results. Inheritance of chromosome 15 only from the mother is associated with Prader-Willi syndrome, in which the patient is typically mentally retarded, short, and obese, with very small hands and feet. By contrast, inheritance of chromosome 15 only from the father is connected with Angelman syndrome, characterized by a large mouth and abnormally red cheeks, as well as by mental retardation with inappropriate laughter and jerky movements. Because of this appearance and behavior, such patients used to be called "happy puppets."

What is the mechanism of genomic imprinting? How is the DNA modified in a reversible way so that the imprint can be erased? The evidence points to DNA methylation as the most probable answer. To begin with, experiments show that genes derived from males and females are methylated differently, and that this methylation can be reversed. More tellingly, several geneticists have inserted foreign genes (transgenes) into developing mice and

observed differences in their behavior in the resulting male and female transgenic mice. In general, these investigations have found that transgenes inherited from the mother are methylated, while those inherited from the father are not. In one strain, a paternally inherited transgene is poorly methylated and is active, whereas the same transgene inherited from the mother is methylated and is inactive. Thus, methylation of a gene can indeed depend on which parent provides it, and the methylation state of a gene can indeed determine whether or not the gene will be active.

But what about erasure? Philip Leder and his colleagues have used transgenic mice to follow the methylated state of a transgene as it moves through gametogenesis (the production of sperm or eggs in male or female mice) and into the developing embryo. These experiments reveal that the methyl groups on the transgene are removed in the early stages of gametogenesis in both males and females. The developing egg then establishes the maternal methylation pattern before the oocyte is completely mature. In the male, some methylation occurs during sperm development, but this methylation pattern is further modified in the developing embryo. Thus, methylation has all the characteristics we expect of an imprinting mechanism: it occurs differently in male and female gametes; it is correlated with gene activity; and it is erased after each generation.

Do any benefits derive from genomic imprinting, or is it just another cause of genetic disorders? David Haig believes that certain imprinting patterns have evolved in response to environmental demands. His favorite example is the insulin-like growth factor (IGF) and its receptor in the mouse. The growth factor tends to make baby mice bigger, but it must interact with its receptor (the type-1 IGF receptor) in order to do so. To complicate the problem, mice have an alternate receptor (a type-2 IGF receptor) that binds IGF but does not pass the growth-promoting signal along. Thus, expression of the IGF gene in developing mice will produce bigger offspring, but expression of the type-2 IGF receptor will sop up the IGF and keep it away from the type-1 receptor, and therefore produce smaller offspring.

Haig points to an inherent biological conflict between the interests of a mother and those of a father of a baby mammal. If the benefits to the mother and father are viewed simply in terms of getting their own genes passed on to their offspring, then the father should favor large offspring, and the mother should favor small ones. The reason is that a large baby is more likely to survive and therefore perpetuate the father's genes. On the other hand, a large baby saps the mother's strength and leaves her fewer resources to provide to other offspring, which could be sired by a different father but still would perpetuate her genes. This is a coldhearted way of looking at parenthood, but it is the sort of thing that can influence evolution. Viewed in this context, it is very interesting that imprinting of male and female gametes in the mouse dictate that the IGF gene provided by a mother mouse is repressed, while that provided by the father is active. On the other hand, the type-2 IGF receptor gene from the father is turned off, whereas that from the mother is active. Both of these phenomena fit with the premise that a male should favor large offspring and a female should favor small ones. We seem to have a battle of the sexes going on at the molecular level, but neither side is winning, because the strategies of each side are canceled by those of the other!

WORKED-OUT PROBLEMS

PROBLEM 1

In *Drosophila melanogaster* a number of loci that affect segment pattern formation have maternal effects. The mutant alleles at five such loci produce the following deviations from normal segmental development:

    (a) *bcd* (bicoid), *exu* (exuperentia) and *swa* (swallow) all cause loss of head and thorax.

    (b) *bic* (bicaudal) produces an abdomen at each end of the embryo. The polarity of the 2 abdomens is reversed.

    (c) *osk* (oskar) prevents the abdomen from forming. Head and thorax develop in normal polarity.

The allele *bcd*⁺ codes for a regulatory protein that forms a gradient by diffusion from the anterior region where maternal *bcd*⁺ mRNA is localized by the *exu*⁺ and *swa*⁺ gene functions. In the following experiments, approximately 1% of cytoplasm and its cells were transferred from the anterior (*a*), middle (*m*) or posterior (*p*) portions of donor early embryos to (*a*) or (*p*) locations of recipient embryos. The results of these transfers are indicated below:

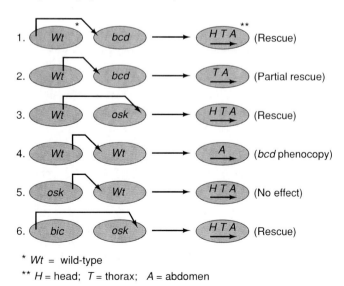

* *Wt* = wild-type

** *H* = head; *T* = thorax; *A* = abdomen

Explain what each of these experimental results tells us about the specification of the normal segmentation pattern in *Drosophila*.

SOLUTION

The transfer of cytoplasm and its cell from donors to recipients in the six transplants reveals the following:

1. *Wt* → *bcd*.   The maternal information from anterior segments that is defective in eggs from *bcd* females is a cytoplasmic component located at the anterior pole of the egg. It can function at quite a distance from the anterior region since it changes the abdomen in *bcd* flies to thorax in the middle of the embryo.

2. *Wt* → *bcd*.   Partial rescue when cytoplasm from the middle portion of the donor embryo is used instead of cytoplasm from the anterior region suggests that the maternal information is distributed in an anterior → posterior gradient.

3. *Wt* → *osk*.   Rescue of *oskar* with cytoplasm from the posterior pole of the *wild-type* embryo indicates a localized, maternally determined *osk* dependent activity necessary for determination (specification) of posterior segments.

4. *Wt* → *Wt*.   The *bcd* phenocopy indicates that a cytoplasmic factor at the posterior end of the embryo inhibits the activity specified by the *bcd* locus at the anterior pole.

5. *osk* → *Wt*.   From the results in 4 we cannot determine the cause of the inhibition. This experiment shows that the inhibition depends on the product of the *osk* gene or something regulated by it.

6. *bic* → *osk*.   On the basis of the results of the previous experiments the *bic* phenotype could be explained by ectopic expression of the *osk* activity at the anterior pole. This hypothesis is supported by the findings in this experiment which indicate that there is *osk*-rescue activity at the anterior pole of *bic* embryos.

PROBLEM 2

The figure at the top of page 413 shows the concentrations (in arbitrary units) of six hypothetical proteins in a hypothetical embryo. The anterior end is on the left (Ant), and the posterior end is on the right (Post). Use this information to answer the following questions.

    (a) If the *erdberen* (*erdb*) gene is active between A concentrations of 40 and 0.5 units, where (in terms of Egg Length) would the *erdb* protein appear?

    (b) If the *pickle* gene (*pkl*) is active between A concentrations of 9 and 1 and inhibited by B concentrations higher than 8, where (in terms of Egg Length) would the *pickle* protein appear?

SOLUTION

    (a) The A concentration never rises to 40, so the *erdb* gene will be active in the very anterior end of the embryo (100% Egg Length), where the A concentration is 15 units. The A concentration falls to 0.5 units at 80% Egg Length, so the

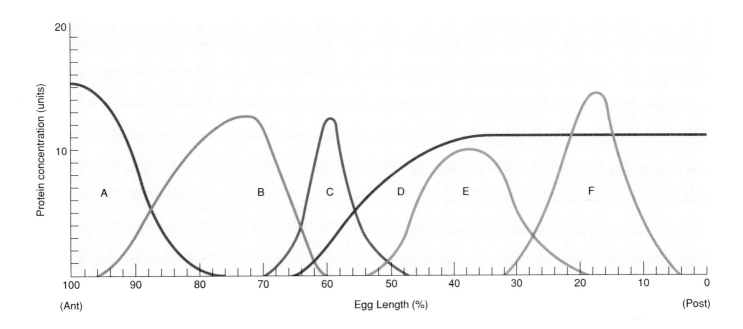

erdb gene will be active throughout the region between 100% and 80% Egg Length, and the *erdb* protein will form a band in this region, as shown below:

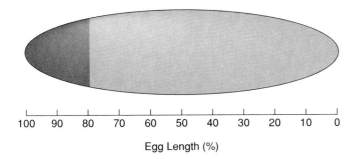

Egg Length (%)

(*b*) The A concentration falls to 9 units at 90% Egg Length, and to 1 at 82% Egg Length, so if A were the only factor involved, the *pkl* gene would be active in a band between 82% and 90% Egg Length. However, the concentration of B rises to 8 units at 84% Egg Length, and that is the threshold above which B inhibits *pkl* activity. Thus, *pkl* protein will be found in a stripe between 84% and 90% Egg Length, as shown below:

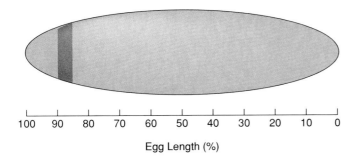

Egg Length (%)

SUMMARY

Multicellular organisms with specialized tissues develop from a single cell, or zygote, by a process called embryogenesis. During embryogenesis, cells differentiate to produce tissues with specialized functions. Differentiation seems to require little if any change in a cell's genetic makeup, since differentiated cells or nuclei from some plants and animals are still totipotent. Instead, the pattern of expression of a common set of genes varies from one kind of differentiated cell to another. Genes are activated during development by hormones and other embryonic factors that interact with enhancers near the genes they control. Even before they differentiate, cells become determined, or irreversibly committed to a given fate.

Cell determination in turn depends on the determination of one or more genes. Brown's model of gene determination invokes tissue-specific transcription factors that associate with the control region of a gene and keep it free of nucleosomes. This allows RNA polymerase and any other effector molecules, such as hormones and their receptors, to bind and cause transcription of the gene, completing the differentiation process. Methylation of cytosines in CG sequences in mammals is generally associated with inactive genes and represents another potential mechanism for controlling determination, at least in mammals.

Embryogenesis in *Drosophila* begins when the zygote nucleus divides several times in the absence of cell division. Cell membranes finally form just before the onset of gastrulation. Later, the embryo develops furrows that indicate segmentation; however, the important developmental unit within the embryo is not the segment but an element called a parasegment, which overlaps parts of two adjacent segments. The pattern of expression of genes during fruit fly development is a cascade in which one class of genes activates (or represses) the next. In general, the order of activation of genes in the cascade is maternal effect genes, gap genes, pair rule genes, segment polarity genes, and selector genes.

Maternal effect genes are active in the mother during oogenesis. Four maternal effect genes, *bicoid, nanos, torso,* and *Toll* establish the fundamental coordinates of the *Drosophila* embryo. The gap genes respond to gradients of maternal effect genes and establish zones of their own products, which are transcription factors that control the expression of other gap genes and the next class of genes in the cascade, the pair rule genes. The concentrations of these transcription factors in adjacent bell-shaped zones allow them to confine the expression of the pair rule genes to narrow stripes between the zones.

The pair rule genes allocate embryonic cells to the fourteen parasegments. Two of these, *even-skipped* and *fushi tarazu,* are expressed in stripes representing every second parasegment. The segment polarity genes, together with the *engrailed* gene, are expressed in every parasegment, and separate each parasegment into an anterior and a posterior compartment. The selector genes encode transcription factors that switch on batteries of genes, many of which also encode transcription factors. The result is the activation of specific sets of genes that cause cells to develop in specific ways—to give rise to a thoracic segment, including legs, for example. As more and more selector genes switch on, the tissue becomes more and more specifically determined. Some master control genes govern development of specific organs; *eyeless,* for example, directs eye development. Once activated, the selector genes remain active, locking cells into a developmental pathway and thereby determining them.

Some gene products activate genes in other cells via signal transduction pathways. A recurring theme in such pathways is protein phosphorylation. In the *sevenless* pathway, the *boss* protein on the surface of photoreceptor R8 binds to the *sevenless* protein on the surface of R7 and activates it. This initiates a series of protein phosphorylations that results in the activation of genes necessary for R7 development.

If one gene is epistatic to another, this suggests strongly that these two genes participate in the same developmental pathway. But to determine which gene is upstream of the other in the pathway, we need to have independent information about the nature of the pathway. In many developmental pathways, downstream genes can be epistatic to upstream genes.

The immune systems of vertebrates can produce millions, if not billions, of different antibodies to react with virtually any foreign substance. These immune systems generate this enormous diversity by three basic mechanisms: (1) assembling genes for antibody light chains and heavy chains from two or three component parts, respectively, each part selected from heterogeneous pools of parts; (2) joining the gene parts by an imprecise mechanism that can delete bases or even add extra bases, thus changing the gene; (3) causing a high rate of somatic mutations, probably during proliferation of a clone of immune cells, thus creating slightly different genes.

2. Is the chicken ovalbumin gene determined, differentiated, or both in the following instances? (*a*) When a young chick's oviduct cells are competent to produce ovalbumin in response to estrogen, but the chick does not yet make estrogen. (*b*) After the chicken matures and begins producing estrogen and ovalbumin.

3. Why does the existence of identical human quintuplets prove that human embryonic cells remain totipotent at least through three cell divisions?

4. Based on the results of experiments with the 5S rRNA gene, what would happen to activity and DNase hypersensitivity if the chicken β^A gene were mixed first with histones, then with extract from fourteen-day embryos?

5. The experiment cited in figure 14.13 examined the location of the control region of the rubisco gene that is light-sensitive. How does having a cloned gene greatly simplify this kind of experiment?

6. In the experiment cited in figure 14.13, why was it important to move the mutated rubisco gene to a new species, tobacco, rather than to reinsert it into a pea plant?

7. If the cloned *Xenopus* 5S rRNA gene is mixed with histones H2A, H2B, H3, and H4, and finally with transcription factors, would you predict the gene to be active or inactive? Why?

8. Why do you think it is important for an early embryo to have a good supply of 5S rRNA?

9. (*a*) Which of the following two genes are determined in the liver of an immature chick that has not begun producing estrogen—the ovalbumin gene and/or the vitellogenin gene? (*b*) Which of these genes are differentiated?

10. If you are interested in demonstrating whether control of expression of a given gene is exerted at the transcriptional level, why is it not sufficient to sample mRNA levels in the cytoplasm before and after that gene's expression is activated?

11. How would you measure the stability of an mRNA?

12. An mRNA for a certain protein has an AUG initiation codon followed by twenty other codons before the first glutamine codon, yet glutamine is the first amino acid in the mature protein product of this mRNA. How can this be?

13. How do mutations in homeotic genes and other homeobox genes show that these genes are important in development?

Answers appear at end of book.

PROBLEMS AND QUESTIONS

1. In what sense is the vertebrate hemoglobin gene a "luxury" to the red blood cell? Why is it not a luxury to the whole organism?

Gene Cloning and Manipulation

Learning Objectives

In this chapter you will learn:

1. How we can produce pure genes in virtually limitless quantities by cloning them in bacteria or higher organisms.

2. How we can obtain the products of cloned genes.

3. How we can use cloned genes as probes to investigate the structure and function of genes in their natural environment.

4. How we can obtain the sequence of bases in a cloned gene.

As we stood in that darkroom looking at those fluorescent orange molecules neatly arranged throughout the gel, we knew that biomedical science would go into warp speed.

Herb Boyer

American geneticist

Imagine that you are a geneticist in the year 1972. You want to investigate the function of eukaryotic genes at the molecular level. In particular, you are curious about the molecular structure and function of the human globin genes: What are the base sequences of these genes? What do their promoters look like? How does RNA polymerase interact with these genes? What changes occur in these genes to cause diseases like thalassemia, where one of the hemoglobin polypeptides is not produced at all?

These questions cannot be answered unless you can purify enough of your gene to study—probably about a milligram's worth. A milligram does not sound like much, but it is an overwhelming amount when you imagine purifying it from whole human DNA. Consider that the DNA involved in one globin gene is less than one part per million in the human genome. This means you will need kilograms of human DNA to start with, which is a very discouraging prospect. And even if you could collect that much material somehow, you would not know how to separate the one gene you are interested in from all the rest of the DNA. In short, you would be stuck.

Gene cloning neatly solves these problems. By linking eukaryotic genes to small bacterial or phage DNAs and inserting these recombinant molecules into bacterial hosts, we can produce large quantities of these genes in pure form. In this chapter we will see how to clone genes in bacteria and in eukaryotes. We will also learn the basic manipulations molecular geneticists use to discover their cloned genes' secrets.

GENE CLONING

The purpose of any cloning experiment is to produce a **clone,** a group of identical organisms. In chapter 14 we saw that some plants can be cloned simply by taking cuttings (Greek: *klon* = twig), and that others can be cloned by growing whole plants from single cells collected from one plant. Even vertebrates can be cloned. John Gurdon produced clones of identical frogs by transplanting nuclei from a single frog embryo to many enucleate eggs (chapter 14). Identical twins also constitute a clone.

The procedure we usually follow in a gene cloning experiment is to place a foreign gene into a bacterial cell, then to grow a clone of these modified bacteria, with each bacterial cell containing the foreign gene. Thus, as long as we ensure that the foreign gene can replicate, we can clone the gene by cloning its bacterial host. Stanley Cohen, Herbert Boyer, and their colleagues performed the first cloning experiment in 1973.

THE ROLE OF RESTRICTION ENDONUCLEASES

Cohen and Boyer's elegant plan depended on invaluable enzymes called **restriction endonucleases.** Stewart Linn and Werner Arber discovered restriction endonucleases in

E. coli in the late 1960s. These enzymes get their name from the fact that they prevent invasion by foreign DNA, such as viral DNA, by cutting it up. Thus, they "restrict" the host range of the virus. Furthermore, they cut at sites within the foreign DNA, rather than chewing it away from the ends, so we call them endonucleases (Greek: *endo* = within) rather than exonucleases (Greek: *exo* = outside). Linn and Arber hoped that their enzymes would cut DNA at specific sites, giving them finely honed molecular knives with which to slice DNA. Unfortunately, these particular enzymes did not fulfill that hope.

However, an enzyme from *Haemophilus influenzae* strain R_d, discovered by Hamilton Smith, did show specificity in cutting DNA. This enzyme is called *Hin*dII (pronounced Hin-dee-two). All restriction enzymes derive the first three letters of their names from the Latin name of the microorganism that produces them. The first letter is the first letter of the genus and the next two letters are the first two letters of the species (hence: *Haemophilus influenzae* yields "*Hin*"). In addition, the strain designation is sometimes included; in this case, the "d" from R_d is used. Finally, if the strain of microorganism produces just one restriction enzyme, the name ends with the Roman numeral I. If more than one enzyme is produced, they are numbered II, III, and so on.

*Hin*dII recognizes this sequence:

and cuts in the middle as shown by the arrows. Py stands for either of the pyrimidines (T or C), and Pu stands for either purine (A or G). Wherever this sequence occurs, and *only* when this sequence occurs, *Hin*dII will make a cut. Happily for molecular geneticists, *Hin*dII turned out to be only one of hundreds of restriction enzymes, each with its own specific recognition sequence. Table 15.1 lists the sources and recognition sequences for several popular restriction enzymes. Note that some of these enzymes recognize four-base sequences instead of the more common six-base sequences. As a result, they cut much more frequently. This is because a given sequence of four bases will occur about once in every $4^4 = 256$ bases, while a sequence of six bases will occur only about once in every $4^6 = 4,096$ bases. Some restriction enzymes, such as *Not*I, recognize eight-base sequences, so they cut much less frequently (once in $4^8 \cong 65,000$ bases); they are therefore called rare cutters. In fact, *Not*I cuts even less frequently than you would expect in mammalian DNA, because its recognition sequence includes two copies of the rare dinucleotide CG (see chapter 11). Notice also that the recognition sequences for *Sma*I and *Xma*I are identical, although the cutting sites are different. We call such enzymes that recognize identical sequences **isoschizomers** (Greek: *iso* = equal; *schizo* = split).

The main advantage of restriction enzymes is their ability to cut a DNA reproducibly in the same places. This property is

Table 15.1   Recognition Sequences and Cutting Sites of Selected Restriction Enzymes★

Enzyme	Recognition Sequence
AluI	A G↓C T
*Bam*HI	G↓G A T C C
*Bgl*II	A↓G A T C T
*Cla*I	A T↓C G A T
*Eco*RI	G↓A A T T C
*Hae*III	G G↓C C
*Hind*II	G T Py↓Pu A C
*Hind*III	A↓A G C T T
*Hpa*II	C↓C G G
*Kpn*I	G G T A C↓C
*Mbo*I	↓G A T C
*Pst*I	C T G C A↓G
*Pvu*I	C G A T↓C G
*Sal*I	G↓T C G A C
*Sma*I	C C C↓G G G
*Xma*I	C↓C C G G G
*Not*I	G C↓G G C C G C

★Only one DNA strand, written 5′ → 3′ left to right, is presented, but restriction enzymes actually cut double-stranded DNA. The cutting site for each enzyme is represented by an arrow.

the basis of many techniques used to analyze genes and their expression. We have seen examples of some of these in previous chapters; in this chapter we will see several more. But this is not the only advantage. Many restriction enzymes make staggered cuts in the two DNA strands (table 15.1), leaving single-stranded overhangs, or **sticky ends,** that can base-pair together briefly. This makes it easier to stitch two different DNA molecules together, as we will see. Note, for example, the complementarity between the ends created by *Eco*RI (pronounced Eeko R-1 or Echo R-1):

$$
\begin{array}{ccc}
\overset{\downarrow}{5'}\text{---GAATTC---}3' & \to & \text{---G}3' \\
3'\text{---CTTAAG---}5' & & \text{---CTTAA}5' \\
& & \underset{\uparrow}{}
\end{array}
\quad + \quad
\begin{array}{c}
5'\text{AATTC---} \\
3'\text{G---}
\end{array}
$$

Note also that *Eco*RI produces 4-base overhangs that protrude from the 5′-ends of the fragments. *Pst*I cuts at the 3′-ends of its recognition sequence, so it leaves 3′-overhangs. *Sma*I cuts in the middle of its sequence, so it produces blunt ends with no overhangs.

Restriction enzymes can make staggered cuts because the sequences they recognize are usually symmetrical. That is, they read the same forward and backward. Thus, *Eco*RI cuts between the G and the A in the top strand (on the left), and between the G and the A in the bottom strand (on the right), as shown by the vertical arrows.

These symmetrical sequences are also called **palindromes.** In ordinary language, palindromes are sentences that read the same forward and backward. Examples are Napoleon's lament: "Able was I ere I saw Elba," or a wart remedy: "Straw? No, too stupid a fad; I put soot on warts," or a statement of preference in

Italian food: "Go hang a salami! I'm a lasagna hog." DNA palindromes also read the same forward and backward, but you have to be careful to read in the same sense (5′ → 3′) in both directions. This means that you read the top strand left to right and the bottom strand right to left.

Cohen and Boyer took advantage of the sticky ends created by a restriction enzyme in their cloning experiment (figure 15.1). They cut two different DNAs with the same restriction enzyme, *Eco*RI. Both DNAs were plasmids. The first, called pSC101, carried a gene that conferred resistance to the antibiotic tetracycline; the other, RSF1010, conferred resistance to both streptomycin and sulfonamide. Both plasmids had just one *Eco*RI **restriction site,** or cutting site for *Eco*RI. Therefore, when *Eco*RI cut these circular DNAs, it converted them to linear molecules and left them with the same sticky ends. These sticky ends then base-paired with one another, at least briefly. Finally, DNA ligase completed the task of joining the two DNAs.

The result was a **recombinant DNA,** two previously separate pieces of DNA linked together. This new, recombinant plasmid was easy to detect. When introduced into bacterial cells, it conferred resistance to both tetracycline, a property of pSC101, and to streptomycin, a property of RSF1010. We have seen several examples of recombinant DNA in this text, but this one differs from most of the others in that it was not created naturally in a cell. Instead, geneticists put it together in a test tube.

Restriction endonucleases recognize specific sequences in DNA molecules and make cuts in both strands. This allows very specific cutting of DNAs. Also, because the cuts in the two strands are frequently staggered, restriction enzymes create sticky ends that help link together two DNAs to form a recombinant DNA in vitro.

VECTORS

Both plasmids in the Cohen and Boyer experiment are capable of replicating in *E. coli*. Thus, both can serve as carriers to allow replication of recombinant DNAs. All gene cloning experiments require such carriers, which we call **vectors,** but a typical experiment involves only one vector, plus a piece of foreign DNA that depends on the vector for its replication. Since the mid-1970s, many vectors have been developed; these fall into two classes: plasmids and phages.

Plasmids as Vectors

In the early years of the cloning era, Boyer and his colleagues developed a set of very popular vectors known as the pBR plasmid series. These old-fashioned vectors are scarcely used anymore, but we describe them here because they provide a simple illustration of cloning methods. One of these plasmids, **pBR322** (figure 15.2), contains genes that confer resistance to two antibiotics, ampicillin and tetracycline. Between these two genes

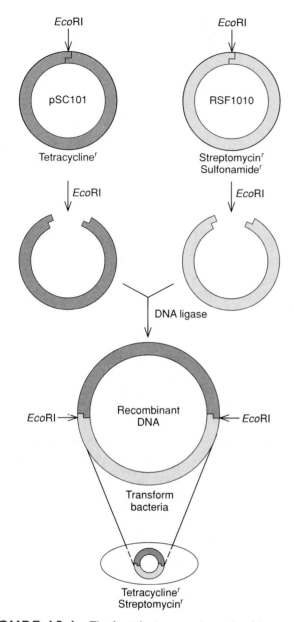

FIGURE 15.1 The first cloning experiment involving a recombinant DNA assembled in vitro. Boyer and Cohen cut two plasmids, pSC101 and RSF1010, with the same restriction endonuclease, EcoRI. This gave the two linear DNAs the same sticky ends, which were then linked in vitro using DNA ligase. The investigators reintroduced the recombinant DNA into E. coli cells by transformation, and screened for clones that were resistant to both tetracycline and streptomycin. These clones were therefore harboring the recombinant plasmid.

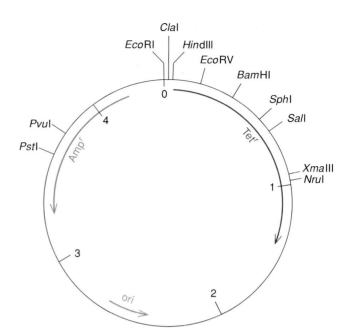

FIGURE 15.2 The plasmid pBR322, showing the locations of eleven unique restriction sites that can be used to insert foreign DNA. The locations of the two antibiotic resistance genes (Amp^r = ampicillin resistance; Tet^r = tetracycline resistance) and the origin of replication (ori) are also shown. Numbers refer to kilobase pairs (kb) from the EcoRI site.

lies the origin of replication. The plasmid was engineered to contain only one cutting site for each of several common restriction enzymes, including EcoRI, BamHI, PstI, HindIII, and SalI. This is convenient because it allows us to use each enzyme to create a site for inserting foreign DNA without losing any of the plasmid DNA.

For example, let us consider cloning a foreign DNA fragment into the PstI site of pBR322 (figure 15.3). First, we would cut the vector with PstI to generate the sticky ends characteristic of that enzyme. In this example, we also cut the foreign DNA with PstI, thus giving it PstI sticky ends. Next, we combine the cut vector with the foreign DNA and incubate them with DNA ligase. As the sticky ends on the vector and on the foreign DNA base-pair momentarily, DNA ligase seals the nicks, attaching the two DNAs together covalently. Once this is done, they cannot come apart again unless they are recut with PstI. Recall that a nick is a broken phosphodiester bond; a restriction enzyme like PstI breaks the bond, and DNA ligase re-forms it, as shown below:

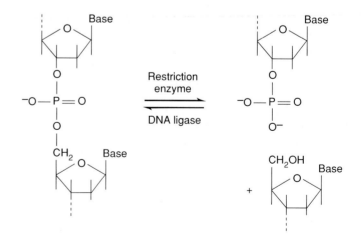

In the following step, we transform E. coli with our DNA mixture. We do this by incubating the bacterial cells in concentrated calcium salt solution to make their membranes leaky, then

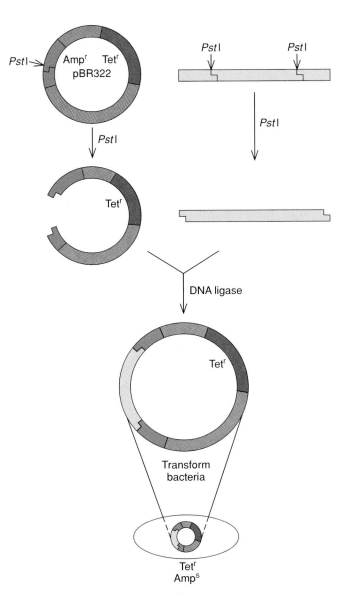

FIGURE 15.3 Cloning foreign DNA using the *Pst*I site of pBR322. We cut both the plasmid and the insert (yellow) with *Pst*I, then join them through these sticky ends with DNA ligase. Next, we transform bacteria with the recombinant DNA and screen for tetracycline-resistant, ampicillin-sensitive cells. The plasmid no longer confers ampicillin resistance because the foreign DNA interrupts that resistance gene (blue).

mixing these permeable cells with the DNA. It would be nice if all the cut DNA had been ligated to plasmids to form recombinant DNAs, but that never happens. Instead, we get a mixture of re-ligated plasmids and re-ligated inserts, along with the recombinants. How do we sort these out? This is where the antibiotic resistance genes of the vector come into play. First we grow cells in the presence of tetracycline, which selects for cells that have taken up either the vector or the vector with inserted DNA. Cells that received no DNA, or that received insert DNA only, will not be tetracycline-resistant and will fail to grow.

Next, we want to find the clones that have received recombinant DNAs. To do this, we screen for clones that are both tetracycline-resistant and ampicillin-sensitive. Figure 15.3 shows

Original (tetracycline)   Replica (ampicillin)

FIGURE 15.4 Screening for inserts in the pBR322 ampicillin resistance gene by replica plating. We perform replica plating as described in chapter 13, or transfer cells to corresponding spots on two petri dishes with toothpicks. The original plate contains tetracycline, so all colonies containing pBR322 will grow. The replica plate contains ampicillin, so colonies bearing pBR322 with inserts in the ampicillin resistance gene will not grow (these colonies are depicted by dotted circles). The corresponding colonies from the original plate can then be picked.

that the *Pst*I site, where we are inserting DNA in this experiment, lies within the ampicillin resistance gene. Therefore, inserting foreign DNA into the *Pst*I site inactivates this gene and leaves the host cell vulnerable to ampicillin.

How do we do the screening? One way is to transfer copies of the clones from the original tetracycline plate to an ampicillin plate. This can be accomplished with a felt transfer tool as illustrated in figure 15.4. We touch the tool lightly to the surface of the tetracycline plate to pick up cells from each clone, then touch the tool to a fresh ampicillin plate. This deposits cells from each original clone in the same relative positions as on the original plate. We look for colonies that do *not* grow on the ampicillin plate and then find the corresponding growing clone on the tetracycline plate. Using that most sophisticated of scientific tools—a sterile toothpick—we transfer cells from this positive clone to fresh medium for storage or immediate use. Notice that we did not call this procedure a **selection,** because it does not remove unwanted clones automatically. Instead, we had to examine each clone individually. We call this more laborious process a **screen** (chapter 13).

Nowadays we can choose from many plasmid cloning vectors besides the pBR plasmids. One useful, though somewhat outdated, class of plasmids is the **pUC** series (figure 15.5). These plasmids are based on pBR322, from which about 40% of the DNA, including the tetracycline resistance gene, has been deleted. Furthermore, the pUC vectors have their cloning sites clustered into one small area called a **multiple cloning site (MCS).** The pUC vectors contain the pBR322 ampicillin resistance gene to allow selection for bacteria that have received a copy of the vector. Moreover, to compensate for the loss of the other antibiotic resistance gene, they have genetic elements that permit a very convenient screen for clones with recombinant DNAs.

Figure 15.5*b* shows the multiple cloning sites of pUC18 and pUC19. Notice that they lie within a DNA sequence coding for the amino terminal portion of β-galactosidase (denoted *lacZ'*). The host bacteria used with the pUC vectors carry a gene fragment that encodes the carboxyl portion of β-galactosidase. By themselves, the β-galactosidase fragments made by these partial genes have no activity. But they can complement each other by intracistronic complementation, which we encountered in chapter 13. In other words, the two partial gene products can cooperate to form an active enzyme. Thus, when pUC18 by

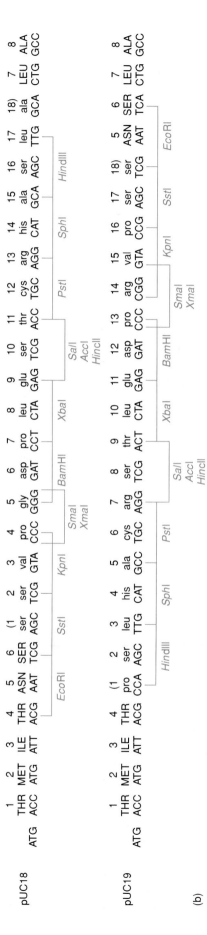

FIGURE 15.5 Architecture of a pUC plasmid. (a) The pUC plasmids retain the ampicillin resistance gene and the origin of replication of pBR322. In addition, they include a multiple cloning site (MCS) inserted into a gene encoding the amino part of β-galactosidase (lacZ'). (b) The MCSs of pUC18 and pUC19. In pUC18, the MCS, containing thirteen restriction sites, is inserted after the sixth codon of the lacZ' gene. In pUC19, it comes after the fourth codon and is reversed. In addition to containing the restriction sites, the MCSs preserve the reading frame of the lacZ' gene, so plasmids with no inserts will still support active expression of this gene. Therefore, clones harboring the vector alone will turn blue in the presence of the synthetic β-galactosidase substrate X-gal, while clones harboring the vector plus an insert will remain white.

itself transforms a bacterial cell carrying the partial β-galactosidase gene, active β-galactosidase is produced. If we plate these clones on medium containing a β-galactosidase indicator, colonies with the pUC plasmid will turn color. The indicator X-gal, for instance, is a synthetic, colorless galactoside; when β-galactosidase cleaves X-gal, it releases galactose plus an indigo dye that stains the bacterial colony blue.

On the other hand, if we have interrupted the plasmid's partial β-galactosidase gene by placing an insert into the multiple cloning site, the gene is usually inactivated. It can no longer make a product that complements the host cell's β-galactosidase fragment, so the X-gal remains colorless. Thus, it is a simple matter to pick the clones with inserts. They are the white ones; all the rest are blue. Notice that, in contrast to screening with pBR322, this is a one-step process. We look simultaneously for a clone that (1) grows on ampicillin and (2) is white in the presence of X-gal. The multiple cloning sites have been carefully constructed to preserve the reading frame of β-galactosidase. Thus, even though the gene is interrupted by eighteen codons (codon numbers in parentheses, with amino acid names in lowercase), a functional protein still results. Further interruption by large inserts, especially those that shift the reading frame, is usually enough to destroy the gene's function.

The multiple cloning site also allows us to cut it with two different restriction enzymes (say, *Eco*RI and *Bam*HI) and then to clone a piece of DNA with one *Eco*RI end and one *Bam*HI end. This is called **directional cloning,** or **forced cloning,** because we force the insert DNA into the vector in only one orientation. (The *Eco*RI and *Bam*HI ends of the insert have to match their counterparts in the vector.) Knowing the orientation of an insert has certain benefits, which we will explore later in this chapter. Forced cloning also has the advantage of preventing the vector from simply re-ligating by itself, since its two restriction sites are incompatible.

Notice that the multiple cloning site of pUC18 is just the reverse of that of pUC19. That is, the restriction sites are in opposite order. This means we can clone our fragment in either orientation simply by shifting from one pUC plasmid to the other. Even more convenient vectors than these are now available. We will discuss some of them later in this chapter.

The first two generations of plasmid cloning vectors were pBR322 and the pUC plasmids. The former has two antibiotic resistance genes and a variety of unique restriction sites into which we can introduce foreign DNA. Most of these sites interrupt one of the antibiotic resistance genes, making selection straightforward. Selection is even easier with the pUC plasmids. These have an ampicillin resistance gene and a multiple cloning site that interrupts a partial β-galactosidase gene. We select for ampicillin-resistant clones that do not make active β-galactosidase and therefore do not turn the indicator, X-gal, blue. The multiple cloning site also makes it convenient to carry out forced cloning into two different restriction sites.

Phages as Vectors

We have already seen how phages serve as natural vectors in transducing bacterial DNA from one cell to another (chapter 13). It was only natural, then, to engineer phages to do the same thing for *all* kinds of DNA.

λ Phage Vectors   Fred Blattner and his colleagues constructed the first phage vectors by modifying the well-known λ phage. They took out the region in the middle of the phage DNA, which codes for proteins needed for lysogeny, but retained the genes needed for lytic infection. These phages are no longer capable of lysogenic infection, but their missing genes can be replaced with foreign DNA. Blattner named these vectors **Charon phages** after Charon, the boatman on the river Styx in classical mythology. Just as Charon carried souls to the underworld, the Charon phages carry foreign DNA into bacterial cells. Charon the boatman is pronounced "Karen," but Charon the phage is usually pronounced "Sharon." A more general term for λ vectors such as Charon 4 (described below) is **replacement vectors,** since λ DNA is removed and replaced with foreign DNA.

One clear advantage of the λ phages over plasmid vectors is that they can accommodate much more foreign DNA. For example, Charon 4 can accept up to about 20 kb of DNA, a limit imposed by the capacity of the λ phage head. When would we need such high capacity? One common use for λ replacement vectors is in constructing **genomic libraries.** Suppose we wanted to clone the entire human genome. This would obviously require a great many clones, but the larger the insert in each clone, the fewer total clones would be needed. In fact, such genomic libraries have been constructed for humans and for a variety of other organisms, and λ replacement vectors have been popular vectors for this purpose.

Aside from their high capacity, some of the λ vectors have the advantage of a minimum size requirement for their inserts. Figure 15.6 illustrates the reason for this requirement: To get the Charon 4 vector ready to accept an insert, we cut it with *Eco*RI. This cuts at two sites near the middle of the phage DNA, yielding two "arms" and two "stuffer" fragments. Next, we purify the arms by ultracentrifugation and throw away the stuffers. The final step is to ligate the arms to our insert, which then takes the place of the discarded stuffers.

At first glance, it may appear that the two arms could simply ligate together without accepting an insert. Indeed, this may happen, but it will not produce a clone, because the two arms constitute too little DNA and will not be packaged into a phage head. The packaging is done in vitro; we simply mix the ligated arms plus inserts with all the components needed to put together a phage particle. Nowadays one can buy the purified λ arms, as well as the packaging extract in cloning kits. This extract has rather stringent requirements as to the size of DNA it will package. It must have at least 12 kb of DNA in addition to λ arms, but no more than 20 kb, or the phage head will overflow.

Since we can be sure that each clone in our genomic library has at least 12 kb of foreign DNA, we know we are not wasting space in our library with clones that contain insignificant

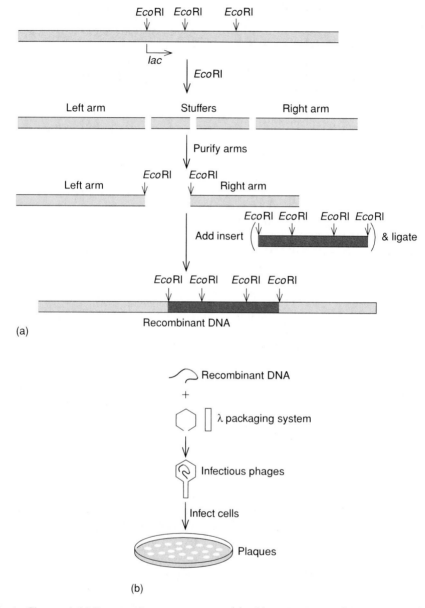

(a)

(b)

FIGURE 15.6   Cloning in Charon 4. (*a*) Forming the recombinant DNA. We cut the vector (yellow) with *Eco*RI to remove the stuffer fragments, and save the arms. Next, we ligate partially digested insert DNA (red) to the arms. (*b*) Packaging and cloning the recombinant DNA. We mix the recombinant DNA from (*a*) with an in vitro packaging extract that contains λ phage head and tail components and all other factors needed to package the recombinant DNA into functional phage particles. Finally, we plate these particles on *E. coli* and collect the plaques that form.

amounts of DNA. This is an important consideration since, even at 12–20 kb per clone, we need about half a million clones to be sure of having each human gene represented at least once. It would be much more difficult to make a human genomic library in pBR322 or a pUC vector, since bacteria selectively take up and reproduce small plasmids. Therefore, most of the clones would contain inserts of a few thousand, or even just a few hundred base pairs. Such a library would have to contain many millions of clones to be complete.

Since *Eco*RI produces fragments whose average size is about 4 kb and yet the vector will not accept any inserts smaller than 12 kb, we cannot cut our DNA completely with *Eco*RI, or most of the fragments will be too small to clone. Furthermore, *Eco*RI, or any other restriction enzyme, cuts in the middle of most eukaryotic genes one or more times, so a complete digest would contain only fragments of most genes. We can avoid these problems by performing an incomplete digestion with *Eco*RI; if the enzyme cuts only about every fourth or fifth site, the average length of the resulting fragments will be about 16–20 kb, just the size the vector will accept and big enough to include the entirety of most eukaryotic genes, introns and all.

Cosmids   Another vector designed especially for cloning large DNA fragments is called a **cosmid.** Cosmids behave both as plasmids and as phages. They contain the cos sites, or cohesive ends, of λ phage DNA, which allow the DNA to be packaged into λ phage heads (hence the "cos" part of the name "cosmid"). They also contain a plasmid origin of replication, so they can replicate as plasmids in bacteria (hence the "mid" part of the name).

Because almost the entire λ genome, except for the cos sites, has been removed from the cosmids, they have room for very large inserts (40–50 kb). Once these inserts are in place, the recombinant cosmids are packaged into phage particles. These particles cannot replicate as phages because they have almost no phage DNA, but they are infectious, so they carry their recombinant DNA into bacterial cells. Once inside, the DNA replicates as a plasmid, using its plasmid origin of replication.

A genomic library is very handy. Once it is established, we can search it for any gene we want. The only problem is that there is no on-line catalog for such a library, so we need some kind of probe to tell us which clone contains the gene of interest. An ideal probe would be a labeled nucleic acid whose sequence matches that of the gene we are trying to find. We would then carry out a **plaque hybridization** procedure in which the DNA from each of the thousands of λ phages from our library is hybridized to the labeled probe. The DNA that forms a labeled hybrid is the right one.

We have encountered hybridization before in several chapters, and we will discuss it again later in this chapter. Figure 15.7 shows how plaque hybridization works. We grow thousands of plaques on each of several petri dishes (only a few plaques are shown here for simplicity). Next, we touch a filter made of a DNA-binding material such as **nitrocellulose** to the surface of the petri dish. This transfers phage DNA from each plaque to the filter. The DNA is then denatured with alkali and hybridized to the labeled probe. When the probe encounters complementary DNA, which should be only the DNA from the clone of interest, it will hybridize, labeling that DNA spot. We then detect this labeled spot with X-ray film. The black spot on the film shows us where to look on the original petri dish for the plaque containing our gene. In practice, the original plate may be so crowded with plaques that it is impossible to pick out the right one, so we pick several plaques from that area, replate at a much lower phage density, and rehybridize to find the positive clone.

M13 Phage Vectors   Another phage frequently used as a cloning vector is the filamentous phage M13. Joachim Messing and his coworkers endowed the phage DNA with the same β-galactosidase gene fragment and multiple cloning sites found in the pUC family of vectors. In fact, the M13 vectors were engineered first; then the useful cloning sites were simply transferred to the pUC plasmids.

What is the advantage of the M13 vectors? The main factor is that the genome of this phage is a single-stranded DNA, so DNA fragments cloned into this vector can be recovered in single-stranded form. As we will see later in this chapter, single-

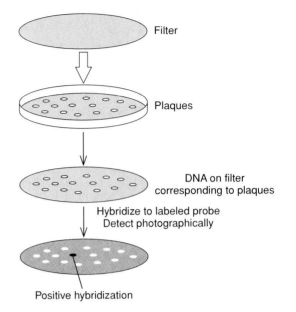

FIGURE 15.7   Selection of positive genomic clones by plaque hybridization. First, we touch a nitrocellulose or similar filter to the surface of the dish containing the Charon 4 plaques from figure 15.6. Phage DNA released naturally from each plaque will stick to the filter. Next, we denature the DNA with alkali and hybridize the filter to a labeled probe for the gene we are studying, then use X-ray film to reveal the position of the label. Cloned DNA from one plaque near the center of the filter has hybridized, as shown by the dark spot on the film.

stranded DNA can be an aid to site-directed mutagenesis, by which we can introduce specific, premeditated alterations into a gene. It also makes it easier to determine the sequence of a piece of DNA.

Figure 15.8 illustrates how we can clone a double-stranded piece of DNA into M13 and harvest a single-stranded DNA product. The DNA in the phage particle itself is single-stranded, but after infecting an *E. coli* cell, it is converted to a double-stranded replicative form (RF). This double-stranded replicative form of the phage DNA is what we use for cloning. After we cut it with one or two restriction enzymes at its multiple cloning site, we can insert foreign DNA with compatible ends. We then use this recombinant DNA to transform host cells, giving rise to progeny phages that bear single-stranded recombinant DNA. The phage DNA, along with phage particles, is secreted from the transformed cells and we can collect it from the growth medium.

Phagemids   Another class of vectors with single-stranded capability has now been developed. These are like the cosmids in that they have characteristics of both phages and plasmids; thus, they are called **phagemids.** One popular variety (figure 15.9) goes by the trade name pBluescript (pBS). It has a multiple cloning site inserted into the *lacZ′* gene, as in the pUC vectors, so clones with inserts can be distinguished by white versus blue staining with X-gal. This vector also has the origin of replication

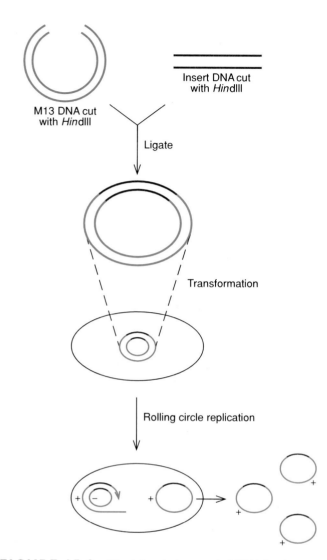

FIGURE 15.8   Obtaining single-stranded DNA by cloning in M13 phage. Foreign DNA (red), cut with *Hind*III, is inserted into the *Hind*III site of the double-stranded phage DNA. The resulting recombinant DNA is used to transform *E. coli* cells, whereupon the DNA replicates by a rolling circle mechanism, producing many single-stranded product DNAs. The product DNAs are called positive (+) strands, by convention. The template DNA is therefore the negative (–) strand.

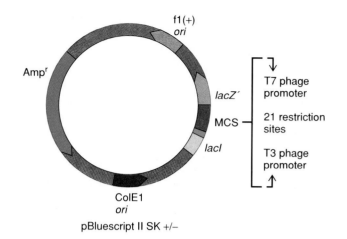

pBluescript II SK +/–

FIGURE 15.9   The pBluescript vector. This plasmid is based on pBR322, and has that vector's ampicillin resistance gene (green) and origin of replication (purple). In addition, it has the phage f1 origin of replication (orange), so it can be secreted in single-stranded form if the cell is infected by an f1 helper phage to provide the replication machinery. The multiple cloning site (MCS, red) contains 21 unique restriction sites situated between two phage promoters (T7 and T3). Thus, any DNA insert can be transcribed in vitro to yield an RNA copy of either strand, depending on which phage RNA polymerase we provide. The MCS is embedded in an *E. coli lacZ′* gene (blue), so the uncut plasmid will produce the β-galactosidase N-terminal fragment when we add an inducer such as IPTG to counteract the repressor made by the *lacI* gene (yellow). Thus, clones bearing the uncut vector will turn blue when we add the indicator X-gal. By contrast, clones bearing recombinant plasmids with inserts in the MCS will have an interrupted *lacZ′* gene, so no functional β-galactosidase will be made. Thus, these clones will remain white.

Two kinds of phages have been especially popular as cloning vectors. The first of these is λ, from which certain nonessential genes have been removed to make room for inserts. Some of these engineered phages can accommodate inserts up to 20 kb, which makes them useful for building genomic libraries, in which it is important to have large pieces of genomic DNA in each clone. Cosmids can accept even larger inserts—up to 50 kb—making them a favorite choice for genomic libraries. The second major class of phage vectors is composed of the M13 phages. These vectors have the convenience of a multiple cloning site and the further advantage of producing single-stranded recombinant DNA, which can be used for DNA sequencing and for site-directed mutagenesis. Plasmids called phagemids have also been engineered to produce single-stranded DNA in the presence of helper phages.

of the single-stranded phage f1, which is related to M13. This means that a cell harboring a recombinant phagemid, if infected by f1 helper phage, will produce and package single-stranded phagemid DNA. A final useful feature of this class of vectors is that the multiple cloning site is flanked by two different phage RNA polymerase promoters. For example, pBS has a T3 promoter on one side and a T7 promoter on the other. This allows us to isolate the double-stranded phagemid DNA and transcribe it in vitro with either of the phage polymerases to produce pure RNA transcripts corresponding to either strand. This is the way, for example, the self-splicing *Tetrahymena* rRNA precursors were produced for in vitro splicing experiments (chapter 9).

Eukaryotic Vectors

Several very useful vectors have been designed for cloning genes into eukaryotic cells. Later in this chapter, we will consider some vectors that are engineered to yield the protein products of genes in eukaryotes. In chapter 16, we will discuss vectors known as yeast artificial chromosomes (YACs) designed for cloning huge pieces of DNA (hundreds of thousands of base pairs) in yeast. We will also introduce vectors based on the Ti plasmid of *Agrobacterium tumefaciens* that can carry genes into plant cells.

IDENTIFYING A SPECIFIC CLONE WITH A SPECIFIC PROBE

We have already mentioned the need for a probe to identify the clone we want among the thousands we do not want. What sort of probe could we employ? Two different kinds are widely used: polynucleotides (or oligonucleotides) and antibodies. Both are molecules able to bind very specifically to other molecules. We will discuss polynucleotide probes here and antibody probes later in this chapter.

Polynucleotide Probes

To probe for the gene we want, we might use the homologous gene from another organism if someone has already managed to clone it. For example, if we were after the human insulin gene and another research group had already cloned the rat insulin gene, we could ask them for their clone to use as a probe. We would hope the two genes have enough similarity in sequence that the rat probe could hybridize to the human gene. This hope is usually fulfilled. However, we generally have to lower the **stringency** of the hybridization conditions so that the hybridization reaction can tolerate some mismatches in base sequence between the probe and the cloned gene.

Researchers use several means to control stringency. High temperature, high organic solvent concentration, and low salt concentration all tend to promote the separation of the two strands in a DNA double helix. We can therefore adjust these conditions until only perfectly matched DNA strands will form a duplex; this is high stringency. By relaxing these conditions (lowering the temperature, for example), we lower the stringency until DNA strands with a few mismatches can hybridize.

Without homologous DNA from another organism, what could we use? There is still a way out if we know at least part of the sequence of the protein product of the gene. We faced a problem just like this in our lab when we cloned the gene for a plant toxin known as ricin. Fortunately, the entire amino acid sequences of both polypeptides of ricin were known. That meant we could examine the amino acid sequence and, using the genetic code, deduce the nucleotide sequence that would code for these amino acids. Then we could construct the nucleotide sequence chemically and use this synthetic probe to find the ricin gene. This sounds easy, but there is a hitch. The genetic code is degenerate, so for most amino acids we would have to consider several different nucleotide sequences.

Fortunately, we were spared much inconvenience because one of the polypeptides of ricin includes this amino acid sequence: Trp-Met-Phe-Lys-Asn-Glu. The first two amino acids in this sequence have only one codon each, and the next three only two each. The sixth gives us two free bases because the degeneracy occurs only in the third base. Thus, we had to make only eight 17-base oligonucleotides (*17-mers*) to be sure of

getting the exact coding sequence for this string of amino acids. This degenerate sequence can be expressed as follows:

		U	G	U		
UGG	AUG	UUC	AAA	AAC	GA	
Trp	Met	Phe	Lys	Asn	Glu	

Using this mixture of eight 17-mers (UGGAUGUUCAAAAA-CGA, UGGAUGUUUAAAAACGA etc.), we quickly identified several ricin-specific clones.

> Specific clones can be identified using polynucleotide probes that bind to the gene itself. Knowing the amino acid sequence of a gene product, one can design an oligonucleotide (a short polynucleotide) that encodes part of this amino acid sequence. This can be one of the quickest and most accurate means of identifying a particular clone.

cDNA CLONING

Molecular geneticists use a variety of techniques to clone cDNAs; here we will consider a fairly simple, yet effective strategy, which is illustrated in figure 15.10. The central part of any cDNA cloning procedure is synthesis of the cDNA from an mRNA template using **reverse transcriptase.** This reverse transcriptase is like any other DNA-synthesizing enzyme in that it cannot initiate DNA synthesis without a primer. To get around this problem, we take advantage of the poly(A) tail at the end of most mRNAs and use oligo(dT) as the primer. The oligo(dT) is complementary to poly(A), so it binds to the poly(A) at the 3′-end of the mRNA and primes DNA synthesis, using the mRNA as the template.

After the mRNA has been copied, yielding a single-stranded DNA (the "first strand"), we remove the mRNA with alkali or **ribonuclease H** (RNase H). This enzyme degrades the RNA part of an RNA/DNA hybrid—just what we need to remove the RNA from our first strand cDNA. Next, we must make a second DNA strand, using the first as the template. Again, we need a primer, and this time we do not have a convenient poly(A) to which to hybridize the primer. Instead, we build an oligo(dC) tail at the 3′-end of the first strand, using the enzyme **terminal transferase** and one of the deoxyribonucleoside triphosphates. In this case, we use dCTP. The enzyme adds dCs, one at a time, to the 3′-end of the first strand. To this tail, we hybridize a short oligo(dG), which primes second strand synthesis. We can use reverse transcriptase again to make the second strand, but DNA polymerase also works. Actually, the most successful enzyme is a fragment of DNA polymerase called the **Klenow fragment.** This piece of enzyme is generated by cleaving *E. coli* DNA polymerase I with a proteolytic enzyme. The Klenow fragment contains the DNA polymerase activity and the 3′ → 5′ exonuclease activity, but it lacks the 5′ → 3′ exonuclease activity normally associated with DNA polymerase I. The latter activity is undesirable because it degrades DNA from the 5′-end, which is damage that DNA polymerase cannot repair.

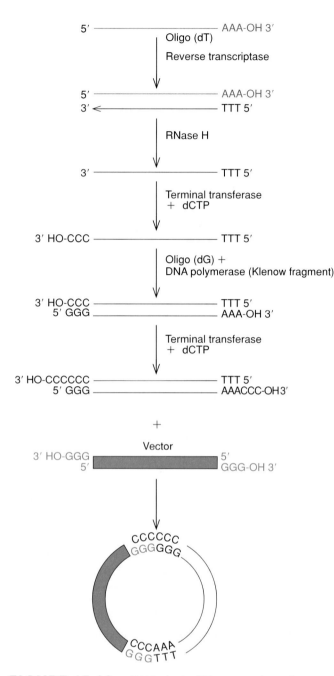

FIGURE 15.10   cDNA cloning. We start with a eukaryotic messenger RNA (red) having poly(A) at its 3′-end. Oligo(dT) hybridizes to the poly(A) and primes reverse transcription, forming the first cDNA strand. We remove the mRNA template, using RNase H, then add an oligo(dC) tail to the 3′-end of the cDNA, using terminal transferase. Oligo(dG) hybridizes to the oligo(dC) and primes second-strand cDNA synthesis by the Klenow fragment of DNA polymerase I. In order to give the double-stranded cDNA sticky ends, we add oligo(dC) with terminal transferase, then anneal these ends to complementary oligo(dG) ends (blue) of a suitable vector (blue). The recombinant DNA can then be used to transform bacterial cells.

Once we have a double-stranded cDNA, we must ligate it to a vector. This was easy with our pieces of genomic DNA, since they had sticky ends, but the cDNA has no sticky ends. This problem is easily solved. We simply tack sticky ends (oligo[dC]) onto the cDNA, again using terminal transferase and dCTP. In the same way, we attach oligo(dG) ends to our vector and allow the oligo(dC)s to anneal to the oligo(dG)s. This brings the vector and cDNA together in a recombinant DNA that can be used directly for transformation. The base pairing between the oligonucleotide tails is strong enough that no ligation is required before transformation. The DNA ligase inside the transformed cells finally performs this task.

What kind of vector should we use? Several choices are available, depending on the way we wish to detect positive clones (those that bear the cDNA we want). We can use a plasmid or phagemid vector such as pUC or pBS; if we do, we usually identify positive clones by **colony hybridization** with a radioactive DNA probe. This procedure is analogous to the plaque hybridization described previously. Or we can use a λ phage, such as λgt11, as a vector. This vector places the cloned cDNA under the control of a *lacZ* promoter, so that transcription and translation of the cloned gene can occur. We can then use an antibody to screen directly for the protein product of the correct gene. We will describe this procedure in more detail later in this chapter.

A cDNA can be synthesized one strand at a time, using mRNA as template for the first strand and this first strand as template for the second. We can endow the double-stranded cDNA with oligonucleotide tails that base-pair with complementary tails on a cloning vector. We can then use this recombinant DNA to transform bacteria. We can detect positive clones by colony hybridization with radioactive DNA probes, or with antibodies if an expression vector such as λgt11 is used.

METHODS OF EXPRESSING CLONED GENES

Why would we want to clone a gene? An obvious reason, suggested at the beginning of this chapter, is that cloning allows us to produce large quantities of pure eukaryotic (or prokaryotic) genes so we can study them in detail. Several specific studies using cloned genes have been described in previous chapters. For example: Cloned genes were manipulated to discover the important control regions of the pea rubisco gene and the frog 5S rRNA genes (chapter 14). The intron-containing region of the *Tetrahymena* rRNA precursor gene was cloned and transcribed to give substrate for the self-splicing reaction (chapter 9). Thus, the gene itself can be a valuable product of gene cloning. Another goal of gene cloning is to make a large quantity of the gene's product, either for investigative purposes or for profit, as we will see in chapter 16.

EXPRESSION VECTORS

The vectors we have examined so far are meant to be used primarily in the first stage of cloning—when we first put a foreign DNA into a bacterium and get it to replicate. By and large, they work well for that purpose, growing readily in *E. coli* and producing high yields of recombinant DNA. Some of them even work as **expression vectors** that can yield the protein products of the cloned genes. For example, the pUC and pBS vectors place inserted DNA under the control of the *lac* promoter, which lies upstream from the multiple cloning site. If an inserted DNA happens to be in the same reading frame as the *lacZ'* gene it interrupts, a **fusion protein** will result. It will have a partial β-galactosidase protein sequence at its amino end and another protein sequence, encoded in the inserted DNA, at its carboxyl end (figure 15.11).

However, if we are interested in high expression of our cloned gene, specialized expression vectors usually work better. These typically have two elements that are required for active gene expression: a strong promoter and a ribosome binding site that includes a Shine-Dalgarno sequence near an initiating ATG codon.

Expression Vectors with Strong Promoters

The main function of an expression vector is to yield the product of a gene—usually, the more product the better. Therefore, expression vectors are ordinarily equipped with very strong promoters; the rationale is that the more mRNA is produced, the more protein product will be made.

One such strong promoter is the *trp* (tryptophan operon) promoter. It forms the basis for several expression vectors, including *ptrpL1*. Figure 15.12 shows how this vector works. It has the *trp* promoter/operator region, followed by a Shine-Dalgarno sequence, and can be used directly as an expression vector by inserting a foreign gene into the *Cla*I site. Alternatively, the *trp* control region can be made "portable" by cutting it out with *Cla*I and *Hind*III and inserting it in front of a gene to be expressed in another vector.

Inducible Expression Vectors

It is usually advantageous to keep a cloned gene turned off until we are ready to express it. One reason is that eukaryotic proteins produced in large quantities in bacteria can be toxic. Even if these proteins are not actually toxic, they can build up to such great levels that they interfere with bacterial growth. In either case, if the cloned gene were allowed to remain turned on constantly, the bacteria bearing the gene would never grow to a great enough concentration to produce meaningful quantities of protein product. The solution is to keep the cloned gene turned off by placing it behind an inducible promoter that can be turned off.

The *lac* promoter is inducible to a certain extent, presumably remaining off until stimulated by the inducer allolactose or by its synthetic analog IPTG. However, the repression

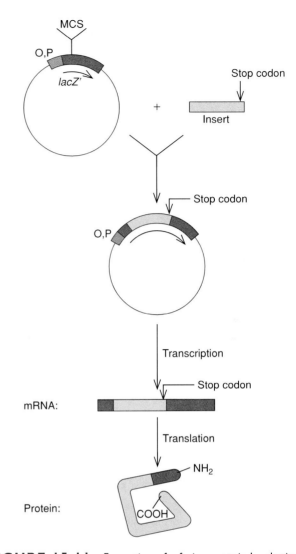

FIGURE 15.11 Formation of a fusion protein by cloning in a pUC plasmid. We insert foreign DNA (yellow) into the multiple cloning site (MCS); transcription from the *lac* promoter (purple) gives a hybrid mRNA beginning with a few *lacZ* codons, changing to insert sequence, then back to *lacZ* (red). This mRNA will be translated to a fusion protein containing a few β-galactosidase amino acids at the beginning (amino end), followed by the insert amino acids for the remainder of the protein. Since the insert contains a translation stop codon, the remaining *lacZ* codons will not be translated.

wrought by the *lac* repressor is incomplete, and some expression of the cloned gene will be observed even in the absence of inducer. One way around this problem is to express our gene in a plasmid or phagemid that carries its own *lacI* gene, as pBS does (figure 15.9). The excess repressor produced by such a vector keeps our cloned gene turned off until we are ready to induce it with IPTG.

Another strategy is to use a tightly controlled promoter such as the λ phage promoter P_L. Expression vectors with this promoter/operator system are cloned into host cells

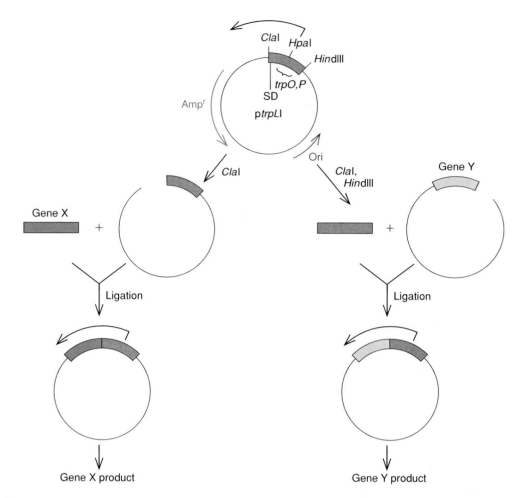

FIGURE 15.12   Two uses of the *ptrpL*1 expression vector. The vector contains a *Cla*I cloning site, preceded by a Shine-Dalgarno ribosome binding site (SD) and the *trp* operator/promoter region (*trpO,P*). Transcription occurs in a counterclockwise direction as shown by the arrow (top). The vector can be used as a traditional expression vector (left) simply by inserting a foreign coding region (X, green) into the unique *Cla*I site. Alternatively (right), the *trp* control region (purple) can be cut out with *Cla*I and *Hind*III and inserted into another plasmid bearing the coding region (Y, yellow) to be expressed.

bearing a temperature-sensitive λ repressor gene (*cI*857). As long as we keep the temperature of these cells relatively low (32° C), the repressor functions, and no expression takes place. However, when we raise the temperature to the non-permissive level (42° C), the temperature-sensitive repressor can no longer function and the cloned gene is induced. Figure 15.13 illustrates this mechanism for the expression vector pKC30.

Expression vectors are designed to yield the protein product of a cloned gene, usually in the greatest amount possible. To optimize expression, these vectors provide strong bacterial promoters and bacterial ribosome binding sites that would be missing on cloned eukaryotic genes. Most cloning vectors are inducible, to avoid premature overproduction of a foreign product that could poison the bacterial host cells.

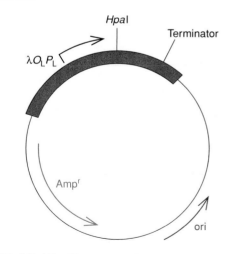

FIGURE 15.13   The inducible expression vector pKC30. A gene to be expressed is inserted into the unique *Hpa*I site, downstream from the λ $O_L P_L$ operator/promoter region. The host cell is a λ lysogen bearing a temperature-sensitive λ repressor gene (*cI*857). To induce expression of the cloned gene, we raise the temperature from 32° to 42° C, which inactivates the temperature-sensitive λ repressor, removing it from O_L and allowing transcription to occur.

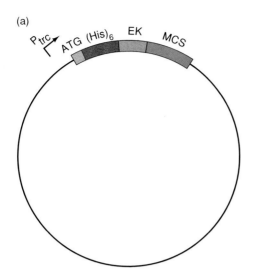

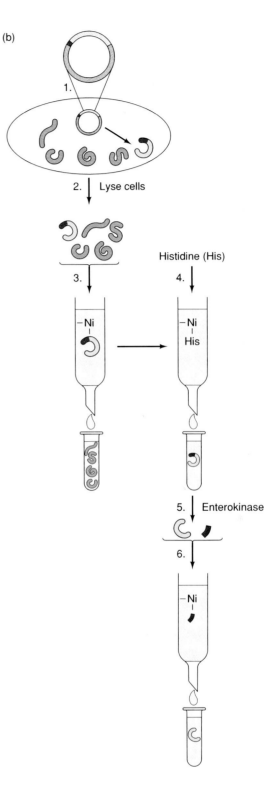

FIGURE 15.14 Using an oligo-histidine expression vector. (a) Map of a generic oligo-histidine vector. Just after the ATG initiation codon (green) lies a coding region (red) encoding six histidines in a row [(His)$_6$]. This is followed by a region (orange) encoding a recognition site for the proteolytic enzyme enterokinase (EK). Finally, the vector has a multiple cloning site (MCS, blue). Usually, the vector comes in three forms with the MCS sites in each of the three reading frames. Thus, we can select the vector that puts our gene in the right reading frame relative to the oligo-histidine. (b) Using the vector. 1. We insert our gene of interest (yellow) into the vector in frame with the oligo-histidine coding region (red), and transform bacterial cells with the recombinant vector. The cells produce the fusion protein (red and yellow), along with other, bacterial proteins (green). 2. We lyse the cells, releasing the mixture of proteins. 3. We pour the cell lysate through a nickel affinity chromatography column, which binds the fusion protein, but not the other proteins. 4. We release the fusion protein from the column with histidine, or a histidine analog, which competes with the oligo-histidine for binding to the nickel. 5. We cleave the fusion protein with enterokinase. 6. We pass the cleaved protein through the nickel column once more to separate the oligo-histidine from the protein we want.

Expression Vectors that Produce Fusion Proteins

When most expression vectors operate, they produce fusion proteins. This might at first seem a disadvantage because the natural product of the inserted gene is not made. However, fusion proteins may be more stable in bacterial cells than normal eukaryotic proteins are, and the extra amino acids at the amino terminus of the fusion protein can be a great help in purifying the protein product.

Consider the oligo-histidine expression vectors, one of which has the trade name pTrcHis (figure 15.14). These have a short sequence right before the multiple cloning site that encodes a stretch of six histidines. Thus, a protein expressed in such a vector will be a fusion protein with six histidines at its amino end. Why would we want to attach six histidines to our protein? Oligo-histidine regions like this have a high affinity

for metals like nickel, so we can purify proteins that have such regions using nickel affinity chromatography. The beauty of this method is its simplicity and speed. After the bacteria have made the fusion protein, we simply lyse them, add the crude bacterial extract to a nickel affinity column and harvest essentially pure fusion protein in only one step. This is possible because very few natural proteins have oligo-histidine regions, so our fusion protein is essentially the only one that binds to the column.

What if we want our protein free of the oligo-histidine tag? The designers of these vectors have thoughtfully provided a way to remove it. Just before the multiple cloning site, there is a coding region for a stretch of amino acids recognized by the proteolytic enzyme enterokinase. So we can use enterokinase to cleave the fusion protein into two parts: the oligo-histidine tag and the protein we want. The site recognized by enterokinase is very rare, and the chance that it exists in our protein is insignificant. Thus, our protein should not be chopped up as we are removing its oligo-histidine tag. If we want, we can run the enterokinase-cleaved protein through the nickel column once more to separate the oligo-histidine fragments from the protein of interest.

λ phages have also served as the basis for expression vectors; one designed specifically for this purpose is λ**gt11**. This phage (figure 15.15) contains the *lac* control region followed by the *lacZ* gene. The cloning sites are located within the *lacZ* gene, so products of a gene inserted into this vector will be fusion proteins with a leader of β-galactosidase.

The expression vector λgt11 has become a popular vehicle for making and screening cDNA libraries. In the examples of screening presented earlier, we looked for the proper DNA sequence by probing with a labeled oligonucleotide or polynucleotide. By contrast, λgt11 allows us to screen a group of clones directly for the expression of the right protein. The main ingredients required for this procedure are a cDNA library in λgt11 and an antiserum directed against the protein of interest.

Figure 15.16 shows how this works. We plate our λ phages with various cDNA inserts and blot the proteins released by each clone onto a support such as nitrocellulose. The nice thing about a λ clone in this regard is that the host cells lyse, forming plaques. In so doing, they release their products, making it easy to transfer proteins from thousands of clones simultaneously, simply by touching a nitrocellulose filter to the surface of a petri dish containing the plaques.

Once we have transferred the proteins from each plaque to nitrocellulose, we probe with our antiserum. Next, we probe for antibody bound to protein from a particular plaque, using labeled protein A from *Staphylococcus aureus*. This protein binds

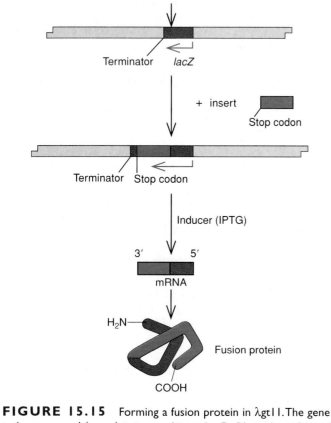

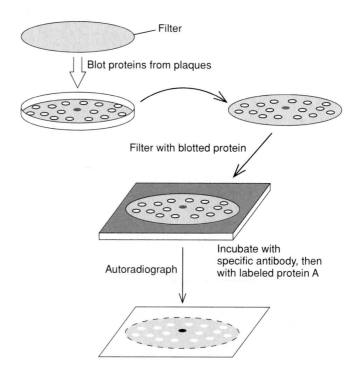

FIGURE 15.15 Forming a fusion protein in λgt11. The gene to be expressed (green) is inserted into the *Eco*RI site near the end of the *lacZ* coding region (red) just before the transcription terminator. Thus, upon induction of the *lacZ* gene by IPTG, a fused mRNA results, containing the inserted coding region just downstream from that of β-galactosidase. This mRNA is translated by the host cell to yield a fusion protein.

FIGURE 15.16 Detecting positive λgt11 clones by measuring expression. A filter is used to blot proteins from phage plaques on a petri dish. One of the clones (red) has produced a plaque containing a fusion protein including β-galactosidase and a part of the protein we are interested in. The filter with its blotted proteins is incubated with an antibody directed against our protein of interest, then with labeled *Staphylococcus* protein A, which binds specifically to antibodies. It will therefore bind only to the antibody-antigen complexes at the spot corresponding to our positive clone. A dark spot on the film placed in contact with the filter reveals the location of our positive clone.

tightly to antibody and labels the corresponding spot on the nitrocellulose. We detect this label photographically, as explained later in this chapter, then go back to our master plate and pick the corresponding plaque. Note that we are detecting a fusion protein, not the cloned protein by itself. Furthermore, it does not matter if we have cloned a whole cDNA or not. Our antiserum is a mixture of antibodies that will react with several different parts of our protein, so even a partial gene will do, as long as its coding region is cloned in the same orientation and reading frame as the leading β-galactosidase coding region.

Expression vectors frequently produce fusion proteins, with the first part of the protein coming from coding sequences in the vector and the latter part from sequences in the cloned gene itself. Fusion proteins have certain advantages: They may be more stable in prokaryotic cells than unmodified eukaryotic proteins, and they may be much simpler to isolate. The λgt11 vector produces fusion proteins that can be detected in plaques with a specific antiserum.

Eukaryotic Expression Systems

Eukaryotic genes are not really "at home" in prokaryotic cells, even when they are expressed under the control of their prokaryotic vectors. One reason is that *E. coli* cells frequently recognize the protein products of cloned eukaryotic genes as outsiders and destroy them. Another is that prokaryotes do not carry out the same kinds of posttranslational modifications as eukaryotes do. For example, a protein that would ordinarily be coupled to sugars in a eukaryotic cell will be expressed as a bare protein when cloned in bacteria. This can affect a protein's activity or stability, or at least its response to antibodies. A more serious problem is that the interior of a bacterial cell is not as conducive to proper folding of eukaryotic proteins as the interior of a eukaryotic cell. Frequently, the result is improperly folded, inactive products of cloned genes. This means that we can frequently express a cloned gene at a stupendously high level in bacteria, but the product forms highly insoluble, inactive granules called inclusion bodies. These are of no use unless we can get the protein to dissolve and regain its activity. On the other hand, it is frequently possible to renature the proteins from inclusion bodies. In that case, the inclusion bodies are an advantage because they can be separated from almost all other proteins by simple centrifugation.

In order to avoid the potential incompatibility between a cloned gene and its host, we can express our gene in a eukaryotic cell. In such cases, we usually do the initial cloning in *E. coli*, using a **shuttle vector** that can replicate in both bacterial and eukaryotic cells. We then transfer the recombinant DNA to the eukaryote of choice by transformation. One eukaryote suited for this purpose is yeast. It shares the advantages of rapid growth and ease of culture with bacteria, yet it is a eukaryote, and thus it carries out the protein folding and glycosylation (adding sugars) expected of a eukaryote. In addition, by splicing our cloned gene to the coding region for a yeast export signal peptide, we can usually ensure that the gene product will be secreted to the growth medium. This is a great advantage in purifying the protein. We simply remove the yeast cells in a centrifuge, leaving relatively pure secreted gene product behind in the medium.

The yeast vectors are based on a plasmid, called the *2-micron plasmid,* that normally inhabits yeast cells. It provides the origin of replication needed by any vector that must replicate in yeast. Yeast-bacterial shuttle vectors also contain the pBR322 origin of replication, so they can also replicate in *E. coli.* In addition, of course, a yeast expression vector must contain a strong yeast promoter.

Another eukaryotic vector that has been remarkably successful is derived from the baculovirus that infects the caterpillar known as the alfalfa looper. Viruses in this class have a rather large circular DNA genome, approximately 130 kb in length. The major viral structural protein, polyhedrin, is made in copious quantities in infected cells. In fact, it has been estimated that when a caterpillar dies of a baculovirus infection, up to 10% of the dry mass of the dead insect is this one protein. This indicates that the polyhedrin gene must be very active, and indeed it is—apparently due to its powerful promoter. Max Summers and his colleagues and Lois Miller and her colleagues, working separately, first developed successful vectors using the polyhedrin promoter in 1983 and 1984, respectively. Since then, many other baculovirus vectors have been constructed using this and other viral promoters. At their best, baculovirus vectors can produce up to half a gram per liter of protein from a cloned gene—a large amount indeed.

MANIPULATING CLONED GENES

Besides simply obtaining the products of cloned genes, we can put them to many uses, as shown in the following three examples: (1) We do not have to be satisfied with the natural product of a cloned gene; once the gene is cloned, we can change it any way we want and collect the correspondingly altered gene product. (2) We can tag cloned genes with radioactive atoms or other markers and use them as probes for many purposes. For example, we can find out how many similar genes exist in an organism's genome or how actively the gene is transcribed in living cells. (3) We can determine the exact base sequence of a cloned gene.

PROTEIN ENGINEERING WITH CLONED GENES

Traditionally, protein biochemists have relied on chemical methods to alter certain amino acids in the proteins they study, so they can observe the effects of these changes on protein activities. But chemicals are rather crude tools for manipulating proteins; it is difficult to be sure that only one amino acid, or even one kind of amino acid, has been altered. Cloned genes make this sort of investigation much more precise, allowing us to perform microsurgery on a protein. By replacing specific bases in a gene, we also replace amino acids at selected spots in the protein product and then we can observe the effects of those changes on that protein's function.

How do we perform such **site-directed mutagenesis?** First, we need a cloned gene whose base sequence is known. (We will see later in this chapter how DNA sequencing is done; for the moment, let us assume that we have already determined the sequence of the gene we want to manipulate.) Then, we need to obtain our gene in single-stranded form. This can be done by cloning it into M13 phage and collecting the single-stranded progeny phage DNA (plus strands) as described in figure 15.8, or by using an analogous procedure with a phagemid (figure 15.9).

Our next task is to change a single codon in this gene. Let us suppose the gene contains the sequence of bases given in figure 15.17, which codes for a sequence of amino acids that includes a tyrosine. The amino acid tyrosine contains a phenolic group:

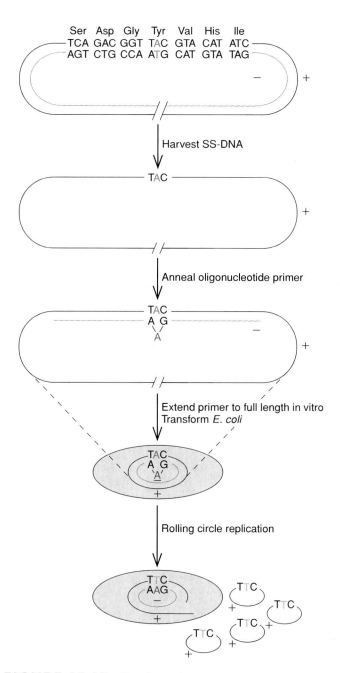

To investigate the importance of this phenolic group, we can change the tyrosine codon to a phenylalanine codon. Phenylalanine is just like tyrosine except that it lacks the phenolic group; instead, it has a simple phenyl group:

If the tyrosine phenolic group is important to a protein's activity, replacing it with phenylalanine's phenyl group should diminish that activity.

We want to change the DNA codon TAC (Tyr) to TTC (Phe). The simplest way is to use an automated DNA synthesizer to make an oligonucleotide (a "25-mer") with the following sequence:

3'-CGAGTCTGCCAA**A**GCATGTATAGTA-5'

This has the same sequence as a piece of the original minus strand, except that the central triplet has been changed from A**T**G to AA**G**, with the altered base boldfaced. Thus, this oligonucleotide will hybridize to the plus strand we harvested from the M13 phage, except for the one base we changed, which will cause an A–A mismatch. The oligonucleotide in this hybrid can serve as a primer for DNA polymerase to complete the negative strand in vitro; then we can transform *E. coli* with the resulting double-stranded replicative form (RF), which still contains the one-base mismatch.

Once inside the host bacterium, the RF replicates by a rolling circle mechanism, using the mutated negative strand as template to produce many copies of the positive strand. These mutated positive strands will contain a TTC phenylalanine codon instead of a TAC tyrosine codon and will be packaged

FIGURE 15.17 Site-directed mutagenesis using an oligonucleotide. We begin with the gene we want to mutagenize, cloned into double-stranded phage M13 DNA. Our goal is to change the tyrosine codon TAC (blue) to the phenylalanine codon TTC. We transform *E. coli* cells with the recombinant phage DNA and harvest single-stranded (plus strand) progeny phage DNA. To this plus strand, we anneal a synthetic primer that is complementary to the region of interest, except for a single base change, an A (red) for a T. This creates an A–A mismatch (blue and red). We extend the primer to form a full-length minus strand in vitro and use this double-stranded DNA to transform *E. coli* cells. The minus strand with the single base change then serves as the template for making many copies of the plus strand, each bearing a TTC phenylalanine codon (pink) in place of the original TAC tyrosine codon.

into phage particles (not shown in figure 15.17), which can be used to infect more bacterial cells. In the final step of the process, we can collect the mutant RF molecules produced by these cells, cut out the mutant gene with restriction enzymes, and clone it into an appropriate expression vector. The protein product will be identical to the wild-type product except for the single change of a tyrosine to a phenylalanine.

Note the precision of this technique compared to traditional chemical mutagenesis, in which we assault organisms with an often lethal dose of mutagen, then examine progeny organisms for the desired mutant characteristics. Before DNA sequencing was possible, we could not be certain of the molecular nature of such a mutation without laborious protein sequencing. Even with modern DNA sequencing techniques, we would have to determine the sequence of an entire mutant gene before we could be sure what specific damage a chemical had caused.

By contrast, site-directed mutagenesis enables us to decide in advance what parts of a protein we want to change and to tailor our mutants accordingly. Nevertheless, in spite of its lack of precision, traditional mutagenesis still plays an important role; one such experiment can quickly and easily create a rich variety of mutants, some of which involve changes we might never think of—or never get around to making by site-directed mutagenesis.

Using cloned genes, we can introduce changes at will, thus altering the amino acid sequences of the protein products. This is most conveniently done with single-stranded cloned DNA from a bacteriophage or phagemid vector and synthetic oligonucleotide primers containing the desired base change.

Using Cloned Genes as Probes

The phenomenon of hybridization—the ability of one single-stranded nucleic acid to form a double helix with another single strand of complementary base sequence—is one of the backbones of modern molecular genetics. Indeed, techniques using hybridization are so pervasive that it would be difficult to overestimate their importance. We have already encountered plaque and colony hybridization earlier in this chapter; we will illustrate here two further examples of hybridization techniques.

Using DNA Probes to Count Genes

Many eukaryotic genes are not found just once per genome, but are parts of families of closely related genes. Suppose we want to know how many genes of a certain type exist in a given organism. If we have cloned a member of that gene family—even a partial cDNA—we can estimate this number.

We begin by using a restriction enzyme to cut genomic DNA that we have isolated from the organism (figure 15.18). It is best to use a restriction enzyme such as *Eco*RI or *Hind*III that recognizes a 6-base-pair cutting site. These enzymes will

produce thousands of fragments of genomic DNA, with an average size of about 4,000 base pairs. Next, we electrophorese these fragments on an agarose gel. The result, if we visualize the bands by staining, will be a blurred streak of thousands of bands, none distinguishable from the others. Eventually, we will want to hybridize a labeled probe to these bands to see how many of them contain coding sequences for the gene of interest. First, however, we must transfer the bands to a medium on which we can conveniently perform our hybridization.

Edward Southern was the pioneer of this technique; he transferred, or blotted, DNA fragments from an agarose gel to nitrocellulose by diffusion, as depicted in figure 15.18. This process has been called **Southern blotting** ever since. Nowadays, blotting is frequently done by electrophoresing the DNA bands out of the gel and onto the blot. Figure 15.18 also illustrates this process. Before blotting, the DNA fragments are denatured with base so that the resulting single-stranded DNA can bind to the nitrocellulose, forming the Southern blot. Media superior to nitrocellulose are now available; some of them use nylon supports and have clever trade names like Gene Screen. Next, we label our cloned DNA by applying DNA polymerase to it in the presence of labeled DNA precursors. Then we denature this labeled probe and hybridize it to the Southern blot. Wherever the probe encounters a complementary DNA sequence, it hybridizes, forming a labeled band corresponding to the band of DNA containing our gene of interest. Finally, we visualize these bands with X-ray film, as explained later in this section.

If we find only one band, the interpretation is relatively easy; there is probably only one gene corresponding to our cDNA. If we find multiple bands, there are probably multiple genes, but it is difficult to tell exactly how many. One gene can give more than one band if it contains one or more cutting sites for the restriction enzyme we used. We can minimize this problem by using a short probe, such as a 100–200-base-pair restriction fragment of the cDNA, for example. Chances are, a restriction enzyme that only cuts on average every 4,000 base pairs will not cut within the 100–200-base-pair region of the genes that hybridize to such a probe.

Using DNA Probes to Measure Gene Activity

Suppose we have cloned a cDNA and want to know how actively the corresponding gene (gene X) is transcribed in a number of different tissues of organism Y. We could answer that question in several ways, but the method we describe here will also tell us the size of the mRNA the gene produces.

We begin by collecting RNA from several tissues of the organism in question. Then we electrophorese these RNAs in an agarose gel and blot them to a suitable support. Since a similar blot of DNA is called a Southern blot, it was natural to name a blot of RNA a **Northern blot.** (By analogy, a blot of protein is called a Western blot.)

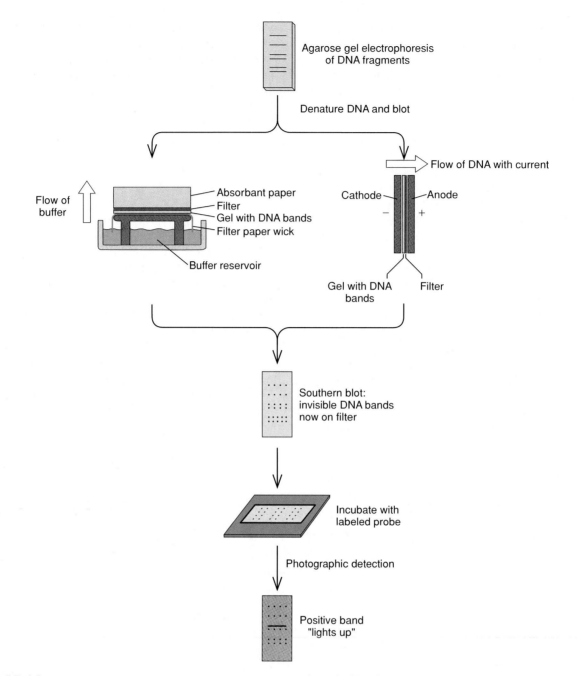

FIGURE 15.18 Southern blotting. First, we electrophorese DNA fragments on an agarose gel. Next, we denature the DNA with base, and transfer the single-stranded DNA fragments from the gel (yellow) to a sheet of nitrocellulose (red) or similar material. This can be done in two ways: by diffusion, in which buffer passes through the gel, carrying the DNA with it (left), or by electrophoresis (right). The blot can then be hybridized to a labeled probe and the labeled bands can be detected by autoradiography or photography.

Next, we hybridize the Northern blot to a labeled cDNA probe. Wherever an mRNA complementary to the probe exists on the blot, hybridization will occur, resulting in a labeled band that we can detect with X-ray film. If we run marker RNAs of known size next to the unknown RNAs, we can tell the sizes of the RNA bands that "light up" when hybridized to the probe.

Furthermore, the Northern blot tells us how actively gene X is transcribed. The more RNA the band contains, the more probe it will bind and the darker the band will be on the film. We can quantify this darkness by measuring the amount of light it absorbs in an instrument called a densitometer. Figure 15.19 shows a Northern blot of RNA from eight different rat tissues, hybridized to a probe for a gene (G3PDH, or glyceraldehyde

3-phosphate dehydrogenase) involved in sugar metabolism. Clearly, this gene is most actively expressed in the heart and skeletal muscle, and least actively expressed in the lung.

Traditionally, the labels used in Southern and Northern blotting experiments have been radioactive. Thus, we incorpo-

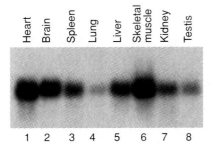

FIGURE 15.19 A Northern blot. Poly(A)⁺ RNA was isolated from the rat tissues indicated at top, then equal amounts of RNA from each tissue were electrophoresed and Northern blotted. The RNAs on the blot were hybridized to a labeled probe for the rat glyceraldehyde-3-phosphate dehydrogenase (G3PDH) gene, and the blot was then exposed to X-ray film. The bands represent the G3PDH mRNA, and their intensities are indicative of the amounts of this mRNA in each tissue.

Courtesy of Clontech.

rate radioactive nucleotides into DNAs to make radioactive probes, which then hybridize to specific bands on the blots, making them radioactive. We can then detect the radioactive bands by autoradiography. This method is still frequently used, but work with radioactive substances requires special training and care in handling and disposal of radioactive waste. For this reason, nonradioactive labels are increasing in popularity.

Figure 15.20 illustrates one way to make and use a nonradioactive labeled probe. First, we replicate the DNA in vitro with DNA polymerase in the presence of nucleotides tagged with the vitamin *biotin*. This produces biotinylated DNA. Next, we incubate the labeled DNA with a blot, where it hybridizes to one or more bands. Next, we react the blot with a complex composed of a protein called *streptavidin,* coupled to the enzyme alkaline phosphatase (AP). Streptavidin binds tightly and specifically to biotin; in fact, it gets part of its name from the fact that it binds biotin avidly. This interaction between the biotinylated probe and the AP-coupled streptavidin tags the reactive bands on the blot with AP. Next, we add an AP substrate such as AMPPD. When AP removes a phosphate group from this molecule, it becomes chemiluminescent, like the tail of a firefly. We can detect this luminescence with X-ray film; dark bands will appear on the film in positions corresponding to the reactive bands on the blot, just as they do on an autoradiograph.

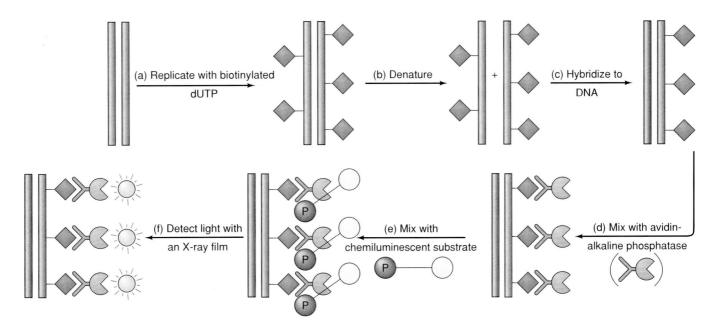

FIGURE 15.20 Detecting nucleic acids with a nonradioactive probe. This sort of technique is usually indirect; we detect a nucleic acid of interest by hybridization to a labeled probe that can in turn be detected because of its ability to produce a colored or light-emitting substance. In this example, we execute the following steps: (*a*) We replicate the probe DNA in the presence of dUTP that is tagged with the vitamin biotin. This generates biotinylated probe DNA. (*b*) We denature this probe and (*c*) hybridize it to the DNA we want to detect (red). (*d*) We mix the hybrids with a bifunctional reagent containing both avidin and the enzyme alkaline phosphatase (green). The avidin binds tightly and specifically to the biotin in the probe DNA. (*e*) We add a phosphorylated compound that will become chemiluminescent as soon as its phosphate group is removed. The alkaline phosphatase enzymes attached to the probe cleave the phosphates from these substrate molecules, making them chemiluminescent (light-emitting). (*f*) We detect the light with an X-ray film.

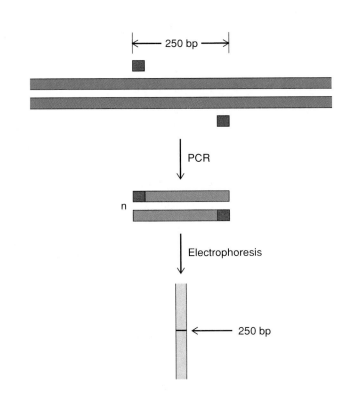

FIGURE 16.9 Sequence tagged sites (STSs). We start with a large cloned piece of DNA, extending indefinitely in either direction. We know the sequences of small areas of this DNA, so we can design primers that will hybridize to these regions and allow PCR to produce double-stranded fragments of predictable lengths. In this example, we have used two PCR primers (red) spaced 250 bp apart. After several cycles of PCR, we have many double-stranded PCR products that are precisely 250 bp long. Electrophoresis of this product allows us to measure its size exactly and confirm that it is the correct one.

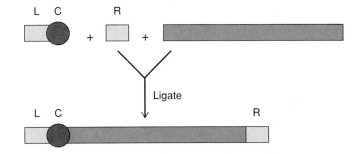

FIGURE 16.10 Cloning in yeast artificial chromosomes (YACs). We begin with two tiny pieces of DNA from the two ends of a yeast chromosome. One of these, the left arm, contains the left telomere (yellow, labeled L) plus the centromere (red). The right arm contains the right telomere (yellow, labeled R). These two arms are ligated to a large piece of foreign DNA (blue)—several hundred kb of human DNA, for example—to form the YAC, which can replicate in yeast cells along with the real chromosomes.

of human (or any other) DNA can be placed in between the centromere and the right telomere, as shown in figure 16.10. The large DNA inserts are prepared by slightly digesting long pieces of human DNA with a restriction enzyme. The YACs, with their human DNA inserts, can then be introduced into yeast cells, where they will replicate just as if they were normal yeast chromosomes.

Once we have a YAC library of the human genome, we need some way to identify the clones that contain the region we want to map. This can be done in several ways. We could hybridize YAC DNA to a labeled DNA probe corresponding to the region of interest, but this is subject to some uncertainty due to possible nonspecific hybridization. A more reliable method is to look for sequence tagged sites (STSs) in the YACs. It is best to screen the YAC library for at least two STSs, spaced hundreds of kb apart, so we will select YACs spanning a long distance.

After we have found a number of positive YACs, we begin mapping by screening them for several additional STSs, so we can line them up in an overlapping fashion as shown in figure 16.11. This set of overlapping YACs is our new contig. We can now begin finer mapping of the contig, and even sequencing the regions we are most interested in.

Using these and other mapping techniques, geneticists have already made great strides in the Human Genome Project. As mentioned earlier in this chapter, we have a genetic map of the whole genome that provides an average resolution of 0.7 cM. We also have relatively high-resolution physical maps of two of the smallest chromosomes, 21 and Y. These maps are especially useful in that they represent contigs of DNA segments cloned in YACs. Thus, if you are interested in a disease gene that maps to one of these chromosomes, you now have a much simplified task. You need only discover two STSs flanking the gene of interest, look on the map to find which YAC or YACs contain these STSs, obtain the YACs, and begin your final search for the gene. Inevitably, we will find that these maps contain mistakes, but they will be corrected as the maps are refined.

Mapping the human genome requires a set of landmarks to which we can relate the positions of genes. Some of these markers are other genes, but many more are anonymous stretches of DNA such as RFLPs, VNTRs, STSs, and microsatellites. The latter two are regions of DNA that can be identified by formation of a predictable length of amplified DNA by PCR with pairs of primers. The large chunks of DNA needed for mapping DNA at the megabase level can be cloned as yeast artificial chromosomes (YACs). Using these tools, geneticists have already produced a map of the whole human genome with an average distance between markers of about 0.7 cM. Two chromosomes, 21 and Y, have been mapped more thoroughly, with each chromosome covered by a contig composed of YACs.

Exon Traps

Once we have a contig stretching over hundreds of kilobases, how do we sort out the genes from the other DNA? One way

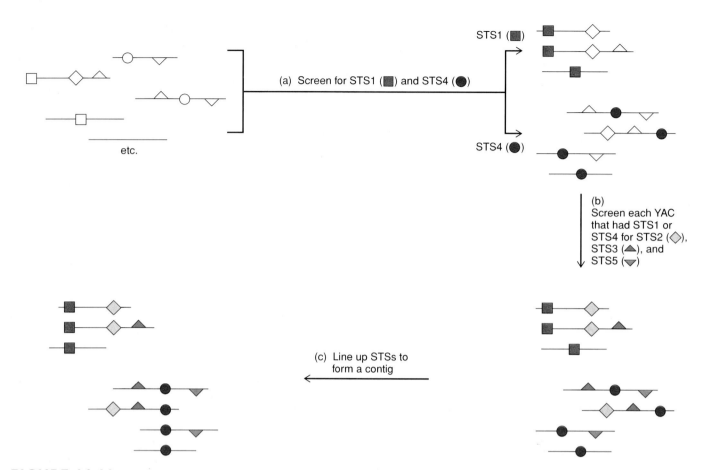

FIGURE 16.11 Mapping with STSs. At top left, several representative YACs are shown, with different symbols representing different STSs placed at specific intervals. In step (*a*) of the mapping procedure, we screen for two or more widely spaced STSs. In this case we screen for STS1 and STS4. All those YACs with either STS1 or 4 are shown at top right. The identified STSs are shown in color.

In step (*b*), each of these positive YACs is further screened for the presence of STS2, STS3, and STS5. The colored symbols on the YACs at bottom right denote the STSs detected in each YAC. In step (*c*), we align the STSs in each YAC to form the contig. Measuring the lengths of the YACs by pulsed-field gel electrophoresis helps to pin down the spacing between pairs of YACs.

is simply to sequence the whole region and look for open reading frames (ORFs), but that is very laborious. Several more efficient methods are available, including a procedure invented by Alan Buckler called exon amplification or **exon trapping.** Figure 16.12 shows how an exon trap works. We begin with a plasmid vector such as pSPL1, which Buckler designed for this purpose. This vector contains a chimeric gene under the control of the SV40 early promoter. The gene was derived from the rabbit β-globin gene by removing its second intron and substituting a foreign intron from the HIV virus, with its own 5′ and 3′ splice sites. We splice human genomic DNA fragments into a restriction site within the intron of this plasmid, then insert the recombinant vector into monkey cells (COS-7 cells) that can transcribe the gene from the SV40 promoter. Now if any of the genomic DNA fragments we placed into the intron are complete exons, with their own 5′ and 3′ splice sites, this exon will become part of the processed transcript in the COS cells. We purify the RNA made by the COS cells, reverse transcribe it to make cDNA, then subject this cDNA to amplification by

PCR, using primers designed to amplify any new exon. Finally, we clone the PCR products, which should represent only exons. Any other piece of DNA inserted into the intron will not have splicing signals; thus, after being transcribed, it will be spliced out along with the surrounding intron and lost.

CG Islands

Another gene-finding technique takes advantage of the fact that active human genes tend to be associated with unmethylated CG sequences, whereas the CG's in inactive regions are almost always methylated (chapter 14). Furthermore, the restriction enzyme *Hpa*II cuts at the sequence CCGG, but only if the second C is unmethylated. In other words, it will cut active genes that have unmethylated CG's within CCGG sites, but it will leave inactive genes (with methylated CCGG's) alone. Thus, geneticists can scan large regions of DNA for "islands" of sites that could be cut with *Hpa*II in a "sea" of other DNA sequences that could not be cut. Such a site is called a **CG island,** or an **HTF island** because it yields *Hpa*II tiny fragments.

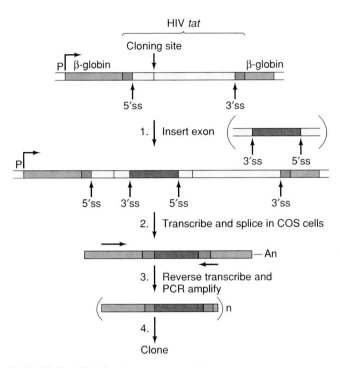

FIGURE 16.12   Exon trapping. We begin with a cloning vector, such as pSPL1, shown here in slightly simplified form. This vector has an SV40 promoter (P), which drives expression of a hybrid gene containing the rabbit β-globin gene (orange), interrupted by part of the HIV *tat* gene, which includes two exon fragments (blue) surrounding an intron (yellow). The exon/intron borders contain 5′ and 3′ splice sites (ss). The *tat* intron contains a cloning site, into which we can insert random DNA fragments. In step 1, we have inserted an exon (red), flanked by parts of its own introns, and its own 5′ and 3′ splice sites. In step 2, we insert this construct into COS cells, where it can be transcribed and then the transcript can be spliced. Note that the foreign exon (red) has been retained in the spliced transcript, because it had its own splice sites. Finally (steps 3 and 4), we subject the transcripts to reverse transcription and PCR amplification, with primers indicated by the arrows. This gives many copies of a DNA fragment containing the foreign exon, which can now be cloned and examined. Note that a non-exon will not have splice sites and will therefore be spliced out of the transcript along with the intron. It will not survive to be amplified in step 3, so we will not waste time studying it.

Several methods are available for identifying the genes in a large contig. One of these is the exon trap, which uses a special vector to help clone exons only. Another is to use methylation-sensitive restriction enzymes to search for CG islands, DNA regions containing unmethylated CG sequences.

KNOCKOUTS

Once geneticists have identified a good candidate for the gene that causes a disease, they have three complementary ways to confirm that they have the right gene. The first is to show that people with the disease have a different form of the gene than people without the disease. The second is to introduce deliberate mutations into the corresponding gene in a lower animal and demonstrate that such mutations cause a disease like the one in humans. The third is to correct the defect in mutant cells by supplying them with a normal copy of the gene. A favorite way of creating deliberate mutations is by targeted disruption of a mouse gene—in other words, making a **knockout mouse.**

Figure 16.13 explains one way to generate a knockout mouse. We begin with cloned DNA containing the mouse gene we want to knock out. We interrupt this gene with another gene that confers resistance to the antibiotic neomycin. Elsewhere in the cloned DNA, outside the target gene, we introduce a thymidine kinase (*tk*) gene. Later, these extra genes will enable us to weed out clones without an interrupted gene.

Next, we mix the engineered mouse DNA with stem cells from an embryonic brown mouse. In a small percentage of these cells, the interrupted gene will find its way into the nucleus and homologous recombination will occur between the altered gene and the resident, intact gene. This recombination places the altered gene into the mouse genome, and removes the *tk* gene. Unfortunately, such recombination events are relatively rare, so many stem cells will experience no recombination and therefore will suffer no interruption of their resident gene. Still other cells will experience nonspecific recombination, in which the interrupted gene will insert randomly into the genome without replacing the intact gene.

The problem now is to eliminate the cells in which homologous recombination did not occur. This is where the extra genes we introduced earlier come in handy. Cells in which no recombination took place will have no neomycin resistance gene. Thus, we can eliminate them by growing the cells in medium containing the neomycin derivative G418. Cells that experienced nonspecific recombination will have incorporated the *tk* gene, along with the interrupted gene, into their genome. We can kill these cells with gangcyclovir, a drug that is lethal to *tk*+ cells. (The stem cells we used are *tk*−.) Treatment with these two drugs leaves us with engineered cells that have undergone homologous recombination and are therefore heterozygous for an interruption in the target gene.

Our next task is to introduce this interrupted gene into a whole mouse (figure 16.14). We do this by injecting our engineered cells into a mouse blastocyst that is destined to develop into a black mouse. Then we place this altered embryo into a surrogate mother, who eventually gives birth to a chimeric mouse. We can recognize this mouse as a chimera by its patchy coat; the black zones come from the original black embryo, and the brown zones result from the transplanted engineered cells.

To get a mouse that is a true heterozygote instead of a chimera, we allow the chimera to mature, then mate it with a black mouse. Since brown (agouti) is dominant, some of the progeny should be brown. In fact, all of the offspring resulting from gametes derived from the engineered stem cells should be brown. But only half of these brown mice will carry the interrupted gene, since the engineered stem cells were heterozygous for the knockout. Southern blots showed that two of the

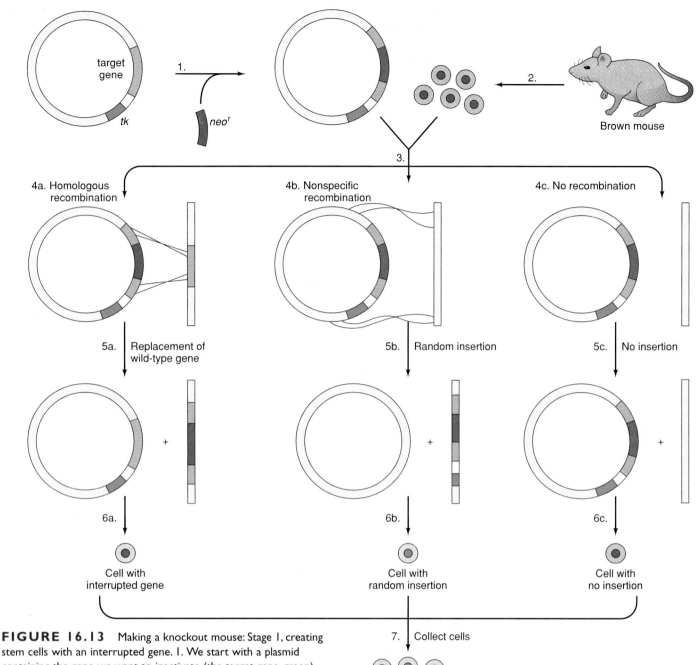

FIGURE 16.13 Making a knockout mouse: Stage 1, creating stem cells with an interrupted gene. 1. We start with a plasmid containing the gene we want to inactivate (the target gene, green) plus a thymidine kinase gene (*tk*). We interrupt the target gene by splicing the neomycin-resistance gene (red) into it. 2. We collect stem cells (tan) from a brown mouse embryo. 3. We transform these cells with the plasmid containing the interrupted target gene. 4 and 5. Three kinds of products result from this transformation: 4a. Homologous recombination between the interrupted target gene in the plasmid and the homologous, wild-type gene causes replacement of the wild-type gene in the cellular genome by the interrupted gene (5a). 4b. Nonspecific recombination with a nonhomologous sequence in the cellular genome results in insertion of the interrupted target gene plus the *tk* gene into the cellular genome (5b). 4c. When no recombination occurs, the interrupted target gene is not integrated into the cellular genome at all (5c). 6. The cells resulting from these three events are color coded as indicated: Homologous recombination yields a cell (red) with an interrupted target gene (6a); nonspecific recombination yields a cell (blue) with the interrupted target gene and the *tk* gene inserted at random (6b); no recombination yields a cell (tan) with no integration of the interrupted gene. 7. We collect the transformed cells, containing all three types (red, blue, and tan). 8. We grow the cells in medium containing the neomycin analog G418 and the drug gangcyclovir. The G418 kills all cells without a neomycin-resistance gene, namely those cells (tan) that did not experience a recombination event. The gangcyclovir kills all cells that have a *tk* gene, namely those cells (blue) that experienced a nonspecific recombination. This leaves only the cells (red) that experienced homologous recombination and therefore have an interrupted target gene.

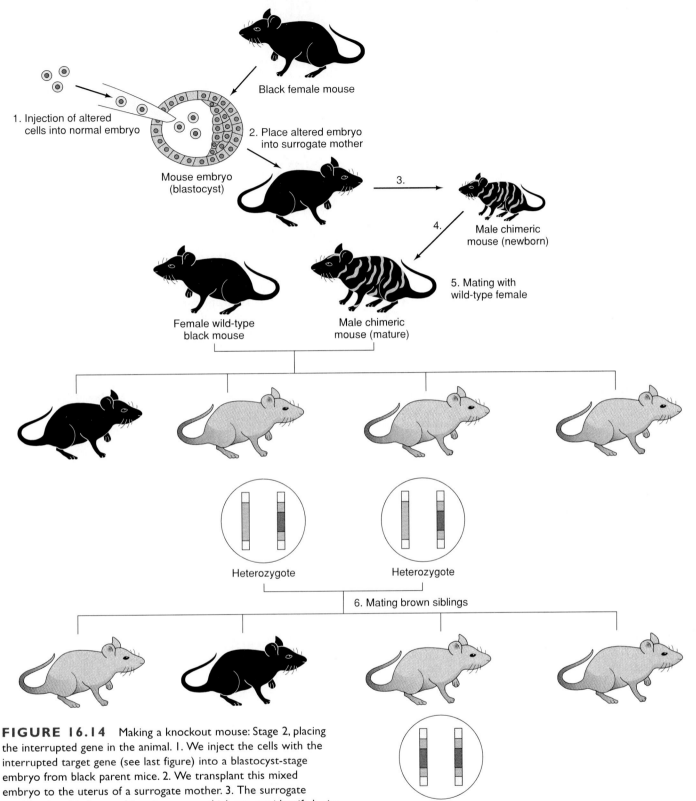

Black female mouse

1. Injection of altered cells into normal embryo

Mouse embryo (blastocyst)

2. Place altered embryo into surrogate mother

3.

4.

Male chimeric mouse (newborn)

5. Mating with wild-type female

Female wild-type black mouse

Male chimeric mouse (mature)

Heterozygote

Heterozygote

6. Mating brown siblings

Homozygote

FIGURE 16.14 Making a knockout mouse: Stage 2, placing the interrupted gene in the animal. 1. We inject the cells with the interrupted target gene (see last figure) into a blastocyst-stage embryo from black parent mice. 2. We transplant this mixed embryo to the uterus of a surrogate mother. 3. The surrogate mother gives birth to a chimeric mouse, which we can identify by its black and brown coat. (Recall that the altered cells came from an agouti [brown] mouse, and they were placed into an embryo from a black mouse.) 4. We allow the chimeric mouse (a male) to mature. 5. We mate it with a wild-type black female. We can discard any black offspring, which must have derived from the wild-type blastocyst; only brown mice could have derived from the transplanted cells. 6. We select a brown brother and sister pair, both of which show evidence of an interrupted target gene (by Southern blot analysis), and mate them. Again we examine the DNA of the brown progeny by Southern blotting. This time, we find one animal that is homozygous for the interrupted target gene. This is our knockout mouse. We can now observe this animal to determine the effects of knocking out the target gene.

brown mice in our example carry the interrupted gene. We mate these and look for progeny that are homozygous for the knockout by direct examination of their DNA. In our example, one of the mice from this mating is a knockout, and now our job is to look for its phenotype. Frequently, as is the case here, this is not obvious. But obvious or not, it can be very instructive.

In other cases, the knockout is lethal and the affected mouse fetuses die before birth. Still other knockouts have intermediate effects. For example, in chapter 17 we will learn about a tumor suppressor gene called *p53*. Humans with defects in this gene are highly susceptible to certain cancers. Mice with their p53 gene knocked out develop normally but are afflicted with tumors at an early age.

> To probe the role of a human gene, geneticists can perform targeted disruption of the corresponding gene in a mouse, and then observe the effect of that mutation on the "knockout mouse."

MAPPING HUMAN DISEASE GENES

Now that we are acquainted with some of the most important tools geneticists use in mapping large regions of DNA, let us see how they have helped us find the genes governing some important human diseases.

HUNTINGTON'S DISEASE

Huntington's disease (HD) is a progressive nerve disorder. It begins almost imperceptibly with small tics and clumsiness. Over a period of years, these symptoms intensify and are accompanied by emotional disturbances. Nancy Wexler, an HD researcher, describes the advanced disease as follows: "The entire body is encompassed by adventitious movements. The trunk is writhing and the face is twisting. The full-fledged Huntington patient is very dramatic to look at." Finally, after ten to twenty years, the patient dies.

Huntington's disease is controlled by a single dominant gene. Therefore, a child of an HD patient has a 50:50 chance of being affected. People who have the disease could avoid passing it on by not having children, except that the first symptoms usually do not appear until after the childbearing years.

Because they did not know the nature of the product of the HD gene (*HD*), geneticists could not look for the gene directly. The next best approach was to look for a gene or other marker that is tightly linked to *HD*. Michael Conneally and his colleagues spent more than a decade trying to find such a linked gene, but with no success.

In their attempt to find a genetic marker linked to *HD*, Wexler, Conneally, and James Gusella turned next to RFLPs. They were fortunate to have a very large family to study. Living around Lake Maracaibo in Venezuela is a family whose members have suffered from HD since the early nineteenth century. The

first member of the family to be so afflicted was a woman whose father, presumably a European, carried the defective gene. So the pedigree of this family can be traced through seven generations, and the number of individuals is unusually large, a typical nuclear family having fifteen to eighteen children.

Gusella and colleagues knew they might have to test hundreds of probes to detect a RFLP linked to *HD*, but they were amazingly lucky. Among the first dozen probes they tried, they found one (called G8) that detected a RFLP that is very tightly linked to *HD* in the Venezuelan family. Figure 16.15 shows the locations of *Hin*dIII sites in the stretch of DNA that hybridizes to the probe. We can see seven sites in all, but only five of these are found in all family members. The other two, marked with asterisks and numbered 1 and 2, may or may not be present. These latter two sites are therefore polymorphic, or variable.

Let us see how the presence or absence of these two restriction sites gives rise to a RFLP. If site 1 is absent, a single fragment 17.5 kb long will be produced. However, if site 1 is present, the 17.5-kb fragment will be cut into two pieces having lengths of 15 kb and 2.5 kb. Only the 15-kb band will show up on the autoradiograph, since the 2.5-kb fragment lies outside the region that hybridizes to the G8 probe. If site 2 is absent, a 4.9-kb fragment will be produced. On the other hand, if site 2 is present, the 4.9-kb fragment will be subdivided into a 3.7-kb fragment and a 1.2-kb fragment.

There are four possible **haplotypes** (clusters of alleles on a single chromosome) with respect to these two polymorphic *Hin*dIII sites, and they have been labeled *A–D*:

Haplotype	Site 1	Site 2	Fragments observed
A	Absent	Present	17.5; 3.7; 1.2
B	Absent	Absent	17.5; 4.9
C	Present	Present	15.0; 3.7; 1.2
D	Present	Absent	15.0; 4.9

The term "haplotype" is a contraction of "haploid genotype," which emphasizes that each member of the family will inherit two haplotypes, one from each parent. For example, an individual might inherit the A haplotype from one parent and the D haplotype from the other. This person would have the AD genotype. Sometimes different genotypes (pairs of haplotypes) can be indistinguishable. For example, a person with the *AD* genotype will have the same RFLP pattern as one with the *BC* genotype, since all five fragments will be present in both cases. However, the true genotype can be deduced by examining the parents' genotypes. Figure 16.16 shows autoradiographs of Southern blots of two families, using the radioactive G8 probe. The 17.5- and 15-kb fragments migrate very close together, so they are difficult to distinguish when both are present, as in the *AC* genotype; nevertheless, the *AA* genotype with only the 17.5-kb fragment is relatively easy to distinguish from the *CC* genotype with only the 15-kb fragment. The *B* haplotype in the first family is obvious because of the presence of the 4.9-kb fragment.

Which haplotype is associated with the disease in the Venezuelan family? Figure 16.17 demonstrates that it is *C*.

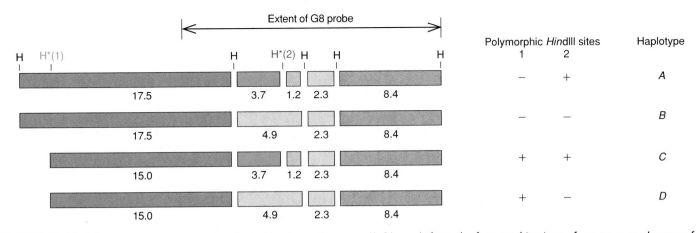

FIGURE 16.15 The RFLP associated with the Huntington's disease gene. The *Hind*III sites in the region covered by the G8 probe are shown. The families studied show polymorphisms in two of these sites, marked with an asterisk and numbered 1 (blue) and 2 (red). Presence of site 1 results in a 15-kb fragment plus a 2.5-kb fragment that is not detected because it lies outside the region that hybridizes to the G8 probe. Absence of this site results in a 17.5-kb fragment. Presence of site 2 results in two fragments of 3.7 and 1.2 kb. Absence of this site results in a 4.9-kb fragment. Four haplotypes

(A–D) result from the four combinations of presence or absence of these two sites. These are listed at right, beside a list of polymorphic *Hind*III sites and a diagram of the *Hind*III restriction fragments detected by the G8 probe for each haplotype. For example, haplotype A lacks site 1 but has site 2. As a result, *Hind*III fragments of 17.5, 3.7, and 1.2 are produced. The 2.3- and 8.4-kb fragments are also detected by the probe, but we ignore them because they are common to all four haplotypes.

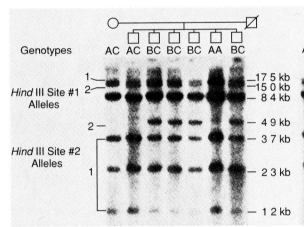
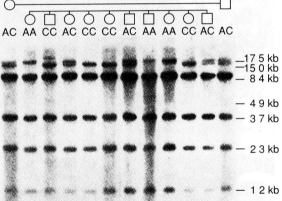

FIGURE 16.16 Southern blots of *Hind*III fragments from members of two families, hybridized to the G8 probe. The bands in the autoradiographs represent DNA fragments whose sizes are listed at the right. The genotypes of all the children and three of the

parents are shown at the top. The fourth parent was deceased, so his genotype could not be determined.

From James F. Gusella, et al., *Nature* 306:236. © 1983 Macmillan Magazines Limited.

Nearly all individuals with this haplotype have the disease. Those who do not will almost certainly develop it later. Equally telling is the fact that no individual lacking the *C* haplotype has the disease. Thus, this is a very accurate way of predicting whether a member of this family is carrying the Huntington's disease gene. A similar study of an American family showed that, in this family, the *A* haplotype was linked with the disease. Therefore, each family varies in the haplotype associated with the disease, but within a family, the linkage between the RFLP site and *HD* is so close that recombination between these sites is very rare. Thus we see that a RFLP can be used as a genetic marker for mapping, just as if it were a gene.

Aside from its use as a genetic counseling tool, what does this analysis reveal? For one thing, it has allowed us to locate *HD* on chromosome 4. Gusella and colleagues did this by making mouse-human hybrid cell lines, each containing only a few human chromosomes, as described in chapter 5. They then prepared DNA from each of these lines and hybridized it to the radioactive G8 probe. Only the cell lines having chromosome 4 hybridized; the presence or absence of all other chromosomes did not matter. Therefore, human chromosome 4 carries *HD*.

At this point, the *HD* mapping team's luck ran out. One long detour arose from a mapping study that indicated the gene

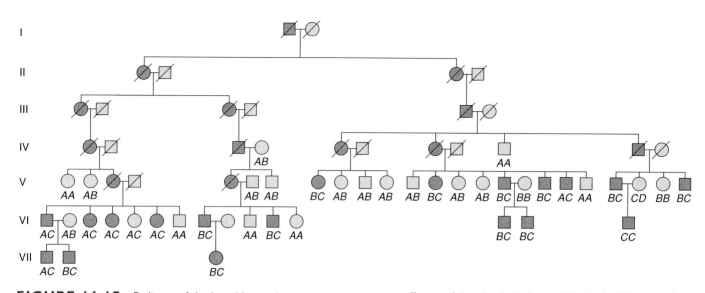

FIGURE 16.17   Pedigree of the large Venezuelan Huntington's disease family. Family members with confirmed disease are represented by purple symbols. Notice that most of the individuals with the *C* haplotype already have the disease, and that no sufferers of the disease lack the *C* haplotype. Thus, the *C* haplotype is strongly associated with the disease, and the corresponding RFLP is tightly linked to the Huntington's disease gene.

lay far out at the end of chromosome 4. This made the search much more difficult because the tip of the chromosome is a genetic wasteland, full of repetitive sequences, and apparently devoid of genes. Finally, after wandering for years in what he calls a genetic "junkyard," Gusella and his group turned their attention to a more promising region. Some mapping work suggested that *HD* resided, not at the tip of the chromosome, but in a 2.2 Mb region several Mb removed from the tip. Over 2 Mb is a tremendous amount of DNA to sift through to find a gene, so Gusella decided to focus on a 500 kb region that was highly conserved among about one-third of HD patients, who seem to have a common ancestor.

To find genes in the 500 kb region, Gusella used an exon trapping strategy and identified a handful of exon clones. He then used these exons to probe a cDNA library to identify the DNA copies of mRNAs transcribed from the target region. One of the clones, called IT15, for "interesting transcript number 15," hybridized to cDNAs that identified a large (10,366 nucleotide) transcript that codes for a large (3,144 amino acid) protein. The presumed protein product did not resemble any known proteins, so that did not provide any evidence that this is indeed *HD*. However, the gene had an intriguing repeat of 23 copies of the triplet CAG (one copy is actually CAA), encoding a stretch of 23 glutamines.

Is this really *HD*? Gusella's comparison of the gene in affected and unaffected individuals in 75 HD families demonstrates that it is. In all unaffected individuals, the number of CAG repeats ranged from 11 to 34, and 98% of these unaffected people had 24 or fewer CAG repeats. In all affected individuals, the number of CAG repeats had expanded to at least 42, up to a high of about 100. Thus, we can predict whether an individual will be affected by the disease by looking at the number of CAG repeats in this gene. Furthermore, the severity, or age of onset of the disease correlates at least roughly with the number of CAG repeats. People with a number of repeats at the low end of the affected range (now known to be 38–40) generally survive well into adulthood before symptoms appear, whereas people with a number of repeats at the high end of the range tend to show symptoms in childhood. In one extreme example, an individual with the highest number of repeats detected (about 100) started showing disease symptoms at the extraordinarily early age of two. Finally, two people were affected, even though their parents were not. In both cases, the affected individuals had expanded CAG repeats, whereas their parents did not. New mutations (expanded CAG repeats), although a rare occurrence in HD, apparently caused both these cases of disease.

Another way of demonstrating that this gene is really *HD* would be to deliberately mutate it and show that the mutation has neurological effects. Obviously, one cannot perform such an experiment in humans, but it would be feasible in mice, if the gene corresponding to *HD* is known. Fortunately, *HD* is conserved in many species, including the mouse, where the gene is known as *Hdh*. In 1995, a team of geneticists led by Michael Hayden created knockout mice with a targeted disruption in exon 5 of *Hdh*. Mice that are homozygous for this mutation die in utero. Heterozygotes are viable, but they show loss of neurons with corresponding lowering of intelligence. This reinforces the notion that *Hdh*, and therefore *HD*, plays an important role in the brain—exactly what we would expect of the gene that causes HD.

How can we put this new knowledge to work? One obvious way is to perform accurate genetic screening to detect people who will be affected by the disease. In fact, by counting the CAG repeats, we may even be able to predict the age of onset of the disease. However, that kind of information is a mixed blessing; it can be psychologically devastating (see Box 16.3). What we really need, of course, is a cure, but that may be a long way off.

16.3 Problems in Genetic Screening

Now that we have identified the genes for Huntington's disease, cystic fibrosis, and other serious maladies, how will we use this information? We are in a position to identify carriers of genetic diseases or people who are at risk for developing these debilitating conditions. This sounds like a good thing; in general, it undoubtedly will be, but there are some ethical problems.

Consider, for example, the actual case of an adoption agency that was trying to place a baby girl whose mother suffered from HD. Understandably, potential adoptive parents wanted to know if the baby had inherited the gene. However, Michael Conneally, the geneticist consulted in the case, decided not to test the child for the following reason: If the results were positive and were later shared with the girl, she would know that she was doomed to get the disease. Many people at risk of developing this terrible disease feel that they cannot cope with knowing for sure that they have the gene, so they elect not to take the test. In this case, the child was too young to choose for herself.

This choice about whether or not to know if one is carrying the *HD* gene has recently become more complex. Now, by counting the number of CAG repeats in the gene, we can predict the age at which the disease will begin. Before, one could cling to the hope that, even if one carried the gene, the disease might not appear until very late in life. Now a genetic test could remove even that hope.

Another dilemma would occur in the following situation: A woman is at risk of developing HD because her father has it. She has decided she cannot bear to know that she has inherited the gene, so she will not submit to a test. Then she gets pregnant. She and her husband want the child, but not if it carries the defective gene. They could check the fetus's genotype by amniocentesis. A negative result would end their worries about the child and allow the mother to cling to hope for her own future, but a positive result would confirm her worst fears, not only for the baby but for herself.

In the case of the autosomal recessive condition cystic fibrosis (CF), a different set of problems arises. Cystic fibrosis is the most common lethal genetic disease afflicting American children. One in twenty Caucasians is a *CF* carrier, and therefore more than one in 2,000 children will be homozygous for *CF* and suffer the disease. Now that the *CF* gene has been identified, one would predict that millions of people would want to be tested to help them with their family planning. Nevertheless, in 1989 the American Society for Human Genetics called for a moratorium on widespread screening, and though small-scale screening has been tried, caution is still the watchword.

There are two main reasons for this reluctance. First, in spite of the expectation that CF would be caused by only a small number of different mutations, the actual number had already risen to 170 by 1992. About 20 of these mutations are common, so we would probably need to test for all 20, and even then we would only discover 85% to 90% of all carriers. Even assuming a 90% detection rate, we would still detect only 81% of all couples at risk ($0.9 \times 0.9 = 0.81$). That seems better than nothing, but missing almost 20% of couples likely to have an affected child is certainly far from perfect, and testing services may fear giving such imperfect advice.

Even if the test were perfect, problems would remain. The major difficulty arises from the sheer volume of the task. Since CF is so common, most prospective parents might want to be tested. But in the United States alone, there are about 150 million people in the reproductive age group. Even assuming we could test that many people, someone has to explain to them what the results mean, and there are not nearly enough professional genetic counselors to do that.

Our experience with screening African-Americans for sickle-cell disease teaches us what can happen to an inadequate program. In the early 1970s, many states passed laws requiring testing of African-Americans for the sickle-cell gene, but did not provide enough counseling to make the program successful. This had several unfortunate outcomes. For one, many carriers of the sickle-cell gene came to the mistaken and distressing conclusion that they actually had the disease. Also, the suggestion to African-American couples who were both carriers that they forego having children, coming as it frequently did from whites, provoked charges of racism. Moreover, in some cases confidentiality was not respected. This led to the denial of health insurance to some carriers. The result of all this was a loss of confidence in the screening process among those it was designed to help. Thus, few people took part, and even fewer used the information they received to help in their family planning.

These cases exemplify the tough decisions that people must make about their own genetic testing and the psychological impact such tests may have on them. Quite a different kind of question arises concerning the mandatory testing of others. Would it be ethical, for example, for insurance companies to demand genetic tests? Those who test positive for such conditions as HD or Alzheimer's disease would either be denied health insurance or forced to pay exorbitant premiums. We have already seen that this happened to carriers of the sickle-cell gene. If it is not fair to burden people because of their genes, over which they have no control, would it be fair to require them to submit to tests before obtaining life insurance? Keep in mind that life insurance companies have traditionally required proof of good health before issuing insurance and that genetic screening could simply be interpreted as an extension of this practice.

Let us take the problem one step further and assume that a life insurance company has paid for genetic testing of an individual and found him likely to develop a debilitating disease such as Alzheimer's. Would it be ethical for the insurance company to share such information with others—the person's employer or prospective employer, for instance? This information would certainly give an indication of the person's long-term fitness for employment, but it would just as certainly place an enormous burden on that person and maybe even render him unemployable. The problem becomes more difficult if the insurer is also the employer, as is the case with some large corporations.

These thorny questions have no easy answers, but the development of more and more tests for genetic diseases will force us to deal with them. Ultimately, if our genetic studies really do lead to effective treatments, these questions should recede in importance.

In order to find a cure for HD, we need to answer at least two questions. First, what is the normal function of the *HD* product (now called **huntingtin**)? To get at that question, geneticists are constructing transgenic mice carrying the mutant form of *HD*. This should shed light on the function of the gene in the brain and in other tissues, where it is also active. Second, we need to know how the expansion of the CAG repeat causes disease. Gusella postulates that the extra glutamines in the protein give it a new function that somehow disrupts brain activity. We now know that huntingtin binds to another protein called huntingtin-associated protein (HAP), so the abnormal huntingtin may associate with HAP in an abnormal way that distorts nerve function.

> Using RFLPs, geneticists mapped the Huntington's disease gene (*HD*) to a region near the end of chromosome 4. Then they used an exon trap to identify the gene itself. The mutation that causes the disease is an expansion of a CAG repeat from the normal range of 11–34 copies, to the abnormal range of at least 38 copies.

CYSTIC FIBROSIS

Cystic fibrosis (CF) is the most common lethal genetic disease afflicting Caucasians. It is caused by an autosomal recessive mutation carried by 1 in 20 people of European descent (over 12 million people in the United States alone). This means that 1 in 400 Caucasian couples will both be carriers, and 1 in 4 of their children will have the disease. In the United States, over one thousand children every year are born with cystic fibrosis.

The disease affects tissues called secretory epithelia, which are responsible for transporting water and salt at the interface between the bloodstream and the external environment (e.g., in the lungs, intestine, and sweat glands). The abnormal secretory epithelia in cystic fibrosis patients fail to carry out this transport properly, which causes buildup of thick mucus in the affected organs. This in turn causes the clinical symptoms of the disease: failure of the pancreas to secrete digestive enzymes into the intestine; bacteria that find a fertile environment for growth in the thick mucus in the lungs; and high levels of salt in the sweat. The latter symptom is not dangerous, but it is a convenient marker that physicians use to diagnose the disease. The lung infections, on the other hand, are very serious. It is primarily because of these that many cystic fibrosis patients do not survive past their twenties.

Since this disease is so prevalent and so devastating, it has been the subject of intense research. Geneticists knew that if they could identify the defective gene and its product, this might suggest new avenues to a treatment for the disease. At least it would provide a test for the mutation so couples would know in advance if they were at risk of having affected children. The test could also be used prenatally to determine whether a fetus was homozygous for the mutation.

As in the case of HD, geneticists trying to find the cystic fibrosis gene did not know the identity of its protein product. They knew that something about salt transport was disrupted, but nothing about the nature of the defective protein. In spite of the enormous difficulties of this kind of searching for a "needle in a haystack," geneticists led by Lap-Chee Tsui, Francis Collins, and John Riordan used a powerful combination of classical and molecular approaches and in 1989 finally located the cystic fibrosis gene (*CF*).

The first step in the search for the gene was to establish its linkage to known markers, especially those whose chromosomal location was known. One of the first markers to be found linked to *CF* was a variability in serum activity of an enzyme called paraoxonase; unfortunately, this marker had not been mapped, and it was not even known what chromosome it resided on. However, using RFLP mapping techniques similar to those employed in the search for the *HD* gene, Tsui and colleagues found linkage between the *CF* gene and a more promising marker—a RFLP called DOCRI–917.

Next, these workers showed that this RFLP is located on chromosome 7. To do this mapping, Tsui and colleagues employed somatic cell hybridization techniques such as those described in chapter 5. They obtained a panel of hamster-human hybrid cells, with known human chromosome content, from another laboratory. Then they screened DNA from these cells for the ability to hybridize to a radioactive probe for DOCRI–917. Whenever the cell contained human chromosome 7, its DNA hybridized to the probe. Whenever the cell did not contain any part of human chromosome 7, its DNA would not hybridize to the probe. Thus, the RFLP is located on chromosome 7, and since *CF* is linked to the RFLP, we know that it is on chromosome 7 as well.

At about the same time, Ray White and coworkers discovered tight linkage between the *CF* gene and a RFLP in a gene called *met*. This was valuable data, since *met* had been mapped to the middle third of the long arm of chromosome 7. This placed *CF* in the same vicinity.

Robert Williamson, Jorg Schmidtke, and colleagues strengthened this conclusion by showing linkage between *CF* and two RFLPs. The first is an anonymous RFLP called pJ3.11; the second is a RFLP in a gene called *TCR*β. Schmidtke and his coworkers used somatic cell hybrid panels to map pJ3.11 to the top (band q22) of the long arm of chromosome 7, and other workers mapped the other RFLP to the bottom (band q3) of the long arm.

Linkage analysis on a number of different markers culminated in the assignment of *CF* to band q31 of chromosome 7, between two closely linked markers, *met* and D7S8 (defined by the pJ3.11 RFLP). The recombination frequency between *met* and *CF* in males was 0.013, while the recombination frequency between *CF* and D7S8 was only 0.009. This placed *met* and D7S8 only about 1–2 million base pairs apart, thus confining the search to a relatively narrow region of the human genome. Still, even a million base pairs is a lot of DNA to sift through looking for a single gene, so creative approaches were needed.

The first step in searching for a gene in a megabase pair region of DNA is to clone at least part of the DNA in question. This is no trivial task, but it has been simplified by techniques such as YAC cloning. In this case, the investigators used another technique called **chromosome walking,** which generates a collection of clones with overlapping DNAs that cover the whole region of interest (figure 16.18). To perform a chromosome walk, from left to right in the example illustrated here, we start with a genomic library and a clone at the left end of the sequence we wish to examine. We cut a piece from the right end of this starting clone and label it; then we use this radioactive fragment to probe the library by plaque hybridization. Any clone that hybridizes will overlap the right end of the starting clone, and is very likely to contain additional DNA to the right. This process is repeated with a probe from the right end of the second clone

to find an overlapping clone farther to the right. This is continued until we have a contig containing overlapping clones representing the whole region. Finally, we can sequence all our clones to obtain the base sequence of the whole region.

Chromosome walking works very well for examining relatively short stretches of a chromosome (100 kb or so), but runs into trouble when we need to canvass hundreds of kb of DNA, as in the present case. The biggest problem is that there are certain regions of DNA, sometimes called poison regions, that are unclonable. Still other regions can be cloned but are unstable and are lost. The reasons for these very frustrating phenomena are not completely clear, but it is known that DNAs with unusual secondary structure or repetitive sequences are frequently lost from bacterial cells.

Collins worked out a clever way around these problems: a modification of chromosome walking called **chromosome jumping.** This technique (figure 16.19) allows the investigator to jump over unclonable regions and start the chromosome walk anew. The key to the procedure is to prepare a library of clones containing about 100 kb of DNA, then to form circles of these DNAs by ligating in a short piece of DNA containing a selectable marker—for example, the *supF* gene, which suppresses amber mutations. This brings together two DNA regions that had been separated by about 100 kb. Next we cut the circular DNA with a restriction enzyme and subclone the fragments into a λ phage with an amber mutation. This ensures that only clones with the *supF* gene will survive, and this gene will mark the place where circle formation took place. We select a clone by plaque hybridization to a probe for a given site in our search (say the *met* locus). Since this cloned DNA contains *met,* and a circle-joining site marked by *supF,* the chances are high that the DNA lying on the other side of the *supF* gene originally lay about 100 kb away from *met.* This can then be used as a new start site for chromosome walking. We have just performed a chromosome jump of 100 kb.

In the end, Riordon, Tsui, Collins, and their colleagues cloned almost 300 kb of DNA, contained within a region of about 500 kb. They suspected that this region contained the *CF* gene, but where was it?

One important clue in this sort of mystery is to find a stretch of DNA that is conserved in several different species. If the DNA is conserved, it probably codes for something and is therefore a gene. To find conserved DNA quickly, the research team blotted fragmented genomic DNA from human, cow, mouse, and chicken and hybridized this "zoo blot" to radioactive probe DNAs from different parts of the cloned human DNA region. Four radioactive probes cross-hybridized to DNA from other species besides human. These probes identified four candidate regions in which to focus the search.

The first candidate region mapped very close to *met,* and far from D7S8, so it could not contain *CF.* The second did not contain any open reading frames, so it was also eliminated from consideration. The third region was used as a probe to look for mRNAs with the same sequence, but none was found. Thus, it appeared that this region is not transcribed and therefore could not contain the *CF* gene.

(a) Chromosome walking

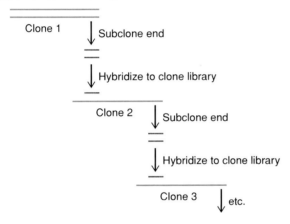

(b) Physical mapping (restriction sites and STSs)

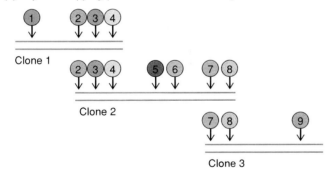

FIGURE 16.18   Mapping by chromosome walking.
(*a*) Chromosome walking. To start the walk, we choose a cloned piece of DNA (clone 1) and subclone one end of it. We then use this small end piece (red) as a probe to identify an overlapping clone (clone 2) in a library. Repeating the process, we subclone the far end of clone 2 to generate a probe to identify yet another overlapping clone (clone 3). We can repeat this cycle as many times as we wish to build a set of overlapping clones spanning large stretches of DNA. (*b*) Physical mapping of restriction sites or STSs in each clone allows us to align the overlapping DNAs and build a map of the whole contig.

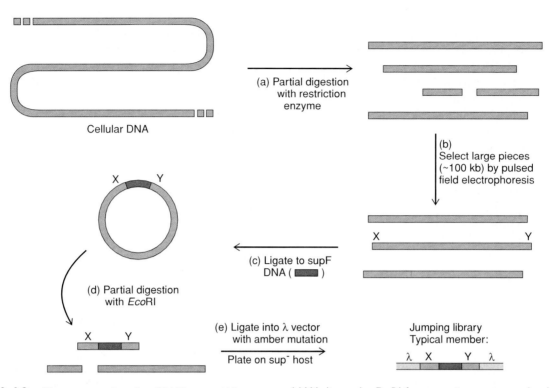

FIGURE 16.19 Chromosome jumping. (*a*) We start with high molecular weight cellular DNA and partially digest it with a restriction enzyme to produce large DNA fragments. (*b*) We subject the fragments to pulsed-field gel electrophoresis and select those that are about 100 kb long. The ends of one of these fragments are labeled X and Y, so we can track them in subsequent steps. (*c*) We ligate these long DNA fragments to a cloned *supF* gene (red) that allows suppression of amber mutations. The *supF* gene inserts between the former free ends (X and Y) of the long DNA fragments. (*d*) We partially digest the circularized DNA with *Eco*RI. Some of the resulting fragments will contain *supF* flanked by X and Y.

(*e*) We ligate the *Eco*RI fragments into a vector that has a lethal amber mutation, package the DNA, and plate the resulting phage particles on a *supF* host. Only the phages with the *supF* gene will replicate, and these constitute our jumping library. Each cloned DNA will have a *supF* gene, flanked by two regions (X and Y) that originally lay about 100 kb apart in the cellular DNA. By adjusting the size fragments we select in step (*b*), we can alter the size of the jump in the jumping library. Assuming that X is a known region encompassed by a previous chromosome walk, we can use Y to begin a new chromosome walk 100 kb away.

Fortunately, the fourth candidate region did contain *CF.* When the research team determined the sequence of this region they found an encouraging sign: the presence of a CG island. It took exhaustive screening of many cDNA libraries with a probe for this region, but finally the researchers were rewarded with a cDNA (clone 10–1) containing a 920-bp piece of DNA encoding part of an mRNA from the sweat gland of a normal (non-CF) individual. This meant that the CG-rich region was transcribed in normal sweat glands, as *CF* ought to be. Clone 10–1 also detected a 6.5-kb transcript in a Northern blot of human RNAs, again demonstrating the expression of this region in human cells. Finally, using cDNAs to hybridize to genomic clones, this group was able to show that *CF* spans approximately 250 kb of DNA and includes at least 24 exons.

How do we know this gene is really *CF*? Several lines of evidence support this conclusion. First, Northern blotting experiments showed that the gene is transcribed in all tissues affected by CF. Second, the base sequence of the gene shows that it encodes a protein with a so-called membrane-spanning domain and so is very likely to be a membrane protein, as the CF protein is predicted to be since it is involved in membrane

transport. Further sequence analysis suggested at first that the product of the *CF* gene was probably not itself a channel for chloride ions, but somehow regulated the transport of these ions across the membrane. This led to renaming the protein product of *CF* as the cystic fibrosis transmembrane conductance regulator **(CFTR).** However, we now know that CFTR is in fact a chloride channel, because it can create such channels when it is the only protein added to an artificial membrane.

A third finding linking this gene with CF is that most CF patients have a 3-bp deletion in this gene, which would result in a mutant protein with one amino acid (a phenylalanine) missing. Work by Tsui and his colleagues showed that about 70% of the first group of CF patients tested had this mutation. The rest presumably had mutations elsewhere in *CF*. Finally, and most tellingly, two teams of workers, including Michael Welsh, as well as Tsui and Collins, have shown that expression of a transplanted wild-type *CF* gene in CF cells cures the defect in chloride ion transport. Thus, the gene encoding CFTR really is *CF,* now usually called the CFTR gene.

Now that we know the identity of the CFTR gene, how can we use this knowledge to find treatments for the disease? One way is to develop an animal model so we can examine in detail the effects of the disease on lungs and other tissues as the disease progresses. The supply of lung tissue from CF patients who die in infancy is small; by the time most patients die, infections have damaged their lungs so badly that their condition in the early stages of the disease is impossible to determine. Two research groups have now bred CFTR knockout mice that should help circumvent these problems. Now we can see precisely what is happening in diseased tissues from birth (or before) until death. And we can try out therapies on these mice before risking them on human patients. One problem, however, is that the knockout mice do not show exactly the same phenotype as human CF patients, so we will have to be careful in interpreting experiments.

Another way knowledge of the CFTR gene can help is in designing therapies for the disease. So far, the genetic findings have suggested several new therapies. The most obvious idea is to supply a wild-type gene to CF patients. We could use any of several viruses or liposomes to carry the CFTR gene to the place where it would do the most good—the lungs—perhaps as simply as by using an aerosol inhaler.

Instead of providing a new gene, we may be able to ameliorate the disease symptoms simply by giving regular doses of the CFTR protein by inhaler. One danger would be that the patients could develop an allergy to the protein, but it appears that they have nonfunctional protein in their lungs already, and the difference between mutant and wild-type proteins (only one amino acid) is so slight that the body may not recognize the wild-type protein as foreign, and so might tolerate it. Another strategy is to stimulate a latent activity of the mutant protein with drugs, and CF specialists are already considering some drugs that may be able to do this.

Using a combination of RFLP mapping and somatic cell genetic techniques, geneticists located CF to a 1–2 Mb region of band q31 of chromosome 7. Then they pinpointed CF by first building a clone library encompassing the DNA region where the gene was presumed to be located. Contiguous clones in this library were identified by chromosome walking and jumping. DNA-DNA hybridization identified four candidate genes that were well conserved among mammalian species. Of these, one had promising characteristics, including an HTF island. This was identified as CF based on: (1) its pattern of expression in organs, such as sweat glands, that are involved in CF; (2) its sequence; (3) the fact that CF patients have mutations in this gene; and (4) the fact that CF cells can be cured of their chloride transport deficiency by introducing the wild-type version of this gene.

DUCHENNE MUSCULAR DYSTROPHY

Geneticists have also found the exact gene responsible for Duchenne muscular dystrophy (DMD), an X-linked muscle-wasting disease. Several methods were employed, including clas-

sical mapping, but a key piece of the puzzle came from studying the rare cases of girls afflicted with the disease. Ronald Worton and coworkers examined the X chromosomes of twelve such patients and found that all had X chromosomes broken in about the same place—band Xp21 near the middle of the chromosome's short arm. This strongly suggested that the break in every case occurred in the DMD gene, inactivating it and thereby causing the disease.

L. M. Kunkel and his colleagues located the DMD gene in this region and found that it is enormous. It contains 61 exons and spans 2 Mb of DNA. Some of its introns are over 100 kb long. The mature mRNA produced from this gene comprises more than 14 kilobases and encodes a gigantic polypeptide having a molecular weight of about 400,000. This protein, called **dystrophin,** is about ten times the size of an average polypeptide. The gene itself is one of the largest yet found in humans or any other organism. It is so huge it may take 24 hours to transcribe.

Dystrophin has been located in human muscle cells using antibodies directed against the product of the very similar mouse DMD gene. It is satisfying that these antibodies react with a protein in normal muscle cells but not in muscle cells from DMD patients. It is also gratifying that this protein has a relative molecular mass of about 400,000. The geneticists studying DMD have thus completed an important part of their job. The next step will be for cell biologists to determine the role dystrophin plays in normal muscle cells. This, in turn, may give clues about how to intervene in the disease. It may even be possible someday to perform gene therapy by providing DMD patients with a functional dystrophin gene.

Geneticists have identified the gene responsible for Duchenne muscular dystrophy. At 2 Mb, it is one of the largest genes ever discovered.

DNA FINGERPRINTING AND DNA TYPING

In 1985, Alec Jeffreys and his colleagues were investigating a DNA fragment from the gene for a human blood protein, α-globin, when they discovered that this fragment contained a sequence of bases repeated several times. As we learned earlier in the chapter, this kind of repeated DNA is called a minisatellite. More interestingly, they found similar minisatellite sequences in other places in the human genome, again repeated several times. This simple finding turned out to have far-reaching consequences, because individuals differ in the pattern of repeats of the basic sequence. In fact, they differ enough that two individuals have only a remote chance of having exactly the same pattern. That means that these patterns are like fingerprints; indeed, they are called DNA fingerprints.

How do we make a DNA fingerprint? We first cut the DNA under study with a restriction enzyme such as *Hae*III.

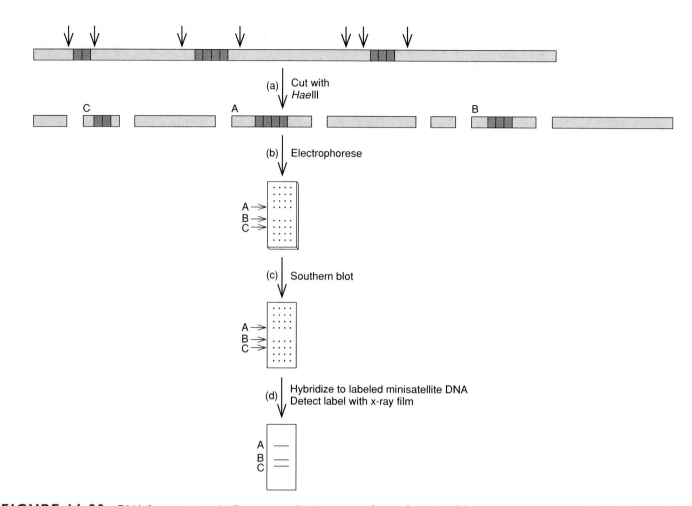

FIGURE 16.20 DNA fingerprinting. (*a*) First, we cut DNA with a restriction enzyme. In this case, the enzyme *Hae*III cuts the DNA in seven places (arrows), generating eight fragments. Only three of these fragments (labeled A, B, and C according to size) contain the minisatellites, represented by blue boxes. The other fragments (yellow) contain unrelated DNA sequences. (*b*) We electrophorese the fragments from part (*a*), which separates them according to their sizes. All eight fragments are present in the electrophoresis gel, but we cannot see any of them. The positions of all fragments, including the three (A, B, and C) with minisatellites are indicated by dotted lines. (*c*) We denature the DNA fragments and Southern blot them. (*d*) We hybridize the DNA fragments on the Southern blot to a labeled DNA with several copies of the minisatellite. This probe will bind to the three fragments containing the minisatellites, but with no others. Finally, we use X-ray film to detect the three labeled bands.

Jeffreys chose this enzyme because the repeated sequence he had found did not contain a *Hae*III recognition site. That means that *Hae*III will cut on either side of the minisatellite regions, but not inside, as shown in figure 16.20*a*. In this case, the DNA has three sets of repeated regions, containing four, three, and two repeats, respectively. Thus, we will get three different-sized fragments bearing these repeated regions.

Next, we electrophorese the fragments we have produced. This separates all the fragments into bands according to their sizes, including the three bearing the minisatellites. Then we separate the two strands of the DNA fragments and blot the single-stranded DNA as just described. At this point, if we have used human DNA, we have millions of fragments, but none of them is visible on the blot; all we can see is a smear of DNA.

We can find the bands with the minisatellites, among millions of irrelevant bands, by hybridizing the blot to a labeled single-stranded DNA containing the minisatellite sequence. This labeled probe DNA will tag any single-stranded fragments on the blot containing the complementary minisatellite sequence. Finally, we can detect the labeled bands with X-ray film. In this case, there are three labeled bands, so we will see three dark bands on the film (figure 16.20*d*).

Of course, real animals have a much more complex genome than the simple piece of DNA in this example, so they will have many more than three fragments that react with the probe. Figure 16.21 shows an example of the DNA fingerprints of several unrelated people and a set of monozygotic twins. As we have already mentioned, this is such a complex pattern of fragments that the patterns for two individuals are extremely unlikely to be identical, unless they come from monozygotic twins. This complexity makes DNA fingerprinting a very powerful identification technique.

FORENSIC USES OF **DNA** FINGERPRINTING AND **DNA** TYPING

A valuable feature of DNA fingerprinting is the fact that, although almost all individuals have different patterns, parts of the pattern (sets of bands) are inherited in a Mendelian fashion. Thus, fingerprints can be used to establish parentage. An immigration case in England illustrates the power of this technique. A Ghanaian boy born in England had moved to Ghana to live with his father. When he wanted to return to England to be with his mother, British authorities questioned whether he was a son or a nephew of the woman. Information from blood group genes was equivocal, but DNA fingerprinting of the boy demonstrated that he was indeed her son.

Scientists studying family relationships in other vertebrate species are finding DNA fingerprinting a very valuable tool. Alec Jeffreys' human probe works surprisingly well with a wide variety of vertebrate species, and probes for other repeated sequences have also been developed.

In addition to testing parentage, DNA fingerprinting has the potential to identify criminals. This is because a person's DNA fingerprint is, in principle, unique, just like a traditional fingerprint. Thus, if a criminal leaves some of his cells (blood, sperm, or hair, for example) at the scene of a crime, the DNA from these cells can identify him. However, as figure 16.21 shows, DNA fingerprints are very complex. They contain dozens of bands, some of which smear together, which can make them hard to interpret.

To solve this problem, geneticists have developed probes that hybridize to single polymorphic DNA loci, rather than to a whole set as in a classical DNA fingerprint. Each probe now gives much simpler patterns, containing only one or a few bands. Of course, each probe by itself is not as powerful an identification tool as a whole DNA fingerprint, but a panel of four or five probes can give enough different bands to be definitive. We sometimes still call such analysis DNA fingerprinting, but a better, more inclusive term is **DNA typing.**

One early, dramatic case of DNA typing involved a man who murdered a man and woman as they slept in a pickup truck, then about forty minutes later went back and raped the woman. This act not only compounded the crime, it also provided forensic scientists with the means to convict the perpetrator. They obtained DNA from the sperm cells he had left behind, typed it, and showed that the pattern matched that of the suspect's DNA. This evidence helped convince the jury to convict the defendant and recommend severe punishment. Figure 16.22 presents an example of DNA typing that was used to identify another rape

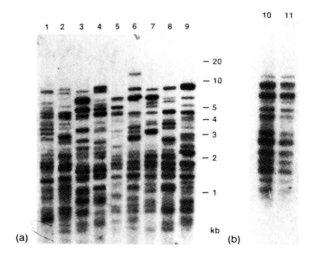

FIGURE 16.21 A DNA fingerprint. (*a*) The nine parallel lanes contain DNA from nine unrelated Caucasians. Note that no two patterns are identical, especially at the upper end. (*b*) The two lanes contain DNA from monozygotic twins, so the patterns are identical.

From G. Vassart, et al. "A Sequence in M13 Phage Detects Hypervariable Minisatellites in Human and Animal DNA." *Science,* 235:683, 6 Feb. 1987. © 1987 American Association for the Advancement of Science.

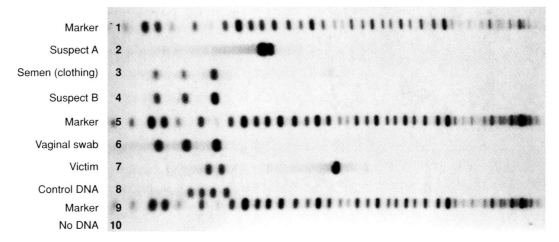

FIGURE 16.22 Use of DNA typing to help identify a rapist. Two suspects have been accused of attacking and raping a young woman, and DNA analyses have been performed on various samples from the suspects and the woman. Lanes 1, 5, and 9 contain marker DNAs. Lane 2 contains DNA from the blood cells of suspect A. Lane 3 contains DNA from a semen sample found on the woman's clothing. Lane 4 contains DNA from the blood cells of suspect B. Lane 6 contains DNA obtained by swabbing the woman's vaginal canal. Lane 7 contains DNA from the woman's blood cells. Lane 8 contains a control DNA. Lane 10 is a control containing no DNA. Partly on the basis of this evidence, suspect B was found guilty of the crime. Note how his DNA fragments in lane 4 match the DNA fragments from the semen in lane 3 and the vaginal swab in lane 6.

Courtesy of Lifecodes Corporation, Stanford, CT.

suspect. The pattern from the suspect clearly matches that from the sperm DNA. This is the result from only one probe. The others also gave patterns that matched the sperm DNA.

One advantage of DNA typing is its extreme sensitivity. Only a few drops of blood or semen are sufficient to perform a test. However, sometimes forensic scientists have even less to go on—a hair pulled out by the victim, for example. While the hair by itself may not be enough for DNA typing, it can be useful if it is accompanied by hair follicle cells. Selected segments of DNA from these cells can be amplified by PCR and typed.

In spite of its potential accuracy, DNA typing has sometimes been effectively challenged in court, most famously in the O. J. Simpson trial in Los Angeles in 1995. Defense lawyers have focused on two problems with DNA typing: First, it is tricky and must be performed very carefully to give meaningful results. Second, there has been controversy about the statistics used in analyzing the data. This second question revolves around the use of the product rule in deciding whether the DNA typing result uniquely identifies a suspect. Let us say that a given probe detects a given allele in one in a hundred people in the general population. Thus, the chance of a match with a given person with this probe is one in a hundred, or 10^{-2}. If we use five probes, and all five alleles match the suspect, we might conclude that the chances of such a match are the product of the chances of a match with each individual probe, or $(10^{-2})^5$, or 10^{-10}. Since there are fewer than 10^{10} (ten billion) people on earth, this would mean this DNA typing would statistically eliminate everyone but the suspect.

When prosecutors first used DNA typing evidence in court, they cited such product rule calculations, which certainly sound impressive. But some geneticists do not believe such calculations are valid. They point out that members of a certain ethnic group may be much more related than members of the general population, so we should use data from within that group, rather than from the whole population. Some have even argued that we should do detailed studies of each ethnic group before any DNA typing is allowed in court. But that could take many years, and few prosecutors would be willing to wait that long to use such a valuable technique. To prevent such a long wait, The National Academy of Sciences established a committee to study the problem and recommend a compromise. In April 1992 they issued their report, which recommends using a "ceiling principle." First, geneticists will analyze the allele frequencies for each probe in 100 individuals from each of the 15 to 20 most prevalent ethnic groups in the United States (German, English, Cuban, Vietnamese, for example). Then, forensic laboratories would be allowed to use the highest frequency in any of these ethnic groups, or 5%, whichever is higher. For example, let us say that the frequencies of the alleles detected in a suspect with five probes are found in the highest frequencies in the Puerto Rican, English, Chinese, Scandinavian, and Russian ethnic groups. These frequencies are 10%, 7%, 3%, 6%, and 5%, respectively. To calculate the chance of finding a match with these five alleles at random in the population, we would first raise the 3% frequency to 5%, then mul-

tiply all these percentages together. The product is 1.05×10^{-6}, or about one in a million. That is a very conservative estimate, which almost all scientists can accept, but it is still very persuasive in court. Finally, we should note that DNA typing cannot only identify criminals, it can just as easily demonstrate a suspect's innocence.

Modern DNA typing uses a battery of DNA probes to detect polymorphic sites in individual animals, including humans. As a forensic tool, DNA typing can be used to test parentage, to identify criminals, or to remove innocent people from suspicion.

WORKED-OUT PROBLEMS

PROBLEM 1

Here is the physical map of a region of DNA you are mapping by RFLP analysis.

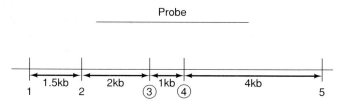

The vertical lines with numbers at the bottom represent restriction sites for the enzyme *Pst*I. Sites 3 and 4 (circled) are polymorphic; the others are not. The distances between *Pst*I sites are given. The extent of a probe you are using to detect the RFLP is indicated by a horizontal line at top. You cut DNA from different individuals with *Pst*I, electrophorese the fragments, blot them to a membrane, and hybridize the blot to the labeled probe.

a. Draw a picture of what the blot will look like when the DNA comes from individuals homozygous for the following haplotypes:

Haplotype	Site 3	Site 4
A	Present	Present
B	Present	Absent
C	Absent	Present
D	Absent	Absent

b. What effect would presence or absence of site 1 have on the results?

SOLUTION

a. Haplotype A: If sites 3 and 4 are both present, all possible fragments will be produced. In order from left to right these are: 1.5 kb, 2 kb, 1 kb, and 4 kb. Of these, the 2-kb,

1-kb, and 4-kb fragments will hybridize to the probe and therefore will be visible. Haplotype B: Site 3 is present, so the 2-kb fragment will still be produced; site 4 is absent, so the 1-kb fragment will be fused to the 4-kb fragment, yielding a 5-kb fragment. Haplotype C: Site 3 is absent, so the 2-kb fragment will be fused to the 1-kb fragment, yielding a 3-kb fragment; site 4 is present, so the 4-kb fragment will be produced. Haplotype D: Sites 3 and 4 are both absent, so the 2-kb, 1-kb, and 4-kb fragments will all be fused together, yielding a 7-kb fragment. The figure below shows a summary of the results you will obtain:

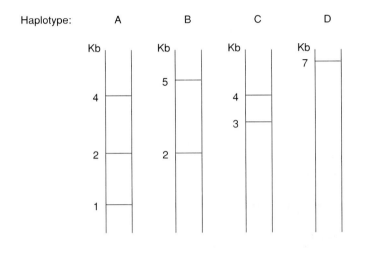

b. The presence or absence of site 1 will make no difference in the results because the probe does not extend to the left of site 2, which is always present. Thus, even though the absence of site 1 will yield a longer fragment to the left of site 2, you will not be able to detect this fragment with your probe.

PROBLEM 2

You are performing DNA typing in a forensic science lab. You use three different probes on a Southern blot of DNA from several suspects and a blood sample collected at the scene of a murder. The bands you observe in the blood sample do not match those of the victim, but they match those of one of the suspects in all six alleles at the three loci. These bands (alleles) are found in the general population with the following frequencies:

Allele	Frequency in general population
1	2%
2	1%
3	0.5%
4	1.5%
5	1.2%
6	3%

The ethnic groups that have these alleles in the highest frequencies, along with those frequencies, are shown in the following table:

Allele	Highest frequency	Ethnic group
1	5%	Russian
2	2%	Cuban
3	1.3%	German
4	6%	Japanese
5	10%	African-American
6	10%	Vietnamese

a. Using the straight product rule, calculate the probability that this combination of alleles occurs in an individual in the general population.
b. Using the ceiling principle, calculate a more conservative probability of finding this combination of alleles.

SOLUTION

a. The simple product rule just requires us to multiply together the probabilities of finding each allele in the general population. Thus,

$$P = 0.02 \times 0.01 \times 0.005 \times 0.015$$
$$\times 0.012 \times 0.03 = 5.4 \times 10^{-12}.$$

b. The ceiling principle requires us to use the frequency for the ethnic group showing the highest frequency for each allele, or 5%, whichever is higher. The second table shows the highest frequencies, two of which, 2% and 1.3%, are less than 5%. Therefore, we must raise these two to 5%, multiply each pair of alleles by 2, and then multiply all these products together. Thus,

$$P = 2(0.05 \times 0.05) \times 2(0.05 \times 0.06)$$
$$\times 2(0.1 \times 0.1) = 6 \times 10^{-7}.$$

(The factor of 2 is a "fudge factor" that has little scientific meaning but is required by many courts.)

PROBLEM 3

You are mapping the gene responsible for a human genetic disease. You find that the gene is linked to a RFLP detected with a probe called B-101. You hybridize labeled B-101 DNA to DNAs from a panel of mouse-human hybrid cells. The table below shows the human chromosomes present in each hybrid cell line, and whether or not the probe hybridized to DNA from each. Which human chromosome carries the disease gene?

Cell line	Human chromosome content	Hybridization to probe
A	1, 2, 17	−
B	3, 5, 9, 12	+
C	5, 18, 22	−
D	1, 3, 9	+
E	2, 9, 18	−
F	2, 3, 7	+

SOLUTION

Look at the human chromosome content in each of the positive cell lines (B, D, and F). These are:

B: 3, 5, 9, 12;
D: 1, 3, 9; and
F: 2, 3, 7.

Now find the one chromosome that is common to all three positive lines. It is chromosome 3. Therefore, the RFLP maps to chromosome 3, and so must the disease gene that is linked to the RFLP.

SUMMARY

The most obvious use of cloned genes is to make valuable proteins. Some products of cloned genes (human insulin, human growth hormone) are already finding uses as pharmaceuticals. In the future, they may also provide safe, effective vaccines. Using a cloned gene directly as a therapeutic tool, geneticists have already been able to transplant a gene to help correct one human genetic disease, immunodeficiency caused by ADA deficiency. They are testing similar techniques to treat other diseases, including acquired diseases like cancer and AIDS. Viruses, especially crippled retroviruses, provide the vectors for these gene transplantation procedures. Geneticists are also transferring genes to plants and animals to give these transgenic organisms more desirable characteristics. Transgenic animals and plants can also be used as factories to give us abundant quantities of the products of their transgenes.

Geneticists have embarked on an ambitious program to map, and finally sequence, the human genome. The first step, to provide a low-resolution linkage map of the whole genome, is finished. Two small chromosomes, 21 and Y, have been mapped more extensively. To map such large regions of DNA, we need to employ special techniques: Yeast artificial chromosomes (YACs) allow large pieces of DNA, up to more than 1 Mb, to be cloned. Several physical markers can be used to identify YACs and map the DNA they contain: RFLPs, VNTRs, STSs, and microsatellites are among them. Microsatellites are especially useful as linkage mapping tools because they are highly polymorphic (variable), and they are spread relatively uniformly throughout the genome.

One benefit of the Human Genome Project is the mapping and identification of genes responsible for human genetic diseases. Already, geneticists have identified the genes that cause several important diseases, including Huntington's disease, cystic fibrosis, and Duchenne muscular dystrophy. Exon trapping and searching for CG islands and for conserved DNA regions have helped identify genes once the search has been narrowed down to a relatively small DNA region. Once the gene has been identified, further confirmation of its role can be obtained by targeted disruption of the corresponding gene in mice and observing the effect of the mutation in these "knockout mice."

DNA typing, including DNA fingerprinting, is a method of identifying individual organisms by the hybridization of specific probes to Southern blots of their DNAs. DNA typing is finding increased use as a forensic tool to identify criminals.

PROBLEMS AND QUESTIONS

1. Why would a vaccine based on a weakened, but live, HIV be risky?

2. Why would a vaccine based on a cloned HIV surface protein not pose the same risk as a live virus vaccine?

3. List two reasons that gene therapists have chosen not to try to treat sickle-cell disease.

4. Why is a disease caused by a defect in a gene governing DNA synthesis potentially easier to treat than an average genetic disease?

5. Fill in the following table that lists the attributes of vectors for gene therapy. Under "insertion of DNA" indicate "+" or "−" if the vector does or does not insert its DNA into the host genome. Under "Cancer risk" and "Capacity" indicate "relatively high" or "relatively low."

Vector	Insertion of DNA	Cancer risk	Capacity
retrovirus			
adenovirus			
AAV			

6. For what two general purposes do geneticists create transgenic animals and plants?

7. List two methods of introducing foreign genes into plants.

8. Below is a physical map of a region you are mapping by RFLP analysis.

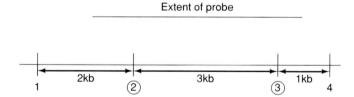

The numbered vertical lines represent restriction sites recognized by *Sma*I. The circled sites (2 and 3) are polymorphic, the others are not. You cut the DNA with *Sma*I, electrophorese the fragments, blot them to a membrane, and probe with a DNA whose extent is shown at top. Give the sizes of bands you will detect in individuals homozygous for the following haplotypes with respect to sites 2 and 3.

Haplotype	Site 2	Site 3
A	Present	Present
B	Present	Absent
C	Absent	Present
D	Absent	Absent

9. Why are microsatellites better tools for linkage mapping than minisatellites?

10. What are the essential elements of a YAC vector?

11. Will the following DNA fragments be detected by an exon trap? Why or why not?

    a. An intron.
    b. Part of an exon.
    c. A whole exon with parts of introns on both sides.
    d. A whole exon with part of an intron on one side.

12. List three ways of detecting a gene in a mega-YAC clone.

13. a. What kind of mutation gave rise to Huntington's disease?
    b. What is the evidence that the gene identified as *HD* is really the gene that causes HD?

14. You are mapping the gene responsible for a human genetic disease. You find that the gene is linked to a RFLP detected with a probe called X-21. You hybridize labeled X-21 DNA to DNAs from a panel of mouse-human hybrid cells. The table below shows the human chromosomes present in each hybrid cell line, and whether or not the probe hybridized to DNA from each. Which human chromosome carries the disease gene?

Cell line	Human chromosome content	Hybridization to X-21
A	1, 5, 21	+
B	6, 7	−
C	1, 22, Y	−
D	4, 5, 18, 21	+
E	8, 21, Y	−
F	2, 5, 6	+

15. a. What kind of *CF* mutation is most prevalent in cystic fibrosis?
    b. What is the evidence that the gene that encodes CFTR is really *CF*?

16. You are performing DNA typing in a forensic science lab. You use three different probes on a Southern blot of DNA from several suspects and a semen sample collected from a rape victim. The bands you observe in the semen sample match those of the suspect in all six alleles at the three loci. These bands (alleles) are found in the general population with the following frequencies:

Allele	Frequency in general population
1	3%
2	2%
3	1%
4	0.5%
5	4%
6	10%

The ethnic groups that have these alleles in the highest frequencies, along with those frequencies, are shown in the following table:

Allele	Highest frequency	Ethnic group
1	6%	Scandinavian
2	3%	Puerto Rican
3	20%	Arab
4	9%	Mexican
5	10%	Korean
6	10%	English

    a. Using the straight product rule, calculate the probability that this combination of alleles occurs in an individual in the general population.
    b. Using the ceiling principle, calculate a more conservative probability of finding this combination of alleles.

Answers appear at end of book.

17

Genes and Cancer

Learning Objectives

In this chapter you will learn:

1. That mutagens, including chemicals and radiation, can cause cancer.

2. That many retroviruses contain oncogenes capable of causing cancer.

3. That normal cells have oncogene counterparts called proto-oncogenes.

4. That proto-oncogenes can be activated by a number of mechanisms, leading to cancer.

5. That many proto-oncogenes code for proteins involved in regulating cell division.

6. That cells contain tumor suppressor genes that keep cell division in check. Cells with defective tumor suppressor genes can become cancerous.

A savage cell which somehow . . . corrupts the forces which normally protect the body, invades the well-ordered society of cells surrounding it, colonizes distant areas and, as a finale to its cannibalistic orgy of flesh consuming flesh, commits suicide by destroying its host.

Pat McGrady
American geneticist

Pat McGrady has defined the cancer cell as savage, and indeed it is. Whereas normal cells are exquisitely sensitive to signals that tell them when to divide and when to rest, cancer cells do not respond to such signals. They divide relentlessly, forming malignant tumors that compress and crowd out surrounding healthy tissue; they consume the body's nutrients voraciously, starving their neighbors. As destructive as such behavior can be, that alone would not be enough to make cancer a deadly disease, at least in most cases. But cancer cells can do more: They can break off from the primary tumor, enter the bloodstream or lymphatic system, travel to remote areas of the body, and establish new tumors there. This process is called **metastasis.** If a tumor has not metastasized, it can usually be cured by surgery. Once it has metastasized to remote locations, it is often fatal.

CHARACTERISTICS OF CANCER CELLS

Ironically, the savage cancer cell is not an invader from the outside; it is directly descended from one of the body's own normal cells. But somewhere along the line, a well-behaved normal cell turned into a renegade. What changed? We certainly do not have the full answer, but examinations of normal and tumor cells in culture show that malignant cells differ in three important respects from their normal counterparts.

First, malignant cells are immortal; they will go on dividing indefinitely. Consider, for example, the HeLa cells cultured from a human cervical carcinoma in 1952 (see chapter 9). These malignant cells have already far outlived their donor, and we have every reason to believe they will continue to prosper as long as people are interested in growing them. This immortality stands in stark contrast to the behavior of normal vertebrate cells. Just as complex multicellular organisms have finite lifetimes, so their cells can divide only a limited number of times (about 60) in culture, before they too "grow old" and die. See chapter 7 for a discussion of the role the enzyme telomerase plays in cell immortality and cancer.

In addition to immortality, cancer cells exhibit a suite of characteristics in contrast to those of normal cells, which we lump together under the name **transformed.** Normal cells need hard surfaces on which to grow and serum to supply growth factors. Also, when they sense that they are crowded, they cease dividing. But transformed cells typically do not show these limitations. They can usually grow in soft agar, as well as on hard surfaces; they usually do not need growth factors; and they continue to divide when they are crowded, even piling up on top of one another until their medium is exhausted.

Figure 17.1 illustrates the difference in appearance between normal cells and transformed cells. Note the elongated shape and parallel order of the normal cells in 17.1*a* and *b,* compared to the rounded shape and disorder of the transformed cells in 17.1*c*

and *d.* An important question arises: Are transformed cells, with their disorganized, relentless growth, the same as malignant cells? The answer is that all malignant cells are transformed, but some transformed cells are not malignant. Transformed nonmalignant cells would not form a malignant tumor if injected into a compatible host animal, but they are clearly not normal; they have traveled most of the way down the road toward malignancy.

Besides immortality and the transformed phenotype, cancer cells usually exhibit a progressive loss of the characteristics of the normal cells from which they derived. Thus, liver tumor cells may at first produce many of the myriad enzymes that normal liver cells do, but they gradually lose those capabilities. Also, cancer cells are usually aneuploid, and this aneuploidy becomes more and more pronounced with time. Thus, tumor cells come to resemble their tissue of origin less and less, both biochemically and genetically.

> Cancer cells differ from normal cells in several respects: They are immortal, they are transformed—capable of growing in physical and chemical environments where normal cells could not survive; and they progressively lose their biochemical and genetic resemblance to their normal ancestors.

THE CAUSES OF CANCER

It is fairly clear that the aneuploidy of cancer cells is an effect, rather than a cause, of malignancy. However, it is also obvious that genes are intimately involved in the malignant transformation of cells. Usually, the genes themselves change; other times only gene expression seems to change. In this chapter we will explore the relationship between genes and **carcinogenesis** (the creation of a cancer cell).

CARCINOGENS AS MUTAGENS

One strong indication of the relationship between genes and cancer is the fact that many agents that cause cancer (**carcinogens**) are also mutagens. Since the late eighteenth century, we have suspected that chemicals can cause cancer. It was then that Sir Percival Pott noticed a high incidence of otherwise rare scrotal cancer in chimney sweeps. Pott concluded that these boys' constant exposure to soot had something to do with the disease. In fact, soot has the same kinds of mutagenic combustion by-products found in cigarette tar, and these very likely caused the trouble.

Chapter 11 explained the mechanism behind most chemical carcinogens: They are electrophilic compounds that attack centers of negative charge in DNA, causing mutations. These mutations, in turn, apparently start the chain of events that leads to cancer. For example, consider benzo[a]pyrene, the most likely

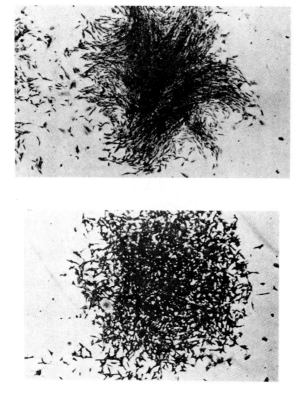

(a)

(c)

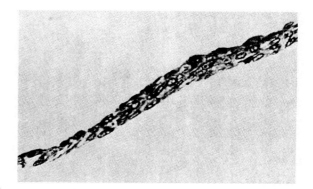

(b)

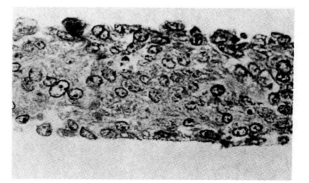

(d)

FIGURE 17.1 Morphological differences between normal and transformed cells. (*a*) A colony of normal baby hamster kidney (BHK) fibroblast cells. Note the regular, parallel arrangement of these cells. (*b*) An enlargement of one section of the colony in (*a*). Note the elongated shape and parallel distribution of these cells. (*c*) A colony of BHK cells transformed by polyoma virus (a tumor virus). Note the disorganized, nonparallel array of cells. (*d*) An enlargement of the colony in (*c*). Note the relatively rounded appearance and disorganized arrangement of these cells.

From John Tooze, *The Molecular Biology of Tumor Viruses*, p. 353, 1973. © Cold Spring Harbor. Photos by G.G.P. Stoker.

carcinogenic culprit in cigarette tar. Benzo[a]pyrene (figure 17.2) belongs to a class of organic compounds called hydrocarbons, which are composed of hydrogen and carbon. Gasoline is primarily a mixture of hydrocarbons, although simpler ones than benzo[a]pyrene. Chemically, hydrocarbons are not very reactive, and they certainly are not electrophilic, so it is at first a little difficult to understand how benzo[a]pyrene could be carcinogenic. But hydrocarbons are also very insoluble in water; thus the body needs to change them chemically in order to excrete them in the urine. This chemical change is what converts the unreactive hydrocarbon to a dangerous, mutagenic carcinogen.

Figure 17.2 shows the enzymatic pathway whereby benzo[a]pyrene is converted to an electrophilic carcinogenic product called the **ultimate carcinogen.** (This is not the most carcinogenic compound in the world, just the most carcinogenic compound to which benzo[a]pyrene is metabolized.) The first step in the pathway is oxidation of the benzo[a]pyrene to an epoxide by an oxygenase. The epoxide then adds the elements of water under the direction of an epoxide hydrase. The product, a dihydrodiol, is oxidized again by the oxygenase to the ultimate carcinogen, a diol epoxide.

Figure 17.3 demonstrates how this diol epoxide can attack an electron-rich center on a DNA base, altering the base, and ultimately the base pair.

Another environmental agent known to cause cancer is radiation. We have seen in chapter 11 that radiation is mutagenic; several lines of evidence show that it is also carcinogenic. For example, consider the fact that people with light complexions, who have relatively little defense against ultraviolet radiation from the sun, are much more likely than dark-skinned people to develop skin cancer. Some people, born with a condition called xeroderma pigmentosum (see chapter 11), have defective DNA repair machinery. As a result, if they dare to venture out in the sun, they develop dozens of skin cancers, including lethal malignant melanomas. This suggests that the unrepaired DNA lesions caused by the ultraviolet radiation are responsible for the cancers. However, the situation may be a bit more complex than that.

Recall from chapter 11 that bacteria have an error-prone DNA repair system. If this system is blocked, the bacteria have a much lower mutation rate. In other words, the improper repair of DNA damage, rather than the damage itself, seems to cause many permanent mutations. If eukaryotes also

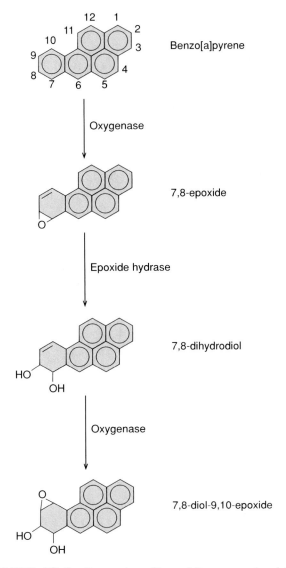

FIGURE 17.2 Conversion of benzo[a]pyrene to the ultimate carcinogenic form.

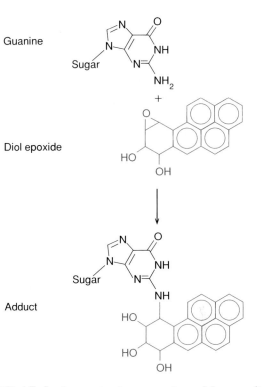

FIGURE 17.3 Interaction between a benzo[a]pyrene diol epoxide (red) and the DNA base guanine.

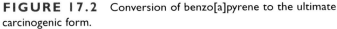

Mutagenic chemicals and radiation can lead to cancer, probably because of the DNA damage they cause. Chemicals can sometimes act directly, but usually they need to be metabolized to an electrophilic ultimate carcinogen that can attack DNA. Direct damage due to chemicals or radiation appears to be carcinogenic, but the inaccurate repair of such damage may also contribute to cancer.

behave this way, the real mutagenic (and carcinogenic) event in some xeroderma pigmentosum patients may not be unrepaired damage but misrepaired damage. The same may be true of the DNA damaged by chemical carcinogens. It may be the cell's inaccurate attempts to repair bases damaged by these chemical agents that cause the permanent mutations leading to cancer.

A related argument may apply to the damage caused by more energetic radiation such as X rays. Remember that X rays cause double-stranded DNA breaks (chromosome breaks). This is obviously a mutagenic event, but it may become a carcinogenic event when the cell attempts to repair the damage by sticking together the severed ends of chromosomes. If more than one chromosome break has occurred, the wrong chromosome segments may be rejoined, resulting in translocations. Such translocations could activate or inactivate genes inappropriately, giving rise to a cancer cell.

VIRUSES AND CANCER

Viruses represent a third kind of agent capable of causing cancer. By studying **oncogenic** (tumor-causing) viruses, molecular virologists have revealed the identities of many of the viral genes **(oncogenes)** that cause tumors. More surprisingly, they have also discovered cellular counterparts of these viral oncogenes, which are now implicated in cancers of nonviral origin. These cellular genes are called cellular oncogenes or, more accurately, cellular **proto-oncogenes.**

The history of tumor virology goes back to 1908, when V. Ellerman and O. Bang transferred leukemia from one chicken to another with an extract that contained viruses. Unfortunately, pathologists of that era did not recognize leukemia as a type of cancer, so they did not appreciate the importance of the experiment.

Only three years later, pioneer tumor virologist Peyton Rous performed similar experiments with chickens using a cell-free extract from a solid tumor. His procedure (figure 17.4) began with a trip to the market near his laboratory at the Rockefeller Institute in New York City, where he gathered **sarcomas** (solid tumors of fibrous tissue) from chickens on sale there. Next, he ground up the sarcomas to produce a homogenate, filtered it to remove anything larger than a virus, and used this filtrate to infect a second chicken. Since the second chicken developed sarcomas that, in turn, produced a filtrable agent that could infect a third chicken, it was clear that the filtrable agent (probably a virus) caused the tumors.

Again the experiments strongly suggested that a virus can cause cancer, and again they were ignored. One reason for the neglect of Rous's work was that chicken tumors were not regarded as suitable models for human cancer. Therefore,

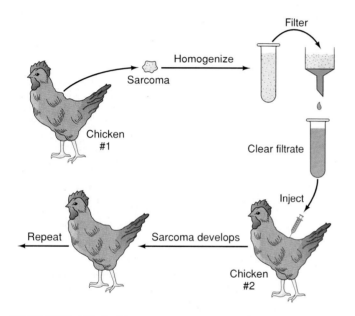

FIGURE 17.4   Viruses can cause tumors. Rous took a sarcoma from one chicken, homogenized it, then filtered the homogenate to yield a clear filtrate from which all substances larger than viruses had been removed. When this filtrate was injected into a second chicken, it caused sarcomas near the injection site. Infectious virus could be obtained from these sarcomas in the same way.

decades passed before mammalian experiments finally demonstrated to everyone's satisfaction that viruses really can cause cancer. Fortunately, Rous lived long enough to see his work win the recognition it deserved, including a Nobel prize in the 1970s. Today, the **Rous sarcoma virus (RSV),** also known as the **avian sarcoma virus (ASV),** is one of the best-studied tumor viruses.

The experiments just described were conducted according to Robert Koch's classic postulates for demonstrating the existence of an infectious agent: (1) Isolate the agent from a diseased animal; (2) use it to infect a second animal; and (3) find the agent in the second diseased animal. This sort of experiment is appropriate for lower animals, but obviously not for humans when cancer is the disease under study. Nevertheless, in two instances in the 1950s, humans were injected with live viruses that were later shown to be tumor viruses (see Box 17.1). Fortunately, no increase in cancer incidence has occurred in either case.

TUMOR VIRUS MOLECULAR GENETICS

Tumor viruses can have genomes made of either DNA or RNA. In fact, we have already discussed both kinds in previous chapters. Adenovirus and SV40 are two DNA tumor viruses that have served as very valuable models for research on gene expression in eukaryotes. For example, introns were first discovered in these two viruses (chapter 9), and adenovirus continues to be a favorite system in which to study splicing. The RNA tumor viruses are all **retroviruses** (chapter 12), which replicate via a double-stranded DNA intermediate. The AIDS virus, HIV (human immunodeficiency virus), is also a retrovirus. Figure 17.5 shows a model of HIV, which also serves to illustrate the structure of an RNA tumor virus.

The ultimate test of a virus's classification as a tumor virus is its **tumorigenicity**—the ability to cause tumors in whole animals. However, testing for this trait is very time-consuming; it may take months for a tumor to develop in a laboratory animal. Fortunately, more convenient systems are available, which may not tell us for certain whether a virus is tumorigenic, but can give a strong hint. The most common of these tests is a cell transformation assay. In this context, the word **transformation**

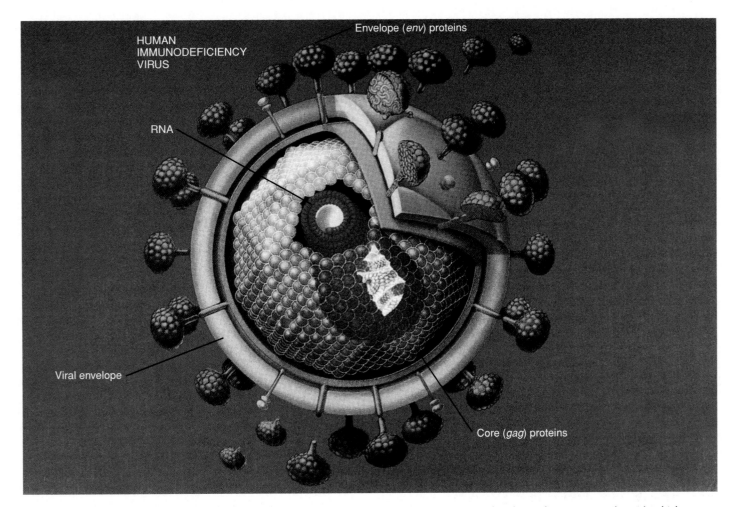

HUMAN
IMMUNODEFICIENCY
VIRUS

Envelope (*env*) proteins

RNA

Viral envelope

Core (*gag*) proteins

FIGURE 17.5 Model of human immunodeficiency virus (HIV), a retrovirus. At the center of the virus is the RNA genome (purple). This is surrounded by a coat made of core proteins (gray), which are products of the *gag* gene. The viral core is enclosed in an envelope (yellow) derived from the cellular plasma membrane. This envelope is decorated with envelope proteins (green), which are products of the viral *env* gene. Homologous genes are found in RNA tumor viruses, and the RNA tumor virus particles can be represented by this same model.

Courtesy Coulter Corporation.

means converting cultured cells from normal behavior to the transformed behavior expected of cancer cells. Thus, instead of observing the development of a tumor in an organism in response to a virus, we simply observe telltale changes in cellular behavior in a culture dish.

Dna and RNA viruses that can cause tumors are classified as tumor viruses. Such viruses have the capacity to transform cells in culture. That is, they can cause normal cells to begin behaving as cancer cells do.

Viral Oncogenes

Since tumor viruses can transform cells, they must contain genes that direct the transformation process. Therefore, identifying and studying these viral oncogenes has been of utmost importance, because it reveals clues about how viruses transform cells. In fact, this quest has been even more rewarding than one might have anticipated, since proto-oncogenes homologous to many viral oncogenes have been found in normal, uninfected human cells. This finding has profound implications: Since we carry genes very similar to viral oncogenes, alterations in the structure or expression of these proto-oncogenes could presumably make them function as true oncogenes—and thus spawn tumors. In fact, as we will discover shortly, the viral oncogenes actually originated as host genes. Retroviruses can pick up these and other genes, which allow them to replicate more successfully, but not all retroviruses have oncogenes.

Because some retroviruses harbor oncogenes that are homologous to cellular proto-oncogenes, let us begin our discussion of oncogenes with a brief examination of the genetics of the retroviruses. Figure 17.6 shows the genetic map of the archetypal transforming retrovirus, the Rous sarcoma virus. This is the DNA form of the viral genome, also called the provirus (see chapter 12 and next paragraph). Three genes are necessary for its replication. The first, called *gag*, codes for a polyprotein that is

FIGURE 17.6 Genetic map of Rous sarcoma virus (DNA, or proviral form of genome). The LTRs on each end of the genome are direct repeats known as long terminal repeats. The genes *gag, pol, env,* and *src* code for the coat proteins, the reverse transcriptase, the membrane proteins, and the transforming protein, respectively.

cleaved to produce several polypeptides that form the viral coat. The second, **pol,** encodes the reverse transcriptase that carries out the replication of the viral RNA via the DNA provirus (chapter 12). The third, **env,** codes for the proteins that cover the outer viral membrane, or **envelope.**

Finally, at the 3′-end, lies **src,** the viral oncogene. This gene is not needed for viral replication—indeed, retroviruses presumably started without oncogenes, and some continue to survive very well without them—but it may confer a competitive advantage on the virus that contains it. The usefulness of the oncogene from the virus's standpoint is that it transforms cells and causes them to grow incessantly. This, in turn, replicates the viral genome, which has been inserted into the host DNA as a **provirus.**

RSV is unusual in that it can replicate on its own. Most transforming retroviruses, in the process of incorporating cellular genes, have lost some of their own, making them **defective.** Some transforming retroviruses contain no intact viral genes at all—only an oncogene and some other cellular DNA sandwiched between two LTRs. These viruses rely on nondefective **helper viruses** to supply the missing functions and allow them to replicate.

As we have suggested, oncogenes apparently did not originate in viruses. Geneticists have now identified dozens of viral oncogenes (table 17.1), and they all seem to have very closely related cellular counterparts called proto-oncogenes. How do we know the viruses picked these genes up from cells, rather than vice versa? The most compelling argument is that the cellular genes typically contain introns, whereas the viral genes do not. This situation is easy to explain by a cell → virus flow of genes (figure 17.7). First a retroviral provirus recombined with a cellular proto-oncogene. Next, the provirus, including the proto-oncogene, was transcribed, then processed to remove the introns. This whole RNA was packaged into a retrovirus particle, which went on to infect another cell. Each of these steps uses well-established genetic mechanisms. By contrast, we know of no precedent for the intron insertion that would have to occur to create a cellular proto-oncogene from a viral oncogene.

In addition to the theoretical argument, ample evidence indicates that a transformation-defective retrovirus can appropriate an oncogene from its host. In one experiment, RSV was deprived of most of its *src* gene (called **v-src** for viral *src* gene) and then injected into chickens. A wild-type RSV with *src* caused tumors near the site of injection within a very short time. Since the defective virus lacked a functional oncogene, it would not be expected to cause tumors at all; however, two months later, some tumors did arise in places remote from the injection site. RSV

particles isolated from these tumors contained functional *src* genes. The simplest explanation is that a cellular *src* gene (**c-src**) recombined with the remnant of the viral oncogene, making it whole again. This mechanism would be similar to the one depicted in figure 17.7, but it would occur much more frequently, because the v-*src* remnant in the provirus would promote homologous recombination with the cellular proto-oncogene.

Slow transforming viruses represent another case in which a retrovirus does not need an oncogene to be a tumor virus. For example, avian leukosis virus (ALV) completely lacks an oncogene, yet can cause tumors because it can activate a cellular proto-oncogene. Several investigators have demonstrated that the ALV provirus inserts next to a cellular proto-oncogene known as c-*myc*. The enhancer embedded in the viral LTR then acts on the nearby c-*myc* gene, activating it and contributing to the transformation process (figure 17.8).

Because of the difficulty of demonstrating that tumor viruses cause human cancer, there are relatively few known human tumor viruses. Nevertheless, there are several examples of viruses that are so strongly associated with certain kinds of cancer that we assume they are involved, even though the evidence is still incomplete. The best studied of these include two DNA viruses and one class of retroviruses. The presumptive human DNA tumor viruses are Epstein-Barr virus, which appears to be involved in Burkitt's lymphoma (see later in this chapter), and human papillomavirus, which is strongly associated with cervical carcinoma. The class of retroviruses includes human T-cell leukemia viruses I and II (**HTLV-I** and **HTLV-II**).

HTLV-I and -II are very similar in their genetic organization to HIV. They even attack the same kind of cells that HIV does: human T lymphocytes. But their effects are very different. Whereas HIV ultimately kills its host cells, HTLV-I and -II transform them. The oncogene in these latter viruses is similar to the HIV *tat* gene, and has a similar name, **tax.** The effect of *tax* is also similar to that of *tat*. It stimulates transcription of the viral provirus to provide a burst of viral gene products just after the infection process begins. But *tax* may also stimulate host genes, perhaps including proto-oncogenes, and thereby stimulate cell division.

All known RNA tumor viruses are retroviruses. Their RNA genomes may contain genes for coat proteins, reverse transcriptase, and envelope surface proteins. They also usually contain an oncogene that can transform host cells. The viral oncogenes appear to have descended from closely related cellular proto-oncogenes. Some retroviruses operate without oncogenes of their own; instead, their provirus forms activate endogenous cellular proto-oncogenes. Most transforming retroviruses lack one or more of the genes required for viral replication. These defective viruses rely on helper viruses for their replication. Two retroviruses strongly implicated in human cancer (T-cell leukemia) are HTLV-I and -II.

Table 17.1   Functions of Cell-Derived Oncogene Products

Class 1—Growth Factors	
sis	PDGF growth factor β-chain
int-2	FGF-related growth factor
hst (KS3)	FGF-related growth factor
FGF-5	FGF-related growth factor
int-1	Growth factor?

Class 2—Receptor and Nonreceptor Protein-Tyrosine Kinases	
src	Membrane-associated nonreceptor protein-tyrosine kinase
yes	Membrane-associated nonreceptor protein-tyrosine kinase
fgr	Membrane-associated nonreceptor protein-tyrosine kinase
lck	Membrane-associated nonreceptor protein-tyrosine kinase
fps/fes	Nonreceptor protein-tyrosine kinase
abl/bcr-abl	Nonreceptor protein-tyrosine kinase
ros	Membrane associated receptor-like protein-tyrosine kinase
erbB	Truncated EGF receptor protein-tyrosine kinase
neu	Receptor-like protein-tyrosine kinase
fms	Mutant CSF-1 receptor protein-tyrosine kinase
met	Soluble truncated receptor-like protein-tyrosine kinase
trk	Soluble truncated receptor-like protein-tyrosine kinase
kit (*W* locus)	Truncated stem cell receptor protein-tyrosine kinase
sea	Membrane-associated truncated receptor-like protein-tyrosine kinase
ret	Truncated receptor-like protein-tyrosine kinase

Class 3—Receptors Lacking Protein Kinase Activity	
mas	Angiotensin receptor

Class 4—Membrane-Associated G Proteins	
H-*ras*	Membrane-associated GTP-binding/GTPase
K-*ras*	Membrane-associated GTP-binding/GTPase
N-*ras*	Membrane-associated GTP-binding/GTPase
gsp	Mutant activated form of $G_s \alpha$
glp	Mutant activated form of $G_i \alpha$

Class 5—Cytoplasmic Protein-Serine Kinases	
raf/mil	Cytoplasmic protein-serine kinase
pim-1	Cytoplasmic protein-serine kinase
mos	Cytoplasmic protein-serine kinase (cytostatic factor)
cot	Cytoplasmic protein-serine kinase?

Class 6—Cytoplasmic Regulators	
crk	SH-2/3 protein that binds to (and regulates?) phosphotyrosine-containing proteins

Class 7—Nuclear Transcription Factors	
myc	Transcription factor
N-*myc*	Transcription factor
L-*myc*	Transcription factor
myb	Transcription factor
lyl-1	Sequence-specific DNA-binding protein
fos	Combines with c-*jun* product to form AP-1 transcription factor
jun	Sequence-specific DNA-binding protein; part of AP-1
erbA	Dominant negative mutant thyroxine (T_3) receptor
rel	Dominant negative mutant NF_κ-B-related protein
vav	Transcription factor?
ets	Sequence-specific DNA-binding protein
ski	Transcription factor?
evi-1	Transcription factor?
gli-1	Transcription factor?
maf	Transcription factor?
pbx	Chimeric E2A-homeobox transcription factor
Hox2.4	Transcription factor?
Unclassified	
dbl	Cytoplasmic truncated cytoskeletal protein?
bcl-2	Plasma membrane signal transducer?

From Hunter, *Cell*, 64:256. Copyright © 1991 Cell Press, Cambridge, MA.

CONVERSION OF CELLULAR PROTO-ONCOGENES TO ONCOGENES

We have seen that viral oncogenes can transform cells, causing malignant tumors, and that their cellular cousins can do the same thing, at least under certain conditions. In fact, we now have abundant evidence that harmless cellular proto-oncogenes can convert to true oncogenes, which at least contribute to **oncogenesis,** the formation of tumors. We even know some of the conditions under which this conversion takes place.

Point Mutations in Proto-Oncogenes

The most straightforward mechanism for converting a cellular proto-oncogene to an oncogene is mutation. Since viral oncogenes resemble their cellular counterparts so closely, it is not difficult to imagine a few well-chosen mutations causing the conversion. In fact, in at least one case—the **Ha-*ras*** oncogene—a single point mutation is sufficient.

The c-Ha-*ras* proto-oncogene is the cellular version of the Harvey rat sarcoma virus oncogene (v-Ha-*ras*). Here is the story of how this gene was implicated in human cancer. First of all,

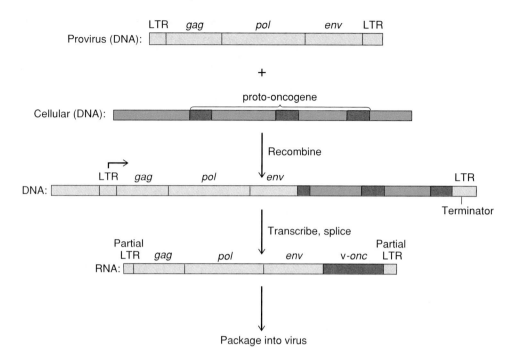

FIGURE 17.7 Hypothetical conversion of a cellular oncogene with introns to a viral oncogene without introns. The provirus, lacking an oncogene, recombines with a cellular proto-oncogene containing two introns (blue). Transcription and splicing of the recombinant provirus yields a viral RNA with the coding sequence of the oncogene (red) without the introns. This RNA can be packaged into virus particles that go on to infect a new cell.

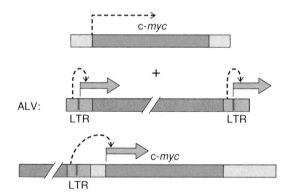

FIGURE 17.8 Activation of a cellular proto-oncogene by an inserted provirus. Before insertion, the cellular proto-oncogene (c-*myc*, blue) is transcribed only weakly, if at all. After the avian leukosis virus (ALV) provirus inserts, c-*myc* transcription is greatly stimulated (---›) by the enhancer (red) in the nearby LTR of the provirus.

Michael Wigler and his colleagues demonstrated that human bladder carcinoma cells contain an oncogene capable of transforming NIH 3T3 cells. Specifically, this team of scientists introduced DNA from a line of human bladder cancer cells into 3T3 cells. The 3T3 cells are mouse cells that are immortal and apparently poised on the brink of transformation. Something in the bladder cancer cell DNA pushed them over into the transformed state (figure 17.9). DNA from normal human cells could not transform them.

Which gene has the transforming activity? Robert Weinberg and his coworkers used radioactive oncogenes to probe Southern blots (chapter 15) of DNA fragments from 3T3 cells transformed by human bladder carcinoma DNA. These experiments revealed that the transformed 3T3 cells have a new DNA band not found in ordinary 3T3 cells, which reacts with a Ha-*ras* probe (figure 17.9). Therefore, the human cancer DNA has introduced a Ha-*ras* oncogene into the 3T3 cells. Furthermore, we know that this new gene is active in the transformed cells, because Northern blots of RNAs made in transformed cells, but not in untransformed 3T3 cells, contain an RNA that reacts with the human Ha-*ras* probe.

Since human bladder cancer cells, but not ordinary bladder cells, contain a transforming agent, and the agent seems to be the Ha-*ras* oncogene, we would like to know how this oncogene differs from the harmless Ha-*ras* proto-oncogene in normal bladder cells. To answer this question, Mariano Barbacid and his colleagues took fragments of the oncogene and proto-oncogene, recombined them in vitro, and used the recombinant genes to try to transform 3T3 cells (figure 17.10). This process narrowed down the important region to a 350-base-pair segment of the gene. Whenever this region came from the cancer cell oncogene, the recombinant gene had transforming activity. When it came from normal cell DNA, it did not transform. A single base change (a G → T transversion) was the only difference between this region in the oncogene and the same region in the proto-oncogene. Therefore, it appears that this subtle change is what converted the proto-oncogene to the oncogene.

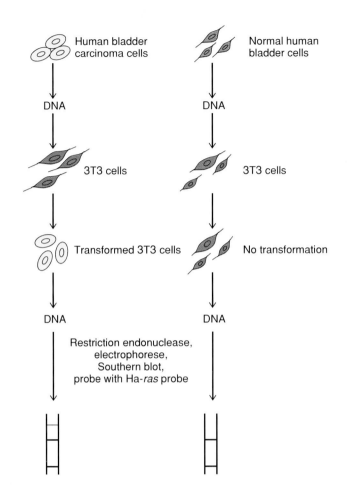

FIGURE 17.9 Human bladder carcinoma cells have DNA with transforming potential. Starting at the top of this figure, DNA is extracted from either human bladder carcinoma cells (left) or normal human bladder cells (right) and introduced into NIH 3T3 cells. The cells that receive the carcinoma DNA become transformed, but those that receive normal cell DNA remain normal. Then, DNA is extracted from both sets of cells that have received exogenous DNA, cut with a restriction endonuclease, electrophoresed, Southern blotted, and hybridized to a radioactive Ha-*ras* DNA probe. Two common bands "light up" in both Southern blots, corresponding to an endogenous mouse *ras* gene of NIH 3T3 cells, but a third band (blue) is found only in the transformed cells. This band corresponds to the human Ha-*ras* gene that transformed these cells.

The Ha-*ras* oncogene belongs to a family of *ras* oncogenes that encode proteins with a molecular weight of 21,000 called **p21** or **Ras.** The G → T transversion that converts the proto-oncogene to the oncogene in the human bladder carcinoma changes a glycine (the twelfth amino acid in Ras) to a valine. When does the change occur? It seems to be a somatic mutation, since it appears in human lung cancer cells but not in normal trachea cells from the same patient. Is this the only cause of the lung cancer? That seems very doubtful, because oncogenesis appears to be a multistep process that takes years to complete. The mutation in *ras* is probably just one step in that process.

Since the pioneering work of Wigler, Weinberg, and others, mutated *ras* genes have been found in a variety of human tumors, including 40% of colon cancers. These mutations are not confined to codon 12, but almost all of them are clustered in three regions of the *ras* gene: codons 12–16; codons 59–63; and codons 116–119. We will discuss the functions of Ras and the effects of these mutations later in this chapter.

Certain human tumor cells contain oncogenes of the *ras* family that can transform mouse 3T3 cells in culture. Normal human DNA lacks this transforming activity. The cellular Ha-*ras* proto oncogene in the first tumor cells studied had suffered a one-base mutation, a G → T transversion in the twelfth codon, which changed a glycine to a valine in the protein product. This was enough to convert the proto-oncogene to an oncogene. Almost all the *ras* mutations found in human tumor cells are clustered in three small regions of the gene.

Proto-Oncogene Translocation

Point mutations are one way to change a proto-oncogene into an oncogene; translocations are another. One of the best-studied examples of a translocation that converts a proto-oncogene to an oncogene concerns the *myc* proto-oncogene in the human cancer **Burkitt's lymphoma.** These cancer cells have typically suffered an 8;14 translocation. In particular, a piece of chromosome 8 bearing *myc* has been translocated to the part of chromosome 14 carrying a cluster of antibody H-chain genes (figure 17.11).

As discussed in chapter 14, antibody genes normally undergo a translocation in which a variable region is juxtaposed with a constant region. The 8;14 translocation in Burkitt's lymphoma seems to be a corruption of this normal process. Instead of a variable region, it is the *myc* gene that winds up next to an H-chain constant region. In the process, two things happen to the *myc* gene that may disturb its normal function. First, the gene loses the first of its three exons. This has no effect on the protein product of the gene, since the first exon does not encode any amino acids, but it may affect the regulation of the gene. Second, the translocated gene has moved close to the very active antibody H-chain genes. This could have an activating effect on *myc*, which is indeed frequently overactive in Burkitt's lymphoma cells.

Thus, the *myc* story provides an example not only of altering a proto-oncogene by translocation, but of activating an oncogene. And that is not all. The translocated *myc* gene may also escape from the mechanisms that normally keep it under control. Which of these factors is important in the transformation process? We do not know the answer yet; it may be that all are important.

Before we leave the topic of *myc* and Burkitt's lymphoma, we should mention that there seems to be another agent at work in this cancer: **Epstein-Barr virus (EBV).** This is the same virus that causes infectious mononucleosis in most places in the world. In fact, about 90% of Americans have been infected by this virus by adulthood, most of them without even knowing it. But this

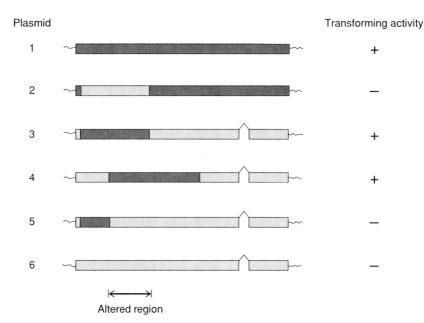

Plasmid   Transforming activity

1   +

2   −

3   +

4   +

5   −

6   −

|←——→|
Altered region

FIGURE 17.10   Pinpointing the difference between the Ha-*ras* oncogene and proto-oncogene. Cloned plasmids containing the oncogene (red) or proto-oncogene (yellow) were cut and rejoined in vitro in different combinations, as shown. In addition, deletions (Λ) were introduced into four of the plasmids. These recombinant plasmids were then tested for transformation activity, and the results are reported at right. Obviously, the deletion did not affect transformation activity, since plasmids 3 and 4, with the deletion, retain full activity. Furthermore, since plasmids 3 and 4 have activity, the alteration in the oncogene must lie in the region of overlap between the oncogene parts of these two plasmids ("altered region" shown at bottom). This conclusion is reinforced by plasmids 2 and 5, which have proto-oncogene sequences in this region and no transformation activity. Further work revealed only one base difference between oncogene and proto-oncogene in this region.

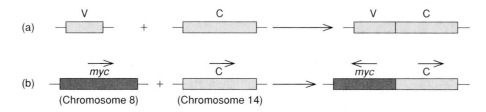

FIGURE 17.11   The 8;14 translocation in Burkitt's lymphoma as a corruption of a normal antibody gene rearrangement. (*a*) Normal antibody gene rearrangement. The variable and constant regions (V and C, respectively) start out as separate entities; recombination brings them together as a functioning antibody heavy chain gene. D and J regions and introns have been omitted throughout this figure for simplicity. (*b*) One kind of abnormal rearrangement involving *myc* from chromosome 8 and an antibody heavy chain constant region from chromosome 14. The *myc* gene, minus its first exon, becomes attached to part of the antibody constant region gene, with the two gene fragments in opposite orientations.

virus is strongly associated with Burkitt's lymphoma in children in central Africa. The difference could be related to the effects of malaria, also endemic to central Africa, on the immune system.

> Some proto-oncogenes suffer translocations that convert them to oncogenes. In Burkitt's lymphoma, the *myc* oncogene is translocated next to a cluster of antibody H-chain genes. In the process, it loses its first (noncoding) exon, becomes activated, and loses its normal control pattern.

Proto-Oncogene Activation

We have already seen two examples of malignancies associated with activation of cellular proto-oncogenes: the activation of the *myc* gene in chickens by the nearby insertion of the strong promoter/enhancer of the avian leukosis virus and the activation of the *myc* gene in humans suffering from Burkitt's lymphoma. Although further examples exist, the phenomenon is by no means universal. Many cancers thrive without obvious activation of proto-oncogenes.

Proto-Oncogene Amplification

Proto-oncogene **amplification** is a special case of activation. More gene product is made, but the primary reason is not

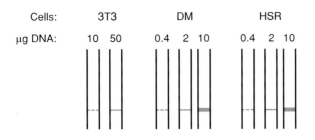

FIGURE 17.12 Amplification of *myc* in tumor cells. Different amounts of DNA from untransformed 3T3 cells, cells showing double minute chromosomes (DM), and cells showing homogeneously staining regions (HSR) were cut with *Eco*RI, electrophoresed, and Southern blotted. The blots were hybridized to radioactive Ki-*ras* DNA, and autoradiographed. The DM and HSR DNAs, respectively, show about 60-fold and 30-fold higher hybridization to the Ki-*ras* probe, demonstrating amplification of that oncogene in these tumor cells.

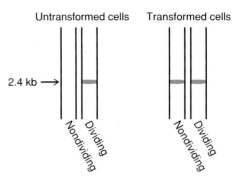

FIGURE 17.13 Loss of control over *myc* in transformed cells. RNA from dividing and nondividing cells that are untransformed or chemically transformed was electrophoresed, Northern blotted, hybridized to radioactive *myc* DNA, and autoradiographed. In dividing untransformed cells, there is a prominent *myc* mRNA of 2.4 kb; this RNA is not produced by nondividing untransformed cells. Thus, the *myc* gene is controlled in untransformed cells. In chemically transformed cells, this control is lost; *myc* mRNA is produced whether the cells are dividing or not.

enhanced expression of the gene. Instead, the gene is copied over and over again (amplified) so that it is finally represented by many more than the usual two copies. Since the cell has an extra dose of proto-oncogene, it will produce more of that gene's product. For example, Harold Varmus and Michael Bishop and their collaborators examined several tumor cell lines for telltale signs of gene amplification. One of these signs is the presence of **double minute chromosomes (DMs),** tiny chromatin bodies lacking centromeres (and containing amplified DNA). Another is the presence of **homogeneously staining regions (HSRs),** parts of chromosomes where the regular, banded staining pattern gives way to an amorphous, uniformly stained zone that contains amplified DNA.

The first two cell lines in which these workers detected probable gene amplifications were from tumors of the mouse adrenal gland. The cell lines are called Y1-DM (with double minute chromosomes) and Y1-HSR (with homogeneously staining regions). Then they looked to see whether a known oncogene had been amplified. Sure enough, the **Ki-*ras*** gene, a cousin of the Ha-*ras* gene, was amplified—about 30-fold and 60-fold, respectively. How did they reach this conclusion? First, they cut various amounts of DNA from the two tumor cell lines with the restriction enzyme *Eco*RI, then electrophoresed the fragments, Southern blotted them, and hybridized the blots to a Ki-*ras* probe. Figure 17.12 shows the results. DNA from the two tumor cell lines gave much stronger hybridization to the probe than did the same amount of DNA from the untransformed mouse cell line, 3T3. Quantifying the amount of radioactivity in the bands tells us how much the Ki-*ras* gene is amplified in the tumor cells.

Loss of Control over Proto-Oncogenes

Arthur Pardee and Gail Sonenshein and their colleagues examined the expression of the *myc* gene in normal and chemically transformed 3T3 cells. They found that the gene was not rearranged, amplified, or over-expressed. However, the gene had lost its response to normal regulation. Using Northern blots

(figure 17.13), these investigators showed that the *myc* gene made no detectable RNA product in quiescent (nondividing cells) normal cells, but a very significant amount in proliferating (dividing) normal cells. On the other hand, after chemical transformation, 3T3 cells made the same amount of *myc* RNA in the quiescent and proliferating states. Thus, when these transformed cells ceased multiplying, they did not turn off the *myc* gene the way normal cells would.

> Oncogenes or proto-oncogenes are frequently activated to abnormally high levels in tumor cells. Sometimes these genes are even amplified, resulting in unusually high production of gene products. Another alteration to a proto-oncogene that may predispose it to participate in oncogenesis is loss of control of the gene's expression.

THE FUNCTIONS OF PROTO-ONCOGENES

The probable functions of several proto-oncogenes have been deduced by examining their base sequences. This frequently reveals close similarity between a proto-oncogene and a gene of known function, so we can guess that the two genes' functions are similar as well. In most cases, proto-oncogenes seem to participate in controlling cell division. Thus, it is easy to see how mutations in these genes could cause a cell to lose control over its growth and become cancerous. We will examine several examples, which fall into four categories: 1) growth factors, 2) growth factor receptors, 3) signal transducers, and 4) transcription factors.

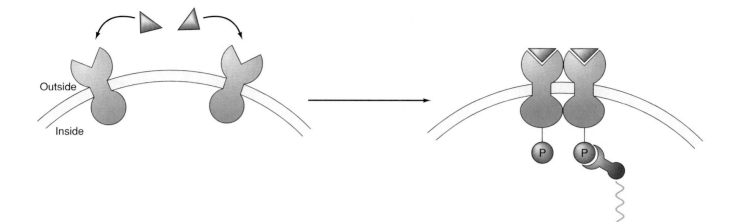

FIGURE 17.14 Function of EGF. Molecules of EGF (red) bind to EGF receptors (blue) that span the cell membrane (yellow). This binding of the growth factor to the receptor domain on the outside of the cell causes the two receptors to come together. Now the protein tyrosine kinase domains of each receptor (round domains inside the cell) are activated to phosphorylate each other. This attracts an adapter molecule (purple) that sends a signal to the cell to divide.

ONCOGENE PRODUCTS AND GROWTH FACTORS

The salient feature of cancer cells is growth. They grow inexorably, as if they sense a constant signal to divide, even though growth is inappropriate. Normal cells grow only when they receive a real signal to do so; one such signal is a growth factor.

Rita Levi-Montalcini and Stanley Cohen shared a Nobel prize in 1986 for their discovery and characterization of the first growth factor, **nerve growth factor (NGF),** in the 1950s. (This is not the same Stanley Cohen who was involved in the earliest gene cloning.) As its name implies, NGF stimulates nerve cells to grow. Two related proteins, **platelet-derived growth factor (PDGF)** and **epidermal growth factor (EGF),** have similar growth promoting properties. Both of these factors exert their effects by binding to specific receptor proteins on the surfaces of cells. The receptor then transmits the growth signal to the interior of the cell.

Cohen and his colleagues have shown that the EGF receptor spans the cell membrane, as depicted in figure 17.14. The part of the protein protruding outside the cell binds the growth factor, which activates the other part of the protein that is in contact with the cytoplasm. Once activated, the inner protein domain expresses its enzymatic activity—protein kinase, which transfers phosphate groups from ATP to proteins. Most protein kinases phosphorylate (add phosphate groups to) the amino acids serine and threonine, but EGF is unusual in that it phosphorylates the amino acid tyrosine.

In light of all this, it is interesting to learn that one oncogene, *sis,* encodes a protein that strongly resembles PDGF and that another, *erb-*B, encodes a protein very much like the protein tyrosine kinase part of the EGF receptor. Two attractive hypotheses for the mechanisms of these oncogenes are easy to imagine. Perhaps a cell that produces too much of the c-*sis* product, or produces it at the wrong time, stimulates *itself* to divide in an uncontrolled way. Or perhaps a cell that overproduces, or inappropriately produces, the *erb-*B product places a sawed-off growth factor receptor into its membrane—a receptor that lacks the growth factor binding part but behaves as if it is always bound to growth factor. This would send a constant growth signal to the interior of the cell.

Growth factors stimulate cells to divide by binding to receptors on cell surfaces and activating a tyrosine protein kinase activity on the cytoplasmic sides of these same receptor proteins. The *sis* oncogene strongly resembles the gene for platelet-derived growth factor (PDGF), and the *erb*-B oncogene is very much like the tyrosine kinase coding region of the epidermal growth factor (EGF) gene. This resemblance to growth promoting substances may explain how oncogene products cause cells to grow uncontrollably.

THE c-*src* PRODUCT: TYROSINE PROTEIN KINASE

The classic example of a viral oncogene with a cellular counterpart is *src.* The products of both the v-*src* and c-*src* genes are protein tyrosine kinases. How do we know this? The first evidence was genetic: Some Rous sarcoma virus mutants are defective in transformation, so they must have mutant *src* genes; these mutants also lack tyrosine kinase activity. Even more convincing are temperature-sensitive mutants whose transforming activity and tyrosine kinase activity both disappear at high temperature. This strongly suggests that the two activities are encoded in the same gene; since we already know that the transforming activity resides in the *src* gene, this means the tyrosine kinase activity does too.

The *src* gene product is itself a target for tyrosine kinase, so it is a phosphoprotein. It is usually called **pp60^src,** which denotes a phosphoprotein (pp) with a molecular weight of 60,000. The

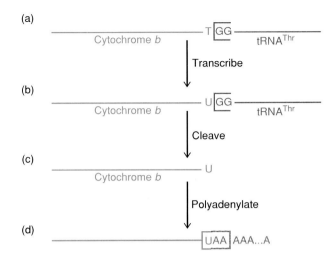

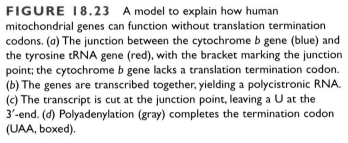

FIGURE 18.23 A model to explain how human mitochondrial genes can function without translation termination codons. (*a*) The junction between the cytochrome *b* gene (blue) and the tyrosine tRNA gene (red), with the bracket marking the junction point; the cytochrome *b* gene lacks a translation termination codon. (*b*) The genes are transcribed together, yielding a polycistronic RNA. (*c*) The transcript is cut at the junction point, leaving a U at the 3'-end. (*d*) Polyadenylation (gray) completes the termination codon (UAA, boxed).

of these open reading frames results in a human genetic disease (see Box 18.1).

Genes reside on both strands of the human mitochondrial DNA, and the mitochondrial RNA polymerase appears to transcribe each strand in one piece, starting from a single promoter on each strand. This means that control of gene expression in mitochondria is relatively primitive and must depend on posttranscriptional mechanisms.

Once the genome-sized RNA is produced, it must be chopped up into usable pieces. We are not sure how this is accomplished, but Giuseppi Attardi has proposed an attractive idea called the **tRNA punctuation hypothesis.** Look again at figure 18.22 and notice that almost all coding regions, rRNA and mRNA alike, are flanked by tRNA genes. Therefore, all that is needed to process the large precursor RNA is a ribonuclease that recognizes the ends of tRNAs and cuts there. This enzyme would free the tRNAs and, in the process, free the larger rRNAs and mRNAs as well.

Clearly, DNA sequencing has given us great insight into the structures of mitochondrial, as well as chloroplast, genomes. However, it is a good idea to reflect on a lesson the open reading frames teach us about both the power and the weakness of DNA sequencing as a genetic mapping tool. Sequencing is extremely powerful in that it reveals the absolute structure of a

---

BOX 18.1 **Mitochondrial Mutations and Human Disease**

Human cells contain only two copies of most chromosomes, so two mutations are all that is needed to cause total inactivation of a given nuclear gene. On the other hand, a typical human cell contains thousands of mitochondria. Does this mean that thousands of independent mitochondrial mutations are required to see a phenotypic effect? If this were so, we would probably never see a human disease that we could attribute to a mitochondrial mutation. However, this must not be the case, because we can trace several human genetic diseases to mutant mitochondrial genes. This is probably because, over time, certain cells will accumulate a preponderance of mutant mitochondria, and even a partial loss of energy production by these mitochondria may be sufficient to interfere with the cell's activity.

This seems to be the case in the best-known example of a disease we can trace to a mitochondrial malfunction: Lebers hereditary optic neuropathy (LHON). This maternally inherited disease is characterized by deterioration of the optic nerve, and consequent loss of vision in early adulthood. The mutation that seems to cause the disease is a G → A transition in ND4, the gene for the fourth subunit of the mitochondrial enzyme NADH dehydrogenase (figure 18.22). In most tissues, this is not a problem, but in optic nerve cells the reduced ability to make ATP is apparently enough to kill the cells and therefore to destroy vision.

Another example of a maternally inherited disease that appears to be due to a mitochondrial mutation is MERRF, which

stands for "myoclonic epilepsy with ragged red fibers." Although geneticists have not yet pinned down the exact cause of this disorder, they have found that the mitochondria in MERRF patients are, to varying degrees, deficient in function. Moreover, the more profound the mitochondrial deficiency, the more severe the symptoms. Another finding that points to mitochondria as the culprits is that the tissues involved in MERRF are those that depend to the greatest extent on the energy that mitochondria furnish.

Several other diseases are not inherited, but seem to derive from somatic mutations involving large deletions of mitochondrial DNA, reminiscent of poky mutations in *Neurospora* and some petites in yeast. These include Kearns-Sayres syndrome (KSS) and chronic progressive external ophthalmoplegia (CPEO), both neuromuscular disorders, and Pearson's syndrome, a fatal malfunction in blood cell production. Of much more general interest is the possibility that a very widespread "disorder"—aging—may be caused in part by progressive inactivation of mitochondria through mutation. According to this hypothesis, time would favor deterioration in two ways: As the body ages, it presumably accumulates more and more mitochondrial mutations; in addition, time would allow for accumulation of a high enough proportion of mutant mitochondria in certain cells to impair their function. Both these trends would lead in turn to the progressive deterioration of the body's tissues that is the hallmark of aging.

gene. However, when the gene we sequence has no known function, its sequence alone can rarely tell us anything about that function.

> The total base sequences of human and bovine mitochondrial DNAs have been determined. This information reveals very little wasted space in these genomes. There are no introns, no untranslated leaders or trailers in the encoded mRNAs, and few if any bases between structural genes. The mammalian mitochondrial genomes encode two rRNAs, twenty-two tRNAs, and thirteen proteins.

Wasted Space in Yeast Mitochondrial Genomes

What accounts for the five-fold difference between the sizes of yeast and mammalian mitochondrial genomes? There are several contributing factors, one being that the yeast genome includes "wasted space" between genes, as figure 18.24 demonstrates. The other differences are exemplified by a single gene, which codes for a mitochondrial protein called cytochrome b. Figure 18.24 depicts the yeast and human cytochrome b genes and their mRNA products. The human gene contains only a coding region; the yeast gene has a very similar coding region, but much more besides. It has an untranslated leader region of about 1,000 bases at its 5′-end and an untranslated trailer region of about 50 bases at its 3′-end. It also has five long introns that expand the total size of the gene to many times the size of the corresponding human gene. By contrast, only after transcription does the human mRNA acquire any untranslated sequences—namely, a poly(A) tail, the first two bases of which constitute the last two

bases of the UAA stop codon. The yeast genome also has at least five independent promoters, so the whole genome is not transcribed in one piece.

How do we relate the map of the yeast mitochondrial structural genes shown in figure 18.22 with the genetic map presented previously (figure 18.11)? Most of the markers discussed earlier are also included in figure 18.22. Note that resistance to the antibiotics erythromycin, chloramphenicol, and spiramycin is conferred by mutations in the gene for the large (21S) rRNA, while resistance to paromomycin is conferred by mutations in the gene for the small (15S) rRNA. These findings fit with the known functions of these antibiotics as blockers of protein synthesis. Finally, resistance to oligomycin is conferred by mutations in the genes for subunits of the enzyme ATPase. Again, this makes sense because oligomycin is known to interfere with the synthesis of ATP in mitochondria, and ATPase is an integral part of the ATP synthesis scheme. This also explains why researchers have found two loci that confer oligomycin resistance: Yeast mitochondrial DNA encodes two subunits of ATPase.

> The yeast mitochondrial DNA is about five times larger than the mammalian. It includes most of the same information found in mammalian genomes, including genes encoding rRNAs, tRNAs, and mRNAs. In addition, it contains introns, long untranslated leaders and trailers at the ends of the encoded mRNAs, plus noncoding spacers between genes.

THE ORIGIN OF MITOCHONDRIA AND CHLOROPLASTS

Where did mitochondria and chloroplasts come from? Why do eukaryotes have them, while prokaryotes do not? Here is the most widely accepted hypothesis: Life started in prokaryotic form; for perhaps a billion years or more, it was the only kind of life on earth. Then, more than a billion years ago, cells with nuclei appeared. These first cells lacked mitochondria and chloroplasts. (Some such primitive eukaryotes persist to this day.) Probably within a short time after the first eukaryotes appeared, one of them engulfed a bacterium. However, instead of treating the bacterial cell as food and digesting it, the larger cell tolerated its guest. The two cells divided more or less together, producing many large progeny cells containing smaller guests. In time, the smaller cells began to specialize in energy production, but they lost many genes, so they could no longer live independent of their host. At the same time, the host lost some of its genes, so it could no longer produce energy efficiently; it became dependent on its guest for energy. The guest had become an ancestral mitochondrion.

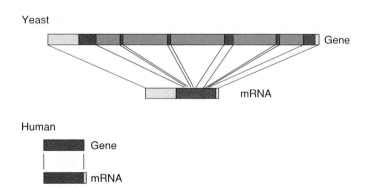

FIGURE 18.24 Contrast between cytochrome b genes from yeast (top) and human (bottom). The two genes contain essentially the same coding region (orange), but this region is interrupted by introns (blue) in yeast. In addition, yeast contains untranslated leaders and trailers (yellow) at the 5′- and 3′-ends, respectively. The mature mRNAs are shown below each gene, with lines to indicate the relationship between regions of the gene and corresponding regions in the mRNA. The untranslated sequence at the end of the human mRNA is poly(A), not encoded in the gene.

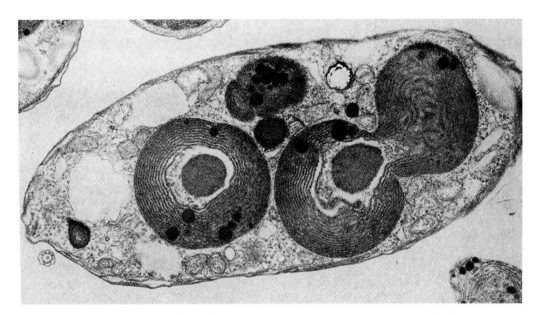

FIGURE 18.25 Cyanelles in the protozoan *Cyanophora paradoxa*. These cyanelles look so much like cyanobacteria that they were originally confused with those prokaryotic organisms.
From William T. Hall and Lynn Margulis, *Scientific American*, 225:53, August 1971.

THE ENDOSYMBIONT HYPOTHESIS

The sort of relationship just described, where two organisms live in a mutually beneficial union, is called **symbiosis.** In this particular case, one partner lives within the other in an **endosymbiotic** relationship, and the guest is therefore called an **endosymbiont.** Thus, this theory of the origin of organelles, formulated by Lynn Margulis, is known as the **endosymbiont hypothesis.** It can also explain the origin of the chloroplast, which apparently evolved from a primitive photosynthesizing cyanobacterium that took up residence in a primitive eukaryote and gradually lost its independence.

What is the evidence for the endosymbiont hypothesis? First, there is the superficial resemblance between modern chloroplasts and cyanobacteria. The protozoan *Cyanophora paradoxa* even provides an apparent intermediate between an endosymbiotic cyanobacterium and a chloroplast (figure 18.25). The green, membrane-rich bodies within this organism were long believed to be guest cyanobacteria representing an early stage in evolution of chloroplasts. In fact, these bodies contain much less DNA than bacteria—too little, in fact, for them to be independent. These *cyanelles* therefore probably represent an advanced intermediate in chloroplast evolution.

Mitochondria do not look much like the purple nonsulfur bacteria from which they presumably derived. Their shapes have apparently diverged considerably over time.

If mitochondria and chloroplasts really evolved from free-living bacteria, we would expect them to have genes and the means to express them, which they do, as this chapter has pointed out. However, they contain many more proteins than structural genes. Where do these proteins come from? They are encoded in the nucleus and made on ribosomes in the cytoplasm, then imported into the organelles.

The resemblance between mitochondria, chloroplasts, and bacteria extends to their chromosomes. The chromosomes of almost all mitochondria and chloroplasts are circles of DNA, just like bacterial chromosomes. The rare exceptions are the linear chromosomes in the mitochondria of certain protozoa and in the chloroplasts of the alga *Acetabularia*. Moreover, the arrangement of the genes in the rRNA gene clusters in certain chloroplasts is strikingly like that in bacteria (figure 18.26), and we

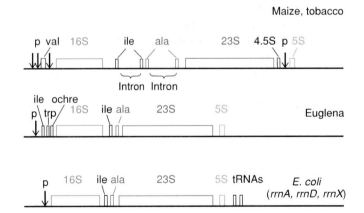

FIGURE 18.26 Similar arrangement of rRNA genes on chloroplasts and bacteria. The rRNA genes are represented by the names of the RNAs (e.g., 16S); tRNA genes are represented by the three-letter names of their amino acids (e.g., ile); ochre codes for a tRNA that suppresses an ochre termination codon; promoters are represented by arrows. Note the constant 16S-ile-ala-23S-5S sequence in all the genomes. The ile and ala tRNA genes in the land plant chloroplasts are interrupted by introns. The arrangement of the nuclear genes for these RNAs is vastly different. This suggests a close evolutionary relationship between prokaryotes and chloroplasts.

have already seen an even more dramatic resemblance between operons in chloroplasts and *E. coli*. And finally, organellar transcription and translation systems resemble those of bacteria much more than they resemble the cytoplasmic transcription and translation systems in the same cells. Several lines of evidence point to this conclusion:

1. Organellar transcription is inhibited by the antibiotic rifampicin, which inhibits prokaryotic, but not eukaryotic, RNA polymerase.

2. The subunit structure of the chloroplast RNA polymerase and the amino acid sequences of these subunits are much more prokaryotic than eukaryotic.

3. The structure of organellar promoters is more like that of prokaryotes than eukaryotes.

4. Organellar protein synthesis is inhibited by the antibiotics chloramphenicol and tetracycline, which inhibit bacterial translation, but not by cycloheximide, a eukaryotic translation inhibitor.

5. The base sequences of mitochondrial and chloroplast rRNAs tend to be similar to those of bacteria, but not to the analogous rRNAs of eukaryotes.

6. The sizes of organellar ribosomes more closely match those of prokaryotic than eukaryotic ribosomes.

7. Organellar protein synthesis initiates with formylmethionine, as does prokaryotic protein synthesis.

Modern mitochondria and chloroplasts appear to have evolved from free-living ancestors of modern purple nonsulfur bacteria and cyanobacteria, respectively. According to the endosymbiont hypothesis, these bacteria invaded larger eukaryotes eons ago and established an endosymbiotic relationship with their hosts. Gradually, the guest organisms lost genes to their hosts and thereby lost their independence as well. At the same time, the host organisms grew dependent on their guests for energy and photosynthesis.

Promiscuous DNA

We have suggested that the free-living prokaryotes that were the ancestors of modern mitochondria and chloroplasts gradually lost most of their genes. In the present day, most of their proteins are encoded in the nucleus, which implies that at least some of the organelles' genes were not simply lost, but transferred to the nucleus. If this is true, we would expect to see modern evidence that genes flow from mitochondria and chloroplasts to nuclei. In fact, we do know of several cases of such "promiscuous DNA." For example, yeast nuclei contain a

region of DNA that matches not one, but parts of several different yeast mitochondrial genes.

This certainly looks like a case of transfer of genes from mitochondrion to nucleus, but how did several mitochondrial genes get scrambled together? The answer is probably that these genes came from a mutant yeast mitochondrion that had lost most of its DNA by internal recombination, leaving only a useless collection of gene fragments. We have seen that such mitochondria exist in petite mutants of yeast.

Another example of promiscuous DNA comes from maize, whose chloroplasts and mitochondria share a 12-kb segment of DNA. In fact, various fragments of plant mitochondrial DNA have been used as probes for similar DNA sequences in chloroplasts of the same species, and positive reactions have been observed in every case. This shows wholesale transfer of genetic material from mitochondrion to chloroplast, or perhaps vice versa, rather than transfer to the nucleus; nevertheless, the mobility of the organellar DNA suggests that it can also move to the nucleus.

One line of evidence underpinning the endosymbiont hypothesis is the presence of "promiscuous DNA" that moves from one locus in the cell to another. Yeast nuclei, for example, have a region of DNA that strongly resembles pieces of several yeast mitochondrial genes. Such transfer of DNA could explain how the ancestors of mitochondria and chloroplasts lost most of their genes to the nucleus.

Summary

Many eukaryotic phenotypes, including leaf color in certain plants, slow growth in yeast (petite) and *Neurospora* (poky), and antibiotic resistance in a wide variety of organisms, are controlled by genes that are inherited in a non-Mendelian fashion. These genes tend to be contributed by only one parent, usually the female. The explanation for this uniparental inheritance is that the relevant genes are carried in organelles (mitochondria or chloroplasts) that reside in the cytoplasm, and the maternal parent usually contributes the bulk of the cytoplasm to the zygote.

Maternal effect is a phenomenon in which the phenotype of the progeny depends, not on the phenotype, but on the genotype of the maternal parent. This can appear to be maternal inheritance in the F_1 generation, but subsequent generations reveal a Mendelian pattern of inheritance.

Linkage in yeast mitochondria has been estimated from recombination frequencies, since both maternal and paternal mitochondria are retained in this organism. It has also been estimated by frequencies of co-deletion from, or co-retention on, diminished petite mitochondrial genomes. Information for mapping yeast mitochondrial genes has been gleaned by observing the extent of hybridization between a petite and a wild-type DNA or between two petite DNAs. Linkage between chloroplast genes has been estimated by subjecting cytohets to recombination and cosegregation analysis.

Chloroplast genomes are almost always circular, and the sizes of these circular double-stranded DNAs range from 57 kb to 195 kb. The chloroplast genome is polyploid, with the number of circles ranging as high as one thousand per chloroplast in young wheat leaves. Physical maps of chloroplast DNAs have been developed by cutting the DNAs with restriction enzymes, identifying the resulting fragments by gel electrophoresis, and hybridizing them together to discover how they overlap. Genes have been assigned to these maps by hybridizing their RNA products to the restriction fragments. Most chloroplast DNAs contain two genes for each rRNA. The two sets of genes are contained in inverted repeats of DNA. In addition, chloroplasts encode a full set of tRNAs and more than eighty proteins. The total base sequences of several chloroplast genomes have been determined, and they have revealed a strikingly prokaryotic nature.

The total base sequences of several mitochondrial DNAs have also been determined. This information reveals that there is very little wasted space in the mitochondrial DNAs of vertebrates. There are no introns, no untranslated leaders or trailers in the encoded mRNAs, and few if any bases between structural genes. The mammalian mitochondrial genomes encode two rRNAs, twenty-two tRNAs, and thirteen proteins. The yeast mitochondrial DNA is about five times larger than the mammalian. It includes most of the same information as found in the mammalian genomes, including rRNAs, tRNAs, and mRNAs. In addition, it has introns, noncoding spacers between genes, and long untranslated leaders and trailers at the ends of the encoded mRNAs.

Modern mitochondria and chloroplasts apparently evolved from free-living ancestors of purple nonsulfur bacteria and cyanobacteria, respectively. According to the endosymbiont hypothesis, these bacteria invaded larger eukaryotes eons ago and established an endosymbiotic relationship with their hosts. Gradually, the guest organisms lost genes to their hosts and therefore lost their independence. By the same token, the host organisms grew dependent on their guests for energy and photosynthesis.

Several lines of evidence support the endosymbiont hypothesis. The most compelling is the striking similarity between the genes and the transcription and translation machinery of organelles and prokaryotes. Another is the presence of "promiscuous DNA" that moves from one locus in the cell to another. Such transfer of DNA could explain how the ancestors of mitochondria and chloroplasts lost most of their genes to the nucleus.

PROBLEMS AND QUESTIONS

1. Why do white strains of the four-o'clock plant live only a short time?

2. A hypothetical mutant phenotype, wrinkled chloroplast, shows a maternal pattern of inheritance in the dandelion. What would be the outcome among first generation offspring resulting from:
   (a) mating male/wrinkled to female/smooth (wild-type)?
   (b) mating male/smooth to female/wrinkled?

3. Present a plausible hypothesis to account for the pattern of inheritance in question 2.

4. When a variegated four-o'clock plant contributes the female gamete to a mating, the progeny exhibit three different phenotypes—green, white, and variegated. Explain this finding.

5. Why can meiosis in *Chlamydomonas* occur only after mating?

6. Why does it make sense to call the mt^+ parent in the *Chlamydomonas* the female?

7. The lengths of the points on a Vulcan's ears depend on those of the mother. That is, if the mother has long points, all her F_1 offspring will have long points, but if she has short points, all her offspring will too. Based on one generation of observation, would we be justified in calling this maternal inheritance?

8. Neutral petite yeast mitochondria have no DNA. Without any genes, how can the mitochondria exist at all?

9. You have two strains of a mold (*Fungus amungus*). One is chloramphenicol-resistant (cap^r), unable to make leucine (leu^-), but able to make biotin (bio^+). The other is chloramphenicol-sensitive (cap^s), able to make leucine (leu^+), but not biotin (bio^-). The *leu* and *bio* genes are known to reside in the nucleus. Describe a heterokaryon experiment you would perform to show whether the cap^r gene is probably carried in the mold's mitochondria.

10. What two other kinds of evidence could you collect to demonstrate clearly whether the *cap* gene is carried in the mitochondria of *Fungus amungus*?

11. You have isolated a novel form of yeast with mitochondrial genes not found in ordinary baker's yeast. You find that this yeast is able to form petites by the usual mechanism, so you subject four of the new genes to petite mapping and make the following observations:

    (1) Whenever *zig* and *twi* are lost or retained together in a petite, *frd* is too. (2) Whenever *frd* and *spa* are lost or retained together, *zig* is too. (3) When *frd* was lost, *zig* was also lost 60% of the time. The *spa* gene was lost together with *zig* only 10% of the time. (4) When *zig* was lost, *twi* was also lost 30% of the time. Draw a linear genetic map of these four genes, starting with *twi* on the left and showing rough distances between them.

12. You isolate two petites of your new yeast and call them A1 and B2. A1 contains *twi*, *frd*, and *zig*, with *twi* and *zig* 40 DNA units apart. B2 contains *twi*, *frd*, and another gene to the left of *twi* called *pap*. A1 hybridizes to 40% of the total yeast mitochondrial genome, while B2 hybridizes to 60%. (A1 therefore includes 40 DNA units, while B2 has 60 DNA units.) The two petite DNAs show an overlap of 25 DNA units. What is the maximum distance in DNA units between *pap* and *twi*?

13. Some chloroplasts have more DNA than an *E. coli* cell has, yet many fewer genes. Explain this apparent contradiction.

14. You cut *Chlamydomonas* chloroplast DNA with the restriction enzyme *Eco*RI, then subject the resulting fragments to electrophoresis and blot them to membrane filters. Next, you probe the blot with radioactive *Bam*HI fragment Ba13. Which

fragments in the blot will "light up" (hybridize to the probe)? Use the map in figure 18.17 to answer this question.

15. What would be the answer to the previous question if the radioactive probe was the Bg5 fragment?

16. How can we use an experiment similar to question 14 to show that the chloroplast rRNAs are located in fragments R07 and R24?

17. You isolate restriction fragment Ba15 from the *Chlamydomonas* chloroplast genome and translate it in vitro. Several mRNAs are produced, and you translate them in vitro to yield a mixture of proteins. You show that one of these proteins binds to an antibody directed against the product of the *psbC* gene. What can you conclude about the location of the *psbC* gene?

18. What are the major factors that make the yeast mitochondrial genome five times larger than that of mammals?

19. How do we know that the cyanelles in *Cyanophor paradoxa* are not cyanobacterial endosymbionts?

20. What does the term "promiscuous DNA" mean in the context of organelle genetics?

Answers appear at end of book.

PART 3

POPULATION GENETICS

Cheetahs, an endangered species with low genetic variation.
© Tim Davis/Tony Stone Images

19

An Introduction to Population Genetics

Learning Objectives

In this chapter you will learn:

1. That there is extensive genetic variation in most natural populations.
2. That genetic variation can be estimated by calculating the proportion of polymorphic loci or the level of heterozygosity in a population.
3. That the Hardy-Weinberg principle shows the relationship between allelic and genotypic frequencies.
4. That inbreeding increases the frequency of homozygotes.
5. That mutation is the original source of genetic variation.
6. That genetic drift may result in loss of genetic variation.
7. That gene flow can introduce new alleles into a population.

T he importance of the great principle of Selection mainly lies in the power of selecting scarcely appreciable differences . . . which can be accumulated until the result is made manifest to the eyes of every beholder.

Charles Darwin

English naturalist

Until now, we have focused on genetic variants that occur rarely in a given group of individuals—for example, mutations that result in inborn errors of metabolism, such as albinism in humans. However, natural populations of most species have alleles in substantial frequencies that affect such traits as body color, red-blood-cell types, or pesticide resistance. Why do these different genetic types exist within a population, and what factors determine their frequencies? Answering these questions necessitates an understanding of **population genetics**, the field that focuses on the extent and pattern of genetic variation in natural populations and the explanations for these observations.

Natural selection has often been considered the major force determining the extent and pattern of genetic variation, an idea that goes back to Charles Darwin's *On the Origin of Species by Means of Natural Selection*, first published in 1859. However, other factors also appear to be important in given situations or in certain species. In this chapter we will introduce those factors: inbreeding, mutation, genetic drift, and gene flow. In chapter 20 we will look at the combined effects of several of these factors in order to understand specific topics in population genetics: namely, the incidence of genetic disease in humans, the evolution of molecules, conservation genetics, agricultural genetics, and pesticide resistance.

We will find Mendelian genetic principles essential to understanding the genetic composition of populations. For example, the principle of segregation (that two alleles in a heterozygous individual are equally represented in the gametes produced by that individual), when applied to all individuals in a population, allows us to understand the overall genetic variation in that population. The genetic constitution of a population is sometimes referred to as its **gene pool**. Let us first describe some examples of genetic variation before discussing the principles of population genetics that explain the basis of this variation and how it may change.

TYPES OF GENETIC VARIATION

Genetic variants can directly affect the morphology or color of an organism in a natural or wild population. A classic example of genetic variation that has fascinated evolutionary biologists in England and France for decades is the color and banding pattern of *Cepaea nemoralis,* a European land snail common in gardens and uncultivated fields. The morphological variation is determined by the alleles at a group of closely linked genes, so that the snails vary strikingly in color and pattern from yellow, unbanded types to yellow, banded types to brown, unbanded ones (figure 19.1). The frequencies of the alleles that determine yellow color and bandedness vary on a large scale from one part of Europe to another and on a small scale from one field to another.

In one detailed study, the frequency of yellow, unbanded types over a 200-meter stretch on an earthen dam in the Netherlands varied from 60% at one end to 40% at the other. Furthermore, these populations remained at the same frequencies over a twelve-year period, suggesting that evolutionary forces were maintaining the differences. For example, it appears from other studies that the different forms confer protective camouflage (or crypsis) from the predation of birds, depending upon the habitat in which they are found. When the snails are resting on grass blades, the bands are vertical, so the banded snails blend in with the grass. But when the snails are in the ground litter, the unbanded brown snails are the most difficult to detect.

In a number of species, variation in chromosome structure has been observed among different individuals. Recently, new techniques developed to examine morphological details of chromosomes (see chapter 4) have uncovered many new chromosomal variants, and it is likely that numerous other small variations in chromosome structure, such as deletions, replications, and inversions, are still to be discovered. One of the most thoroughly studied examples of chromosomal variation was initiated

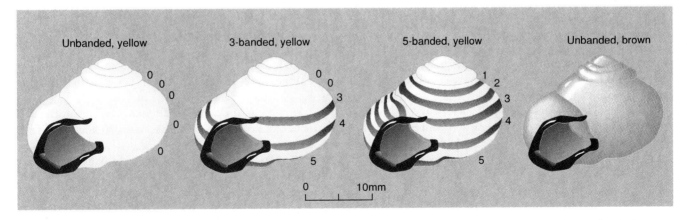

FIGURE 19.1 Several shell banding patterns observed in the snail *Cepaea nemoralis* where the non-zero numbers indicate the presence of particular bands.

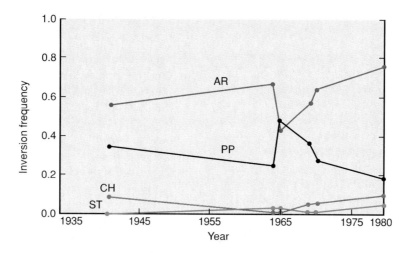

FIGURE 19.2   Frequencies of inversions on the third chromosome in *Drosophila pseudoobscura* that have taken place over a period of three decades in the Capitan area of New Mexico. (AR = Arrowhead; PP = Pikes Peak; CH = Chiricahua; ST = Standard.)

by Theodosius Dobzhansky, a Russian geneticist who moved to the United States in the 1920s. He and his students analyzed the inversion variants on the third chromosome of the fruit fly *Drosophila pseudoobscura,* a widespread species in the mountains of western North America (see figure 4.25).

In the 1930s, Dobzhansky began a long-term study of variation in this inversion. Some of these data, showing the frequencies of four inversion types over thirty years of sampling at a site in New Mexico, are given in figure 19.2. Initially, the inversion type AR (Arrowhead) was most common, but in 1965 type PP (Pikes Peak) increased in frequency, only to decline in later years. Two other inversion types, CH (Chiricahua) and ST (Standard), were always present, but at low frequencies in the samples from this population.

These inversions, although inherited as single units like alleles (see chapter 4), generally contain hundreds of genes. Because different inversions may have different alleles at these loci, they can substantially affect the survival of individual flies. In fact, some of the changes observed over time may have been caused by differential survival of various inversions to DDT, an insecticide that was first applied in the 1940s.

Most alleles that cause genetic diseases in humans occur in very low frequencies. However, some disease alleles have higher frequencies in certain human populations than in others. For example, the frequency of the sickle-cell allele at the β-globin locus is quite high in many African populations and in populations with African ancestry, such as African-Americans. The sickle-cell allele also occurs in some Mediterranean and Asian populations (figure 19.3).

The presence of the sickle-cell allele can be ascertained by biochemical examination of the β-globin molecule (see a photo of normal and sickle cells, figure 11.6). Variant forms of a protein can be detected by **gel electrophoresis** (see also chapter 6), a technique that allows the separation of different proteins extracted from blood, tissues, or whole organisms. The process is carried out by imposing an electric field across a gelatinous supporting medium (such as a polyacrylamide or starch gel) into which the protein has been placed. The proteins are allowed to migrate for a specific amount of time and are then stained with various protein-specific chemicals, resulting in bands on the gel that permit the relative mobility of a specific protein to be determined (figure 19.4). The relative mobility is generally a function of the charge, size, and shape of the molecule. If two proteins that are the products of different alleles have different amino acid sequences, they often have different mobilities as well, because the differences in sequence result in a change in charge, size, or shape of the molecule.

The three β-globin genotypes (homozygotes for the normal and sickle-cell allele plus a heterozygote) have different banding patterns when they are run on an electrophoretic gel, as shown in figure 19.5. The normal A_1A_1 genotype has only one band that migrates relatively fast, as in the first column, while individuals with sickle-cell disease, genotype A_2A_2 in column 3, also have only one band, but it migrates more slowly. The heterozygous individuals, A_1A_2, who have the sickle-cell trait (they are carriers) have two bands (see column 2), each band representing the protein synthesized from the respective allele. Sometimes these genotypes at the β-globin locus are represented as *AA, AS,* and *SS,* instead of A_1A_1, A_1A_2, A_2A_2.

The reason for the high frequency of this disease allele is that heterozygotes (those who carry this allele and the normal allele) have a relatively high resistance to malaria. The relative immunity of sickle-cell carriers to malaria is thought to occur because the red blood cells of heterozygotes are sickle-shaped, a state that inhibits infection by the malarial parasite. When we compare the frequency of the sickle-cell allele and the past incidence of malaria on a geographic scale (before the extensive eradication of the mosquito that carries the malarial parasite), we find an extremely high concordance (see figure 19.3).

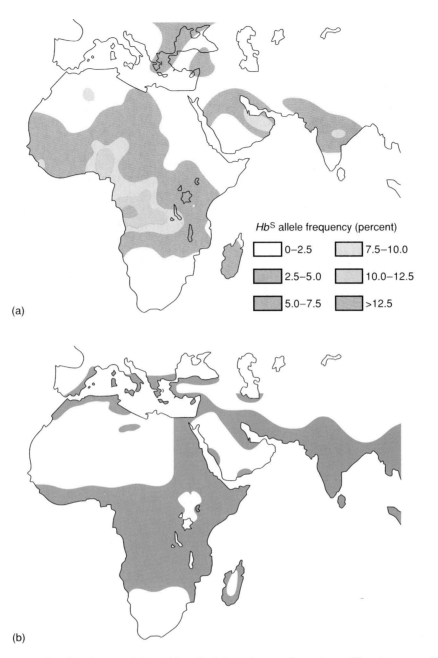

Extensive genetic variation occurs in most populations at several different levels, including variants affecting color, chromosomal structure, and protein characteristics.

MEASURING GENETIC VARIATION

How much genetic variation exists in a given population? How can we quantify this variation in a standardized manner? To

properly answer these questions, we need to introduce some quantitative measures and the symbols and statistics that accompany them. However, let us first define a **population** as a group of interbreeding individuals of the same species that exist together in time and space. For example, from a genetic viewpoint, all the breeding largemouth bass in a small lake in a given year constitute a population.

To determine the frequencies of the genotypes or alleles at a given locus in a population, we must first count the number of individuals having different genotypes. For example, at the β-globin gene, let N_{11}, N_{12}, and N_{22} indicate the number of

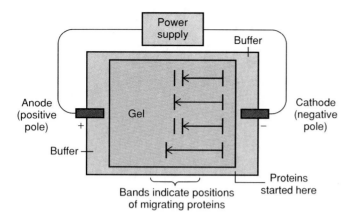

FIGURE 19.4 Diagram of a gel electrophoresis apparatus. The buffers are used to conduct electricity and ensure a given pH.

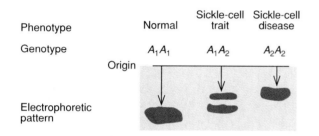

FIGURE 19.5 The phenotypes, genotypes, and electrophoretic patterns of the β-globin gene.

individuals counted with genotypes A_1A_1, A_1A_2, and A_2A_2, respectively. If there are N total individuals in the sample, the estimated frequencies of these three genotypes in the population are:

$$P = \frac{N_{11}}{N} \quad \text{for} \quad A_1A_1$$

$$H = \frac{N_{12}}{N} \quad \text{for} \quad A_1A_2$$

$$Q = \frac{N_{22}}{N} \quad \text{for} \quad A_2A_2$$

Because there are three genotypes, the sum of the frequencies of the three genotypes must be one; that is, P + H + Q = 1.

Notice that all the alleles in the homozygote A_1A_1 are A_1 and that all the alleles in the homozygote A_2A_2 are A_2. In addition, half of the alleles in the heterozygotes are A_1 and half are A_2. Therefore, we can calculate the frequency of allele A_2 (designated by q) as the sum of half the frequency of the heterozygote plus the frequency of the A_2A_2 homozygote. In other words:

$$q = \tfrac{1}{2}H + Q$$

If we substitute the numbers of different genotypes given in the first formulas, then:

$$q = \frac{\tfrac{1}{2}N_{12}}{N} + \frac{N_{22}}{N}$$

$$= \frac{\tfrac{1}{2}N_{12} + N_{22}}{N}$$

If we designate the frequency of the other allele, A_1, as p, then using the same logic:

$$p = P + \tfrac{1}{2}H$$

and

$$p = \frac{N_{11} + \tfrac{1}{2}N_{12}}{N}$$

Because there are only two alleles, the sum of their frequencies must be one, or p + q = 1. Furthermore, if we subtract q from both sides of this equation, we get the useful relationship p = 1 − q.

Let us illustrate the use of these formulas by a numerical example. In parts of Africa, the frequency of the sickle-cell allele is quite high, considering that it causes a severe disease when homozygous. In one sample of 400 individuals from western Africa, 320 were normal (genotype A_1A_1), 72 were carriers of the sickle-cell trait (A_1A_2), and 8 (mostly young individuals) actually had sickle-cell disease (A_2A_2). In other words, N_{11}, N_{12}, and N_{22} were 320, 72, and 8, respectively. The frequencies of the three genotypes in this population would be expressed as:

$$P = {}^{320}\!/\!_{400} = 0.80$$
$$H = {}^{72}\!/\!_{400} = 0.18$$
$$Q = {}^{8}\!/\!_{400} = 0.02$$

This means that 2% of the individuals have sickle-cell disease and 18% are carriers of the sickle-cell allele. Using the formula given for the frequency of allele A_2, the frequency of the sickle-cell allele is:

$$q = \tfrac{1}{2}(0.18) + 0.02$$
$$= 0.11$$

The frequency of the normal allele, A_1, is p = 1 − q = 0.89.

Now that we have introduced these formulas, we can discuss two useful ways to quantify the extent of genetic variation in a population. Using the allelic frequencies calculated for a given gene, we can define a gene as **polymorphic** (having many forms; that is, having two or more alleles in substantial frequency) or **monomorphic** (having only one allele in high frequency). The simplest measure of genetic variation is to categorize loci as polymorphic or monomorphic and then to calculate the proportion of all polymorphic loci. By general (although somewhat arbitrary) definition, if the frequency of the most common allele is 0.99 or greater (or if all other alleles combined have a frequency of 0.01 or less), the gene is considered monomorphic; otherwise, it is polymorphic. For example, the sickle-cell gene in the example just given is polymorphic for the two alleles A_1 and A_2 because the frequency of A_1 is less than 0.99.

A second approach for measuring genetic variation in a population is to calculate the average heterozygosity over all loci. The measure of heterozygosity is useful because it takes into account all levels of genetic variation rather than just classifying loci into two categories as does the first method. The average heterozygosity for a sample of n different loci is:

$$H = \tfrac{1}{n}(H_A + H_B + H_C \ldots)$$

where H_A is the heterozygosity for gene A, H_B for gene B, and so on. As a simple example, if the heterozygosities at five loci are $H_A = 0.5$, $H_B = 0.0$, $H_C = 0.1$, $H_D = 0.0$, and $H_E = 0.2$, then $n = 5$, and the average heterozygosity is:

$$H = (\tfrac{1}{5})(0.5 + 0.0 + 0.1 + 0.0 + 0.2)$$
$$= 0.16$$

Note that in this case three loci, A, C, and E, are polymorphic, making the proportion of polymorphic loci $\tfrac{3}{5} = 0.6$.

The first attempts to quantify the overall extent of genetic variation in a given species were made in the mid-1960s by Richard Lewontin in *D. pseudoobscura* and by Harry Harris in humans. Both surveys used protein electrophoresis and demonstrated substantial genetic variation from a sample of genes. As an illustration, some of Harris's data from a Caucasian population in England is given in table 19.1 for seventy-one different genes. Twenty of these loci are polymorphic; that is, they have the most common allele with a frequency of 0.99 or less. Therefore, the proportion of polymorphic loci in this sample is $\tfrac{20}{71} = 0.282$. Notice that the level of heterozygosity ranges from 0.00 for the monomorphic loci to around 0.5 for five polymorphic loci, most of which have multiple alleles. The average heterozygosity over all seventy-one loci is 0.067. Another way to interpret this value is that, on the average, about 6.7% of the loci in a given individual are heterozygous for such biochemical variants.

Table 19.1   Heterozygosity for Seventy-One Human Loci

Locus	Heterozygosity (H)
51 monomorphic loci	0.00
Peptidase C	0.02
Peptidase D	0.02
Glutamate-oxaloacetate transaminase	0.03
Leukocyte hexokinase	0.05
6-phosphogluconate dehydrogenase	0.05
Alcohol dehydrogenase-2	0.07
Adenylate kinase	0.09
Pancreatic amylase	0.09
Adenosine deaminase	0.11
Galactose-1-phosphate uridyl transferase	0.11
Acetylcholinesterase	0.23
Mitochondrial malic enzyme	0.30
Phosphoglucomutase-1	0.36
Peptidase A	0.37
Phosphoglucomutase-3	0.38
Pepsinogen	0.47
Alcohol dehydrogenase-3	0.48
Glutamate-pyruvate transaminase	0.50
RBC acid phosphatase	0.52
Placental alkaline phosphatase	0.53

From Harris and Hopkinson, *Annals of Human Genetics,* 36:9–20, 1972. Copyright © 1972 Cambridge University Press, New York, NY.

A relatively recent discovery in population genetics (and one that caused a radical reexamination of its foundation) is the high degree of polymorphism and heterozygosity that was found in nearly all natural populations once modern biochemical techniques were applied. At first thought, this level of genetic variation is surprising, because little variation appears in the wild-type morphology of most populations. However, the success of plant and animal breeding programs in changing morphological traits such as size (see chapters 3 and 20) suggests that there is extensive background genetic variation.

Surveys of the genetic variation in various species demonstrate that up to 20% of the genes examined by electrophoresis are polymorphic and that 10% of these genes are heterozygous in a given individual. Table 19.2 lists a number of species with different levels of polymorphism. Generally, the highest levels of genetic variation occur in invertebrates and in some plants, while vertebrates generally have lower levels of variation. Several organisms are of particular interest. The horseshoe crab, an animal that has not changed morphologically in millions of years according to fossil records, has a high level of genetic variation. Thus, **stasis** (no change) in morphology over geological time does not necessarily indicate lack of genetic variation on a biochemical level. On the other hand, the Northern elephant seal, which was nearly hunted to extinction in the nineteenth century, and the cheetah, also an endangered species, show no genetic variation in these electrophoretic surveys. In both cases, small population numbers appear to have caused a loss of genetic variation via genetic drift, a phenomenon that will be discussed later in this chapter and in chapter 20. Finally, barley is a plant with approximately 99% self-fertilization, and thus it has almost no heterozygotes (see discussion of inbreeding), yet it has a substantial number of polymorphic loci when plants are examined over several populations.

B y counting the numbers of different genotypes in a population, we can calculate genotypic frequencies, allelic frequencies, and heterozygosity. Two measures of genetic variation, the proportion of polymorphic loci and the level of heterozygosity, are used to quantify the extent of genetic variation. In many species, the amount of genetic variation observed for genes coding for proteins is substantial.

THE HARDY-WEINBERG PRINCIPLE

The description of the genetic variation in a population is often greatly simplified when allelic frequencies, rather than genotypic frequencies, can be used. In 1908, shortly after the rediscovery of Mendel's work, an English mathematician, Godfrey Hardy, and a German physician, Wilhelm Weinberg, showed that a simple relationship often exists between allelic frequencies and genotypic frequencies. (Actually, William Castle, an American

Table 19.2   Average Proportions of Polymorphic Loci and Average Heterozygosity for Various Organisms

Organism	Number of Genes Examined	Proportion Polymorphic	Heterozygosity
Humans	71	0.28	0.067
Northern elephant seal	24	0.0	0.0
Elephant	32	0.29	0.089
Cheetah	52	0.0	0.0
Tree frog	27	0.41	0.074
Drosophila pseudoobscura	24	0.42	0.12
Horseshoe crab	25	0.25	0.057
Barley	28	0.30	0.003

geneticist, had published a special case of this relationship in 1903.) This relationship, generally known today as the **Hardy-Weinberg principle,** states that the genotypic frequencies for a gene with two different alleles are a binomial function of the allelic frequencies.

To understand the Hardy-Weinberg principle, assume that a population is segregating for two alleles, A_1 and A_2, at gene A, with frequencies of p and q (remember p + q = 1), respectively. We will assume that the male and female gametes unite at random to form zygotes; that is, the product rule of probability discussed in chapter 2 is applicable. In addition, we should assume that the factors that we will discuss in some detail later—mutation, genetic drift, gene flow, and selection—are not affecting genetic variation.

The random union of gametes can be illustrated graphically, as in figure 19.6. The unit length across the top of the square in this figure is divided into proportions p and q, representing the frequencies of the female gametes, A_1 and A_2, respectively. Likewise, the side of the square is divided into proportions representing the frequencies of the male gametes. (The frequencies of the alleles in female and male gametes are assumed to be identical.) To envision random union of gametes, one must imagine a pool composed of both female and male gametes, p with A_1 alleles and q with A_2 alleles—and then assume that zygote formation occurs by chance collision in the gametic pool.

The areas within the unit square represent the probabilities (or proportions) of different progeny zygotes. For example, the upper left square has an area of p^2, the frequency of genotype A_1A_1, and the lower right square has an area of q^2, the frequency of A_2A_2. The other two gametic combinations both have frequencies of pq and result in heterozygotes, making the total proportion of heterozygotes 2pq. Overall, therefore, the three progeny zygotes (A_1A_1, A_1A_2, A_2A_2) are formed in proportions (p^2, 2pq, q^2). If we square the sum of the allelic frequencies (that is, $[p + q]^2 = p^2 + 2pq + q^2$) = 1, we can see that the Hardy-Weinberg proportions are really binomial proportions and that genotypic frequencies sum to unity.

Of course, different populations may have different allelic frequencies. To illustrate the effect of different allelic frequencies on genotypic frequencies, figure 19.7 gives the frequencies of

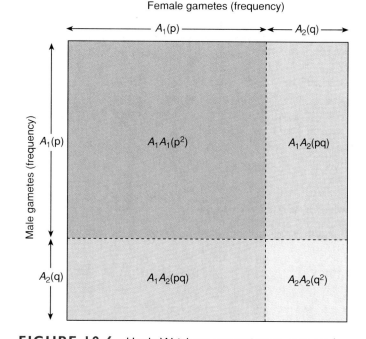

FIGURE 19.6   Hardy-Weinberg proportions as generated from the random union of gametes, using a unit square.

the three different genotypes, assuming Hardy-Weinberg proportions for the total range of allelic frequencies. Notice that the heterozygote is the most common genotype for intermediate allelic frequencies, while one of the homozygotes is the most common genotype for nonintermediate frequencies. The maximum frequency of the heterozygote occurs when p = q = 0.5. In this case, half of the individuals are heterozygotes. When the locus is monomorphic (p > 0.99 or q > 0.99), the amount of heterozygosity is very low.

In most organisms, random union of gametes does not occur. Instead, the parental genotypes pair and mate; then these mated pairs produce gametes that unite independently of the alleles they contain. Let us illustrate this situation when **random mating** occurs between different female and male genotypes. We will consider a diploid organism and assume that the three possible genotypes, A_1A_1, A_1A_2, and A_2A_2, are present in the population in frequencies P, H, and Q, respectively (remember that P + H + Q = 1). Given the three genotypes in each sex, nine combinations of matings between the male and female genotypes are possible, as shown in table 19.3. The frequency of a particular mating type, given random mating and using the product rule of probability again, is equal to the product of the frequencies of the genotypes that constitute a particular mating type. For example, the frequency of mating type $A_1A_1 \times A_1A_1$ is P^2. Only six mating types will be distinguished here, because reciprocal matings—for example, $A_1A_1 \times A_1A_2$ and $A_1A_2 \times A_1A_1$, where the first genotype is the female—have the same genetic consequences.

The mating types, their frequencies, and the expected frequencies of their offspring genotypes, using the principle of segregation, are given in table 19.4. For example, the mating $A_1A_1 \times A_1A_1$ produces only A_1A_1 progeny; the mating $A_1A_1 \times A_1A_2$ produces $\frac{1}{2}A_1A_1$ and $\frac{1}{2}A_1A_2$; and so on. If the frequency of A_1A_1 progeny contributed by each mating type is summed (adding

down the first progeny column), we find that $P^2 + PH + \frac{1}{4}H^2$ $= (P + \frac{1}{2}H)^2$ of the progeny are A_1A_1. From the expression for the frequency of A_1 given earlier, we know that p = P + $\frac{1}{2}$H and, therefore, that p^2 of the progeny are genotype A_1A_1. Similarly, the frequencies of A_1A_2 and A_2A_2 progeny are 2(P + $\frac{1}{2}$H) (Q + $\frac{1}{2}$H) = 2pq and (Q + $\frac{1}{2}$H)^2 = q^2, respectively. In other words, the Hardy-Weinberg principle is also true for random mating in the population, an assumption that is entirely reasonable for most genes.

The crucial point of the Hardy-Weinberg principle is that, given any set of initial genotypic frequencies (P, H, Q) after one generation of random mating, the genotypic frequencies in the progeny are in the proportions p^2, 2pq, and q^2. For example, given the initial genotypic frequencies (0.2, 0.4, 0.4) in which p = 0.4 and q = 0.6, after one generation, the genotypic frequencies become [(0.4)^2, 2(0.4)(0.6), (0.6)^2 = 0.16, 0.48, 0.36]. Furthermore, the genotypic frequencies will stay in these exact proportions generation after generation, given continued random mating and the absence of factors that change allelic or genotypic frequencies. (After all this discussion of p's and q's, one might think that it is the source of the phrase "Mind your p's and q's." A more probable origin, however, is the tradition of British bartenders who tell their customers, "Mind your pints and quarts"—p's and q's for short.)

In the allelic frequency estimates given earlier in this chapter, all alleles were assumed to be expressed (observable) in heterozygotes. However, in many cases—for example, metabolic

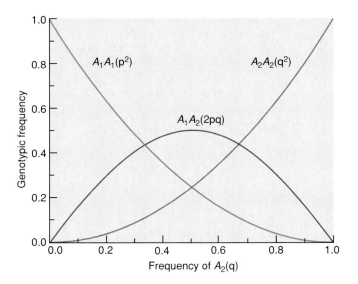

FIGURE 19.7 The allelic frequencies of three different genotypes for a population in Hardy-Weinberg proportions illustrating that heterozygotes are the most common genotype when allelic frequencies are intermediate.

Table 19.3 Frequencies of Different Mating Types for Two Alleles with Random Mating

Female Genotypes (Frequencies)	Male Genotypes (Frequencies)		
	A_1A_1(P)	A_1A_2(H)	A_2A_2(Q)
A_1A_1(P)	p^2	PH	PQ
A_1A_2(H)	PH	H^2	HQ
A_2A_2(Q)	PQ	HQ	Q^2

Table 19.4 The Hardy-Weinberg Principle, Assuming Random Mating and Normal Segregation

Parents		Progeny		
Mating Type	Frequency	A_1A_1	A_1A_2	A_2A_2
$A_1A_1 \times A_1A_1$	p^2	p^2	—	—
$A_1A_1 \times A_1A_2$	2PH	PH	PH	—
$A_1A_1 \times A_2A_2$	2PQ	—	2PQ	—
$A_1A_2 \times A_1A_2$	H^2	$\frac{1}{4}H^2$	$\frac{1}{2}H^2$	$\frac{1}{4}H^2$
$A_1A_2 \times A_2A_2$	2HQ	—	HQ	HQ
$A_2A_2 \times A_2A_2$	Q^2	—	—	Q^2
Total	1	$(P + \frac{1}{2}H)^2 = p^2$	$2(P + \frac{1}{2}H)(Q + \frac{1}{2}H) = 2pq$	$(Q + \frac{1}{2}H)^2 = q^2$

disorders in humans and color polymorphisms in various other organisms—the heterozygote is generally indistinguishable from one of the homozygotes because one of the alleles is completely dominant. In these instances, the population is usually assumed to be in Hardy-Weinberg proportions, to allow the estimation of allelic frequencies. For example, with two alleles and complete dominance, it is assumed that the proportion of A_2A_2 genotypes in the population is:

$$q^2 = \frac{N_{22}}{N}$$

By taking the square root of both sides of this equation, an estimate of the frequency of A_2 is:

$$q = \left(\frac{N_{22}}{N}\right)^{\frac{1}{2}}$$

(The superscript ½ means square root.)

This approach is applied in many human metabolic diseases where it is often important to know the proportion of heterozygotes (carriers) in a population—individuals who usually cannot be distinguished from homozygotes for the normal allele. An estimate of the proportion of carriers in a population is:

$$H = 2pq$$

where q is the estimate of allelic frequency and $p = 1 - q$. For example, assuming that only 1 in 40,000 individuals at birth has a particular rare recessive disease such as albinism (or is an A_2A_2 homozygote), then from the equation, $q = (1/40,000)^{\frac{1}{2}} = 0.005$. Therefore, $p = 0.995$, and the frequency of carriers is $H = 2(0.005)(0.995) = 0.00995$. In other words, about 1% of such a population carries the disease allele. People who are not geneticists are often surprised at how many individuals are carriers of recessive disease alleles when the actual frequency of the disease itself is so low.

THE CHI-SQUARE TEST

The observed genotypic numbers are often quite close to the expected genotypic numbers calculated from the Hardy-Weinberg principle. But in some organisms, such as self-fertilizing plants, the observed and expected genotypic frequencies in a population may differ greatly. Therefore a statistical test is necessary to determine whether the fit is sufficiently close to expected Hardy-Weinberg proportions. The common test employed for this purpose is the chi-square (χ^2) test used in chapter 2 to examine Mendel's principles. To determine whether the observed numbers are consistent with Hardy-Weinberg predictions, a chi-square value can be calculated as:

$$\chi^2 = \Sigma \frac{(O - E)^2}{E}$$

O and E are the observed and expected *numbers* of a particular genotype (remember to use numbers and not proportions), and Σ indicates that it is summed over all genotypic classes. In other words, this chi-square value is the squared difference of the observed and expected genotypic numbers, divided by the expected genotypic numbers, and summed over all genotypic classes. Knowing the calculated value of chi-square and the degrees of freedom, the probability that the observed numbers will deviate from the expected numbers by chance can be obtained from a chi-square table (see table 2.5).

Using the Hardy-Weinberg expected frequencies for the two-allele case, where all genotypes are distinguishable, the chi-square expression becomes:

$$\chi^2 = \frac{(N_{11} - p^2N)^2}{p^2N} + \frac{(N_{12} - 2pqN)^2}{2pqN}$$
$$+ \frac{(N_{22} - q^2N)^2}{q^2N}$$

The three terms represent the squared, standardized deviations from the expected numbers for genotypes A_1A_1, A_1A_2, and A_2A_2, respectively. For example, N_{11} is the observed number of genotype A_1A_1, and p^2N is the expected number of A_1A_1.

As an example of this procedure, let us use the hemoglobin data given previously. The observed numbers of the three genotypes were: $N_{11} = 320$, $N_{12} = 72$, and $N_{22} = 8$, giving $p = 0.89$ and $q = 0.11$. Therefore, the expected numbers of the three genotypes are $p^2N = (0.89)^2 400 = 316.8$, $2pqN = 2(0.89)(0.11)400 = 78.3$, and $q^2N = (0.11)^2 400 = 4.8$. Using the chi-square formula:

$$\chi^2 = \frac{(320 - 316.8)^2}{316.8} + \frac{(72 - 78.3)^2}{78.3}$$
$$+ \frac{(8 - 4.8)^2}{4.8}$$
$$= 0.03 + 0.51 + 2.13$$
$$= 2.67$$

To determine whether this value is statistically significant, we first need to know the degrees of freedom. As in chapter 2, we start by counting the number of categories (or genotypes)—three in this case. Next, we subtract 1 from this total because the numbers of all genotypes sum to N. Furthermore, in this application of the chi-square, we subtract an additional 1 because we had to estimate the allelic frequency (in this case only q because $p = 1 - q$) in order to calculate the expected numbers of the genotypes. Overall then, the remaining degrees of freedom are $3 - 1 - 1 = 1$.

Now we can look at table 2.5 to determine whether a chi-square value of 2.67 with 1 degree of freedom is significant. Because 2.67 is less than 3.84 (the lowest level of statistical significance for 1 degree of freedom), we can conclude that the observed numbers in the sample are consistent with those expected under the Hardy-Weinberg principle.

MORE THAN TWO ALLELES

Many genes have more than two alleles, as we discussed in chapter 3. For example, the β-globin gene containing the sickle-cell allele actually has more than one hundred variants, most of which are rare. Also, many genes examined by protein electrophoresis have more than two alleles. Further, when DNA sequences for genes are considered, many alleles are generally present in a population.

In the case of multiple alleles, the frequency of an allele can be calculated as the sum of the frequency of its homozygote plus half the frequency of each heterozygote that carries the allele in question. To illustrate, assume that a sample of N individuals is categorized for the genotypes from three alleles. As discussed in chapter 3, six genotypes are possible: three homozygotes (A_1A_1, A_2A_2, and A_3A_3) and three heterozygotes (A_1A_2, A_1A_3, and A_2A_3). The frequency of allele A_1 is calculated in the same manner as for two alleles, except that A_1 alleles are in two different heterozygotes, A_1A_2 and A_1A_3. Therefore, the frequency of A_1 is:

$$p = \frac{N_{11} + \frac{1}{2}N_{12} + \frac{1}{2}N_{13}}{N}$$

where N_{13} is the number of A_1A_3 heterozygotes.

The Hardy-Weinberg principle can also be extended to more than two alleles. Let us illustrate this with the human ABO blood group system, which has three alleles: I^A, I^B, and I^O. Table 19.5 gives the frequencies of these alleles in gametes p, q, and r, respectively, and the frequencies of the resulting genotypes and phenotypes. As for two alleles, the genotypic frequencies can be calculated as the product of particular male and female gametic frequencies. For example, the frequency of genotype I^AI^A is (p)(p) = p^2. Overall, the genotypic frequencies are equal to the square of the following trinomial (because there are three alleles):

$$(p + q + r)^2 = p^2 + 2pq + q^2 + 2pr + 2qr + r^2$$

Check table 19.5 to see that you understand the genotypes associated with these frequencies.

Note that because of the recessivity of allele I^O, two genotypes have the A phenotype (I^AI^A and I^AI^O) and two have the B phenotype (I^BI^B and I^BI^O). The frequency of the O phenotype, using the Hardy-Weinberg principle, is the square of the frequency of the recessive allele I^O, that is r^2; the frequencies of the A and B phenotypes are $p^2 + 2pr$ and $q^2 + 2qr$, respectively; and the frequency of the AB phenotype is 2pq. In the United States Caucasian population, the frequencies of alleles I^A, I^B, and I^O are approximately 0.28, 0.06, and 0.66, resulting in phenotypic frequencies of 0.45, 0.08, 0.03, and 0.44 for ABO blood group types A, B, AB, and O, respectively. Luckily, the rarest phenotype, type AB, is also the universal recipient, so people with this phenotype can receive transfusions from individuals of any ABO blood group type.

An important consequence of having many alleles at a locus is that, in general, most individuals in a population are heterozygotes. For example, at the *HLA-A* and *HLA-B* loci, over 90% of the individuals are heterozygotes and less than 10% are homozygotes (see chapter 3). To illustrate this calculation, the Hardy-Weinberg heterozygosity for the ABO system is:

$$\begin{aligned} H &= 2pq + 2pr + 2qr \\ &= 2(0.28)(0.06) + 2(0.28)(0.66) + 2(0.06)(0.66) \\ &= 0.482 \end{aligned}$$

When there are many alleles (which means many more heterozygotes than homozygotes), the simplest approach is to calculate the Hardy-Weinberg homozygosity and subtract it from unity. The remainder is the Hardy-Weinberg heterozygosity.

X-LINKED GENES

A number of human diseases, such as hemophilia and muscular dystrophy, are determined by recessive alleles on the X chromosome. These diseases are much more common in males than in females, a fact that can be explained by the Hardy-Weinberg principle. For example, table 19.6 gives the expected frequencies of different genotypes, assuming random union of gametes for an X-linked trait. The frequency of affected males, A_2Y, is q, while the frequency of affected females, A_2A_2, is q^2. If q is low, as it is for most disease alleles (say, 0.005), the frequencies of affected males and females are 0.005 and 0.000025, respectively. In other words, the ratio of affected males to females is $q:q^2$ or 1:q. Thus, if q = 0.005, the ratio is 200:1. Both because of the high number of males affected by X-linked traits and the relatively high frequency of the hemophilia allele for a disease allele, a high proportion of individuals with recessive genetic diseases (nearly one individual in four in some populations) are male hemophiliacs. On the other hand, very few females have hemophilia.

Table 19.5   The Hardy-Weinberg Principle Applied to the ABO Blood Group System

Female Gametes (Frequencies)	Male Gametes (Frequencies)		
	I^A(p)	I^B(q)	I^O(r)
I^A(p)	I^AI^A(p)^2 A	I^AI^B(pq) AB	I^AI^O(pr) A
I^B(q)	I^AI^B(pq) AB	I^BI^B(q^2) B	I^BI^O(qr) B
I^O(r)	I^AI^O(pr) A	I^BI^O(qr) B	I^OI^O(r^2) O

Table 19.6   Frequencies of Gametes and Genotypes for an X-linked Gene

Female Gametes (Frequencies)	Male Gametes (Frequencies)		
	Female Progeny		Male Progeny
	A_1(p)	A_2(q)	Y(1)
A_1(p)	A_1A_1(p^2)	A_1A_2(pq)	A_1Y(p)
A_2(q)	A_1A_2(pq)	A_2A_2(q^2)	A_2Y(q)

The Hardy-Weinberg principle states that genotypic frequencies are a binomial function of allelic frequencies after one generation of random mating. Genotypic numbers can be examined statistically, using the chi-square test to determine whether they are consistent with Hardy-Weinberg expectations. The Hardy-Weinberg principle and estimation of allelic frequencies can also be applied to multiple alleles and X-linked genes.

Table 19.7 Positive Assortative Mating in Studies of Five Human Traits

Trait	Number of Studies Showing Correlation Coefficients of:			
	<0.1	0.1–0.2	0.2–0.3	>0.3
Height	7	8	7	5
Weight	1	2	3	1
Cephalic index	14	5	3	—
Hair color	—	2	2	1
Eye color	2	1	1	1

INBREEDING

So far in this chapter, we have considered populations in which mating is random, but there are obvious exceptions to this assumption in some species, such as self-fertilizing plants or invertebrates. **Inbreeding,** nonrandom mating that results from mating of close relatives, is of particular importance in humans, because many individuals who have recessive diseases are the products of matings between relatives. Furthermore, inbreeding is used in both plant and animal breeding to develop varieties or lines that have certain desired characteristics. In chapter 20 we will discuss the joint effects of inbreeding and selection that result in the phenomenon of inbreeding depression.

Nonrandom mating with respect to genotype occurs in populations where the mating individuals are either more closely or less closely related than mates drawn by chance (at random) from the population. The results of these two types of matings are called inbreeding and **outbreeding,** respectively. As we will see, neither inbreeding nor outbreeding can by itself cause a change in allelic frequency, but both do cause a change in genotypic frequencies. In an inbred population, the frequency of homozygotes is increased and the frequency of heterozygotes is reduced relative to random mating (or Hardy-Weinberg) proportions. In outbreeding, the opposite occurs, with the frequency of heterozygotes increased and the frequency of homozygotes reduced relative to random mating proportions. These genotypic changes affect all loci in the genome.

Nonrandom mating based on phenotypes rather than on genotypes may also occur. If the mated pairs in a population are composed of individuals with the same phenotype more often than would be expected by chance, **positive assortative mating** has occurred. By the same token, if mated pairs share the same phenotype less often than expected, we say that **negative assortative mating** has occurred. For example, positive assortative mating occurs in humans for traits such as height and eye color (table 19.7), although even for these traits it is absent in some populations. From these studies, it is obvious that cephalic index, a measure of head shape, is unimportant in selecting mates! Assortative mating generally affects the genotypic frequencies of only those loci involved in determining the phenotypes used in mate selection. Because positive assortative mating has effects similar to those of inbreeding, except that it generally affects only one or a few loci

rather than all loci, we will consider only the effects of inbreeding in the following discussion.

SELF-FERTILIZATION

In general, the most extreme type of inbreeding is self-fertilization, or "selfing," where both pollen and egg (or sperm and egg) are produced by the same individual. With complete self-fertilization, a population is divided into a series of inbred lines that quickly become highly homozygous. For example, the lines of pea plants used by Mendel were homozygous for different alleles that affected morphology or color because they were allowed to self-fertilize.

To illustrate how such high homozygosity occurs, assume that the frequencies of the three genotypes in the parents are P, H, and Q for A_1A_1, A_1A_2, and A_2A_2, respectively (table 19.8). Notice that A_1A_1 and A_2A_2 are true-breeding genotypes; that is, all selfed progeny produced by them are A_1A_1 and A_2A_2, respectively. However, only half the progeny of A_1A_2 are A_1A_2, with the other half split equally between homozygotes A_1A_1 and A_2A_2 due to Mendelian segregation.

If we sum the columns of table 19.8 for the different progeny types, we find that the frequency of heterozygotes is decreased by one-half compared to the parents, while the frequencies of the two homozygotes are each increased. For example, if the initial frequencies were 0.25, 0.5, and 0.25 for genotypes A_1A_1, A_1A_2, and A_2A_2, these frequencies would be 0.375, 0.25, and 0.375 after one generation of complete selfing.

To understand how inbreeding changes the genotypic proportions, let us examine the frequency of heterozygotes when complete self-fertilization occurs over several generations. The relationship for heterozygosity in two succeeding generations is a general one; thus, after one generation the heterozygosity is halved, and after two generations it is halved again, so that the result is one-fourth the initial value. After t generations of complete selfing, heterozygosity is halved t times, so that it is $(½)^t$ of its initial value. For example, if H was initially 0.5, after five generations, the expected heterozygosity would be $(½)^5\ 0.5 = 0.016$. This is a reduction of heterozygosity from 50% to less than 2% in five generations, showing that

Table 19.8 Frequency of Parents and Progeny with Complete Self-Fertilization

	Parent	Progeny		
	Frequency	A_1A_1	A_1A_2	A_2A_2
A_1A_1	P	P	—	—
A_1A_2	H	¼H	½H	¼H
A_2A_2	Q	—	—	Q
	1.0	P + ¼H	½H	Q + ¼H

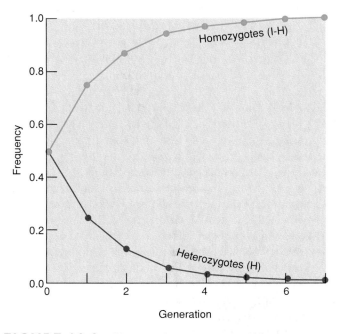

FIGURE 19.8 Change in heterozygosity and homozygosity over time with self-fertilization.

the reduction in heterozygosity through selfing is very rapid. The change in heterozygosity and homozygosity (1 − H) is given in figure 19.8.

One of the most important aspects of inbreeding is that, even though genotypic frequencies may be greatly altered, allelic frequencies remain unchanged overall. To illustrate this, let us indicate frequencies of the alleles and genotypes in the progeny generation by primes (′). Therefore, the frequency of allele A_1 in the next generation, p′, is:

$$p' = P' + \tfrac{1}{2}H'$$

where P′ and H′ are the frequencies of genotypes A_1A_1 and A_1A_2 in the next generation. By substituting the values of P′ and H′ from table 19.8 (the sums of columns 3 and 4), then:

$$p' = P + \tfrac{1}{4}H + \tfrac{1}{4}H$$
$$= p$$

In other words, as stated previously, no change in allelic frequency occurs from generation to generation due to self-fertilization.

THE INBREEDING COEFFICIENT

In order to describe the effect of inbreeding on genotypic frequencies in general, we will use a measure called the **inbreeding coefficient** (f). This value is defined as the probability that the two alleles at a gene in an individual are identical by descent. (Alleles are identical by descent when the two alleles in a diploid individual are derived from one particular allele in their ancestry, a situation we will illustrate later using pedigrees.) The largest value f can take is 1, when all genotypes in a population are homozygotes containing alleles identical by descent, as in the pure-breeding lines of Mendel's peas. The lowest value f can take is 0, when no alleles are identical by descent. In large, random mating populations, it is generally assumed that f is close to 0, because any inbreeding that may occur is between very distant relatives and thus will have little effect on the inbreeding coefficient.

A general formulation of the proportion of the three genotypes A_1A_1, A_1A_2, and A_2A_2, using the inbreeding coefficient and the allelic frequencies, can be derived as shown in table 19.9. Here it is assumed that the population can be divided into an inbred proportion (f) and a random mated proportion (1 − f). In the inbred proportion, the frequencies of A_1A_1, A_1A_2, and A_2A_2 are p, 0, and q, respectively. These frequencies are the proportions of lines expected for each of the genotypes if, for example, complete selfing continues. When the inbred and random mated proportions are added together (using the relationship q = 1 − p), the genotypic frequencies become:

$$P = p^2 + fpq$$
$$H = 2pq - 2fpq$$
$$Q = q^2 + fpq$$

The first term in each equation is the Hardy-Weinberg proportion of the genotypic frequencies, and the second term is the deviation from that value. Note that an individual may be homozygous either because the two alleles are identical by descent (the fpq terms) or because the two alleles are identical in kind through random mating (the p^2 and q^2 terms in the equations for P and Q, respectively). For example, A_2A_2 homozygotes that occur unrelated to inbreeding contain alleles identical in kind, while A_2A_2 homozygotes resulting from the same ancestral allele are termed identical by descent. The size of the coefficient of inbreeding reflects the genotypes' deviation from Hardy-Weinberg proportions such that, when f is 0, the zygotes are in Hardy-Weinberg proportions, and when f is positive, as from inbreeding, a deficiency of heterozygotes and an excess of homozygotes occur.

CALCULATING THE INBREEDING COEFFICIENT

We will discuss two ways the inbreeding coefficient can be estimated—from genotypic frequencies or from pedigrees. By

Table 19.9 Frequency of Different Genotypes with Inbreeding

Genotype	Inbred (f)	Random Mated (1 − f)	Total	
A_1A_1	fp	$(1 - f)p^2$	$fp + (1 - f)p^2 = p^2 + fpq$	
A_1A_2	—	$(1 - f)2pq$	$(1 - f)2pq = 2pq - 2fpq$	
A_2A_2	fq	$(1 - f)q^2$	$fq + (1 - f)q^2 = q^2 + fpq$	
	f	1 − f	1	1

the first method, we estimate the inbreeding coefficient in a natural population using the expression for the frequency of heterozygotes given previously. Thus, we can solve the expression for f as:

$$H = 2pq - 2fpq$$

$$1 - f = \frac{H}{2pg}$$

$$f = 1 - \frac{H}{2pg}$$

From the bottom equation, it is clear that the inbreeding coefficient (f) is a function of the ratio of the observed heterozygosity (H) over the Hardy-Weinberg heterozygosity (2pq). With inbreeding, H is less than 2pq, so f is greater than 0. As expected, if no heterozygotes are observed (H = 0), the breeding coefficient is 1.

Many species of plants have mating systems that include both self-fertilization and random mating (often called outcrossing) with other individuals. If the proportion of self-fertilization is high, nearly all individuals in the population should be homozygotes. An example is *Avena fatua,* a highly self-fertilizing species of wild oats, which is widespread in many areas around the Mediterranean Sea and was introduced into California where it is responsible for the "golden" hills in summer. A sample from the plant revealed that the frequencies of the three genotypes at an enzyme locus were P = 0.67, H = 0.06, and Q = 0.27. In order to estimate the inbreeding coefficient, we must first calculate the frequencies of the alleles. As before, the frequency of A_2 (q) is:

$$q = \frac{1}{2}H + Q$$
$$= (\frac{1}{2})(0.06) + 0.27$$
$$= 0.3$$

the value of p is therefore 1 − q = 0.7. Using the expression given previously:

$$f = 1 - \frac{0.06}{2(0.3)(0.7)}$$
$$= 0.86$$

This is an extremely high inbreeding coefficient, not far from 1, the maximum value possible. This high value suggests that most of the population reproduces by self-fertilization and that very little mating is random.

As stated earlier, the inbreeding coefficient is known as the probability of identity by descent, which implies that homozygosity is the result of the two alleles in an individual having descended from the same ancestral allele. The probability of identity by descent varies, depending upon the relationship of the parents of the individual being examined. For example, if the parents are unrelated (or not closely related), there is no possibility (or a very low one) that the individual will be homozygous by descent. The other extreme (given one generation of sexual reproduction) occurs when the same individual produces both gametes; that is, when self-fertilization occurs. In this case, the probability of an offspring having identical alleles by descent is 0.5. To illustrate, assume that the parent has the genotype A_1A_2. Progeny are produced in the proportions $\frac{1}{4}A_1A_1, \frac{1}{2}A_1A_2$, and $\frac{1}{4}A_2A_2$; thus, half the progeny, the A_1A_1 and the A_2A_2, will have alleles identical by descent. We assumed the parent was heterozygous so that we could identify the alleles in the progeny. However, even if the parent were homozygous for alleles identical in kind, we could arbitrarily designate them as different in order to calculate the probability of identity by descent.

A second way to obtain the inbreeding coefficient for a progeny is from a pedigree in which a **consanguineous mating** (a mating between relatives) has occurred. We use the pedigree to calculate the probability of identity by descent in the progeny. As an example, let us calculate the inbreeding coefficient for an offspring of two half sibs, individuals who share one common parent. Figure 19.9*a* gives the pedigree for this type of mating, where X and Y are two half sibs having the same mother but different fathers. The two fathers are indicated as open squares because they do not contribute to the inbreeding coefficient. The mother of X and Y is indicated as the common ancestor (CA). In figure 19.9*b*, the same pedigree appears in a different form, with the fathers omitted and diamonds symbolizing all individuals, because sex is not important in determining the inbreeding coefficient here. The arrows in this figure indicate the direction of parent to offspring transmission.

Let us assume that the mother (CA) has the genotype A_1A_2. To calculate the inbreeding coefficient, we need to know the probability that her grandchild (Z) is either A_1A_1 or A_2A_2—that is, identical by descent for either of her alleles. The first alternative, that Z is A_1A_1, can occur only if X and Y each contribute to Z a gamete containing A_1. The probability of an A_1 allele in X is the probability that an A_1 allele came from CA, or 0.5. Because the probability of transmission of A_1 to Z from X is also

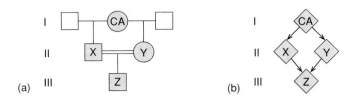

FIGURE 19.9 A pedigree illustrating a mating between two half sibs, X and Y. (*a*) With all individuals; (*b*) without the father. (CA = common ancestor, and the double line indicates a consanguineous mating.)

0.5, the joint probability of both events (that is, A_1 from CA to X and then to Z) is $0.5 \times 0.5 = 0.25$, using the product rule. Likewise, the probability of A_1 to Z through Y is 0.25. Therefore, the probability of an A_1A_1 progeny, who has received an A_1 from X and Y each, is $(0.25)(0.25) = 0.0625$. Using the same approach, the probability of an A_2A_2 progeny is 0.0625. The overall probability of identity by descent in Z is then $0.0625 + 0.0625 = 0.125$.

A more straightforward way to calculate the inbreeding coefficient from a pedigree is the **chain-counting technique.** A chain for a given common ancestor starts with one parent of the inbred individual, goes up the pedigree to the common ancestor, and comes back down to the other parent. For the example in figure 19.9, the chain is quite simple and is expressed X-CA-Y. The number of individuals in the chain (N) can be used in the following formula to calculate the inbreeding coefficient as:

$$f = (\tfrac{1}{2})^N$$

For example, for the pedigree in figure 19.9, the inbreeding coefficient would be $(\tfrac{1}{2})^3 = 0.125$, the same result we obtained by the longer approach, using the definition of the inbreeding coefficient.

> Inbreeding increases the frequency of homozygotes and reduces the frequency of heterozygotes, without changing allelic frequencies. The inbreeding coefficient can be measured either from genotypic frequencies or from pedigrees.

MUTATION

Mutation is a particularly important process in population genetics and evolution because it is the major source of new genetic variation in a species or population. Mutation can take many forms (see chapter 11). For example, it may involve only one DNA base, several bases, the major part of a chromosome, whole chromosomes, or sets of chromosomes. The immediate cause of a particular mutation might be a mistake in DNA replication, the physical breakage of the chromosome by a mutagenic agent, the insertion of a transposable element, or a failure in disjunction of meiosis. Our discussion will center on **spontaneous**

mutations, which appear without apparent explanation, rather than on **induced mutations,** which are caused by some mutagenic agent.

In order to examine the effect of mutation on genetic variation in a population, let us assume that the rate of mutation from wild-type alleles (A_1) to deleterious (or detrimental) alleles (A_2) for a given gamete per generation is u. This is known as the **forward mutation rate.** We will call the rate of mutation from deleterious alleles to wild-type alleles v, or the **backward mutation rate.** Because only A_1 alleles can mutate to A_2 and the A_1 allele has a frequency of p, the increase in the frequency of A_2 from forward mutation is $u \times p$. Using the same logic, the frequency of A_2 alleles can be decreased by backward mutation by the amount $v \times q$. Overall then, the change per generation in the frequency of A_2 (Δq) due to mutation is:

$$\Delta q = up - vq$$

The maximum positive value for this change is u, when $p = 1$ and $q = 0$ (all alleles are wild-type). The maximum negative value is v, when $p = 0$ and $q = 1$. However, because the mutation rates u and v are generally small (see chapter 11), the expected change due to mutation is also quite small. For example, if $u = 10^{-5}$, $v = 10^{-6}$, and $q = 0.0$, then:

$$\Delta q = (0.00001)(1.0) - (0.000001)(0.0)$$
$$= 0.00001$$

This is obviously quite a small change in allelic frequency.

Even so, mutation is of fundamental importance in determining the incidence of rare genetic diseases (see chapter 20). In fact, the balance between mutation (increasing the frequency of the disease allele) and selection (reducing the frequency of the disease allele) can explain the observed incidence of diseases such as albinism. In addition, mutation in concert with genetic drift provides a reasonable explanation for the large amounts of molecular variation recently observed in many species (see chapter 20).

> Mutation is the original source of genetic variation. It can cause only a small change in allelic frequency per generation.

GENETIC DRIFT

We have assumed until now that the population under study is large. But in fact, many populations are small, such as the number of individuals in an endangered species or in a species colonizing a new habitat. As a result of small population numbers, chance effects, called **genetic drift,** which are introduced by sampling gametes from generation to generation in a small population, can change allelic frequencies. In a large population, usually only a small chance change in the allelic frequency will be due to genetic drift. But in a small population, allelic frequency can undergo large fluctuations in different generations in a seemingly unpredictable pattern. Genetic drift is fundamental to efforts toward genetic conservation and understanding molecular evolution (see chapter 20).

An example will illustrate the type of allelic frequency change that can be expected in a small population. Assume that a diploid population has five individuals (ten copies of the gene, or ten gametes) and that, initially, half of these are A_1 and half are A_2. The ten gametes for the next generation are randomly chosen from the gametic pool, a process analogous to flipping a coin ten times and counting the number of heads and tails. Sometimes the proportion of A_1 alleles is greater than the proportion of A_2, sometimes the proportions are equal, and sometimes the proportion of A_2 is greater than the proportion of A_1. If many small populations, each containing ten gametes, are randomly chosen from a population with equal numbers of A_1 and A_2, the proportions of populations with different frequencies of A_2 will be as shown in figure 19.10. For example, a frequency of 0.4 indicates that four of the ten gametes were A_2. Notice that in most of the populations, the frequencies are close to the initial frequency of 0.5, but that significant variation occurs around that value. For example, there is a greater than 1% chance that the allele frequency will change from 0.5 to 0.1 (the far left end of the histogram).

To illustrate the potential for cumulative genetic change over time due to genetic drift, let us follow four particular populations of size twenty (forty gametes per population) over a number of generations (figure 19.11). These values were obtained using computer simulation in which random numbers represented the chance segregation that took place from generation to generation. The solid lines indicate the four replicates, and the broken line is the mean frequency of allele A_2 over the four replicates. All of the replicates were initiated with the frequency of A_2 equal to 0.5. After one generation of genetic drift,

the frequencies were 0.625, 0.55, 0.55, and 0.475. By generation 15, they were even more widely dispersed—0.8, 0.5, 0.475, and 0.3. In one of the replicates, the frequency of A_2 went to 1.0 in generation 19, meaning that all of the alleles in the population were A_2. In another replicate, the A_2 allele frequency went to 0.0 in generation 28. The other two replicates were still segregating for both alleles at the end of thirty generations. The mean of the four replicates varied from 0.625 in generation 19 to 0.475 in generation 30, but was generally near the initial frequency of 0.5. (In cases where there are enough replicate populations, no change occurs in the mean allelic frequency over all populations, because increases in allelic frequency in some populations are cancelled by reductions in allelic frequency in others.)

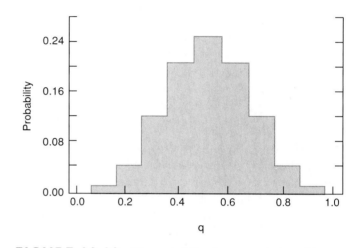

FIGURE 19.10 The probability that a population with ten gametes will have a given allelic frequency after one generation of genetic drift.

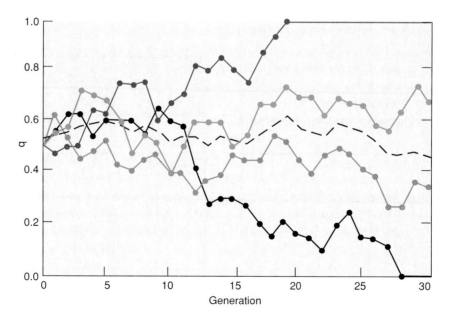

FIGURE 19.11 The allelic frequencies over thirty generations for four replicate populations of size twenty. The mean is represented by the broken line. When a population goes to fixation (q = 1.0) or loss (q = 0.0), then there is no change in allelic frequency.

FIGURE 19.12   The Northern elephant seal.
© Richard Humbert/Biological Photo Service.

From this example, it is obvious that genetic drift can cause large and erratic changes in the allelic frequencies of individual populations in a rather short time and can lead to differentiation among populations. Within a small population, there is a high probability that a gene will become fixed (i.e., go to a frequency of 1.0) for one allele or the other over a period of time. But in a large population over the same time span, there would be little change in allelic frequency due to genetic drift, and consequently, little cumulative differentiation among populations.

One prediction of genetic drift—that a small population will become highly homozygous over time—is supported by the findings in several endangered species. For example, the Northern elephant seal was hunted to near-extinction at the turn of the century, until fewer than one hundred individuals survived (figure 19.12). Although the population size has rebounded to over 70,000 individuals in recent years, it was low for a number of generations. Apparently due to this "bottleneck," genetic variation was lost, resulting in an estimated zero heterozygosity for the twenty-four genes mentioned in table 19.2.

Another example of a seemingly genetically impoverished species is the Pere David's deer. This species was discovered in the hunting compound of a Chinese emperor in 1856, presumably having been sequestered there for three hundred years. Approximately four hundred of the deer are living today, all descended from eleven animals taken from China. A survey of this species has also shown no heterozygosity for the enzyme genes examined.

We can identify the effects of genetic drift in several ways. First, populations eventually go either to fixation for A_2 ($q = 1$) or to loss of A_2 ($q = 0.0$). In both cases, the heterozygosity in the population is zero. The probability that a population will eventually become fixed for A_2 is equal to the initial frequency of A_2. For example, if q is initially 0.2, the probability that the population will be fixed for A_2 is 0.2, and the probability that it will lose the A_2 allele is 0.8. We can understand this point by realizing that, for A_2 to become fixed, it must change by chance from 0.2 to 1.0, a much bigger change than for the loss of A_2 (from 0.2 to 0.0).

Second, because the mean frequency over replicate populations does not change and the distribution of the allelic frequencies over replicate populations does, the overall effect of drift can be understood by examining the variation of the allelic frequency over replicate populations. One statistic used to measure the extent of variation is the variance (see chapter 3). The expected variance (V) in allelic frequency over replicate populations, due to one generation of genetic drift, is:

$$V = \frac{pq}{2N}$$

Where N is the population size in a large population (large N), V is small, indicating only a minor effect from genetic drift. The impact of genetic drift can also be understood by examining the rate of loss of heterozygosity. In chapter 20, we will use the decay of heterozygosity to measure the effectiveness of genetic conservation in endangered species.

To establish a general yardstick for measuring the effect of genetic drift on allelic frequency relative to other factors, we note that the square root of the variance (the standard deviation, see chapter 3) is approximately equal to the average absolute value (indicated by parallel vertical lines) of the allelic frequency change, or:

$$|\Delta q| = V^{1/2}$$

For example, if p = q = 0.5 and N = 50, then:

$$V = \frac{0.5 \times 0.5}{2 \times 50} = 0.0025$$

Taking the positive square root of V, then:

$$|\Delta q| = 0.05$$

Obviously, if N is relatively small, the effect of genetic drift on allelic frequency may be substantial, much larger than that of mutation and comparable to that of selection or gene flow, as we will see later.

When N is large, as in most species, the expected per-generation change in allelic frequency from genetic drift is small. However, the cumulative change over many generations may be quite important, as we will find out when we discuss molecular evolution in chapter 20.

Genetic drift in small populations can result in large chance changes in allelic frequency. Genetic drift may be the cause of the low heterozygosity observed in some endangered species.

GENE FLOW

Previously, we have considered only a single large population in which random mating occurred throughout. Many species, however, have smaller populations that are divided because of ecological or behavioral factors. For example, the populations of fish in tide pools, deciduous trees in farmlands, and insects on host plants are separated from each other because the suitable habitat for the species is not continuous. With this type of population structure, differing amounts of genetic connectedness can exist among the separated populations. The extent of this genetic connection depends primarily on the amount of **gene flow**—that is, genetically effective migration—among the populations. As a result of gene flow, genetic variants can be introduced from one population to another.

For example, the genetic variants that confer resistance to insecticides like DDT generally are initially present in only a limited number of populations (see chapter 20). However, the movement of individuals between populations introduces resistant alleles to populations that did not have them previously. As a result, DDT resistance can spread rapidly throughout mobile species such as mosquitoes or houseflies.

The simplest type of gene flow takes place when an allele enters a single population from an outside source. For example, such unidirectional gene flow occurs in an island population that receives migrants from a continental population. This description basically applies to species having populations on islands and nearby large landmasses, to aquatic species that live in ponds and have lakes as a source of migration, and to peripheral populations that are constantly replenished by individuals from the main part of the species range.

To examine the effect of this type of population structure, let us assume that an island population receives migrants from a large source (continent) population, as shown in figure 19.13. Although reciprocal gene flow may occur, we will assume it is so low that it has a negligible effect on the allelic frequency in the source population. Let the proportion of migrants moving into the island population each generation be m, so that the proportion of nonmigrants (or residents) in the island population is $1 - m$. If the frequency of A_2 in the migrants is q_m and the frequency of A_2 on the island before migration is q, the allelic frequency of A_2 after migration, q', is:

$$q' = (1 - m)q + mq_m$$
$$= q - m(q - q_m)$$

The change in allelic frequency after one generation of gene flow is the difference between the frequencies after gene flow and before gene flow, or:

$$\Delta q = q' - q$$

If we substitute q' from above into this expression, then:

$$\Delta q = -m(q - q_m)$$

From these expressions, it is obvious that no change in allelic frequency will occur (that is, $\Delta q = 0$) if $m = 0$ or if $q = q_m$. Only the second alternative is of interest here, because a value of $m = 0$ indicates no gene flow. Note that q_m and m are assumed to be constant over time and that they have values between zero and unity. If q is smaller than q_m, the frequency of A_2 increases on the island because of the gene flow; that is, Δq is positive. If q is greater than q_m, the frequency of A_2 will decrease, and Δq is negative. As a result of these two effects, there is a **stable equilibrium frequency** of A_2 on the island at $q_m = q$, as indicated by the solid circles in figure 19.14. (A stable equilibrium occurs where the population frequency returns to a given value after being perturbed either above or below that value.)

The effect of different allelic frequencies in the migrants on Δq when $m = 0.1$ (10% of the individuals of each generation are migrants) can be seen in figure 19.14. Note that the allelic frequency changes linearly as the frequency moves away from the equilibrium value and that it reaches an absolute maximum at either zero or unity, depending upon the value of q_m. From the values in this figure, it is obvious that the change in allelic frequency due to gene flow can be important. It may equal the size of changes caused by genetic drift in small populations and may be much larger than changes caused by mutation.

Many populations are composed of individuals who have descended from different populations. If we assume that only

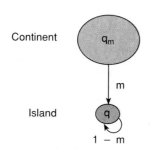

FIGURE 19.13 A model of the gene flow from a continental population to an island population, where m is the rate of gene flow and q_m is the allelic frequency in the migrants.

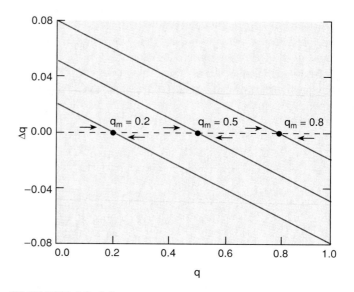

FIGURE 19.14 The expected change in allelic frequency in one generation due to gene flow when $m = 0.1$ and $q_m = 0.2, 0.5,$ or 0.8. The arrows indicate the direction of changes in allelic frequency.

two ancestral populations have contributed to the gene pool in a given population (for clarity, we will call this population the **hybrid population**), we can measure the amount of **admixture,** the proportion of the gene pool descended from one of the ancestral populations. We achieve this estimate by rearranging the first equation of this section so that it reads:

$$q' = (1 - m)q + mq_m$$

Then, we solve for m so that:

$$m = \frac{q - q'}{q - q_m}$$

Now, so we won't confuse the estimate for the amount of admixture (M) with the rate of gene flow (m), let us substitute M for m in the equation above, so that:

$$M = \frac{q - q'}{q - q_m}$$

Such admixture estimates have been made for human populations that have descended from different racial groups. For example, the African-American population is primarily of African origin, mixed with some more recent Caucasian ancestry. How can we estimate the proportion of Caucasian ancestry in African-American populations? One approach is to use a gene that has quite different allelic frequencies in the two ancestral populations, such as an allele at the Duffy blood group locus, another gene that codes for red-blood-cell antigens. Table 19.10 gives estimates of the ancestral frequencies of this allele and estimates of admixture in two U.S. cities. For example, to calculate M in the Oakland sample:

$$M = \frac{0.0 - 0.094}{0.0 - 0.429}$$
$$= 0.220$$

Although the estimate of admixture is relatively low in both samples, notice that the estimate of admixture is much higher (0.22) in the northern locality (Oakland) and only 0.037 in Charleston, indicating that certain southern African-American populations may have very little Caucasian admixture.

	African-Americans (q′)	African (q)	Caucasian (q_m)	M
Locality				
Oakland, CA	0.094	0.0	0.429	0.220
Charleston, SC	0.016	0.0	0.429	0.037

Table 19.10   Estimates of Caucasian Admixture (M) in African-Americans

Gene flow between populations can introduce new alleles and, in some cases, can rapidly change allelic frequencies. The extent of past gene flow between populations can be estimated.

NATURAL SELECTION

Charles Darwin is often considered the chief architect of the theory of natural selection. He provided strong evidence for it from his collection of plants and animals gathered on an around-the-world voyage on the H.M.S. *Beagle*. Darwin's theory held that variation among individuals of a species was the raw material through which selection could eventually produce individuals better adapted to their environment. Darwin was primarily concerned with the evolution of quantitative traits (discussed in chapter 3), but the same general principles operate regarding traits determined by single genes.

Selection can occur in several different ways (affecting fertility or mating success, for example), but here we will assume for simplicity that it results from differential survival of genotypes. To illustrate, assume 50% of the haploid genotype A_1 survive to adulthood, while only 40% of A_2 survive to adulthood, making the survival rate of A_1 higher than that of A_2. Intuitively, one would expect the frequency of A_1 to increase and the

THE FAR SIDE   By GARY LARSON

Natural selection at work

frequency of A_2 to decrease over time in a population initially composed of these two genotypes.

Selection has two basic effects on the amount of genetic variation. First, selection favoring a particular allele may lead to the reduction of genetic variation and, consequently, to homozygosity for the favored allele, as for DDT resistance alleles in mosquitoes exposed to DDT for many generations. Second, selection may result in the maintenance of two or more alleles in a population, such as for the banding polymorphism in snails or for the sickle-cell allele in humans.

In order to understand the effects of selection on genetic variation, we must consider the relative fitnesses of the different genotypes. **Relative fitness** can be defined as the relative ability of different genotypes to pass on their alleles to future generations. If, as in the example given previously, the relative fitnesses of two genotypes are based solely on their viabilities, these values can be used to calculate relative fitness values. Generally, the highest relative fitness value is assumed to be 1.0, and the others are standardized by dividing by this value. (As we will see, this is done to simplify the algebra.) Therefore, in the previous example, the relative fitness of genotype A_1 is $^{0.5}\!/_{0.5} = 1.0$ and that of genotype A_2 is $^{0.4}\!/_{0.5} = 0.8$.

SELECTION AGAINST A HOMOZYGOTE

We will begin by considering the situation in which selection operates against a homozygote, with the other homozygote and the heterozygote having equal relative fitnesses. This is sometimes called purifying selection because it purifies the gene pool of detrimental alleles. Let the relative fitnesses of the three possible genotypes, A_1A_1, A_1A_2, and A_2A_2, be designated as 1, 1, and $1 - s$, respectively, where the fitness of genotype A_2A_2 is smaller by an amount s, called the **selective disadvantage** (or **selection coefficient**) of the homozygote. If s = 1, then A_2 is a recessive lethal like a number of disease alleles in humans. If the relative fitness of A_2A_2 is 0.8, the selective disadvantage is s = 1 − 0.8 = 0.2.

Because we are considering only selection resulting from different viabilities of the genotypes, the relative fitness of a genotype is its relative probability of surviving from the newly formed zygote stage until reproduction. We can give the genotypic fre-

quencies before selection in terms of allelic frequencies, assuming Hardy-Weinberg proportions. The weighted contribution of the three genotypes to the next generation is the product of the frequency of the genotypes before selection and their relative fitnesses (table 19.11). The **mean fitness** of the population (w) is the sum of the weighted contributions of the different genotypes, or:

$$w = p^2(1) + 2pq(1) + q^2(1 - s)$$
$$= p^2 + 2pq + q^2 - sq^2$$

Because $p^2 + 2pq + q^2 = 1$ (that is, the sum of the frequencies of all the genotypes is 1), the mean fitness is:

$$w = 1 - sq^2$$

If the relative contributions are standardized by the mean fitness, the frequencies of the genotypes after selection can be obtained as shown in the bottom line of table 19.11. The frequency of A_2 after selection (q′) is equal to half the frequency of the heterozygote A_1A_2 (because half the genes in A_1A_2 are A_2 alleles) plus the frequency of the homozygote A_2A_2, or:

$$q' = (\tfrac{1}{2})\left(\frac{2pq}{1 - sq^2}\right) + \frac{q^2(1 - s)}{1 - sq^2}$$
$$= \frac{pq + q^2(1 - s)}{1 - sq^2}$$
$$= \frac{q(p + q - sq)}{1 - sq^2}$$
$$= \frac{q(1 - sq)}{1 - sq^2}$$

Therefore, the frequency of A_2 after selection is a function of the frequency before selection and of the selection coefficient.

As an example, let us assume that the homozygote A_2A_2 is inviable; in other words, that A_2 is a recessive lethal. In order to make the fitness of A_2A_2 zero, then 1 − s = 0, or s = 1.0. Let us now calculate the change in the frequency of A_2 over several generations. If the initial frequency of A_2 is 0.5, then from the formula just given:

$$q' = \frac{0.5[1 - (1.0)(0.5)]}{1 - (1.0)(0.5)^2}$$
$$= 0.333$$

Table 19.11 Frequency of Genotypes Before and After Selection, Assuming Hardy-Weinberg Proportions Before Selection

	Genotype			Total
	A_1A_1	A_1A_2	A_2A_2	
Relative fitness	1	1	$1 - s$	—
Frequency before selection	p^2	$2pq$	q^2	1.0
Weighted contribution	p^2	$2pq$	$q^2(1 - s)$	$1 - sq^2$
Frequency after selection	$\dfrac{p^2}{1 - sq^2}$	$\dfrac{2pq}{1 - sq^2}$	$\dfrac{q^2(1 - s)}{1 - sq^2}$	1.0

The same formula can be used to calculate the frequency in the next generation, assuming that q = 0.333 before selection, so that:

$$q' = \frac{0.333[1 - (1.0)(0.333)]}{1 - (1.0)(0.333)^2}$$
$$= 0.250$$

This process can be repeated over subsequent generations to calculate the allelic frequency. After twenty generations of selection against A_2A_2, the frequency of A_2 has been reduced from 0.5 to approximately 0.045 (see Box 19.1).

Many of the principles of population genetics have been investigated using *Drosophila* laboratory populations as a model. For example, the *Glued* mutant is a lethal homozygote that also

19.1 Industrial Melanism

he peppered moth, *Biston betularia,* is generally a mottled light color, called typical, but in a number of areas of England, a dark, melanic form is common (figure B19.1). The melanics were quite rare 150 years ago, but now they constitute nearly 100% of some urban populations. The melanic moths have increased because their dark color serves as protective camouflage from bird predation in the polluted areas they inhabit. This phenomenon, which has occurred in a number of other species, is called **industrial melanism.** A further confirmation of this hypothesis is the increasing frequency of the nonmelanic forms in areas where the air has become cleaner due to pollution control. In other words, there are both spatial and temporal associations between air pollution and the frequency of melanic peppered moths.

To measure the extent of selection between the two moth phenotypes, British biologists Cedric Clarke and Phillip Sheppard pinned dead moths of the two types on dark and light backgrounds and measured their survival from bird predation. On the dark background, fifty-eight of seventy melanic moths

survived, while only thirty-nine of seventy typical moths survived (table B19.1a). As a result, the relative fitness of the typical moths was only 0.67, compared to 1.0 for the melanics in this environment. On the other hand, on the light background (table B19.1b), the relative fitness of typicals was highest (or 1) and that of melanics was only 0.75.

Let us use the relative fitnesses estimated on the dark background to calculate the expected change in the typical allele. In this case, the selective disadvantage against typical homozygotes is s = 0.33, because 1 − s = 0.67. If we assume that the frequency of the typical allele is 0.2, then using the formula:

$$q' = \frac{0.2[1 - (0.33)(0.2)]}{1 - (0.33)(0.2)^2}$$
$$= 0.189$$

In other words, the allelic frequency will decline from 0.2 to 0.189, a significant but relatively small change.

(a)

(b)

FIGURE B19.1   Melanic and typical forms of the peppered moth *Biston betularia* (a) on a soot-covered trunk and (b) on a lichen-coated tree. Both photos © Michael Tweedie/Photo Researchers, Inc.

Table B19.1   Survival of Melanic and Typical Moths on Dark and Light Backgrounds

(a) Dark Background

	Melanic		Typical	
	Exposed	*Survived*	*Exposed*	*Survived*
Number	70	58	70	39
Survival		$\frac{58}{70} = 0.83$		$\frac{39}{70} = 0.56$
Relative fitness		$\frac{0.83}{0.83} = 1$		$\frac{0.56}{0.83} = 0.67$

(b) Light Background

	Melanic		Typical	
	Exposed	*Survived*	*Exposed*	*Survived*
Number	40	24	40	32
Survival		$\frac{24}{40} = 0.6$		$\frac{32}{40} = 0.8$
Relative fitness		$\frac{0.6}{0.8} = 0.75$		$\frac{0.8}{0.8} = 1$

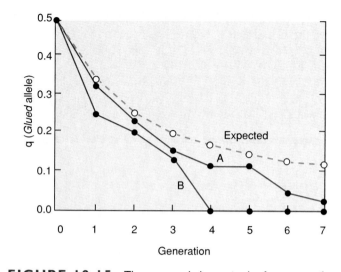

FIGURE 19.15 The expected change in the frequency of the lethal allele *Glued* (blue line) and the observed change in the replicate populations (red lines).

Source: Data from Clegg, et al., *Genetics*, 83:793–810, 1976.

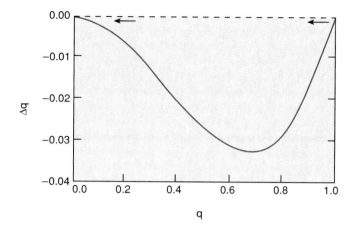

FIGURE 19.16 The change in the frequency of A_2 in one generation for different initial frequencies (q) when there is selection against the recessive genotype A_2A_2. The arrows at top indicate the direction of allele frequency change.

reduces eye size and affects eye appearance in heterozygotes. Michael Clegg and several coworkers followed the decline in the frequency of the *Glued* mutant in artificial populations initially composed of all heterozygotes, that is, those having an initial frequency of 0.5. Figure 19.15 gives the expected change (broken line) just as we calculated above and the observed change in two replicates, A and B. Notice that the general trend observed is close to that expected for the first few generations but that there is a difference in the later generations. Further investigation also showed selection against the heterozygotes, explaining why the observed frequency declined faster than expected.

The amount of allelic frequency change in one generation due to selection is defined as:

$$\Delta q = q' - q$$

By substituting q' given previously, then:

$$\Delta q = \frac{q(1 - sq)}{1 - sq^2} - q$$

$$= \frac{q(1 - sq) - q(1 - sq^2)}{1 - sq^2}$$

$$= \frac{q(1 - sq - 1 + sq^2)}{1 - sq^2}$$

$$= \frac{-sqq^2}{1 - sq^2}$$

Notice that the change in allelic frequency is negative, indicating that selection in this case is always reducing the frequency of A_2.

To illustrate the connection between this formula and the one given for q', let us assume again that q = 0.5 and that s = 1.0. Putting these values in the formula for change in allelic frequency:

$$\Delta q = \frac{-(1.0)(0.5)(0.5)^2}{1 - (1.0)(0.5)^2}$$

$$= -0.167$$

In other words, the allelic frequency has been reduced from 0.5 to 0.333, as we found earlier.

The change in allelic frequency is a function of allelic frequency and selective disadvantage. Figure 19.16 illustrates how Δq changes for different initial frequencies when the selective disadvantage is 0.2. The change in frequency is 0.0 when either p = 0 or q = 0, because the population is monomorphic for either the A_2 or the A_1 alleles. The change in frequency is greatest for an intermediate allelic frequency and becomes quite low when q approaches 0.0. This occurs because most of the A_2 alleles are in heterozygotes and are not subject to selection.

HETEROZYGOUS ADVANTAGE

The mode of selection just described, that is, selection against homozygotes, leads to the eventual fixation of one allele and, as a result, reduces the genetic variation in the population. However, when the heterozygote has a higher fitness than both homozygotes, two alleles can be maintained in the population. A classic example of this type of selection is sickle-cell disease, mentioned earlier; another example is resistance to the rodenticide warfarin in rats, which will be discussed in chapter 20.

To investigate the heterozygous advantage model, we will use a fitness array where the heterozygote, A_1A_2, has the maximum relative fitness (1) and where the homozygotes' fitnesses are $1 - s_1$ and $1 - s_2$ (s_1 and s_2 being the selective disadvantages, or selection coefficients, of the homozygotes A_1A_1 and A_2A_2, respectively). If we follow the same procedure used before, the change in allelic frequency due to selection is:

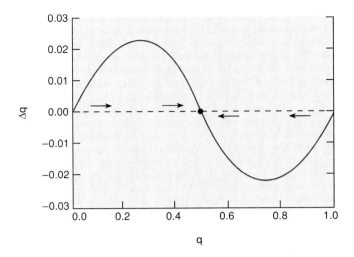

FIGURE 19.17 The change in the frequency of A_2 in one generation for different initial frequencies when there is a symmetric heterozygote advantage ($s_1 = s_2$). The arrows indicate the direction of allele frequency change, and the solid circle indicates the equilibrium.

$$\Delta q = \frac{pq(s_1p - s_2q)}{w}$$

In order to maintain both alleles in the population, Δq must be zero for some value of q between 0 and 1. The point where this occurs is called the stable equilibrium frequency of allele A_2—that is, the allelic frequency for which the change is zero. We found earlier that gene flow could also cause a stable equilibrium.

The value of q, when this expression equals zero and both alleles are present, occurs when the term inside the parentheses in the equation is zero, or:

$$s_1p - s_2q = 0$$

By solving this equation for q, one can show that an equilibrium occurs when:

$$q_e = \frac{s_1}{s_1 + s_2}$$

where q_e signifies the equilibrium frequency of allele A_2. In other words, for this allelic frequency, there is no change, and both alleles are maintained in the population. Notice that the equilibrium is a function only of the selection coefficients for the two homozygotes.

The properties of the equilibrium can be seen in another way by rewriting Δq using the equilibrium frequency so that:

$$\Delta q = \frac{-pq(s_1 + s_2)(q - q_e)}{w}$$

From this formulation, one can see that when q is greater than q_e, Δq is negative; thus, the allelic frequency will decrease toward the equilibrium. Likewise, when q is less than q_e, Δq is positive and the allelic frequency will increase toward the equilibrium. In

Table 19.12 Number of Eggs Laid and Hatched by Pigeons of Three Different Transferrin Genotypes

	Female Genotype		
	$Tf^A Tf^A$	$Tf^A Tf^B$	$Tf^B Tf^B$
Number of eggs laid	128	267	144
Number of eggs hatched	59	180	75
Proportion hatched	0.46	0.67	0.52
Relative fitness	0.69	1.0	0.78

other words, as in the gene flow example discussed earlier, the equilibrium is stable because after a perturbation away from it, the allelic frequency will change back toward the equilibrium frequency again.

The change in allelic frequency is a function of the allelic frequency and the selection coefficients. An example is given in figure 19.17, where the fitnesses of the three genotypes are 0.8, 1.0, and 0.8, so that $s_1 = s_2 = 0.2$. As a result of these fitness values, the equilibrium frequency is $q_e = 0.2/(0.2 + 0.2) = 0.5$. If q is smaller than 0.5, the change in allelic frequency is positive, with a maximum value just above q = 0.2. When the value of q is greater than 0.5, the change in allelic frequency is negative, with a maximum value just below 0.8.

We mentioned earlier that heterozygote advantage is the basis of the maintenance of the sickle-cell allele at the β-globin gene in humans. A number of studies have shown that heterozygotes have a higher viability than normal homozygotes in areas where malaria is prevalent. Heterozygotes are apparently resistant to malaria because their abnormal red blood cells, sickle-shaped rather than donut-shaped, are hard for the parasite to penetrate. If we let the relative fitness of heterozygotes be 1, that of normal homozygotes should be about 0.8, making $s_1 = 0.2$. In malarial areas, sickle-cell disease is generally fatal before individuals can reproduce, so that the fitness of genotype A_2A_2 is 0, making $s_2 = 1$. Putting these selection coefficients into the formula given previously, the predicted equilibrium is $q_e = 0.2/(0.2 + 1.0) = 0.17$. In fact, the observed frequency of the sickle-cell allele is between 0.15 and 0.2 in many regions of Africa where malaria is endemic, making these calculations consistent with observations in human populations (see figure 19.3).

Transferrin is an iron-binding protein that is polymorphic in most vertebrate species. In pigeons there are two common alleles, called Tf^A and Tf^B, which vary in frequency from about 0.38 to 0.59 in different breeds. Jeffrey Frelinger found that the eggs from female pigeons heterozygous for these two alleles were better able to inhibit microbial growth than were the eggs from homozygous females. His data, given in table 19.12, show that 67% of the eggs from heterozygous females survived infection while only 46% and 52% of the two homozygotes survived. If we standardize these values to calculate relative fitnesses, then $s_1 = 0.31$ and $s_2 = 0.22$, and the predicted equilibrium is 0.584, a value that is at the upper end of the range observed in different pigeon breeds.

Natural selection can result in the elimination of detrimental alleles by selection against homozygotes or in the maintenance of genetic variation by selection for heterozygotes.

WORKED-OUT PROBLEMS

PROBLEM 1

One of the most common types of consanguineous matings in humans is that between two first cousins. In this case, the mating individuals have the same set of grandparents. Draw a pedigree showing this type of mating. In this pedigree there are two common ancestors, the two grandparents. Write out the chains for the two different common ancestors and calculate the inbreeding coefficient (f) that results from each. The total inbreeding coefficient is the total of these two values. What is it?

SOLUTION

The pedigree for this mating is given below. CA_1 and CA_2 are the two grandparents who are common ancestors, and X and Y are the mating individuals.

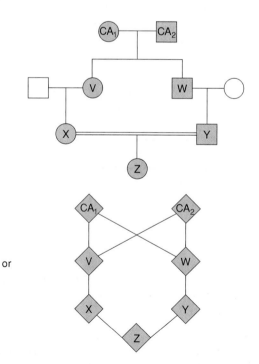

The chains are Z-X-V-CA_1-W-Y-Z and Z-X-V-CA_2-W-Y-Z. The f value for CA_1 is $(\frac{1}{2})^7$, and the f value for CA_2 is also $(\frac{1}{2})^7 = 0.03125$. Therefore, the total inbreeding coefficient is the sum of these or $0.03125 + 0.03125 = 0.0625$ or $\frac{1}{16}$.

PROBLEM 2

A certain plant population has 289 individuals, some bearing white flowers, others red flowers. The red allele is dominant over the white allele. Of the 289 plants, 246 have red flowers. Assuming the population is in equilibrium, (a) What is the frequency of the white allele? (b) How many of the red plants are expected to be heterozygous?

SOLUTION

(a) Several important relationships must be considered to solve this problem. First, the sum of the frequency of both alleles is unity: $p + q = 1$ (where p is the frequency of one allele and q is the frequency of the other allele). Second, the frequency of each of the genotypes also adds up to unity: $p^2 + 2pq + q^2 = 1$ (where p^2 is the frequency of one homozygous genotype, q^2 is the frequency of the other homozygous genotype, and 2pq is the frequency of the heterozygotes).

In problems of this sort, it is best to make q^2 the frequency of the homozygous recessive genotype. Hence, q is equal to the frequency of the recessive allele. For this problem, you know that 246 plants have red flowers. The remaining 43 plants have white flowers (289 – 246 = 43). To determine the frequency of white-flowered plants, divide 43 by 289. This gives 0.149. Thus, $q^2 = 0.149$. The square root of q^2 is q. So, $q = \sqrt{0.149} = 0.386$. The frequency of the white allele is 0.386.

To find the number of heterozygous red plants in the population (b), you need the frequency of the heterozygotes. First, find p (the frequency of the red allele), since the frequency of the heterozygotes is 2pq. To find p, substitute into the equation $p + q = 1$. Thus, $p = 1 - q = 1 - 0.386 = 0.614$. The frequency of heterozygotes is determined by substituting values in $2pq = 2(0.614)(0.386) = 0.474$. The frequency of heterozygotes is therefore 0.474. To find the number of heterozygotes, multiply the total number of plants in the population by the frequency. In this case, the number of heterozygous red-flowered plants is $(289)(0.474) = 136.986$ or 137.

SUMMARY

Extensive genetic variation occurs in most populations at several different levels, including variants affecting color, chromosomal structure, and protein characteristics. By counting the numbers of different genotypes in a population, we can calculate genotypic frequencies, allelic frequencies, and heterozygosity. We can use two measures of genetic variation, the proportion of polymorphic loci and the level of heterozygosity, to quantify the extent of genetic variation. In many species, the amount of genetic variation observed for genes coding for proteins is substantial. The Hardy-Weinberg principle states that genotypic frequencies are a binomial function of allelic frequencies after one generation of random mating. Genotypic numbers can be examined statistically using the chi-square test to determine whether they are consistent with Hardy-Weinberg expectations. The Hardy-Weinberg principle and the methods of estimating allelic frequencies also apply to multiple alleles and X-linked genes.

Inbreeding increases the frequency of homozygotes and reduces the frequency of heterozygotes, without changing allelic frequencies. The inbreeding coefficient can be measured either from genotypic frequencies or from pedigrees. Mutation is the original source of genetic variation, but is can cause only a small change in allelic frequency per generation. Genetic drift in small populations can result in larger chance changes in allelic frequency. Genetic drift may be the cause of the low heterozygosity observed in some endangered species. Gene flow between populations introduces new alleles, and in some cases rapidly changes allelic frequencies. The extent of past gene flow between populations can be estimated. Natural selection eliminates detrimental alleles by selection against homozygotes, or maintains genetic variation by selection for heterozygotes.

PROBLEMS AND QUESTIONS

1. Calculate the frequency of alleles A_1 and A_2 for the following populations:

Population	N_{11}	N_{12}	N_{22}
a	20	40	40
b	30	8	12
c	72	122	91

What are the Hardy-Weinberg frequencies and expected genotypic numbers for the three populations? Calculate the chi-square values for these populations. Are these populations statistically different from Hardy-Weinberg proportions?

2. Assume that the frequencies of genotypes A_1A_1, A_1A_2, and A_2A_2 are 0.1, 0.4, and 0.5, respectively. Use these values and table 19.3 to illustrate that random mating will produce Hardy-Weinberg proportions in one generation. Using a chi-square test and assuming that there are 200 individuals, is this population statistically different from Hardy-Weinberg proportions?

3. If q = 0.3 and there are Hardy-Weinberg proportions, what is the most common genotype and what is its frequency? What is the least frequent genotype and its frequency?

4. If A_2 is recessive and 4 out of 400 individuals are A_2A_2 (the rest are the dominant phenotype), what is the estimated frequency of A_2? What proportion of the population is expected to be heterozygous?

5. A survey of 50 *Drosophila* by Dobzhansky and his colleagues in New Mexico (see figure 19.2) found the following numbers of the various inversion genotypes: AR AR, 18; AR PP, 19; AR ST, 5; PP PP, 5; PP ST, 3; and ST ST, 0. (*a*) Treating the inversions AR, PP, and ST as if they are three different alleles, calculate the frequency of the three inversion types. (*b*) Using these frequencies, calculate the expected numbers of the six genotypes if they were in Hardy-Weinberg proportions.

6. Calculate the frequency of alleles A_1, A_2, and A_3 in the following populations:

Population	N_{11}	N_{12}	N_{13}	N_{22}	N_{23}	N_{33}
a	10	20	30	10	20	10
b	40	—	—	40	—	120

What are the expected genotypic frequencies for these two populations? Calculate the chi-square values for these populations. Are these populations statistically different from Hardy-Weinberg proportions? (There are 3 degrees of freedom in this case.)

7. In some populations, approximately 8% of the males have a particular type of color blindness (an X-linked recessive trait). What proportion of females would you expect to have this trait? What proportion of females would you expect to be carriers for this allele?

8. If the initial genotypic frequencies are 0.04, 0.32, and 0.64 for A_1A_1, A_1A_2, and A_2A_2, what are the frequencies after one generation of complete self-fertilization? What is the frequency of heterozygotes after five generations of complete self-fertilization?

9. Calculate the frequency of the three genotypes, A_1A_1, A_1A_2, and A_2A_2, given f = 0.2 and q = 0.4. Assuming that the frequencies of the three genotypes are 0.35, 0.1, and 0.55, what is f?

10. A survey of three populations of wild oats in northern California found that the observed frequencies of heterozygotes were 0.05, 0.07, and 0.03, while the expected frequencies of heterozygotes were 0.41, 0.44, and 0.32, respectively. (*a*) What are the estimated inbreeding coefficients in these three populations? (*b*) What do you think accounts for this level of inbreeding in this species?

11. (*a*) Draw the pedigree for a mating between two half-first cousins who share only one grandparent. What is the expected inbreeding coefficient? (*b*) Draw a pedigree for a mating between full sibs. What is the expected inbreeding coefficient? (There are two chains in this case because there are two common ancestors. To calculate the overall inbreeding coefficient, you need to add the two f values together.)

12. If u = 10^{-6} and v = 10^{-7}, what is Δq for q = 0.0, 0.25, 0.5, 0.75, and 1.0? When is Δq largest? Explain why.

13. What is the effect of a small population size on the amount of genetic variation in that population over a period of time? Discuss the effects on fixation of alleles and the level of homozygosity in the population.

14. If the initial frequency in a small population is 0.3 for A_2, what is the probability that the population will become fixed for A_2? For A_1? Explain your answers.

15. In a population of 100 individuals with q = 0.4, what are V and the approximate value of $|\Delta q|$ after one generation? How does this value of Δq compare to that for mutation? (For example, see problem 12.)

16. What is the effect of gene flow on the amount of genetic variation in a population? How does it depend upon the level of gene flow and the allele frequency in the migrants?

17. Assume that m = 0.05 and q_m = 0.1. (*a*) Draw a Δq curve (as in figure 19.14) with these values for different levels of q. (*b*) What is the expected equilibrium frequency?

18. In two ancestral populations, the frequency of an allele is 0.0 and 0.6. In a hybrid population, the frequency of the allele is 0.08. What is the estimate of admixture? What kinds of assumptions need to be made for such estimates of gene flow?

19. (a) For a recessive with s = 0.4, what is Δq when q = 0.0, 0.25, 0.5, 0.75, and 1.0? (b) Graph these results on a Δq curve. (c) For a recessive with s = 0.1, what is the allelic frequency after one and two generations when the initial frequency is 0.2?

20. In another experiment like that described in box table 19.1a, 120 melanic moths were exposed and 96 survived, while only 74 of 120 typicals survived. Estimate the relative fitness of the two morphs. What is the expected frequency of the melanic allele after one generation, given that its initial frequency is 0.4?

21. If the relative fitnesses of genotypes A_1A_1, A_1A_2, and A_2A_2, are 0.8, 1.0, and 0.2, respectively, what is the expected equilibrium frequency of A_2? Calculate Δq for q = 0.1 and q = 0.8. What does this tell you about the expected change in frequency?

22. As discussed in the text, it appears that the two common transferrin alleles are maintained in pigeon populations by heterozygote advantage. Three other breeds of pigeons have frequencies of 0.32, 0.81, and 1.0, respectively. After a number of generations, what would you expect the frequencies in these breeds to become? Why do you think they were different from these values initially?

23. Compare in general the expected change in allelic frequencies that result from mutation, gene flow, genetic drift, and selection. Which factors will have the largest effect on Δq, and when? In what ways is the effect on allelic frequency from genetic drift different from that due to mutation, selection, and gene flow?

24. Inbreeding affects genetic variation in a different way than do other factors. Describe its influence, and tell how you could detect whether high inbreeding is present in a population. How might inbreeding be used in plant or animal breeding?

25. In the study of snails mentioned at the beginning of this chapter, it was suggested that differential selection via protective camouflage is important in determining the frequencies of different snail types. Design an experiment that would illustrate this directly. Some people suggest that genetic drift is important in determining snail types. How would you experimentally measure the importance of genetic drift? (In this experiment, you can remove snails, add snails, look at different sites, etc.)

Answers appear at end of book.

20

Extensions and Applications of Population Genetics

I t is the triumph of scientific men . . . to desire tests by which the value of beliefs may be ascertained, and to feel sufficiently masters of themselves to discard contemptuously whatever may be found untrue.

Sir Francis Galton

British scientist

Learning Objectives

In this chapter you will learn:

1. That the incidence of many human genetic diseases can be explained by a mutation-selection balance.

2. That selection, inbreeding, and genetic drift can also influence the incidence of genetic diseases.

3. That the theory of neutrality is consistent with many findings in molecular evolution.

4. How the molecular clock and phylogenetic trees are used to understand the relationships between species.

5. That conservation genetics is concerned with the preservation of genetic variation in agricultural plants and animals as well as in rare and endangered species.

6. That the primary goals of conservation genetics are to avoid inbreeding depression and loss of genetic variation.

7. That pesticide resistance occurs in many pest species.

8. That pesticide resistance can provide excellent case studies of evolutionary change.

7.

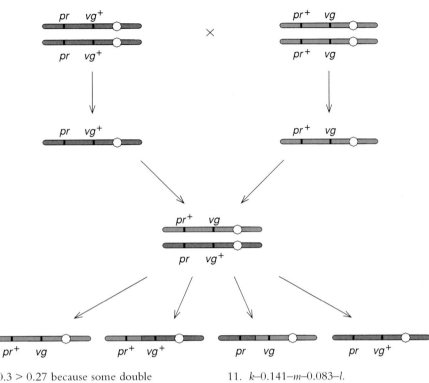

8. The gene order is *a–c–b*. 0.3 > 0.27 because some double crossovers have occurred.

9. The gene order would be *a–c–d–b*. *bcd/+ + +×bcd/bcd*.

10. The use of three-point crosses confirmed the linearity of genes on chromosomes predicted from two-locus crosses. In addition, one three-point cross can give as much information as three crosses involving two genes each.

11. *k*–0.141–*m*–0.083–*l*.

12. 0.0117. I = 1–0.00636/0.0117 = 0.46. There is some interference because the observed number of double recombinants is about half the proportion of doubles expected.

13.

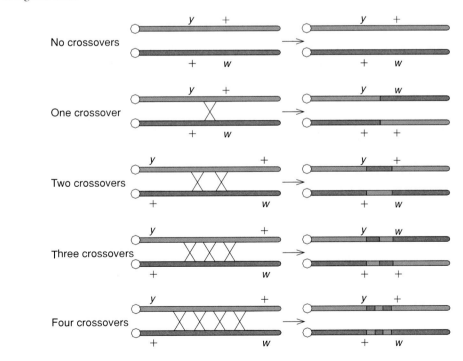

14. In humans, one cannot obviously make specific crosses planned to determine the linkage of two genes. Further, the progeny number in humans is much smaller than in these other organisms in which a single mating may produce over a hundred progeny.

15. 2 *h* C/Y, 2 *H* *c*/Y, 1 *H* C/Y; 0.2.

16. 0.214. Families having no recombinant male offspring are omitted here, causing a bias in the data.

17. 0.5 A_1A_2, 0.5 A_2—; all would be A_1A_2.

18. Chromosome 11.

19. In four-stranded crossing over, only two of the four present chromatids are involved in any recombination event, while in two-stranded crossing over all (or both) are involved. As a result, four-stranded crossing over generally will give both parental and nonparental gametes in the progeny, while two-stranded crossing over will give only parental (100%), when there is no crossing over, or only nonparental (100%), when there is crossing over.

20. X^+Y, $X^W$$X^+$Y; wild-type male and female; white is a recessive allele.

21. *DDDDdddd; DDddDDdd, DDddddDD, ddDDDDdd,* or *ddDDddDD.*

22. Centromere–*F*–*E*.

23. 0.028; 0.0045.

24. Recombination may be localized into certain regions.

25. Mendel might not have formulated the principle of independent assortment correctly at all.

CHAPTER 6

1. The critical observation was that heat-killed cells of a virulent strain of *Streptococcus pneumoniae* were able to change (transform) cells of an avirulent strain into cells that were virulent. Since genes control heredity, the heritable nature of this transformation made it very likely that the transforming agent was itself a gene or genes.

2. The Avery and colleagues approach offers the stronger argument for DNA as the genetic material. These workers carefully purified their DNA before using it to transform cells. In the Hershey-Chase approach, 10% of the phage protein was not accounted for and therefore could have entered the host cells along with the phage DNA. In principle, this protein could have been the genetic material.

3. No. Both protein and DNA contain C and H. By contrast, the radioactive tracers ^{32}P and ^{35}S used by Hershey and Chase are specific for DNA and protein, respectively, therefore allowing for a more unequivocal identification of each cellular component.

4. 5′-CGAATGCT-3′; 3′-TCGTAAGC-5′.

5. ATGTCCAGTCG
   TACAGGTCAGC

6. Thymine, 30%; cytosine, 20%. The thymine and adenine contents are identical, and so are the cytosine and guanine contents.

7. (*a*) 12,000 helical turns, (120,000 base pairs/10 base pairs/turn); (*b*) 40.8 microns, (12,000 turns × 34 × 10^{-4} microns/turn); (*c*) 79.2 × 10^6, (1.2 × 10^5 base pairs × 660/base pair); 240,000 phosphorus atoms, (120,000 in each strand).

8. Since the $C_0t_{1/2}$ of the phage DNA is 200-fold less than that of the *E. coli* DNA, its size should also be 200-fold less, or 21,000 base pairs.

9. The phage DNA is ½₀ the size of the bacterial DNA, so its half-C_0t is also ½₀ that of the bacterial DNA.

10. 550 bp.

11. Avery and colleagues used cell-free extracts of virulent (S) cells as the transforming agent. Since no virulent cells were mixed with the avirulent ones, there was no possibility of resurrecting any virulent cells.

12. About 53%. Because the T3 mRNAs are transcribed from the T3 DNA, and are therefore complementary to one of the T3 DNA strands, their GC content should closely resemble the GC content of the T3 DNA template. Consider this example:

    DNA: GAGCTACGGC
           CTCGATGCCG
                   ↓ Transcription
    mRNA: GAGCUACGGC

    The GC content of the DNA is 70% (14 out of 20 bases), and so is the GC content of the mRNA (7 out of 10 bases).

13. No, you cannot predict the purine content of a transcript knowing only the purine content of the double-stranded DNA template. The purine content of the transcript will depend on which strand is transcribed. For example, if the strand (GGGCTA) is transcribed, the mRNA product will be CCCGAU (⅓, or 33.3% purine). Transcription of the complementary DNA strand yields GGGCUA, with a purine content of 66.7%.

14. (*a*)
    ════════════════
    5′ - - - - - - - - →3′

    (*b*) 5′ ──────────── 3′
         ════════════════
         - - - - - - - →

    (*c*)
    ════════════════
    3′ ──────────── 5′
    - - - - - - - →

15. (*a*) 5′GTCAATCGTAGCGGCCATAT3′
        3′CAGTTAGCATCGCCGGTATA5′
    (*b*) 5′AUAUGGCCGCUACGAUUGAC3′

16. 100 proteins (120,000/1,200 = 100).

CHAPTER 7

1. (a) Dispersive:

   (b) Conservative:

   (c) Semiconservative:

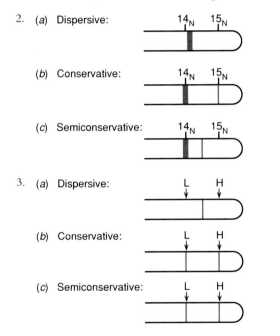

Note: The 8 DNA duplexes after 3 generations of dispersive replication would all be heterogeneous mixtures of ^{15}N and ^{14}N labeled DNA, with ⅞ of the label being ^{14}N.

2. (a) Dispersive:

   (b) Conservative:

   (c) Semiconservative:

3. (a) Dispersive:

   (b) Conservative:

   (c) Semiconservative:

4. The central space represents DNA that replicated before the radioactive thymidine was added. The dotted lines represent incorporation of low specific activity radioactive thymidine. The solid lines represent incorporation of high specific activity thymidine.

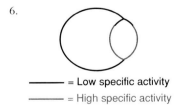

5. There are two classes of replicons; one begins replicating before the other. The late-starting replicons produce the pairs of streaks with the large gap in the middle.

6.

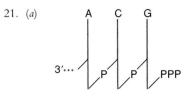

——— = Low specific activity

——— = High specific activity

7. (a) The gradient centrifugation of the pulse-labeled DNA would show only a peak of large DNA. No Okazaki fragments would appear, no matter how short the labeling period. (b) The DNA

polymerase must be able to make DNA in both the 5′ → 3′ and 3′ → 5′ directions.

8. Helicases separate the two DNA strands in advance of the replicating fork. This process requires energy, which is provided by the cleavage of ATP by the ATPase.

9. A helicase unwinds DNA, while a topoisomerase changes the topology of the DNA by increasing or decreasing its number of supercoils. Helicases do not break DNA strands, whereas transient strand breakage is a necessary part of a topoisomerase's function.

10. SSB binds cooperatively to single-stranded, but not double-stranded, DNA. This means that once one molecule of SSB binds to single-stranded DNA, it greatly facilitates the binding of other SSB molecules; such cooperative binding does not occur on double-stranded DNA.

11. Positive supercoils.

12. The primosome binds to the template for the lagging strand. This makes sense because the lagging strand needs a new primer for each new Okazaki fragment, whereas the leading strand replicates more or less continuously and rarely needs a new primer.

13. A replisome with two polymerase centers is able to replicate both strands more or less at the same time and do so processively in the direction of the replicating fork, without having to waste time dissociating from and reassociating with the DNA.

14. Primer synthesis cannot require a base-paired nucleotide to add to. Therefore, the proofreading function cannot operate. If the wrong nucleotide is incorporated into a primer, it cannot be removed by proofreading.

15. RNA genomes are replicated by RNA-synthesizing enzymes that do not require primers and therefore do not permit proofreading. Large RNA genomes would thus accumulate too many mistakes. Furthermore, RNA is inherently much less stable than DNA; therefore, large RNA genomes would tend to be degraded.

16. The DNA between the repeats is deleted.

17. The DNA between the repeats is inverted.

18. (a) 6 A's, 2 a's; (b) 4 A's, 4 a's.

19. Assuming a replication rate of 1,000 nucleotides per second at each fork and no pauses, it should take 2.1×10^3 seconds, or 35 minutes; that is $\dfrac{4.2 \times 10^6}{1 \times 10^3} \times \dfrac{1}{2}$, since the DNA replicates at the rate of 1,000 nucleotides per second and is bidirectional.

20. 40.8 microns. Assuming a rate of 1,000 bp/sec in each direction, the two replisomes will replicate 60 × 2,000 = 120,000 bp in one minute. Thus, 120,000 bp × 3.4×10^{-4} microns/bp = 40.8 microns.

21. (a)

(b)

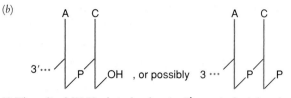

3′···   P   OH , or possibly   3 ···   P   P

(c) The edited DNA chain has lost its 5′-terminal triphosphate. Since phosphodiester bond formation in earthly DNA synthesis requires such a triphosphate, this would present a barrier to continued DNA synthesis until the triphosphate was restored. Presumably, Zorbian organisms have a mechanism to deal with this problem.

CHAPTER 8

1. RNA polymerase holoenzyme transcribes un-nicked T4 DNA actively *in vitro*. The transcription is specific (immediate early genes only) and asymmetric. The core enzyme transcribes such a template only weakly. What transcription occurs is nonspecific and symmetric.

2. (a) Up. A C → T mutation in the first C would create a −10 box with this sequence: TATAGT. This is closer to the consensus sequence. (b) Down. A T → A mutation in the last position would create CATAGA. This is less like the consensus −10 box.

3. The genes in the operon are all transcribed together, from a single promoter. Therefore, controlling transcription from that promoter controls all the genes simultaneously.

4. (a) The *lac* operon would always be turned on, because repressor cannot turn it off unless it can bind to the *lac* operator. (b) Same as (a). (c) The operon would be uninducible. Repressor would remain bound to operator even in the presence of inducer. (d) The operon would be transcribed only weakly. Even in the absence of glucose, CAP plus cAMP would be unable to facilitate polymerase binding to the *lac* promoter.

5.
Phenotype for β-galactosidase Production		
No Inducer	**Inducer**	**Reasoning**
−	+	Wild-type
−	+	Z^+ is dominant
−	+	I^+ is dominant
−	−	I^s is dominant
+	+	O^c is dominant
−	+	O^c is only *cis*-dominant.
+	+	O^c is dominant, even in the presence of I^s repressor because the O^c operator cannot bind the I^s repressor.

6. (a) Since the mutant's phenotype is constitutive, its genotype must carry a mutation. Therefore, I^x must be I^-.
(b) Since the mutation in I prevents *lac* function, and is dominant, I^x must be I^s.

(c) Since the partial diploid behaves as wild-type, I^x is either I^+ or I^- (I^- is recessive).
(d) Since the mutant is constitutive, I^x must be I^-.
(e) Since the mutant is constitutive, O^x must be O^c.
(f) Since the phenotype of the partial diploid is wild-type, O^x must be O^+.
(g) Since the partial diploid fails to make β-galactosidase, even with wild-type forms of the control loci (I^+O^+), Z^x must be Z^-.

7.
	No Inducer	**Inducer**	
(a)	−	+	P^+ is dominant. The *lac* operon with the wild-type promoter will still function, even if the other does not. The phenotype is therefore wild-type.
(b)	−	−	With neither promoter able to bind RNA polymerase, neither *lac* operon can function.
(c)	−	−	The operon with the wild-type Z gene has a nonfunctional promoter, and the operon with a wild-type promoter has a nonfunctional Z gene.
(d)	−	−	Even in the absence of active repressor, neither *lac* operon can be productive without a functional promotor.

8. Two of the enzymes contain two different polypeptides. Therefore, the five genes code for the five different polypeptides contained in the three enzymes.

9. (a) 700-fold; 70-fold from derepression and 10-fold from relief of attenuation. (b) 70-fold; the 10-fold enhancement from relief of attenuation is lost because a ribosome traversing the leader will fall off the ribosome at the new UGA (stop) codon. (c) 70-fold; the 10-fold enhancement from relief of attenuation is lost because the tryptophan codons are missing, so the ribosome will not stall in the absence of the amino acid. (d) 70-fold; attenuation would not work because the string of G's would not be recognized as part of a proper terminator.

10. (a) The early genes only, because the gene 28 product is needed for the switch to middle transcription. (b and c) The early and middle genes only, because the products of both genes 33 and 34 are needed for the switch to late transcription.

11. Only the early (class I) genes would be transcribed, since gene 1 encodes the T7 RNA polymerase that transcribes the later genes.

12. The cells would not be able to sporulate, since σ^C and σ^E are needed to switch the transcriptional specificity of *B. subtilis* RNA polymerase from vegetative to sporulation genes.

13. The repressor gene (*cI*).

14. All *but* the repressor gene.

15. Transcripts made in the delayed early phase of infection all initiate at the immediate early promoters. Therefore, immediate

early products, including pN, continue to be made throughout the delayed early phase.

16. It cannot result in *cro* expression because the nontemplate strand of the *cro* gene is transcribed, yielding an "antisense" RNA.

17. Transcription from P_{RE} depends on products of the delayed early genes *cII* and *cIII,* while transcription from P_{RM} depends on repressor itself. Therefore, neither of these promoters can be active before other λ genes have been expressed.

18. Positive. The expression of the delayed early genes is turned off unless pN acts positively to turn it on.

19. (*a*) Some cells within the plaque are lysogenized and thus are immune to superinfection. These continue to grow and cloud the plaque. (*b*) No, a repressor mutant would be unable to lysogenize cells, so its plaques would be clear.

20. Neither. Since active pN is not made, no antitermination will occur, so the phage will be stuck in the immediate early phase. Since no cII is made, activation of the *cI* gene will not occur, so a lysogen cannot form.

21. The recognition helix.

CHAPTER 9

1. A classical polymerase III promoter lies within the coding region of the gene that it controls. Any change in the promoter will therefore alter the sequence of bases within the coding region of the gene, and therefore alter its product.

2. First, nucleosomes are found in all eukaryotes that have been studied. Second, most of the histones are extremely well conserved. If they were not so vitally important, they would have evolved more rapidly.

3. The 11-nm fiber is essentially a string of nucleosomes. The nucleosomes contain all five kinds of histones, but H1 is bound only to the outside of each nucleosome. Since H1 is not necessary for forming nucleosomes, its absence would not perturb the 11-nm fiber, beyond reducing its average diameter slightly. On the other hand, histone H1 plays an essential role in coiling the nucleosome chain into the 25-nm fiber, or solenoid. Removing H1 would prevent formation of this chromatin structure.

4. The euchromatic chromosome, because the other X chromosome, which is heterochromatic, would contain highly condensed DNA whose genes would be inaccessible to transcription and would therefore be inactive.

5. Similarity: Both stimulate transcription of genes. Difference: σ causes holoenzyme to bind tightly to promoters, but does not bind to them by itself. Most eukaryotic transcription factors, on the other hand, bind to promoters independently of RNA polymerase.

6.

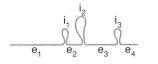

7.

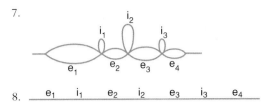

8.   e₁   i₁   e₂   i₂   e₃   i₃   e₄

9. 27 bases downstream from the second A of the TATA box (TATAAAT) lies the sequence TT**AG**CT. Within this sequence there are two purines (A and G, boldfaced) at 29 and 30 bases downstream from the middle of the TATA box. These are the most likely cap sites.

10.

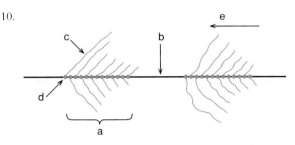

11. GpppXpY . . .

12. To show that poly(A) is present in the mRNAs, degrade them with RNase A and T1, then electrophorese the remaining RNA to see if a poly(A) has survived. (RNase A cuts RNA after C's and U's, and RNase T1 cuts after G's, so only strings of A's, like poly[A], will be left.) To show that the poly(A) is at the 3′-end of the mRNA, digest the mRNA briefly with a 3′-specific RNase to see if the poly(A) is rapidly lost. Only if the poly(A) is at the 3′-end will it be degraded immediately.

13. First, clone a gene containing the GAGA box, then introduce mutations into the GAGA sequence. Put the mutated gene into an in vitro or in vivo transcription system from the algae and measure the effect of the mutations on transcription. If the GAGA box behaves like the SV40 TATA box, mutations in this region should cause an increase in heterogeneity of transcription start sites.

14. It would be part of the intron. In fact, that phosphate participates in the branch of the lariat.

15. It would stay with the exon. In fact, the phosphate becomes part of the phosphodiester bond linking the two exons in the mature mRNA.

CHAPTER 10

1. The tRNAs would bind together much more tightly than free trinucleotides because their anticodons are bent into a helical shape that facilitates base pairing.

2. The alanine would be incorporated where a methionine is supposed to be inserted because it is the anticodon, not the amino acid, that determines the response to a given codon.

3. The aminoacyl-tRNA synthetases are crucial to fidelity of translation; if they make mistakes, the translation machinery will incorporate incorrect amino acids.

4. First reaction: aminoacyl-AMP + pyrophosphate. Second reaction: aminoacyl-tRNA.

5. (a) Fully overlapping, 10 (AUG, UGG, GGC etc.); nonoverlapping, 4 (AUG, GCA, GUG, CCA). (b) Fully overlapping, 3 (UGC, GCC, CCA); nonoverlapping, 1 (CCA).

6. (a) The reading frame would shift by two bases, and the gene would therefore not function. (b) The reading frame would not change, and the gene would therefore be functional, unless the extra three bases coded for an amino acid that disrupted the function of the gene product. (c) The reading frame would not change, and the gene would therefore be functional, unless the changes in the two adjacent codons were deleterious.

7. A repeating dipeptide. The two codons would be UUCGUU and CGUUCG, assuming that translation starts at the beginning of the RNA.

8. The proteins would be made with arginines where phenylalanines should have been. Most or all of these proteins, because of their altered primary structures, would function abnormally or not at all.

9. It would eliminate initiation at that point. A subsequent AUG in the middle of the message might then function as the start codon, but the resulting protein product would be incomplete—truncated at the amino terminus.

10. (a) fMet-Phe-Lys-Met-Val-Thr-Trp. The next to last codon (UAA) causes termination, so the last codon (AUC) is not translated. (b) fMet-Val-Thr-Trp. Note that the base change destroyed the first start codon, so the fourth codon (AUG) would serve as the initiating codon, resulting in a shortened peptide only 4 amino acids long. (c) No change in amino acid sequence; UUU and UUC both code for Phe. (d) fMet-Leu-Lys-Met-Val-Thr-Trp. (e) fMet-Phe-Lys-Met-Val-Thr. The change converted the Trp codon to a stop signal.

11. AUG AAC CAC GAC UUA UAG(stop). The alternative, AUG UUC UGA(stop), starts closer to the 5′-end but is much shorter.

12. (a) Leu, Ser, and Arg. (b) Trp and Met.

13. 256 (4^4).

14. CUU and CUC.

15. UAI.

16. Cloacin DF13 should profoundly inhibit translation, because it removes the part of the 16S rRNA that base-pairs with Shine-Dalgarno sequences on mRNAs, and binds the mRNAs to the ribosome. Thus, without strong ribosome-mRNA binding, translation would be inhibited.

17. This is a very poorly translated mRNA. The main reason is that it has no leader, and therefore no Shine-Dalgarno sequence to which ribosomes would readily bind.

18. Messenger RNA (b) would be translated most actively. (a) would not bind the 40S ribosome because of the hairpin near the cap. (c) would not be translated as well as (b) because of its shorter leader. (d) would not be translated well because of the large, stable hairpin in its leader. (e) would not be translated as well as

(b) because it lacks a hairpin just downstream from the initiation codon.

19. (a) Translocation; (b) aminoacyl-tRNA transfer to the ribosome; (c) peptide bond formation.

20. fMet-tRNA. This aminoacyl-tRNA must move to the P site to make room in the A site for the next incoming aminoacyl-tRNA.

21. 402. If this does not make sense, consider a dipeptide with only two amino acids. There is only one peptide bond linking the two amino acids, so only one peptidyl transferase reaction was needed to synthesize the dipeptide. Thus, the number of peptidyl transferase reactions required to make a polypeptide is one less than the number of amino acids in that polypeptide.

22. All QT proteins made in this suppressor strain would have an extra tyrosine at the carboxyl end, because the suppressor tRNA would recognize the first stop codon (UAG) in every gene as a tyrosine codon.

23. Wild-type Gln anticodon: 3′-GUU-5′ (recognizes the Gln codon 5′-CAA-3′). Ochre suppressor anticodon: 3′-AUU-5′ (recognizes the ochre codon 5′-UAA-3′). The change was a G → A transition in the 3′ terminal base of the anticodon.

24. Show that the mutant will grow on *c* but not on *b*. This shows that the *b* → *c* enzyme is defective. Therefore, its gene is defective.

CHAPTER 11

1. A somatic mutation. If it were a germ-line mutation, the color would be uniform, or at least in a pattern of some sort. The unique spot of white was probably caused by a mutation inactivating a gene responsible for color in a single somatic cell. This mutation was passed on to all descendents of this one cell.

2. We know the yellow color is dominant, because two-thirds of the progeny are yellow; thus, the heterozygotes are yellow. A lethal gene must be involved since one-quarter of the progeny are inviable; the lethality of the yellow gene must be recessive, because only the homozygotes die.

3. No. Mutations that inactivate a gene encoding a subunit of DNA polymerase III would surely be lethal, since they would block a function crucial to life: replication of DNA.

4. Conditional lethal mutations would be appropriate. Bacteria could be tested for the trait by attempting to grow them at normal temperature, but they could be maintained alive at lower temperature.

5. No, the product of a temperature-sensitive gene, not the gene itself, is temperature-sensitive.

6. After the patch of fur was shaved, the skin was relatively cool, so the temperature-sensitive, color-producing enzyme was active. This led to a patch of dark fur. The new fur warmed the skin, inactivating the color-producing enzyme, so the dark patch gradually turned white as new white hairs replaced dark ones.

7. Transitions. Tautomerization of T causes it to base-pair like C, resulting in T → C transitions.

8. (*a*) Pyrimidine (especially thymine) dimers. (*b*) If this damage goes unrepaired, it can either stall DNA replication or cause error-prone repair, which can induce mutations.

9. (*a*) DNA strand breaks. (*b*) Double-strand breaks are difficult to repair properly and therefore constitute lasting damage.

10. The mRNA becomes 5′ AUGGCCUAAAGAGG 3′; the frameshift mutation has shifted the reading frame one base to the right. Now the mRNA codes for the truncated peptide fMet–Ala since the third codon, UAA, is a stop codon.

11. Add a base, preferably early in the mRNA, to compensate for the lost base. This will restore the reading frame.

12. Photoreactivating enzyme (DNA photolyase) and O^6-methylguanine methyl transferase.

13. An exonuclease degrades a polynucleotide (DNA or RNA) from the end, one base at a time; an endonuclease cuts within a polynucleotide.

14. A long gap in one of the strands. Long gaps do occur during repair in *E. coli,* suggesting that this bacterium uses the second, stepwise method.

15. Because the damage remains, even though DNA replication has occurred.

16. (*a*) The wild-type sequence is TGG (the only Trp codon), and the mutant sequence could be either TGT or TGC (both Cys codons). (*b*) Since the sense of the codon changed from Trp to Cys, the change was a missense mutation (specifically, a transversion).

17. (*a*) A C (or a U) was inserted after the second base, yielding

    Asp   Ala   Stop
    GAC   GCA   UAA   GC

(*b*) The fourth base (a C) was deleted, yielding

    Glu   Ile   Ser
    GAG   AUA   AGC

18.

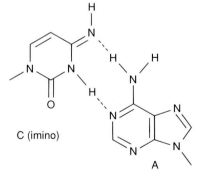

19.

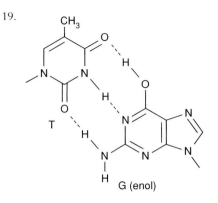

20. (*a*) AUCGU AUG CAU GCA UCG UGA C. In this and in other mRNA sequences below, the codons, beginning with the first AUG, are set off from one another by spaces.

(*b*) fMet–His–Ala–Ser

(*c*) DNA: GTATCGTAT**A**CATGCATCGTGAC
             CATAGCATA**T**GTACGTAGCACTG
   RNA: AUCGUAU**AC** AUG CAU CGU GAC
   Protein: fMet–His–Arg–Asp

(*d*) DNA: GTAC—The rest of the sequence remains
            CAT**G** the same.
   RNA: **AC**CGU AUG CAU GCA UCG UGA
   Protein: fMet–His–Ala–Ser. Since the mutation occurred outside the coding region, no change in protein sequence results.

(*e*) DNA: GT**A**TCGTATGCATGCA**C**CGTGAC
            CATAGCATACGTACGT**G**GCACTG
   RNA: AUCGU AUG CAU GCA **C**CG UGA
   Protein: fMet–His–Ala–Pro

(*f*) DNA: GTATCGTATGCATGCATC**A**TGAC
            CATAGCATACGTACGTAG**T**ACTG
   RNA: AUCGU AUG CAU GCA UC**A** UGA C
   Protein: fMet–His–Ala–Ser. Note that the mutation did not change the meaning of the codon, so no change in the peptide sequence occurs.

(*g*) DNA: GTATCGTATG**CT**GCATCGTGAC
            CATAGCATAC**GA**CGTAGCACTG
   RNA: AUCGU AUG C**U**G CAU CGU GAC
   Protein: fMet–Leu–His–Arg–Asp

CHAPTER 12

1. (*a*) Genes for transposase and terminal inverted repeats that the transposase recognizes. (*b*) Terminal inverted repeats that the other transposon's transposase recognizes.

2. Transposase is necessary for transposition.

3. The DNA does not always jump. In replicative transposition, the transposon does not even leave its original location; instead, it replicates and one of the copies goes to the new location.

4. (*a*) The inserted transposon will be surrounded by 5-bp direct repeats of host DNA. (*b*) See figure 12.3; imagine the cuts 5 bp apart, rather than 9 bp.

5. After allowing a certain time for transposition to occur, isolate the target plasmid and use it to transform cells with no inherent antibiotic resistance. Plate the transformed cells on medium containing both chloramphenicol and ampicillin. Only those cells that received a plasmid with both antibiotic resistance genes will survive; therefore, the number of survivors is related to the rate at which Tn3 transposed to the target plasmid.

6. Fewer. Homologous recombination requires some homology between the transposon and its target DNA, which automatically limits the number of target sites; illegitimate recombination requires little if any homology, so a transposon using this mechanism is free to transpose anywhere in the target DNA.

7. (*a*) Replicon fusion (cointegrate formation) and resolution. (*b*) A cointegrate.

8. (*a*) No product; the transposase is needed to form the cointegrate intermediate. (*b*) A cointegrate.

9. (1) The Mu DNA integrates into the host DNA for lytic replication, and (2) the DNA replicates by transposition.

10. Several transposons conferring resistance to a variety of antibiotics can transpose to the same plasmid in a pathogenic bacterium. This bacterium will be resistant to all these antibiotics, and it can even share the plasmid with other dangerous bacteria, making them antibiotic-resistant. When these bacteria cause disease, they are very difficult to kill because they have built-in resistance to our drugs.

11. Two. At the boundaries between the two plasmids.

12. No, Tn3 transposes by replicating itself, so it would not leave the *C* gene during transposition, and reversion would not be nearly as frequent.

13. (*a*) Inhibitors of double-stranded DNA replication would block transposition of Tn3, but not Ty, because Tn3 transposition depends on this process, but the retrotransposon Ty does not. (*b* and *c*) Inhibitors of transcription and reverse transcription would block transposition of Ty, but not Tn3, because Ty transposition, just like retrovirus replication, requires these processes. Tn3 does not have to be transcribed, because all the gene products needed for its transposition are already present in the extract. (*d*) Inhibitors of translation will have no effect on either transposon; all proteins needed for transposition are supplied in the extract.

14. The intron would remain in the transposon. Since Tn3 transposition does not involve an RNA intermediate, there would be no opportunity for splicing.

15. The intron would be spliced out.

CHAPTER 13

1. Genotypes and phenotypes (respectively) are: (*a*) *lac⁻*, Lac⁻; (*b*) *leu⁺*, Leu⁺; (*c*) *thi⁻*, Thi⁻; (*d*) *str^r*, Str^r; (*e*) *pur^s*, Pur^s; (*f*) *bio⁺*, Bio⁺.

2. Prototrophs: (*a*), (*b*), (*d*), (*e*), and (*f*). Auxotroph; (*c*). The *thi⁻* strain is the only one that cannot make a substance (thiamine) needed for growth on minimal medium containing glucose.

3. Yes, he was justified. If single mutants had been used, their reversion rate (one in a million) would have produced ten times as many prototrophs as their recombination rate (one in ten million). Therefore, the investigators would have observed almost as many prototrophs with the two mutant strains grown separately (one in a million) as when they were grown together (1.1 in a million); thus, they would not have known whether they were observing recombination plus reversion, or just reversion.

4. It usually does not contain a full set of genes from the donor cell, so it is not a true diploid. Before recombination, it is an unstable merodiploid; after recombination, it is haploid.

5. The Ton^r phenotype is caused by a mutation. If it were due to an adaptation, all cultures would be about equal in resistance to the phage. (These were the actual results obtained by Luria and Delbrück in 1943.)

6. Integration of the F plasmid into the host chromosome. This mobilizes the host chromosome so that it behaves as the F plasmid normally would and allows the chromosome to be transferred (usually only in part) to a recipient cell.

7. Usually, the connection between conjugating cells breaks before the whole chromosome has had a chance to transfer.

8.
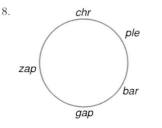

9. Conjugation is probably interrupted, just as it is in earthly bacteria. This lowers the likelihood that genes near the end of the chromosome will be transferred; conjugation usually ceases before such transfer can take place.

10. 25 minutes. The *chr* marker would be transferred first, 50 minutes after conjugation begins; *ple* would be transferred next, 75 minutes after conjugation begins. The difference in transfer times is therefore 25 minutes.

11. *gap-bar-ple-chr-zap*

12. By recombination, place one mutant *zap* operon on an F–*zap* plasmid that is *hisB⁺*. Place this F–*zap* plasmid in an *X. gigantii* strain that is sensitive to lethalmycin (*let^s*). Conjugate this strain with a recipient F⁻ strain that is *zap⁻* because it carries the other mutant *zap* operon, and is also *hisB⁻* and *let^r*. Select for cells that are *hisB⁺* and *let^r* by growing them in the absence of histidine and in the presence of lethalmycin. This will yield cells that have undergone conjugation, with transfer of the F–*zap* plasmid to the F⁻ cells. This places both mutant *zap* operons in the same cell— one in the chromosome, the other in the F′ plasmid. Test these cells for Zap phenotype. If the mutations are in different genes,

they will complement each other and give rise to a large number of Zap⁺ colonies. If the mutations are in the same gene, very few, if any, Zap⁺ colonies will arise.

13. They could be due to recombination between two DNAs with mutations in the same gene (either *zapA* or *zapB*). Alternatively, they could be due to intracistronic complementation, such as that observed in the *E. coli lacZ* gene. However, this mechanism should be relatively efficient, so the first possibility (recombination) is probably correct.

14. λ phage heads can only accommodate about 10% extra DNA. Thus, the λ DNA will be able to pick up host genes from one side (*gal*) or the other (*bio*) of the prophage, but not from both.

15. Yes, they would cotransduce, because they are less than 2.4% of the *E. coli* genome apart, and this is the amount of DNA that can fit into a P1 phage head. Using the expression: frequency = $(1-d/L)^3$, we would predict a cotransduction frequency of 2%, not too far from the actual frequency of 1%.

16. First, infect the first Zip⁻ strain (genotype *zap⁺ zip₁⁻*) with Q2 and collect the progeny phage. Use these to infect the second Zip⁻ strain (genotype *zap⁻ zip₂⁻*) and select for Zap⁺ transductants. Since these transductants have received the *zap* gene, there is a good chance that they have also received the closely linked *zip* gene. Therefore, the Zap⁺ transductants can be screened for Zip⁺ transductants that result from recombination between the two *zip⁻* mutations (*zip₁⁻* and *zip₂⁻*). The farther apart the mutations in the two damaged *zip* genes, the higher the frequency of these Zip⁺ transductants.

17. *zip zap bop*. If the order were *zip bop zap,* Bop⁺ Zip⁻ Zap⁻ transductants would be less common than Bop⁺ Zip⁺ Zap⁺.

18. The marker order is *awk nrd kat.* The Awk⁺ Kat⁺ Nrd⁻ class of transductants is the least frequent (1 in 1,063); these would be the least frequent, resulting from 4 cross-over events, only if the order of the three markers is as indicated. The cotransduction frequency between *awk* and *kat* is (1 + 115)/1,063 = 0.11. The cotransduction frequency between *awk* and *nrd* is (309 + 115)/1,063 = 0.40.

19. (250 plaques/0.05 ml) × 10⁶ dilution factor = 5 × 10⁹ pfu/ml.

20. The curve will start to rise sooner and the rise will be more gradual. This is because infectious phages appear in infected cells before lysis occurs. These will be detected by breaking the cells open.

21. K M L (or L M K).

CHAPTER 14

1. The red blood cell itself can live without hemoglobin, but the whole organism would die without this product, which is needed to carry oxygen to all its tissues.

2. (*a*) Determined only; (*b*) determined and differentiated.

3. After two cell divisions, there are only four cells, enough for identical quadruplets if the cells separate instead of staying together in one embryo, but not enough to produce quintuplets.

One more cell division would produce eight blastomeres. This is more than enough for identical quintuplets, provided the cells are all totipotent. If they have lost any of their genetic potential, they cannot give rise to a whole person.

4. You would observe neither DNase hypersensitivity nor activity. This is because the histones, which you added first, would bind throughout the gene and its 5′-flanking region, tying it up in nucleosomes. The factors in the fourteen-day extract would not be expected to remove these nucleosomes.

5. With a cloned gene, deletions of precisely defined length and position can be introduced, and the altered genes can then be tested for light sensitivity. Without a cloned gene, the deletions would have to be introduced by hit-and-miss genetic means, and their locations and extents would not be as well defined.

6. Expression of the gene with the mutated 5′-flanking region must be distinguishable from the unmutated gene. If the mutated pea rubisco gene were reintroduced into a pea plant, its expression would not be as easily distinguishable from expression of the resident pea rubisco gene.

7. The gene will be active because no histone H1 is present to cross-link the nucleosomes.

8. Early embryonic cells divide rapidly, thus they must require and produce a great deal of protein. 5S rRNA is one of the components of the ribosomes that must make this protein.

9. (*a*) The vitellogenin gene (ovalbumin is not produced in the liver); (*b*) neither.

10. Messenger RNA levels in the cytoplasm reflect much more than just transcription. They are also potentially influenced by capping, polyadenylation, splicing, transport, and RNA degradation.

11. Pulse-label cells with radioactive RNA precursor, then chase with unlabeled precursor. After various durations of chase, hybridize the labeled total mRNA (polyadenylated RNA) to a cDNA that is complementary to the mRNA you are studying. The more labeled RNA that can still hybridize after a given time, the more stable the mRNA.

12. The first twenty-one amino acids in this protein constitute a signal peptide that is removed during maturation of the protein.

13. Individuals with mutations in these genes experience abnormal development.

CHAPTER 15

1. *Xgi*I (*Xenobacterium giganticus*).

2. GATATC

3. Every 4⁵ = 1,024 bases.

4. *Bam*HI and *Bgl*II. 5′-GATC-3′.

5. No; after ligating the two DNAs together, the site (GGATCT) is a hybrid—*Bam*HI-like on one end and *Bgl*II-like on the other. Therefore, neither enzyme will recognize this site.

6. You would see a uniform lawn of bacteria instead of individual colonies. This is because *all* cells will grow, not just those containing plasmids with antibiotic resistance genes.

7. Tetracycline. Inserts in the *Bam*HI site of pBR322 will inactivate the tetracycline gene.

8. No. The *Eco*RI site of pBR322 lies outside both the ampicillin and tetracycline resistance genes.

9. No. *E. coli* cells selectively replicate pUC plasmids with small inserts. Therefore, it would take millions of pUC clones to build a complete genomic library from a complex genome.

10. Yes. The pUC plasmids are very convenient, and since the total amount of DNA we need to clone is not large, we do not have to worry about the fact that each clone will have only a limited amount of DNA.

11. Since the tobacco probe and the pea gene will almost certainly have different sequences, you should use a lower temperature (lower stringency) so that a stable hybrid can form between the two DNAs of similar, but not identical, sequence.

12. (*a*) Tyr-Met-Cys-Trp-Ile. (Only the first two bases of the Ile codon would be used.) (*b*) $2 \times 1 \times 2 \times 1 \times 1 = 4$. (*c*) $6 \times 2 \times 1 \times 2 \times 1 = 24$.

13. When the vector and insert have tails long enough to form stable hybrids (approximately 12 bases). Then the host cell can complete the ligation step after transformation.

14. It prevents premature synthesis of the protein product of the cloned gene, which may be harmful to the host cell.

15. One in six; one-half will be in the right orientation and one-third of these will be in the right reading frame, since there are three possible reading frames; ($\frac{1}{2} \times \frac{1}{3} = \frac{1}{6}$).

16. Thr   Asp   Ser   Val   Gly
   ACC   GAT   **TCA**   GTG   GGC

   Boldface type denotes the base change that alters the Ala codon to a Ser codon.

17. Cut the viral DNA with *Eco*RI, electrophorese the resulting five fragments, and Southern blot the fragments. Hybridize the Southern blot to a radioactive probe made from the cloned cDNA, and autoradiograph to see which of the five DNA fragments "light(s) up."

18. Collect mRNAs from all five tissues, electrophorese them, then Northern blot. Hybridize a radioactive probe (made from the cloned petunia gene) to the Northern blot and autoradiograph. The tissue that contains the highest concentration of the specific mRNA will hybridize best and will give the most intense band on the autoradiograph.

19. The size(s) of the mRNA(s) made from this gene.

20. ATGAGTTGGCCAGCAGAGCGAGCATGGATGTAA

21.

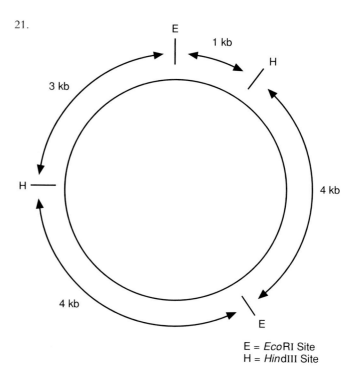

E = *Eco*RI Site
H = *Hind*III Site

22. (*a*) *Pst*I. (*b*) The two *Pst*I fragments would be 1.5 and 6 kb if the insert has the correct orientation. (*c*) The two *Pst*I fragments would be 3 and 4.5 kb if the insert has the incorrect orientation.

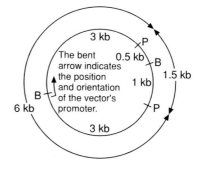

The bent arrow indicates the position and orientation of the vector's promoter.

23. (*a*) 9 and 6 kb. (*b*) 9 kb.

24. (*a*) 12 and 8 kb; (*b*) 7 and 3 kb; (*c*) 12 and 7 kb; (*d*) 8 kb.

25.
(*a*) ---AGCT--- *Alu*I ---AG + CT---
      ---TCGA--- → ---TC   GA--- (blunt)

(*b*) ---CGATCG--- *Pvu*I ---CGAT3' +   CG---
      ---GCTAGC--- → ---GC   3'TAGC---
                                (2-base 3'-overhang)

(c)
$$
\begin{array}{ccc}
\text{---GATC---} & \textit{MboI} & \text{---} \\
\text{---CTAG---} & \rightarrow & \text{---CTAG5'} \\
\end{array}
+
\begin{array}{c}
\text{5'GATC---} \\
\text{---}
\end{array}
$$
(4-base 5'-overhang)

26. (a) and (c) are palindromes. The center of symmetry is through the central AT pair in (c). (b) and (d) are mirror images.

CHAPTER 16

1. If any of the live virus had not been weakened enough, it could cause AIDS, which would probably be fatal.

2. The protein would come from cells, such as bacteria or yeast, carrying the cloned gene, not from the HIV itself, so it could not possibly cause AIDS.

3. (1) The β-globin gene is ordinarily tightly controlled, and there is no way of guaranteeing such control in an implanted gene. (2) In order to cure the disease, we must be able to transform stem cells, and that has been difficult so far.

4. The corrected cells with a healthy gene should be able to synthesize DNA and therefore replicate more rapidly than uncorrected cells. Thus, the corrected cells should outgrow the resident, mutant cells.

5.

Vector	Insertion of DNA	Cancer Risk	Capacity
retrovirus	+	relatively high	relatively high
adenovirus	−	relatively low	relatively high
AAV	+	relatively low	relatively low

6. (1) To give the organisms more desirable characteristics. (2) To harvest the products of the transgenes.

7. (1) Clone the gene into a Ti plasmid, insert it into *Agrobacterium tumefaciens,* and infect a plant with the engineered bacterium. (2) Coat metal pellets with the gene and "shoot" them into plant cells.

8.

Haplotype	Fragment Sizes (kb)
A	2, 3, and 1
B	2 and 4
C	5 and 1
D	6

9. Microsatellites are spread relatively uniformly throughout the genome, whereas minisatellites tend to cluster at the ends of chromosomes.

10. Two telomeres (one at each end) and a centromere, and, of course, at least one restriction site into which foreign DNA can be inserted.

11. (a) No, only exons can be trapped.
(b) No, only complete exons can be trapped.

(c) Yes, an exon with introns on both sides will have the splicing signals to ensure that it is spliced into the mature transcript made in the COS cells.
(d) No, even though the exon is complete, it will be missing the splicing signal on the side with no intron, so it will not be spliced into the mature transcript.

12. (1) Use an exon trap. (2) Look for CG islands. (3) Look for a conserved DNA region; that is, one that is homologous to DNAs from other organisms.

13. (a) The mutation is an expansion of a repeat of the sequence CAG, from a range of 11–34 copies to a range of 38 or more copies. (b) First, the *HD* genes of affected people always have CAG repeat regions that have expanded beyond the normal range. Second, the degree of expansion correlates with the age of onset of the disease. Finally, two individuals are affected by HD even though their parents are not. The affected individuals have expanded CAG repeats, while their parents do not.

14. Chromosome 5 must carry the disease gene. Chromosome 5 is the only one present in all three cell lines whose DNA hybridizes to the disease-linked probe.

15. (a) The most prevalent *CF* mutation is a three-bp deletion, which results in deletion of a single amino acid (phenylalanine) from CFTR. (b) The following evidence shows that the gene encoding CFTR is *CF:* (1) Northern blots show that this gene is transcribed in all tissues affected by CF. (2) We know that CFTR is a chloride ion channel, since it can generate such channels when added to an artificial membrane. (3) CF patients all have mutations in this gene. (4) A wild-type gene encoding CFTR can cure the defect in chloride ion transport in CF cells.

16. (a) 1.2×10^{-10} ($0.03 \times 0.02 \times 0.01 \times 0.005 \times 0.04 \times 0.1$).
(b) 4.32×10^{-6} ($2[0.06 \times 0.05\star] \times 2[0.2 \times 0.09] \times 2[0.1 \times 0.1]$).
$\star$Remember to raise 4% to 5% before computing the product.

CHAPTER 17

1. Xeroderma pigmentosum patients have defective DNA repair systems that cannot repair ultraviolet damage properly. As a result, these people respond to ultraviolet exposure by developing multiple skin cancers.

2. Chemicals, radiation, and viruses.

3. He filtered the sarcoma extract to remove anything as large as (or larger than) bacteria. The agent that passed through this filter was still able to generate sarcomas when injected into other chickens.

4. Transformation means the change in a cell's behavior from normal to tumor cell-like.

5. (a) Yes; (b) no.

6. The primary product of the *gag* gene is a single polypeptide. Only after this polypeptide is completed is it cut into the smaller pieces that are the final products of the gene.

7. Both are viral DNA genomes inserted into their hosts' DNA.

8. Oncogenes probably developed first in cells. Cellular oncogenes usually have introns, whereas viral oncogenes do not. It is easier to explain loss of introns through transcription, splicing, reverse transcription, and incorporation into a virus, than to explain gain of introns through some unknown mechanism.

9. Their proviruses can insert near cellular oncogenes and activate them because of the strong enhancers in their LTRs.

10. NIH 3T3 cells are not normal. They are already immortal and poised on the brink of transformation, needing just one final nudge to push them over. The mutant Ha-*ras* gene provides that nudge.

11. It is a somatic mutation. We know this because tumor cells have a mutant Ha-*ras* gene, while normal cells from the same patient have only the normal gene.

12. Extract DNAs from human lung cancer cells and normal human cells, cut them with a restriction enzyme, electrophorese the fragments, Southern blot the fragments and hybridize the blots to a radioactive *aal* probe (v-*aal* DNA from the aardvark leukemia virus). Perform autoradiography and compare the intensities of the bands in the cancer cell and normal cell lanes. If the bands from the cancer cells are more intense, the c-*aal* oncogene has been amplified.

13. Extract RNAs from human liver cancer cells and normal human liver cells, electrophorese them, Northern blot, and hybridize the blots to a radioactive *bas* probe (v-*bas* DNA from the bat sarcoma virus). Perform autoradiography and compare the intensities of the bands in the cancer cell and normal cell lanes. If the band in the cancer cell lane is more intense, the c-*bas* oncogene is excessively transcribed in these cells.

14. The *was* gene encodes the tyrosine protein kinase.

15. (*a*) A mutation that enhanced the GTPase activity of Ras plus GAP would tend to slow growth of the cell because the mutant p21 would get rid of the stimulatory GTP more rapidly—by cleaving it to GDP. (*b*) A mutation that reduced the GTPase activity of Ras plus GAP would tend to accelerate growth of the cell, perhaps even help transform the cell, because the mutant p21 would keep the stimulatory GTP around longer before it would cleave it to GDP.

16. Normal, because the normal cell would provide the regulatory factor missing in the retinoblastoma cell, thus bringing cell division under control.

17. A human born with one normal and one retinoblastoma gene is usually initially free of malignancy—just like the hybrid cells with one normal and one malignant "parent." But the person develops the malignancy later, once a normal retinal cell gene is damaged by a somatic mutation. Given enough time, the hybrid cells would probably accumulate such mutations and also become malignant. Thus, the main difference in these experiments is time.

18.

Normal Cell	Tumor Cell
Limited lifespan	Immortal
Grow only on solid surfaces	Grow in soft medium
Need serum factors	Serum-independent
Contact inhibited	No contact inhibition
Irregular shape (usually)	Rounded
Ordered growth	Disordered growth
Resemble tissue of origin	Lose resemblance to tissue of origin
Diploid	Usually aneuploid
Not metastasizing	Metastasizing

CHAPTER 18

1. Because they lack functional chloroplasts and therefore cannot carry out photosynthesis, which is essential for the survival of higher plants.

2. (*a*) Smooth; (*b*) wrinkled. In maternal inheritance only the maternal genotype matters.

3. The gene for a smooth or wrinkled chloroplast is carried in the chloroplast itself. Since the gamete of the female parent contributes the chloroplasts to the zygote, only the genotype of that parent matters in determining the chloroplast type in the offspring.

4. Ova from the variegated plant may contain: all white chloroplasts, in which case the progeny will be white; all green chloroplasts, in which case the progeny will be green; or a mixture of green and white, in which case the progeny will be variegated.

5. Because the vegetative algal cells are haploid. These must fuse to form a diploid zygote before reductional division can take place.

6. Because the *mt*⁺ parent contributes the chloroplast genes to the progeny, just as the maternal parent does in higher plants.

7. No, it could be maternal effect. We would have to carry out our observations to the third generation to see if inheritance is Mendelian (maternal effect) or non-Mendelian (maternal inheritance).

8. Most of the proteins in a mitochondrion are not products of mitochondrial genes. Instead, they are encoded in nuclear genes and imported into the mitochondria. These proteins are still made in neutral petites.

9. Fuse the *capᵣ leu⁻ bio⁺* hyphae with the *capˢ leu⁺ bio⁻* hyphae and later look for sections of hyphae that are *capᵣ leu⁺ bio⁺*. If these are found, they demonstrate that the known nuclear genes *leu⁺* and *bio⁻* are assorting independently of *capᵣ*. This means *capᵣ* cannot be encoded in the nucleus and is likely to be encoded in the mitochondria.

10. (*a*) Show that the pattern of inheritance of *capᵣ* is maternal. (*b*) Show that injection of mitochondria from a *capᵣ* mold into a *capˢ* mold renders the latter *capᵣ*.

11. *twi*——*frd*—*zig*————*spa*. Rationale: You know that *frd* is between *twi* and *zig* because they are deleted together; you know that *zig* is between *frd* and *spa* because they are deleted together. You can estimate the distances between genes from co-deletion frequencies. For example, *zig* is lost together with *frd* much more frequently than it is lost with *spa*; therefore, *zig* is much closer to *frd* than to *spa*.

12. 35 DNA units. Rationale: *twi* and *zig* lie 40 units apart on a 40-unit fragment; therefore, they lie at the extreme ends of A1, with *twi* at the left end. Furthermore, since A1 and B2 overlap by 25 DNA units and the total length of B2 is 60 units, the maximum distance between *pap* and *twi* is 60 − 25, or 35 units.

13. Chloroplasts are polyploid: They have many identical copies of one rather small genome. Therefore, they have more total DNA, but many fewer *different* genes than the *E. coli* genome, which has very few repeated genes.

14. R24′, R01, and R25.

15. R25.

16. Hybridize radioactive rRNA to the *Eco*RI blot and show that R07 and R24 "light up."

17. It lies within the restriction fragment Ba15.

18. Much more space between genes in yeast mitochondria, much longer noncoding regions in yeast mitochondrial mRNAs, and the presence of introns in yeast mitochondria.

19. If they were true endosymbionts, they would have enough genes to be independent. But they have far too little DNA for this to be true.

20. DNA that has moved from nucleus to organelle, or vice versa, or even from one organelle to another.

CHAPTER 19

1.

	p	q	p^2	2pq	q^2
a	0.4	0.6	0.16	0.48	0.36
b	0.68	0.32	0.462	0.435	0.102
c	0.467	0.533	0.218	0.498	0.284
	p^2N	$2pqN$	q^2N	χ^2	
a	16	48	36	2.78	
b	23.1	21.8	5.1	20.1	
c	62.1	141.9	80.9	5.63	

With 1 degree of freedom, *a* is not significant, *b* is significant at the 0.01 level, and *c* is significant at the 0.05 level.

2. The expected numbers are 18, 84, and 98 for genotypes A_1A_1, A_1A_2, and A_2A_2. 0.45—with 1 degree of freedom, it is consistent.

3. A_1A_1, 0.49; A_2A_2, 0.09.

4. 0.1; 0.18.

5. (*a*) AR: 0.6, PP: 0.32, and ST: 0.08. (*b*) AR AR: 18, AR PP: 19.2, AR ST: 4.8, PP PP: 5.12, PP ST: 2.56, and ST ST: 0.32.

6.

	A_1	A_2	A_3	χ^2
a	0.35	0.3	0.35	2.27
b	0.2	0.2	0.6	400

Expected genotypic frequencies

	A_1A_1	A_1A_2	A_1A_3	A_2A_2	A_2A_3	A_3A_3
a	0.1225	0.21	0.245	0.09	0.210	0.1225
b	0.04	0.08	0.24	0.04	0.24	0.36

With 3 degrees of freedom, *a* is not significant and *b* is significant at the 0.01 level.

7. 0.0064; 0.147.

8. 0.12, 0.16, 0.72; 0.01.

9. 0.408, 0.384, 0.208; f = 0.792.

10. (*a*) 0.878, 0.841, and 0.906. (*b*) This species has a very high level of self-fertilization and a low level of random mating so that the frequency of heterozygotes is quite low.

11. (*a*)

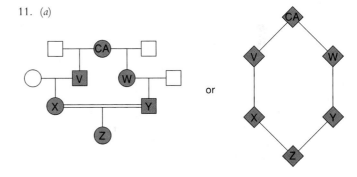

Inbreeding coefficient = 0.03125

(*b*)

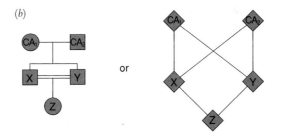

Inbreeding coefficient = 0.25

12. 10^{-6}, 7.25×10^{-7}, 4.5×10^{-7}, 1.75×10^{-7}, 10^{-7}. Δq is largest when p = 1 and q = 0 because all alleles are wild-type.

13. A small population size results in genetic drift, a factor that increases the fixation of alleles in a population and thereby reduces the heterozygosity.

14. 0.3; 0.7. The probability of fixation of an allele is based on its initial frequency. Hence, the probability of fixation is lower for the allele in lowest frequency because it is less likely to increase all the way to 1.0.

15. 0.0012, 0.035. The expected is much larger.

16. Gene flow may have several effects on the genetic variation in a population. For example, if an allele is missing from a population, gene flow from a polymorphic population can introduce it to the population and therefore increase the genetic variation. On the other hand, if the population is polymorphic, then gene flow from a monomorphic population can reduce the level of genetic variation. A high level of gene flow can speed up these changes, while a low level of gene flow may cause the changes to occur over many generations.

17. (a)

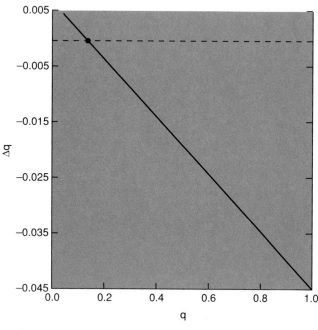

(b) 0.1.

18. 0.13. We assume that there has been no selection or genetic drift and that the allelic frequency estimates are correct.

19. (a) 0.0, −0.019, −0.056, −0.073, 0.0.
    (b)

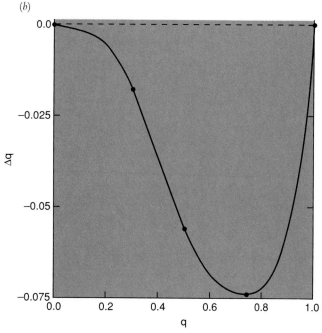

(c) 0.1968, 0.1937.

20. Melanic 1, typical 0.771; 0.4.

21. 0.2, 0.011, −0.2; there is a stable equilibrium at q = 0.2.

22. 0.585, 0.585, and 1.0; assuming that the level of selection is the same in the three breeds, the observed differences could be due to chance, i.e., genetic drift.

23. In general, selection and gene flow have the largest effects on Δq, and mutation has the smallest. Genetic drift can have a large effect in small populations. Genetic drift can either increase or decrease allelic frequency, while the other factors in particular situations will change allelic frequencies only in one direction.

24. Inbreeding increases the frequency of homozygotes and decreases the frequency of heterozygotes. Examine genotypic frequencies compared to Hardy-Weinberg expectations. Inbreeding in concert with selection may fix populations for favorable homozygotes.

25. Snails could be placed on different backgrounds to determine the rate of predation. Look at changes in allelic frequency from generation to generation to see if they vary more in small populations than in large.

CHAPTER 20

1. 0.0000167, 0.0041; new frequencies 0.0000333, 0.0058. When u is doubled, the equilibrium genotypic frequency is essentially doubled. This is not true of the equilibrium allelic frequencies as indicated by the value differences; .0000166 difference in genotypic frequency and 0.0017 difference in allelic frequency.

2. 0.000001; 0.00002.

3. Malaria increased the frequency to equilibrium in Africa; gene flow and selection are reducing the frequency today. Mutation may be a limiting factor.

4. u = sq² = 0.0004. This value is much higher than known mutation rates, which are about 10^{-5} to 10^{-7}. Genetic drift, heterozygote advantage, or past selection have been offered as explanations.

5. Populations that are started with a very few founders are most likely to have some disease alleles that are higher in frequency than in other populations. This is an extreme form of genetic drift in which in one generation the frequency of an allele changes dramatically because of the restricted number of individuals.

6. Neutrality considers genetic variation a function of mutation and genetic drift, predicting alleles to be transient and fewer in small populations. Selection predicts constant alleles and more variation in small populations. One experiment would involve studying the change in frequency of a polymorphic allele in *D. melanogaster* when the flies are exposed to different environmental or nutritional conditions (see pp. 561–563).

7. The molecular clock assumes a regular replacement of molecular variants over time. The rate of amino acid substitution is the change in amino acid sequence per unit time. Nucleotide substitution would be faster because there are fewer constraints and the third position changes fast.

8. (*a*) The population size in the first population is larger so there is less genetic drift reducing the amount of heterozygosity.
(*b*) Population sizes (N) of 2,778 and 1,316.

9. 0.444, 0.25; there are more physical constraints on the second protein.

10. Species a and c; species d.

11. If we assume that the molecular information gives us the true relationship, we can examine morphology and physiology, knowing these patterns and the time since divergence.

12. Selection is haploid, no dominance. Genetic drift is more likely because of only one copy versus four for diploid (male and female).

13. 0.00124; 0.03125. The second is larger because more recessive homozygotes are exposed to selection.

14. 0.478, 0.475; 0.456, 0.433; the reduction in heterozygosity is very low for the larger population and not too great for the smaller one.

15. 167.

16. Look at morphological, economic, and disease and pest resistance traits. Grow separately in their natural habitats, avoiding unnatural selection if possible.

17. 0.000312; 0.000064; nearly five times as many progeny from half-first cousins are affected.

18. Avoid inbreeding, keep a large population, and avoid unnatural selection.

19. 0.337. Gene flow or mutation has not introduced the allele, and/or selection for resistance has not occurred. (See chapter 19.)

20. 0.81; 0.518. (See chapter 19.)

21. Mosquitoes have a high reproductive capacity, short generation length, and high dispersal. Furthermore, they adapt to DDT in several different ways.

22. All the new crops may be identical, resulting in loss of genetic variation and susceptibility to particular diseases.

Appendix B: Selected Readings

Chapter 1

Creighton, H. B., and B. McClintock. 1931. A correlation of cytological and genetical crossing-over in *Zea mays. Proceedings of the National Academy of Sciences* 17:492–97.

Mirsky, A. E. 1968. The discovery of DNA. *Scientific American* 218(June):78–88.

Morgan, T. H. 1910. Sex-limited inheritance in *Drosophila. Science* 32:120–22.

Sturtevant, A. H. 1913. The linear arrangement of six sex-linked factors in *Drosophila,* as shown by their mode of association. *Journal of Experimental Zoology* 14:43–59.

Chapter 2

Bodmer, W. F., and L. L. Cavalli-Sforza. 1986. *Genetics, evolution, and man.* San Francisco: W. H. Freeman.

Hartl, D. L., and V. Orel. 1992. What did Gregor Mendel think he discovered? *Genetics* 131:245–53.

Lewis, R. 1997. *Human Genetics,* 2d ed. Dubuque, IA: Wm. C. Brown.

McKusick, V. 1990. *Mendelian inheritance in man,* 9th ed. Baltimore: Johns Hopkins University Press.

Olby, R. C. 1966. *Origins of Mendelism.* London: Constable.

Peters, J. A. 1962. *Classic papers in genetics.* Englewood Cliffs, N.J.: Prentice-Hall.

Piergorsch, W. W. 1990. Fisher's contributions to genetics and heredity, with special emphasis on the Gregor Mendel controversy. *Biometrics* 46:915–24.

Stern, C., and E. R. Sherwood. 1966. *The origin of genetics: A Mendel source book.* San Francisco: W. H. Freeman and Co.

Sutton, H. E. 1988. *An introduction to human genetics,* 4th ed. New York: Harcourt, Brace, Jovanovich.

Todd, N. B. 1977. Cats and commerce. *Scientific American* 237 (November):100–107.

Vogel, F., and A. G. Motulsky. 1979. *Human genetics.* New York: Springer.

Chapter 3

Bodmer, W. F., and L. L. Cavalli-Sforza. 1970. Intelligence and race. *Scientific American* 223(October):19–29.

Bouchard, T. J., D. T. Lykken, M. McGue, N. L. Segal, and A. Tellegen. 1990. Sources of human psychological differences: The Minnesota study of twins reared apart. *Science* 250:223–28.

Clausen, J., D. D. Keck, and W. H. Heisey. 1941. Regional differences in differentiation in plant species. *American Naturalist* 75:231–50.

Falconer, D. S. *Quantitative genetics,* 3rd ed. London: Halstead Press.

Hedrick, P. W. 1984. *Population biology.* Boston: Jones and Bartlett Publishers.

Kettlewell, B. 1973. *The evolution of melanism.* Oxford: Oxford University Press.

Nakamura, Y., M. Leppert, P. O'Connell, R. Wolff, T. Holm, M. Culver, C. Martin, E. Fujimoto, M. Hoff, E. Kumlin, and R. White. 1987. Variable number of tandem repeat (VNTR) markers for human gene mapping. *Science* 235:1616–22.

Searle, A. G. 1968. *Comparative genetics of coat colour in mammals.* London: Logos Press.

Silvers, W. K. 1979. *The coat colors of mice: A model for mammalian gene action and interaction.* New York: Springer.

Singer, S. 1978. *Human genetics.* San Francisco: W. H. Freeman.

Sokal, R. R., and F. J. Rohlf. 1981. *Biometry.* San Francisco: W. H. Freeman.

Chapter 4

Borgaonkar, D. S. 1991. *Chromosomal variation in man.* New York: Wiley-Liss.

Bridges, C. 1931. Nondisjunction as proof of the chromosome theory of heredity. *Genetics* 1:1–52; 107–63.

Cherfas, J. 1991. Sex and the single gene. *Science* 252:782.

Hassold, T. J. 1986. Chromosome abnormalities in human reproductive wastage. *Trends in Genetics* 3(April):105–10.

Hassold, T. J., and P. A. Jacobs. 1984. Trisomy in man. *Annual Review of Genetics* 18:69–98.

Mcintosh, J. R., and K. L. McDonald. 1989. The mitotic spindle. *Scientific American* 261(October):48–56.

McLeish, J., and B. Snoad. 1958. *Looking at chromosomes.* New York: Macmillan.

Swanson, C. P., T. Merz, and W. J. Young. 1981. *Cytogenetics: The chromosome in division, inheritance, and evolution.* Englewood Cliffs, N.J.: Prentice-Hall.

Varma, R. S., and A. Babu. 1989. *Human chromosomes: A manual of basic techniques.* New York: Pergamon Press.

Von Wettstein, D., et al. 1984. The synaptonemal complex in genetic segregation. *Annual Review of Genetics* 18:331–414.

CHAPTER 5

Beeman, R. W. 1987. A homoeotic gene cluster in the red flour beetle. *Nature* 327:247–49.

Culliton, B. J. 1990. Mapping terra incognita (humani corporis). *Science* 250:210–12.

Finchman, J. R. S., P. R. Day, and A. Radford. 1979. *Fungal genetics.* London: Blackwell.

Lindsley, D. L., and G. G. Zimm. 1992. *The genome of* Drosophila melanogaster. New York: Academic Press.

Morton, N. E. 1991. Parameters of the human genome. *Proceedings of the National Academy of Sciences* 88:7474–76.

Ott, J. 1985. *Analysis of human genetic linkage.* Baltimore: Johns Hopkins University Press.

Ruddle, F. H. 1974. Hybrid cells and human genes. *Scientific American* 231(July):36–44.

Stephens, J. C., M. L. Cavanaugh, M. I. Gradie, M. L. Mador, and K. K. Kidd. 1990. Mapping the human genome: Current status. *Science* 250:237–44.

CHAPTER 6

Adams, R. L. P., R. H. Burdon, A. M. Campbell, and R. M. S. Smellie, eds. 1976. *Davidson's The Biochemistry of the Nucleic Acids,* 8th ed. The structure of DNA, chapter 5. New York: Academic Press.

Avery, O. T., C. M. McLeod, and M. McCarty. 1944. Studies on the chemical nature of the substance-inducing transformation of pneumococcal types. *Journal of Experimental Medicine* 79:137–58.

Chargaff, E. 1950. Chemical specificity of nucleic acids and their enzymatic degradation. *Experimentia* 6:201–209.

Dickerson, R. E. 1983. The DNA helix and how it is read. *Scientific American* 249(December):94–111.

Hershey, A. D., and M. Chase. 1952. Independent functions of viral protein and nucleic acid in growth of bacteriophage. *Journal of General Physiology* 36:39–56.

Watson, J. D., and F. H. C. Crick. 1953. Molecular structure of nucleic acids: A structure for deoxyribose nucleic acid. *Nature* 171:737–38.

———. 1953. Genetical implications of the structure of deoxyribonucleic acid. *Nature* 171:964–67.

CHAPTER 7

Cairns, J. 1963. The chromosome of *Escherichia coli. Cold Spring Harbor Symposia on Quantitative Biology* 28:43–46.

Callan, H. G. 1973. DNA replication in the chromosomes of eukaryotes. *Cold Spring Harbor Symposia on Quantitative Biology* 38:195–203.

Cox, M. M., and I. R. Lehman. 1987. Enzymes of general recombination. *Annual Review of Biochemistry* 56:229–62.

Dressler, D., and H. Potter. 1982. Molecular mechanism of genetic recombination. *Annual Review of Biochemistry* 51:727–62.

Greider, C. W. and Blackburn, E. H. 1996. Telomeres, telomerase, and cancer. *Scientific American* 274(Feb.):92–97.

Herendeen, D. R., and Kelly, T. J. 1996. DNA polymerase III: Running rings around the fork. *Cell* 84:5–8.

Kornberg, A., and T. A. Baker. 1992. *DNA replication,* 2nd ed. New York: W. H. Freeman.

Kunkel, T. A. 1992. DNA replication fidelity. *Journal of Biological Chemistry* 267:18251–54.

McHenry, C. S. 1988. DNA polymerase III holoenzyme of *Escherichia coli. Annual Review of Biochemistry* 57:519–50.

Meselson, M., and F. W. Stahl. 1958. The replication of DNA in *Escherichia coli. Proceedings of the National Academy of Sciences* 44:671–82.

Meselson, M. S., and C. M. Radding. 1975. A general model for recombination. *Proceedings of the National Academy of Sciences* 72:358–61.

Radman, M., and R. Wagner. 1988. The high fidelity of DNA replication. *Scientific American* 259(August):24–30.

Smith, G. R. 1987. Mechanism and control of homologous recombination in *Escherichia coli. Annual Review of Genetics* 21:179–201.

Wang, J. C. 1982. DNA topoisomerases. *Scientific American* 247(July):94–109.

Zakian, V. A. 1995. Telomeres: Beginning to understand the end. *Science* 270:1601–6.

CHAPTER 8

Adhya, S., and S. Garges. 1990. Positive control. *Journal of Biological Chemistry* 265:10797–10800.

Beadle, G. W., and E. L. Tatum. 1941. Genetic control of biochemical reactions in *Neurospora. Proceedings of the National Academy of Sciences* 27:499–506.

Beckwith, J. R., and D. Zipser, eds. 1970. The lactose operon. Cold Spring Harbor Laboratory.

Conceptual foundations of genetics: Selected readings. 1976. Edited by H. O. Corwin and J. B. Jenkins. Boston: Houghton Mifflin Co.

Friedman, D. I., M. J. Imperiale, and S. L. Adhya. 1987. RNA 3′-end formation in the control of gene expression. *Annual Review of Genetics* 21:453–88.

Helmann, J. D., and M. J. Chamberlin. 1988. Structure and function of bacterial sigma factors. *Annual Review of Biochemistry* 57:839–72.

Jacob, F. 1966. Genetics of the bacterial cell (Nobel lecture). *Science* 152:1470–78.

Jacob, F., and J. Monod. 1961. Genetic regulatory mechanisms in the synthesis of proteins. *Journal of Molecular Biology* 3:318–56.

Losick, R., and M. Chamberlin. 1976. RNA polymerase. Cold Spring Harbor Laboratory.

Miller, J. H., and W. S. Reznikoff, eds. 1978. The operon. Cold Spring Harbor Laboratory.

Monod, J. 1966. From enzymatic adaptation to allosteric transitions (Nobel lecture). *Science* 154:475–83.

Ptashne, M. 1989. How gene activators work. *Scientific American* 260(January):24–31.

Ptashne, M. 1992. *A genetic switch.* Cambridge, MA: Cell Press.

Ptashne, M., and W. Gilbert. 1970. Genetic repressors. *Scientific American* 222(June):36–44.

Schleif, R. 1987. DNA binding proteins. *Science* 241:1182–87.

Zubay, G., D. Schwartz, and J. Beckwith. 1970. Mechanism of activation of catabolite-sensitive genes: A positive control system. *Proceedings of the National Academy of Sciences* 66:104–10.

CHAPTER 9

Buratkowski, S. 1994. The basics of basal transcription by RNA polymerase II. *Cell* 77:1–3.

Burlingame, R. W., W. E. Love, B.-C. Wang, R. Hamlin, N.-H. Xuong, and E. N. Moudrianskis. 1985. Crystallographic structure of the octameric histone core of the nucleosome at a resolution of 3.3A. *Science* 228:546–53.

Chambon, P. 1981. Split genes. *Scientific American* 244(May):60–71.

Conaway, J. W., and R. C. Conaway. 1991. Initiation of eukaryotic messenger RNA synthesis. *Journal of Biological Chemistry* 266:17721–24.

Friedman, D. E., M. J. Imperiale, and S. L. Adhya. 1987. RNA 3'-end formation in the control of gene expression. *Annual Review of Genetics* 21:453–88.

Geiduschek, E. P., and G. P. Tocchini-Valentini. 1988. Transcription by RNA polymerase III. *Annual Review of Biochemistry* 57:873–914.

Kornberg, R. D., and A. Klug. 1981. The nucleosome. *Scientific American* 244(February):52–64.

McKnight, S. L. 1991. Molecular zippers in gene regulation. *Scientific American* 264(April):54.

Miller, O. L. 1973. The visualization of genes in action. *Scientific American* 228(March):34–42.

Proudfoot, N. 1991. Poly(A) signals. *Cell* 64:671.

Ptashne, M. 1989. How gene activators work. *Scientific American* 260(January):24–31.

Rhodes, D., and A. Klug. 1993. Zinc fingers. *Scientific American* 268(February):56–65.

Sharp, P. A. 1987. Splicing of messenger RNA precursors. *Science* 235:766–71.

Steitz, J. A. 1988. "Snurps." *Scientific American* 258(June):36–41.

Struhl, K. 1994. Duality of TBP, the universal transcription factor. *Science* 263:1103–4.

Tjian, R. 1995. Molecular machines that control genes. *Scientific American* 272(February):54–61.

CHAPTER 10

Bjork, G. R., et al. 1987. Transfer RNA modification. *Annual Review of Biochemistry* 56:263–87.

Brenner, S., F. Jacob, and M. Meselson. 1961. An unstable intermediate carrying information from genes to ribosomes for protein synthesis. *Nature* 190:576–81.

Crick, F. H. C. 1958. On protein synthesis. *Symposium of the Society for Experimental Biology* 12:138–63.

Crick, F. H. C. 1966. The genetic code. *Scientific American* 215(October):55–62.

Crick, F. H. C., L. Barnett, S. Brenner, and R. J. Watts-Tobin. 1962. General nature of the genetic code for proteins. *Nature* 192:1227–32.

Fox, T. D. 1987. Natural variation in the genetic code. *Annual Review of Genetics* 21:67–91.

Khorana, H. G. 1968. Synthesis in the study of nucleic acids. *Biochemical Journal* 109:709–25.

Kozak, M. 1991. Structural features in eukaryotic mRNAs that modulate the initiation of translation. *Journal of Biological Chemistry* 266:19867–70.

Lake, J. A. 1981. The ribosome. *Scientific American* 245(August):84–97.

Leder, P., and M. Nirenberg. 1964. RNA codewords and protein synthesis, II. Nucleotide sequence of a valine RNA codeword. *Proceedings of the National Academy of Sciences* 52:420–27.

Moore, P. B. 1988. The ribosome returns. *Nature* 331:223–27.

Nomura, M. 1984. The control of ribosome synthesis. *Scientific American* 250(January):102–14.

Quigley, G. J., and A. Rich. 1976. Structural domains of transfer RNA molecules. *Science* 194:796–806.

Rich, A., and S. H. Kim. 1978. The three-dimensional structure of transfer RNA. *Scientific American* 238(January):52–62.

CHAPTER 11

Brenner, S., L. Barnett, F. H. C. Crick, and A. Orgel. 1961. The theory of mutagenesis. *Journal of Molecular Biology* 3:121–24.

Little, J. W. 1982. The SOS regulatory system of *Escherichia coli*. *Cell* 29:11–12.

Loeb, L. A. 1985. Apurinic sites as mutagenic intermediates. *Cell* 40:483–84.

Miller, J. H. 1983. Mutational specificity in bacteria. *Annual Review of Genetics* 17:215–38.

Radman, M., and R. Wagner. 1986. Mismatch repair in *Escherichia coli*. *Annual Review of Genetics* 20:523–38.

Rennie, J. 1993. DNA's new twists. *Scientific American* 266(March):122–32.

Sancar, A., and G. B. Sancar. 1988. DNA repair enzymes. *Annual Review of Biochemistry* 57:29–67.

Suzuki, D. T. 1970. Temperature-sensitive mutations in *Drosophila melanogaster*. *Science* 170:695–706.

Walker, G. C. 1985. Inducible DNA repair systems. *Annual Review of Biochemistry* 54:425–58.

CHAPTER 12

Baltimore, D. 1985. Retroviruses and retrotransposons: The role of reverse transcription in shaping the eukaryotic genome. *Cell* 40:481–82.

Cohen, S. N., and J. A. Shapiro. 1980. Transposable genetic elements. *Scientific American* 242(February):40–49.

Craig, N. L. 1985. Site-specific inversion: Enhancers, recombination proteins, and mechanism. *Cell* 41:649–50.

Craigie, R. 1996. Quality control in Mu DNA transposition. *Cell* 85:137–40.

Doering, H.-P., and P. Starlinger. 1984. Barbara McClintock's controlling elements: Now at the DNA level. *Cell* 39:253–59.

Engels, W. R. 1983. The P family of transposable elements in *Drosophila*. *Annual Review of Genetics* 17:315–44.

Federoff, N. V. 1984. Transposable genetic elements in maize. *Scientific American* 250(June):84–99.

Garfinkel, D. J., J. D. Boeke, and G. R. Fink. 1985. Ty element transposition: Reverse transcriptase and virus-like particles. *Cell* 42:507–17.

Grindley, N. G. F. and Leschziner, A. E. 1995. DNA transposition: From a black box to a color monitor. *Cell* 83:1063–6.

Mizuuchi, K., and R. Craigie. 1986. Mechanism of bacteriophage Mu transposition. *Annual Review of Genetics* 20:385–429.

Shapiro, J. A. 1983. *Mobile genetic elements.* New York: Academic Press.

Wessler, S. R. 1988. Phenotypic diversity mediated by the maize transposable elements Ac and Spm. *Science* 242:399–405.

CHAPTER 13

Benzer, S. 1961. Genetic fine structure. In *Harvey Lectures,* Series 56. New York: Academic Press.

Hayes, W. 1968. *The genetics of bacteria and their viruses,* 2nd ed. New York: Wiley.

Ippen-Ihler, K. A., and E. G. Minkley. 1986. The conjugation system of F, the fertility factor of *Escherichia coli*. *Annual Review of Genetics* 20:593–624.

Jacob, F., and E. L. Wollman. 1961. *Sexuality and the genetics of bacteria.* New York: Academic Press.

Morse, M. L., E. M. Lederberg, and J. Lederberg. 1956. Transduction in *Escherichia coli* K–12. *Genetics* 41:142–56.

Stent, G. S., and R. Calendar. 1978. *Molecular genetics: An introductory narrative,* chapters 9–13. San Francisco: W. H. Freeman.

Wollman, E. L., F. Jacob, and W. Hayes. 1956. Conjugation and genetic recombination in *Escherichia coli* K–12. *Cold Spring Harbor Symposia on Quantitative Biology* 21:141–62.

CHAPTER 14

Barinaga, M. 1991. Dimers direct development. *Science* 251:1176.

Borst, P., and D. R. Greaves. 1987. Programmed gene rearrangements altering gene expression. *Science* 235:658–67.

Brown, D. D. 1981. Gene expression in eukaryotes. *Science* 211:667–74.

Coulter, D., and E. Wieschaus. 1986. Segmentation genes and the distributions of transcripts. *Nature* 321:472–74.

DeRobertis, E. M., and J. B. Gurdon. 1979. Gene transplantation and the analysis of development. *Scientific American* 241(December):74–82.

DeRobertis, E. M., G. Oliver, and C. V. E. Wright. 1990. Homeobox genes and the vertebrate body plan. *Scientific American* 263(July):46.

Gurdon, J. B. 1968. Transplanted nuclei and cellular differentiation. *Scientific American* 219(December):24–35.

Gurdon, J. B. 1992. The generation of diversity and pattern in animal development. *Cell* 68:185–99.

Hoffman, M. 1991. How parents make their mark on genes. *Science* 252:1250.

Holliday, R. 1987. The inheritance of epigenetic defects. *Science* 238:163–70.

———. 1989. A different kind of inheritance. *Scientific American* 260(June):40–48.

Lawrence, P. A. 1992. *The making of a fly.* Blackwell Scientific Publications, London.

Levine, M., and T. Hoey. 1988. Homeobox proteins as sequence-specific transcription factors. *Cell* 55:537–40.

McGinnis, W. and Kuziora, M. 1994. The molecular architects of body design. *Scientific American* 270(February)58–66.

Moses, P. B., and N.-H. Chua. 1988. Light switches for plant genes. *Scientific American* 258(April):64–69.

Nüsslein-Volhard, C. 1991. Determination of the embryonic axes of *Drosophila*. *Development Supplement* 1:1–10.

Sapienza, C. 1990. Parental imprinting of genes. *Scientific American* 263(October):52–60.

Thomas, B. J., and R. Rothstein. 1991. Sex, maps, and imprinting. *Cell* 64:1.

Tonegawa, S. 1985. The molecules of the immune system. *Scientific American* 253(October):122–31.

Wolffe, A. P., and D. D. Brown. 1988. Developmental regulation of two 5S ribosomal RNA genes. *Science* 241:1626–32.

CHAPTER 15

Cohen, S. 1975. The manipulation of genes. *Scientific American* 233(July):24–33.

Cohen, S., A. Chang, H. Boyer, and R. Helling. 1973. Construction of biologically functional bacterial plasmids *in vitro*. *Proceedings of the National Academy of Sciences* 70:3240–44.

Nathans, D., and H. O. Smith. 1975. Restriction endonucleases in the analysis and restructuring of DNA molecules. *Annual Review of Biochemistry* 44:273–93.

Sanger, F., G. M. Air, B. G. Barrell, N. L. Brown, A. R. Coulson, J. C. Fiddes, C. A. Hutchison, P. M. Slocombe, and M. Smith. 1977. Nucleotide sequence of the φX174 DNA. *Nature* 265:687–95.

Watson, J. D., J. Tooze, and D. T. Kurtz. 1983. *Recombinant DNA: A Short Course.* New York: W. H. Freeman.

CHAPTER 16

Anderson, W. F. 1995. Gene therapy. *Scientific American* 273(September):124–8.

Beardsley, T. 1996. Vital Data (Human Genome). *Scientific American* 274(March):100–105.

Capecchi, N. R. 1994. Targeted gene replacement. *Scientific American* 270(March):52–59.

Chilton, M.-D. 1983. A vector for introducing new genes into plants. *Scientific American* 248(June):50–59.

Gasser, C. S., and R. T. Fraley. 1992. Transgenic crops. *Scientific American* 266(June):62–69.

Gilbert, W., and L. Villa-Komaroff. 1980. Useful proteins from recombinant bacteria. *Scientific American* 242(April):74–94.

Gusella, J. F., et al. 1983. A polymorphic DNA marker genetically linked to Huntington's disease. *Nature* 306:234–38.

Landegren, U., R. Kaiser, C. T. Caskey, and L. Hood. 1988. DNA diagnostics: Molecular techniques and automation. *Science* 242:229–37.

Morell, V. 1993. Huntington's disease gene finally found. *Science* 260:28–30.

Murray, T. H. 1991. Ethical issues in human genome research. *FASEB Journal* 5:55–60.

Neufeld, P. J., and N. Colman. 1990. When science takes the witness stand. *Scientific American* 262(May):18–25.

Verma, I. M. 1990. Gene therapy. *Scientific American* 263(November):68.

Welch, M. J., and Smith, A. E. 1995. Cystic fibrosis. *Scientific American* 273(December):52–9.

White, R., M. Leppert, D. T. Bishop, D. Barker, J. Berkowitz, C. Brown, P. Callahan, T. Holm, and L. Jerominski. 1985. Construction of linkage maps with DNA markers for human chromosomes. *Nature* 313:101–104.

White, R., and J.-M. Lalouel. 1988. Chromosome mapping with DNA markers. *Scientific American* 258(February):20–29.

CHAPTER 17

Barbacid, M. 1987. *ras* genes. *Annual Review of Biochemistry* 56:779–827.

———. 1982. Oncogenes. *Scientific American* 246(March):80–92.

———. 1987. The molecular genetics of cancer. *Science* 235:305–11.

———. 1991. Molecular themes in oncogenesis. *Cell* 64:235.

Cavenee, W. K., and White, R. L. 1995. The genetic basis of cancer. *Scientific American* 272(March):72–79.

Chantley, et al. 1991. Oncogenes and signal transduction. *Cell* 64:281.

Croce, C. M., and G. Klein. Chromosomal translocations and human cancer. *Scientific American* 252(March):54–60.

Haber, D. A., and D. E. Housman. 1991. Rate-limiting steps: The genetics of pediatric cancers. *Cell* 64:5.

Hunter, T. 1984. The proteins of oncogenes. *Scientific American* 251(August):70–79.

Marx, J. 1991. Possible new colon cancer gene found. *Science* 251:1317.

Marx, J. 1994. Oncogenes reach a milestone. *Science* 266:1942–4.

Modrich, P. 1994. Mismatch repair, genetic stability, and cancer. *Science* 266:1959–60.

Weinberg, R. A. 1983. The molecular basis of cancer. *Scientific American* 249(November):126–42.

———. 1988. Finding the anti-oncogene. *Scientific American* (September):44–51.

Weiss, R., N. Teich, H. Varmus, and J. Coffin, eds. 1984. *The molecular biology of tumor viruses,* 2d ed. Cold Spring Harbor Laboratory.

DeDuve, C. 1966. The birth of complex cells. *Scientific American* 274(April):50–57.

Gillham, N. W. G. 1978. *Organelle heredity.* New York: Raven Press.

Grivell, L. 1983. Mitochondrial DNA. *Scientific American* 248(March):78–89.

Margulis, L. 1971. Symbiosis and evolution. *Scientific American* 225(August):49–57.

Palmer, J. D. 1985. Comparative organization of chloroplast genomes. *Annual Review of Genetics* 19:325–54.

Sager, R. 1965. Genes outside chromosomes. *Scientific American* 212(January):70–79.

———. 1972. *Cytoplasmic inheritance.* New York: Academic Press.

Warren, G., and Wickner, W. 1996. Organelle inheritance. *Cell* 84:395–400.

CHAPTER 18

Beale, G., and J. Knowles. 1978. *Extranuclear genetics.* London: Edward Arnold Publishers.

APPENDIX C: EXPLORATIONS IN CELL BIOLOGY AND GENETICS CD-ROM

Here are descriptions and questions to accompany the CD-ROM modules referenced in the text.

MODULE 10—EXPLORING MEIOSIS: DOWN SYNDROME

This interactive exercise allows the user to explore how the failure of chromosomes to separate during meiosis can result in Down syndrome, a congenital condition involving serious mental retardation. The user evaluates the influence of the age of the mother and, separately, of the father, on the incidence of Down syndrome babies, creating in each case a graph showing the incidence as a function of age. The way in which primary nondisjunction in meiosis I produces Down children can be explored by comparing an animation of normal meiosis with one in which primary nondisjunction of chromosome 21 occurs.

Questions to Explore:

1. Is there a difference between the sexes in the influence of parental age on the incidence of Down syndrome? If so, to what would you attribute the difference?

2. When in meiosis does the event occur that results in Down syndrome? What is the nature of the event? Can you think of a reason why sex might affect the probability of this event?

3. At what parental age does the probability of having a Down syndrome child become appreciable? Why at that age?

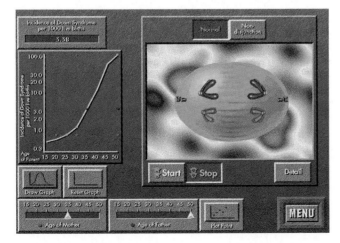

MODULE 11—CONSTRUCTING A GENETIC MAP

This interactive exercise allows students to explore the influence of physical separation upon genetic recombination by moving the relative position of genes on a chromosome. The exercise analyzes the behavior of three alleles in a dihybrid cross. The symbols identifying the alleles are specified by the user. If recombination frequencies are entered for the eight recombinant types, the program will draw a genetic map of the three loci. If instead the user specifies the relative locations of the three genes on the chromosome, and the number of F_2 progeny to be analyzed, the program will indicate the expected number of each recombinant type among the F_2 progeny.

Questions to Explore:

1. In a 3-point cross (that is, a dihybrid cross involving three alleles), why does the sum of the two short lengths not add up to the long one (that is, for the map *a-b-c,* why do the recombination frequencies *a-b* plus *b-c* not equal *a-c*)?

2. Can you do a 3-point cross in which one parent is recessive for one gene (*a-B-C*) and the other parent recessive for the other two genes (*A-b-c*)?

3. In a 3-point cross in which two genes are close together and the third gene far away, how can you tell what side of the two genes the third gene is on (*a-b—c,* or *c—a-b*)?

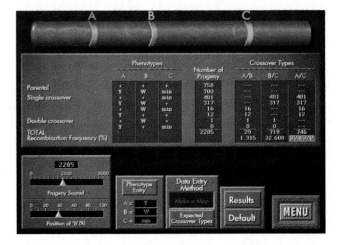

MODULE 12—HEREDITY IN FAMILIES

This interactive exercise allows students to explore how the character and location of a genetic trait influence its heritability within families. The key variables are dominance/recessiveness of the inherited allele and X-linkage versus autosomal location of the gene encoding the allele. From a bank of 30 actual pedigrees, one is selected at random, and the two variables are analyzed. The program scores the answer selected and then presents another pedigree from the bank. As the 30 pedigrees are examined in random order, the program keeps score of the analyses. These pedigrees are often all a human geneticist has to work with in attempting to assess the dominance and chromosomal location of a trait.

Questions to Explore:

1. In analyzing family pedigrees, why are sex-linked alleles expressed so much more frequently in male offspring than in females?

2. Why do you imagine some genetic disorders like sickle-cell anemia and hemophilia are common, while others are rare?

3. In a pedigree, what effect does sex linkage have on your ability to determine if a trait is being influenced by a single gene?

4. What is the minimal evidence you would accept that a trait appearing in a family is indeed hereditary (that is, caused by an allele), rather than environmentally induced (that is, not due to an allele)?

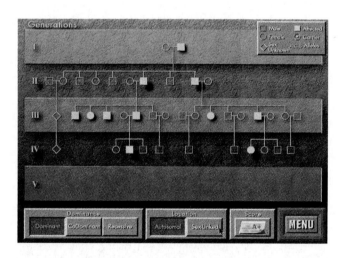

MODULE 13—GENE SEGREGATION WITHIN FAMILIES

This interactive exercise allows the user to explore how heredity affects individual families. Among the most common practical questions that arise about human heredity are ones concerning the potential makeup of families. In a family with three children, what is the likelihood all three will be girls? This interactive exercise tests the ability of the user to make such predictions. The key tool in determining what a potential family is likely to consist of is a simply calculated array of probabilities, the so-called binomial distribution. Using them, it is possible to assess the probability that a family of a given size will have so many girls, or, for parents heterozygous for a recessive trait, that so many of the offspring will be double-recessive.

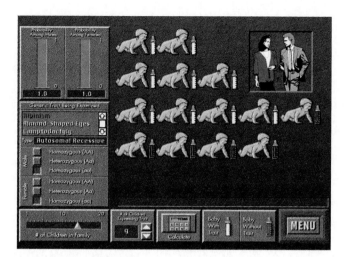

Questions to Explore:

1. Assuming the probability of having boys and girls to be equal, how many four-child families out of a hundred would be expected to have exactly two boys?

2. Imagine that two parents both heterozygous for albinism have five children. What is the probability that three of them will be albino?

MODULE 14—DNA FINGERPRINTING: YOU BE THE JUDGE

In this interactive exercise, the user analyzes the DNA evidence presented in real courtroom trials, attempting to ascertain the guilt or innocence of the suspect in each instance. The exercise selects a case at random from its library of real cases, and presents the user with the physical evidence (DNA samples from victim, perpetrator, and suspect) and a variety of DNA probes (RFLPs) to use as tools in the analysis. Selecting a probe, the user treats the DNA samples and examines the gel banding patterns that result. The user notes any differences between suspect and perpetrator (a difference would indicate innocence of the suspect), and, if no differences occur, uses the frequency of matches to estimate the probability that two randomly selected people would match that well. Then, by selecting a second and a third enzyme, the user can repeat the analysis to gain a clearer statistical picture.

Questions to Explore:

1. What is the relationship between the certainty with which guilt can be ascertained and the number of enzymes employed in the analysis? At what point is it reasonable to stop and say the conclusion is certain?

2. In how many of the cases that you investigated did you declare an innocent person to be guilty? Are there any circumstances where a guilty person may be indicated to be innocent by this DNA fingerprinting analysis?

3. What sources of bias might exist in the analysis? Are the "mistakes" that you make in analyzing this series of cases (that is, declaring a guilty person innocent, or an innocent person guilty) biased toward convicting the innocent or freeing the guilty?

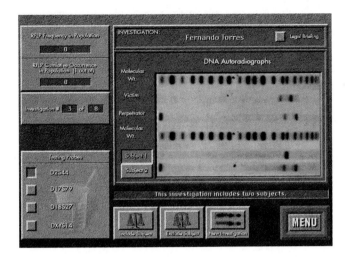

MODULE 16—GENE REGULATION

In this interactive exercise, the user explores the various strategies employed by organisms to regulate the transcription of genes. Two strategies are explored: (1) bacterial gene regulation, with the focus on rapid adaptation to environmental changes, and (2) eukaryotic gene regulation, with the focus on complex, hard-wired programs dictating fixed patterns of gene activity. In the bacterial simulation, the user will design a regulatory mechanism for a sugar-utilizing enzyme, selecting elements from among activator and repressor proteins, and locating their binding sites on the DNA. The user then varies the level of sugar in the environment and assesses the success of the proposed regulatory mechanism in optimizing use of the sugar. In the eukaryotic simulation, the user will have a broader choice of regulatory tools, including transcription factors, and the ability to locate regulatory genes at far distant sites. The challenge is to design a regulatory mechanism that will permit the various mammalian globin genes to be expressed at different specific times in development.

Questions to Explore:

1. What difference can be seen between bacteria and eukaryotes in the need for proximity of regulatory sites to the gene they are regulating? What do you suppose accounts for this difference?

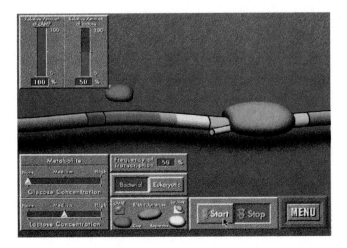

2. Can you design a bacterial regulatory mechanism that allows expression of a particular gene to be controlled by many regulatory genes simultaneously?

3. Transcription is only the first stage in a long process of gene expression. At what other levels might you expect regulation to occur? Do any of these have any advantages over transcriptional control?

MODULE 17—MAKING A RESTRICTION MAP

In this interactive exercise the student constructs a restriction map by entering measured band position data from a set of electrophoresis gels. The data may be supplied by the user from real lab experiments, or the user may choose to analyze one of several data sets provided by the interactive exercise. Maps such as these, based on determining the relative positions of overlapping DNA restriction fragments, are the principal way in which today's geneticists construct physical maps of genes.

Questions to Explore:

1. Do you believe restriction maps such as these have greater resolution than maps prepared by scoring the frequency with which recombinants appear among the progeny of genetic crosses? Why?

2. What great advantage do restriction maps have over genetic maps? What great advantage do genetic maps have over restriction maps?

3. In preparing your restriction map, do you find that there is any particular advantage to employing enzymes with different size restriction sites? Why do you think this is so?

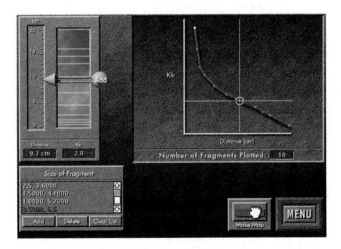

GLOSSARY

A

amber codon   UAG, coding for termination.

amber mutation   *See* nonsense mutation.

amber suppressor   A tRNA bearing an anticodon that can recognize the amber codon (UAG) and thereby suppress amber mutations.

Ames test   A test for mutagens developed by Bruce Ames, in which the rate of reversion of auxotrophic bacterial strains (*Salmonella*) to prototrophy upon exposure to a chemical is taken as a measure of the chemical's mutagenicity.

amino acid   The building block of proteins.

aminoacyl-tRNA synthetase   The enzyme that links a tRNA to its cognate amino acid.

amino tautomer   The normal tautomer of adenine or cytosine found in nucleic acids.

amino terminus   The end of a polypeptide with a free amino group. The end at which protein synthesis begins.

amplification   Selective replication of a gene to produce more than the normal single copy in a haploid genome.

anabolic metabolism   The process of building up substances from relatively simple precursors. The *trp* operon encodes anabolic enzymes that build the amino acid tryptophan.

anaphase   A short stage of nuclear division in which the chromosomes move to the poles.

aneuploid   Not having the normal diploid number of chromosomes.

annealing of DNA   The process of bringing back together the two separate strands of denatured DNA to re-form a double helix.

Antennapedia **complex (ANT-C)**   A large genetic locus in *Drosophila* that contains several homeotic genes, including *Antennapedia*.

Antennapedia (*Antp*) **mutants**   Mutants of *Drosophila* in which a leg appears where an antenna ought to be.

antibody   A protein with the ability to recognize and bind to a substance, usually another protein, with great specificity. Helps the body's immune system recognize and trigger an attack on invading agents.

anticoding strand   The complement of the coding strand.

anticodon   A three-base sequence in a tRNA that base-pairs with a specific codon.

anticodon loop   The loop, conventionally drawn at the bottom of tRNA molecule, that contains the anticodon.

antigen   A substance recognized and bound by an antibody.

anti-oncogene   A gene that acts to keep cell division in check. Inactivation of an anti-oncogene can lead to malignant transformation of the cell.

antiparallel   The relative polarities of the two strands in a DNA double helix; if one strand goes 5′–3′, top to bottom, the other goes 3′–5′. The same antiparallel relationship applies to any double-stranded polynucleotide or oligonucleotide, including the RNAs in a codon-anticodon pair.

antisense strand   The complement of the sense strand. *See* anticoding strand.

antiterminator   A protein, such as the λ *N* gene product, that overrides a terminator and allows transcription to continue.

apoptosis   Programmed cell death. A process by which cells kill themselves according to a controlled pathway, usually for the good of the entire organism.

aporepressor   A repressor in an inactive form, without its corepressor.

apurinic site (AP site)   A deoxyribose in a DNA strand that has lost its purine base.

assembly map   A scheme showing the order of addition of ribosomal proteins during self-assembly of a ribosomal particle in vitro.

assortative mating   Mating that is nonrandom with respect to the phenotypes in a population—that is, where the probability of individuals of two given phenotypes mating is not equal to the product of their frequencies in the population.

asymmetric transcription   Transcription of only one strand of a given region of a double-stranded polynucleotide.

ATPase   An enzyme that cleaves ATP, releasing energy for other cellular activities.

attenuation   A mechanism of transcription control that involves premature transcription termination.

attenuator   A region of DNA upstream from one or more structural genes, where premature transcription termination (attenuation) can occur.

att **sites**   Sites on phage and host DNA where recombination occurs, allowing integration of the phage DNA into the host genome as a prophage.

autoradiography   A technique in which a radioactive sample is allowed to expose a photographic emulsion, thus "taking a picture of itself."

autosome   Any chromosome except the sex chromosomes.

auxotroph   An organism that requires one or more substances in addition to minimal medium.

A-type particles Intracellular, noninfectious particles that resemble retroviruses. May be vehicles for transposition of retrotransposons.

B

back mutation *See* reversion.

bacteriophage *See* phage.

Barr body Structure formed when the inactivated X chromosome is heterochromatinized.

base pair A pair of bases (A–T or G–C), one in each strand, that occur opposite each other in a double-stranded DNA or RNA.

base A cyclic, nitrogen-containing compound linked to deoxyribose in DNA and to ribose in RNA.

B-DNA The standard Watson-Crick model of DNA favored at high relative humidity and in solution.

β-galactosidase An enzyme that breaks the bond between the two constituent sugars of lactose.

bidirectional DNA replication Replication that occurs in both directions at the same time from a common starting point, or origin of replication. Requires two active replicating forks.

binomial formula The probability that in a group of a given size, a certain number of individuals will be one type and a certain number will be another type.

Bithorax complex (BX-C) A large genetic locus in *Drosophila* containing several homeotic genes that control segmentation in the fly's thorax and abdomen.

bivalent The structure formed when two homologous chromosomes are paired in meiosis I.

β-lactamase An enzyme that breaks down ampicillin and related antibiotics and renders a bacterium resistant to the antibiotic.

blastomere An early embryonic cell formed by cleavage of the egg.

blastula The hollow ball stage of animal embryogenesis.

branch migration Lateral motion of the branch of a chi structure during crossing over.

bromodeoxyuridine (BrdU) An analogue of thymidine that can be incorporated into DNA in place of thymidine, and can then cause mutations by forming an alternate tautomer that base-pairs with guanine instead of adenine.

Burkitt's lymphoma A human cancer in which the *myc* oncogene in the tumor cells is translocated next to an immunoglobulin H chain gene.

burst size The difference between the final and initial phage concentrations in a cell; the number of progeny phage produced by a single cell.

C

cap A methylated guanosine bound through a 5′–5′ triphosphate linkage to the 5′-end of an mRNA, an hnRNA or an snRNA.

CAP (catabolite activator protein) A protein which, together with cAMP, activates operons that are subject to catabolite repression. Also known as CRP.

cap binding protein (CBP) A protein that associates with the cap on a eukaryotic mRNA and allows the mRNA to bind to a ribosome. Also known as eIF4F.

carboxyl terminus The end of a polypeptide with a free carboxyl group.

carcinogen A substance that causes cancer.

carcinogenesis The sequence of events that convert a normal cell to a cancer cell.

catabolic metabolism The process of breaking substances down into simpler components. The *lac* operon encodes catabolic enzymes that break down lactose into its component parts, galactose and glucose.

catabolite activator protein *See* CAP.

catabolite repression The repression of a gene or operon by glucose or, more likely, by a catabolite, or breakdown product of glucose.

CCAAT box An upstream motif, having the sequence CCAAT, found in many eukaryotic promoters recognized by RNA polymerase II.

cDNA A DNA copy of an RNA, made by reverse transcription.

centimorgan (cM) *See* map unit.

centromere Constricted region on the chromosome where spindle fibers are attached during cell division.

charging Coupling a tRNA with its cognate amino acid.

Charon phages A set of cloning vectors based on λ phage.

chiasma (chiasmata, pl.) The structure formed between nonsister chromatids during meiosis; physical evidence of recombination.

chi-square test A statistical test to determine if the observed numbers deviate from those expected under a particular hypothesis. Often expressed as χ^2 test.

chi structure The branched DNA structure formed by the first strand exchange during crossing over.

chloramphenicol An antibiotic that kills bacteria by inhibiting the peptidyl transferase reaction catalyzed by 50S ribosomes.

chloroplast The photosynthetic organelle of green plants and other photosynthetic eukaryotes.

chromatids Copies of a chromosome produced in cell division.

chromatin The material of chromosomes, composed of DNA and chromosomal proteins.

chromosome The physical structure, composed largely of DNA, that contains the genes of an organism.

chromosome theory of inheritance The theory that genes are contained in chromosomes.

***cis*-acting** A term that describes a genetic element, such as an enhancer, a promoter, or an operator, that must be on the same chromosome in order to influence a gene's activity.

***cis*-dominant** Dominant, but only with respect to genes on the same piece of DNA. For example, an operator constitutive mutation in one copy of the *lac* operon of a merodiploid *E. coli* cell is dominant with respect to that copy of the *lac* operon, but not to the other. That is because the operator controls the operon that is directly attached to it, but it cannot control an unattached operon.

***cis-trans* test** A test to see whether two mutations are true alleles (located in the same gene) or pseudoalleles (located in different genes). When the mutations are in *cis* position, wild-type behavior is always observed. When the mutations are in *trans* position, wild-type behavior is observed if the mutations are pseudoalleles, but mutant behavior is observed if they are alleles.

cistron A genetic unit defined by the *cis-trans* test. For all practical purposes, it is synonymous with the word *gene*.

clastogen An agent that causes DNA strand breaks.

clones Individuals formed by an asexual process so that they are genetically identical to the original individual.

closed promoter complex The complex formed by relatively loose binding between RNA polymerase and a prokaryotic promoter. It is "closed" in the sense that the DNA duplex remains intact, with no "opening up," or melting of base pairs.

coding strand The polynucleotide strand that has the same polarity and "sense" as the mRNA transcript.

codominant Two alleles whose phenotypic effects are both expressed in the heterozygote.

codon A three-base sequence in mRNA that causes the insertion of a specific amino acid into protein, or causes termination of translation.

cointegrate An intermediate in transposition of a transposon such as Tn3 from one replicon to another. The transposon replicates, and the cointegrate contains the two replicons joined through the two transposon copies.

colinearity A relationship between a gene and its protein product in which the distance between any two mutations is directly proportional to the distance between the corresponding altered amino acids.

colony hybridization A procedure for selecting a bacterial clone containing a gene of interest. DNAs from a large number of clones are simultaneously tested with a labeled probe that hybridizes to the gene of interest.

common garden experiment Growing organisms of different types in the same environment to determine whether the differences are genetic or environmental.

complementary polynucleotide strands Two strands of DNA or RNA that have complementary sequences; that is, wherever one has an adenine the other has a thymine, and wherever one has a guanine the other has a cytosine.

complementation group A group of mutants that cannot complement each other, thus demonstrating that their mutations all lie in the same gene.

complementation test *See cis-trans* test.

composite transposon A bacterial transposon composed of two types of parts: two arms containing IS or IS-like elements, and a central region comprised of the genes for transposition and one or more antibiotic resistance genes.

concatemers DNAs of multiple genome length.

conditional lethal A mutation that is lethal under certain circumstances, but not under others (e.g., a temperature-sensitive mutation).

conditional probability The probability of an event occurring, contingent upon given circumstances.

conjugation (of bacteria) The movement of genetic material from an F^+ to an F^- bacterium.

consanguineous mating A mating between related individuals.

consensus sequence The average of several similar sequences. For example, the consensus sequence of the −10 box of an *E. coli* promoter is TATAAT. This means that if you examine a number of such sequences, T is most likely to be found in the first position, A in the second, and so forth.

conservation genetics An effort to maintain genetic variation in a species by managing natural or seminatural populations or by collecting and storing diverse genetic types of a species.

conservative replication DNA (or RNA) replication in which both parental strands remain together, producing a progeny duplex both of whose strands are new.

conservative transposition Transposition in which both strands of the transposon DNA are conserved as they leave their original location and move to the new site.

constant region The region of an antibody that is more or less the same from one antibody to the next.

constitutive Always turned on.

constitutive mutant An organism containing a constitutive mutation.

constitutive mutation A mutation that causes a gene to be expressed at all times, regardless of normal controls.

contig A group of cloned DNAs containing contiguous or overlapping sequences.

copia A transposable element found in *Drosophila* cells.

core element An element of the eukaryotic promoter recognized by RNA polymerase I. Includes the bases surrounding the transcription start site.

core polymerase *See* RNA polymerase core.

corepressor A substance that associates with an aporepressor to form an active repressor (e.g., tryptophan is the corepressor of the *trp* operon).

cos The cohesive ends of the linear λ phage DNA.

cosmid A vector designed for cloning large DNA fragments. A cosmid contains the cos sites of λ phage, so it can be packaged into λ heads, and a plasmid origin of replication, so it can replicate as a plasmid.

$C_0 t$ The product of DNA concentration and time of annealing in a DNA renaturation experiment.

cotransduction Transduction of two or more genetic markers in the same virus.

Cro The product of the lambda *cro* gene. A repressor that binds preferentially to $O_R 3$ and turns off the lambda repressor gene (*cI*).

crossing over Physical exchange between homologous chromosomes that takes place in meiosis I. Also called recombination.

crossover The joining of corresponding DNA strands on homologous chromosomes by strand breakage and rejoining.

CRP Cyclic AMP receptor protein. *See* CAP.

C value The amount of DNA in picograms (trillionths of a gram) in a haploid genome of a given species.

C value paradox The fact that the C value of a given species is not necessarily related to the genetic complexity of that species.

cyanobacteria (blue-green algae) Photosynthetic bacteria. The ancestors of modern cyanobacteria are thought to have invaded eukaryotic cells and evolved into chloroplasts.

cyclic AMP (cAMP) An adenine nucleotide with a cyclic phosphodiester linkage between the 3′ and 5′ carbons. Implicated in a variety of control mechanisms in prokaryotes and eukaryotes.

cytidine A nucleoside containing the base cytosine.

cytohet A cell with cytoplasm, and therefore mitochondria and/or chloroplasts, from both parents.

cytoplasmic ribosomes The ribosomes that translate the mRNAs encoded in the nucleus.

cytosine (C) The pyrimidine base that pairs with guanosine in DNA.

D

deamination of DNA   The removal of an amino group (NH_2) from a cytosine or adenine in DNA, in which the amino group is replaced by a carbonyl group (C = O). This converts cytosine to uracil and adenine to hypoxanthine.

defective virus   A virus that is unable to replicate without a helper virus.

degenerate code   A genetic code, such as the one employed by all life on Earth, in which more than one codon can stand for a single amino acid.

degrees of freedom   An integer used to determine whether a chi-square value is statistically significant.

delayed early genes   Phage genes whose expression begins after that of the immediate early genes. Their transcription depends on at least one phage protein.

deleterious   An allele that causes some reduction in viability.

deletion   A mutation involving a loss of one or more base pairs; the case in which a chromosomal segment or gene is missing. Also called *deficiency.*

deletion mapping   Genetic mapping that employs deletion mutants. For example, one of the mating partners may contain a well-defined deletion, and the other an unknown point mutation. If recombination occurs, we know the point mutation lies outside the deleted area.

denaturation (DNA)   Separation of the two strands of DNA.

denaturation (protein)   Disruption of the three-dimensional structure of a protein without breaking any covalent bonds.

deoxyribose   The sugar in DNA.

determinant   A substance in an embryonic cell that contributes to determination of that cell.

determination   The irreversible commitment of a cell to differentiate along a given pathway.

dideoxyribonucleotide   A nucleotide, deoxy at both 2′ and 3′ positions, used to stop DNA chain elongation in DNA sequencing.

differentiated cell   A cell that has undergone differentiation.

differentiation   Expression of the specialized characteristics of a given cell type.

dihybrid cross   A cross between two individuals that are heterozygous at two genes; for example, *AaBb* × *AaBb.*

dihydrouracil loop   *See* D-loop.

dimer (protein)   A complex of two polypeptides. These can be the same (in a homodimer), or different (in a heterodimer).

diploid   The chromosomal number of the zygote and other cells (except the gametes). Symbolized as 2n.

directional cloning   Insertion of foreign DNA into two different restriction sites of a vector, such that the orientation of the insert can be predetermined.

dispersive replication   A hypothetical mechanism in which the DNA becomes fragmented so that new and old DNA coexist in the same strand after replication.

D-loop   A loop formed when a free DNA end "invades" a double helix, base-pairing with one of the strands and forcing the other to "loop out."

DNA (deoxyribonucleic acid)   A polymer composed of deoxyribonucleotides linked together by phosphodiester bonds. The stuff of which most genes are made.

DNA fingerprints   The use of highly variable regions of DNA to identify particular individuals.

DNA glycosylase   An enzyme that breaks the glycosidic bond between a damaged base and its sugar.

DNA gyrase   A topoisomerase that pumps negative superhelical turns into DNA. Relaxes the positive superhelical strain created by unwinding the *E. coli* DNA during replication.

DNA ligase   An enzyme that joins two double-stranded DNAs end to end.

DNA melting   *See* denaturation (DNA).

DNA photolyase   The enzyme that carries out light repair by breaking thymine dimers.

DNA polymerase I   One of three different DNA-synthesizing enzymes in *E. coli;* used primarily in DNA repair.

DNA polymerase III holoenzyme   The enzyme within the replisome, which actually makes DNA during replication.

DNase   Deoxyribonuclease, an enzyme that degrades DNA.

DNase footprinting   A method of detecting the binding site for a protein on DNA by observing the DNA region this protein protects from degradation by DNase.

DNase hypersensitive sites   Regions of chromatin that are about a hundred times more susceptible to attack by DNase I than bulk chromatin. These usually lie in the 5′-flanking regions of active or potentially active genes.

DNA typing   The use of molecular techniques, especially Southern blotting, to identify a particular individual.

dominant   An allele or trait that expresses its phenotype when heterozygous with a recessive allele; for example, *A* is dominant over *a* because the phenotypes of *AA* and *Aa* are the same.

double helix   The shape two complementary DNA strands assume in a chromosome.

down mutation   A mutation, usually in a promoter, that results in less expression of the gene.

Drosophila melanogaster   A species of fruit fly used widely by geneticists.

Ds   A defective transposable element found in corn, which relies on an *Ac* element for transposition.

duplication   The situation in which a chromosomal segment or gene is represented twice.

E

editing   The process of checking each nucleotide for complementarity with its base-pairing partner as it is inserted during DNA replication.

electrophile   A molecule that seeks centers of negative charge in other molecules and attacks them there.

electrophoresis   A procedure in which voltage is applied to charged molecules, inducing them to migrate. This technique can be used to separate DNA fragments. It can also be used to separate and identify proteins or enzymes resulting from the presence of different alleles.

elongation factor   A protein that is necessary for either the aminoacyl-tRNA binding or the translocation step in the elongation phase of translation.

embryogenesis   Embryonic development.

encode   To contain the information for making an RNA or polypeptide. A gene can encode an RNA or a polypeptide.

endonuclease   An enzyme that makes cuts within a polynucleotide strand.

endoplasmic reticulum (ER)   Literally, "cellular network"; a network of membranes in the cell upon which proteins destined for export from the cell are synthesized.

endospores   Dormant spores formed within a cell, as in *Bacillus subtilis.*

endosymbiont hypothesis   A theory of the origin of organelles, which holds that mitochondria and chloroplasts began as free-living prokaryotes that invaded and initiated a mutually beneficial relationship with primitive eukaryotic cells.

enhancer   A DNA element that strongly stimulates transcription of a gene or genes. Enhancers are usually found upstream from the genes they influence, but they can also function if inverted or moved hundreds or even thousands of base pairs away.

enol tautomer   An abnormal tautomer of uracil, thymine, or guanine, found only rarely in nucleic acids. Base-pairs abnormally and so can cause mutations.

enucleate egg   An egg without a nucleus.

env   The gene in a retrovirus that encodes the viral envelope (membrane) protein.

enzyme   A molecule—usually a protein, but sometimes an RNA—that catalyzes, or accelerates and directs, a biochemical reaction.

epigenetic change   A change in the expression of a gene, not in the structure of the gene itself.

epistasis   The situation in which the alleles at one gene cover up or alter the expression of alleles at another gene.

equilibrium   A state at which there is no change in the allelic frequencies of the population.

error-prone repair   A mechanism used by *E. coli* cells to replicate DNA that contains thymine dimers. DNA is replicated across from a dimer even though no base pairing is possible. This usually leads to mistakes.

Escherichia coli (E. coli)   An intestinal bacterium; the favorite subject for bacterial genetics.

euchromatin   Chromatin that is extended, accessible to RNA polymerase, and at least potentially active. These regions stain either lightly or normally and are thought to contain most of the genes.

eukaryote   An organism whose cells have nuclei.

euploid   *n.* A polyploid organism whose chromosome number is an integral multiple of the chromosome number of the organism from which it derived. *adj.* Having the normal number of chromosomes for the organism in question.

excinuclease   A contraction of the term *excision nuclease.* This is the name given to the enzyme system that catalyzes excision repair of damaged nucleotides in DNA.

excision repair   A DNA repair mechanism that removes damaged nucleotides, then replaces them with normal ones.

exon   A region of a gene that is ultimately represented in that gene's mature transcript. The word refers to both the DNA and its RNA product.

exon shuffling   A hypothetical process in which exons from different genes can combine to form new genes over evolutionary time.

exonuclease   An enzyme that degrades a polynucleotide from the end inward.

expression site   A locus on a chromosome where a gene can be moved to be expressed efficiently. For example, the expression site in trypanosomes is at the end (telomere) of a chromosome.

expression vector   A cloning vector that allows expression of the cloned gene.

expressivity   The level of phenotypic expression for a particular genotype.

F

F′   An F plasmid that has picked up a piece of host DNA.

F₁   The progeny of a cross between two parental types that differ at one or more genes; the first filial generation.

F₂   The progeny of a cross between two F₁ individuals or the progeny of a self-fertilized F₁; the second filial generation.

fate map   A diagram that shows which parts of an embryo are destined to differentiate into each segment of the adult.

F-duction   The transfer of bacterial genes from one cell to another on an F′ plasmid.

F plasmid   The fertility plasmid that allows a donor bacterium to conjugate with a recipient.

fertility plasmid   *See* F plasmid.

fine structure mapping   Extensive genetic mapping within a gene, ideally down to the nucleotide level.

fingerprint (protein)   The specific pattern of peptide spots formed when a protein is cut into pieces (peptides) with an enzyme (e.g., trypsin), and then the peptides are separated by chromotography.

FISH   Fluorescence in situ hybridization. A means of hybridizing a fluorescent probe to whole chromosomes to determine the location of a gene or other DNA sequence within a chromosome.

5′-end   The end of a polynucleotide with a free (or phosphorylated or capped) 5′-hydroxyl group.

flagellin   The protein of which a flagellum is composed.

flagellum   A whiplike structure on the surface of a cell, used for propulsion (Latin: *flagellum,* little whip; plural, flagella).

fluctuation test   A test to determine whether a given microbial trait is caused by a mutation. If it is, individual cultures should vary widely in the number of cells showing the trait, since mutation is a rare and random event and may occur early in the growth of some cultures and later in others. The test can also be used to measure mutation rates.

fMet   *See* N-formyl methionine.

F pilus   The thin projection, carried on the surface of an F⁺ bacterium, which allows bacterial cells to join during conjugation.

forced cloning   *See* directional cloning.

forked-line approach   A technique used to determine progeny types for multiple genes, based on independent assortment between the genes.

forward mutation   A mutation away from wild-type.

founder effect   The chance changes in allelic frequency resulting from the initiation of a population by a small number of individuals.

frameshift mutation   An insertion or deletion of one or two bases in the coding region of a gene, which changes the reading frame of the corresponding mRNA.

fushi tarazu (ftz)   A homeobox-containing gene in *Drosophila* that helps define segmentation in the early embryo.

fusion protein   A protein resulting from the expression of a recombinant DNA containing two open reading frames (ORFs) fused together. One or both of the ORFs can be incomplete.

G

gag   The gene in a retrovirus that encodes the viral coat proteins.

galactoside permease   An enzyme encoded in the *E. coli lac* operon that transports lactose into the cell.

GAL4   A transcription factor that activates the galactose utilization (GAL) genes of yeast by binding to an upstream control element (UAS_G).

gamete   A haploid sex cell.

gamma rays   Very high energy radiation that ionizes cellular components. The ions then can cause chromosome breaks.

GC box   A hexamer having the sequence GGGCGG on one strand, which occurs in a number of mammalian structural gene promoters. The binding site for the transcription factor Sp1.

gene   The basic unit of heredity. Contains the information for making one RNA and, in most cases, one polypeptide.

gene cloning   Generating many copies of a gene by inserting it into an organism, such as a bacterium, where it can replicate along with the host.

gene cluster   A group of related genes grouped together on a eukaryotic chromosome.

gene conversion   The conversion of one gene's sequence to that of another.

gene expression   The process by which gene products are made.

gene flow   Movement of individuals from one population to another that results in the introduction of migrant alleles through mating and subsequent reproduction.

gene mutation   A mutation confined to a single gene.

gene pool   The total variety and number of alleles within a population or species.

generalized recombination   *See* homologous recombination.

generalized transduction   Use of P1 or other phage heads to carry bacterial genes from one cell to another.

genetic code   The set of 64 codons and the amino acids (or terminations) they stand for.

genetic counseling   Counseling based on the risks of producing genetically defective infants, usually given where some familial history of a genetic disease exists.

genetic drift   The chance changes in allelic frequency that result from the sampling of gametes from generation to generation.

genetic mapping   Determining the linear order of genes and the distances between them.

genetic marker   A mutant gene or other peculiarity in a genome that can be used to "mark" a spot in a genome for mapping purposes.

genetic recombination   *See* recombination.

genetic variance   The variation of a phenotype in a population that results from genetic causes, symbolized by V_G. The extent of the additive genetic variance, symbolized by V_A, is important in determining the rate and amount of response to directional selection.

genome   One complete set of genetic information from a genetic system; e.g., the single, circular chromosome of a bacterium is its genome.

genomic library   A set of clones containing DNA fragments derived directly from a genome, rather than from mRNA.

genotype   The allelic constitution of a given individual. The genotypes at locus *A* in a diploid individual may be *AA*, *Aa*, or *aa*.

genotype–environment interaction   The effect on a phenotype value that results from a specific genotype and a specific environment and that is not predictable from either separately. Symbolized by GE_{ij}.

germinal mutation   *See* germ-line mutation.

germ-line mutation   A mutation that affects the sex cells, enabling it to be passed on to progeny.

glucose   A simple, six-carbon sugar used by many forms of life as an energy source.

glycosidic bond (in a nucleoside)   The bond linking the base to the sugar (ribose or deoxyribose) in RNA or DNA.

Golgi apparatus   A membranous organelle that packages newly synthesized proteins for export from the cell.

G protein   A protein that binds GTP, which activates the protein to perform a function. When the G protein cleaves the GTP to GDP, it becomes inactivated until a new molecule of GTP binds.

group I introns   Self-splicing introns in which splicing is initiated by a free guanosine or guanosine nucleotide.

group II introns   Self-splicing introns in which splicing is initiated by formation of a lariat-shaped intermediate.

guanine (G)   The purine base that pairs with cytosine in DNA.

guanosine   A nucleoside containing the base guanine.

H

hairpin   A structure resembling a hairpin (bobby pin), formed by intramolecular base pairing in an inverted repeat of a single-stranded DNA or RNA.

half chiasma   *See* chi structure.

half-C_0t   The C_0t at which half of the DNA molecules in an annealing solution will be renatured.

half-life   The time it takes for half of a population of molecules to disappear.

haploid   The chromosomal number in the gamete. Symbolized as n.

haplotype   A set of linked genes or DNA sequences that tend to be inherited together.

Hardy-Weinberg principle   The principle stating that after one generation of random mating, single-locus genotypic frequencies can be represented as a function of the allelic frequencies.

helicase   An enzyme that unwinds a polynucleotide double helix.

helix-turn-helix   A structural motif in certain DNA-binding proteins, especially those from prokaryotes, that fits into the DNA wide groove and gives the protein its binding capacity and specificity.

helper virus (or phage)   A virus that supplies functions lacking in a defective virus, allowing the latter to replicate.

hemoglobin   The red, oxygen-carrying protein in the red blood cells.

heritability   The proportion of phenotypic variance that is the additive genetic component. Symbolized as H^2. Realized heritability is an estimate of H^2 from a selection experiment.

hermaphrodite   An animal that has both male and female reproductive organs.

heterochromatin   Chromatin that is condensed and inactive.

heteroduplex   A double-stranded polynucleotide whose two strands are not completely complementary.

heterogeneous nuclear RNA (hnRNA)   A class of large, heterogeneous-sized RNAs found in the nucleus, including unspliced mRNA precursors.

heterokaryon A cell containing two or more nuclei of different origins.

heterokaryotype Individuals that are heterozygous for two chromosomal types.

heterozygosity The proportion of heterozygotes in a population.

heterozygote A diploid genotype in which the two alleles for a given gene are different; for example, A_1A_2.

heterozygote advantage The situation in which the heterozygote has a higher fitness than either homozygote, leading to a stable equilibrium.

Hfr A strain of *E. coli* that experiences a high frequency of recombination because of integration of the F plasmid into the bacterial chromosome, which mobilizes the chromosome.

histones A class of five small, basic proteins intimately associated with DNA in most eukaryotic chromosomes.

hnRNA *See* heterogeneous nuclear RNA.

homeobox A sequence of about 180 base pairs found in homeotic genes and other development-controlling genes in eukaryotes. Encodes a DNA-binding region (homeodomain) of a protein.

homeodomain A 60-amino acid domain of a DNA binding protein that allows the protein to bind tightly to a specific DNA region.

homeotic gene A gene in which a mutation causes the transformation of one body part into another.

homologous chromosomes Chromosomes that are identical in size, shape, and, except for allelic differences, genetic composition.

homologous recombination Recombination that requires extensive sequence similarity between the recombining DNAs.

homozygote A diploid genotype in which both alleles for a given gene are identical; for example, A_1A_1 or *aa*.

host-range The set of hosts that can be productively infected by a virus (or phage).

hot spots Highly mutable sites within a genome—sites that accumulate more than the average number of mutations.

housekeeping genes Genes that code for proteins used by all kinds of cells.

human immunodeficiency virus (HIV) The agent that causes acquired immune deficiency syndrome (AIDS).

hybrid cell A cell created by fusing two dissimilar cells. First, a heterokaryon forms; then, after nuclear fusion, a hybrid cell.

hybrid dysgenesis A phenomenon observed in *Drosophila* in which the hybrid offspring of two certain parental strains suffer so much chromosomal damage that they are sterile, or dysgenic.

hybridization (of polynucleotides) Forming a double-stranded structure from two polynucleotide strands (either DNA or RNA) from different sources.

hybrid polynucleotide The product of polynucleotide hybridization.

hyperchromic shift The increase in a DNA solution's absorbance of 260 nm light upon denaturation.

hyphae Strandlike chains of fungal cells.

I

ICR *See* internal control region.

illegitimate recombination Recombination that requires little if any sequence similarity between the recombining DNAs and is therefore not site-specific.

imaginal disk A group of insect larval cells that give rise to a body part, such as a wing, leg, or antenna.

imino tautomer An abnormal tautomer of adenine or cytosine, found only rarely in nucleic acids. Base-pairs abnormally and so can cause mutations.

immediate early genes Phage genes that are expressed immediately upon infection. They are transcribed even if phage protein synthesis is blocked, so their transcription depends only on host proteins.

immunity region The control region of a λ or λ-like phage, containing the gene for the repressor as well as the operators recognized by this repressor.

immunoglobulin (antibody) A protein that binds very specifically to an invading substance and alerts the body's immune defenses to destroy the invader.

imprinting A phenomenon in which an autosome inherited from the female parent produces a different phenotype than the same autosome, with the same genotype, inherited from the male parent.

inbreeding Nonrandom mating in which the mating individuals are more closely related than individuals drawn by chance from the population.

inbreeding coefficient The probability that the two homologous alleles in an individual are identical by descent. Symbolized as f.

inbreeding depression The reduction of population fitness due to inbreeding.

in cis A condition in which two genes are located on the same chromosome.

incision Nicking a DNA strand with an endonuclease.

independent assortment A principle discovered by Mendel, which states that genes on different chromosomes behave independently.

inducer A substance that releases negative control of an operon.

inducer (embryonic) A substance that one embryonic cell makes to influence the development of another cell.

initiation factor A protein that helps catalyze the initiation of translation.

inosine (I) A nucleoside containing the base hypoxanthine, which base-pairs with cytosine.

insertion sequence (IS) A simple type of transposon found in bacteria, containing only inverted terminal repeats and the genes needed for transposition.

interference The difference between the observed number of double recombinants and the number expected when recombination in one area does not influence that in another area.

internal control region (ICR) The part of a eukaryotic promoter recognized by RNA polymerase III that lies inside the gene's coding region.

interphase The stage of the cell cycle during which DNA is synthesized but the chromosomes are not visible.

intervening sequence (IVS) *See* intron.

intracistronic complementation Complementation of two mutations in the same gene. Can occur by cooperation among different defective monomers to form an active oligomeric protein.

in trans A condition in which two genes are located on separate chromosomes.

intron A region that interrupts the transcribed part of a gene. An intron is transcribed, but is removed by splicing during maturation of the transcript. The word refers to the intervening sequence in both the DNA and its RNA product.

inversion   An alteration in the sequence of genes in a chromosome. A pericentric inversion includes the centromere, while a paracentric inversion does not.

inverted repeat   A symmetrical sequence of DNA, reading the same forward on one strand and backward on the opposite strand. For example:

GGATCC
CCTAGG

isoschizomers   Two or more restriction endonucleases that recognize and cut the same restriction site.

J

joining region (J)   The segment of an immunoglobulin gene encoding the last thirteen amino acids of the variable region. One of several joining regions is joined by a chromosomal rearrangement to the rest of the variable region, introducing extra variability into the gene.

K

karyotype   A pictorial or photographic representation of all the chromosomes in a given individual.

keto tautomer   The normal tautomer of uracil, thymine, or guanine found in nucleic acids.

kilobase pair   One thousand base pairs.

Klenow fragment   A fragment of DNA polymerase I, created by cleaving with a protease, that lacks the $5' \rightarrow 3'$ exonuclease activity of the parent enzyme.

L

lacA   The *E. coli* gene that encodes galactoside transacetylase.

lacI   The *E. coli* gene that encodes the *lac* repressor.

lac **operon**   The operon encoding enzymes that permit a cell to metabolize the milk sugar lactose.

lac **repressor**   A protein, the product of the *E. coli lacI* gene, that forms a tetramer which binds to the *lac* operator and thereby represses the *lac* operon.

lactose   A two-part sugar, or disaccharide, composed of two simple sugars, galactose and glucose.

lacY   The *E. coli* gene that encodes galactoside permease.

lacZ   The *E. coli* gene that encodes β-galactosidase.

lagging strand   The strand that is made discontinuously in semidiscontinuous DNA replication.

lambda (λ) phage   A temperate phage of *E. coli*. Can replicate lytically or lysogenically.

lambda (λ) repressor   The protein that forms a dimer and binds to the lambda operators O_R and O_L thus repressing all other phage genes except the repressor gene itself.

lariat   The name given the lasso-shaped intermediate in certain splicing reactions.

late genes   The last phage genes to be expressed. By definition, their transcription begins after the onset of phage DNA replication. Phage structural proteins are generally encoded by late genes.

latent period   The time between infection of a bacterium by a phage and the appearance of progeny phages in the medium. Also called the lag period.

leader   A sequence of untranslated bases at the 5′-end of an mRNA.

leading strand   The strand that is made continuously in semidiscontinuous DNA replication.

lethal   An allele that causes mortality. A recessive lethal causes mortality only when homozygous.

leucine zipper   A domain in a DNA binding protein that includes several leucines spaced at regular intervals. Appears to permit formation of a dimer with another leucine zipper protein. The dimer is then empowered to bind to DNA.

LexA   The product of the *E. coli lexA* gene. A repressor that represses, among other things, the *umuDC* operon.

light repair   Direct repair of a thymine dimer by the enzyme DNA photolyase.

linkage   The physical association of genes on the same chromosome.

locus (loci, pl.)   The position of a gene on a chromosome, used synonymously with the term gene in many instances.

long terminal repeats (LTRs)   Regions of several hundred base-pairs of DNA found at both ends of the provirus of a retrovirus.

luxury genes   Genes that code for specialized cell products.

Lyonized chromosome   An X chromosome in a female mammal that is composed entirely of heterochromatin and is genetically inactive.

lysis   Rupturing the membrane of a cell, as by a virulent phage.

lysogen   A bacterium harboring a prophage.

M

MAP kinase (mitogen activated protein kinase)   A group of protein kinases that phosphorylate proteins, some of which are transcription factors, in response to mitogens (mitosis-inducing substances). The net result of this protein phosphorylation is usually cell division.

map unit (centimorgan)   The distance separating two genetic loci that recombine with a frequency of 1%.

marker   A gene or mutation that serves as a signpost at a known location in the genome.

maternal effect   A phenotype governed by the nuclear genotype, not the phenotype, of the maternal parent, and unaffected by genes from the paternal parent.

maternal effect genes   Nuclear genes active in an oocyte, which define a phenotype that is unaffected by genes from the male gamete.

maternal inheritance   A non-Mendelian pattern of inheritance that always depends only on the maternal parent. The paternal parent does not influence the F_1 or any succeeding generation.

maternal message   An mRNA made in great quantity during oogenesis, then stored, complexed with protein in the egg cytoplasm.

mean   The arithmetic average.

mean fitness The sum of the product of the relative fitness and the frequency of the genotypes in the population. Symbolized as w.

megabase pair One million base pairs.

meiosis Cell division that produces gametes (or spores) having half the number of chromosomes of the parental cell.

Mendelian genetics *See* transmission genetics.

merodiploid A bacterium that is only partially diploid—that is, diploid with respect to only some of its genes.

merozygote *See* merodiploid.

mesoderm The middle layer of embryonic cells, which develops into connective tissue, muscle, and blood, among other tissues.

message *See* mRNA.

messenger RNA *See* mRNA.

metacentric Chromosome with the centromere in the middle.

metal finger *See* zinc finger.

metaphase Intermediate stage of nuclear division in which the chromosomes move to the equatorial plane.

metastasis The spread of cancer cells through the blood or lymph system to establish secondary (metastatic) tumors.

microconidium The small, male gamete of the mold *Neurospora*.

microsatellite A short DNA sequence (usually 2–4 bp) repeated many times in tandem. A given microsatellite is found in varying lengths, scattered around a eukaryotic genome.

microsatellite instability The tendency of a given copy of a microsatellite to change size from parent to offspring, or even within an organism's lifetime.

minimal medium A growth medium for microbes that contains only the simple substances needed for growth of prototrophs.

minus ten box (–10 box) *See* Pribnow box.

minus thirty-five box (–35 box) An *E. coli* promoter element centered about 35 bases upstream from the start of transcription.

mismatch repair The correction of a mismatched base incorporated by accident—in spite of the editing system—into a newly synthesized DNA.

missense mutation A change in a codon that results in an amino acid change in the corresponding protein.

mitosis Cell division that produces two daughter cells having nuclei identical to the parental cell.

modifier gene A gene that modifies the phenotype of another gene.

molecular clock A hypothesis suggesting that a regular turnover or replacement of molecular variants (amino acid or nucleotide substitution) takes place over time.

molecular genetics The study of the structure and activities of genes at the molecular level.

monohybrid cross A cross between two individuals that are heterozygous at one gene; for example, *Aa* × *Aa*.

monomorphic The situation in which all the individuals in a population are the same genetic type or have the same allele.

morphogen A substance that helps determine shape in a developing embryo.

morphological mutation A mutation that affects the morphology, or appearance, of an organism.

mosaic embryo An embryo in which the early blastomeres are different from one another.

mRNA (messenger RNA) A transcript that bears the information for making one or more proteins.

multiple cloning site (MCS) A region in certain cloning vectors, such as the pUC plasmids and M13 phage DNAs, that contains several restriction sites in tandem. Any of these can be used for inserting foreign DNA.

multiplicity of infection (m.o.i.) The ratio of infectious virus (or phage) particles to infectable cells in an experiment.

mutagen A mutation-causing agent.

mutant An organism (or genetic system) that has suffered at least one mutation.

mutation The original source of genetic variation caused, for example, by a change in a DNA base or a chromosome. Spontaneous mutations are those that appear without explanation, while induced mutations are those attributed to a particular mutagenic agent.

mutation rate The rate of mutational change from one allelic form to another. Symbolized by u or v.

mutation–selection balance The equilibrium that results when mutation introduces detrimental alleles into a population and selection eliminates them.

mutator strain A strain of bacteria that makes more than the usual number of mistakes during DNA replication, yielding a higher than normal mutation rate.

mutually exclusive events Events such that if one occurs, the other cannot; for example, heads and tails on a coin.

myc A cellular oncogene, encoding a nuclear protein, that is rearranged in human Burkitt's lymphoma.

N

negative control A control system in which gene expression is turned off unless a controlling element (e.g., a repressor) is removed.

Neurospora crassa A common bread mold, developed by Beadle and Tatum as a subject for genetic investigation.

neutrality theory The theory used to explain the existence of molecular variation when there is no differential selection (alleles are neutral with respect to each other) and new variants are introduced into the population by mutation and eliminated by genetic drift.

N-formyl methionine (fMet) The initiating amino acid in prokaryotic translation.

nick A single-stranded break in DNA.

nitrocellulose A type of paper that has been changed chemically so it binds single-stranded DNA. Commonly used for blotting DNA prior to hybridizing with a labeled probe.

nondisjunction Abnormal separation of chromosomes in meiosis (or mitosis).

nonpermissive conditions Those conditions under which a conditional mutant gene cannot function.

nonsense codons UAG, UAA, and UGA These codons tell the ribosome to stop protein synthesis.

nonsense mutation A mutation that creates a premature stop codon within a gene's coding region. Includes amber mutations (UAG), ochre mutations (UAA), and opal mutations (UGA).

nontemplate DNA strand The strand complementary to the template strand. Sometimes called the coding strand, or sense strand.

norm of reaction The phenotypic pattern for a given genotype produced under different environmental conditions.

Northern blotting Transferring RNA fragments to a support medium (*see* Southern blotting).

nucleic acid   A chainlike molecule (DNA or RNA) composed of nucleotide links.

nucleocapsid   A structure containing a viral genome (DNA or RNA) with a coat of protein.

nucleoid   A nonmembrane-bound area of a bacterial cell in which the chromosome resides.

nucleolus   A cell organelle found in the nucleus that disappears during part of cell division. Contains the rRNA genes.

nucleoside   A base bound to a sugar—either ribose or deoxyribose.

nucleosome   A repeating structural element in eukaryotic chromosomes, composed of a core of eight histone molecules with about 200 base pairs of DNA wrapped around the outside.

nucleotide   The subunit, or chain-link, in DNA and RNA, composed of a sugar, a base, and at least one phosphate group.

nutritional mutation   A mutation that makes an organism dependent on a substance, such as an amino acid or vitamin, that it previously could make for itself.

O

ochre codon   UAA, coding for termination.

ochre mutation   *See* nonsense mutation.

ochre suppressor   A tRNA bearing an anticodon that can recognize the ochre codon (UAA) and thereby suppress ochre mutations.

Okazaki fragment   Small DNA fragments, 1,000–2,000 bases long, created by discontinuous synthesis of the lagging strand.

oligomeric protein   A protein that contains more than one polypeptide subunit.

oligonucleotide   A short piece of RNA or DNA.

oncogene   A gene that participates in causing tumors.

oncogenesis   The formation of tumors.

oncogenic   Tumor-causing.

one gene-one polypeptide hypothesis   The hypothesis, now generally regarded as valid, that one gene codes for one polypeptide.

oocyte 5S rRNA genes   The 5S rRNA genes (haploid number about 19,500 in *Xenopus laevis*) that are expressed only in oocytes.

oogenesis   Development of an egg.

opal codon   UGA, coding for termination.

opal mutation   *See* nonsense mutation.

opal suppressor   A tRNA bearing an anticodon that can recognize the opal codon (UGA) and thereby suppress opal mutations.

open promoter complex   The complex formed by tight binding between RNA polymerase and a prokaryotic promoter. It is "open" in the sense that approximately ten base pairs of the DNA duplex open up, or separate.

open reading frame (ORF)   A reading frame that is uninterrupted by translation stop codons.

operator   A DNA element found in prokaryotes that binds tightly to a specific repressor and thereby regulates the expression of adjoining genes.

operon   A group of genes coordinately controlled by an operator.

ORF   *See* open reading frame.

origin of replication   The unique spot in a replicon where replication begins.

P

P_{RE} The lambda promoter from which transcription of the repressor gene occurs during the establishment of the lysogenic state.

P_{RM} The lambda promoter from which transcription of the repressor gene occurs during maintenance of the lysogenic state.

pair-rule genes   Genes that, when mutated, give rise to insects missing segment-sized body parts at two-segment intervals.

palindrome   *See* inverted repeat.

parthenogenesis   Reproduction in which organisms produce offspring identical to themselves without fertilization.

paternity exclusion   The process of using genetic markers to exclude given men as fathers of a particular child.

pBR322   One of the original plasmid vectors for gene cloning.

PCNA   Proliferating cell nuclear antigen. A eukaryotic protein that confers processivity on DNA polymerase δ during leading strand synthesis.

PCR   *See* polymerase chain reaction.

pedigree   A family tree illustrating the inheritance of particular genotypes or phenotypes.

P element   A transposable element of *Drosophila,* responsible for hybrid dysgenesis. Can be used to mutagenize *Drosophila* deliberately.

penetrance   The proportion of individuals with a given genotype that express it at the phenotypic level.

peptide bond   The bond linking amino acids in a protein.

peptidyl transferase   An enzyme that is an integral part of the large ribosomal subunit and catalyzes the formation of peptide bonds during protein synthesis.

permissive conditions   Those conditions under which a conditional mutant gene can function.

petite   A slow-growing strain of yeast with defective mitochondria.

phage   A bacterial virus.

phage Mu   A temperate phage of *E. coli* that replicates lytically by transposition.

phage P1   A lytic phage of *E. coli* used in generalized transduction.

phagemid   A plasmid cloning vector with the origin of replication of a single-stranded phage, which gives it the ability to produce single-stranded cloned DNA upon phage infection.

phase variation   The replacement of one type of protein in a bacterium's flagella with another type.

phenocopy   Occurs when environmental factors induce a particular abnormal phenotype that resembles a genetically determined phenotype.

phenotype   The morphological, biochemical, behavioral, or other properties of an organism. Often only a particular trait of interest, such as weight, is considered.

phosphodiester bond   The sugar-phosphate bond that links the nucleotides in a nucleic acid.

photoreactivating enzyme   *See* DNA photolyase.

photoreactivation   *See* light repair.

phylogenetic tree   A diagram that organizes the relationship between species or populations, putatively indicating ancestral relationships.

physical map   A genetic map based on physical characteristics of the DNA, such as restriction sites, rather than on locations of genes.

plaque   A hole that a virus makes on a layer of host cells by infecting and either killing the cells or slowing their growth.

plaque assay An assay for virus (or phage) concentration in which the number of plaques produced by a given dilution of virus is determined.

plaque-forming unit (pfu) A virus capable of forming a plaque in a plaque assay.

plaque hybridization A procedure for selecting a phage clone that contains a gene of interest. DNAs from a large number of phage plaques are simultaneously tested with a labeled probe that hybridizes to the gene of interest.

plasmid A circular DNA that replicates independently of the cell's chromosome.

pleiotropic mutation A mutation that affects the expression of several other genes.

pleiotropy The situation in which a single gene affects two or more characteristics.

point mutation An alteration of one, or a very small number, of contiguous bases.

pol The gene in a retrovirus that encodes the viral reverse transcriptase and RNase H.

poly(A) Polyadenylic acid. The string of about two hundred A's added to the end of a typical eukaryotic mRNA.

polyadenylation Addition of poly(A) to the 3′-end of an RNA.

poly(A) polymerase The enzyme that adds poly(A) to an mRNA, or to its precursor.

polycistronic message An mRNA bearing information from more than one gene.

polygenic trait A phenotypic trait determined by a number of genes.

polymerase chain reaction (PCR) Amplification of a region of DNA using primers that flank the region and repeated cycles of DNA polymerase action.

polymorphism The existence of two or more genetically determined forms (alleles) in a population in substantial frequency. In practice, a polymorphic gene is one at which the frequency of the most common allele is less than 0.99.

polynucleotide A polymer composed of nucleotide subunits; DNA or RNA.

polypeptide A single protein chain.

polyploids Organisms with three or more sets of chromosomes. Autopolyploids receive all their chromosomes from the same species, while allopolyploids obtain their chromosomes from two (or more) species.

polyribosome *See* polysome.

polysome A messenger RNA attached to (and presumably being translated by) several ribosomes.

polytene chromosomes Large chromosomes in diptera that contain many parallel replicates tightly held together.

population A group of individuals that exist together in time and space and can interbreed.

population genetics The study of the variation of genes between and within populations.

positive control A control system in which gene expression depends on the presence of a positive effector such as CAP (and cAMP).

positive strand The strand of a simple viral genome with the same sense as the viral mRNAs.

post-replication repair *See* recombination repair.

posttranscriptional control Control of gene expression that occurs during the posttranscriptional phase when transcripts are processed by splicing, clipping, and modification.

posttranslational modification The set of changes that occur in a protein after it is synthesized.

pp60^{src} The product of the *src* gene, a tyrosine protein kinase.

Pribnow box (−10 box) An *E. coli* promoter element centered about ten bases upstream from the start of transcription.

primary structure The sequence of amino acids in a polypeptide, or of nucleotides in a DNA or RNA.

primary transcript The initial, unprocessed RNA product of a gene.

primase The enzyme within the primosome that actually makes the primer.

primer A small piece of RNA that provides the free end needed for DNA replication to begin.

primosome A complex of about twenty polypeptides, which makes primers for *E. coli* DNA replication.

probability The proportion of the time that a particular event occurs.

probe (nucleic acid) A piece of nucleic acid, labeled with a tracer (usually radioactive) that allows an experimenter to track the hybridization of the probe to an unknown DNA. For example, a radioactive probe can be used to identify an unknown DNA band after electrophoresis.

processed pseudogene A pseudogene that has apparently arisen by retrotransposon-like activity: transcription of a normal gene, processing of the transcript, reverse transcription, and reinsertion into the genome.

processing (of RNA) The group of cuts that occur in RNA precursors during maturation, including splicing, 5′- or 3′-end clipping, or cutting rRNAs out of a large precursor.

processivity The tendency of an enzyme to remain bound to one or more of its substrates during repetitions of the catalytic process. Thus, the longer a DNA or RNA polymerase continues making its product without dissociating from its template, the more processive it is.

product rule A rule stating that the probability of both of two independent events occurring is the product of their individual probabilities.

prokaryotes Microorganisms that lack nuclei. Synonymous with bacteria, including cyanobacteria (blue-green algae).

programmed cell death *See* apoptosis.

promoter A DNA sequence to which RNA polymerase binds prior to initiation of transcription—usually found just upstream from the transcription start site of a gene.

proofreading The process the cell uses to check the accuracy of DNA replication as it occurs, and to replace a mis-paired base with the right one.

prophage A phage genome integrated into the host's chromosome.

prophase Early stage of nuclear division in which chromosomes coil, condense, and become visible.

protein A polymer, or polypeptide, composed of amino acid subunits. Sometimes the term *protein* denotes a functional collection of more than one polypeptide (e.g., the hemoglobin protein consists of four polypeptide chains).

protein sequencing Determining the sequence of amino acids in a protein.

proteolytic processing Cleavage of a protein into pieces.

proto-oncogene A cellular counterpart of a viral oncogene. Proto-oncogenes do not ordinarily participate in oncogenesis unless they are over-expressed or mutated.

protoperithecium   The large, female gamete of the mold *Neurospora*.

prototroph   An organism that can grow on defined minimal medium.

provirus   A double-stranded DNA copy of a retroviral RNA, which inserts into the host genome.

pseudogene   A nonallelic copy of a normal gene, which is mutated so that it cannot function.

pulse-chase   The process of giving a short period, or "pulse," of radioactive precursor so that a substance such as RNA becomes radioactive, then adding an excess of unlabeled precursor to "chase" the radioactivity out of the substance.

pulse labeling   Providing a radioactive precursor for only a short time. For example, DNA can be pulse labeled by incubating cells for a short time in radioactive thymidine.

pure-breeding line   Identical individuals that always produce progeny like themselves.

purine   The parent base of guanine and adenine.

puromycin   An antibiotic that resembles an aminoacyl-tRNA and kills bacteria by forming a peptide bond with a growing polypeptide and then releasing the incomplete polypeptide from the ribosome.

pyrimidine   The parent base of cytosine, thymine, and uracil.

Q

quantitative trait   A trait that generally has a continuous distribution in a population and is usually affected by many genes and many environmental factors.

quaternary structure   The way two or more polypeptides interact in a complex protein.

quenching   Quickly chilling heat-denatured DNA to keep it denatured.

R

random-mating population   A group of individuals in which the probability of members mating with individuals of particular types is equal to their frequency in the population.

ras   A family of oncogenes, found in several human tumors, that encode a G protein called p21.

rate of allelic substitution   The rate of replacement of alleles in a population. If amino acid or nucleotide sequences are known, the rate of amino acid or nucleotide substitution can be calculated.

reading frame   One of three possible ways the triplet codons in an mRNA can be translated. For example, the message CAGUGCUCGAC has three possible reading frames, depending on where translation begins: (1) CAG UGC UCG; (2) AGU GCU CGA; (3) GUG CUC GAC. A natural mRNA generally has only one correct reading frame.

recA   The *E. coli* gene that encodes the RecA protein, which is involved in homologous recombination, and also functions as a protease during the SOS response.

recessive   An allele or trait that does not express its phenotype when heterozygous with a dominant allele; for example, *a* is recessive to *A* because the phenotype for *Aa* is like *AA* and not like *aa*.

reciprocal matings   Crosses that are identical genetically except that the male and female parents are switched.

recognition helix   The α-helix in a DNA-binding motif of a DNA-binding protein that fits into the wide groove of its DNA target and makes sequence-specific contacts that define the specificity of the protein. In effect, the recognition helix recognizes the specific sequence of its DNA target.

recombinant DNA   The product of recombination between two (or more) fragments of DNA. Can occur naturally in a cell, or be fashioned by geneticists *in vitro*.

recombination   Reassortment of genes or alleles in new combinations. Technically, recombination can occur by either independent assortment of chromosomes or by crossing over between chromosomes. However, the term usually refers to recombination by crossing over.

recombination repair   A mechanism that *E. coli* cells use to replicate DNA containing thymine dimers. First, the two strands are replicated, leaving a gap across from the dimer. Next, recombination between the progeny duplexes places the gap across from normal DNA. Finally, the gap is filled in using the normal DNA as template.

recurrence risk   The probability that a given genetic disease will recur in a family, given that one individual (or more) in the family has already exhibited the disease.

regulative embryo   An embryo in which the early blastomeres are indistinguishable from one another.

regulatory gene   A gene that encodes a protein that regulates other genes.

relative fitness   The relative ability of different genotypes to pass on their alleles to future generations.

release factor   A protein that causes termination of translation at stop codons.

renaturation of DNA   *See* annealing of DNA.

repetitive DNA   DNA sequences that are repeated many times in a haploid genome.

rep helicase   The product of the *E. coli rep* gene.

replacement vector   A cloning vector derived from λ phage, in which a significant part of the phage DNA is removed, and must be replaced by a similar segment of foreign DNA.

replicating fork   The point where the two parental DNA strands separate to allow replication.

replicative transposition   Transposition in which the transposon DNA replicates, so one copy remains in the original location as another copy moves to the new site.

replicon   All the DNA replicated from one origin of replication.

replisome   The large complex of polypeptides, including the primosome, which replicates DNA in *E. coli*.

repressed   Turned off. When an operon is repressed, it is turned off, or inactive.

resolution   The second step in transposition through a cointegrate intermediate; it involves separation of the cointegrate into its two component replicons, each with its own copy of the transposon.

resolvase   The enzyme that catalyzes resolution of a cointegrate.

res **sites**   Sites on the two copies of a transposon in a cointegrate, between which recombination occurs to accomplish resolution.

restriction endonuclease   An enzyme that recognizes specific base sequences in DNA and cuts at or near those sites.

restriction fragment   A piece of DNA cut from a larger DNA by a restriction endonuclease.

restriction fragment length polymorphism (RFLP)   A variation from one individual to the next in the number of cut sites for a given restriction endonuclease in a given locus. This results in a variation in the lengths of restriction fragments generated by that enzyme cutting within that locus. These variable lengths, which are really alleles, can be treated like any other genetic marker and used in mapping studies.

restriction map   A map that shows the locations of restriction sites in a region of DNA.

restriction site   A sequence of bases recognized and cut by a restriction endonuclease.

restrictive conditions   *See* nonpermissive conditions.

retinoblastoma   A malignant tumor of immature retina cells.

retrotransposon   Transposable elements such as *copia* and Ty that transpose via a retrovirus-like mechanism.

retrovirus   An RNA virus whose replication depends on formation of a provirus by reverse transcription.

reverse transcriptase   RNA-dependent DNA polymerase; the enzyme, commonly found in retroviruses, that catalyzes reverse transcription.

reverse transcription   Synthesis of a DNA using an RNA template.

reversion   A mutation that cancels the effects of an earlier mutation in the same gene.

RF (replicative form)   The circular double-stranded form of the genome of a single-stranded DNA phage such as ϕX174. The DNA assumes this form in preparation for rolling circle replication.

RFLP   *See* restriction fragment length polymorphism.

rho (ρ)   A protein that is needed for transcription termination at certain terminators in *E. coli* and its phages.

ribonuclease H (RNase H)   An enzyme that degrades the RNA part of an RNA-DNA hybrid.

ribonucleoside triphosphates   The building blocks of RNA: ATP, CTP, GTP, and UTP.

ribose   The sugar in RNA.

ribosomal RNA   *See* rRNA.

ribosome   An RNA-protein particle that translates mRNAs to produce proteins.

rise period   The time during which phage-infected cells are lysing, releasing progeny phages to the medium.

RNA-dependent DNA polymerase   *See* reverse transcriptase.

RNA polymerase   The enzyme that directs transcription, or synthesis of RNA.

RNA polymerase core   The collection of subunits of a prokaryotic RNA polymerase having basic RNA chain elongation capacity, but no specificity of initiation; all the RNA polymerase subunits except the σ factor.

RNA (ribonucleic acid)   A polymer composed of ribonucleotides linked together by phosphodiester bonds.

RNA splicing   The process of removing introns from a primary transcript and attaching the exons to each other.

rolling circle replication   A mechanism of replication in which one strand of a double-stranded circular DNA remains intact and serves as the template for elongation of the other strand at a nick.

rRNA (ribosomal RNA)   The RNA molecules contained in ribosomes.

S

Saccharomyces cerevisiae   Baker's yeast.

sarcoma   A malignant tumor of connective tissue.

screen   A genetic sorting procedure that allows one to distinguish desired organisms from unwanted ones, but does not automatically remove the latter.

secondary structure   The local folding of a polypeptide or RNA. In the latter case, the secondary structure is defined by intramolecular base pairing.

second-site reversion   A mutation in a site different from that of an original mutation, which reverses the effect of the original mutation.

sedimentation coefficient   A measure of the rate at which a molecule or particle travels toward the bottom of a centrifuge tube under the influence of a centrifugal force.

segmentation genes   Genes that establish the segmentation pattern of an animal, especially an insect.

segregation   A principle discovered by Mendel, which states that heterozygotes produce equal numbers of gametes having the two different alleles.

selection   A genetic sorting procedure that eliminates unwanted organisms, usually by preventing their growth or by killing them.

selective disadvantage   The reduction in relative fitness of genotype A_2A_2 as compared with that of A_1A_1; also known as the *selection coefficient,* symbolized by s.

selector gene   A gene that controls the fate of cells in a given compartment of a developing embryo, causing that part of the organism's body to be different from another part.

self-fertilization   Fertilization of eggs with sperm or pollen from the same individual. Sometimes called *selfing.*

semiconservative replication   DNA replication in which the two strands of the parental duplex separate completely and pair with new progeny strands. One parental strand is therefore conserved in each progeny duplex.

semidiscontinuous replication   A mechanism of DNA replication in which one strand is made continuously and the other is made discontinuously.

sense strand   *See* coding strand.

sequence tagged site (STS)   A short stretch of DNA that can be identified by using PCR with defined primers to amplify it.

sequencing   Determining the amino acid sequence of a protein, or the base sequence of a DNA or RNA.

sex chromosome   A chromosome involved in determining the sex of an organism. In humans, the X and Y chromosomes are the sex chromosomes.

sex-duction   *See* F-duction.

sex-influenced   Traits that are expressed differently in the two sexes.

sex-limited   Traits that are expressed only in one sex.

sex-linked   Alleles that are on the sex chromosomes.

Shine-Dalgarno (SD) sequence   A G-rich sequence (consensus = AGGAGGU) that is complementary to a sequence at the 3′-end of *E. coli* 16S rRNA. Base pairing between these two sequences helps the ribosome bind an mRNA.

shuttle vector   A cloning vector that can replicate in two or more different hosts, allowing the recombinant DNA to shuttle back and forth between hosts.

sickle-cell disease   A genetic disease in which abnormal β-globin is produced. Because of a single amino acid change, this blood protein tends to aggregate under low oxygen conditions, distorting red blood cells into a sickle shape.

sigma (σ)   The prokaryotic RNA polymerase subunit that confers specificity of transcription—that is, ability to recognize specific promoters.

signal peptide   A stretch of about twenty amino acids, usually at the amino terminus of a polypeptide, that helps to anchor the nascent polypeptide and its ribosome in the endoplasmic reticulum. Polypeptides with a signal peptide are destined for packaging in the Golgi apparatus and are usually exported from the cell.

silencer   A DNA element that can act at a distance to decrease transcription from a eukaryotic gene.

silent mutations   Mutations that cause no detectable change in an organism, even in a haploid organism or in homozygous condition.

single-copy DNA   DNA sequences that are present once, or only a few times, in a haploid genome.

single-strand-binding protein   *See* SSB.

site-directed mutagenesis   A method for introducing specific, predetermined alterations into a cloned gene.

site-specific recombination   Recombination that always occurs in the same place and depends on limited sequence similarity between the recombining DNAs.

small nuclear RNAs (snRNAs)   A class of nuclear RNAs a few hundred nucleotides in length. These RNAs, together with tightly associated proteins, make up **snRNPs (small nuclear ribonucleoproteins)** that participate in splicing, polyadenylation, and 3′-end maturation of transcripts.

solenoid   The 25nm diameter hollow tube formed by the coiling of a string of nucleosomes. Represents the second order of chromatin folding in eukaryotes (Greek: *solenoeides,* pipe-shaped).

somatic 5S rRNA genes   The 5S rRNA genes (haploid number about 400 in *Xenopus laevis*) that are expressed in both somatic cells and oocytes.

somatic cell genetics   Genetic manipulations involving fusing somatic cells instead of mating organisms.

somatic cell hybridization   Fusion of cells from two different species; used to determine chromosomal location of genes.

somatic cells   Nonsex cells.

somatic mutation   A mutation that affects only somatic cells, so it cannot be passed on to progeny.

SOS response   The activation of a group of genes, including *recA,* that helps *E. coli* cells to respond to environmental insults such as chemical mutagens or radiation.

Southern blotting   Transferring DNA fragments separated by gel electrophoresis to a suitable support medium such as nitrocellulose, in preparation for hybridization to a labeled probe.

spacer DNA   DNA sequences found between, and sometimes within, repeated genes such as rRNA genes.

specialized transduction   Use of λ or a similar phage to carry bacterial genes from one cell to another.

spliceosome   The large RNA-protein body upon which splicing of nuclear mRNA precursors occurs.

splicing   The process of linking together two RNA exons while removing the intron that lies between them.

spore   (1) A specialized haploid cell formed sexually by plants or fungi, or asexually by fungi. The latter can either serve as a gamete, or germinate to produce a new haploid cell. (2) A specialized cell formed asexually by certain bacteria in response to adverse conditions. Such a spore is relatively inert and resistant to environmental stress.

sporulation   Formation of spores.

src   The oncogene of Rous sarcoma virus (RSV), which encodes a tyrosine protein kinase.

SSB   Single-strand-binding protein, used during DNA replication. Binds to single-stranded DNA and keeps it from base-pairing with a complementary strand.

standard deviation   The square root of the variance.

statistically significant   A chi-square value (or other statistic) large enough to establish that the results are not consistent with the hypothesis used.

sticky ends   Single-stranded ends of double-stranded DNAs that are complementary and can therefore base-pair and stick together.

stop codon   One of three codons (UAG, UAA, and UGA) that code for termination of translation.

streptomycin   An antibiotic that kills bacteria by causing their ribosomes to misread mRNAs.

stringency (of hybridization)   The combination of factors (temperature, salt, and organic solvent concentration) that influence the ability of two polynucleotide strands to hybridize. At high stringency, only perfectly complementary strands will hybridize. At reduced stringency, some mismatches can be tolerated.

structural genes   Genes that code for proteins.

STS   *See* sequence tagged site.

sum rule   A rule stating that the combined probability of two (or more) mutually exclusive events occurring is the sum of their individual probabilities.

supercoil   *See* superhelix.

superhelix   A form of circular double-stranded DNA in which the double helix coils around itself like a twisted rubber band.

suppression   Compensation by one mutation for the effects of another.

suppressor mutation   A mutation that reverses the effects of a mutation in the same or another gene.

SV40   Simian virus 40; a DNA tumor virus with a small circular genome, capable of causing tumors in certain rodents.

T

T antigen   The major product of the early region of the DNA tumor virus SV40. A DNA binding protein; has the ability to transform cells and thereby cause tumors.

TATA box   An element with the consensus sequence TATAAAA, found about 25 base pairs upstream from the start of transcription in most eukaryotic promoters recognized by RNA polymerase II.

tautomeric shift   The reversible change of one isomer of a DNA base to another by shifting the locations of hydrogen atoms and double bonds.

tautomerization   The shift of a base from one tautomeric form to the other. Also called tautomeric shift.

tautomers   Two forms of the same molecule that differ only in the location of a proton and a double bond. For example, two tautomers of adenine are possible, an amine form and an imine form.

TΨC loop The loop in a tRNA molecule, conventionally drawn on the right, that contains the nearly invariant sequence TΨC, where Ψ is pseudouridine.

T-DNA The tumor-inducing part of the Ti plasmid.

telocentric Chromosome with the centromere very near one end.

telomerase An enzyme that can extend the very ends of telomeres after DNA replication.

telomere A structure at the end of a eukaryotic chromosome, containing tandem repeats of a short DNA sequence.

telophase The last stage of nuclear division in which the nuclear membrane forms and encloses the chromosomes in the daughter cells.

temperate phage A phage that can enter a lysogenic phase in which a prophage is formed.

temperature-sensitive mutation A mutation that causes a product to be made that is defective at high temperature (the restrictive temperature) but functional at low temperature (the permissive temperature).

template A polynucleotide (RNA or DNA) that serves as the guide for making a complementary polynucleotide. For example, a DNA strand serves as the template for ordinary transcription.

template DNA strand The DNA strand of a gene that is complementary to the RNA product of that gene; that is, the strand that served as the template for making the RNA. Also called the anticoding strand, or antisense strand.

terminal transferase An enzyme that adds deoxyribonucleotides, one at a time, to the 3′-end of a DNA.

terminator *See* transcription terminator.

tertiary structure The overall three-dimensional shape of a polypeptide or RNA.

testcross A cross between an F_1 individual (or an individual of unknown genotype) and the recessive parent of the F_1 (the tester).

tetrad analysis The use of the four meiotic products from a single meiosis (a tetrad) to determine the behavior of genes in meiosis.

tetramer (protein) A complex of four polypeptides.

TFIIIA and C Transcription factors that help activate the vertebrate 5S rRNA genes by facilitating binding of TFIIIB.

TFIIIB The transcription initiation factor that activates genes transcribed by RNA polymerase III.

thalassemia A genetic disease marked by failure to produce a functional mRNA for one of the two major adult hemoglobin proteins, α-globin or β-globin.

3′-end The end of a polynucleotide with a free (or phosphorylated) 3′-hydroxyl group.

three-point cross (eukaryotic) A testcross in which one parent is heterozygous for three genes. (prokaryotic) A cross in which the two parental genotypes differ at three genes.

thymidine A nucleoside containing the base thymine.

thymine (T) The pyrimidine base that pairs with adenine in DNA.

thymine dimer Two adjacent thymines in one DNA strand linked covalently, interrupting their base pairing with adenines in the opposite strand. This is the main genetic damage caused by ultraviolet light.

Ti plasmid The tumor-inducing plasmid from Agrobacterium tumefaciens. Used as a vector to carry foreign genes into plant cells.

topoisomerase An enzyme that changes a DNA's superhelical form, or topology.

totipotent cell A cell that retains full genetic capability, including the ability, under appropriate circumstances, to give rise to a complete organism.

***trans*-acting** A term that describes a genetic element, such as a repressor gene or transcription factor gene, that can be on a separate chromosome and still influence another gene. These *trans*-acting genes function by producing a diffusible substance that can act at a distance.

transcript An RNA copy of a gene.

transcription The process by which an RNA copy of a gene is made.

transcription factor A protein that stimulates transcription of a eukaryotic gene by binding to a promoter or enhancer element.

transcription terminator A specific DNA sequence that signals transcription to terminate.

transducing virus (or phage) A virus that can carry host genetic information from one cell to another.

transductants Cells that have received genetic information from a transducing virus.

transduction The use of a phage (or virus) to carry host genes from one cell to another cell of different genotype.

transfer RNA *See* tRNA.

transformation (genetic) An alteration in a cell's genetic makeup caused by introducing exogenous DNA.

transformation (malignant) The process by which a normal cell begins behaving like a tumor cell.

transformed cell An immortal cell that displays many of the characteristics of a cancer cell, but that may or may not be tumorigenic.

transgene A foreign gene transplanted into an organism, making the recipient a transgenic organism.

transgenic organism An organism into which a new gene or set of genes has been transferred.

transition A mutation in which a pyrimidine replaces a pyrimidine, or a purine replaces a purine.

translation The process by which ribosomes use the information in mRNAs to synthesize proteins.

translocation The translation elongation step, following the peptidyl transferase reaction, that involves moving an mRNA one codon's length through the ribosome and bringing a new codon into the ribosome's A site.

translocation (chromosomal) Movement of a chromosome segment from one chromosome to another, nonhomologous chromosome.

transmission genetics The study of the transmission of genes from one generation to the next.

transposable element A DNA element that can move from one genomic location to another.

transposase The name for the collection of proteins, encoded by a transposon, that catalyze transposition.

transposon *See* transposable element.

transversion A mutation in which a pyrimidine replaces a purine, or vice versa.

trisomy Diploids having one extra chromosome (2n + 1).

tRNA (transfer RNA) Relatively small RNA molecules that bind amino acids at one end and "read" mRNA codons at the other, thus serving as the "adapters" that translate the mRNA code into a sequence of amino acids.

$tRNA_f^{Met}$ The tRNA responsible for initiating protein synthesis.

trp **operon** The operon that encodes the enzymes needed to make the amino acid tryptophan.

trypanosomes Protozoa that parasitize both mammals and tsetse flies; the latter spread the disease by biting mammals.

tumorigenicity The ability to cause tumors in whole animals.

tumor suppressor gene *See* anti-oncogene.

Ty A yeast transposon that transposes via a retrovirus-like mechanism.

U

UCE *See* upstream control element.

ultimate carcinogen The most carcinogenic form to which a given substance can be metabolized.

ultraviolet radiation Relatively low energy radiation found in sunlight. Causes thymine dimers in DNA.

unassigned reading frame (URF) An open reading frame that codes for a polypeptide of unknown identity.

undermethylated region A region of a gene or its flank that is relatively poor in, or devoid of, methyl groups.

unidirectional DNA replication Replication that occurs in one direction, with only one active replicating fork.

uniparental inheritance A pattern of inheritance in which only one parent provides genes to the progeny.

unique DNA *See* single-copy DNA.

up mutation A mutation, usually in a promoter, that results in more expression of a gene.

upstream control element (UCE) An element of eukaryotic promoters recognized by RNA polymerase I. Includes bases between positions approximately −100 to −150.

uracil (U) The pyrimidine base that replaces thymine in RNA.

uridine A nucleoside containing the base uracil.

V

variable loop The loop, or stem, in a tRNA molecule lying between the anticodon and TΨC loops.

variable region The region of an antibody that binds specifically to a foreign substance, or antigen. As its name implies, it varies considerably from one kind of antibody to another.

variance A measure of the dispersion of the mean value in a population.

vector A piece of DNA (a plasmid or a phage DNA) that serves as a carrier in gene cloning experiments.

vegetative cell A cell that is reproducing by division, rather than sporulating or reproducing sexually.

viroid An infectious agent (of plants) containing only a small circular RNA.

virulent phage A phage that lyses its host.

virusoid A viroid-like RNA that depends on a virus for its replication.

visible mutations *See* morphological mutations.

W

wobble The ability of the third base of a codon to shift slightly to form a non-Watson-Crick base pair with the first base of an anticodon, thus allowing a tRNA to translate more than one codon.

wobble hypothesis Francis Crick's hypothesis that invoked wobble to explain how one anticodon could decode more than one codon.

wobble position The third base of a codon, where wobble base-pairing is permitted.

X

xeroderma pigmentosum A disease characterized by extreme sensitivity to sunlight. Even mild exposure leads to many skin cancers. Caused by a defect in DNA repair.

X rays High energy radiation that ionizes cellular components. The ions then can cause chromosome breaks.

Y

YAC *See* yeast artificial chromosome.

yeast artificial chromosome A high-capacity cloning vector consisting of yeast left and right telomeres and a centromere. DNA placed between the centromere and one telomere becomes part of the YAC and will replicate in yeast cells.

Z

Z-DNA A left-handed helical form of double-stranded DNA whose backbone has a zigzag appearance. This form is stabilized by stretches of alternating purines and pyrimidines.

zinc finger A finger-shaped protein structural motif found in DNA-binding proteins and characterized by two cysteines and two histidines bound to a zinc ion. By inserting into the major groove of the DNA, these fingers help the proteins bind DNA.

zygote A fertilized egg.

CREDITS

Chapter 2

Figure 2.14a: From Robert H. Tamarin, *Principles of Genetics,* 5th ed. Copyright © 1996 Times Mirror Higher Education Group, Inc., Dubuque, Iowa. Reprinted by permission. All Rights Reserved.

Chapter 4

Figure 4.40: From J. J. Yunis and O. Prakash, *Science,* 251:1525–30, 19 March 1982. Copyright © 1982 by the American Association for the Advancement of Science, Washington, DC.

Chapter 5

Figure 5.22: From C. B. Bridges, *Science,* 83(2148):210, 1936. Copyright © 1936 by the American Association for the Advancement of Science, Washington, DC.

Chapter 6

Figure 6.14b: From *Nature* 171:737, 1953.
Figure 6.14c: From *Nature* 175:834, 1955.
Figure 6.18: From Paul Doty, *Harvey Lectures* 55:121, 1961. Copyright © 1961 Academic Press, Orlando, FL. Reprinted by permission.

Chapter 7

Figure 7.23: Modified from Herendeen and Kelly, *Cell* 84:5, 1996.
Figure 7.27: From *Nature* 337:336, 1989. Copyright © 1989 Macmillan Magazines Ltd. Reprinted by permission.

Chapter 8

Figure 8.33a: From *Nature* 327:593, 1987. Copyright © 1987 Macmillan Magazines Ltd. Reprinted by permission.

Chapter 9

Figure 9.6: From Widom and Klug, *Cell* 43:210, 1985. Copyright © 1985 Cell Press, Cambridge, MA. Reprinted by permission.

Figure 9.22: From S. McKnight and R. Tjian, *Cell* 46:796, 1986. Copyright © Cell Press, Cambridge, MA. Reprinted by permission.
Figure 9.29: From E. K. O'Shea, J. D. Klemm, P. S. Kim, and T. Alber, *Science,* 254:541, 1991. Copyright © 1991 by the American Association for the Advancement of Science, Washington, DC.
Figure 9.30: From *Nature* 373:258, 1995. Copyright © 1995 Macmillan Magazines Ltd. Reprinted by permission.
Figure 9.32b: From Susan M. Berget, Claire Moore, and Phillip A. Sharp, "Spliced Segments at the 5′ Terminus of Adenovirus 2 Late mRNA," *Proceedings of the National Academy of Science* 74(8):3717–3175, August 1977.
Figure 9.34: From Shirley M. Tilghman et al., "The Intervening Sequence of a Mouse β-globin Gene Is Transcribed Within the 15S B-globin mRNA Precursor," *Proceedings of the National Academy of Science* 75(3):1309–1313, March 1978.

Chapter 10

Figure 10.3: Reprinted from Linus Pauling: *The Nature of the Chemical Bond, Third Edition.* Copyright © 1960 by Cornell University. Used by permission of the publisher, Cornell University Press.
Figure 10.5: From R. F. Dickerson, *The Proteins,* 2d edition. Copyright © 1964 Academic Press, Orlando, FL. Reprinted by permission.
Figure 10.7: From *Nature* 329:509, 1987. Copyright © 1987 Macmillan Magazines Ltd. Reprinted by permission.
Figure 10.12: From *Nature* 331:225, 1988. Copyright © 1988 Macmillan Magazines Ltd. Reprinted by permission.
Figure 10.19a: From S.-H. Kim et al., *Science,* 185:436, 1974. Copyright © 1974 by the American Association for the Advancement of Science, Washington, DC.
Figure 10.39a: From *Nature* 237:84, 1972. Copyright © 1972 Macmillan Magazines Ltd. Reprinted by permission.

Chapter 11

Quote on page 295: "Look Through My Window" Words and Music by John Phillips. © Copyright 1966, 1967 MCA Music Publishing, a Division of MCA Inc. Copyright renewed. All Rights Reserved. International Copyright Secured.

Chapter 13

Figure 13.31: From: *Molecular Genetics* by Stent and Calendar. Copyright © 1978 by W. H. Freeman and Company. Used with permission.

Chapter 14

Figure 14.9: From Peter A. Lawrence, *The Making of a Fly.* Copyright © 1992 Blackwell Scientific Publications. Reprinted by permission of Blackwell Science Ltd.
Figure 14.10: From P. Walter et al., *Cell* 38:6, 1984. Copyright © 1984 Cell Press, Cambridge MA. Reprinted by permission.
Figure 14.22: From Peter A. Lawrence, *The Making of a Fly.* Copyright © 1992 Blackwell Scientific Publications. Reprinted by permission of Blackwell Science Ltd.
Figure 14.28: From Peter A. Lawrence, *The Making of a Fly.* Copyright © 1992 Blackwell Scientific Publications. Reprinted by permission of Blackwell Science Ltd.
Figure 14.29: From Peter A. Lawrence, *The Making of a Fly.* Copyright © 1992 Blackwell Scientific Publications. Reprinted by permission of Blackwell Science Ltd.
Figure 14.31: From Small et al., *Genes and Development* 5:829, 1991.
Figure 14.32: From Peter A. Lawrence, *The Making of a Fly.* Copyright © 1992 Blackwell Scientific Publications. Reprinted by permission of Blackwell Science Ltd.
Figure 14.36: From Scott, *Cell* 71:551, 1992. Copyright © 1992 Cell Press, Cambridge, MA. Reprinted by permission.
Figure 14.37: From Lawrence et al., *Cell* 78:183, 1994. Copyright © 1994 Cell Press, Cambridge, MA. Reprinted by permission.

Figure 14.40: From Peter A. Lawrence, *The Making of a Fly.* Copyright © 1992 Blackwell Scientific Publications. Reprinted by permission of Blackwell Science Ltd.

Figure 14.46: From J. E. Sulston and H. R. Horvitz, *Developmental Biology* 56:111, 1977. Copyright © 1977 Academic Press, Orlando, FL. Reprinted by permission.

Text Art on page 412: From John Kuspira and Ramesh Bhambhani, *Compendium of Problems in Genetics.* Copyright © 1994 Times Mirror Higher Education Group, Inc., Dubuque, Iowa. Reprinted by permission. All Rights Reserved.

Chapter 15

Figure 15.5: Courtesy of Life Technologies, Inc.

Chapter 18

Figure 18.10: From: *An Introduction to Genetic Analysis* 3/E by Suzuki et al. Copyright © 1986 by W. H. Freeman and Company. Used with permission.

Appendix C

Appendix C: From Peter H. Raven and George B. Johnson, *Biology,* 4th ed. Copyright © 1996 Times Mirror Higher Education Group, Inc., Dubuque, Iowa. Reprinted by permission. All Rights Reserved.

Electronic Illustrators Group

1.3, 1.4, 2.4, 2.5, 2.6, 2.7, 2.9, 2.11, 2.12, 2.13, 2.14, 2.15, 2.16, 2.19, TA 2.3, TA 2.4, 3.1, 3.2, 3.4, 3.5, 3.7, 3.9, 3.10, 3.12, 3.14, 3.15, 3.17, 3.19, 3.20, 3.21, 3.22, 3.23, 3.25, 3.26, 3.27, 3.28, 3.29, 3.30, 3.32, 3.33, 3.34, 3.35, 3.36A, 3.37, 3.38, BX 3.1–2, 4.2, 4.3, 4.4, 4.6, 4.7, 4.9, 4.10, 4.11, 4.13, 4.14, 4.15, 4.16, 4.17, 4.18, 4.19, 4.20, 4.21, 4.22, 4.23, 4.24, 4.26, 4.27, 4.29, 4.30, 4.31, 4.323, 4.33, 4.35, 4.39, 4.40, 4.41, 5.2, 5.3, 5.4, 5.5, 5.6, 5.7, 5.8, 5.9, 5.10, 5.11, 5.12, 5.13, 5.14, 5.15, 5.16, 5.17, 5.18, 5.19, 5.20, 5.21, 5.22, 5.23, BX 5.1–2, 5.24, 6.4, 6.5, 6.6, 6.7, 6.8, 6.9, 6.10, 6.11, 6.13, 6.14, 6.15, 6.16, 6.17, 6.18, 6.19, 6.20, 6.22, 6.24A, 6.25B, 6.27, 6.28, TA 6.6, TA 6.7, 7.1, 7.2, 7.3, 7.4, 7.5, 7.6, 7.7, 7.8, 7.10, 7.11, 7.12, 7.13, 7.14B–C, 7.15B, 7.16B, 7.18, 7.19, 7.20, 7.21, 7.22, 7.23, 7.24, 7.25, 7.26, 7.27, 7.28, 7.29, 7.30, 7.31, 7.33, 7.34, TA 7.1, TA 7.2, TA 7.3, TA 7.4, TA 7.5, 8.1, 8.2, 8.3, 8.4, 8.5, 8.6, 8.7, 8.8, 8.9A–D, 8.10, 8.11, 8.12, 8.13, 8.14, 8.15, 8.16, 8.17, 8.18, 8.19, 8.20, 8.21, 8.23, 8.24, 8.25, 8.26, 8.27, 8.28, 8.29, 8.30, 8.31, 8.33B, 9.4, 9.6, 9.8, 9.11, 9.13, 9.14, 9.15, 9.16, 9.17, 9.18, 9.19, 9.20, 9.21, 9.22, 9.23, 9.24, 9.25A, 9.26, 9.27B, 9.28, 9.29B, 9.30, 9.31, 9.32B–C, 9.33, 9.34, 9.35, 9.37, 9.38, 9.39, 9.40, 9.42, 9.43A, 9.44, 9.45, 9.46, 9.47, TA 9.1, TA 9.2, TA 9.3, TA 9.4, TA 9.5, TA 9.6, TA 9.7, TA 9.8, TA 9.9, 10.1, 10.2, 10.3, 10.4, 10.5, 10.6, 10.7, 10.8, 10.9, 10.10, 10.12, 10.15A, 10.17, 10.18, 10.19, 10.20, 10.21, 10.22, 10.24, 10.25, 10.26, 10.27, 10.28, 10.29, 10.30, 10.31, 10.32, 10.33, 10.34, 10.35, 10.36, 10.37, 10.38, 10.39, 10.40, 10.41, 10.42, TA 10.1, 11.2, 11.7, 11.9, 11.10, 11.11, 11.12, 11.13, 11.14, 11.15, 11.16, 11.17, 11.18, 11.19, 11.20, 11.21, 11.22, 11.23, 11.24, 11.25, 11.26, 11.27, 11.28, 11.29, 11.30, 11.31, 11.33, 12.1, 12.2A, 12.3, 12.4, 12.5, 12.6, 12.7, 12.8, BX 12.1, 12.9, 12.10, 12.11, 12.12, 12.13, 12.15, 12.16, 12.17, 12.20, 13.1, 13.3, 13.4, 13.8, 13.9, 13.10, 13.11, 13.12, 13.13, 13.14, 13.15, 13.16, 13.17, 13.18, 13.19, 13.20, 13.21, 13.23, 13.26, 13.28, 13.29, 13.30, 13.31, 13.32, TA 13.1, TA 13.2, 14.3, 14.5, 14.6, 14.7, 14.8, 14.9, 14.10, 14.11, 14.12, 14.13, 14.16, 14.17, 14.18, 14.19, 14.21, 14.22, 14.25, 14.28, 14.29, 14.31, 14.32, 14.33, 14.36, 14.37, 14.40, BX 14.1–1, 14.41, 14.43, 14.44, 14.45, 14.47, 14.48, 14.49, TA 14.1, TA 14.2, TA 14.3, 15.1, 15.2, 15.3, 15.4, 15.5, 15.6, 15.7, 15.8, 15.9, 15.10, 15.11, 15.12, 15.13, 15.14, 15.15, 15.16, 15.17, TA 15.1, TA 15.2, 15.18, 15.20, 15.22, 15.23, 15.24A–B, 15.25, 15.26, 15.27, 15.28, TA 15.3, 16.1, 16.2, 16.3A, 16.4, 16.7, 16.8, 16.9, 16.10, 16.11, 16.12, 16.13, 16.14, 16.15, 16.17, 16.18, 16.19, 16.20, TA 16.1, TA 16.2, TA 16.3, 17.2, 17.3, 17.4, 17.6, 17.7, 17.8, 17.9, 17.10, 17.11, 17.12, 17.13, 17.14, 17.15, 17.16, 17.17, 17.18, 17.20, 17.21, 18.4, 18.6, 18.7, 18.8, 18.9, 18.11, 18.12, 18.13, 18.14, 18.17, 18.19(lineart), 18.20, 18.21, 18.22, 18.23, 18.24, 18.26, 19.1, 19.2, 19.3, 19.4, 19.5, 19.6, 19.7, 19.8, 19.9, 19.10, 19.11, 19.13, 19.14, 19.15, 19.16, 19.17, TA 19.1, 20.1, 20.2, 20.3, 20.5, 20.6, 20.7, 20.8, 20.9, 20.10, 20.15, 20.16, 20.17, 20.18, 20.19, 20.20, 20.21, 20.22, 20.23, 20.24.

Illustrious Inc.

BX 1.1–1, BX 1.2A, BX 1.2B, BX 1.3

Carlyn Iverson

2.1, 2.2, 2.3, 3.6, 3.11, 4.34, 14.4, 14.20, 18.5, 18.10

INDEX